Equations of Conics in Standard Position

Circle: $x^2 + y^2 = r^2$
Parabola: $y^2 = 4px$, $x^2 = 4py$
Ellipse: $x^2/a^2 + y^2/b^2 = 1$
Hyperbola: $x^2/a^2 - y^2/b^2 = 1$

Functions

Function notation: $y = f(x)$
Composite function: $(f \circ g)(x) = f(g(x))$
Inverse function notation: $y = f^{-1}(x)$
The basic property: $(f \circ f^{-1})(x) = (f^{-1} \circ f)(x) = x$

Exponential Functions

$f(x) = a^x$, $a > 0$, $a \neq 1$

Domain: the set of all real numbers
Range: the set of all positive real numbers

Properties

$a^x a^y = a^{x+y}$
$(a^x)^y = a^{xy}$
$a^{-x} = 1/a^x$

If $a > 1$ and $x < y$, then $a^x < a^y$ (increasing)
If $0 < a < 1$ and $x < y$, then $a^x > a^y$ (decreasing)

The Natural Exponential Function

$f(x) = e^x$, $e \simeq 2.71828\ldots$

Logarithmic Functions

$y = \log_a x$ if and only if $x = a^y$, $a > 0$, $a \neq 1$.

Domain: the set of all positive real numbers
Range: the set of all real numbers.

Properties of Logarithms

$\log_a(xy) = \log_a x + \log_a y$
$\log_a(x/y) = \log_a x - \log_a y$
$\log_a x^m = m \log_a x$

Change of base formula: $\log_a x = \log_b x/\log_b a$
Common logarithms: $y = \log x$ (base 10)
Natural logarithms: $y = \ln x = \log_e x$ (base e)

Right Triangle Trig

$\sin \alpha = A/C \qquad \csc \alpha = C/A$
$\cos \alpha = B/C \qquad \sec \alpha = C/B$
$\tan \alpha = A/B \qquad \cot \alpha = B/A$

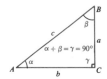

Complementary Angles

$\sin \alpha = \cos(90° - \alpha)$
$\cos \alpha = \sin(90° - \alpha)$
$\tan \alpha = \cot(90° - \alpha)$
$\sin((\pi/2) - \alpha) = \cos \alpha$
$\cos((\pi/2) - \alpha) = \sin \alpha$
$\tan((\pi/2) - \alpha) = \cot \alpha$

Special Angles

$\sin 45° = \cos 45° = \sqrt{2}/2$

$\sin 30° = \cos 60° = 1/2$
$\sin 60° = \cos 30° = \sqrt{3}/2$

Circular Functions

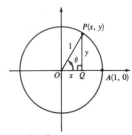

$y = \sin \theta$
$x = \cos \theta$

COLLEGE
ALGEBRA
AND TRIGONOMETRY
WITH
APPLICATIONS

SECOND EDITION

■

COLLEGE ALGEBRA AND TRIGONOMETRY WITH APPLICATIONS

JOSÉ BARROS-NETO
Rutgers University

WEST PUBLISHING COMPANY

St. Paul ▪ New York ▪ Los Angeles ▪ San Francisco

■ **To the student** A Solutions Manual containing worked-out solutions for all the odd-numbered exercises in this book is available from West Publishing Company. Please ask your instructor or college bookstore for details.

This text was set in Avant-Garde and Times Roman by the Interactive Composition Corporation. Susan Gerstein was the copyeditor. Nancy Blodget designed the interior; Delor Erickson designed the cover and provided the original cover art. New art for this edition was prepared by ANCO/Boston and Rolin Graphics.

Copyright © 1985   By WEST PUBLISHING COMPANY

Copyright © 1988   By WEST PUBLISHING COMPANY
50 W. Kellogg Boulevard
P.O. Box 64526
St. Paul, MN 55164-1003

96 95 94 93 92 91 90 89   8 7 6 5 4 3 2 1

Library of Congress Cataloging-in-Publication Data

Barros-Neto, José
College algebra and trigonometry with applications.

Includes index.
1. Algebra. 2. Trigonometry, Plane. I. Title.
QA154.2.B384 1988   512'.13   87-23202
ISBN 0-314-59695-X

CONTENTS

Success in the study of calculus or mathematically oriented courses in fields such as economics, business, chemistry, or biology requires solid preparation in algebra, a knowledge of the elementary functions and their properties, and some exposure to trigonometry and the trigonometric functions.

The second edition of *College Algebra and Trigonometry with Applications* is designed for students who need additional preparation for the study of calculus or mathematically oriented courses. In this new edition, I've kept the same organization as the first edition, but I've added to it several important pedagogical features. The first three chapters contain a thorough review of algebra, the study of linear and quadratic equations and inequalities, and coordinate geometry. These chapters are flexible enough so that they can be used in courses of different levels. An instructor teaching a course designed for students with limited preparation, and consequently moving at a slower pace, will find in these chapters enough basic material for these students. For a class of well-prepared students, most of the review material may be omitted or covered briefly so that the instructor can advance more quickly to functions and trigonometry.

The core of the book consists of six chapters: two chapters on functions and their basic properties, one chapter on exponential and logarithmic functions, and three chapters on trigonometry. All these chapters include a large variety of applications to economics, biology, geology, chemistry, physics, and engineering. The first of the three chapters on trigonometry covers right triangle trigonometry and discusses problems involving right triangles. Trigonometric functions and their properties, graphs of trigonometric functions, oscillatory motion, and inverses of trigonometric functions are the subject of the second chapter. Trigonometric identities and applications of trigonometry make up the third chapter on trigonometry. This second edition contains two additional sections: vectors in the plane and polar coordinates.

The last three chapters of the book contain supplementary material. There is a chapter on linear systems of equations, matrices, determinants, Cramer's rule, nonlinear systems of equations, systems of inequalities, and linear programming. Another chapter discusses polynomials and contains several applications that students may need in a calculus course. Finally, the last chapter contains the principle of mathematical induction, arithmetic and geometric sequences and series, permutations and combinations, and elementary probability.

BOOK ORGANIZATION

The book is written in a concise and informal style. Lists of axioms or formal statements of theorems are avoided as much as possible. Occasionally, elementary proofs are presented, but only in situations where the learning of such proofs may improve the students' understanding of the subject matter. Explanations always proceed from the particular to the general case. Selected examples, often taken from the applied sciences, are used to motivate and prepare students for new concepts and definitions. The various topics discussed throughout the book are never presented as isolated abstract mathematical entities. Whenever possible, I show how concrete ideas studied and developed in the natural sciences evolve into mathematical concepts.

A typical chapter or section of the book starts with one or more examples taken from the applied sciences, or with a relevant historical note. This is done as a motivation, and to show the interplay between mathematics and the applied sciences.

Examples, Practice Exercises, and Exercises

Each section of the book contains a number of examples that illustrate the ideas and techniques currently under discussion. Each example is followed by a practice exercise of the same level of difficulty. By solving the practice exercises, students will have the opportunity to check their knowledge immediately, before proceeding to the next subject.

In this second edition, I've extensively revised the exercises at the end of each section. The lists of exercises are now more diversified and complete. The level of difficulty of the exercises ranges from routine, to average, to challenging. Exercises that are conceptually more difficult and/or may require lengthier calculations or algebraic manipulations are denoted by the symbol $\square$.

Chapter Summaries

At the end of each chapter, a detailed chapter summary contains a review of the main definitions, results, and formulas discussed in the chapter. Students will find the chapter summaries particularly helpful when working the exercises of the chapter tests or solving the review exercises.

Chapter Tests

Immediately after each chapter summary, there is a chapter test containing ten problems of average difficulty and designed to test the students' learning progress. The answer to these tests are given at the back of the book together with the answers to odd-numbered exercises.

Review Exercises

Following each chapter summary and chapter test, students and instructors will find a comprehensive list of review exercises with varying levels of difficulty.

Historical Notes and Mathematical Vignettes

Interspersed throughout the text in separate boxes, historical notes describe facts that are relevant to the subject matter and give the students a historical perspective on the development of mathematical concepts. The mathematical vignettes, a new feature in this second edition, deal with present day applications of mathematics and how they relate to the applied sciences.

Calculators

Calculators are important computational tools, and their use should be encouraged. They expedite calculations and improve accuracy. Because instructors may object to the use of calculators at this level of instruction, the book includes a large number of examples, practice exercises, and exercises that do not need a calculator. However, they may require the use of tables such as the ones found in the appendix. All examples, practice exercises, and exercises requiring the use of calculators are indicated by the symbol ▣ .

ACKNOWLEDGMENTS

A book at this level personifies the teaching experience, taste, and pedagogic ideas of the author. However, it would have been impossible to have written this book without the multitude of comments, criticisms, and suggestions of several reviewers. I'm indebted to the following individuals who reviewed all or parts of this second edition at various stages of preparation:

Annette Blackwelder, Florida State University
Yungchen Chen, Southwest Missouri State University
H. Joan Dykes, Edison Community College
John Findeis, Armstrong State College
Ken Goldstein, Miami Dade Community College
D. W. Hall, Michigan State University
Lyman Holden, Southern Illinois University
Ruth Hunt, University of Missouri at Columbia
Lynda Morton, University of Missouri at Columbia
Glenn Prigge, University of North Dakota
Eugenie V. Sturgis, Clemson University
George L. Szoke, University of Akron
Michael D. Taylor, University of Central Florida
John Toby, Northshore Community College
V. C. Varadachari, Lakewood Community College

I'm also very thankful to the staff of West Publishing Company for the care, support, and assistance they provided in the preparation of this book.

Finally, my gratitude goes to my wife, Iva, for her help, patience, and encouragement.

José Barros-Neto
November 1987

CHAPTER 1

FUNDAMENTALS

OF

ALGEBRA

Throughout this book, we shall mostly consider real numbers. We will begin the chapter with an informal discussion about *natural numbers, integers, rational numbers,* and *irrational numbers;* these numbers together form the *real number system.* It is important to pay special attention to the distinction between a rational number and an irrational number. After we introduce the real number system, we will show how real numbers can be represented by the points of a line and will discuss the basic concepts of algebra. We will then define polynomials, briefly review some of the algebraic properties of polynomials, and discuss special products and the factoring of polynomials. Next we will introduce rational expressions, which are formal quotients of polynomials. Their relationship to polynomials is analogous to that of fractions to integers. We will review in some detail the algebraic operations on rational expressions. We next discuss radicals and fractional powers of real numbers, and then we end the chapter with a section on complex numbers and their properties.

 ## 1.1 NATURAL NUMBERS, INTEGERS, RATIONAL AND IRRATIONAL NUMBERS

Natural Numbers

The most common number system consists of the *natural numbers*

$$1, 2, 3, \ldots, n, \ldots$$

also called *counting numbers.*

Two fundamental operations are defined on natural numbers: *addition* (or *sum*) and *multiplication* (or *product*). The symbol $+$ is used to indicate addition, while the symbols $\times$ and $\cdot$ are used to indicate multiplication. Very often multiplication is indicated by juxtaposition of the factors: we may write, for example, $2n$ instead of $2 \cdot n$, and apq to represent the product of a, p, and q.

The natural numbers are *closed* under the addition or multiplication operations. This means that every time we add or multiply two natural numbers, the result is a natural number.

Prime Numbers

The natural number 60 can be written as the product of 4 and 15, that is, $60 = 4 \cdot 15$. The numbers 4 and 15 are called *factors* of 60, and 60 is said to be a *multiple* of 4 and 15.

If the only factors of a natural number greater than 1 are the number itself and 1, then the natural number is said to be a *prime number*. Some examples of prime numbers are 2, 3, 5, 7, 11, 13, 17, and 19. It can be shown that there are infinitely many prime numbers. Going back to the number 60, since $4 = 2 \cdot 2$ and $15 = 3 \cdot 5$, we can write $60 = 2 \cdot 2 \cdot 3 \cdot 5$; the numbers 2, 3, and 5 are the *prime factors* of 60.

Every natural number greater than 1 that is not a prime number is called a *composite number*. Some composite numbers are 4, 6, 8, 9, 10, 15, 30, and 60. The following theorem about composite numbers is important in many applications.

Fundamental Theorem of Arithmetic

Every composite number can be written in a unique way, except for the order of the factors, as a product of prime factors.

When a number is written as a product of all its prime factors, we obtain the *prime factorization* of the number.

EXAMPLE 1

Find the prime factorizations of 180 and 210.

Solution: First find as many factors 2 as possible, then factors 3, factors 5, and so on.

$$180 = 2 \cdot 90 = 2 \cdot 2 \cdot 45 = 2 \cdot 2 \cdot 3 \cdot 15 = 2 \cdot 2 \cdot 3 \cdot 3 \cdot 5$$
$$210 = 2 \cdot 105 = 2 \cdot 3 \cdot 35 = 2 \cdot 3 \cdot 5 \cdot 7$$

Practice Exercise 1

Write 360 as a product of prime numbers.

Answer: $360 = 2 \cdot 2 \cdot 2 \cdot 3 \cdot 3 \cdot 5$

Integers

Subtraction (denoted by the symbol $-$) is the opposite of the addition operation. For example, $5 - 3 = 2$, because $5 = 3 + 2$; the number 2 is the *difference* of 5 and 3.

The natural numbers are not closed under subtraction. For instance, since there is no natural number that can be added to 6 to give 4, the difference $4 - 6$ cannot be defined as a natural number. Similarly, the difference $1 - 1$ is not a natural number.

In order to have a number system in which the difference of any two numbers is always defined, the *integers* are introduced; they consist of the number 0 and the *signed numbers* $\pm n$, where n is a natural number. An integer $+n$ (with a plus sign) is called *positive*, and an integer $-n$ (with a minus sign) is called *negative*. We also have $0 = +0 = -0$. Every natural number n is identified with a positive integer $+n$; thus natural numbers are *positive integers*. The addition and multiplication operations extend to the integers, and the integers are closed under the operations of addition, subtraction, and multiplication.

Sets

Throughout this book we shall use some set notation. A *set* is a collection of objects, which are called the *elements* of the set. Sets are usually denoted by uppercase letters, such as A, B, S, X, and Y, while the elements of a set are denoted by lowercase letters, such as a, b, s, x, and y. The *empty* or *null set* $\varnothing$ is the set that contains no element.

A pair of braces $\{\ \}$ used with words or symbols can describe a set. For example, the set of natural numbers is denoted by

$$\mathbb{N} = \{1, 2, 3, \ldots, n, \ldots\},$$

and the set of integers is denoted by

$$\mathbb{Z} = \{\ldots, -n, \ldots, -2, -1, 0, 1, 2, \ldots, n, \ldots\}.$$

If a is an element of set A, we write $a \in A$; the notation $a \notin A$ means that a is not an element of set A.

Given two sets A and B, if every element of A is an element of B, then A is said to be a *subset* of B, and we write $A \subset B$. Set B is said to *contain* set A, or set A is said to be *contained* in set B. As an example, the set $\mathbb{N}$ of natural numbers is a subset of the set $\mathbb{Z}$ of integers, and we write $\mathbb{N} \subset \mathbb{Z}$.

Rational Numbers

Division (indicated with the symbol $\div$) is the inverse of multiplication. For example, $12 \div 3 = 4$ because $12 = 3 \cdot 4$. The number 12 is called the *dividend*, 3 is called the *divisor*, and 4 is called the *quotient*. Instead of writing $12 \div 3$ to denote the quotient of 12 by 3, we may also write $12/3$ or $\frac{12}{3}$.

The integers are not closed under the operation of division. For example, since there is no integer that when multiplied by 3 gives 2, the quotient $2 \div 3$ cannot be defined as an integer. Similarly, $-\frac{4}{5}$ and $\frac{8}{21}$ are not integers. In order to have them defined, the set of integers is enlarged, this time by introducing *rational numbers*.

If m and n denote integers, with $n \neq 0$, then m/n is said to represent a *rational number*. The number m/n is also called a *fraction*, where m is the *numerator* and n the *denominator*. It is necessary to assume $n \neq 0$, because division by zero is undefined.

We denote the set of all rational numbers by

$$\mathbb{Q} = \left\{ \frac{m}{n} : m, n \in \mathbb{Z} \text{ with } n \neq 0 \right\}.$$

(The colon is read "such that.")

Two rational numbers m/n and p/q are said to be *equal* exactly when

$$mq = np. \tag{1.1}$$

For example, $2/3 = 6/9$ because $2 \cdot 9 = 18 = 3 \cdot 6$; and $-5/7 = 5/-7$ because $(-5)(-7) = 35 = 7 \cdot 5$. However, $3/5 \neq 7/8$.

From this definition it follows that when the numerator and denominator of a fraction have a common factor, we can *simplify* the fraction by cancelling the common factor. For example,

$$\frac{18}{30} = \frac{3 \cdot 6}{5 \cdot 6} = \frac{3}{5}.$$

If a rational number m/n, with $n > 0$, is such that m and n have no common factor other than 1, then m/n is said to be in *lowest terms*. For example, $3/5$ is in lowest terms for $18/30$, and $2/3$ is in lowest terms for $30/45$. Notice that the lowest terms for $12/-24$ is $-1/2$ and not $1/-2$.

By identifying every integer m with the rational number $m/1$, we make $\mathbb{Z}$, the set of integers, a subset of $\mathbb{Q}$, the set of rational numbers. Thus, we have the following inclusions among the three different systems of numbers: $\mathbb{N} \subset \mathbb{Z} \subset \mathbb{Q}$.

Any two rational numbers can be written in equivalent forms with the same denominator. For example, $3/4$ and $5/6$ can be written $9/12$ and $10/12$, two fractions with the same denominator. This method is used to add or subtract fractions.

The four arithmetic operations extend to the rational number as follows. If m/n and p/q are rational numbers, then their *sum* is defined by

$$\frac{m}{n} + \frac{p}{q} = \frac{mq}{nq} + \frac{pn}{nq} = \frac{mq + pn}{nq}. \tag{1.2}$$

The *difference* is defined by

$$\frac{m}{n} - \frac{p}{q} = \frac{mq}{nq} - \frac{pn}{nq} = \frac{mq - pn}{nq}. \tag{1.3}$$

The *product* is defined by

$$\frac{m}{n} \times \frac{p}{q} = \frac{mp}{nq}. \tag{1.4}$$

Finally, the *quotient* is defined by

$$\frac{m}{n} \div \frac{p}{q} = \frac{m}{n} \times \frac{q}{p} = \frac{mq}{np}. \tag{1.5}$$

The set of rational numbers is closed under the operations of addition, subtraction, multiplication, and division. The basic properties of these operations will be reviewed in detail after we introduce real numbers.

EXAMPLE 2   Perform the indicated operations and simplify.

(a) $\dfrac{2}{3} + \dfrac{3}{4}$ (b) $\dfrac{3}{4} - \dfrac{2}{3}$ (c) $\dfrac{2}{3} \times \dfrac{3}{4}$ (d) $\dfrac{2}{3} \div \dfrac{3}{4}$

Solution: We have the following.

(a) $\dfrac{2}{3} + \dfrac{3}{4} = \dfrac{8}{12} + \dfrac{9}{12} = \dfrac{8+9}{12} = \dfrac{17}{12}$ (b) $\dfrac{3}{4} - \dfrac{2}{3} = \dfrac{9}{12} - \dfrac{8}{12} = \dfrac{9-8}{12} = \dfrac{1}{12}$

(c) $\dfrac{2}{3} \times \dfrac{3}{4} = \dfrac{2 \cdot 3}{3 \cdot 4} = \dfrac{6}{12} = \dfrac{1}{2}$ (d) $\dfrac{2}{3} \div \dfrac{3}{4} = \dfrac{2}{3} \times \dfrac{4}{3} = \dfrac{8}{9}$

Practice Exercise 2   Perform the following operations.

(a) $\dfrac{5}{6} + \dfrac{3}{8}$ (b) $\dfrac{5}{6} - \dfrac{3}{8}$ (c) $\dfrac{5}{6} \times \dfrac{3}{8}$ (d) $\dfrac{5}{6} \div \dfrac{3}{8}$

Answer: (a) $\dfrac{29}{24}$ (b) $\dfrac{11}{24}$ (c) $\dfrac{5}{16}$ (d) $\dfrac{20}{9}$

EXAMPLE 3   Calculate the sum $\dfrac{7}{12} + \dfrac{5}{20}$.

Solution: According to formula (1.2), we have

$$\frac{7}{12} + \frac{5}{20} = \frac{140}{240} + \frac{60}{240} = \frac{140 + 60}{240} = \frac{200}{240} = \frac{5}{6}.$$

Notice that when adding or subtracting fractions we can always use the product of the denominators as the common denominator. But the best choice of common denominator is the *least common multiple* (LCM) of the denominators. Recall that the least common multiple of several numbers is the *smallest natural number* that is a multiple of all of them. To find the LCM of two (or even more) integers, write their prime factorizations and then form a new integer whose prime factorization contains each of the primes from the given integers the *largest* number of times that it occurs in any of the given integers. For example,

$$12 = 2 \cdot 2 \cdot 3 \quad \text{and} \quad 20 = 2 \cdot 2 \cdot 5$$

so their LCM is $2 \cdot 2 \cdot 3 \cdot 5 = 60$. We now have

$$\frac{7}{12} + \frac{5}{20} = \frac{35}{60} + \frac{15}{60} = \frac{50}{60} = \frac{5}{6}.$$

Practice Exercise 3   Calculate the difference $\dfrac{17}{12} - \dfrac{3}{4}$. Give your answer in simplified form.

Answer: $\dfrac{2}{3}$

Properties of Equality

The equality between rational numbers (also integers and rational numbers) satisfies two basic properties, which we now describe.

> If x and y are rational numbers such that
> $$x = y,$$
> then
> $$x + z = y + z \quad \text{and} \quad xz = yz$$
> for every rational number z.

In other words, if we add the same number to both sides of an equality or multiply both sides of an equality by the same number, the equality is still true.

Decimal Numbers

A rational number m/n denotes the division of m by n. By performing a long division we obtain the *decimal representation* of the rational number. For example,

$$\frac{1}{2} = 0.5 \quad \text{and} \quad \frac{3}{4} = 0.75.$$

Also,

$$\frac{1}{3} = 0.333 \ldots, \quad \frac{2}{11} = 0.1818 \ldots, \quad \frac{1}{6} = 0.1666 \ldots,$$

where the three dots mean a digit or a group of digits repeats without end. These examples show that decimal representations of rational numbers are of two types:

(1) *terminating* or *finite decimals,* such as $1/2 = 0.5$, $3/4 = 0.75$, and $1/8 = 0.125$;

(2) *repeating* or *periodic decimals,* such as $1/3 = 0.333 \ldots$, $2/11 = 0.1818 \ldots$, and $1/6 = 0.1666 \ldots$

Repeating decimals are called *nonterminating* or *infinite* decimals. The digit or group of digits that repeats indefinitely is called the *period* of the repeating decimal. Repeating decimals can be written as follows: $1/3 = 0.\overline{3}$, $2/11 = 0.\overline{18}$, and $1/6 = 0.1\overline{6}$, where the bar indicates the period.

Every rational number can be represented by a terminating or a repeating decimal number. If the only factors of the denominator of a rational number are 2s and 5s, then its decimal representation is a terminating decimal. Indeed, we can always multiply the numerator and the denominator by 2s and 5s so that the denominator becomes a power of 10. For example,

$$\frac{1}{2} = \frac{1 \cdot 5}{2 \cdot 5} = \frac{5}{10} = 0.5,$$

$$\frac{21}{20} = \frac{21}{2 \cdot 2 \cdot 5} = \frac{21 \cdot 5}{2 \cdot 2 \cdot 5 \cdot 5} = \frac{105}{100} = 1.05,$$

$$\frac{1}{8} = \frac{1}{2 \cdot 2 \cdot 2} = \frac{5 \cdot 5 \cdot 5}{2 \cdot 2 \cdot 2 \cdot 5 \cdot 5 \cdot 5} = \frac{125}{1000} = 0.125.$$

However, if the denominator of a rational number reduced to its simplest form contains at least one prime factor other than 2 or 5, then the decimal representation is a repeating decimal. For example, $4/9 = 0.\overline{4}$, $7/6 = 1.1\overline{6}$, and $50/11 = 4.\overline{54}$.

Conversely, *every terminating or repeating decimal represents a rational number,* as shown in the examples that follow.

EXAMPLE 4   Write 0.375 as a rational number.

Solution: We have

$$0.375 = \frac{3}{10} + \frac{7}{100} + \frac{5}{1000}$$

$$= \frac{300}{1000} + \frac{70}{1000} + \frac{5}{1000} = \frac{375}{1000} = \frac{3}{8}.$$

Practice Exercise 4   Write 0.675 as a fraction.

Answer: $\dfrac{27}{40}$

EXAMPLE 5   Transform $0.\overline{3}$ into a fraction.

Solution: Set $r = 0.333\ldots$. If we multiply both sides of this equality by 10, the equality is still true (the second property of equality, as stated previously), and we obtain

$$10r = 3.333\ldots$$

The reason we have multiplied by 10 is because the period of the given decimal number contains only one digit. As a consequence, the decimal parts of r and $10r$ are the same. Now, if we add $-r = -0.333\ldots$ to both sides of the last equality, we still have equality (the first property of equality, as stated previously) and we get

$$
\begin{array}{rl}
10r = & 3.333\ldots \\
\underline{-r = } & \underline{-0.333\ldots} \\
9r = & 3
\end{array}
$$

Finally, multiplying both sides by 1/9, we obtain

$$r = \frac{3}{9} = \frac{1}{3}.$$

Practice Exercise 5   What fraction has decimal representation $0.666\ldots$?

Answer: $\dfrac{2}{3}$

EXAMPLE 6   Find a fraction whose decimal representation is $0.1\overline{36}$.

Solution: Set $r = 0.13636\ldots$ and multiply both sides by 10 and 1000 to get

$$10r = 1.36\ldots \quad \text{and} \quad 1000r = 136.36\ldots.$$

The reason we multiply both sides of the original equality by 10 and also 1000 is to obtain two numbers with the same decimal part. Subtract $10r$ from $1000r$ to cancel out the decimal part and get

$$1000r = 136.36\ldots$$
$$\underline{-10r = -1.36\ldots}$$
$$990r = 135$$
$$r = \frac{135}{990} = \frac{3}{22}.$$

Practice Exercise 6   Write $0.31818\ldots$ as a fraction.

Answer: $\dfrac{7}{22}$

Irrational Numbers

We have seen that every rational number has a terminating or a repeating decimal representation and vice versa. Thus, if we can produce a decimal number that is *nonterminating* and *nonrepeating,* such a number cannot represent a rational number. The infinite decimal number

$$0.101001000100001\ldots$$

is nonterminating and nonrepeating. Notice that it contains only the digits 0 and 1 distributed according to the following pattern: one zero follows the first digit 1, two zeros follow the second digit 1, three zeros follow the third digit 1, and so on. By using different digits or creating new patterns of distribution of digits, it is possible to write many examples of nonterminating and nonrepeating decimals, such as

$$0.010110111011110\ldots,$$
$$0.120120012000120000012\ldots,$$
$$0.10110111011110\ldots.$$

None of these decimal numbers can be a rational number, because they are all nonterminating and nonrepeating decimals. Such decimals are called *irrational numbers*.

In order to give examples of irrational numbers, it is not necessary to rely on the decimal notation as we did above. Irrational numbers have been known for more than 2,000 years, while the decimal notation, as it is used today, was introduced in the 16th century by the Dutch mathematician Simon Stevin (1548–1620).

To describe other examples of irrational numbers, we recall the Pythagorean theorem. Among the many mathematical discoveries of the early Greek mathematicians, the Pythagorean theorem stands out as one of the simplest and most beautiful results in geometry. The theorem, leading to the discovery of irrational

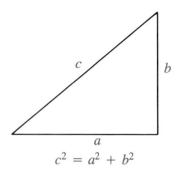

$$c^2 = a^2 + b^2$$

Figure 1.1

---

numbers, was destined to shatter the Greek concept of numbers based entirely on natural and rational numbers.

The Pythagorean Theorem

In a right triangle, the square of the length c of the hypotenuse is equal to the sum of the squares of the lengths a and b of the other two sides ($c^2 = a^2 + b^2$).

If $a = b = 1$, then by the Pythagorean theorem we have

$$c^2 = 1^2 + 1^2$$

or

$$c^2 = 2.$$

HISTORICAL NOTE

THE NUMBER π

The Babylonians used the fraction 25/8 as an approximation for π.

That is, the length c is a number that multiplied by itself is equal to 2. The Greek mathematicians, whose number system contained only natural and rational numbers, believed that c was a rational number and tried, without success, to find such a number. Later, they realized that *there is no rational number whose square can be equal to 2*. This led to the invention of $\sqrt{2}$, an irrational number.

Other examples of irrational numbers are $\sqrt{3}$, $\sqrt{5}$, $\sqrt{7}$, $\sqrt{10}$, $\sqrt{11}$, $\sqrt{12}$, the number π—which gives the ratio between the length of a circle and its diameter—and the number e, which is used in the definition of *natural logarithms*.

To decide whether a given number is irrational may not be an easy task. Sometimes, quite involved proofs are required, as in the cases of the numbers π and e. Such proofs are beyond the scope of this book and will not be discussed here.

Every irrational number can be represented as a decimal number. Some common examples are, $\sqrt{2} = 1.41421\ 35624 \ldots$, $\sqrt{3} = 1.73205\ 08076 \ldots$, $\pi = 3.14159\ 26535 \ldots$, and $e = 2.71828\ 18284 \ldots$. The decimal form of an irrational number is always nonterminating and nonrepeating.

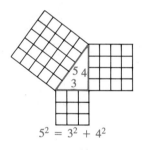

$$5^2 = 3^2 + 4^2$$

Figure 1.2

EXERCISES 1.1

Find the prime factorization of each of the following integers.

1. 240

2. 480

3. 432

4. 567

5. 2310

6. 4725

Calculate each of the following quantities and simplify your answer.

7. $\dfrac{25}{120} + \dfrac{42}{80}$

8. $\dfrac{18}{45} - \dfrac{7}{120}$

9. $\dfrac{21}{60} \times \dfrac{24}{84}$

10. $\dfrac{7}{15} \div \dfrac{35}{45}$

Find the decimal representation of the given fractions.

11. $\dfrac{3}{7}$

12. $\dfrac{5}{8}$

13. $\dfrac{21}{40}$

14. $\dfrac{4}{15}$

15. $\dfrac{3}{11}$

16. $\dfrac{12}{25}$

17. $\dfrac{15}{120}$

18. $\dfrac{18}{240}$

Write the following decimal numbers as fractions.

19. 0.15

20. 1.25

21. $0.\overline{123}$

22. $0.\overline{321}$

23. $0.4\overline{18}$

24. $0.05\overline{3}$

25. $1.1\overline{06}$

26. $2.0\overline{5}$

Write the following sums as repeating decimals.

27. $0.\overline{4} + 0.\overline{15}$

28. $0.\overline{3} + 0.\overline{21}$

29. $0.\overline{12} + 0.\overline{203}$

30. $0.\overline{25} + 0.\overline{123}$

1.2   REAL NUMBERS

A *real number* is either a rational or an irrational number. According to Section 1.1, every real number has a decimal representation. If the real number is rational, then its decimal representation is terminating or repeating; if the real number is irrational, then its decimal representation is nonterminating and nonrepeating.

The set of all real numbers, denoted by $\mathbb{R}$, contains as subsets the natural numbers, the integers, the rational and irrational numbers. We have the following inclusions:

$$\mathbb{N} \subset \mathbb{Z} \subset \mathbb{Q} \subset \mathbb{R}.$$

The real number system and its properties form the basis of calculus. Throughout this book, unless otherwise stated, the word *number* will always signify *real number*.

The operations of addition and multiplication defined for rational numbers extend to real numbers. Also, the equality between real numbers satisfies the same properties that hold for the equality between rational numbers (see page 6). The list of six properties of $\mathbb{R}$ on page 11 provides, in principle, everything we need to know about the arithmetic in $\mathbb{R}$.

From properties I through VI we can derive the rules for the arithmetic of real numbers, as well as the rules used to simplify algebraic expressions whose terms represent numbers.

Difference

The unique number y, in property V, such that $x + y = 0$ is called the *additive inverse* or *opposite* of x and is denoted by $y = -x$. The existence of the additive inverse allows us to define the *difference* of any two real numbers. If x and y are numbers, their difference is defined by

$$x - y = x + (-y). \tag{1.6}$$

Properties of Real Numbers

Let $\mathbb{R}$ denote the set of all real numbers and let x, y, and z be arbitrary elements of $\mathbb{R}$.

I.   *Closure.*
     The set $\mathbb{R}$ is closed under the addition and multiplication operations. That is, $x + y$ and xy are both real numbers.

II. *Commutative properties.*
$$x + y = y + x \quad \text{and} \quad xy = yx.$$

III. *Associative properties.*
$$(x + y) + z = x + (y + z) \quad \text{and} \quad x(yz) = (xy)z.$$

IV. *Existence of identities.*
     The number 0, called the *additive identity,* is the unique number such that $x + 0 = x$, for every $x \in \mathbb{R}$.
     The number 1, called the *multiplicative identity,* is the unique number such that $x \cdot 1 = x$, for every $x \in \mathbb{R}$.

V.   *Existence of additive and multiplicative inverses.*
     For every $x \in \mathbb{R}$, there is a unique $y \in \mathbb{R}$ such that $x + y = 0$.
     For every $x \in \mathbb{R}$, with $x \neq 0$, there is a unique $y \in \mathbb{R}$ such that $xy = 1$.

VI. *Distributive property of the product over the sum.*
$$x(y + z) = xy + xz.$$

Quotient

When $x \neq 0$, the unique number y, in property V, such that $xy = 1$ is called the *multiplicative inverse* or *reciprocal* of x. The multiplicative inverse of x is denoted by $\frac{1}{x}$, or $1/x$, or x^{-1}, and satisfies the relations

$$xx^{-1} = x^{-1}x = 1.$$

The division operation is now defined as multiplication by a reciprocal. If x and y are numbers with $y \neq 0$, then the *quotient* $\frac{x}{y}$ (also denoted by x/y or $x \div y$) is defined by

$$\frac{x}{y} = x\left(\frac{1}{y}\right) = x \cdot y^{-1}. \tag{1.7}$$

The quotient x/y is also called a *fraction.* By using properties I through VI, you can check that no matter what the real numbers x, y, z, and w are (as long as y and w are nonzero), the fractions formed from them will obey the following rules.

These rules are easy to remember; they are analogous to the rules of arithmetic of the rational numbers given on page 4.

Operations on Fractions

$$\frac{x}{y} + \frac{z}{w} = \frac{xw}{yw} + \frac{yz}{yw} = \frac{xw + yz}{yw}$$

$$\frac{x}{y} - \frac{z}{w} = \frac{xw}{yw} - \frac{yz}{yw} = \frac{xw - yz}{yw}$$

$$\frac{x}{y} \cdot \frac{z}{w} = \frac{xz}{yw}$$

$$\frac{x}{y} \div \frac{z}{w} = \frac{x}{y} \cdot \frac{w}{z} = \frac{xw}{yz} \quad \text{provided that } z \neq 0$$

The number zero (additive identity) has two very important properties, called *multiplicative properties of zero,* that can be derived from properties I through VI.

Multiplicative Properties of Zero

$$x \cdot 0 = 0 \cdot x = 0$$

If $xy = 0$, then either $x = 0$ or $y = 0$.

EXAMPLE 1

What properties are being illustrated in each of the following examples?

(a) $\dfrac{3}{5} + \left(15 + \dfrac{4}{9}\right) = \left(\dfrac{3}{5} + 15\right) + \dfrac{4}{9}$

(b) $a(7 + 8) = a \cdot 7 + a \cdot 8 = 7a + 8a$

(c) $a(3 - \pi) = a(3 + (-\pi)) = a \cdot 3 + a \cdot (-\pi)$

Solution: **(a)** Associativity of the sum
(b) Distributivity of the product over the sum and commutativity of the product
(c) The definition of difference and the distributive property

Practice Exercise 1

What properties are being illustrated in each of the following?
(a) $(a - 1.36) + 5.12 = 5.12 + (a - 1.36)$
(b) $\left(\sqrt{2} + \sqrt{3}\right) + 0 = 0 + \left(\sqrt{2} + \sqrt{3}\right) = \sqrt{2} + \sqrt{3}$
(c) $1(r + 2) = r + 2$

Answer: **(a)** Commutativity of the sum   **(b)** Commutativity of the sum; the fact that 0 is the additive identity   **(c)** The fact that 1 is the multiplicative identity

EXAMPLE 2

Compute the following expression.

$$\frac{\dfrac{2}{3} - \dfrac{3}{4} + \dfrac{5}{12}}{\dfrac{5}{9} - \dfrac{1}{4}}$$

Solution: We have

$$\frac{\dfrac{2}{3} - \dfrac{3}{4} + \dfrac{5}{12}}{\dfrac{5}{9} - \dfrac{1}{4}} = \frac{\dfrac{8}{12} - \dfrac{9}{12} + \dfrac{5}{12}}{\dfrac{20}{36} - \dfrac{9}{36}} = \frac{\dfrac{4}{12}}{\dfrac{11}{36}} = \frac{4}{12} \times \frac{36}{11} = \frac{12}{11}.$$

Practice Exercise 2

Compute and give your answer in simplified form.

$$\frac{\dfrac{4}{5} + \dfrac{7}{10} - \dfrac{3}{4}}{\dfrac{2}{5} + \dfrac{1}{2}}$$

Answer: $\dfrac{5}{6}$

The Coordinate Line

Real numbers can be put into a *one-to-one correspondence* with the points of a line. That is, each real number corresponds to a unique point on a line, and each point on a line corresponds to a unique real number.

On a line (Figure 1.3), fix an arbitrary point O, the *origin,* and a *unit segment* OU. The point O corresponds to the number 0, and the point U corresponds to the number 1. By repeating the unit segment, moving from left to right, obtain the points corresponding to the positive integers 1, 2, 3, ... Starting at the origin and repeating the unit segment, moving now from right to left, obtain the points corresponding to the negative integers -1, -2, -3, ...

To find the point on the line corresponding to a fraction m/n, with n a natural number, divide the unit segment OU in n equal parts. Starting at the origin, repeat the first subdivision m times to the right if m is a positive integer, or m times to the left of the origin if m is a negative integer. Figure 1.3 shows the location of 5/4, 5/2, and $-3/2$.

Thus, each rational number is the *coordinate* of a unique point on the line. However, not every point on the line has a rational coordinate. There are points that do not correspond to rational numbers. Such points correspond to irrational numbers.

Points on the line associated with irrational numbers can, in certain cases, be located by geometric construction with a ruler and a compass. In Figure 1.4, the triangle OUA is a right isosceles triangle whose sides OU and UA both measure one unit of length. According to the Pythagorean theorem, the length of the hypotenuse OA is equal to $\sqrt{2}$ units. If we draw an arc of a circle with center at the origin and radius OA, the arc intersects the line at a point P that has coordinate $\sqrt{2}$.

A similar procedure can be used to locate the points corresponding to irrational numbers such as $\sqrt{3}$, $\sqrt{5}$, and $\sqrt{7}$ (see Figure 1.5). However, the points associated with π and e cannot be located with a ruler and a compass.

In establishing the one-to-one correspondence between real numbers and points of a line, we can say that every point P to the right of the origin corresponds

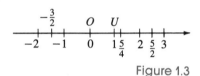

Figure 1.3

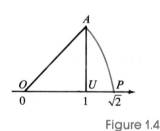

Figure 1.4

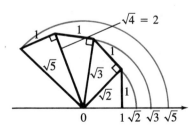

Figure 1.5

Locating $\sqrt{2}$, $\sqrt{3}$, and $\sqrt{5}$ with the help of a ruler and a compass.

to a unique *positive real number r*. Points to the left of the origin correspond to *negative real numbers*. The number *r* is called the *coordinate* or *abscissa* of the point *P*. A *coordinate line* or *coordinate axis* (or simply *axis*) is a line on which a coordinate system has been defined.

EXERCISES 1.2

Indicate what properties of the sum and product are being illustrated in each of Exercises 1–8.

1. $\left(a + \dfrac{3}{8}\right) + b = a + \left(\dfrac{3}{8} + b\right)$

2. $r + \dfrac{3}{11} = \dfrac{3}{11} + r$

3. $\dfrac{1}{5} \cdot (-x) = (-x) \cdot \dfrac{1}{5}$

4. $\left(\dfrac{7}{8} - a\right) \cdot \dfrac{9}{15} = \dfrac{9}{15} \cdot \left(\dfrac{7}{8} - a\right)$

5. $\dfrac{3}{4} \cdot \left(\dfrac{1}{2} + \dfrac{10}{11}\right) = \dfrac{3}{4} \cdot \dfrac{1}{2} + \dfrac{3}{4} \cdot \dfrac{10}{11}$

6. $\dfrac{8}{12} \cdot \left(\dfrac{3}{5} + 0\right) = \dfrac{8}{12} \cdot \dfrac{3}{5}$

7. $(a + b) \cdot \dfrac{3}{10} = \dfrac{3}{10}\,a + \dfrac{3}{10}\,b$

8. $x \cdot \left(\dfrac{2}{3} - y\right) = \dfrac{2}{3}\,x - xy$

In Exercises 9–14, compute the given expressions in two ways: (a) using the distributive property; and (b) first performing the operations inside the parentheses. Give your answer in simplified form.

9. $\dfrac{3}{10}\left(\dfrac{2}{3} + \dfrac{1}{4}\right)$

10. $\dfrac{4}{5}\left(\dfrac{3}{7} + \dfrac{2}{9}\right)$

11. $\dfrac{5}{12}\left(\dfrac{-4}{5} + \dfrac{8}{12}\right)$

12. $\dfrac{6}{7}\left(\dfrac{1}{2} - \dfrac{8}{11}\right)$

13. $\left(-\dfrac{4}{9}\right)\left(\dfrac{1}{3} - \dfrac{2}{5}\right)$

14. $(-5)\left(\dfrac{4}{7} + \dfrac{5}{9}\right)$

In Exercises 15–28, compute and give your answers in simplified form.

15. $\dfrac{2 + \dfrac{1}{2}}{2 - \dfrac{1}{2}}$

16. $\dfrac{5 - \dfrac{1}{3}}{2 + \dfrac{4}{5}}$

17. $\dfrac{\dfrac{2}{3} + \dfrac{1}{2}}{\dfrac{7}{9} - \dfrac{2}{3}}$

18. $\dfrac{\dfrac{4}{5} - \dfrac{2}{3}}{\dfrac{3}{4} + \dfrac{2}{5}}$

19. $\dfrac{\dfrac{1}{3} - \dfrac{3}{4}}{\dfrac{1}{2} + \dfrac{2}{3}}$

20. $\dfrac{\dfrac{3}{5} - \dfrac{1}{4}}{\dfrac{3}{8} + \dfrac{2}{3}}$

21. $\dfrac{\dfrac{2}{3} + \dfrac{3}{6} - \dfrac{3}{12}}{\dfrac{7}{8} - \dfrac{4}{3}}$

22. $\dfrac{\dfrac{3}{4} - \dfrac{4}{10} + \dfrac{2}{5}}{\dfrac{1}{3} - \dfrac{4}{5}}$

23. $\dfrac{\dfrac{4}{9} - \dfrac{1}{6} + \dfrac{5}{18}}{\dfrac{2}{3} + \dfrac{5}{6}}$

24. $\dfrac{\dfrac{3}{8} + \dfrac{5}{6} - \dfrac{1}{4}}{\dfrac{3}{4} + \dfrac{5}{9}}$

25. $\dfrac{\dfrac{3}{5} \times \left(\dfrac{2}{3} - \dfrac{3}{4}\right)}{\dfrac{4}{6} + \dfrac{2}{3} - \dfrac{2}{5}}$

26. $\dfrac{\dfrac{3}{8} \times \left(\dfrac{3}{5} - \dfrac{2}{3}\right)}{\dfrac{5}{6} - \dfrac{1}{2} + \dfrac{3}{4}}$

27. $\dfrac{\dfrac{3}{5} - \dfrac{3}{4} \times \dfrac{6}{8}}{\dfrac{1}{3} \times \dfrac{3}{7} + \dfrac{5}{4}}$

28. $\dfrac{\left(\dfrac{5}{6} - \dfrac{1}{4}\right)\left(\dfrac{2}{3} + \dfrac{1}{8}\right)}{\dfrac{3}{4} \times \dfrac{2}{5} - \dfrac{5}{6}}$

☐ **29.** The sum of two rational numbers is necessarily a rational number. Is the sum of two irrational numbers necessarily an irrational number?

☐ **30.** Is the product of two irrational numbers necessarily an irrational number?

☐ **31.** Show that $\sqrt{2} + \dfrac{5}{6}$ is irrational. (*Hint:* If $x = \sqrt{2} + \dfrac{5}{6}$, then $x - \dfrac{5}{6} = \sqrt{2}$. Can x be rational?)

☐ **32.** Show that $\dfrac{4}{5}\sqrt{2}$ is irrational.

☐ **33.** Show that if a is a rational number and b is an irrational number, then $a + b$ is irrational.

34. Show that if $a \neq 0$ is a rational number and b is an irrational number, then ab is irrational.

35. Which of the following are rational numbers? Irrational numbers?

    (a) $(2\sqrt{3})\sqrt{3}$ (b) $4 + \sqrt{2}$

    (c) $(2 + 3\sqrt{2}) + (2 - 3\sqrt{2})$

    (d) $0.23023002300023\ldots$

36. Which of the following are rational numbers? Irrational numbers?

    (a) $\sqrt{12/27}$ (b) $\dfrac{3}{4}\sqrt{2} + 1$ (c) $(0.5)\sqrt{5}$

    (d) $0.\overline{23}$

37. The fraction 22/7 was used by Archimedes as an approximation of π. Is 22/7 greater than or less than π?

38. The fractions 41/29 and 99/70 are approximations of $\sqrt{2}$. Are they greater than or less than $\sqrt{2}$?

39. Show that if $\dfrac{a}{b} = \dfrac{c}{d}$, then $\dfrac{a + b}{b} = \dfrac{c + d}{d}$.

40. Show that if $\dfrac{a}{b} = \dfrac{c}{d}$, then $\dfrac{a + b}{a - b} = \dfrac{c + d}{c - d}$, provided $a \neq b$ and $c \neq d$.

1.3   INTEGER EXPONENTS

The multiplication operation leads to the following definition of *powers of a real number*.

Positive Exponents

If a is a real number and n is a natural number, then we define

$$a^n = \underbrace{a \cdot a \cdots a}_{n \text{ factors}}. \qquad (1.8)$$

For example,

$$2^5 = 2 \cdot 2 \cdot 2 \cdot 2 \cdot 2 = 32,$$

$$\left(-\frac{1}{5}\right)^2 = \left(-\frac{1}{5}\right)\left(-\frac{1}{5}\right) = \frac{1}{25},$$

$$\left(\frac{3}{4}\right)^3 = \frac{3}{4} \cdot \frac{3}{4} \cdot \frac{3}{4} = \frac{27}{64},$$

$$(-1)^5 = (-1)(-1)(-1)(-1)(-1) = -1.$$

In the expression a^n, called a *power* or an *exponential*, a is the *base* and n is the *exponent*. We read it "nth power of a," "a to the nth power," or simply, "a to the n."

Powers of a real number have a few simple properties. First, consider multiplication of exponentials with the same base. The product of 3^2 by 3^4 is

$$3^2 \cdot 3^4 = \underbrace{(3 \cdot 3)}_{\substack{2 \\ \text{factors}}}\underbrace{(3 \cdot 3 \cdot 3 \cdot 3)}_{\substack{4 \\ \text{factors}}}$$

$$= \underbrace{3 \cdot 3 \cdot 3 \cdot 3 \cdot 3 \cdot 3}_{6 \text{ factors}} = 3^6$$

This suggests that to multiply powers of 3, we should keep the same base and *add* the exponents. The same is true for any other base such as $4/5$, 8, π, or e. In general, the product of a^n by a^m contains a total of $n + m$ factors equal to a; thus

$$a^n \cdot a^m = a^{n+m}.$$

For example, $x^6 \cdot x^{10} = x^{16}$. The last property can be generalized further:

$$a^n \cdot a^m \cdot a^p = a^{n+m+p}.$$

For example, $y^3 \cdot y^8 \cdot y^7 = y^{18}$.

    Sometimes we have to take a power of an exponential. For example, suppose that we want to raise 2^3 to the fourth power. We have

$$(2^3)^4 = 2^3 \cdot 2^3 \cdot 2^3 \cdot 2^3$$
$$= 2^{3+3+3+3}$$
$$= 2^{3 \cdot 4} = 2^{12}.$$

This indicates that to raise a power to a power, we should keep the same base and *multiply* the exponents. Symbolically:

$$(a^n)^m = a^{nm}.$$

For example, $(x^5)^3 = x^{15}$. This property can also be generalized:

$$((a^n)^m)^p = a^{nmp}.$$

For example, $((y^2)^4)^3 = y^{24}$.

    Next, consider division of powers. The quotients $3^4/3^4$, $5^8/5^6$, and $7^2/7^5$ can be simplified as follows:

$$\frac{3^4}{3^4} = 1,$$

$$\frac{5^8}{5^6} = \frac{5^6 \cdot 5^2}{5^6} = 5^2,$$

and

$$\frac{7^2}{7^5} = \frac{7^2}{7^2 \cdot 7^3} = \frac{1}{7^3}.$$

HISTORICAL NOTE

POWERS OF A NUMBER

The notation x^n, denoting an integer power of a variable, originated with René Descartes (1596–1650), a French mathematician. Prior to its introduction, notations used for a variable and powers of a variable were somewhat clumsy. For example, Francois Viète (1540–1603), a French lawyer and mathematician, introduced letters to denote constants and variables. He would have written the powers of a number A as follows: Aq (square of A), Ac (cube of A), Aqq (fourth power of A), and so on.

These examples illustrate the following property of exponents.

If $a \neq 0$, then

$$\frac{a^n}{a^m} = \begin{cases} a^{n-m} & \text{if } n > m \\ 1 & \text{if } n = m \\ \dfrac{1}{a^{m-n}} & \text{if } n < m. \end{cases}$$

In many instances, it is necessary to take powers of products or quotients, such as

$$(2a)^4 = (2a)(2a)(2a)(2a)$$
$$= (2 \cdot 2 \cdot 2 \cdot 2)(a \cdot a \cdot a \cdot a)$$
$$= 16a^4,$$

and

$$\left(\frac{x}{4}\right)^3 = \frac{x}{4} \cdot \frac{x}{4} \cdot \frac{x}{4} = \frac{x^3}{64}.$$

These two examples illustrate the following properties of exponents:

$$(ab)^n = a^n b^n$$

and

$$\left(\frac{a}{b}\right)^n = \frac{a^n}{b^n} \quad \text{if} \quad b \neq 0.$$

Up to now we have considered only exponents that are natural numbers (i.e., positive integers). It is necessary to extend the definition to exponents that are negative integers or 0.

Zero or Negative Exponents

If a is a real number and $a \neq 0$ then by definition,

$$a^0 = 1 \tag{1.9}$$

and

$$a^{-n} = \frac{1}{a^n}, \quad \text{for all natural numbers } n. \tag{1.10}$$

Note that the restriction $a \neq 0$ is essential. Otherwise, the right-hand side of (1.10) is not defined. Also, the symbol 0^0 is not defined.

Definitions (1.9) and (1.10) can be justified as follows. If n and m are natural numbers with $m > n$, we have

$$\frac{a^m}{a^n} = a^{m-n}. \tag{1.11}$$

If we set $n = m$, then the left-hand side of (1.11) becomes 1 while the right-hand side equals a^0. Thus, $a^0 = 1$. On the other hand, if we set $m = 0$ in (1.11) we get

$$\frac{1}{a^n} = \frac{a^0}{a^n} = a^{0-n} = a^{-n},$$

which is formula (1.10).

Some numerical examples are:

$$7^0 = 1, \quad \left(-\frac{2}{3}\right)^0 = 1, \quad (12.5)^0 = 1,$$

$$3^{-4} = \frac{1}{3^4} = \frac{1}{81},$$

$$\left(\frac{3}{5}\right)^{-2} = \frac{1}{\left(\frac{3}{5}\right)^2} = \frac{1}{\frac{9}{25}} = \frac{25}{9},$$

and

$$\left(-\frac{1}{2}\right)^{-3} = \frac{1}{\left(-\frac{1}{2}\right)^3} = \frac{1}{-\frac{1}{8}} = -8.$$

Properties of Exponents

If a and b are numbers and if n and m are integers, then the following are true.

1. $a^n \cdot a^m = a^{n+m}$

2. $(a^n)^m = a^{nm}$

3. $\dfrac{a^n}{a^m} = a^{n-m}, \quad a \neq 0$

4. $(ab)^n = a^n b^n$

5. $\left(\dfrac{a}{b}\right)^n = \dfrac{a^n}{b^n}, \quad b \neq 0$

6. $a^0 = 1, \quad a \neq 0$

7. $a^{-n} = \dfrac{1}{a^n}, \quad a \neq 0$

The reason for definitions (1.9) and (1.10) is that the properties we have described for powers in which the exponents are positive integers are true when the exponents are any integers. For future reference we have listed the properties of integer exponents on page 18.

EXAMPLE 1   Use the properties of exponents to simplify each expression.

(a) $(2ax^2)(5ay^4)$

(b) $\left(\dfrac{3a^2}{5bc^3}\right)^2\left(\dfrac{5ab^2}{c}\right)$

Solution: We have

(a) $(2ax^2)(5ay^4) = 2 \cdot 5 \cdot a \cdot a \cdot x^2 \cdot y^4 = 10a^2x^2y^4,$

(b) $\left(\dfrac{3a^2}{5bc^3}\right)^2 \cdot \left(\dfrac{5ab^2}{c}\right) = \dfrac{9a^4}{25b^2c^6} \cdot \dfrac{5ab^2}{c} = \dfrac{45a^5b^2}{25b^2c^7} = \dfrac{9a^5}{5c^7}.$

Practice Exercise 1   Simplify each of the following expressions.

(a) $(3ax^2y^3)^2$

(b) $\left(\dfrac{3x^2}{yz}\right)^4$

Answer: (a) $9a^2x^4y^6$ (b) $\dfrac{81x^8}{y^4z^4}$

EXAMPLE 2   Write each of the following expressions with positive exponents only and simplify.

(a) $(4a^{-2}x^3)^{-2}$

(b) $\dfrac{6a^2b^{-3}}{2a^{-3}b}$

Solution: (a) According to the definition (1.10) and properties of exponents, we get

$$(4a^{-2}x^3)^{-2} = \frac{1}{(4a^{-2}x^3)^2} = \frac{1}{4^2(a^{-2})^2(x^3)^2}$$

$$= \frac{1}{16a^{-4}x^6} = \frac{1}{16x^6} \cdot \frac{1}{a^{-4}} = \frac{a^4}{16x^6}.$$

(b) Similarly,

$$\frac{6a^2b^{-3}}{2a^{-3}b} = \frac{6}{2} \cdot \frac{a^2}{a^{-3}} \cdot \frac{b^{-3}}{b} = 3 \cdot a^2 \cdot a^3 \cdot \frac{1}{b \cdot b^3}$$

$$= \frac{3a^5}{b^4}.$$

Practice Exercise 2   Write with positive exponents only and simplify.

(a) $(5b^2y^{-4})^3$ (b) $\dfrac{12u^3v^4}{4u^{-1}v^2}$

Answer: (a) $\dfrac{125b^6}{y^{12}}$ (b) $3u^4v^2$

Scientific Notation

In the applied sciences we often work with extremely large or small numbers. For example, one light-year (the distance traveled by a ray of light in one year) is equal to

$$9{,}460{,}000{,}000{,}000 \text{ kilometers;}$$

the mass of a proton is equal to

$$0.\,000\ 000\ 000\ 000\ 000\ 000\ 000\ 000\ 001\ 673 \text{ kilograms.}$$

In order to simplify notation and calculations, such numbers are usually expressed in scientific notation.

A number is said to be written in *scientific notation* when it is written in the form

$$a \times 10^n,$$

where a is a decimal number such that $1 \le a < 10$ and n is an integer.

For example, in scientific notation one light-year is written as 9.46×10^{12} kilometers and the mass of a proton as 1.673×10^{-27} kilograms.

EXAMPLE **3**   Write the following numbers in scientific notation.
(a) 25×10^{-7} (b) 0.112×10^6

Solution:
(a) $25 \times 10^{-7} = (2.5 \times 10) \times 10^{-7} = 2.5 \times (10^1 \times 10^{-7}) = 2.5 \times 10^{-6}$
(b) $0.112 \times 10^6 = (1.12 \times 10^{-1}) \times 10^6 = 1.12 \times (10^{-1} \times 10^6)$
         $= 1.12 \times 10^5$

Practice Exercise 3   Rewrite in scientific notation.

(a) 0.034×10^8 (b) 416.3×10^{-10}

Answer: (a) 3.4×10^6 (b) 4.163×10^{-8}

As the following example shows, some calculations are greatly simplified by using scientific notation and the properties of exponents.

EXAMPLE 4   Evaluate the expression

$$\frac{(1.21 \times 10^5)(8.1 \times 10^{-8})}{(3.3 \times 10^7)^2}$$

and give your answer in scientific notation.

Solution: Using the associativity of the product and the properties of exponents, we write

$$\frac{(1.21 \times 10^5)(8.1 \times 10^{-8})}{(3.3 \times 10^7)^2} = \frac{(1.21)(8.1)(10^5 \times 10^{-8})}{(3.3)^2(10^7)^2}$$

$$= \frac{(1.21)(8.1)(10^5 \times 10^{-8})}{(10.89)(10^{14})}$$

$$= \frac{(1.21)(8.1)}{10.89} \times 10^{5-8-14}$$

$$= \frac{9.801}{10.89} \times 10^{-17} = 0.9 \times 10^{-17}$$

$$= 9.0 \times 10^{-18}$$

Practice Exercise 4   Perform the indicated operations and give your answer in scientific notation.

$$\frac{(9.12 \times 10^{-7})(2.3 \times 10^8)}{(1.15 \times 10^{-12})(2.4 \times 10^4)}$$

Answer: 7.6×10^9

Significant Digits

The result of a measurement is a number. Since measurements are frequently inaccurate, numbers obtained by measurement are only approximations to some exact value that we really cannot determine.

There is a standard language that is used to describe the accuracy of measurements. Let us give an example using the meter as unit of length. If we say that 12 meters is the measurement of a pole *to the nearest meter*, this really means that the pole has length between 11.5 meters and 12.5 meters. That is, the *exact value* of the length is no less than 11.5 meters and no more than 12.5 meters. If the measurement is done more accurately and we find that 12.4 meters is the measurement of the pole *to the nearest tenth of a meter*, then the length of the pole is really between 12.35 meters and 12.45 meters. Also, if we say that the pole measures 12.00 meters, then we mean that the pole measures 12 meters *to the nearest hundredth of a meter* and, in this case, the measurement of the pole is between 11.995 meters and 12.005 meters.

The first measurement of 12 meters is said to have two significant digits of accuracy; the second of 12.4 meters has three significant digits of accuracy; and the third of 12.00 meters has four significant digits of accuracy.

The number of *significant digits* of accuracy in a measurement is the number of digits from the leftmost nonzero digit to the rightmost digit zero or nonzero.

The following table shows some numbers, how many significant digits there are in each number, and the measurement range of each number.

Number	Number of significant digits	Range of measurement		
0.2016	4	0.20155	to	0.20165
1.36	3	1.355	to	1.365
0.04	1	0.035	to	0.045
0.0012	2	0.00115	to	0.00125

Final zeros of a number may or may not be significant digits. For example, suppose that a road measures 12,000 meters, to the nearest meter; then all the zeros are significant digits. If the road measures 12,000 meters to the nearest ten meters, then the first two zeros are significant digits, but the last one is not. Finally, if the road measures 12,000 meters to the nearest thousand meters, then none of the three zeros are significant digits. To avoid ambiguity, we write the three measurements in scientific notation, as follows: 1.2000×10^4 meters represents a measurement to the nearest meter (five significant digits); 1.200×10^4 meters is the measurement to the nearest ten meters (four significant digits); and 1.2×10^4 meters is the measurement to the nearest thousand meters (two significant digits).

In future calculations we shall use the following rules concerning significant digits:

(1) When adding or subtracting numbers obtained by measurements, round off the answer so that it will show no more decimal places than the measurement with fewest decimal places.

(2) When multiplying or dividing numbers representing measurements, round off the answer so that it has the same number of significant digits as the number with the fewest significant digits.

(3) When computing powers or roots, round off the answer so that it has the same number of significant digits as the number whose power or root we are computing.

EXAMPLE **5**   In physics, the formula

$$F = G\frac{mm'}{r^2}$$

gives the magnitude of the force of attraction between two masses m and m' placed at a distance r from each other, where G is a physical constant. Find F, knowing that $m = 3.9 \times 10^{20}$ units of mass, $m' = 1.5 \times 10^{20}$ units of mass, and $r = 1.0 \times 10^6$ units of distance, assuming that $G = 1$. Give your answer in scientific notation.

Solution: We have

$$F = \frac{(3.9 \times 10^{20})(1.5 \times 10^{20})}{(1.0 \times 10^6)^2}$$

$$= \frac{(3.9)(1.5)(10^{20} \times 10^{20})}{10^{12}}$$

$$= (3.9)(1.5) \times 10^{28}$$

$$= 5.85 \times 10^{28}$$

$$\simeq 5.9 \times 10^{28} \text{ units of force.}$$

Practice Exercise 5   Two spherical lead balls of masses 4.50 kg and 5.00 kg are placed so that their centers are 0.500 m apart. Find the magnitude of the force of attraction F between these masses if the constant G is equal to $6.67 \times 10^{-11} \text{N} \cdot \text{m}^2/\text{kg}^2$. (Here N is the abbreviation of Newton, a unit of force.)

Answer: $F = 6.00 \times 10^{-9} \text{ N}$

Using Your Calculator

Scientific calculators allow the entry of a number in scientific notation. Calculations can then be quickly performed, and the answer is displayed in scientific notation. There are two principal standard ways (called *logic modes*) to enter data into calculators: the *algebraic* mode and the *reverse Polish notation* (RPN) mode. Algebraic logic uses parentheses and follows the rules of ordinary algebra. The RPN logic avoids parentheses and, perhaps, is a little more difficult to learn. All the sample computations shown in this book were performed on a calculator that operates in the algebraic mode. When we list a sequence of keystrokes, we will be using the algebraic mode. The same sequence of keystrokes will probably work (perhaps after a few minor changes) on most calculators using the algebraic mode. However, you should consult the owner's manual for details on how to operate your calculator.

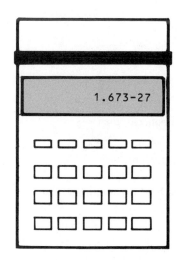

Figure 1.6

C   Here are the keystrokes you should perform in order to enter the number 1.673×10^{-27}.

$\boxed{\text{ON/C}}$ 1.673 $\boxed{\text{EE}}$ 27 $\boxed{+/-}$

Calculators have a key $\boxed{y^x}$ or $\boxed{x^y}$ to compute powers of a number. For example, $20^8 = 2.56 \times 10^{10}$. The keystrokes 20 $\boxed{y^x}$ 8 $\boxed{=}$ will give you the answer in scientific notation.

Some Fundamental Constants

Mass of an electron	9.107×10^{-31} kg
Charge of an electron	4.802×10^{-10} statcoulomb
Mass of a proton	1.673×10^{-27} kg
Mass of the sun	1.999×10^{30} kg
Avogadro's number	6.022×10^{23}
Earth-moon mean distance	3.843×10^{8} m
Earth-sun mean distance	1.496×10^{11} m

EXERCISES 1.3

In Exercises 1–16, simplify each expression and write all answers without exponents.

1. 11^4 **2.** 25^3

3. 5^{-3} **4.** 3^{-4}

5. $\left(\dfrac{3}{4}\right)^{-2}$ **6.** $\left(\dfrac{3}{4}\right)^{-3}$

7. $(-8)^3$ **8.** $(-5)^4$

9. $(-7)^{-3}$ **10.** $(-9)^{-3}$

11. $3(-4)^{-3}$ **12.** $2\left(\dfrac{1}{2}\right)^{-4}$

13. $3^{-2} + 4^{-1}$ **14.** $2^{-4} + 5^{-2}$

15. $\left(\dfrac{2}{3}\right)^{-1} + 3^{-1}$ **16.** $\left(\dfrac{3}{4}\right)^{-2} + 9^{-1}$

In Exercises 17–40, write each expression with positive exponents only.

17. $(2xy^3)^{-2}$ **18.** $(3a^4b^{-1})^3$

19. $\dfrac{x^9}{x^{21}}$ **20.** $\dfrac{u^{15}}{u^8}$

21. $x^{-2} + \dfrac{1}{x^{-2}}$ **22.** $\dfrac{1}{a^{-3}} - a^{-1}$

23. $\dfrac{(x+y)^{10}}{(x+y)^{13}}$ **24.** $\dfrac{(u-v)^{15}}{(u-v)^9}$

25. $\dfrac{3x^2y^{-4}}{z^{-5}}$ **26.** $\dfrac{3c^{-1}}{a^{-2}b}$

27. $(4x^{-8}y^{-3})x^2y^2(3y)$ **28.** $3r^2s^2(4r^{-5}s^{-7})2s$

29. $(3x^2y^{-1})^2$ **30.** $(5x^{-2}y^3)^3$

31. $\left(\dfrac{4ab^2}{c^2}\right)^{-4}$ **32.** $\left(\dfrac{5a^2x^3}{b}\right)^{-3}$

33. $\left(\dfrac{x^{-2}y^{-3}}{2}\right)^2$ **34.** $\left(\dfrac{2}{a^{-3}b^{-2}}\right)^3$

35. $\dfrac{3x^2y^{-1}}{(xy^2)^{-2}}$ **36.** $\dfrac{a^2b^{-5}}{(a^4b^{-2})^{-3}}$

37. $\dfrac{a^{-2}b^{-3}c^{-4}}{3ab}$ **38.** $\dfrac{xyz^{-2}}{x^2yz}$

39. $\left[\left(\dfrac{a^{-2}x}{y^2}\right)^2\right]^2$ **40.** $\left(\dfrac{x^{-1}y^4}{z^3}\right)^{-2}$

☐ **41.** Show that if a and b are nonzero, then $\dfrac{a^{-n}}{b^{-n}} = \dfrac{b^n}{a^n}$.

☐ **42.** Show that if a and b are nonzero, then $\left(\dfrac{a}{b}\right)^{-1} = \dfrac{b}{a}$.

In Exercises 43–52, write each of the following numbers in scientific notation.

43. $5,150,000,000$ **44.** $12,015,000,000$

45. $0.000\,000\,001\,8$ **46.** $0.000\,000\,000\,025$

47. $0.000\,000\,000\,514\,2$ **48.** $0.000\,000\,010\,342$

49. 0.213×10^7 **50.** 0.136×10^{-6}

51. 405.6×10^{-8} **52.** 512.7×10^{11}

In Exercises 53–58, the given numbers represent approximate measurements. Find the range of each measurement.

53. 5.6 ft   **54.** 7.8 m
55. 15.76 km   **56.** 105.7 mi
57. 21.315 kg   **58.** 13.27 lb

State the number of significant digits in each of the following numbers.

59. 31.6   **60.** 6.05
61. 56.012   **62.** 0.5014
63. 0.01506   **64.** 0.00015
65. 3.2×10^{-8} **66.** 1.83×10^{21}

In Exercises 67–70, perform the indicated operations and give your answers in scientific notation. Do not use a calculator.

67. $\dfrac{(3.2 \times 10^{-3})(2.5 \times 10^{-2})}{(0.8 \times 10^{4})(0.5 \times 10^{-5})}$

68. $\dfrac{(4.8 \times 10^{5})(1.3 \times 10^{-3})}{(2.6 \times 10^{-7})(0.6 \times 10^{4})}$

69. $\dfrac{(1.2 \times 10^{-8})(4.9 \times 10^{6})}{(1.5 \times 10^{-9})(7.0 \times 10^{3})}$

70. $\dfrac{(2.3 \times 10^{7})(1.5 \times 10^{-6})}{(2.5 \times 10^{-8})(4.6 \times 10^{3})}$

[c] In Exercises 71–74, perform the indicated operations and give your answers in scientific notation.

71. $\dfrac{3.1 \times 10^{-12}}{9.0 \times 10^{6}}$ **72.** $\dfrac{1.25 \times 10^{11}}{0.60 \times 10^{5}}$

73. $\dfrac{(4.0 \times 10^{12})(3.5 \times 10^{-9})}{(2.0 \times 10^{-4})^{2}}$

74. $\dfrac{(8.15 \times 10^{-6})(3.2 \times 10^{-3})}{(1.2 \times 10^{5})(2.71 \times 10^{8})}$

[c] In Exercises 75–78, convert each of the numbers to scientific notation, simplify, and give your answer in scientific notation.

75. $\dfrac{(0.000\ 000\ 012)(41000)}{0.000\ 000\ 001\ 5}$

76. $\dfrac{(560\ 000\ 000)(0.000\ 002)}{0.000\ 000\ 021}$

77. $\dfrac{(3210)(0.000\ 000\ 000\ 18)}{(321\ 000\ 000)(0.000\ 516)}$

78. $\dfrac{(1\ 200\ 000\ 000)(0.000\ 25)}{(0.000\ 000\ 000\ 004)(136\ 000)}$

79. The speed of light is 300,000 kilometers per second. How many kilometers does a light signal travel in one hour? Give your answer in scientific notation.

80. A light ray travels 186,000 miles per second. How many miles does it travel in 100 minutes?

81. The mass of the earth is 5.983×10^{24} kilograms (kg). A metric ton is 10^{6} grams (g). Express the mass of the earth in metric tons.

82. The mass of the sun is 1.999×10^{27} metric tons. Express the mass of the sun in grams.

[c] **83.** The mass of the moon is 1.228×10^{-2} times the mass of the earth. Find the mass of the moon in kilograms. (See Exercise 81 for the mass of the earth.)

[c] **84.** The mass of an electron is 9.107×10^{-31} kg. If the mass of a proton is 1.837×10^{3} times the mass of an electron, find the mass of the proton.

[c] **85.** Find the force of attraction between the earth and the moon, knowing that the mass of the earth is 5.983×10^{24} kg, the mass of the moon is 7.337×10^{22} kg, the mean distance between the earth and the moon is 3.843×10^{8} m, and the gravitational constant G is 6.672×10^{-11} N $\cdot$ m^2/kg^2.

[c] **86.** Find the force of attraction between the earth and the sun. (Mass of the earth: 5.983×10^{24} kg, mass of the sun: 1.999×10^{30} kg, earth–sun mean distance: 1.496×10^{11} m, $G = 6.672 \times 10^{-11}$ N $\cdot$ m^2/kg^2)

[c] **87.** The speed of light (in a vacuum) is 2.998×10^{8} meters per second. If 1 meter is equal to 6.214×10^{-4} miles, find the speed of light in miles per second.

[c] **88.** The speed of light is 186,000 miles per second. Find **(a)** the distance traveled by a ray of light in one year; **(b)** the distance it travels in four years.

[c] **89.** At standard temperature and pressure, 1 g mol (or 32 g) of oxygen, O_2, occupies a volume of 22.4 liters = 22.4×10^{-3} m^3. The number of molecules of oxygen in 1 g mol is constant and equal to 6.02×10^{23} (Avogadro's number). Find the volume of each oxygen molecule in liters and in m^3.

[c] **90.** In the liquid state, 1 g mol of oxygen is found to have a volume smaller than the original gaseous volume by a factor of 700. Use Exercise 89 to find the volume, in m^3, of each molecule of liquid oxygen.

1.4 **POLYNOMIALS**

When a letter, such as x, is used to denote a real number, without any particular number being assigned to it, we say that x is a *real variable*. Other letters, such as

t, u, v, w, y, z, and so on, can also be used to denote real variables.

Let x be a variable in $\mathbb{R}$ and consider the powers

$$x^0 = 1, \quad x^1 = x, \quad x^2, \ldots, \quad x^n, \ldots$$

where the exponents are *nonnegative integers*. These powers of x are the basic elements used to define polynomials.

Polynomial in One Variable

A *polynomial* in a single variable x over $\mathbb{R}$ is an expression of the form

$$a_n x^n + a_{n-1} x^{n-1} + \cdots + a_1 x + a_0. \qquad (1.12)$$

where a_0, a_1, . . . , a_n are real numbers, called *coefficients* of the polynomial, and n is a nonnegative integer.

If $a_n \neq 0$, then the *degree* of the polynomial is n. If $a_n = a_{n-1} = \cdots = a_1 = 0$, but $a_0 \neq 0$, then the polynomial has degree 0. Every nonzero real number defines a polynomial of degree 0. When *all* coefficients of a polynomial are zero, the polynomial is said to be the *zero polynomial*. By convention, no degree is assigned to the zero polynomial.

The term $a_k x^k$ in a given polynomial is called the *term of degree k* or kth-*degree term*; for example, the term

$$-x^2 \quad \text{in the polynomial} \quad 4x^3 - x^2 + 5x + 7$$

is called its *second-degree term*. If $a_n \neq 0$, then $a_n x^n$, the term of degree n, is called the *leading term* and a_n the *leading coefficient* of the polynomial (1.12). The term a_0 is called the *constant term*. For example,

$$4x^3 - x^2 + 5x + 7$$

is a polynomial of degree 3 with leading coefficient 4 and constant term 7. In a polynomial, terms with zero coefficients are frequently left unwritten, for example,

$$x^3 - 3x + 1 \quad \text{is short for} \quad x^3 + 0x^2 - 3x + 1.$$

A polynomial containing exactly one nonzero term, such as $5x^2$, is called a *monomial*. A *binomial* is a polynomial containing exactly two terms: $3x^6 - 8$ is a binomial of degree 6. A polynomial containing exactly three nonzero terms is a *trinomial*. For example, $5x^2 - 2x + 3$ is a trinomial of degree 2 with leading coefficient 5 and constant term 3.

A *linear polynomial* is a polynomial of degree 1:

$$a_1 x + a_0, \quad \text{with } a_1 \neq 0.$$

A polynomial of degree 2,

$$a_2 x^2 + a_1 x + a_0, \quad \text{with } a_2 \neq 0,$$

is called a *quadratic polynomial*. Polynomials of degree 3 are called *cubic polynomials*.

Polynomials can be added, subtracted, and multiplied by using the associative and commutative properties for the sum and product of numbers, the distributive property of the product over the sum, and the properties of exponents.

EXAMPLE 1

Add the polynomials $x^3 + 4x + 2$ and $3x^2 - 2x + 1$.

Solution: We have

$(x^3 + 4x + 2) + (3x^2 - 2x + 1)$

$\quad = x^3 + 3x^2 + (4x - 2x) + (2 + 1)$ Associative and commutative properties

$\quad = x^3 + 3x^2 + (4 - 2)x + 3$ Distributive property

$\quad = x^3 + 3x^2 + 2x + 3.$

Sometimes it is convenient to use the following scheme to add polynomials.

$$\begin{array}{r} x^3 \qquad + 4x + 2 \\ 3x^2 - 2x + 1 \\ \hline x^3 + 3x^2 + 2x + 3 \end{array}$$

Practice Exercise 1

Add the polynomials $x^4 - 2x^3 + 5x^2 - x + 4$ and $2x^3 - x^2 - 5$.

Answer: $x^4 + 4x^2 - x - 1$

EXAMPLE 2

Find the difference $(x^4 - 5x^3 + x^2 - 3x + 1) - (2x^3 - 5x^2 + 2x - 4)$.

Solution: The polynomial $-(2x^3 - 5x^2 + 2x - 4) = -2x^3 + 5x^2 - 2x + 4$ is called the additive inverse of the polynomial $2x^3 - 5x^2 + 2x - 4$. To subtract the two polynomials, you may add the first polynomial to the additive inverse of the second one:

$(x^4 - 5x^3 + x^2 - 3x + 1) - (2x^3 - 5x^2 + 2x - 4)$

$\quad = (x^4 - 5x^3 + x^2 - 3x + 1) + (-2x^3 + 5x^2 - 2x + 4)$

$\quad = x^4 + (-5 - 2)x^3 + (1 + 5)x^2 + (-3 - 2)x + (1 + 4)$

$\quad = x^4 - 7x^3 + 6x^2 - 5x + 5.$

You should also try subtracting the given polynomials in an arrangement similar to the addition scheme used in Example 1.

Practice Exercise 2

Subtract $x^4 - x^3 + 5x - 1$ from $4x^3 - 5x^2 + 6x - 2$.

Answer: $-x^4 + 5x^3 - 5x^2 + x - 1$

Note that to *add* or *subtract* polynomials we *add* or *subtract the coefficients of the terms of the same degree.*

EXAMPLE 3

Multiply the polynomials $4x + 2$ and $2x^2 + 3x - 5$.

Solution: We have

$(4x + 2)(2x^2 + 3x - 5)$

$\quad = 4x(2x^2 + 3x - 5) + 2(2x^2 + 3x - 5)$ Distributive property

$$\begin{aligned}
&= (4x)(2x^2) + (4x)(3x) + (4x)(-5) \\
&\quad + 2(2x^2) + 2(3x) + 2(-5) \qquad \text{Distributive property} \\
&= 8x^3 + 12x^2 - 20x + 4x^2 + 6x - 10 \qquad \text{Computing the products} \\
&= 8x^3 + 16x^2 - 14x - 10.
\end{aligned}$$

When multiplying polynomials, the following scheme may be used.

$$\begin{array}{r}
2x^2 + 3x - 5 \\
4x + 2 \\
\hline
4x^2 + 6x - 10 \\
8x^3 + 12x^2 - 20x \\
\hline
8x^3 + 16x^2 - 14x - 10
\end{array}$$

Notice that the degree of the product of polynomials is equal to the sum of the degrees of the polynomials; the leading term of the product is the *product* of the leading terms, and the constant term of the product is the *product* of the constant terms.

Practice Exercise 3   Compute the product

$$(4x^3 - 3x^2 + 5)(x^2 - 2).$$

Answer: $4x^5 - 3x^4 - 8x^3 + 11x^2 - 10$

Polynomials in More than One Variable

We may also consider polynomials in two or more variables. A *polynomial in two variables x and y over* $\mathbb{R}$ is a finite sum of terms of the form $ax^n y^m$, where n and m are nonnegative integers and a is a real number. For example, $x + y - 1$, $x^2 - y^2$, and $3x^2 + y^2 + 4xy - 3x + 5y - 6$ are polynomials in x and y. Similarly, there are polynomials in three or more variables.

The sum of polynomials in two variables is found by adding the coefficients of terms of the same degree in x and in y.

EXAMPLE **4**   Add the polynomials $x^2 + xy + y^2 + 5x - y + 3$ and $3x^2 + 6y^2 - 7x + 9y + 11$.

Solution: We may arrange our work as follows.

$$\begin{array}{r}
x^2 + xy + y^2 + 5x - y + 3 \\
3x^2 \qquad\;\; + 6y^2 - 7x + 9y + 11 \\
\hline
4x^2 + xy + 7y^2 - 2x + 8y + 14
\end{array}$$

Practice Exercise 4   Find the sum of $3u^2 - 2uv + v^2 - 2u + 3v - 1$ and $u^2 - 3v^2 - 4v + 3$.

Answer: $4u^2 - 2uv - 2v^2 - 2u - v + 2$

Polynomials in two or more variables are multiplied in the same way as are polynomials of one variable.

EXAMPLE 5   Multiply $x + y$ and $x^2 - xy + y^2$.

Solution: We have

$$(x + y)(x^2 - xy + y^2) = x(x^2 - xy + y^2) + y(x^2 - xy + y^2)$$
$$= x^3 - x^2y + xy^2 + x^2y - xy^2 + y^3$$
$$= x^3 + y^3.$$

The following scheme may also be used:

$$
\begin{array}{r}
x^2 - xy + y^2 \\
x + y \\
\hline
x^2y - xy^2 + y^3 \\
x^3 - x^2y + xy^2 \\
\hline
x^3 \qquad\qquad + y^3
\end{array}
$$

Practice Exercise 5   Find the product of $x^3 + x^2y + xy^2 + y^3$ and $x - y$.

Answer: $x^4 - y^4$

Now, we briefly consider the division of a polynomial by a monomial. If the degree (in each individual variable) of the monomial is less than or equal to the degrees (in each individual variable) of each term of the polynomial, then the division is a simple operation: just divide each term of the polynomial by the monomial.

EXAMPLE 6   Divide $6x^4y^2 - 8x^3y^3 + 2x^2y^4$ by $2x^2y$.

Solution: We write

$$\frac{6x^4y^2 - 8x^3y^3 + 2x^2y^4}{2x^2y} = \frac{6x^4y^2}{2x^2y} - \frac{8x^3y^3}{2x^2y} + \frac{2x^2y^4}{2x^2y}$$
$$= 3x^2y - 4xy^2 + y^3.$$

Practice Exercise 6   Perform the division $\dfrac{6x^8 + 12x^6 + 15x^4}{3x^2}$.

Answer: $2x^6 + 4x^4 + 5x^2$

Division of one polynomial by another will be covered in detail in Chapter 11.

EXERCISES 1.4

In Exercises 1–28, perform the indicated operations and simplify as completely as possible.

1. $(2x^2 - 3x + 1) + (-x^2 + x - 4)$

2. $(x^3 - 5x + 2) + (x^2 + 7x + 8)$

3. $\left(\frac{1}{2}x^4 + 2x^2 - \frac{1}{5}\right) - \left(\frac{1}{3}x^4 - 3x^2 + \frac{1}{2}\right)$

4. $3x^2 + 5x + 4) - (x^4 + 3x^2 - 4)$

5. $(x^7 + 5x^5 + 3x^3 + x + 1) + (4x^4 + 2x^2 + 1)$

6. $(2t^3 + 3t^2 + 4t + 1) + (-5t^2 - 6t - 4)$

7. $(t^5 - t^4 + t^3 - t^2 - 1) + (2 + 3t + 4t^2 + 5t^3)$

8. $(1 - t^2 + t^3 - t^4 + t^5) - (2t^3 + 3t^2 - 4t + 5)$

9. $(x^4 - 1) + (x^4 - 3x^2) - (-x^4 - 2x^3 + x^2 + x - 5)$

10. $(5x^2 - 6x + 1) + (3x - 8) - (x^3 - 2x^2 + 4x - 1)$

11. $(5x^3 - 4x^2y + 3xy^2 - 4y^3 + xy + 1)$
$\qquad - (3x^2y - 6xy^2 + 4x^2 - 6xy + x - 4)$

12. $(7x^3 - 5x^2y + 10xy^2 - 6y^3 + 3x^2$
$\qquad\qquad - 5xy - 3y^2 + 3x - 4y + 7)$
$\qquad + (8x^2y - 6xy^2 + x^2 - 6xy + 7y^2 - 5x + 8y - 3)$

13. $(2x^2 - 3x + 1)(4x - 2)$

14. $(5t^3 - 3t)(t^2 - t + 1)$

15. $(3x^2 - 2x + 1)(x^2 - 2x + 3)$

16. $(x^2 + x + 1)(x^2 + x - 1)$

17. $(x^3 - 4x^2 + x)(x^4 - 2x^2)$

18. $(x^4 - 3x^2 + 1)(x^3 + x)$

19. $(x^5 - x^3 + 1)(x^2 + x + 1)$

20. $(y^2 - y + 1)(y^6 + y^4 - 1)$

21. $(a + b)(a^2 - ab + b^2)$

22. $(x - y)(x^2 + xy + y^2)$

23. $(x + y + 1)(x + y - 1)$

24. $(a - b + 1)(a - b - 1)$

25. $(2x - 3y + 4)(2x + 3y - 4)$

26. $(3u - v - 2)(3u - v + 2)$

27. $(x^2 + y^2 + xy)(x^2 - y^2 - xy)$

28. $(2x^2 - y^2 + 4x)(2x^2 - 3y^2 - 6x)$

In Exercises 29–38, you are asked to multiply three or more polynomials. Multiply two of them, then multiply the result by the third polynomial, and so on.

29. $(x - 1)(x - 2)(x - 3)$

30. $(u + 1)(u - 2)(u + 3)$

31. $(3x - 1)(3x - 2)(3x - 3)$

32. $(2x + 1)(2x + 2)(2x + 3)$

33. $(2x + 1)(2x + 1)(2x + 1)$

34. $(x - 1)(x - 1)(x - 1)$

35. $(a + b)(a - b)(a^2 + b^2)$

36. $(2x - y)(2x + y)(4x^2 + y^2)$

37. $(3x + 2)(3x - 2)(x + 1)(x - 1)$

38. $(4x - 1)(4x + 1)(x - 3)(x + 3)$

In Exercises 39–50, perform the indicated divisions.

39. $(6x^4 + 8x^2 - 10x) \div 2x$

40. $(12x^3 - 4x^2 - 8x) \div 4x$

41. $(9x^7 - 12x^5 - 3x^4 + 6x^3) \div 3x^2$

42. $(10x^8 - 15x^6 + 20x^4) \div 5x^2$

43. $\dfrac{16x^9 - 20x^7 + 12x^5}{4x^3}$

44. $\dfrac{15x^8 - 10x^7 + 5x^5 + 25x^3}{5x^3}$

45. $(10x^3y^2 - 6x^2y^3) \div 2xy^2$

46. $(9u^3v^3 - 6u^2v^4 + 12u^4v^2) \div 3u^2v$

47. $\dfrac{3x^2yz - 6xyz^3 + 9x^3y^2z^3}{3xyz}$

48. $\dfrac{4ax^2y^3 + 6a^2xy^4 - 10ax^3y^2}{2axy}$

49. $\dfrac{16a^2u^2v^4 - 8a^3u^3v^3 + 12a^4u^4v^2}{4a^2uv}$

50. $\dfrac{10m^4x^6y^2 + 15m^3x^5y^3 + 20m^2x^4y^4}{5m^2x^2y}$

HISTORICAL NOTE

ALGEBRA

The word *algebra* was part of the title of an Arabic manuscript from about A.D. 800 describing certain rules to solve equations. A literal translation of the word *al-jabr* (algebra) is *the restoration*, meaning the operations that are necessary in order to "restore" or "balance" an equation. Today, the word is used to describe the branch of mathematics dealing with statements of relations that utilize letters and other symbols to represent numbers or values.

1.5 SPECIAL PRODUCTS; FACTORING

In our dealings with polynomials, we frequently encounter certain products called *special products*. They deserve special attention, and you should readily recognize them because of their numerous applications. Here is a list of some of the most common ones.

Special Products

I. *Square of a binomial*

$$(x + y)^2 = x^2 + 2xy + y^2$$

II. *Product of a sum by a difference*

$$(x + y)(x - y) = x^2 - y^2$$

III. *Product of two binomials*

$$(x + a)(x + b) = x^2 + (a + b)x + ab$$

IV. *Sum of cubes*

$$x^3 + y^3 = (x + y)(x^2 - xy + y^2)$$

V. *Difference of cubes*

$$x^3 - y^3 = (x - y)(x^2 + xy + y^2)$$

VI. *Product of two binomials*

$$(mx + n)(px + q) = mpx^2 + (mq + np)x + nq$$

You should check the validity of formulas I through VI by computing the indicated products. Formulas I, II, IV, and V definitely need to be memorized in view of the many applications. Formulas III and VI are similar to patterns that follow and don't require memorization. Also, notice that formula III is a particular case of formula VI, when $m = p = 1$, $n = a$, and $q = b$.

EXAMPLE 1   Find each of the following products.

(a) $(2a + 7)^2$ **(b)** $(3u^2 + 4)(3u^2 - 4)$
(c) $(3a + 2)(9a^2 - 6a + 4)$ **(d)** $(3x + 1)(4x + 3)$

Solution: **(a)** By formula I, $(x + y)^2 = x^2 + 2xy + y^2$, with $x = 2a$ and $y = 7$;

$$(2a + 7)^2 = (2a)^2 + 2(2a) \cdot 7 + 7^2$$
$$= 4a^2 + 28a + 49.$$

(b) This is a product of a sum by a difference: $(x + y)(x - y) = x^2 - y^2$. So

$$(3u^2 + 4)(3u^2 - 4) = (3u^2)^2 - 4^2$$
$$= 9u^4 - 16.$$

(c) We can use formula IV, $(x + y)(x^2 - xy + y^2) = x^3 + y^3$:

$$(3a + 2)(9a^2 - 6a + 4) = (3a + 2)[(3a)^2 - (3a) \cdot 2 + 2^2]$$
$$= (3a)^3 + 2^3$$
$$= 27a^3 + 8.$$

(d) This is a product of two binomials, $(mx + n)(px + q) = mpx^2 + (mq + np)x + nq$. Hence

$$(3x + 1)(4x + 3) = 3 \cdot 4 \cdot x^2 + (3 \cdot 3 + 1 \cdot 4)x + 1 \cdot 3$$
$$= 12x^2 + 13x + 3.$$

Practice Exercise 1   Use the special product formulas to perform each multiplication.
(a) $(2b - 7)^2$ **(b)** $(2v + 5)(2v - 5)$ **(c)** $(4u - 3)(16u^2 + 12u + 9)$
(d) $(3x + 2)(x + 6)$

Answer: **(a)** $4b^2 - 28b + 49$ **(b)** $4v^2 - 25$ **(c)** $64u^3 - 27$
(d) $3x^2 + 20x + 12$

Factoring Polynomials

In Section 1.1, we mentioned that each natural number can be written as a product of prime factors. For example, $60 = 2 \cdot 2 \cdot 3 \cdot 5$, where 2, 3, and 5 are the prime factors of 60.

The situation is analogous for polynomials. For example, direct multiplication shows that the quadratic polynomial $2x^2 + 7x + 3$ can be written as the following product:

$$2x^2 + 7x + 3 = (2x + 1)(x + 3).$$

In the product $(2x + 1)(x + 3)$, the polynomials $2x + 1$ and $x + 3$ are called *factors* of the product. The process of finding the factors of a polynomial is called *factoring*.

A polynomial is said to be *prime* or *irreducible* if it cannot be written as a product of two polynomials of positive degrees. For example, $2x + 1$ and $x + 3$ are prime polynomials.

A polynomial is said to be in *completely factored* form when it is written as a product of prime polynomials.

When factoring polynomials, it is necessary to specify *over* what set of numbers we want to carry out the factorization. For example, the polynomial $x^2 - 4$ with *integer* coefficients can be written as a product of polynomials with *integer* coefficients:

$$x^2 - 4 = (x - 2)(x + 2).$$

We say that $x^2 - 4$ has been factored *over the integers*. On the other hand, it is not possible to write $x^2 - 2$ as a product of two polynomials of degree 1 with integer coefficients. Thus, $x^2 - 2$ is said to be *irreducible over the integers*. However, since

$$x^2 - 2 = (x - \sqrt{2})(x + \sqrt{2}),$$

the polynomial $x^2 - 2$ can be factored *over the real numbers*.

There are polynomials that cannot be factored over the real numbers, such as

$$x^2 + 1, \quad x^2 + 4, \quad 3x^2 + 1, \quad \text{and} \quad x^2 - 2x + 5.$$

These polynomials are said to be *irreducible over the real numbers*. Later, after introducing the concept of *complex numbers*, we shall see that these polynomials can be factored *over the complex numbers*.

In this chapter, polynomials with integer coefficients will be factored so that the factors contain only integer coefficients.

Now we will discuss several examples. If all terms of a polynomial contain a *common factor*, then we can factor the polynomial by using the distributive property.

EXAMPLE 2

Factor each polynomial.
(a) $6ax^3 + 4ax^2$ (b) $2a(x + y) - 5b(x + y)$

Solution: (a) We look for the "largest" common factor of the two monomials. Since $6ax^3 = (2ax^2)3x$ and $4ax^2 = (2ax^2)2$, it follows that $2ax^2$ is a common factor. Thus

$$6ax^3 + 4ax^2 = (2ax^2)3x + (2ax^2)2$$
$$= 2ax^2(3x + 2).$$

(b) In this case, $x + y$ is the largest common factor of $2a(x + y)$ and $5b(x + y)$. Thus

$$2a(x + y) - 5b(x + y) = (x + y)(2a - 5b)$$

Practice Exercise 2

Factor each polynomial.
(a) $4u^2x^2 - 8u^2x$ (b) $(u - v)5x + 3(u - v)$

Answer: (a) $4u^2x(x - 2)$ (b) $(u - v)(5x + 3)$

Most polynomials that can be factored at all are factored by recognizing them as special products.

EXAMPLE 3

Factor the polynomial $4y^2 - 25$.

Solution: Since

$$4y^2 - 25 = (2y)^2 - 5^2,$$

we see that the given polynomial is a *difference of two squares*. According to formula II, we write

$$4y^2 - 25 = (2y)^2 - 5^2,$$
$$= (2y + 5)(2y - 5).$$

Practice Exercise 3   Factor the expression $64x^2 - 49$.

Answer: $(8x + 7)(8x - 7)$

EXAMPLE 4   Factor the polynomial $2x^3 + 16$.

Solution: First, we notice that 2 is a common factor of both coefficients. Using the distributive property, we write

$$2x^3 + 16 = 2(x^3 + 8).$$

Since $8 = 2^3$, we can apply formula IV to obtain

$$2x^3 + 16 = 2(x^3 + 8)$$
$$= 2(x + 2)(x^2 - 2x + 4).$$

Practice Exercise 4   Factor $3x^3 - 24$.

Answer: $3(x - 2)(x^2 + 2x + 4)$

EXAMPLE 5   Factor the quadratic polynomial $2x^2 + 10x + 12$.

Solution: Since 2 is a common factor, we can write

$$2x^2 + 10x + 12 = 2(x^2 + 5x + 6).$$

Next, we must look for two numbers a and b such that $a + b = 5$ and $ab = 6$. Since 2 and 3 satisfy these requirements, we write

$$2x^2 + 10x + 12 = 2(x^2 + 5x + 6)$$
$$= 2(x + 2)(x + 3).$$

Practice Exercise 5   Factor $3x^2 + 24x + 45$.

Answer: $3(x + 3)(x + 5)$

EXAMPLE 6   Factor the polynomial $2x^2 - 5x - 3$.

Solution: Using formula VI, we first look for two integers m and p so that $mp = 2$. Since the only factors of 2 are 1 and 2, the factorization of the given polynomial is of the form

$$2x^2 - 5x - 3 = (2x + \quad)(x + \quad),$$

where the blank spaces will be occupied by two integers n and q such that $nq = -3$ and $mq + np = -5$. Since the integral factors of -3 are -1 and 3 or 1 and -3, trial and error lead us to the factorization

$$2x^2 - 5x - 3 = (2x + 1)(x - 3).$$

Inspection of the other products,

$$(2x - 1)(x + 3)$$
$$(2x + 3)(x - 1)$$
$$(2x - 3)(x + 1),$$

shows that none of them equals $2x^2 - 5x - 3$. As an exercise, you should check that in each of these three products, the term of highest degree is $2x^2$ and the constant term is -3. However, the term of degree 1 fails to be $-5x$.

Practice Exercise 6   Write $3x^2 + 11x - 4$ as a product of two linear polynomials.

Answer: $(3x - 1)(x + 4)$

EXAMPLE 7   Factor completely the expression $81a^4 - 1$.

Solution: The given expression is a difference of two squares,

$$81a^4 - 1 = (9a^2)^2 - 1^2.$$

Thus, using formula II, we can write

$$81a^4 - 1 = (9a^2)^2 - 1$$
$$= (9a^2 + 1)(9a^2 - 1).$$

But $9a^2 - 1 = (3a)^2 - 1^2$ is also the difference of two squares. Repeating the process, we obtain the complete factorization of the given expression:

$$81a^4 - 1 = (9a^2 + 1)(9a^2 - 1)$$
$$= (9a^2 + 1)[(3a)^2 - 1^2]$$
$$= (9a^2 + 1)(3a + 1)(3a - 1).$$

Practice Exercise 7   Factor completely $16u^4 - 1$.

Answer: $(4u^2 + 1)(2u + 1)(2u - 1)$

Factoring by Grouping

To factor certain expressions, it is sometimes necessary to group some of their terms, to factor out any common factors that may be present, and then to use the distributive property. This technique, illustrated in the next example, is called *factoring by grouping*.

EXAMPLE 8   Factor the following expressions.
(a) $6xy + 2ay - 3bx - ab$ **(b)** $2ax^2 - 2ay - x^2b + by$.

Solution: **(a)** We arrange the terms in two groups and factor each group as follows:

$$6xy + 2ay - 3bx - ab = (6xy - 3bx) + (2ay - ab)$$
$$= 3x(2y - b) + a(2y - b).$$

Since $2y - b$ is a common factor of the last two terms, we obtain

$$6xy + 2ay - 3bx - ab = (2y - b)(3x + a).$$

(b) Analogously,

$$2ax^2 - 2ay - x^2b + by = (2ax^2 - 2ay) - (x^2b - by)$$
$$= 2a(x^2 - y) - b(x^2 - y)$$
$$= (x^2 - y)(2a - b).$$

Practice Exercise 8   Factor each of the following expressions.
(a) $2au^2 + bu^2 - 4av - 2bv$ **(b)** $3mn - 6mq - pn + 2pq$

Answer: **(a)** $(u^2 - 2v)(2a + b)$ **(b)** $(3m - p)(n - 2q)$

We conclude this section with the following hints that may help you to factor a polynomial.

Factoring Hints

1. Look for a common factor.
2. See if you can use one of the special products.
3. Try factoring by grouping.
4. Look to see if further factoring can be performed.
5. Remember that a *completely factored* polynomial must be written as a product of *prime* factors.
6. Check your factorization by multiplying the factors back out.

EXERCISES 1.5

In Exercises 1–16, use the special products formulas to expand each of the following products.

1. $(2x + 1)(2x + 3)$
2. $(2x - 4)(2x + 3)$
3. $(2x + 3a)(2x - a)$
4. $(3y + b)(3y + 2b)$
5. $(2ax - 5b)^2$
6. $(3r + as)^2$
7. $(4y - 3b)^2$
8. $(3au - 4b)^2$
9. $(x^2 + 5)(x^2 - 5)$
10. $(5a^2 + 3)(5a^2 - 3)$
11. $(x + \sqrt{2})(x - \sqrt{2})$
12. $(t - \sqrt{5})(t + \sqrt{5})$
13. $(\sqrt{a} + \sqrt{b})(\sqrt{a} - \sqrt{b})$
14. $(\sqrt{2}\,x + 1)(\sqrt{2}\,x - 1)$
☐ **15.** $(x - y + a)^2$
☐ **16.** $(u - v + 2b)^2$

In Exercises 17–68, factor completely each of the following polynomials

17. $x^2 + 6x + 8$
18. $y^2 - 6y + 5$
19. $x^2 - 11x + 24$
20. $x^2 - 11x + 28$
21. $x^2 + 2x - 35$
22. $x^2 + 5x - 36$
23. $y^2 - 4y - 12$
24. $z^2 - 4z - 21$
25. $p^2 - 11p + 30$
26. $q^2 - 13q + 36$
27. $4x^2 + x - 3$
28. $3x^2 + 7x - 6$
29. $2x^2 + 16x + 30$
30. $3x^2 - 6x - 45$
31. $5x^2 - 20$
32. $4x^2 - 36$
33. $3ax^2 + 6ax - 9a$
34. $2mu^2 - 10mu + 8m$

35. $64a^2 - 9b^2$

36. $b^2 - 81c^2$

37. $16x^4 - 81$

38. $81y^4 - 625$

39. $x^3 - 27$

40. $r^3 - 64$

41. $8x^3 + 64$

42. $64x^3 + 27$

43. $16ax^3 + 2a$

44. $3x^3 - 24$

45. $2x^2 + 5x - 3$

46. $2x^2 - 7x + 6$

47. $4x^2 - 4x + 1$

48. $4x^2 - 20x + 25$

49. $6x^2 - 5x + 1$

50. $6x^2 + 7x + 2$

51. $3x^2 + 11x + 6$

52. $5x^2 + 9x - 2$

53. $2ac - 6ad + bc - 3bd$

54. $3mu - 6nu - 5mv + 10nv$

55. $6ax - 9ay - 2bx + 3by$

56. $8rt - 2pt + 12rq - 3pq$

57. $x^2y + 2ay - 3bx^2 - 6ab$

58. $2uy^2 + 6bu - 5y^2 - 15b$

59. $(a + b)^3 - 8$

60. $(x + y)^4 - 16$

61. $x^6 - 64$

62. $y^6 - 729$

☐ **63.** $(x + 2y)^2 - 4z^2$

☐ **64.** $(2a - b)^2 - 16c^2$

☐ **65.** $(2x - 1)^3 + 8$

☐ **66.** $(2x + 1)^3 - 8$

☐ **67.** $(a + b)^2 - (a - b)^2$

☐ **68.** $(a - b)^2 - (a + b)^2$

In Exercises 69–74, the given polynomials have rational coefficients. Factor each polynomial over the reals. [*Hint:* Write $\dfrac{u^2}{2} - 8 = \dfrac{1}{2}(u^2 - 16)$ and complete the factorization.]

69. $\dfrac{u^2}{2} - 8$

70. $\dfrac{v^2}{5} - 5$

71. $x^2 - \dfrac{1}{6}x - \dfrac{1}{6}$

72. $x^2 + \dfrac{7x}{12} + \dfrac{1}{12}$

73. $\dfrac{x^3}{2} + 4$

74. $\dfrac{x^3}{3} - 9$

75. Without using a calculator, compute the difference $(321.4)^2 - (320.4)^2$. (*Hint:* Use special product formula II.)

76. Without a calculator, find the difference $(162.25)^2 - (160.25)^2$.

☐ **77.** Show that any odd number (that is, a number of the form $2n + 1$, with n a natural number) can be written as the difference of the squares of two consecutive natural numbers.

☐ **78.** Show that if you square an integer and subtract 1, the result equals the product of the integer preceding and the integer following that integer.

☐ **79.** Show that one-fourth of the difference between the squares of the integer following and the integer preceding a given integer is the integer itself.

☐ **80.** Show that if you take a number, multiply it by the number four units larger, add four units to the product, take the square root and subtract 2 from the result, you get the number you started with.

1.6 RATIONAL EXPRESSIONS

A *rational expression* is the quotient of two polynomials. For example,

$$\frac{3x + 1}{5}, \quad \frac{x + 5}{2x^2 - 3x - 1}, \quad \text{and} \quad \frac{x^4 + x + 2}{x^2 + 5x + 1}$$

are rational expressions. More generally, if $a_nx^n + a_{n-1}x^{n-1} + \cdots + a_1x + a_0$ is a polynomial of degree n and $b_mx^m + b_{m-1}x^{m-1} + \cdots + b_1x + b_0$ is a polynomial of degree m, we may form the rational expression

$$\frac{a_nx^n + a_{n-1}x^{n-1} + \cdots + a_1x + a_0}{b_mx^m + b_{m-1}x^{m-1} + \cdots + b_1x + b_0}.$$

Since x denotes an arbitrary real number in this expression while the coefficients $a_0, a_1, \ldots, a_n$ and $b_0, b_1, \ldots, b_n$ denote fixed real numbers, the rational expression represents the quotient of two real numbers, *except for those values of x for which the denominator is equal to zero*. Thus, when we deal with rational expressions, it will always be understood that *we are excluding the values of x for which the denominator is equal to zero.* For example, $(2x + 1)/(x - 3)$ has a meaning for all real numbers x such that $x \neq 3$.

Rational expressions can be added, subtracted, multiplied, divided, and simplified in the same manner as fractions [Section 1.2].

Simplifying Rational Expressions

To simplify a rational expression, we divide both numerator and denominator by any factor they have in common. In analogy with rational numbers (or fractions), rational expressions whose numerator and denominator have no common factors are said to be in *lowest terms.*

EXAMPLE 1

Simplify $\dfrac{x^2 - 4}{2x^2 - 3x - 2}$.

Solution: We factor both numerator and denominator and cancel the common factor:

$$\frac{x^2 - 4}{2x^2 - 3x - 2} = \frac{(x + 2)(x - 2)}{(2x + 1)(x - 2)} = \frac{x + 2}{2x + 1}.$$

Note that the equality

$$\frac{x^2 - 4}{2x^2 - 3x - 2} = \frac{x + 2}{2x + 1}$$

is valid only for values of x *different* from 2 and $-\frac{1}{2}$, because if $x = 2$, the denominator of the first rational expression equals zero, and if $x = \frac{1}{2}$, both denominators are zero. From now on, we shall always assume such restrictions when simplifying rational expressions.

Practice Exercise 1

Reduce $\dfrac{2x^2 + 5x - 3}{x^2 + x - 6}$ to lowest terms.

Answer: $\dfrac{2x - 1}{x - 2}$.

Addition and Subtraction

To add or subtract two rational expressions with the same denominator, we add or subtract the numerators and keep the same denominator.

EXAMPLE 2

Perform the indicated sum and simplify:

$$\frac{3x^2 + 2x - 6}{x^2 - 1} + \frac{2 - x}{x^2 - 1}.$$

Solution: We have

$$\frac{3x^2 + 2x - 6}{x^2 - 1} + \frac{2 - x}{x^2 - 1} = \frac{3x^2 + 2x - 6 + 2 - x}{x^2 - 1}$$

$$= \frac{3x^2 + x - 4}{x^2 - 1}$$

$$= \frac{(3x + 4)(x - 1)}{(x + 1)(x - 1)}$$

$$= \frac{3x + 4}{x + 1}.$$

Practice Exercise 2   Subtract as indicated and simplify:

$$\frac{2x^2 + 2x - 5}{x^2 - 4} - \frac{3x + 5}{x^2 - 4}$$

Answer: $\dfrac{2x - 5}{x - 2}$

   In order to add or subtract rational expressions with different denominators, we have to replace them by equivalent rational expressions with the same denominator, just as we do with ordinary fractions. We can always choose as common denominator the product of the denominators. However, computation is simpler if we use the *least common denominator* (LCD) of the rational expressions.

Least Common Denominator

To find the LCD of several rational expressions, proceed as follows:

1. Factor completely each of the denominators.
2. Multiply all prime factors, each one raised to the highest power that factor occurs in any one factorization.

EXAMPLE 3   Subtract

$$\frac{3x - 1}{x - 1} - \frac{2x}{2x + 3}.$$

Solution: Take the product $(x - 1)(2x + 3)$ of the denominators as the common denominator and write

$$\frac{3x - 1}{x - 1} - \frac{2x}{2x + 3} = \frac{(3x - 1)(2x + 3)}{(x - 1)(2x + 3)} - \frac{2x(x - 1)}{(x - 1)(2x + 3)}$$

$$= \frac{6x^2 + 7x - 3}{(x - 1)(2x + 3)} - \frac{2x^2 - 2x}{(x - 1)(2x + 3)}$$

$$= \frac{6x^2 + 7x - 3 - 2x^2 + 2x}{(x - 1)(2x + 3)}$$

$$= \frac{4x^2 + 9x - 3}{(x - 1)(2x + 3)}.$$

Practice Exercise 3   Add

$$\frac{4x}{3x + 1} + \frac{x + 1}{x - 2}.$$

Answer: $\dfrac{7x^2 - 4x + 1}{(3x + 1)(x - 2)}$

EXAMPLE 4   Perform the indicated operations and simplify.

$$\frac{2}{x^2 + 2x} + \frac{6}{2x + 4} - \frac{3x + 1}{x^2}$$

Solution: First, we find the LCD of the denominators according to the following scheme.

$$x^2 + 2x = x(x + 2)$$
$$2x + 4 = 2(x + 2)$$
$$\underline{ x^2 = x^2 }$$
$$\text{LCD} = 2(x + 2)x^2$$

Observe that there are three prime factors: 2, x, and $x + 2$. The highest exponent of the factors 2 and $x + 2$ is 1, and the highest exponent of the factor x is 2. Thus, the LCD is the product of 2, $x + 2$, and x^2. Next, we write

$$\frac{2}{x^2 + 2x} + \frac{6}{2x + 4} - \frac{3x + 1}{x^2} = \frac{2}{x(x + 2)} + \frac{6}{2(x + 2)} - \frac{3x + 1}{x^2}$$

$$= \frac{2(2x)}{2(x + 2)x^2} + \frac{6(x^2)}{2(x + 2)x^2} - \frac{2(x + 2)(3x + 1)}{2(x + 2)x^2}$$

$$= \frac{4x}{2(x + 2)x^2} + \frac{6x^2}{2(x + 2)x^2} - \frac{2(3x^2 + 7x + 2)}{2(x + 2)x^2}$$

$$= \frac{4x + 6x^2 - 6x^2 - 14x - 4}{2(x + 2)x^2}$$

$$= \frac{-10x - 4}{2(x + 2)x^2}$$

$$= -\frac{\cancel{2}(5x + 2)}{\cancel{2}(x + 2)x^2}$$

$$= -\frac{5x + 2}{(x + 2)x^2}.$$

Practice Exercise 4   Add and simplify:

$$\frac{2x - 1}{x^2 + 2x + 1} + \frac{3x - 4}{x^2 - 1}.$$

Answer: $\dfrac{5x^2 - 4x - 3}{(x + 1)^2(x - 1)}$

Multiplication and Division

The product of rational expressions is obtained by multiplying their numerators and multiplying their denominators. In order to simplify computations, it is advisable first to factor the polynomials and to cancel out common factors that appear in the numerator and denominator.

EXAMPLE **5**   Multiply $\dfrac{x^2 - 6x + 9}{2x + 10} \cdot \dfrac{2x + 6}{x^2 - 9}$.

Solution: First, factor completely the polynomials appearing in both expressions and cancel out common factors.

$$\frac{x^2 - 6x + 9}{2x + 10} \cdot \frac{2x + 6}{x^2 - 9} = \frac{(x - 3)^2}{2(x + 5)} \cdot \frac{2(x + 3)}{(x + 3)(x - 3)}$$

$$= \frac{2(x - 3)^2(x + 3)}{2(x + 5)(x + 3)(x - 3)}$$

$$= \frac{x - 3}{x + 5}.$$

Practice Exercise 5   Multiply $\dfrac{3x - 6}{2x^2 - 5x + 2} \cdot \dfrac{x^2 + 6x + 8}{3x + 6}$.

Answer: $\dfrac{x + 4}{2x - 1}$

The pattern for dividing one rational expression by another is the same as it is for fractions: *invert the divisor and multiply*. Just as in any other multiplication, cancel out any common factors that may appear in the (new) numerators and denominators before actually carrying out the multiplication.

EXAMPLE **6**   Divide $\dfrac{2x^2 + 5x - 3}{x^2 + 10x + 25} \div \dfrac{2x - 1}{x^2 + 5x}$. Give your answer in simplified form.

Solution: We have

$$\frac{2x^2 + 5x - 3}{x^2 + 10x + 25} \div \frac{2x - 1}{x^2 + 5x}$$

$$= \frac{2x^2 + 5x - 3}{x^2 + 10x + 25} \cdot \frac{x^2 + 5x}{2x - 1} \qquad \text{Invert the divisor}$$

Factor numerator and denominator

$$= \frac{(2x - 1)(x + 3)x(x + 5)}{(x + 5)^2(2x - 1)}$$

Cancel out common factors

$$= \frac{x(x + 3)}{x + 5}.$$

Practice Exercise 6   Divide and simplify: $\dfrac{2x^2 - 5x + 3}{x^2 + 2x} \div \dfrac{x^2 - 5x + 4}{x + 2}$.

Answer: $\dfrac{2x - 3}{x(x - 4)}$

Compound Expressions

On many occasions, we have to deal with quotients in which the numerator and denominator are neither polynomials nor rational expressions but can be reduced to rational expressions.

EXAMPLE 7   Simplify the expression $\dfrac{1 - \dfrac{7}{x^2 - 9}}{1 - \dfrac{1}{x - 3}}$.

Solution: First write the numerator and denominator as rational expressions, and then divide.

$$\frac{1 - \dfrac{7}{x^2 - 9}}{1 - \dfrac{1}{x - 3}} = \frac{\dfrac{x^2 - 9 - 7}{x^2 - 9}}{\dfrac{x - 3 - 1}{x - 3}}$$

$$= \frac{\dfrac{x^2 - 16}{x^2 - 9}}{\dfrac{x - 4}{x - 3}} = \frac{x^2 - 16}{x^2 - 9} \cdot \frac{x - 3}{x - 4}$$

$$= \frac{(x + 4)(x - 4)(x - 3)}{(x + 3)(x - 3)(x - 4)} = \frac{x + 4}{x + 3}$$

Practice Exercise 7   Simplify the compound expression $\dfrac{\dfrac{2}{x + 3} + 1}{1 - \dfrac{2}{x + 7}}$.

Answer: $\dfrac{x + 7}{x + 3}$

EXERCISES 1.6

In Exercises 1–16, simplify each of the following rational expressions.

1. $\dfrac{15a^2x^4}{3ax^2}$

2. $\dfrac{18m^3n^6}{9m^2n^2}$

3. $\dfrac{10x^6 + 15x^4 - 5x^2}{5x}$

4. $\dfrac{18a^4 + 27a^5}{9a^3}$

5. $\dfrac{6x - 18}{6x - 24}$

6. $\dfrac{9x - 45}{9x + 27}$

7. $\dfrac{12z^2 + 6z}{9z^2 - 3z}$

8. $\dfrac{12y^2 - 8y}{4y^2 - 16y}$

9. $\dfrac{x + 1}{x^2 + 9x + 8}$

10. $\dfrac{x^2 - 1}{x^2 + 5x + 4}$

11. $\dfrac{u^2 - 9}{u^2 + 6u + 9}$

12. $\dfrac{z^2 - 16}{2z^2 + 5z - 12}$

13. $\dfrac{2x^2 + 10x}{2x^2 + 9x - 5}$

14. $\dfrac{2x^2 + x - 3}{2x^2 + 5x + 3}$

15. $\dfrac{4x^2 + 7x + 4}{4x^2 + 11x + 6}$

16. $\dfrac{10x^2 + 26x - 12}{5x^2 + 18x - 8}$

In Exercises 17–34, perform the operations and simplify.

17. $3 + \dfrac{x - 1}{x + 2}$

18. $5 - \dfrac{x - 3}{x - 4}$

19. $x - 2 - \dfrac{x - 2}{x - 3}$

20. $x - 5 + \dfrac{x - 1}{x + 4}$

21. $\dfrac{3}{2x - 1} - \dfrac{x}{x + 4}$

22. $\dfrac{5}{3x + 1} - \dfrac{x}{x + 2}$

23. $\dfrac{x + 4}{x - 4} - \dfrac{x - 1}{x + 4}$

24. $\dfrac{x + 2}{x - 1} + \dfrac{x + 3}{x - 5}$

25. $\dfrac{5x + 10}{x^2 + 6x + 8} + \dfrac{3}{x + 4}$

26. $\dfrac{3x + 6}{x^2 - 4} + \dfrac{5x}{x - 2}$

27. $\dfrac{2x - 1}{x^2 + 6x + 9} + \dfrac{x}{x + 3}$

28. $\dfrac{x - 1}{x^2 + 5x + 6} + \dfrac{x - 2}{x^2 + 4x + 4}$

29. $\dfrac{x - 2}{2x^2 - x - 1} + \dfrac{x + 2}{2x^2 + 3x + 1}$

30. $\dfrac{2x}{3x^2 + x - 2} - \dfrac{2x - 8}{3x^2 - 8x + 4}$

31. $\dfrac{1}{x - 1} - \dfrac{1}{3x - 1} + \dfrac{1}{4 - 4x}$

32. $\dfrac{x^2}{(x + 1)^2} + \dfrac{1}{x - 1} - \dfrac{1}{2}$

33. $\dfrac{3}{x^2 + 3x + 2} + \dfrac{2}{x^2 - x - 6} - \dfrac{1}{x^2 - 2x - 3}$

34. $\dfrac{2}{x - 5} - \dfrac{2}{x + 5} - \dfrac{15}{x^2 - 25}$

In Exercises 35–46, perform the multiplications and simplify.

35. $\dfrac{x + 1}{x^2 + x} \cdot \dfrac{x}{x - 1}$

36. $\dfrac{x^4 - 4x^2}{x + 2} \cdot \dfrac{x - 2}{x^2 - 2x}$

37. $\dfrac{4y - 8}{4(y + 2)} \cdot \dfrac{y}{2y - 4}$

38. $\dfrac{x^2 - 16}{x^2 - 9} \cdot \dfrac{3 - x}{2x + 8}$

39. $\dfrac{5}{8b^2} \cdot \dfrac{4b + 6}{10b + 15}$

40. $\dfrac{7a^2}{4} \cdot \dfrac{6a - 8}{21a - 28}$

41. $\dfrac{3x + 2}{x^2 - 1} \cdot \dfrac{4x + 4}{9x + 6}$

42. $\dfrac{2x - 5}{x^2 - 4} \cdot \dfrac{3x - 6}{4x - 10}$

43. $\dfrac{2x^2 + 5x + 2}{x^2 + 5x - 6} \cdot \dfrac{x^2 + 7x + 6}{4x^2 + 4x + 1}$

44. $\dfrac{2x^2 - 3x - 2}{x^2 + x - 2} \cdot \dfrac{x^2 + 3x + 2}{4x^2 - 1}$

45. $\dfrac{x^3 - x^2 - 2x}{x^2 - 1} \cdot \dfrac{2x + 1}{x^2 - 2x}$

46. $\dfrac{x^3 - 9x}{x^2 - x - 2} \cdot \dfrac{x^2 - 1}{x^2 + 2x - 3}$

In Exercises 47–70, perform the divisions and simplify.

47. $\dfrac{\dfrac{2x}{x^3}}{\dfrac{4x}{x^5}}$

48. $\dfrac{\dfrac{3y^2}{4y}}{\dfrac{9y}{2y^6}}$

49. $\dfrac{\dfrac{2x - 1}{6}}{\dfrac{2x + 1}{12}}$

50. $\dfrac{\dfrac{x + 5}{8}}{\dfrac{x - 5}{16}}$

51. $\dfrac{\dfrac{x + 4}{3x + 12}}{2x}$

52. $\dfrac{\dfrac{x - 6}{2x - 12}}{3x}$

53. $\dfrac{\dfrac{x^2 - 1}{x}}{\dfrac{x + 1}{x^2}}$

54. $\dfrac{\dfrac{x^2 - 9}{2x^3}}{\dfrac{x - 3}{4x^2}}$

55. $\dfrac{\dfrac{5x^2}{4x^2 - 1}}{\dfrac{10x}{2x + 1}}$

56. $\dfrac{\dfrac{9a^2 - 1}{4 - a}}{\dfrac{3a + 1}{a^2 - 16}}$

57. $\dfrac{\dfrac{2x^2 - 7x + 3}{2x + 1}}{\dfrac{x^2 - 9}{10x + 5}}$

58. $\dfrac{\dfrac{x - 5}{x^2 + 8x + 15}}{\dfrac{x - 2}{x^2 + 6x + 9}}$

59. $\dfrac{\dfrac{3x^2 - 8x - 3}{4x^2 - 1}}{\dfrac{x^2 + x - 12}{2x^2 + 3x + 1}}$

60. $\dfrac{\dfrac{2}{3}(x^2 + 2x - 15)}{\dfrac{5}{6}(x^2 - 9)}$

61. $\dfrac{\dfrac{1}{x + 1} + \dfrac{1}{x - 1}}{\dfrac{2}{x}}$

62. $\dfrac{1 + \dfrac{1}{2x}}{\dfrac{2}{3x} - 1}$

63. $\dfrac{\dfrac{3}{x-2}-1}{3-\dfrac{1}{x-2}}$

64. $\dfrac{\dfrac{x}{x+1}-2}{3-\dfrac{x}{x+1}}$

67. $\dfrac{1-\dfrac{5}{x^2-4}}{\dfrac{x-3}{2-x}}$

68. $\dfrac{\dfrac{5}{x^2-9}+1}{\dfrac{5}{x-3}+1}$

65. $\dfrac{\dfrac{2x}{x-1}+\dfrac{3}{x+1}}{\dfrac{4x}{x+1}-\dfrac{2}{x-1}}$

66. $\dfrac{\dfrac{x}{x+2}-\dfrac{2}{x-3}}{\dfrac{3}{x+2}-\dfrac{x}{x-3}}$

69. $\dfrac{\dfrac{7}{x^2-9}-1}{\dfrac{1}{x-3}-1}$

70. $1-\dfrac{x-\dfrac{1}{x}}{1-\dfrac{1}{x}}$

1.7 RADICALS

Let a be a real number and let n be a natural number. Can we find a real number x such that the statement

$$x^n = a \qquad (1.13)$$

is true? Before answering this question, let us discuss a few examples.

If $a = 4$ and $n = 2$, then both $x = 2$ and $x = -2$ satisfy equation (1.13), because $2^2 = 4$ and $(-2)^2 = 4$.

If $a = -27$ and $n = 3$, then $x = -3$ is the unique real number such that $(-3)^3 = -27$.

Finally, if $a = -1$ and $n = 2$, then there is *no* real number x such that $x^2 = -1$.

These examples indicate that the following general results are true.

n even	$a > 0$	There are *two* real numbers with opposite signs satisfying (1.13).
n odd	$a > 0$ or $a < 0$	There is a *unique* real number satisfying (1.13).
n even	$a < 0$	*No* real number satisfies (1.13).

We now make the following definition.

---

*n*th Root of a Real Number

A real number x satisfying $x^n = a$ is called an *n*th root of a.

For example, 2 and -2 are *second* (or *square*) *roots* of 4; the number -3 is the *third* (or *cube*) *root* of -27; and 5 and -5 are *fourth roots* of 625.

The Principal *n*th Root of a Real Number

The principal *n*th root of a real number *a* is the *positive n*th root of *a* if $a > 0$ and *n* is even, or it is the *unique n*th root of *a* if *n* is odd.

For example, 2 is the *principal square root* of 4; the number 3 is the *principal cube root* of 27; and 5 is the *principal fourth root* of 625.

Notation

The principal *n*th root of *a* is denoted $\sqrt[n]{a}$.

In this notation, the symbol $\sqrt{\ }$ is called a *radical,* the natural number *n* is the *index* of the radical, and *a* is the *radicand*.

When $n = 2$, it is customary to omit the index from the radical. We write $\sqrt{a}$ and read "the principal square root of *a*" or simply "square root of *a*." When $n = 3$, the number $\sqrt[3]{a}$ is called the *principal third root* of *a* or simply the *cube root* of *a*.

From the definition of principal *n*th root, it follows that:

$$\sqrt[n]{a} = b \quad \text{implies} \quad a = b^n.$$

EXAMPLE 1

Compute the following radicals.

(a) $\sqrt{64}$ (b) $\sqrt[5]{-\dfrac{32}{243}}$

Solution:

(a) The integers 8 and -8 are such that $8^2 = (-8)^2 = 64$. Since $\sqrt{64}$ denotes the principal square root of 64, it follows that $\sqrt{64} = 8$.

(b) We have $\sqrt[5]{-\dfrac{32}{243}} = -\dfrac{2}{3}$, because $\left(-\dfrac{2}{3}\right)^5 = -\dfrac{32}{243}$.

Practice Exercise 1

Compute the following radicals.

(a) $\sqrt[3]{-64}$ (b) $\sqrt[4]{625}$

Answer: (a) -4 (b) 5

Now we list all the properties of radicals that are necessary to work with radicals. We assume that all of the numbers are such that the expressions are defined.

Properties of Radicals

1. $\left(\sqrt[n]{a}\right)^n = a$

2. $\sqrt[n]{ab} = \sqrt[n]{a}\sqrt[n]{b}$

3. $\sqrt[n]{\dfrac{a}{b}} = \dfrac{\sqrt[n]{a}}{\sqrt[n]{b}}$ $(b \neq 0)$

4. $\sqrt[m]{\sqrt[n]{a}} = \sqrt[mn]{a}$

Property 1 is a restatement of the definition of the principal nth root of a. We prove property 2, leaving the others as exercises. Let $x = \sqrt[n]{a}$ and $y = \sqrt[n]{b}$. By the definition of nth root,

$$x^n = a \quad \text{and} \quad y^b = b.$$

Multiplying these two expressions, we have

$$x^n \cdot y^n = (xy)^n = ab.$$

Thus, xy is the nth root of ab, that is, $xy = \sqrt[n]{ab}$. Substituting $\sqrt[n]{a}$ for x and $\sqrt[n]{b}$ for y, we obtain

$$\sqrt[n]{a} \cdot \sqrt[n]{b} = \sqrt[n]{ab}.$$

Simplifying Radicals

A radical is said to be in *simplified form* when

1. all possible factors have been eliminated from under the radical, and
2. the index of the radical is the smallest possible.

EXAMPLE 2   Use the properties of radicals to simplify each expression. Assume that all letters represent positive numbers.

(a) $\sqrt[3]{216}$ **(b)** $\sqrt{\dfrac{15}{144}}$ **(c)** $\sqrt{\dfrac{8a^4}{b^6}}$ **(d)** $\sqrt[3]{\sqrt{64x^7}}$

Solution: **(a)** Since $216 = 8 \cdot 27$, we write

$$\sqrt[3]{216} = \sqrt[3]{8 \cdot 27} = \sqrt[3]{8} \cdot \sqrt[3]{27} = 2 \cdot 3 = 6.$$

(b) Using property 3, we obtain

$$\sqrt{\dfrac{15}{144}} = \dfrac{\sqrt{15}}{\sqrt{144}} = \dfrac{\sqrt{15}}{12}.$$

(c) Using properties 2 and 3, we write

$$\sqrt{\dfrac{8a^4}{b^6}} = \dfrac{\sqrt{8a^4}}{\sqrt{b^6}} = \dfrac{\sqrt{2 \cdot 4 \cdot a^4}}{\sqrt{b^6}} = \dfrac{2a^2\sqrt{2}}{b^3}.$$

(d) Here we use properties 4 and 2:

$$\sqrt[3]{\sqrt{64x^7}} = \sqrt[6]{64x^7} = \sqrt[6]{2^6 x^6 x} = 2x\sqrt[6]{x}.$$

Practice Exercise 2   Use the properties of radicals to simplify each radical.

$$\text{(a)} \ \sqrt{\frac{32}{81}} \quad \text{(b)} \ \sqrt[3]{-512} \quad \text{(c)} \ \sqrt{\frac{16x^3}{y^4}} \quad \text{(d)} \ \sqrt{\sqrt{\sqrt{81a^4b^5}}}$$

Answer: (a) $4\sqrt{2}/9$ (b) -8 (c) $\dfrac{4x\sqrt{x}}{y^2}$ (d) $3ab\sqrt[4]{b}$

Sum and Difference of Radicals

The distributive property of the product over the sum allows us to combine terms of a sum or a difference of radicals with the *same index* and the *same radicand*.

EXAMPLE 3   Perform the following operations and simplify.

$$\text{(a)} \ 3\sqrt{2} + 4\sqrt{18} - \sqrt{32} \quad \text{(b)} \ 5\sqrt[3]{x^4} - 3\sqrt[3]{x^7} + \sqrt[3]{8x^4}$$

Solution: (a) We have

$$
\begin{aligned}
3\sqrt{2} + 4\sqrt{18} - \sqrt{32} &= 3\sqrt{2} + 4\sqrt{2 \cdot 9} - \sqrt{2 \cdot 16} \\
&= 3\sqrt{2} + 4 \cdot 3\sqrt{2} - 4\sqrt{2} \\
&= 3\sqrt{2} + 12\sqrt{2} - 4\sqrt{2} \\
&= (3 + 12 - 4)\sqrt{2} \quad \text{\small Distributive property} \\
&= 11\sqrt{2}.
\end{aligned}
$$

(b) First, we remove all third powers from under each radical; then we use the distributive property.

$$
\begin{aligned}
5\sqrt[3]{x^4} - 3\sqrt[3]{x^7} + \sqrt[3]{8x^4} &= 5\sqrt[3]{x^3 x} - 3\sqrt[3]{x^6 x} + \sqrt[3]{2^3 x^3 x} \\
&= 5x\sqrt[3]{x} - 3x^2\sqrt[3]{x} + 2x\sqrt[3]{x} \\
&= (5x - 3x^2 + 2x)\sqrt[3]{x} \\
&= (7x - 3x^2)\sqrt[3]{x}
\end{aligned}
$$

Practice Exercise 3   Perform the indicated operations and simplify.

$$\text{(a)} \ 2\sqrt[3]{16} - 5\sqrt[3]{2} + 2\sqrt[3]{128} \quad \text{(b)} \ \sqrt{8x^5} + \sqrt{18x^3} - \sqrt{2x^5}$$

Answer: (a) $7\sqrt[3]{2}$ (b) $(x^2 + 3x)\sqrt{2x}$

Rationalizing Denominators

Property 3 of radicals tells us that the radical of a quotient is the quotient of radicals. For instance,

$$\sqrt{\frac{15}{2}} = \frac{\sqrt{15}}{\sqrt{2}},$$

and we have a fraction whose denominator contains a radical. In most computations, it is preferable to replace such a fraction by an equivalent one whose denominator does not contain a radical. In this example, we can achieve this by multiplying both numerator and denominator by $\sqrt{2}$. Thus,

$$\sqrt{\frac{15}{2}} = \frac{\sqrt{15}}{\sqrt{2}} = \frac{\sqrt{15} \cdot \sqrt{2}}{\sqrt{2} \cdot \sqrt{2}} = \frac{\sqrt{30}}{\sqrt{4}} = \frac{\sqrt{30}}{2}.$$

This process is known as *rationalizing the denominator*.

EXAMPLE 4   Rationalize the denominator in each of the following expressions.

(a) $\dfrac{\sqrt{2} + \sqrt{3}}{\sqrt{5}}$ (b) $\dfrac{\sqrt{x} - \sqrt{y}}{\sqrt{x}}$ (c) $\dfrac{\sqrt[3]{3}}{\sqrt[3]{2}}$

Solution: (a) We multiply both numerator and denominator by $\sqrt{5}$, obtaining

$$\frac{\sqrt{2} + \sqrt{3}}{\sqrt{5}} = \frac{(\sqrt{2} + \sqrt{3})\sqrt{5}}{\sqrt{5} \cdot \sqrt{5}} = \frac{\sqrt{10} + \sqrt{15}}{5}.$$

(b) Multiplying the numerator and denominator by $\sqrt{x}$ gives us

$$\frac{\sqrt{x} - \sqrt{y}}{\sqrt{x}} = \frac{(\sqrt{x} - \sqrt{y})\sqrt{x}}{\sqrt{x}\sqrt{x}} = \frac{x - \sqrt{xy}}{x}.$$

(c) Since $\sqrt[3]{2} \cdot \sqrt[3]{2^2} = \sqrt[3]{2^3} = 2$, we multiply both numerator and denominator by $\sqrt[3]{2^2}$ to obtain

$$\frac{\sqrt[3]{3}}{\sqrt[3]{2}} = \frac{\sqrt[3]{3} \cdot \sqrt[3]{2^2}}{\sqrt[3]{2} \cdot \sqrt[3]{2^2}} = \frac{\sqrt[3]{3 \cdot 2^2}}{\sqrt[3]{2^3}} = \frac{\sqrt[3]{12}}{2}.$$

Practice Exercise 4   Rationalize the denominators of the following expressions.

(a) $\dfrac{\sqrt{2} - \sqrt{5}}{\sqrt{3}}$ (b) $\dfrac{\sqrt{a} + \sqrt{b}}{\sqrt{c}}$ (c) $\dfrac{\sqrt[4]{2}}{\sqrt[4]{3}}$

Answer: (a) $\dfrac{\sqrt{6} - \sqrt{15}}{3}$ (b) $\dfrac{\sqrt{ac} + \sqrt{bc}}{c}$ (c) $\dfrac{\sqrt[4]{54}}{3}$

EXAMPLE 5   Rationalize the denominators of the following expressions.

(a) $\dfrac{5 - \sqrt{2}}{3 + \sqrt{2}}$ (b) $\dfrac{1}{\sqrt{x} + \sqrt{y}}$

Solution: (a) Recalling the special product $(x + y)(x - y) = x^2 - y^2$, we see that $(3 + \sqrt{2})(3 - \sqrt{2}) = 9 - 2 = 7$. Thus, multiplying both numerator and denominator by $3 - \sqrt{2}$, we obtain

$$\frac{5 - \sqrt{2}}{3 + \sqrt{2}} = \frac{(5 - \sqrt{2})(3 - \sqrt{2})}{(3 + \sqrt{2})(3 - \sqrt{2})}$$

$$= \frac{5 \cdot 3 - 5\sqrt{2} - 3\sqrt{2} + \sqrt{2} \cdot \sqrt{2}}{3^2 - (\sqrt{2})^2}$$

$$= \frac{15 - 5\sqrt{2} - 3\sqrt{2} + 2}{9 - 2}$$

$$= \frac{17 - 8\sqrt{2}}{7}.$$

The expression $3 - \sqrt{2}$ is called the *conjugate* of $3 + \sqrt{2}$.

(b) In this case $\sqrt{x} - \sqrt{y}$ is the conjugate of $\sqrt{x} + \sqrt{y}$. Multiplying the numerator and denominator by $\sqrt{x} - \sqrt{y}$ gives us

$$\frac{1}{\sqrt{x} + \sqrt{y}} = \frac{\sqrt{x} - \sqrt{y}}{(\sqrt{x} + \sqrt{y})(\sqrt{x} - \sqrt{y})} = \frac{\sqrt{x} - \sqrt{y}}{x - y}.$$

Practice Exercise 5   Rationalize the denominators of the following expressions.

(a) $\dfrac{1}{\sqrt{5} - \sqrt{2}}$ **(b)** $\dfrac{\sqrt{a} + \sqrt{b}}{\sqrt{a} - \sqrt{b}}$

Answer: **(a)** $\dfrac{\sqrt{5} + \sqrt{2}}{3}$ **(b)** $\dfrac{a + b + 2\sqrt{ab}}{a - b}$

Simplifying Expressions Containing Radicals

An expression containing radicals is in *simplified form* if

1. all radicals have been simplified,
2. all radicals have been removed from denominators, and
3. all possible operations have been performed.

EXERCISES 1.7

In Exercises 1–50, simplify each expression. Assume that all letters represent nonnegative real numbers and that all denominators are different from zero.

1. $\sqrt{81}$ **2.** $\sqrt[3]{125}$

3. $\sqrt[5]{-32}$ **4.** $\sqrt[7]{128}$

5. $\sqrt{\dfrac{16}{25}}$ **6.** $\sqrt[4]{\dfrac{81}{625}}$

7. $(\sqrt{3} + 2)(\sqrt{3} - 2)$ **8.** $(\sqrt{5} + 3)(\sqrt{5} - 3)$

9. $(3\sqrt{5} + 4)(3\sqrt{5} - 4)$

10. $(6 + 2\sqrt{3})(6 - 2\sqrt{3})$

11. $(\sqrt{3} - \sqrt{7})^2$ **12.** $(\sqrt{8} + \sqrt{2})^2$

13. $(2\sqrt{3} + \sqrt{2})(3\sqrt{2} - \sqrt{3})$

14. $(3\sqrt{15} - \sqrt{3})(5\sqrt{3} + \sqrt{5})$

15. $(4\sqrt{3} - 2)(6 + \sqrt{3})$ **16.** $(9\sqrt{2} + 3)(6 - \sqrt{2})$

17. $(3\sqrt{3} + 1)(2\sqrt{3} - 3)$

18. $(5\sqrt{2} + 4)(3\sqrt{2} - 1)$

19. $(\sqrt[3]{5^2} + \sqrt[3]{5} + 1)(\sqrt[3]{5} - 1)$

20. $(\sqrt[3]{7} - 1)(\sqrt[3]{7^2} + \sqrt[3]{7} + 1)$

21. $(\sqrt[3]{11} + 1)(\sqrt[3]{11^2} - \sqrt[3]{11} + 1)$

22. $(\sqrt[3]{13^2} - \sqrt[3]{13} + 1)(\sqrt[3]{13} + 1)$

23. $\sqrt[3]{8a^6}$ **24.** $\sqrt[3]{216x^3y^6}$

25. $\sqrt[3]{-27a^5}$ **26.** $\sqrt[4]{32p^5q^6}$

27. $\sqrt{2x}\sqrt{8x^3}$ **28.** $\sqrt{3y^2}\sqrt{27y^4}$

29. $\sqrt{3ab^3}\sqrt{6a^3b}$ **30.** $\sqrt{4u^3v^5}\sqrt{5v^5u^3}$

31. $\sqrt{2ax^2}\sqrt{8a^3x^4}$ **32.** $\sqrt{5u^2v^3}\sqrt{10u^4v^7}$

33. $\sqrt[3]{2xy^2}\sqrt[3]{4x^5y^7}$ **34.** $\sqrt[5]{9a^3b}\sqrt[5]{27a^2b^9}$

35. $\sqrt[6]{4a^2bc^3}\sqrt[6]{16a^4b^5c^9}$ **36.** $\sqrt[3]{3r^2s}\sqrt[3]{-9r}$

37. $\sqrt{\dfrac{16a^2b^4}{c^6}}$ **38.** $\sqrt[3]{\dfrac{27ax^4}{8b^6}}$

39. $\sqrt[5]{\dfrac{32a^6b^7}{c^5}}$

40. $\sqrt{\dfrac{50m^2n^4}{p^3}}$

41. $\sqrt{\dfrac{1}{2a^3b}}$

42. $\sqrt[3]{\dfrac{3b^3x}{2a}}$

43. $\sqrt[3]{\dfrac{4am^2}{9n}}$

44. $\sqrt{\dfrac{3x^2}{yz}}$

45. $\sqrt{\sqrt[3]{128a^7b^8x^8}}$

46. $\sqrt{\sqrt{16m^2p^4q^6}}$

47. $\sqrt{(m+n)^2}$

48. $\left(\sqrt[4]{2ab^2x^3}\right)^4$

49. $\sqrt{25a\sqrt{25a^3}}$

50. $\sqrt[3]{27m^2\sqrt{9m^5}}$

In Exercises 51–62, perform the indicated operations and simplify.

51. $3\sqrt{2} + 5\sqrt{8}$

52. $5\sqrt{3} - 4\sqrt{12}$

53. $2\sqrt{3} - 4\sqrt{27}$

54. $6\sqrt{5} - 3\sqrt{20}$

55. $\sqrt{2} + \sqrt{5} - 6\sqrt{8} + 3\sqrt{45}$

56. $5\sqrt{2} - 4\sqrt{27} + 6\sqrt{3} + 3\sqrt{8}$

57. $\sqrt[3]{16} + 2\sqrt[3]{54}$

58. $\sqrt[3]{40} - 2\sqrt[3]{135}$

59. $\sqrt{4x} - \sqrt{16x} + \sqrt{25x}$

60. $\sqrt{8a} + \sqrt{18a} + \sqrt{50a}$

61. $\sqrt{8a^3} + \sqrt{18a^3} + \sqrt{50a^3}$

62. $\sqrt[3]{8x^4} - \sqrt[3]{27x^7} + \sqrt[3]{64x^4}$

In Exercises 63–86, rationalize the denominators.

63. $\dfrac{1 + \sqrt{3}}{\sqrt{2}}$

64. $\dfrac{2 - \sqrt{2}}{\sqrt{3}}$

65. $\dfrac{a + \sqrt{b}}{\sqrt{c}}$

66. $\dfrac{m - \sqrt{n}}{\sqrt{p}}$

67. $\dfrac{3}{1 - \sqrt{2}}$

68. $\dfrac{4}{2 + \sqrt{2}}$

69. $\dfrac{\sqrt{5} + \sqrt{2}}{\sqrt{3}}$

70. $\dfrac{\sqrt{7} - \sqrt{3}}{\sqrt{6}}$

71. $\dfrac{\sqrt{a} - \sqrt{c}}{\sqrt{c}}$

72. $\dfrac{\sqrt{2x} - \sqrt{y}}{\sqrt{xy}}$

73. $\dfrac{\sqrt{x^3} + \sqrt{x^5}}{\sqrt{x^3}}$

74. $\dfrac{2\sqrt{a^3} + \sqrt{a^5}}{\sqrt{a^3}}$

75. $\dfrac{\sqrt{6} - 4}{2 + \sqrt{5}}$

76. $\dfrac{6}{2 + 3\sqrt{2}}$

77. $\dfrac{a}{\sqrt{a} + 2}$

78. $\dfrac{3}{2 - \sqrt{x}}$

79. $\dfrac{\sqrt{x}}{\sqrt{x} + \sqrt{y}}$

80. $\dfrac{2a}{\sqrt{a} + \sqrt{b}}$

81. $\dfrac{\sqrt{x - 1}}{1 - \sqrt{x - 1}}$

82. $\dfrac{2}{\sqrt{x + 1} - \sqrt{x}}$

83. $\dfrac{1}{a + \sqrt{b}}$

84. $\dfrac{1}{\sqrt{a + b} + c}$

☐ **85.** $\dfrac{1}{\sqrt{x^2 + 1} + x}$

☐ **86.** $\dfrac{h}{\sqrt{2(x + h) + 1} - \sqrt{2x + 1}}$

☐ **87.** Show that $\sqrt{3} + \sqrt{5} = \sqrt{8 + 2\sqrt{15}}$.

☐ **88.** Show that $\sqrt{3} - \sqrt{2} = \sqrt{5 - 2\sqrt{6}}$.

☐ **89.** Prove that $\sqrt[n]{\dfrac{a}{b}} = \dfrac{\sqrt[n]{a}}{\sqrt[n]{b}}$.

☐ **90.** Prove that $\sqrt[m]{\sqrt[n]{a}} = \sqrt[mn]{a}$.

1.8   RATIONAL EXPONENTS

In this section, we extend the definition of powers to *rational* values of the exponents.

> ### Rational Exponents
>
> If m and n are *natural numbers* and a is a real number such that $\sqrt[n]{a}$ exists, then
>
> $$a^{m/n} = \left(\sqrt[n]{a}\right)^m = \sqrt[n]{a^m}. \tag{1.14}$$

Formula (1.14) tells us that a rational power $a^{m/n}$ can be computed in two different ways: we may *raise the nth root of a to the mth power,* or we can *take the nth root of the mth power of a.*

EXAMPLE 1

Find the value of each expression.
(a) $16^{3/4}$ (b) $(8a^6)^{2/3}$

Solution: (a) $16^{3/4} = \left(\sqrt[4]{16}\right)^3 = 2^3 = 8$

or

$$16^{3/4} = \sqrt[4]{16^3} = \sqrt[4]{4096} = 8$$

(b) $(8a^6)^{2/3} = \left(\sqrt[3]{8a^6}\right)^2 = (2a^2)^2 = 4a^4$

or

$$(8a^6)^{2/3} = \sqrt[3]{(8a^6)^2} = \sqrt[3]{64a^{12}} = 4a^4$$

Notice that both expressions, $\left(\sqrt[4]{16}\right)^3$ and $\left(\sqrt[3]{8a^6}\right)^2$, are easier to evaluate than their counterparts.

Practice Exercise 1

Compute each of the following expressions.
(a) $8^{2/3}$ (b) $(81a^8)^{3/4}$

Answer: (a) 4   (b) $27a^6$

Setting $m = 1$ in the definition of rational exponents gives us the formula

$$a^{1/n} = \sqrt[n]{a}. \qquad \textbf{(1.15)}$$

That is, the principal *n*th root of a number can be viewed as a rational exponent. Combining formulas (1.14) and (1.15), we may write

$$a^{m/n} = (a^{1/n})^m = (a^m)^{1/n}.$$

Negative Exponents

The definition of rational exponents can be extended to the case of *negative* exponents by defining

$$a^{-m/n} = \frac{1}{a^{m/n}}.$$

EXAMPLE 2

(a) Evaluate $8^{-2/3}$.   (b) Simplify $(25a^6)^{-1/2}$.

Solution: We use the definition of negative rational exponents for both parts.

(a) $8^{-2/3} = \dfrac{1}{8^{2/3}} = \dfrac{1}{(\sqrt[3]{8})^2} = \dfrac{1}{4}$

(b) $(25a^6)^{-1/2} = \dfrac{1}{(25a^6)^{1/2}} = \dfrac{1}{\sqrt{25a^6}} = \dfrac{1}{5a^3}$

Practice Exercise 2   **(a)** Evaluate $81^{-3/4}$.   **(b)** Simplify $(8a^6)^{-1/3}$.

Answer: **(a)** $1/27$ **(b)** $1/(2a^2)$

Properties 1 through 5 of integer exponents (Section 1.3) are valid for rational exponents, too. For future reference, we list them here.

Properties of Rational Exponents

Let a and b be real numbers and let r and s be rational numbers. The following properties hold whenever the indicated powers are defined.

1. $a^r \cdot a^s = a^{r+s}$

2. $(a^r)^s = a^{rs}$

3. $\dfrac{a^r}{a^s} = a^{r-s}$, $a \neq 0$

4. $(ab)^r = a^r \cdot b^r$

5. $\left(\dfrac{a}{b}\right)^r = \dfrac{a^r}{b^r}$, $b \neq 0$

6. $a^0 = 1$, $a \neq 0$

7. $a^{-r} = \dfrac{1}{a^r}$, $a \neq 0$

EXAMPLE 3   Simplify and write your answers using only positive exponents. Assume that all letters represent positive numbers.

(a) $(2a^{1/3})(4a^{1/2})$ **(b)** $\left(\dfrac{4x^{-2}}{y^{-4}}\right)^{1/2}$

Solution: **(a)** $(2a^{1/3})(4a^{1/2}) = 8a^{1/3+1/2} = 8a^{5/6}$

(b) $\left(\dfrac{4x^{-2}}{y^{-4}}\right)^{1/2} = \left(\dfrac{4y^4}{x^2}\right)^{1/2} = \dfrac{\sqrt{4y^4}}{\sqrt{x^2}} = \dfrac{2y^2}{x}$

Practice Exercise 3   Assuming that all variables represent positive real numbers, simplify each expression and write your answers with positive exponents only.

(a) $(3a^{1/4}b^{-3/4})^4$ **(b)** $\left(\dfrac{2u^{1/2}}{u^{1/3}}\right)^3$

Answer: **(a)** $\dfrac{81a}{b^3}$ **(b)** $8u^{1/2}$

Many expressions involving radicals can best be simplified by changing the radicals to rational exponents, performing the operations, and changing back to radicals.

EXAMPLE 4   Express the following quantities as one radical in simplest form.

(a) $\sqrt{2} \cdot \sqrt[3]{4}$ (b) $\dfrac{\sqrt[3]{4a^2}}{\sqrt{2a}}$

Solution:

(a) $\sqrt{2} \cdot \sqrt[3]{4} = \sqrt{2} \cdot \sqrt[3]{2^2} = 2^{1/2} \cdot 2^{2/3} = 2^{1/2+2/3} = 2^{7/6} = 2 \cdot 2^{1/6} = 2\sqrt[6]{2}$

(b) $\dfrac{\sqrt[3]{4a^2}}{\sqrt{2a}} = \dfrac{\sqrt[3]{(2a)^2}}{\sqrt{2a}} = \dfrac{(2a)^{2/3}}{(2a)^{1/2}} = (2a)^{2/3-1/2} = (2a)^{1/6} = \sqrt[6]{2a}$

↝ **Practice Exercise 4**   Write as one radical in simplest form.

(a) $\dfrac{\sqrt[4]{8}}{\sqrt{2}}$ (b) $\sqrt{2x^2} \cdot \sqrt[4]{8x}$

Answer: (a) $\sqrt[4]{2}$ (b) $2x\sqrt[4]{2x}$

EXERCISES 1.8

In Exercises 1–8, write each of the following using radicals instead of rational exponents. Assume that all variables represent positive real numbers.

1. $8^{3/2}$ **2.** $27^{4/3}$
3. $64^{-1/4}$ **4.** $81^{-3/2}$
5. $(a^2b^3)^{3/5}$ **6.** $(4xy^2)^{2/3}$
7. $(x^2 + y^2)^{1/2}$ **8.** $(x^4 + xy)^{1/4}$

In Exercises 9–16, write each of the following using rational exponents instead of radicals.

9. $\sqrt[3]{5^2}$ **10.** $\sqrt[6]{7^5}$
11. $\sqrt[4]{x^3}$ **12.** $\sqrt[5]{y^2}$
13. $a^2\sqrt{a}$ **14.** $b\sqrt{b^3}$
15. $a^3\sqrt[4]{a^5}$ **16.** $x^2\sqrt[3]{x^5}$

In Exercises 17–32, simplify each of the following expressions. All variables represent positive real numbers.

17. $(-8)^{2/3}$ **18.** $(-27)^{4/3}$
19. $81^{-1/2}$ **20.** $9^{-3/2}$
21. $(-243)^{3/5}$ **22.** $128^{-5/7}$
23. $(5u^{1/2})(3u^{3/2})$ **24.** $2ax^{1/6}(3ax^{1/3})$
25. $(-125x^6)^{2/3}$ **26.** $(16a^2x^4)^{3/2}$
27. $(m^{-2}n^{-3})^{-1/6}$ **28.** $(8a^{-3}b^{-6})^{-2/3}$
29. $\left(\dfrac{5a^{1/4}}{b^{1/2}}\right)^2$ **30.** $\left(\dfrac{a^{1/3}}{b^{3/2}}\right)^6$
31. $\left(\dfrac{81x^{-8}}{y^4}\right)^{3/4}$ **32.** $\left(\dfrac{-125a^3b^6}{c^{-9}}\right)^{2/3}$

In Exercises 33–46, simplify the given expressions and write the answers with positive exponents only. All variables represent positive numbers.

33. $(36a^8b^2)^{-1/2}$ **34.** $(125a^3)^{-2/3}$
35. $(64a^2b^3)^{-2/3}$ **36.** $(32x^8y^4)^{1/4}$
37. $\left(\dfrac{36x^2y^4}{z^3}\right)^{-1/2}$ **38.** $\left(\dfrac{a^8b^{12}}{c^3}\right)^{-3/4}$
39. $(a^{-2/3}b^{-1/2})^{-6}$ **40.** $(u^{-3}v^2)^{-1/6}$
41. $\left(\dfrac{16a^4x^{-1}}{25a^{-2}x^3}\right)^{1/2}$ **42.** $\left(\dfrac{8a^{-1}b^2}{27a^2b^{-4}}\right)^{1/3}$
43. $\left(\dfrac{4}{9}a^{-1/3}x^4\right)^{-3/2}$ **44.** $\left(\dfrac{8}{27}b^6y^{-3/2}\right)^{2/3}$
45. $\left(\dfrac{a^2}{b^{1/4}}\right)\left(\dfrac{b^{-1/2}}{a^{3/2}}\right)$ **46.** $\dfrac{(a^{-6}x)^{-1/3}}{(a^2x^4)^{-1/2}}$

In Exercises 47–60, simplify and write each expression as a radical with least positive index. Assume that all variables represent positive real numbers.

47. $\sqrt{2} \cdot \sqrt[3]{3}$ **48.** $\sqrt{3} \cdot \sqrt[4]{9}$
49. $\sqrt{a} \cdot \sqrt[4]{2a}$ **50.** $\sqrt[3]{2x} \cdot \sqrt[6]{2x}$
51. $\dfrac{\sqrt[4]{25}}{\sqrt{5}}$ **52.** $\dfrac{\sqrt[3]{4}}{\sqrt{2}}$
53. $\dfrac{\sqrt{2a^2}}{\sqrt[4]{2a}}$ **54.** $\dfrac{\sqrt{2a}}{\sqrt[3]{a}}$
55. $\dfrac{\sqrt{4x^2y^2}}{\sqrt[6]{2xy}}$ **56.** $\dfrac{\sqrt{4a^3b^3}}{\sqrt[3]{2ab}}$

57. $\dfrac{\sqrt[3]{m^2 v}}{\sqrt[6]{m^3 v}}$ **58.** $\dfrac{\sqrt{2ax^3}}{\sqrt[3]{4a^2 x}}$ **59.** $\sqrt[3]{2a\sqrt{2a}}$ **60.** $\sqrt{2a\sqrt[3]{2a}}$

HISTORICAL NOTE

COMPLEX NUMBERS

The concept of complex numbers was first proposed by Hieronimo Cardano (1501–1576), an Italian mathematician, in his "Ars Magna," a treatise on the solution of cubic and quartic equations. Cardano's ideas were ignored until 1799, when Carl Friedrich Gauss (1777–1855), a German mathematician, revived them and used complex numbers in several of his works.

1.9   COMPLEX NUMBERS

Although the real number system is more than adequate for most of the applications discussed in this book, the concept of complex numbers is needed in a few instances, such as solving quadratic equations. Moreover, complex numbers are indispensable in certain branches of mathematics and are often used in physics and engineering.

In order to understand why we need complex numbers, notice that the square x^2 of an arbitrary real number x is *always nonnegative*. Thus, there is *no* real number x such that $x^2 = -1$. In other words, it is impossible to find a real number satisfying the statement $x^2 + 1 = 0$. In view of this impossibility, complex numbers were invented.

The Imaginary Unit

The *imaginary unit i* is defined by requiring that

$$i^2 = -1.$$

We also write i as $\sqrt{-1}$. If the imaginary unit is combined with two real numbers by addition and multiplication, a complex number is obtained.

Complex Numbers

A *complex number* is a number of the form

$$a + bi,$$

where a and b are real numbers. The set of all complex numbers is denoted by $\mathbb{C}$.

If $a = 0$, then the complex number $0 + bi$ is called a *pure imaginary* number. If $b = 0$, then the complex number is equal to the real number a. Conversely, every real number can be viewed as a complex number by writing $a = a + 0i$. In this way, the set $\mathbb{R}$ of real numbers becomes a subset of the set $\mathbb{C}$ of complex numbers.

We have described several systems of numbers: *natural numbers, integers, rational, real,* and *complex numbers.* The set $\mathbb{N}$ of natural numbers is contained in the set $\mathbb{Z}$ of integers. These form a subset of $\mathbb{Q}$, the set of rational numbers. Real numbers consist of rational and irrational numbers; thus the set $\mathbb{R}$ contains the set $\mathbb{Q}$. Finally, $\mathbb{C}$, the largest set, contains $\mathbb{R}$. In set notation, the inclusion among the different systems of numbers can be written

$$\mathbb{N} \subset \mathbb{Z} \subset \mathbb{Q} \subset \mathbb{R} \subset \mathbb{C}.$$

Before extending the operations of addition and multiplication to complex numbers, we introduce some definitions and terminology.

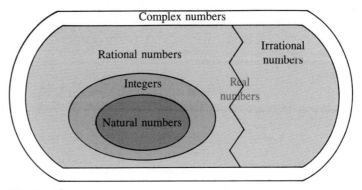

Figure 1.7

Equality

Two complex numbers $a + bi$ and $c + di$ are said to be *equal*, written

$$a + bi = c + di,$$

if and only if

$$a = c \quad \text{and} \quad b = d.$$

Real and Imaginary Parts of a Complex Number

Given any complex number $z = a + bi$, we call a the *real part* of z and b the *imaginary part* of z. Thus, two complex numbers are equal if and only if their real parts are equal and their imaginary parts are equal.

Addition and Multiplication

If $a + bi$ and $c + di$ are complex numbers, we define

$$(a + bi) + (c + di) = (a + c) + (b + d)i$$

and

$$(a + bi)(c + di) = (ac - bd) + (bc + ad)i.$$

The right-hand sides of the sum and product of two complex numbers are obtained by using the associative, commutative, and distributive properties together with the relation $i^2 = -1$. This is because we want the same arithmetic rules used for real numbers to hold for complex numbers.

EXAMPLE **1**

Write each of the following numbers in the form $a + bi$.
(a) $(2 + 3i) + (5 - 4i)$ **(b)** $(3 + 5i)(2 + 4i)$

Solution: **(a)** We have

$$(2 + 3i) + (5 - 4i) = 2 + 3i + 5 - 4i$$
$$= (2 + 5) + (3 - 4)i$$
$$= 7 - i.$$

(b) Using the distributive property, we obtain

$$(3 + 5i)(2 + 4i) = 3(2 + 4i) + 5i(2 + 4i)$$
$$= 3 \cdot 2 + 3(4i) + (5i)2 + (5i)(4i)$$
$$= 6 + 12i + 10i + 20i^2.$$

Since $i^2 = -1$, we get

$$(3 + 5i)(2 + 4i) = 6 + 22i + 20(-1)$$
$$= 6 + 22i - 20$$
$$= -14 + 22i.$$

Practice Exercise 1

Find the sum and product of the complex numbers $2 - 3i$ and $5 + 4i$.

Answer: $7 + i$ and $22 - 7i$

Complex Conjugate

If $z = a + bi$ is a complex number, then the number $\bar{z} = a - bi$ is called the *complex conjugate* of z.

Complex conjugate numbers have the same real part, but their imaginary parts have the opposite sign. Notice that the complex conjugate of $\bar{z}$ is z.

EXAMPLE **2**

Find the product $(2 + 3i)(2 - 3i)$.

Solution: Using the distributive property, we have

$$(2 + 3i)(2 - 3i) = 2 \cdot 2 + 2(-3i) + (3i)2 + (3i)(-3i)$$
$$= 4 - 6i + 6i - 9i^2$$
$$= 4 - 9(-1)$$
$$= 13.$$

Thus, the product of the complex number $2 + 3i$ by its own complex conjugate is the real number 13.

Practice Exercise 2   Find the product $(3 + 5i)(3 - 5i)$.

Answer: 34

In general, *the product of the complex number by its own complex conjugate is a real number*. Indeed, if $z = a + bi$, then

$$
\begin{aligned}
z\bar{z} &= (a + bi)(a - bi) \\
&= a \cdot a + a(-bi) + (bi)a + (bi)(-bi) \quad \text{Distributive} \\
&= a \cdot a - a \cdot bi + a \cdot bi - (b \cdot b)i^2 \quad \text{property} \\
&= a^2 + b^2. \quad\quad\quad\quad\quad\quad\quad\quad\quad i^2 = -1
\end{aligned}
$$

Thus, the product $z\bar{z}$ is a *nonnegative real number*. Furthermore, if $z = a + bi \neq 0$ (that is, if at least one of a and b is not zero), then $z\bar{z} > 0$.

The set $\mathbb{C}$ of complex numbers is closed under the operations of addition and multiplication, and these operations satisfy the commutative, associative, and distributive properties. Furthermore, $0 = 0 + 0i$ is the additive identity, and $1 = 1 + 0i$ is the multiplicative identity.

Additive Inverses

If $z = a + bi$ is a complex number, then the *additive inverse* of z is the number $-z = (-a) + (-b)i$.

A quick calculation shows that

$$
\begin{aligned}
z + (-z) &= (a + bi) + ((-a) + (-b)i) \\
&= (a + (-a)) + (b + (-b))i \\
&= 0 + 0i \\
&= 0.
\end{aligned}
$$

EXAMPLE 3   Find the difference $(2 + 3i) - (4 + 5i)$.

Solution: To find the difference, we add to $2 + 3i$ the additive inverse of $4 + 5i$:

$$
\begin{aligned}
(2 + 3i) - (4 + 5i) &= (2 + 3i) + ((-4) + (-5)i) \\
&= (2 + (-4)) + (3 + (-5))i \\
&= (-2) + (-2)i \\
&= -2 - 2i.
\end{aligned}
$$

Practice Exercise 3   Write $(7 + 2i) - (5 + 8i)$ in the form $a + bi$.

Answer: $2 - 6i$

Multiplicative Inverses

Every complex number $z \neq 0$ has a multiplicative inverse, that is, a complex number w such that $zw = 1$. If $z = a + bi$, then the real and imaginary parts of w can be determined as follows. Multiplying both sides of the equality

$$zw = 1$$

by $\bar{z}$ yields

$$\bar{z}(zw) = \bar{z}$$

or

$$(\bar{z}z)w = \bar{z}$$

But $\bar{z}z = z\bar{z} = a^2 + b^2$, so

$$(a^2 + b^2)w = \bar{z},$$

hence

$$w = \frac{\bar{z}}{a^2 + b^2}.$$

Recalling that $\bar{z} = a - bi$, we have

$$w = \frac{a - bi}{a^2 + b^2}$$

or

$$w = \frac{a}{a^2 + b^2} - \frac{b}{a^2 + b^2}\, i,$$

which is the multiplicative inverse of $z = a + bi \neq 0$.

The multiplicative inverse (also called *reciprocal*) of a complex number $z = a + bi$ is denoted by z^{-1}, $1/z$, $(a + bi)^{-1}$, or $1/(a + bi)$.

Quotient

If z and w are complex numbers, with $w \neq 0$, then the *quotient* of z by w is defined by

$$\frac{z}{w} = zw^{-1}.$$

EXAMPLE 4   Write each of the following numbers in the form $a + bi$.

(a) $\dfrac{1}{3 + 4i}$ (b) $\dfrac{3 + 5i}{2 - 4i}$

Solution: (a) According to the formula for the reciprocal of a complex number, we can write

$$\frac{1}{3 + 4i} = \frac{3}{25} - \frac{4}{25}\,i.$$

However, there is no need to memorize the formula. To find the reciprocal of z, just multiply the numerator and denominator of the expression $1/z$ by $\bar{z}$, as follows:

$$\frac{1}{3 + 4i} = \frac{1 \cdot (3 - 4i)}{(3 + 4i)(3 - 4i)} = \frac{3 - 4i}{9 + 16} = \frac{3}{25} - \frac{4}{25}\,i.$$

(b) Multiplying the numerator and denominator of the given complex fraction by $2 + 4i$, which is the complex conjugate of $2 - 4i$, we obtain

$$\frac{3 + 5i}{2 - 4i} = \frac{(3 + 5i)(2 + 4i)}{(2 - 4i)(2 + 4i)} = \frac{-14 + 22i}{4 + 16}$$

$$= \frac{14}{20} + \frac{22}{20}\,i = -\frac{7}{10} + \frac{11}{10}\,i.$$

Practice Exercise 4   Write each of the following numbers in the form $a + bi$.

(a) $\dfrac{1}{2 - 3i}$ (b) $\dfrac{3 - 2i}{3 + 4i}$

Answer: (a) $\dfrac{2}{13} + \dfrac{3}{13}\,i$ (b) $\dfrac{1}{25} - \dfrac{18}{25}\,i$

EXERCISES 1.9

In each of Exercises 1–36, write the given complex numbers in the form $a + bi$.

1. $(3 + 2i) + (-4 + 3i)$

2. $(-5 + 4i)$ $+ (-3 - 2i)$

3. $(5 - 6i) + (3 - 5i)$

4. $(12 + 5i) - (6 - 8i)$

5. $-(3 + 4i) + (5 - 9i)$

6. $(8 - 5i) - (-3 - 4i)$

7. $10 - (8 + 5i)$

8. $-9 + (-5 + 12i)$

9. $(7 - 6i) - 3i$

10. $2i - (4 - 6i)$

11. $(4 - 3i)(3 + 4i)$

12. $(5 - 5i)(3 - 8i)$

13. $(-8 - i)(4 + i)$

14. $(7 + i)(8 + 5i)$

15. $\left(\dfrac{1}{2} + \dfrac{3}{2}i\right)(3 - 4i)$

16. $\left(\dfrac{3}{4} - 5i\right)\left(1 - \dfrac{1}{4}i\right)$

17. $(3 + 5i)(3 - 5i)$

18. $(-2 + 7i)(-2 - 7i)$

19. $\left(-\dfrac{2}{3} + \dfrac{3}{4}i\right)\left(-\dfrac{2}{3} - \dfrac{3}{4}i\right)$

20. $\left(\dfrac{1}{5} + \dfrac{3}{8}i\right)\left(\dfrac{1}{5} - \dfrac{3}{8}i\right)$

21. $(2 + 5i)(2 + 5i)$

22. $(3 - 4i)(3 - 4i)$

23. $i(3 - 2i)(3 + 5i)$

24. $(1 - i)(3 - 4i)(5i)$

25. $3i(1 - 2i)(3 - 4i)$

26. $2i\left(\frac{1}{2} + \frac{3}{4}i\right)\left(\frac{1}{2} - \frac{3}{4}i\right)$

27. $\dfrac{1}{1 + i}$

28. $\dfrac{1}{5 - 3i}$

29. $\dfrac{3 - 3i}{4 + 4i}$

30. $\dfrac{1 + i}{3 + i}$

31. $\dfrac{2 + 5i}{2 - 5i}$

32. $\dfrac{5 - 3i}{5 + 3i}$

33. $\dfrac{3 - i}{2 + 4i}$

34. $\dfrac{4 - 5i}{3 - 2i}$

35. $\dfrac{2 - 3i}{3 + 6i}$

36. $\dfrac{3 + 7i}{4 - 3i}$

In Exercises 37–40, let $z = a + bi$ and $w = c + di$. Prove each of the following relations.

37. $\overline{z + w} = \bar{z} + \bar{w}$

38. $\overline{zw} = \bar{z}\,\bar{w}$

39. $\overline{z^{-1}} = (\bar{z})^{-1}$

40. $\overline{(-z)} = -\bar{z}.$

CHAPTER SUMMARY

Real numbers form the basis of a college algebra and trigonometry course. They consist of the *natural numbers*, the *integers*, the *rational numbers*, and the *irrational numbers*. Rational numbers, also called *fractions*, are ratios of integers.

Real numbers can be represented by *decimals*. If the decimal is *terminating* or *repeating*, the number is a rational one. *Nonterminating* and *nonrepeating* decimals correspond to irrational numbers.

The set of all real numbers is closed under the addition and multiplication operations. These operations satisfy the following important properties:

> *Commutativity*
> $$x + y = y + x \quad \text{and} \quad xy = yx$$
> *Associativity*
> $$(x + y) + z = x + (y + z) \quad \text{and} \quad (xy)z = x(yz)$$
> The number 0 is the *additive identity*
> $$x + 0 = x$$
> The number 1 is the *multiplicative identity*
> $$x \cdot 1 = x$$
> *Additive inverse*
> $$x + (-x) = 0$$
> *Multiplicative inverse*
> $$x\left(\frac{1}{x}\right) = 1 \quad (\text{for } x \neq 0)$$
> *Distributivity of the product over the sum*
> $$x(y + z) = xy + xz$$

Real numbers can be represented by points on a *coordinate line* and points on a coordinate line are labeled by real numbers.

Positive numbers correspond to points located to the right of the *origin* (zero), while *negative numbers* are located to the left of the origin.

The multiplication operation leads to the definition of *integer powers* of a real number, and generalizing this concept leads to the definitions of *radicals* and *rational powers* of a real number.

The powers

$$x^0 = 1, \quad x^1 = x, \quad x^2, \ldots, \quad x^n, \ldots$$

where x is a real number and the exponents are *nonnegative integers*, are the basic elements used to define *polynomials*. A *rational expression* is the quotient of two polynomials. *Factoring*, the process of determining the *prime factors* of a polynomial, is aided by the following *special products:*

> I. *Square of a binomial*
> $$(x + y)^2 = x^2 + 2xy + y^2$$
> II. *Product of a sum by a difference*
> $$(x + y)(x - y) = x^2 - y^2$$
> III. *Product of two binomials*
> $$(x + a)(x + b) = x^2 + (a + b)x + ab$$
> IV. *Sum of cubes*
> $$x^3 + y^3 = (x + y)(x^2 - xy + y^2)$$
> V. *Difference of cubes*
> $$x^3 - y^3 = (x - y)(x^2 + xy + y^2)$$
> VI. *Product of two binomials*
> $$(mx + n)(px + q) = mpx^2 + (mq + np)x + nq$$

Most of the time we shall deal with real numbers. Nonetheless, the concept of *complex numbers* is needed when solving quadratic equations and, more generally, when discussing properties of zeros of polynomials. Complex numbers are introduced by defining the *imaginary unit, i,* which satisfies the relation $i^2 = 1$. A complex number z can be written as follows:

$$z = a + bi,$$

where a is the *real part* and b the *imaginary part* of z. The operations of addition and multiplication can be extended to complex numbers, and the set of all complex numbers is closed under these operations. Moreover, they satisfy the commutative, associative, and distributive properties. The numbers 0 and 1 are, respectively, the additive and multiplicative identities. Every complex number $z = a + bi$ has a complex conjugate

$$\bar{z} = a - bi.$$

The product $z\bar{z}$ of a complex number and its complex conjugate is always a nonnegative real number:

$$z\bar{z} = a^2 + b^2.$$

Every complex number $z = a + bi \neq 0$ has a *multiplicative inverse,* that is, a complex number z^{-1} such that $zz^{-1} = 1$. The multiplicative inverse z^{-1} can be written

$$z^{-1} = \frac{a}{a^2 + b^2} - \frac{b}{a^2 + b^2}\, i.$$

CHAPTER TEST

1. Which of the following are rational numbers? Irrational numbers?
   (a) $\left(5 + \sqrt{5}\right)\left(5 - \sqrt{5}\right)$ (b) $\left(-1 - 7\sqrt{3}\right)^2$

2. Simplify the expression $\dfrac{(p\, q^2 r^3)^{-1}\,(p^2 q^3 r)^2}{(p^3 q\, r^2)^{-1}}$.

3. Simplify $(4a^2 + 4a - 2) - (-4a^2 - 8a) + (-3a + 2)$ as completely as possible.

4. Perform the operations and simplify:
   $$\frac{y + 1}{y^2 + 5y + 6} - \frac{y - 2}{y^2 - y - 6} .$$

5. Factor $48u^2 - 75$ as completely as possible.

6. Rationalize the denominator: $\dfrac{\sqrt{5} - 5}{\sqrt{2} - 2}$.

7. Factor the polynomial $20z^2 + 7bz - 6b^2$.

8. Perform the division: $\dfrac{-30a^2 x^5 + 15a^4 x^5 + 25a^3 x^4}{-5a^2 x^4}$.

9. Factor completely the expression $16nv - 12qv - 12nw + 9qw$.

10. Write $\dfrac{5}{4 + i}$ in the form $a + bi$.

REVIEW EXERCISES

In Exercises 1–6, write each fraction as a decimal number.

1. $\dfrac{7}{16}$ 2. $\dfrac{32}{80}$

3. $\dfrac{9}{11}$ 4. $\dfrac{8}{15}$

5. $\dfrac{15}{64}$ 6. $\dfrac{14}{125}$

In Exercises 7–12, write each decimal number as a fraction in lowest terms.

7. 5.18 8. 3.45

9. $0.\overline{141}$ 10. $0.\overline{45}$

11. $1.3\overline{18}$ 12. $2.21\overline{6}$

In Exercises 13–16, write each sum as a repeating decimal.

13. $0.\overline{6} + 0.\overline{7}$

14. $0.\overline{3} + 0.\overline{8}$

15. $0.\overline{5} + 0.\overline{13}$

16. $0.\overline{4} + 0.\overline{18}$

In Exercises 17–22, compute each expression and give your answer in simplified form.

17. $\dfrac{\dfrac{1}{2} + \dfrac{3}{4}}{\dfrac{5}{6} - \dfrac{2}{3}}$

18. $\dfrac{\dfrac{4}{5} - \dfrac{2}{3}}{\dfrac{1}{2} + \dfrac{3}{7}}$

19. $\dfrac{\dfrac{2}{5}\left(\dfrac{3}{8} - \dfrac{1}{2}\right)}{\dfrac{4}{10} - \dfrac{3}{5}}$

20. $\dfrac{\left(\dfrac{3}{7} + \dfrac{2}{3}\right)\left(\dfrac{5}{12} - \dfrac{3}{4}\right)}{\dfrac{5}{6} - \dfrac{1}{8}}$

21. $\dfrac{\dfrac{5}{8} \div \left(\dfrac{7}{12} + 1\right)}{\dfrac{3}{14} - \dfrac{3}{7}}$

22. $\dfrac{\dfrac{3}{5} + \dfrac{5}{6}}{\dfrac{3}{4} \times \left(\dfrac{4}{5} - \dfrac{3}{8}\right)}$

23. The difference of two rational numbers is a rational number. Is the difference of two irrational numbers necessarily an irrational number?

24. Is the quotient of two irrational numbers always an irrational number?

25. Which of the following numbers are rational? Which are irrational numbers?

(a) $\left(3 + 5\sqrt{2}\right)\left(3 - 5\sqrt{2}\right)$ (b) $\sqrt{3} - 5\sqrt{3}$

(c) $0.150150015000150000015\ldots$ (d) $\sqrt{8}/\sqrt{2}$

26. Which of the following numbers are rational? Which are irrational?

(a) $0.\overline{45}$ (b) $\left(2 - 3\sqrt{2}\right)^2$

(c) $\dfrac{5 + 2\sqrt{3}}{5 - 2\sqrt{3}}$ (d) $\sqrt{54}\,\sqrt{24}$

In Exercises 27–44, simplify and write the answers with positive exponents only.

27. $(3a^{-5}x^2)(4ay^{-3})$

28. $(6x^2y^5)(4xy^{-3})^{-2}$

29. $\dfrac{6a^2b^{-5}}{4c^{-4}}$

30. $\dfrac{5au}{a^{-2}b^{-1}u^{-4}}$

31. $\left(\dfrac{3my^{-4}}{2y}\right)$

32. $\left(\dfrac{3a^2u^5}{b^{-1}}\right)^2$

33. $\left(\dfrac{m^3p^2q^5}{2mp}\right)^3$

34. $\left(\dfrac{5a^3t^6}{2s^2}\right)^2$

35. $\left(\dfrac{a^2x^{-3}}{b}\right)^{-2}$

36. $\left(\dfrac{3a^4x^2}{y^{-3}}\right)^{-3}$

37. $(25a^6x^4)^{-1/2}$

38. $(64b^3x^6y^{-9})^{-2/3}$

39. $(x^{-1/4}y^{-2/4})^{-2}$

40. $(a^{-3}m^2n^6)^{-1/6}$

41. $\left(\dfrac{8a^3x^{-1}}{27a^{-4}x^3}\right)^{1/2}$

42. $\left(\dfrac{8}{27}x^5y^{-2/3}\right)^{2/3}$

43. $\dfrac{(b^{-8}y)^{-1/4}}{(b^3y^6)^{-1/6}}$

44. $\left(\dfrac{x^4}{b^{1/2}}\right)\left(\dfrac{b^{-1/2}}{x^{3/2}}\right)^4$

Without using a calculator, perform the operations in Exercises 45–48 and give your answer in scientific notation. Simplify as much as possible.

45. $\dfrac{(3.6 \times 10^{-4})(3.5 \times 10^{-2})}{(0.7 \times 10^5)(0.6 \times 10^{-7})}$

46. $\dfrac{(2.4 \times 10^{-5})(4.5 \times 10^4)}{(0.5 \times 10^2)(0.3 \times 10^{-3})}$

47. $\dfrac{(3.1 \times 10^{-3})(4.2 \times 10^{-5})}{(2.0 \times 10^5)(7.0 \times 10^4)}$

48. $\dfrac{(1.2 \times 10^6)(8.2 \times 10^{-4})}{(4.1 \times 10^{-7})(4.8 \times 10^{-5})}$

C Perform the indicated operations in Exercises 49–52 and give the answers in scientific notation.

49. $\dfrac{(1.313 \times 10^{-6})(2.11 \times 10^9)}{(3.12 \times 10^3)^2(1.4 \times 10^{-8})}$

50. $\dfrac{(2.123 \times 10^{-6})(2.15 \times 10^{-5})}{(4.17 \times 10^6)^2(5.123 \times 10^8)}$

51. $\dfrac{(2.134 \times 10^{-8})(3.15 \times 10^{-3})}{(5.07 \times 10^4)(2.312 \times 10^6)}$

52. $\dfrac{(5.11 \times 10^{16})(6.12 \times 10^{-4})}{(3.25 \times 10^{-6})^3}$

53. The speed of light is approximately 300,000 kilometers per second. How many kilometers does a light signal travel in 4×10^5 seconds?

54. The speed of light is approximately 186,000 miles per second. How many miles does a light signal travel in 3×10^{-7} seconds?

C **55.** Let $V = xyz$. Find V if $x = 1.516 \times 10^3$, $y = 1.217 \times 10^4$, and $z = 1.605 \times 10^5$.

C **56.** Let $F = mn/r^2$. If $m = 3.21 \times 10^{-3}$, $n = 1.32 \times 10^{-5,}$ and $r = 5 \times 10^{-3}$, find F.

C **57.** The speed of light is approximately 300,000 kilometers per second and there are approximately 31,560,000 seconds in a mean solar year. How many kilometers does a light-year represent?

C **58.** How many miles does a light ray travel in a mean solar year? (See Exercise 54 for the speed of light in miles per second and Exercise 57 for the length of a mean solar year.)

C **59.** Light emanating from the sun takes 8 minutes to reach the earth. Find the distance from the earth to the sun in kilometers.

C **60.** In Exercise 59, find the distance from the earth to the sun in miles.

c **61.** One gallon is approximately equal to 3.785 liters. How many liters are there in 170,000 gallons?

c **62.** One millimeter is about 3.9×10^{-2} inches. How many inches are there in 50,000 millimeters?

c **63.** One kilogram is approximately 2.203 lb. Express the weight of the earth in pounds. (The earth's mass is 5.983×10^{24} kg.)

c **64.** The sun's mass is 1.999×10^{30} kg. Find its weight in pounds. (1 kg = 2.205 lb)

In Exercises 65–74, use special products to find each of the following products.

65. $(u - 5)(u + 2)$

66. $(v - 1)(v + 10)$

67. $(x^2 - 2)(x^2 + 3)$

68. $(a^2 + 3)(a^2 + 5)$

69. $(3a^2 + b^2)(3a^2 - b^2)$

70. $(6y^2 - z^2)(6y^2 + z^2)$

71. $\left(x + \sqrt{a}\right)\left(x - \sqrt{a}\right)$

72. $\left(2x + \sqrt{b}\right)\left(2x - \sqrt{b}\right)$

73. $\left(\sqrt{x} + 2\right)\left(\sqrt{x} - 2\right)$

74. $\left(\sqrt{3a} + b\right)\left(\sqrt{3a} - b\right)$

In Exercises 75–84, factor as completely as possible each of the following polynomials.

75. $v^2 - 10v + 25$

76. $a^2 - 8a + 16$

77. $3x^2 + 10x - 8$

78. $3x^2 - 2x - 8$

79. $2ax + 5bx - 8ay - 20by$

80. $2a^3 + 10a^2 + 3a + 15$

81. $2bu - 3bv + 2au - 3av$

82. $15ax + 3bx - 20ay - 4by$

83. $8x^3 - 18x(a^2 + 2a + 1)$

84. $9(4x^2 - 4xy + y^2) - 16a^2$

In Exercises 85–98, perform the indicated operations and simplify.

85. $\dfrac{5}{x - 6} + \dfrac{2}{x - 3}$

86. $\dfrac{8}{x - 5} - \dfrac{4}{x + 2}$

87. $\dfrac{5x}{x^2 - 4} + \dfrac{1}{x - 2}$

88. $\dfrac{5}{x - 3} - \dfrac{x}{x^2 - 6x + 9}$

89. $\dfrac{4}{x^2 - 2x + 1} - \dfrac{3}{x^2 - 1}$

90. $\dfrac{1}{x^2 - 25} + \dfrac{1}{x^2 + 10x + 25}$

91. $\dfrac{xy}{x^3 - y^3} - \dfrac{y}{x^2 + xy + y^2}$

92. $\dfrac{1}{x + a} - \dfrac{a}{x^2 - ax + a^2}$

93. $\dfrac{1}{3x - 3} + \dfrac{1}{2x + 2} - \dfrac{1}{1 - x^2}$

94. $\dfrac{x}{x + y} + \dfrac{y}{x - y} - \dfrac{2xy}{x^2 - y^2}$

95. $\dfrac{x - \dfrac{x^2}{x - 5}}{1 + \dfrac{25}{x^2 - 25}}$

96. $\dfrac{3 - \dfrac{9}{3 - x}}{1 + \dfrac{x^2}{9 - x^2}}$

97. $\dfrac{x + \dfrac{1}{1 - (1/x)}}{x - \dfrac{1}{1 + (1/x)}}$

98. $\dfrac{x + \dfrac{-1}{x - (1/x)}}{x - \dfrac{1}{x + (1/x)}}$

In Exercises 99–110, simplify each of the following radicals. Assume that all radicals are defined and all denominators are different from zero.

99. $\sqrt{48x^3y^6}$

100. $\sqrt{81ab^5c^7}$

101. $\sqrt[4]{625a^5b^7}$

102. $\sqrt[3]{343a^6x^5y^4}$

103. $\sqrt{3ab^3}\,\sqrt{2a^2b^5}$

104. $\sqrt[3]{4r^4s^2}\,\sqrt[3]{6r^3s^5}$

105. $\sqrt[5]{\dfrac{32ab^3}{c^6}}$

106. $\sqrt{\dfrac{18a^3x^4}{75b^5}}$

107. $\sqrt{\sqrt[3]{64a^6b^8c^9}}$

108. $\sqrt[3]{\sqrt{25m^2x^4y^6}}$

109. $\left(\sqrt{3ax^2y}\right)^4$

110. $\left(\sqrt[3]{2a^2bx^4}\right)^2$

In Exercises 111–122, rationalize the denominators.

111. $\dfrac{4 + \sqrt{5}}{\sqrt{3}}$

112. $\dfrac{\sqrt{5} - 2}{\sqrt{7}}$

113. $\dfrac{\sqrt{a} + \sqrt{b}}{\sqrt{a}}$

114. $\dfrac{\sqrt{m} - \sqrt{n}}{\sqrt{n}}$

115. $\dfrac{\sqrt{2} + \sqrt{5}}{\sqrt{3} - \sqrt{2}}$

116. $\dfrac{\sqrt{6} - \sqrt{8}}{\sqrt{5} - 1}$

117. $\dfrac{2\sqrt{x} + \sqrt{y}}{\sqrt{x} - \sqrt{y}}$

118. $\dfrac{\sqrt{x} + 3\sqrt{y}}{3\sqrt{x} + \sqrt{y}}$

119. $\dfrac{\sqrt{x + 1} + \sqrt{x}}{\sqrt{x - 1} - \sqrt{x}}$

120. $\dfrac{\sqrt{a + b} - \sqrt{b}}{\sqrt{a + b} + \sqrt{b}}$

121. $\dfrac{\sqrt{x^2 + 1} - \sqrt{x^2 - 1}}{\sqrt{x^2 + 1} - \sqrt{x^2 - 1}}$

122. $\dfrac{\sqrt{1 + a^2} + \sqrt{1 - a^2}}{\sqrt{1 + a^2} - \sqrt{1 - a^2}}$

In Exercises 123–136, write each of the following complex numbers in the form $a + bi$.

123. $-(-3 + 4i) + (5 - 9i)$

124. $(8 - 5i) - (-3 - 4i)$

125. $(-8 - i)(4 + i)$

126. $(7 + i)(8 + 5i)$

127. $(-3 - 2i)(-3 + 2i)$

128. $(10 - 4i)(10 + 4i)$

129. $i(5 - 3i)(5 + 3i)$

130. $(5i)(3 - 6i)(-5i)$

131. $3i(-2 - 6i)(2 + 6i)$

132. $(4i)(-3 - 8i)(3 + 8i)$

133. $\dfrac{1}{4 - 5i}$

134. $\dfrac{1}{-3 - 2i}$

135. $\dfrac{i}{5 + 8i}$

136. $\dfrac{3 + 6i}{i}$

In Exercises 137–140, $z = a + bi$ and $w = c + di$. Prove each of the following relations.

☐ **137.** $\overline{\overline{z}} = z$

☐ **138.** $\overline{z - w} = \overline{z} - \overline{w}$

☐ **139.** $\overline{z/w} = \overline{z}/\overline{w}$

☐ **140.** $a = \dfrac{z + \overline{z}}{2}, \ b = \dfrac{z - \overline{z}}{2i}$

CHAPTER 2

LINEAR AND QUADRATIC EQUATIONS AND INEQUALITIES

The basic idea in solving equations and inequalities is to carry out a sequence of admissible operations that transforms them into simpler, but equivalent, equations and inequalities so we can easily solve them. In this chapter, we explain how to solve linear and quadratic equations and inequalities, and then we discuss several applications. One of the most important mathematical skills is the ability to translate problems that occur in real-life or work situations into mathematical equations. We will devote a good part of this chapter to the study of word problems that lead to linear or quadratic equations or inequalities.

 2.1 SOLVING LINEAR EQUATIONS

Expressions such as

$$x + 5 = 7, \quad x^2 + 2x - 3 = 0, \quad \text{or} \quad \frac{1}{x + 1} = \frac{3}{x - 2}$$

are called *equations* in x. The letter x is called a *variable*. Given an equation, every number that, when substituted for x, makes the equation a true statement is called a *solution* or a *root* of the equation. For example, 2 is the only number that can be substituted for x in the equation $x + 5 = 7$ to make it a true statement. Thus 2 is the unique solution of this equation. On the other hand, since

$$1^2 + 2 \cdot 1 - 3 = 0 \quad \text{and} \quad (-3)^2 + 2(-3) - 3 = 0,$$

the numbers 1 and -3 are both solutions of the equation $x^2 + 2x - 3 = 0$. If a number is a solution of an equation, we say that the number *satisfies* the equation.

Two equations are said to be *equivalent* if they have exactly the same solutions. For example, $x + 3 = 4$ and $2x + 1 = 3$ are equivalent, because each has the number 1 as its unique solution.

If an equation is satisfied for all possible values assigned to the variable x, then the equation is called an *identity*. For example, $x^2 - 4 = (x - 2)(x + 2)$ is an identity, since every number is a solution of this equation.

Equations that are satisfied by some values of x but not by all values of x are sometimes called *conditional equations*. For example, $x + 5 = 7$ and $x^2 + 2x - 3 = 0$ are conditional equations. Most of the time in this book, the term "equation" will mean conditional equation.

Linear Equations

A *linear equation in one variable* is an expression of the form

$$ax + b = 0, \qquad (2.1)$$

where a and b are real numbers with $a \neq 0$. This equation has a unique solution

$$x = -\frac{b}{a}. \qquad (2.2)$$

To show that (2.2) is the unique solution, we first assume that equation (2.1) *has* a solution. By adding $-b$ to both sides of the equation and then multiplying by $1/a$, we obtain

$$(ax + b) + (-b) = 0 + (-b)$$
$$ax = -b$$
$$\frac{1}{a}(ax) = \frac{1}{a}(-b)$$
$$x = -\frac{b}{a}.$$

This shows that *if* equation (2.1) has a solution, then it *must be* $-b/a$. On the other hand, to verify that $-b/a$ *is* a solution, we substitute $-b/a$ for x and obtain

$$a\left(-\frac{b}{a}\right) + b = (-b) + b = 0.$$

EXAMPLE 1   Solve the equation $(3/5)x + 2 = 0$.

Solution: We have

$$\frac{3}{5}x + 2 = 0$$
$$\frac{3}{5}x = -2$$
$$x = \left(\frac{5}{3}\right)(-2)$$
$$x = -\frac{10}{3}.$$

Practice Exercise 1   Solve the equation $\frac{5}{8}x - 15 = 0$.

Answer: $x = 24$

In some cases it is necessary to use algebraic operations to reduce the given equation to the form (2.1) before solving it.

EXAMPLE 2   Solve the equation $\frac{(x-2)}{2} = \frac{x}{5} - \frac{1}{4}$.

Solution: Multiplying both sides by 20 (the LCM of 2, 4, and 5) to eliminate the denominators, we obtain

$$\frac{20(x-2)}{2} = 20\left(\frac{x}{5}\right) - 20\left(\frac{1}{4}\right),$$
$$10(x-2) = 4x - 5,$$
$$10x - 20 - 4x + 5 = 0.$$

Hence

$$6x - 15 = 0,$$

which is a linear equation of the form (2.1). The solution is $x = 15/6 = 5/2$. To check it, we substitute $5/2$ for x in the given equation and obtain

$$\frac{(5/2) - 2}{2} = \frac{5/2}{5} - \frac{1}{4}$$
$$\frac{1/2}{2} = \frac{1}{2} - \frac{1}{4}$$
$$\frac{1}{4} = \frac{1}{4},$$

a true statement. Therefore, $x = 5/2$ is the solution of the given equation.

Practice Exercise 2   Find the solution of the equation $\frac{x}{3} = \frac{x-1}{2} + \frac{4}{9}$.

Answer: $x = \frac{1}{3}$

EXAMPLE 3   Solve the equation $(x+1)(x-2) = (x-1)(x+3) + 4$.

Solution: Computing the products and simplifying, we obtain

$$x^2 - x - 2 = x^2 + 2x - 3 + 4$$
$$-x - 2 = 2x + 1$$
$$-3x = 3$$
$$x = -1.$$

You can check this result by substituting -1 for x in the original equation.

Practice Exercise 3   Solve $(2x+1)(x-3) = (x-4)(2x-3) + 3$.

Answer: $x = 3$

Some equations that contain the variable in one or more denominators may eventually be reduced to linear equations. But we must exclude values of the variable for which denominators are equal to zero because for such values the equation has no meaning.

EXAMPLE 4   Solve $\dfrac{1}{x + 1} = \dfrac{3}{x - 2}$.

Solution: We assume that $x \neq -1$ and $x \neq 2$, so that both denominators, $x + 1$ and $x - 2$, are different from zero. In order to eliminate the denominators, we multiply both sides of the given equation by the product $(x + 1)(x - 2)$ and obtain

$$\frac{(x + 1)(x - 2)}{x + 1} = \frac{3(x + 1)(x - 2)}{x - 2}$$
$$x - 2 = 3(x + 1)$$
$$x - 2 = 3x + 3$$
$$-2x = 5$$
$$x = -\frac{5}{2}.$$

We leave it to you to check that $-5/2$ is the solution of the given equation.

Practice Exercise 4   Solve $\dfrac{2}{x - 3} = \dfrac{4}{x - 1}$.

Answer: $x = 5$

The next example shows why we must avoid values of the variable that make a denominator equal to zero.

EXAMPLE 5   Solve $\dfrac{4x}{x - 1} = 1 + \dfrac{4}{x - 1}$.

Solution: Multiplying both sides by $x - 1$ and simplifying, we obtain

$$\left[\frac{4x}{x - 1}\right](x - 1) = (x - 1) + \left[\frac{4}{x - 1}\right](x - 1)$$
$$4x = x + 3$$
$$3x = 3$$
$$x = 1.$$

But 1 cannot be a solution because the denominator $x - 1$ equals 0 when $x = 1$. Therefore, the given equation has no solution.

Practice Exercise 5   Find the solution of $\dfrac{2 - x}{1 - 2x} = 1 + \dfrac{3x}{1 - 2x}$.

Answer: No solution. (Can you explain why?)

Literal Equations

In applied situations, we often encounter equations involving several variables, and we are asked to solve a given equation for one of the variables in terms of the remaining variables. Such an equation, whose solution is an expression involving variables with unspecified values, is called a *literal equation*.

EXAMPLE 6

In optics, the simple lens equation is $1/f = 1/d_1 + 1/d_2$, where f is the focal length, d_1 is the object distance, and d_2 is the image distance. Solve the equation for f.

Solution: Multiplying both sides of the equation by $fd_1 d_2$, the product of all the denominators, we obtain

$$d_1 d_2 = fd_2 + fd_1$$
$$d_1 d_2 = f(d_2 + d_1),$$

or

$$f(d_1 + d_2) = d_1 d_2.$$

Hence

$$f = \frac{d_1 d_2}{d_1 + d_2}.$$

Practice Exercise 6

Solve the equation $\dfrac{1}{t} - \dfrac{1}{s} = \dfrac{1}{a}$ for s.

Answer: $s = \dfrac{at}{a - t}$

EXAMPLE 7

The thermal expansion formula

$$l = l_0(1 + at)$$

gives the length l of a rod at a temperature of t degrees, where l_0 is the length of the rod at $0°$ and a is the linear coefficient of thermal expansion. Solve it for t.

Solution: The letter t is the unknown variable, while the other letters represent known constants. By carrying out the multiplication on the right-hand side, we have

$$l = l_0 + l_0at.$$

Adding $-l_0$ to both sides of the last equation allows us to isolate the term containing t:

$$l - l_0 = l_0at,$$

or

$$l_0at = l - l_0.$$

Multiplying both sides by $1/l_0a$, we obtain the solution $t = (l - l_0)/l_0a$.

Practice Exercise 7

The area A of a trapezoid with bases b_1 and b_2 and height h is given by

$$A = \frac{1}{2}h(b_1 + b_2).$$

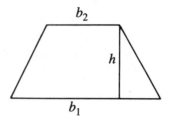

Solve this formula for b_1.

Answer: $b_1 = \dfrac{2A}{h} - b_2$

EXERCISES 2.1

Solve each of the following equations in Exercises 1–36.

1. $\dfrac{3}{2}x - 5 = 0$

2. $\dfrac{4}{5}x + 6 = 0$

3. $\dfrac{2}{3}x + \dfrac{3}{4} = 0$

4. $\dfrac{3}{4}x - \dfrac{5}{6} = 0$

5. $\sqrt{2}x - 3 = 0$

6. $2x - \sqrt{3} = 0$

7. $5(x + 2) = 2(x - 1)$

8. $4(x - 3) = 5x + 4$

9. $3(x - 4) + 6 = 2(x + 1) - 7$

10. $3x + 5 = 2(x - 4) + 7$

11. $0.1(t - 4) + 0.2t = 0.5$

12. $0.2m - 0.3(m - 1) = m$

13. $\sqrt{3}\,x - 4 = 5 - \sqrt{3}\,x$

14. $\sqrt{5}\,x + 2 = x - 4$

15. $\sqrt{3}\,y - 2 = \sqrt{2}\,y + 6$

16. $\sqrt{2}\,y - 6 = \sqrt{5} - 3$

17. $\dfrac{1}{x + 1} = \dfrac{2}{x + 3}$

18. $\dfrac{2}{x - 3} = \dfrac{5}{x + 4}$

19. $\dfrac{2}{x - 3} = \dfrac{3}{x - 4}$

20. $\dfrac{3}{x - 2} = \dfrac{5}{x + 2}$

21. $\dfrac{1}{t} - \dfrac{1}{4} = \dfrac{2}{3} - \dfrac{1}{2t}$

22. $\dfrac{2}{3t} = \dfrac{3}{2t} - \dfrac{1}{5}$

23. $\dfrac{x}{5} - \dfrac{x + 1}{4} = \dfrac{3x + 1}{2} - 3$

24. $\dfrac{2x - 1}{6} - \dfrac{1}{3} = \dfrac{x + 2}{4} - \dfrac{x}{2}$

25. $\dfrac{4x}{x - 3} + \dfrac{12}{3 - x} = \dfrac{1}{5}$

26. $\dfrac{4x}{2x + 1} - \dfrac{3}{8} = -\dfrac{2}{2x + 1}$

27. $(x - 1)(x + 3) = x(x - 2) + 4$

28. $(x - 2)(x - 3) = 3 + (x - 1)(x + 2)$

29. $(3x - 1)(x - 2) = 3x(x - 4) + 5$

30. $x(x + 2) - 4 = x(x - 5)$

C 31. $2.105x - 3.14 = 1.276x + 0.317$

C 32. $1.41x + 1.73 = 0.005x + 2.413$

C 33. $\dfrac{2.516}{x - 3.018} = \dfrac{5.101}{x + 4.021}$

C 34. $\dfrac{1}{0.05t} - 0.126 = 1.306 - \dfrac{1}{1.08t}$

C 35. $(5.3 \times 10^8)x - 3.15 \times 10^{-5}$
$= (1.21 \times 10^7)x + 1.07 \times 10^{-6}$

C 36. $(1.316 \times 10^{-7})x + 1.105 \times 10^{-8}$
$= 3.06 \times 10^{-3} - (6.72 \times 10^{-5})x$

In Exercises 37–46, equations in two variables are given. Solve each equation for the indicated variable.

37. $2xy + 3y = x + 1$, for y

38. $3uv - 5v = 2u - 4$, for v

39. $3uv - 5v = 2u - 4$, for u

40. $2xy + 3y = x + 1$, for x

41. $x(2y - 2) = y(4x - 3) + 1$, for y

42. $x(2y - 2) = y(4x - 3) + 1$, for x

43. $y = \dfrac{x - 5}{3x + 2}$, for x

44. $y = \dfrac{x + 3}{2x - 1}$, for x

45. $u = \dfrac{2v + 3}{v - 4}$, for v

46. $r = \dfrac{3s - 1}{2s + 1}$, for s

In Exercises 47–60, formulas that occur in mathematics and the applied sciences are given. Solve for the indicated variable in terms of the others.

47. The relationship between a temperature measurement in degrees Fahrenheit (F) and in degrees Celsius (C) is $F = (9/5)C + 32$. Solve for C.

48. Solve the equation $1/f = 1/d_1 + 1/d_2$ for d_1.

49. Boyle's law for gases states that $PV/T = C$, where P is the pressure, V is the volume, T is the temperature, and C is a constant. **(a)** Solve for T. **(b)** Solve for V.

50. The formula $A = (1/2)bh$ gives the area of a triangle of base b and height h. **(a)** Solve for b. **(b)** Find the length of the base if the area is equal to 150 cm² and the height is 35 cm.

51. The perimeter P of a rectangle with sides x and y is given by the formula $P = 2x + 2y$. **(a)** Solve it for y. **(b)** If $x = 50$ cm and $P = 300$ cm, find y.

52. The surface area of a parallelepiped with sides a, b, and c is given by $S = 2ab + 2ac + 2bc$. Solve for a.

53. Solve the equation $p = p_0(1 + rt)$ first for p_0 and then for r.

54. In an electric circuit containing two capacitors C_1 and C_2 in series, the total capacitance C_T is given by

$$\frac{1}{C_T} = \frac{1}{C_1} + \frac{1}{C_2}.$$

Solve for C_1.

55. The lateral surface area of a right circular cylinder (think of a soda can *without* top and bottom) is equal to $S = 2\pi rh$, where r is the radius and h is the height of the cylinder. Solve this equation for r.

56. The total surface area of a right circular cylinder (think of a soda can *with* top and bottom) of radius r and height h is $S = 2\pi r^2 + 2\pi rh$. Solve the equation for h.

57. Let $F = G(m_1 m_2)/r^2$. Solve for m_1.

58. One of Newton's laws says that the force exerted on an object is the product of the object's mass by its acceleration: $F = ma$. Solve for a.

59. Ohm's law states that in an electrical circuit $V = RI$, where V is the potential (in volts), I the current (in amperes), and R the resistance (in ohms). Solve the equation for R.

60. The intelligence quotient (IQ) of an individual is obtained by dividing his/her mental age (MA) by his/her chronological age (CA) and multiplying the result by 100. Thus, IQ is given by the formula

$$IQ = (MA/CA)100.$$

Solve this equation for MA.

2.2 WORD PROBLEMS INVOLVING LINEAR EQUATIONS

In mathematics and the applied sciences, we encounter a large variety of problems leading to linear equations. In such problems, known quantities are related to unknown quantities. By correctly expressing the relationship among them, we obtain an equation translating the problem into mathematical terms. The solution of the equation will give us the solution of the problem.

There are no fixed rules that lead us from a specific problem to a mathematical equation associated with the problem. Each problem has its own characteristics, and different problems may give rise to different equations. However, on many occasions, we are faced with similar problems, and knowing how to solve one of them might help us solve others.

Here are some suggestions for solving these problems:

Guidelines for Solving Word Problems

1. Read each problem carefully several times, looking for all relevant information.
2. Whenever possible, draw a diagram to reorganize the information.
3. Identify known and unknown quantities.
4. Assign letters to unknown quantities.
5. Write down an equation relating known and unknown quantities according to the available information.
6. Solve the equation.
7. Check the solution in the original problem.

EXAMPLE 1   Find three consecutive integers whose sum is 243.

Solution: If x denotes the first number, then $x + 1$ denotes the second, and $x + 2$ denotes the third. The mathematical equation corresponding to this problem is then

$$x + (x + 1) + (x + 2) = 243.$$

Solving for x, we have

$$3x + 3 = 243$$
$$3x = 240$$
$$x = 80.$$

Thus, the three consecutive integers are 80, 81, and 82. We now check our answer: $80 + 81 + 82 = 243$.

Practice Exercise 1   The sum of three consecutive integers is -45. Find the integers.

Answer: $-14, -15, -16$

EXAMPLE 2   A girl has 43 nickels and quarters in her piggy bank, totaling $5.75. How many nickels and how many quarters does she have?

Solution: Let x denote the number of nickels. Since the number of nickels and quarters is 43, it follows that the number of quarters is $43 - x$. Now, we can organize our work according to the following table:

	Number	Value per coin	Total value
Nickels	x	5¢	$5x$¢
Quarters	$43 - x$	25¢	$25(43 - x)$¢

Since the piggy bank total is the total value of nickels plus the total value of quarters, we obtain the equation

$$5x + 25(43 - x) = 575.$$

Solving for x, we get

$$5x + 1075 - 25x = 575$$
$$20x = 500$$
$$x = 25.$$

The girl has 25 nickels and $43 - 25 = 18$ quarters.
    Check: $25 \times 5 + 18 \times 25 = 125 + 450 = 575$¢ $= \$5.75$

Practice Exercise 2   Sharon spent $5.90 buying pencils and notebooks. Each pencil cost 25¢ and each notebook cost $1.10. She bought 2 more pencils than notebooks. How many pencils and notebooks did she buy?

Answer: 6 pencils and 4 notebooks

Percent Problems

The word *percent* is used to indicate a fraction whose denominator is 100. For example, $25/100 = 0.25$ is written as 25% (and read "25 percent"). To find $r\%$ of a number, multiply the number by $r/100$. For example, 25% of 120 is equal to

$$\frac{25}{100} \times 120 = 0.25 \times 120 = 30.$$

EXAMPLE 3   If 30% of a number is 210, find the number.

Solution: Let x be the number. If 30% of x is 210, we obtain the equation

$$0.3x = 210,$$

which, when solved for x, gives us

$$x = 700.$$

Check: 30% of 700 $= 0.3 \times 700 = 210$

Practice Exercise 3   If $x\%$ of 350 is 84, find x.

Answer: $x = 24$

EXAMPLE 4   Among 3200 college freshmen, 2080 are taking a calculus course. What percentage of the freshmen are taking calculus?

Solution: The fraction of college freshmen taking calculus is

$$\frac{2080}{3200} = \frac{13}{20}.$$

Converted to a decimal, this is

$$\frac{13}{20} = 0.65.$$

Therefore, 65% of the freshmen are taking calculus.

Practice Exercise 4   In a small town, 210 people live in brick structures and 290 live in wooden structures. What percentage of the whole population lives in brick structures?

Answer: 42%.

Simple Interest Problems

If a sum of money p_0 (the *initial principal*) is invested at a *simple interest rate* of $100r$ percent per year, the *amount of interest* at the end of t years is given by

$$i = p_0 rt. \tag{2.3}$$

The *amount of principal* at the end of t years is equal to the initial principal p_0 plus the amount of interest:

$$p = p_0 + p_0 rt$$

or

$$p = p_0(1 + rt). \tag{2.4}$$

EXAMPLE 5

If \$1,500 is invested at 6% annual simple interest, find **(a)** the amount of interest at the end of one year, and **(b)** the amount of principal at the end of two years.

Solution: The interest rate is 6% per year, so the variable r appearing in formulas (2.3) and (2.4) is $r = 6/100 = 0.06$.
(a) According to (2.3), the amount of interest at the end of one year ($t = 1$) is

$$i = 1{,}500 \times 0.06 \times 1 = \$90.$$

(b) At the end of two years the amount of principal is

$$\begin{aligned}
p &= 1{,}500(1 + 0.06 \times 2) \\
&= 1{,}500(1.12) \\
&= \$1{,}680.
\end{aligned}$$

Practice Exercise 5

How much money invested at a simple interest rate of 8% per year earns \$192 in a year? What is the principal at the end of four years?

Answer: \$2,400 and \$3,168

Mixture Problems

EXAMPLE 6

A flask contains 120 milliliters (ml) of a saline solution whose salt concentration is 8%. How many ml of water should be added to decrease the salt concentration to 6%?

Solution: Let x be the number of ml of water to be added to the 120 ml of saline solution. The volume of the new solution will be $(120 + x)$ ml.

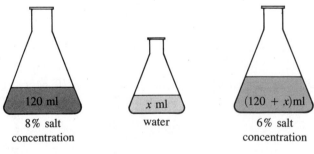

Figure 2.1

     The volumes of the original and the new solutions are different, but the amount of salt remains the same. In the original solution the amount of salt is 8% of 120 ml, that is,

$$(0.08) \cdot 120 = 9.6 \text{ ml},$$

while in the new solution the amount of salt is 6% of $(120 + x)$ ml, that is,

$$(0.06) \cdot (120 + x) \text{ ml}.$$

Since these two quantities are the same, we obtain the equation

$$(0.06)(120 + x) = (0.08) \cdot 120.$$

Solving for x, we have

$$7.2 + 0.06x = 9.6$$
$$0.06x = 2.4$$
$$6x = 240$$
$$x = 40 \text{ ml.}$$

Therefore, 40 ml of water should be added to decrease the salt concentration to 6%.

Practice Exercise 6

How many kilograms of water must be evaporated from 80 kg of a 6% saline solution to obtain a 15% solution?

Answer: 48 kg

Investment Problems

Investors often mix stocks and bonds to improve the performance of their financial holdings.

EXAMPLE 7

An investor owns $5,000 of stock A yielding 6% a year. He wants to buy shares of stock B yielding 8% per year. How much should he invest so that his total investment earns $6\frac{3}{4}\%$ per year?

Solution: If x denotes the dollar amount used to buy shares of stock B, then the total investment is $5000 + x$. The yield from this amount is

$$6\tfrac{3}{4}\% \text{ of } (5000 + x) = 0.0675(5000 + x) \text{ dollars.}$$

The yield from the original $5,000 invested in stock A is

$$6\% \text{ of } 5000 = 0.06 \times 5000 = 300 \text{ dollars,}$$

and the yield from x dollars invested in stock B is

$$8\% \text{ of } x = 0.08x \text{ dollars}$$

Thus, we obtain the equation

$$0.08x + 300 = 0.0675(5000 + x)$$
$$0.08x + 300 = 337.50 + 0.0675x$$
$$0.0125x = 37.50$$
$$125x = 375000$$
$$x = \$3,000.$$

Practice Exercise 7

Darlene invested part of $10,000 in a stock yielding 8% a year and the rest in another yielding 11% a year. If after one year she earned $938, how much did she invest in each stock?

Answer: $5,400 at 8% and $4,600 at 11%

Rate Problems

If an object moves along a line with a constant *rate* (or *speed*) r and covers a *distance* d in *time* t, then the quantities are related by the equation

$$d = rt. \tag{2.5}$$

EXAMPLE 8

On a river, a boat traveled upstream from a point A to a point B at a rate of 15 miles per hour (mi/h). It returned to A, traveling downstream, at a rate of 20 mi/h. Knowing that the total traveling time was $2\frac{1}{3}$ h, find the distance from A to B.

Solution: Let d be the distance from A to B. According to (2.5), if t_1 is the time it took the boat to travel upstream, then $d = 15t_1$, so $t_1 = d/15$. Similarly, if t_2 denotes the time it took to travel downstream, then $d = 20t_2$, so $t_2 = d/20$. With these facts we can make the following table:

	Distance	Rate	Time
Upstream	d	15	$d/15$
Downstream	d	20	$d/20$

Since the total traveling time was $2\frac{1}{3}$ hr, we have the equation

$$\frac{d}{15} + \frac{d}{20} = 2\frac{1}{3}.$$

Solving for d, we obtain

$$\frac{d}{15} + \frac{d}{20} = \frac{7}{3}$$
$$4d + 3d = 140$$
$$7d = 140$$
$$d = 20 \text{ miles.}$$

Practice Exercise 8

A single-engine plane flies from one city to another at a rate of 120 km/h. Flying back at a rate of 100 km/h, it takes the plane 1/2 hour longer to complete the trip. What is the distance between the two cities?

Answer: 300 km

EXAMPLE 9

A boy was in a town 18.5 miles from his home, and he got a car ride for all but the last 3.5 miles, which he had to walk. The average speed of the car was 45 mi/h, and the whole trip took 1 hour and 30 minutes. Find the boy's walking rate.

Solution: Again, this is a problem where we have to use formula (2.5). Let r be the boy's walking rate. Since he traveled 15 miles by car at a rate of 45 mi/h and walked 3.5 miles at a rate of r mi/h, we can use this information to complete the following table:

	Distance	Rate	Time (h)
By car	15	45	$15/45 = 1/3$
Walking	3.5	r	$3.5/r$

The whole trip took 1 hour and 30 minutes ($= 3/2$ h), so we can write the equation

$$\frac{1}{3} + \frac{3.5}{r} = \frac{3}{2}.$$

Solving for r, we obtain

$$2r + 21 = 9r$$
$$-7r = -21$$
$$r = \frac{21}{7} = 3 \text{ mi/h}.$$

To check our solution, note that the car ride lasted $1/3$ h $= 20$ minutes, while the walking time was $3.5/3$ h $= 7/6$ h $= 1$ hour and 10 minutes. Thus, the total time was 1 hour and 30 minutes.

Practice Exercise 9   Janet can ride her bicycle 4 miles per hour faster than Peter. If Peter rides 12 miles in the same time that Janet rides 15 miles, find the rates of Janet and Peter.

Answer: Janet: 20 mi/h, Peter: 16 mi/h

Work Problems

EXAMPLE 10   A water pipe fills a tank in 2 hours and another one fills it in 4 hours. How long will it take to fill the tank if both pipes are used?

Solution: In one hour the first pipe fills $1/2$ of the tank, while in that time the second pipe fills $1/4$ of the tank. Together, they fill $(1/2) + (1/4) = 3/4$ of the tank in one hour.

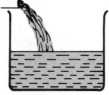

First pipe fills ½ of the tank in 1 hr

Second pipe fills ¼ of the tank in 1 hr

Both pipes fill ½ + ¼ = ¾ of the tank in 1 hr

Figure 2.2

If t denotes the time it takes for both pipes to fill the tank, then in one hour they fill $1/t$ of the tank. We thus obtain the following equation:

$$\frac{1}{t} = \frac{3}{4},$$

so

$$t = \frac{4}{3}h = 1 \text{ h } 20 \text{ min}.$$

Practice Exercise 10   Each working alone, Ed can paint a garage in 2 days and Liz can paint it in 3 days. How long will it take if both work together?

Answer: 6/5 days = 1.2 days

 ## EXERCISES 2.2

Solve the following problems.

1. Find four consecutive integers whose sum is 450.
2. The sum of four consecutive integers is -330. Find the integers.
3. If the sum of three consecutive even numbers is 480, find the numbers.
4. Find three consecutive even numbers whose sum is 612.
5. If the difference of the squares of two consecutive natural numbers is 49, find the two numbers.
6. Find two consecutive even numbers, knowing that the difference of their squares is 116.
7. Three consecutive even numbers are such that the sum of the first and third numbers equals the second number plus 64. Find the numbers.
8. Find four consecutive numbers so that the sum of the last three equals 218 plus twice the first number.
9. If the perimeter of a rectangle is 98 cm and its length is 5 cm less than twice its width, find its dimensions.
10. The length of a rectangle is three times its width. If the length is reduced by 4 m and the width is increased by 1 m, then the area is decreased by 12 m². Find the dimensions of the original rectangle.
11. In an algebra course, a student took three tests and ended up with an average of 85. Knowing that her first and third test scores were 90 and 87, find her second test score.
12. A student has test scores of 77, 81, and 79. What should his next test score be in order to bring his average up to 80?
13. A toymaker makes dolls at a cost of $15 each. Her fixed weekly costs are $585. How many dolls has she made in a week if her total costs were $4335?
14. A manufacturer makes bicycles at a cost of $115 each. His fixed monthly costs are $2,575. How many bicycles has he made in a month if his total costs were $11,200?
15. The price tag of a tape recorder was reduced by $15. If this reduction corresponds to 12% of the original price, find the original price.
16. After getting a 20% discount, Peter bought a camera for $150. Find the original price of the camera.
17. What is the wholesale price of a tape recorder whose retail price of $116 is 45% above its wholesale price?
18. What is the retail price of a TV set whose wholesale price of $217 is 30% below its retail price?
19. In a poll, 858 people preferred a new product as opposed to 35% who preferred another brand. How many people were polled?
20. In a town of 8000 people, 4400 voted in the last election. What percentage of the people voted?
21. A girl has $4.55 in nickels, dimes, and quarters. If she has four fewer dimes than quarters and seven more nickels than dimes, find how many of each type of coins she has.
22. A boy has 72 coins consisting of nickels and dimes and totaling $4.90. How many nickels and dimes does he have?
23. Chris has $2,400 earning simple interest at a rate of 8% per year. What is the amount of interest at the end of one year? How much money will she have at the end of the year?
24. Suppose that you invest $1,200 at a simple interest annual rate of $6\frac{1}{4}\%$ per year. How much money will you have at the end of the year?
25. A man puts 5.5% of his salary into a retirement plan. If his yearly salary is $21,520, how much will he put into the fund in one year?
26. Joan deposited $2,500 in a bank account earning simple interest. At the end of four years she had $3,175 in her account. Find the yearly rate of interest.
27. A businessman wishes to invest $50,000 in a tax-free fund and in a money market fund with annual yields of 7.1% and 14.5%, respectively. How much should he invest in each fund in order to realize a gain of $5,622 in one year?
28. A woman has $5,000 more invested at a simple interest rate of 8.5% a year than she has invested at 7%. If her total yearly interest is $2,285, how much money is invested at each rate?

29. A lab technician wants to prepare 720 grams of a compound by mixing 5 parts of chemical A with 3 parts of chemical B. How many grams of each chemical should be used?

30. An alloy contains 5 parts of copper, 3 parts of zinc, and 2 parts of nickel. How much of each metal is needed to make 300 kg of the alloy?

31. To make a certain amount of concrete, a builder mixes 3 parts of cement, 2 parts of sand, and 2 parts of stone. How many cubic feet of each ingredient should he use to prepare 875 cubic feet of concrete?

32. Analysis of a certain brand of dog food shows that it is 5 parts protein, 3 parts fiber, and 1 part fat. How many grams of each substance will be found in 540 grams of the dog food?

33. How many cubic centimeters (cm^3) of water should be added to 480 cm^3 of a 2% saline solution so as to decrease the salt concentration to 1.5%?

34. What volume of water should be evaporated from 600 liters of a 15% saline solution so that the remaining is a 25% saline solution?

35. A chemical manufacturer has two acid solutions, A and B. The first one is 20% acid and the second one is 40% acid. How many gallons of each should she mix to obtain 160 gallons of a 35% acid solution?

36. A lab assistant has two solutions of sulfuric acid: a 30% solution and a 70% solution. How many milliliters of each should he use to prepare 20 ml of a 60% solution?

37. On the moon's surface, a lunar rover traveled from the base of the lunar module to a nearby hill at a rate of 8 mi/h and then back at a rate of 12 mi/h. If the total traveling time was $1\frac{1}{4}$ h, find the total distance traveled.

38. At a river resort, you rent a motor boat for a sightseeing trip. You travel upstream at 9 mi/h and then return, traveling downstream at 12 mi/h. The whole trip lasts 3.5 h. How far from the resort were you when you turned back?

39. An airplane doing meteorological observation flies against a headwind, traveling from A to B at 250 mi/h. It flies back from B to A, with the wind, at 300 mi/h. If the whole trip lasted $5\frac{1}{2}$ h, find **(a)** the distance from A to B and **(b)** the time it took to fly from A to B.

40. An airplane traveled from one city to another at a cruising speed of 150 mi/h. On the return trip, traveling at 200 mi/h, the elapsed time was 1 hour less. What is the distance between the two cities?

41. On a trip to his summer cottage in the mountains, a man takes a ferryboat across a lake and then drives 180 mi at an average of 45 mi/h. The distance from the ferry dock to the cottage is 225 mi. If the total traveling time is 5 hours and 15 minutes, find the average cruising speed of the ferryboat.

42. A boy gets a ride in a motorboat across a lake and then walks the rest of the way home. The lake is 10 miles across and the boat travels at an average of 30 mi/h. The boy travels 20 miles altogether. If the entire trip takes 1 hour and 40 minutes, what is his average speed while walking?

43. An automobile leaves a city at 9:00 AM, traveling on a highway at a rate of 50 mi/h. Half an hour later, another automobile leaves the same city traveling in the same direction at a rate of 60 mi/h. How long will it take for the second automobile to overtake the first one? At what time will it overtake the first?

44. A highway 200 miles long connects two cities, A and B. A car leaves A traveling toward B at a uniform rate of 55 mi/h. At the same time, another car leaves B traveling toward A at a rate of 45 mi/h. How long will it take for the cars to pass each other? How far will they be from the respective cities from which they departed?

45. A swimming pool can be filled in 6 hours using water from a pipe. Another pipe fills the pool in 4 hours. If both pipes are used, how long will it take to fill the swimming pool?

46. A construction crew takes 30 days to rebuild a road. A second crew is able to rebuild it in 20 days. Working together, how long will it take both crews to complete the job?

47. Pipes A and B can fill a tank in 2 h and 3 h, respectively. Pipe C can empty it in 4 h. If the three pipes are open simultaneously, how long will it take to fill the tank?

48. Three pipes are connected to a storage tank. The first one fills the tank in 3 hours, the second in 4 hours, and the third in 6 hours. How long will it take to fill the tank if all three pipes are used?

49. A machine can print 125 labels per minute while another one can print 200 labels per minute. Working together, how long will it take them to print 6500 labels?

50. An electronic card sorter operates at a rate of 250 cards/min and a second one at a rate of 225 cards/min. Operating together, how long will it take them to sort a batch of 5700 cards?

2.3   **THE ORDER RELATION IN ℝ**

In Section 1.2, we described several properties of the real number system. It was mentioned that the set of all real numbers is closed under addition and multiplication and that these operations satisfy the commutative, associative, and distributive properties. Moreover, real numbers can be represented by points of a line, called a *coordinate* or *number line*, and vice versa: points on a coordinate line are labeled by real numbers. Points to the right of the origin on the number line correspond to positive numbers, while points to the left of the origin correspond to negative numbers. Besides these properties, the real number system has another important feature, which we now describe.

Order Relation

Given two real numbers a and b, the number a is said to be *greater than b*, and we write $a > b$, if $a - b$ is *positive*. If the difference $a - b$ is *negative*, then a is said to be *smaller than b* and we write $a < b$. These statements define an *order relation* among real numbers.

For example, $9 > 4$, because $9 - 4 = 5$, a positive number, and $-8 < -6$, because $(-8) - (-6) = -8 + 6 = -2$, a negative number.

If $a > b$, then the point corresponding to a on a coordinate line will be to the right of the point corresponding to b (Figure 2.3). For short, we say that a is *to the right* of b.

Similarly, if $a < b$ then a is *to the left* of b.

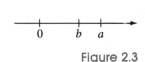

Figure 2.3

EXAMPLE 1   Write in ascending numerical order the following numbers: **(a)** -2 and -5, **(b)** $\sqrt{2}$ and 1.41.

Solution: **(a)** Since $(-5) - (-2) = (-5) + 2 = -3$, then $-5 < -2$.
(b) $\sqrt{2}$ is an irrational number and 1.4142136 is a decimal approximation of $\sqrt{2}$ (obtained with the help of a calculator). Comparing this approximation with the rational number 1.41, we see that $1.41 < \sqrt{2}$.

Practice Exercise 1   Locate on a coordinate line the following pairs of numbers: **(a)** 1/2 and 1/3, **(b)** $\sqrt{5}$ and 2.24.

Answer:

Inequalities

Given two real numbers a and b, we write $a \leq b$ to indicate that a is *less than or equal to b*. Similarly, $a \geq b$ means that a is *greater than or equal to b*. Both statements are, sometimes, called *weak inequalities*.

The statements $a < b$ or $a > b$ as previously defined are called *strict inequalities*.

The term "inequality" will be used for any of the statements $a < b$, $b > a$, $a \leq b$, or $a \geq b$.

Observe that $a \leq b$ consists of two *mutually exclusive* statements: *either $a < b$ or $a = b$*. For example, $2 \leq 5$ is a true statement because $2 < 5$. Also $5 \leq 5$ is a true statement because $5 = 5$.

Now we list the basic properties of inequalities. Together with properties I through VI of real numbers, discussed in Section 1.2, they will be used in the next section to solve inequalities.

Properties of Inequalities

1. If x and y are real numbers, then exactly one of the following statements is true: $x < y$, $x = y$, or $x > y$. This is the *trichotomy property*.
2. If $x \leq y$ and $y \leq z$, then $x \leq z$. This is the *transitive property*.
3. If $x \leq y$ and z is a real number, then $x + z \leq y + z$.
4. If $x \leq y$ and $z > 0$, then $xz \leq yz$. However, if $z < 0$, then $xz \geq yz$.

Properties 1, 2, 3, and 4 are statements about strict inequalities as well as equalities. Property 3 asserts that a strict inequality or an equality is preserved if we add *any* number to both sides of the inequality. For example:

  (a) since $-3 < 5$, it follows that $-3 + 4 < 5 + 4$, or $1 < 9$.
  (b) since $-2 < 3$, it follows that $-2 + (-4) < 3 + (-4)$, or $-6 < -1$,
  (c) since $8 = 8$, it follows that $8 + (-2) = 8 + (-2)$, or $6 = 6$.

Property 4 says that if we multiply both sides of an inequality by a *positive* number, then the inequality is preserved. However, multiplication by a *negative* number reverses the sense of the inequality. Notice that an equality is always preserved if it is multiplied by either a positive or a negative number. For example:

  (a) since $-4 < 3$, it follows that $(-4) \cdot 2 < 3 \cdot 2$, or $-8 < 6$,
  (b) since $-4 < 3$, it follows that $(-4) \cdot (-2) > 3 \cdot (-2)$, or $8 > -6$,
  (c) since $-5 = -5$, it follows that $(-5) \cdot (-1) = (-5) \cdot (-1)$, or $5 = 5$.

Intervals

The order relation $\leq$ and $<$ on real numbers allows us to define the notion of *intervals*.

Closed Intervals

If a and b are real numbers, with $a \leq b$, the closed interval $[a, b]$ is defined as the set of all real numbers x such that $a \leq x \leq b$.

In set notation, the closed interval $[a, b]$ is denoted by

$$[a, b] = \{x \in \mathbb{R} : a \leq x \leq b\}.$$

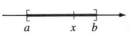

Figure 2.4

Figure 2.4 shows the geometric interpretation of $[a, b]$ on the real line: it is a line segment with left endpoint a and right endpoint b. *Closed* means that both endpoints are included in the interval. Observe that if $a = b$, then the closed interval reduces to a single point.

Open Intervals

If a and b are real numbers, with $a < b$, then the *open interval* (a, b) is defined as the set of all real numbers x such that $a < x < b$.

Figure 2.5

In set notation, the open interval (a, b) is denoted by

$$(a, b) = \{x \in \mathbb{R}: a < x < b\};$$

it is illustrated on a number line in Figure 2.5. Again the interval is a line segment. *Open* means that the interval does not include its endpoints.

Half-Open Intervals

If $a < b$, we define

$$[a, b) = \{x \in \mathbb{R}: a \le x < b\} \quad \text{and} \quad (a, b] = \{x \in \mathbb{R}: a < x \le b\}.$$

These intervals, illustrated in Figure 2.6, are called *half-open intervals*.

Note that "open" and "closed" as used in this special technical way are not really antonyms. Half-open intervals are not open, but they are not closed either.

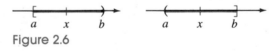

Figure 2.6

Unbounded Intervals

The intervals we just defined are called *bounded* intervals, in contrast to *unbounded* intervals, which we will define next. A point P on a line divides the line into two parts, each of which is called a *half-line*. If the abscissa of P is a, then the set of all numbers x such that $x \ge a$ is said to be an *unbounded* (or *infinite*) *closed interval* (Figure 2.7). The set of all numbers x such that $x \le a$ is also an unbounded closed interval. The set of all numbers x such that $x < a$ is said to be an *unbounded open interval* (Figure 2.8), as is the set of all numbers x such that $x > a$.

To represent unbounded intervals we use the *symbols* $+\infty$ and $-\infty$. These are *not* real numbers but just notational devices. In set notation, the interval illustrated in Figure 2.7 is denoted by

$$[a, +\infty) = \{x \in \mathbb{R}: x \ge a\};$$

and the interval shown in Figure 2.8 is denoted by

$$(-\infty, a) = \{x \in \mathbb{R}: x < a\}.$$

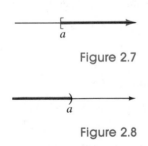

Figure 2.7

Figure 2.8

Analogously, we have the infinite intervals

$$(a, +\infty) = \{x \in \mathbb{R}: x > a\} \quad \text{and} \quad (-\infty, a] = \{x \in \mathbb{R}: x \leq a\}.$$

The real line itself can be viewed as an unbounded interval, denoted by $(-\infty, +\infty)$.

Absolute Values

If x is a real number, then its *absolute value* $|x|$ is defined by

$$|x| = \begin{cases} x \text{ if } x \geq 0, \\ -x \text{ if } x < 0. \end{cases}$$

This means that if a number is nonnegative, its absolute value is the number itself. If a number is negative, then its absolute value is the additive inverse of the number. For example, $|5| = 5$ and $|-8| = -(-8) = 8$. Loosely speaking, the absolute value of a signed number is *the number without the sign*.

EXAMPLE 2   Determine the set of numbers satisfying each of the following relations.
(a) $|x| = |-4|$ **(b)** $|2x| = 6$ **(c)** $|x - 5| = 3$ **(d)** $|1 - 3x| = 11$

Solution: **(a)** Since $|-4| = 4$, we can write the given relation as $|x| = 4$. According to the definition of absolute value, the last relation is the same as $x = 4$ or $-x = 4$. Hence, $x = 4$ and $x = -4$ are the two numbers satisfying the given relation.
(b) Using the definition of absolute value, we write $|2x| = 6$ as

$$2x = 6 \quad \text{or} \quad -2x = 6.$$

Thus,

$$x = 3 \quad \text{or} \quad x = -3.$$

(c) If $|x - 5| = 3$, then

$$x - 5 = 3 \quad \text{or} \quad -(x - 5) = 3.$$

Solving for x in each equation, we obtain $x = 8$ or $x = 2$.
(d) The relation $|1 - 3x| = 11$ is equivalent to

$$1 - 3x = 11 \quad \text{or} \quad -(1 - 3x) = 11.$$

Solving for x, we get

$$x = -\frac{10}{3} \quad \text{or} \quad x = 4.$$

In Example 2 we solved *equations involving absolute values*.

Practice Exercise 2   Solve each of the following equations.
(a) $|x| = |8|$ **(b)** $|3x| = 12$ **(c)** $|4 - x| = |-2|$ **(d)** $|2x - 3| = 7$

Answer: **(a)** -8 or 8 **(b)** -4 or 4 **(c)** 6 or 2 **(d)** -2 or 5

EXAMPLE 3

Determine whether each expression inside the absolute value bars is nonnegative or positive in order to find its absolute value.

(a) $|-7 + 3|$ **(b)** $|4 - \sqrt{7}|$
(c) $|\pi - (22/7)|$ **(d)** $|1 - a|$ if $a > 1$

Solution: **(a)** $|-7 + 3| = |-4| = 4$
(b) Since $\sqrt{7} < 4$ (Why?), we know that $4 - \sqrt{7} > 0$, so $|4 - \sqrt{7}| = 4 - \sqrt{7}$.
(c) An approximation for π is 3.1416. Next, check that 3.1428 is an approximation for $22/7$. Thus, $\pi < 22/7$ or $\pi - (22/7) < 0$. Hence,

$$|\pi - (22/7)| = -(\pi - (22/7)) = 22/7 - \pi.$$

(d) If $a > 1$, then $a - 1 > 0$ and $1 - a < 0$, so $|1 - a| = -(1 - a) = a - 1$.

Practice Exercise 3

Find the absolute value of each expression.
(a) $|-4 + 9|$ **(b)** $|2 - \pi|$ **(c)** $|\sqrt{2} - (7/5)|$ **(d)** $|a - 1|$ if $a > 1$
Answer: **(a)** 5   **(b)** $\pi - 2$ **(c)** $\sqrt{2} - 7/5$ **(d)** $a - 1$

Properties of the Absolute Value

The absolute value of a number is always nonnegative and satisfies the following properties:

$$|x| = |-x|,$$
$$|xy| = |x||y|,$$
$$-|x| \le x \le |x|.$$

For example, if $x = -15$, then $|-15| = |15| = 15$. If $x = -7$ and $y = 3$, then $xy = (-7) \cdot 3 = -21$ and $|-21| = |-7| \cdot |3|$. If $x = -4$, then $|-4| = 4$ and we have $-|-4| = -4 < 4$.

Using the notion of absolute value, we can define the *distance* between two points on a coordinate line.

The Distance between Two Points on an Axis

Let A and B be two points with coordinates a and b, respectively. The *distance between A and B*, denoted by $d(A, B)$ or AB, is defined by

$$d(A, B) = |b - a|.$$

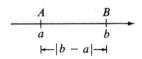

Figure 2.9

It follows that the distance $d(A, B)$ is always a nonnegative number. (Can you explain why?) Also, $d(A, B)$ is called the *length of the segment AB,* as shown in Figure 2.9.

The Distance from a Point to the Origin

If P is a point with abscissa x, the distance between P and the origin is

$$d(P, O) = |x - 0| = |x|.$$

Thus, *the absolute value $|x|$ represents the distance from P to the origin.*

Properties of Distance

From the definition of absolute value, we can derive the following properties for the distance between two points:

$$d(A, B) = 0 \quad \text{if and only if} \quad A = B;$$
$$d(A, B) = d(B, A).$$

EXAMPLE 4   Let A, B, C, and D be points with abscissas 5, 8, -4, and -7, respectively. Find $d(A, B)$, $d(C, A)$, and $d(D, C)$.

Solution: See Figure 2.10.

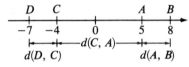

$$d(A, B) = |8 - 5| = 3$$
$$d(C, A) = |5 - (-4)| = |5 + 4| = 9$$
$$d(D, C) = |-4 - (-7)| = |-4 + 7| = 3$$

Figure 2.10   Note that $d(B, A) = |5 - 8| = |-3| = 3$, so $d(A, B) = d(B, A)$.

Practice Exercise 4   Let E, F, G, and H be points with abscissas -8, -5, -2, and 4, respectively. Find $d(E, G)$, $d(F, H)$, and $d(H, E)$.

Answer: 6, 9, and 12

 ## EXERCISES 2.3

In Exercises 1–10, write the given numbers in ascending numerical order.

1. $-2, 2, \sqrt{3}$

2. $-3, 1, \sqrt{5}$

3. $-\dfrac{2}{3}, \dfrac{1}{2}, -\sqrt{2}$

4. $\dfrac{3}{5}, -\dfrac{2}{3}, -\sqrt{3}$

5. $\dfrac{1}{4}, \sqrt{5}, \sqrt{3}, 4$

6. $\dfrac{2}{3}, \sqrt{6}, \sqrt{2}, 3$

7. $-\sqrt{2}, 1, -\dfrac{5}{2}, \sqrt{6}$

8. $-\sqrt{5}, 2, \dfrac{6}{5}, \sqrt{7}$

9. $-2.45, -\sqrt{6}, \sqrt{2}, \dfrac{17}{12}$

10. $-\sqrt{2}, -\dfrac{41}{29}, \sqrt{7}, 2.63$

In Exercises 11–16, write each of the given intervals in set notation and sketch the interval on a number line.

11. $[-2, 3]$

12. $[-5, 0]$

13. $[2, 8)$

14. $(-2, 7)$

15. $(-\infty, 5)$

16. $[0, +\infty)$

In Exercises 17–22, write each inequality in interval notation and sketch the interval.

17. $-2 < x < 7$

18. $3 < x < 11$

19. $-3 < x < 3$

20. $-2 \geq x \geq -6$

21. $x \geq -2$

22. $7 > x > -7$

In Exercises 23–44, solve each equation.

23. $|x| = |-5|$ **24.** $|-x| = 4$

25. $|2x| = |-7|$ **26.** $|-3x| = |-2|$

27. $-x = |-2|$ **28.** $-x = |-8|$

29. $|x - 1| = 3$ **30.** $|2 - x| = |-5|$

31. $|3 - 2x| = 4$ **32.** $|2x - 1| = 7$

33. $|3x - 4| = |-2|$ **34.** $|4x - 1| = |-2|$

35. $|3x - 1| + 4 = 7$ **36.** $|4x + 2| - 3 = 6$

37. $|2 - 5x| - 1 = 9$ **38.** $|3 - 6x| - 10 = 12$

39. $|3 - 4x| - 10 = -4$ **40.** $|4 - 2x| - 15 = -9$

41. $|3x - 4| = |2x + 5|$ **42.** $|5y - 4| = |y - 6|$

43. $|m + 3| = |2 - m|$ **44.** $|2p - 4| = |3 - 2p|$

In Exercises 45–58, find the value (without absolute value bars) of each of the following expressions.

45. $|-15|$ **46.** $-|-9|$

47. $|-9 + 4|$ **48.** $|5 - 12|$

49. $|-6 - 8|$ **50.** $|-10 - 3|$

51. $|\pi - (25/8)|$ **52.** $|\sqrt{10} - 4|$

53. $|\sqrt{2} - \sqrt{3}|$ **54.** $|\sqrt{3} - \sqrt{5}|$

55. $|a - 3|$ if $a < 3$ **56.** $|5 - b|$ if $b > 5$

57. $|2x - 10|$ if $x < 5$ **58.** $|3x - 9|$ if $x > 3$

In Exercises 59–64, the abscissas of A, B, and C are given. Find $d(A, B)$, $d(A, C)$, and $d(B, C)$.

59. $-6, -4, -1$ **60.** $-8, -2, 5$

61. $-\dfrac{5}{2}, 0, \dfrac{7}{2}$ **62.** $-5, -\dfrac{1}{5}, 0$

63. $0.5, 1.26, 0.08$ **64.** $1.25, 2.11, 0.75$

Prove each of the following properties of the absolute value.

65. $|x| = |-x|$

66. $|xy| = |x||y|$ (*Hint:* First consider the case when both x and y have the same sign; then prove the case when x and y have opposite signs.)

67. If $x \neq 0$, then $|1/x| = 1/|x|$. (*Hint:* Use the relation $x(1/x) = 1$.)

68. If $y \neq 0$, then $|x/y| = |x|/|y|$. (*Hint:* Use Exercises 66 and 67.)

69. $d(A, B) = 0$ if and only if $A = B$.

70. $d(A, B) = d(B, A)$

2.4   LINEAR INEQUALITIES

> A *linear inequality in one variable* is an expression of the form
> $$ax + b \leq 0, \tag{2.6}$$
> where $a, b \in \mathbb{R}$ and $a \neq 0$.

To *solve* this inequality means to find all real numbers that, when substituted for x, make (2.6) a true statement. The set of all such numbers is the *solution set* of the inequality. In order to find the solution set of (2.6), we apply the properties of inequalities listed in the previous section and replace (2.6) by a chain of equivalent but simpler inequalities, ending up with an inequality whose form and solution set are as simple as possible.

EXAMPLE 1   Solve the inequality $(3/5)x + 2 \leq 0$ and sketch its solution set.

Solution: According to property 3 of inequalities (Section 2.3), adding -2 to both sides of the given inequality yields the equivalent inequality

$$\frac{3}{5}x \leq -2.$$

Multiplying both sides by 5/3 (a positive number) preserves the sense of the inequality (property 4), giving us

$$x \le \frac{5}{3}(-2)$$

or

$$x \le -\frac{10}{3}.$$

Figure 2.11

The solution set, illustrated in Figure 2.11, consists of all real numbers that are less than or equal to $-10/3$.

It is important to notice how the solution set of the inequality $(3/5)x + 2 \le 0$ is related to the solution of the equation $(3/5)x + 2 = 0$. The latter has a *unique* solution, $x = -10/3$. The point with abscissa $-10/3$ splits the real line into two half-lines. The inequality $(3/5)x + 2 \le 0$ is satisfied for all points on the half-line to the left of and including $-10/3$. (What inequality describes the points with abscissas greater than $-10/3$?)

Practice Exercise 1   Find and sketch the solution set of the inequality $-4x + (2/3) \le 0$.

Answer: $\left\{ x \in \mathbb{R} : x \ge \frac{1}{6} \right\}$

We can also consider inequalities of the form

$$ax + b \ge 0, \qquad ax + b < 0, \qquad \text{or} \qquad ax + b > 0.$$

Inequalities such as $ax + b < 0$ or $ax + b > 0$ are called *strict inequalities*. Notice that multiplication by -1 changes the inequality $ax + b \ge 0$ into an equivalent inequality of the form (2.6).

EXAMPLE 2   Solve the inequality $1 - 2x > 7$ and graph the results.

Solution: We have

$$
\begin{aligned}
1 - 2x &> 7 \\
-2x &> 6 && \text{Add } -1 \text{ to both sides} \\
x &< \left(-\frac{1}{2}\right)6 && \text{Multiply both sides by} \\
& && -1/2, \text{ reversing the sense of} \\
& && \text{the inequality} \\
x &< -3.
\end{aligned}
$$

Thus, the solution set is $\{x \in \mathbb{R} : x < -3\}$.

Practice Exercise 2   Find the solution set of the inequality $2 - 3x < 9$ and graph it.

Answer: $\left\{ x \in \mathbb{R} : x > -\frac{7}{3} \right\}$

In many cases, we must perform algebraic operations in order to transform a given inequality into a linear inequality.

EXAMPLE 3 Solve $\dfrac{x-2}{2} \geq \dfrac{x}{5} - \dfrac{1}{4}$.

Solution: Multiplying both sides of the inequality by 20 eliminates the denominators:

$$10(x-2) \geq 4x - 5$$
$$10x - 20 \geq 4x - 5$$
$$6x \geq 15.$$

Hence

$$x \geq \frac{15}{6} = \frac{5}{2},$$

and the solution set is $\{x \in \mathbb{R}: x \geq 5/2\}$.

Practice Exercise 3 Solve the inequality

$$\frac{x}{3} < \frac{5}{6} - \frac{x-1}{4}.$$

Answer: $x < \dfrac{13}{7}$

EXAMPLE 4 Find the solution set of the inequality $(2x + 1)(x - 2) \geq (x - 1)(2x + 3) + 5$.

Solution: First, computing the products, we obtain

$$2x^2 - 3x - 2 \geq 2x^2 + x - 3 + 5.$$

Next, adding $-2x^2$ to both sides, we obtain the equivalent inequality

$$-3x - 2 \geq x + 2.$$

Solving for x, we get

$$-4x \geq 4.$$

We multiply by $-1/4$ to reverse the sense of the inequality,

$$x \leq -1,$$

so the solution set is $\{x \in \mathbb{R}: x \leq -1\}$.

Practice Exercise 4 Solve the inequality $(x + 4)(3x - 1) < 3(x + 1)^2 - 7$.

Answer: $x < 0$

EXAMPLE 5 Solve the inequality

$$\frac{3x}{6x - 1} > \frac{1}{2}.$$

Solution: *Do not* multiply both sides by $2(6x - 1)$ in an attempt to eliminate the denominators! If $x < 1/6$, then $2(6x - 1) < 0$ (can you see why?), and multiplication by a negative number reverses the sense of the inequality. Instead, we add $-1/2$ to both sides of the given inequality to obtain the equivalent inequality

$$\frac{3x}{6x - 1} - \frac{1}{2} > 0.$$

Next, we rewrite the two terms on the left-hand side with the same denominator:

$$\frac{6x}{2(6x - 1)} - \frac{6x - 1}{2(6x - 1)} > 0$$

$$\frac{6x - (6x - 1)}{2(6x - 1)} > 0$$

$$\frac{1}{2(6x - 1)} > 0.$$

Since the numerator is positive, in order for this rational expression to be positive, the denominator $2(6x - 1)$ must be positive. Thus, the last inequality is equivalent to

$$2(6x - 1) > 0,$$

or

$$6x - 1 > 0;$$

hence

$$x > \frac{1}{6}.$$

We conclude that the solution set of the given inequality is $\{x \in \mathbb{R}: x > 1/6\}$.

Practice Exercise 5   Find the solution set of the inequality $\dfrac{x}{4x - 3} < \dfrac{1}{4}$.

Answer: $\left\{x \in \mathbb{R}: x < \dfrac{3}{4}\right\}$

Inequalities Involving Absolute Values

EXAMPLE 6   Describe and sketch the sets of numbers such that **(a)** $|x| < 5$, **(b)** $|x| \geq 4$.

Solution: **(a)** As we already know (Section 2.3), $|x|$ represents the distance between a point with abscissa x and the origin. Since -5 and 5 are the coordinates of the only two points that are 5 units away from the origin, then every number x such that $|x| < 5$ must lie between -5 and 5, and vice versa (Figure 2.12). In other words, the statement $|x| < 5$ is equivalent to the statement $-5 < x < 5$. It follows that the set of all numbers satisfying the inequality $|x| < 5$ is the same as the open interval $(-5, 5)$.

(b) If $|x| \geq 4$, then the distance between the point with abscissa x and the origin is greater than or equal to 4. Such points lie outside the open interval $(-4, 4)$ (Figure 2.13). We can also say that the statement $|x| \geq 4$ is equivalent to the statement $x \leq -4$ or $x \geq 4$.

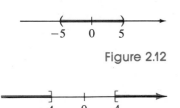

Figure 2.12

Figure 2.13

Thus the set of numbers such that $|x| \leqq 4$ is the *union* of the intervals $(-\infty, -4]$ and $[4, +\infty)$, and can be represented in set notation as follows

$$(-\infty, -4] \cup [4, +\infty).$$

Practice Exercise 6   Determine all numbers satisfying the inequalities **(a)** $|x| > 2$, **(b)** $|x| \leq 3$.

Answer: **(a)** $\{x \in \mathbb{R}: x < -2 \text{ or } x > 2\}$ **(b)** $\{x \in \mathbb{R}: -3 \leq x \leq 3\}$

Example 6 and Practice Exercise 6 illustrate two important properties of the absolute value that we now state in general terms.

The Two Basic Absolute Value Inequalities

If $b > 0$, then

$$|x| \leq b \qquad \text{if and only if} \qquad -b \leq x \leq b \tag{2.7}$$

and

$$|x| \geq b \qquad \text{if and only if} \qquad x \leq -b \text{ or } x \geq b. \tag{2.8}$$

EXAMPLE **7**   Solve the inequality $|x - 2| < 7$.

Solution: According to (2.7) with $b = 7$ and x replaced by $x - 2$, we have

$$|x - 2| < 7 \qquad \text{if and only if} \qquad -7 < x - 2 < 7.$$

The last inequality can be written as a double inequality:

$$-7 < x - 2 \qquad and \qquad x - 2 < 7.$$

Solving both inequalities, we obtain

$$-5 < x \qquad and \qquad x < 9,$$

that is,

$$-5 < x < 9.$$

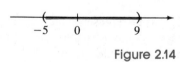

Figure 2.14

Thus, x satisfies $|x - 2| < 7$ if and only if $-5 < x < 9$. In other words, the solution set of the given inequality is the open interval $(-5, 9)$.

In Example 7, we could have worked as well with the inequality $-7 < x - 2 < 7$ instead of splitting it into two inequalities. By adding 2 to each of the three terms in the last inequality, we obtain

$$-7 + 2 < x - 2 + 2 < 7 + 2$$

or

$$-5 < x < 9.$$

Practice Exercise 7   Find the solution set of the inequality $|x + 3| \leq 5$.

Answer: $[-8, 2]$

EXAMPLE **8**   Find the solution set of the inequality $|2x + 1| \geq 5$.

Solution: According to (2.8) with $b = 5$ and x replaced by $2x + 1$, we may write

$$|2x + 1| \geq 5 \qquad \text{if and only if} \qquad 2x + 1 \leq -5 \text{ or } 2x + 1 \geq 5.$$

Solving each of these inequalities, we obtain

$$2x \leq -6 \qquad or \qquad 2x \geq 4.$$

Hence

$$x \leq -3 \qquad or \qquad x \geq 2.$$

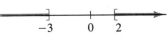

Figure 2.15

Thus, the solution set of $|2x + 1| \geq 5$ (Figure 2.15) consists of all real numbers lying outside the interval $(-3, 2)$. Or, in set notation

$$(-\infty, -3] \cup [2, +\infty).$$

Practice Exercise 8   Solve the inequality $|3x - 1| > 4$.

Answer: $x < -1$ or $x > \dfrac{5}{3}$

Applications

EXAMPLE **9**   In a chemical experiment, the temperature must be kept between 104°F and 122°F. Find the corresponding range of temperature in the Celsius scale.

Solution: The equation expressing Fahrenheit degrees in terms of Celsius degrees is

$$F = \left(\frac{9}{5}\right)C + 32.$$

Thus, we have to solve for C the inequalities

$$104 \leq \left(\frac{9}{5}\right)C + 32 \leq 122,$$

$$72 \leq \left(\frac{9}{5}\right)C \leq 90,$$

$$72\left(\frac{5}{9}\right) \leq C \leq 90\left(\frac{5}{9}\right);$$

hence

$$40° \leq C \leq 50°.$$

Practice Exercise 9

Assume that in Ohm's law, $V = RI$, the resistance R is kept constant and equal to 15 ohms. If the voltage V varies from 120 to 210 volts, find the range of variation of the intensity I of the current.

Answer: $8 \le I \le 14$

EXAMPLE 10

The owner of a movie theater estimates that his daily operating cost C is given by the equation $C = 2.5x + 100$, while his daily revenue R is given by $R = 3x$, where x is the number of tickets sold in a day. How many tickets a day must he sell to realize a profit?

Solution: To make a profit the revenue R must be greater than the cost C, from which we get the inequality

$$3x > 2.5x + 100.$$

Solving for x, we obtain

$$0.5x > 100,$$

or

$$x > 200.$$

Thus, he must sell more than 200 tickets a day to realize a profit. This number is called the *break-even* point; that is, by selling exactly 200 tickets, his revenue equals his cost, and there is no profit:

$$R = 3 \times 200 = 600 \quad \text{and} \quad C = 2.5 \times 200 + 100 = 600.$$

Practice Exercise 10

The daily cost of manufacturing x dozen pencils is given by $C = 0.5x + 150$, and the revenue from selling them is given by $R = 3.5x$. How many dozen pencils must be sold in order to realize a profit?

Answer: $x > 50$

EXERCISES 2.4

In Exercises 1–10, find and sketch the solution set of each inequality.

1. $6x - 3 > 8$

2. $3x - 4 < 10$

3. $5 - 2x \le 7$

4. $4 - 5x \ge -8$

5. $\dfrac{3}{4}x - 2 \ge \dfrac{1}{3}$

6. $\dfrac{2}{3}x + \dfrac{1}{2} \ge \dfrac{1}{4}$

7. $3x - 2 \ge 4 - 2x$

8. $5x - 2 < 3 - 2x$

9. $\dfrac{2x - 3}{4} \ge \dfrac{3x}{2} - \dfrac{1}{3}$

10. $3 - \dfrac{x}{5} \le \dfrac{x + 1}{2} - 2$

Solve each of the following inequalities in Exercises 11–30.

11. $3 \le 4x - 1 \le 5$

12. $-2 \le 3x + 1 \le 4$

13. $1 \le \dfrac{2x - 5}{6} < 3$

14. $\dfrac{1}{3} \le \dfrac{3x + 1}{2} \le 2$

15. $5 > 2 - \dfrac{x}{3} \ge 0$

16. $3x \le 4 - \dfrac{2x}{5} < 6$

17. $(3x + 2)(x - 4) < (x + 2)(3x - 1)$

18. $6x(2x - 8) \ge (4x - 1)(3x + 2)$

19. $(x + 1)^2 > (x - 2)(x - 1)$

20. $(2x - 3)(2x + 5) \leq (2x + 1)^2$

21. $\dfrac{2}{x + 3} > 0$ **22.** $\dfrac{3}{x + 5} < 0$

23. $\dfrac{-3}{4 - x} < 0$ **24.** $\dfrac{-2}{5 - x} > 0$

25. $\dfrac{x}{x + 4} > 1$ **26.** $\dfrac{2x}{x - 5} < 2$

27. $\dfrac{3x}{3x - 1} - 1 < 0$ **28.** $\dfrac{2x}{3 - 2x} + 1 > 0$

29. $\dfrac{2x}{4x - 1} > \dfrac{1}{2}$ **30.** $\dfrac{2x}{2 - 3x} < -\dfrac{2}{3}$

In Exercises 31–54, solve each absolute value inequality and sketch the solution set.

31. $|2x| > 8$ **32.** $|3x| < 6$

33. $|3x| - 1 \leq 5$ **34.** $|2x| + 3 \geq 11$

35. $|x - 4| > 1$ **36.** $|x + 2| < 6$

37. $|3x + 1| < 9$ **38.** $|4x - 5| \leq 1$

39. $|3 - 6x| > 2$ **40.** $|1 - 2x| \leq 4$

41. $|2x - 1| \geq |-3|$ **42.** $|6x - 3| \geq |-4|$

43. $\left|\dfrac{3 - x}{4}\right| < 1$ **44.** $\left|\dfrac{5 - 2x}{3}\right| \geq 2$

45. $\left|\dfrac{2x - 5}{3}\right| < 3$ **46.** $\left|\dfrac{3x - 2}{2}\right| \leq 2$

47. $\left|2x - \dfrac{1}{3}\right| < 4$ **48.** $\left|3x + \dfrac{2}{4}\right| \geq 1$

49. $\left|\dfrac{x}{3} - \dfrac{1}{2}\right| \geq 1$ **50.** $\left|\dfrac{3}{4} - \dfrac{5x}{2}\right| < 1$

51. $\left|\dfrac{3x}{4} - \dfrac{1}{3}\right| \geq 1$ **52.** $\left|\dfrac{3x}{5} - \dfrac{1}{2}\right| < 3$

53. $|2x + 6| > 0$ **54.** $|3x - 6| > 0$

Solve the following problems.

55. If the temperature of a room must be kept between 20° and 25°C, find the range of temperatures in degrees Fahrenheit.

56. The temperature of an oven is kept between 284° and 347°F. What is the range of temperatures in degrees Celsius?

57. According to Ohm's law, $V = RI$. If V is kept constant and equal to 200 volts, find the range in resistance (measured in ohms) if the intensity I of the current varies from 10 to 20 amperes.

58. One of Newton's laws states that force equals mass times acceleration: $F = ma$. If $F = 120$ units and the mass m varies from 15 to 48 units of mass, find the range for the acceleration.

59. Peter's scores on his first three algebra tests were 85, 90, and 95. What range of scores on the fourth test will give him an average between 80 and 85, inclusive?

60. Ann wants to keep her test average above 75. What is the minimum score that she must get on her fourth test if her previous scores are 70, 76, and 80?

61. Assume that in Boyle's law, $PV/T = C$, the temperature T is kept at 100° and $C = 1$. If the pressure P varies in the interval $25 \leq P \leq 75$, find the interval of variation for V.

62. Keeping the same assumptions as in Exercise 61, if the volume is no smaller than 50 and no greater than 80, what is the interval of variation for the pressure?

63. The formula $v = 32t$ gives the velocity v, in feet per second, of an object in free fall, after t seconds. Over what time interval will the velocity be no smaller than 144 feet per second but no greater than 208 feet per second?

64. Suppose that the force F (in pounds) required to stretch a spring x inches beyond its natural length is given by $F = 2.5x$. Find the range of F if the range of x is $10 \leq x \leq 12$.

65. Recall that the intelligence quotient, IQ, of an individual is given by

$$\text{IQ} = 100\,\frac{\text{MA}}{\text{CA}},$$

where MA is the mental age and CA is the chronological age of the subject. If the range of the intelligence quotient of a group of 15-year-old children is $70 \leq \text{IQ} \leq 140$, find the range of their mental ages.

66. In a group of children, the intelligence quotient varies from 80 to 120. If MA = 12, find the range for CA. [Use the formula in Exercise 65.]

67. The driver of a car wants to travel 150 mi in a time interval of 2.5 to 3 hours. Use the formula $d = rt$ to find the corresponding range for the speed r.

68. Using same assumptions as in Exercise 67, over what time interval will the speed be greater than 40 miles per hour but smaller than 60 miles per hour?

69. The daily cost equation for a jeans manufacturer is $C = 775 + 2.5x$ and the daily revenue equation is $R = 15x$, where x is the number of pairs of jeans sold per day. Find how many pairs of jeans they have to sell in a day to realize a profit.

70. A publisher estimates that to print a certain magazine, the fixed costs are $3,000 and the variable costs are 35 cents per copy. How many copies must they sell at 60 cents per copy to realize a profit?

2.5   SOLVING QUADRATIC EQUATIONS

> A *quadratic equation* is an expression of the form
>
> $$ax^2 + bx + c = 0, \qquad (2.9)$$
>
> where a, b, and c are real numbers with $a \neq 0$. A *solution* or *root* of the quadratic equation is a number r that, when substituted for x, makes (2.9) a true statement:
>
> $$ar^2 + br + c = 0. \qquad (2.10)$$

In Section 2.1 we discussed linear equations and showed how the solution of a linear equation can easily be obtained by using properties of the real numbers system. The situation is different for quadratic equations.

First of all, a quadratic equation with real coefficients may have:

(a) *two distinct real roots;*

(b) *one real root* (called a *double root*)*; or*

(c) *no real roots.*

For example, the numbers 1 and 2 are two distinct roots of $x^2 - 3x + 2 = 0$. This can be seen if you substitute 1 and 2 for x into the given equation:

$$1^2 - 3 \cdot 1 + 2 = 0 \quad \text{and} \quad 2^2 - 3 \cdot 2 + 2 = 0.$$

The only solution (or double root) of $x^2 - 6x + 9 = 0$ is the number 3. Finally, the equation $x^2 + 1 = 0$ has no real roots.

Second, the solutions of a quadratic equation are not obtained as easily as the solution of a linear equation. Either we have to perform some algebraic operations to transform the quadratic equation into an equivalent but simpler equation, or we have to use the quadratic formula.

If a quadratic equation does not have real roots, we say that the equation is *not solvable over the real numbers* $\mathbb{R}$. As we shall see, quadratic equations that are not solvable over $\mathbb{R}$ always have *complex roots*, so although they are not solvable over $\mathbb{R}$ they are solvable over $\mathbb{C}$.

There are essentially two methods for solving a quadratic equation, by *factoring* and by *completing the square*. When the latter method is applied to the general equation (2.9), it yields the *quadratic formula*.

Solving by Factoring

The solutions of a quadratic equation $ax^2 + bx + c = 0$ are easily obtained whenever we can factor the quadratic polynomial $ax^2 + bx + c$. The method uses the multiplicative property of zero (Section 1.2): if r and s are real numbers such that $rs = 0$, then either $r = 0$ or $s = 0$ (or both).

EXAMPLE **1**   Solve the quadratic equation $2x^2 - 7x - 4 = 0$ by factoring.

Solution: After factoring the quadratic polynomial, we can write the given equation in an equivalent form:

$$(2x + 1)(x - 4) = 0.$$

Since the product is zero, it follows that either $2x + 1 = 0$ or $x - 4 = 0$. The first equation has solution $x = -1/2$ and the second has solution $x = 4$. To check that both numbers are solutions of the quadratic equation, we substitute $-1/2$ and 4 for x into the equation:

$$2\left(-\frac{1}{2}\right)^2 - 7\left(-\frac{1}{2}\right) - 4 = \frac{1}{2} + \frac{7}{2} - 4 = 0;$$
$$2(4)^2 - 7(4) - 4 = 32 - 28 - 4 = 0.$$

Practice Exercise 1   Find the solutions of the equation $3x^2 + 11x - 4 = 0$.

Answer: $-4, \dfrac{1}{3}$

EXAMPLE 2   Solve $x^2 - 6x + 9 = 0$ by factoring.

Solution: The quadratic polynomial $x^2 - 6x + 9$ is a *perfect square*, that is, $x^2 - 6x + 9 = (x - 3)^2$. Thus the given equation is equivalent to

$$(x - 3)^2 = 0.$$

The latter has a unique solution $x = 3$, called a *double root*.

Practice Exercise 2   Solve the quadratic equation $4x^2 - 4x + 1 = 0$.

Answer: $\dfrac{1}{2}$ (double root)

Solving by Completing the Square

Suppose first that we have a quadratic equation of the special type

$$(x + p)^2 = q, \tag{2.11}$$

with $p, q \in \mathbb{R}$ and $q > 0$. To solve this equation, take square roots of both sides to obtain

$$x + p = \pm\sqrt{q}.$$

Next, add $-p$ to both sides to get

$$x = -p \pm \sqrt{q}.$$

Thus, the two solutions of (2.11) are $x_1 = -p + \sqrt{q}$ and $x_2 = -p - \sqrt{q}$.

EXAMPLE 3   Solve the equation $(x - 2)^2 - 8 = 0$.

Solution: We have

$$(x - 2)^2 - 8 = 0$$
$$(x - 2)^2 = 8$$
$$x - 2 = \pm\sqrt{8} = \pm2\sqrt{2}$$
$$x = 2 \pm 2\sqrt{2}.$$

Hence the solutions are $x_1 = 2 + 2\sqrt{2}$ and $x_2 = 2 - 2\sqrt{2}$.

Practice Exercise 3　Find the solutions of $(2x + 1)^2 - 6 = 0$.

Answer:　$\dfrac{-1 - \sqrt{6}}{2}, \dfrac{-1 + \sqrt{6}}{2}$

The method of solving a quadratic equation by completing the square consists of transforming the given equation into an equation of the form (2.11) before solving it.

EXAMPLE 4　Solve $x^2 + 6x + 8 = 0$ by completing the square.

Solution:　First, observe that in a perfect square such as $(x + m)^2 = x^2 + 2mx + m^2$, the term m^2 is equal to *the square of one-half the coefficient of* x. Next, write the given equation with its constant term on the right-hand side:

$$x^2 + 2 \cdot (3)x = -8.$$

(Note that we have written the coefficient of x as $2 \cdot 3$ so that it appears in the form $2 \cdot m$, with $m = 3$.) If we add $9 = 3^2$ to the left-hand side of the equation, then it will become a perfect square: $x^2 + 2 \cdot (3)x + 9 = (x + 3)^2$. In order to balance the equation, we must add 9 to the right-hand side of it, so that we obtain

$$x^2 + 2 \cdot (3)x + 9 = -8 + 9,$$

or

$$(x + 3)^2 = 1,$$

which is an equation of the type (2.11). Taking square roots and solving for x yields

$$x + 3 = \pm\sqrt{1} = \pm1,$$
$$x = -3 \pm 1.$$

Hence, the solutions are $x = -2$ and $x = -4$.

Practice Exercise 4　Use the method of completing the square to solve the equation $x^2 + 4x - 5 = 0$.

Answer:　$-5, 1$

EXAMPLE 5　Solve $2x^2 - 12x + 9 = 0$ by completing the square.

Solution:　Since $2x^2 - 12x + 9 = 2(x^2 - 6x + 9/2) = 0$, the given equation is equivalent to

$$x^2 - 6x + \frac{9}{2} = 0,$$

where the leading coefficient is now 1. Next, we complete the square by arranging our work as follows:

$$x^2 - 6x \qquad\qquad = -\frac{9}{2},$$

$$x^2 + 2 \cdot (-3)x \qquad\qquad = -\frac{9}{2},$$

$$x^2 + 2 \cdot (-3)x + (-3)^2 = -\frac{9}{2} + (-3)^2.$$

Observe that, in order to complete the square, we have to add to both sides *the square of one-half the coefficient of x*, which is $(-3)^2 = 9$. Now we write

$$x^2 - 6x + 9 = -\frac{9}{2} + 9$$

$$(x - 3)^2 = \frac{9}{2}$$

$$x - 3 = \pm\sqrt{\frac{9}{2}} = \pm\frac{3\sqrt{2}}{2}$$

$$x = 3 \pm \frac{3\sqrt{2}}{2} = \frac{6 \pm 3\sqrt{2}}{2}.$$

Thus, the solutions are

$$x_1 = \frac{6 + 3\sqrt{2}}{2} \quad \text{and} \quad x_2 = \frac{6 - 3\sqrt{2}}{2}.$$

Practice Exercise 5   Find, by the method of completing the square, the solutions of $3x^2 - 6x - 3 = 0$.
Answer: $1 - \sqrt{2}, 1 + \sqrt{2}$

The Quadratic Formula

We now use the method of completing the square to derive the quadratic formula, with which we can solve any quadratic equation. Consider the quadratic equation

$$ax^2 + bx + c = 0,$$

where $a, b, c \in \mathbb{R}$ and $a \neq 0$. By dividing both sides by a, we obtain the equivalent equation

$$x^2 + \left(\frac{b}{a}\right)x + \frac{c}{a} = 0.$$

Shifting the constant term c/a to the right-hand side and writing b/a, the coefficient of x, as $2(b/2a)$, we get

$$x^2 + 2\left(\frac{b}{2a}\right)x = -\frac{c}{a}.$$

Now we complete the square by adding $b^2/4a^2$, the square of $b/2a$, to both sides of the equation,

$$x^2 + 2\left(\frac{b}{2a}\right)x + \frac{b^2}{4a^2} = -\frac{c}{a} + \frac{b^2}{4a^2}.$$

We rewrite the last expression as follows:

$$\left(x + \frac{b}{2a}\right)^2 = \frac{b^2 - 4ac}{4a^2}.$$

If the quantity $b^2 - 4ac$ is *nonnegative*, then we can take square roots of both sides and obtain

$$x + \frac{b}{2a} = \pm\sqrt{\frac{b^2 - 4ac}{4a^2}} = \pm\frac{\sqrt{b^2 - 4ac}}{2a}$$

$$x = -\frac{b}{2a} \pm \frac{\sqrt{b^2 - 4ac}}{2a}$$

$$= \frac{-b \pm \sqrt{b^2 - 4ac}}{2a}.$$

The Quadratic Formula

Let $ax^2 + bx + c = 0$, with $a, b, c \in \mathbb{R}$ and $a \neq 0$, be a quadratic equation. If the quantity $b^2 - 4ac$ is nonnegative, then the quadratic equation has *real roots* given by

$$x = \frac{-b \pm \sqrt{b^2 - 4ac}}{2a}. \qquad (2.12)$$

If $b^2 - 4ac > 0$, then the equation has *two distinct real roots*. If $b^2 - 4ac = 0$, then the equation has *one real root* (called a *double* root).

EXAMPLE 6

Use the quadratic formula to solve the equation $2x^2 - 5x - 3 = 0$.

Solution: Applying formula (2.12) with $a = 2$, $b = -5$, and $c = -3$ gives us

$$x = \frac{-(-5) \pm \sqrt{(-5)^2 - 4 \cdot 2 \cdot (-3)}}{(2 \cdot 2)} = \frac{5 \pm \sqrt{25 + 24}}{4}$$

$$= \frac{5 \pm \sqrt{49}}{4} = \frac{5 \pm 7}{4}.$$

Hence, the solutions are

$$x_1 = \frac{5 + 7}{4} = 3 \quad \text{and} \quad x_2 = \frac{5 - 7}{4} = -\frac{2}{4} = -\frac{1}{2}.$$

Practice Exercise 6

Solve the equation $3x^2 - x - 1 = 0$.

Answer: $\dfrac{1 - \sqrt{13}}{6}, \dfrac{1 + \sqrt{13}}{6}$

Complex Roots

When the quantity $b^2 - 4ac$ is *negative*, the quadratic equation has *no real roots*. The solutions are complex numbers that are not real, as we now explain.

The Principal Square Root of a Negative Number

Let $q > 0$ be a real number. The *principal square root of* $-q$ is the complex number $i\sqrt{q}$, where $\sqrt{q}$ is the principal square root of q. We write

$$\sqrt{-q} = i\sqrt{q}.$$

For example, $\sqrt{-4} = 2i$, $\sqrt{-9} = 3i$, and $\sqrt{-5} = i\sqrt{5}$.

Now, consider a quadratic equation of the form

$$x^2 + q = 0,$$

where $q > 0$. This equation is equivalent to

$$x^2 = -q.$$

From the definition of the principal square root of a negative number, we conclude that the last equation has two complex solutions,

$$x_1 = i\sqrt{q} \quad \text{and} \quad x_2 = -i\sqrt{q}.$$

This can be checked by substituting $i\sqrt{q}$ and $-i\sqrt{q}$ for x into the equation and recalling that $i^2 = -1$ (Section 1.9). We have

$$\left(i\sqrt{q}\right)^2 + q = i^2\left(\sqrt{q}\right)^2 + q = -q + q = 0$$

and

$$\left(-i\sqrt{q}\right)^2 + q = (-i)^2\left(\sqrt{q}\right)^2 + q = -q + q = 0.$$

Reasoning in a similar way, it is easy to check that the solutions of a quadratic equation of the form

$$(x + p)^2 + q = 0,$$

where $p, q \in \mathbb{R}$ and $q > 0$, are the complex numbers

$$x_1 = -p + i\sqrt{q} \quad \text{and} \quad x_2 = -p - i\sqrt{q}.$$

EXAMPLE 7   Solve the equation $(x - 2)^2 + 27 = 0$.

Solution: We have

$$(x - 2)^2 + 27 = 0$$
$$(x - 2)^2 = -27$$
$$x - 2 = \pm i\sqrt{27} = \pm 3i\sqrt{3}$$
$$x = 2 \pm 3i\sqrt{3}$$

Hence $x_1 = 2 + 3i\sqrt{3}$ and $x_2 = 2 - 3i\sqrt{3}$ are the solutions of the equation.

Check: $\left[\left(2 + 3i\sqrt{3}\right) - 2\right]^2 + 27 = \left(3i\sqrt{3}\right)^2 + 27 = -27 + 27 = 0,$
$\left[\left(2 - 3i\sqrt{3}\right) - 2\right]^2 + 27 = \left(-3i\sqrt{3}\right)^2 + 27 = -27 + 27 = 0.$

Practice Exercise 7

Solve the equation $(x + 1)^2 + 16 = 0$.

Answer: $-1 - 4i, -1 + 4i$

EXAMPLE 8

Solve $x^2 - 6x + 13 = 0$ by completing the square.

Solution: We have

$$x^2 - 6x + 13 = 0$$
$$x^2 + 2(-3)x + (-3)^2 = -13 + (-3)^2$$
$$x^2 + 2(-3)x + (-3)^2 = -13 + 9$$
$$(x - 3)^2 = -4.$$

Taking square roots, we get

$$x - 3 = \pm\sqrt{-4} = \pm 2i$$

or

$$x = 3 \pm 2i.$$

Thus, the given equation has two complex roots, $x = 3 + 2i$ and $x = 3 - 2i$.

Practice Exercise 8

By completing the square, find the solutions of $x^2 - 4x + 13 = 0$.

Answer: $2 - 3i, 2 + 3i$

If $b^2 - 4ac < 0$, then formula (2.12), together with the definition of the principal square root of a negative number, gives us two complex roots.

EXAMPLE 9

Solve the quadratic equation $3x^2 - x + 1 = 0$.

Solution: Apply the quadratic formula (2.12) with $a = 3$, $b = -1$, and $c = 1$:

$$x = \frac{-(-1) \pm \sqrt{(-1)^2 - 4 \cdot 3 \cdot 1}}{2 \cdot 3}$$
$$= \frac{1 \pm \sqrt{1 - 12}}{6}$$
$$= \frac{1 \pm \sqrt{-11}}{6}$$
$$= \frac{1 \pm i\sqrt{11}}{6}.$$

The two complex solutions are $x_1 = \left(1 + i\sqrt{11}\right)/6$ and $x_2 = \left(1 - i\sqrt{11}\right)/6$.

Practice Exercise 9

Use the quadratic formula to solve the equation $2x^2 - 3x + 2 = 0$.

Answer: $\dfrac{3 - i\sqrt{7}}{4}, \dfrac{3 + i\sqrt{7}}{4}$

Now, we summarize the results of this section.

The Discriminant

The quantity $b^2 - 4ac$ that appears under the radical sign in formula (2.12) is called the *discriminant* of the quadratic polynomial $ax^2 + bx + c = 0$, and it is denoted by

$$\Delta = b^2 - 4ac. \qquad (2.13)$$

The discriminant plays an important role in the study of quadratic polynomials and equations. If all the coefficients of a quadratic equation are integers and the discriminant is a perfect square, then the roots are rational numbers. If the discriminant is a positive number but not a perfect square, then the roots are irrational numbers. If the discriminant is equal to zero, then the quadratic equation has a double root. Finally, if the discriminant is negative, then the roots are complex numbers.

In general, we can say the following:

(a) If $\Delta > 0$, then the quadratic equation has *two distinct real roots*:

$$x_1 = \frac{-b + \sqrt{\Delta}}{2a} \quad \text{and} \quad x_2 = \frac{-b - \sqrt{\Delta}}{2a}. \qquad (2.14)$$

(b) If $\Delta = 0$, then the quadratic equation has a *double root*:

$$x_1 = x_2 = -\frac{b}{2a}. \qquad (2.15)$$

(c) If $\Delta < 0$, then the quadratic equation has *two complex roots:*

$$x_1 = \frac{-b + i\sqrt{-\Delta}}{2a} \quad \text{and} \quad x_2 = \frac{-b - i\sqrt{-\Delta}}{2a}. \qquad (2.16)$$

Note that since $\Delta < 0$, then $-\Delta > 0$, and we have $\sqrt{\Delta} = \sqrt{-(-\Delta)} = i\sqrt{-\Delta}$. Also, observe that the complex roots of a quadratic equation with *real coefficients* are always complex conjugates of each other. According to formulas (2.16), the real and imaginary parts of the complex roots are, respectively,

$$-\frac{b}{2a} \quad \text{and} \quad \pm\frac{\sqrt{-\Delta}}{2a}.$$

Factoring Quadratic Polynomials

We have seen that whenever we can factor a quadratic polynomial as a product of two linear factors, then we can readily obtain the solutions of the corresponding quadratic equation: they are the solutions of the linear factors. Conversely, if we know the solutions of a quadratic equation, then we can express the quadratic polynomial as a product of two linear factors.

EXAMPLE 10

Solve each quadratic equation and use the solutions to factor the corresponding quadratic polynomial: **(a)** $x^2 + 6x + 8 = 0$, **(b)** $2x^2 - 5x - 3 = 0$.

MATHEMATICAL VIGNETTE

ALGORITHMS

An *algorithm* is a procedure for solving a certain class of problems with a specified set of mathematical tools. For example, the tools of algebra are addition, subtraction, multiplication, division, and root extraction. Using these tools, we can derive a familiar algorithm, namely, the quadratic formula, which can solve any quadratic equation.

Solution: **(a)** We have seen (Example 4) that the solutions of the equation $x^2 + 6x + 8 = 0$ are $x = -2$ and $x = -4$. From these we obtain the linear factors $x + 2$ and $x + 4$. Next, it is easy to check that these are factors of the quadratic polynomial:

$$x^2 + 6x + 8 = (x + 2)(x + 4).$$

(b) The solutions of $2x^2 - 5x - 3 = 0$ are $x = -1/2$ and $x = 3$ (see Example 6). From these we obtain the linear factors $x + 1/2$ and $x - 3$. However, multiplication of these two factors *will not* reproduce the original polynomial: the leading coefficient of $2x^2 - 5x - 3$ is 2, while the leading coefficient of the product $(x + 1/2)(x - 3)$ is 1. To obtain the original polynomial, we have to multiply this product by 2. Thus

$$2x^2 - 5x - 3 = 2(x + 1/2)(x - 3).$$

Since $2(x + 1/2) = 2x + 1$, the last expression can also be written

$$2x^2 - 5x - 3 = (2x + 1)(x - 3).$$

Practice Exercise 10   Solve the quadratic equations **(a)** $x^2 + 2x - 3 = 0$ and **(b)** $3x^2 + x - 2 = 0$. Use the solutions to factor each polynomial.

Answer: **(a)** $(x - 1)(x + 3)$,   **(b)** $(3x - 2)(x + 1)$

Factoring Quadratic Polynomials with Known Roots

It can be proved that if x_1 and x_2 are the roots (real or complex) of the quadratic equation

$$ax^2 + bx + c = 0,$$

then the following factorization holds:

$$ax^2 + bx + c = a(x - x_1)(x - x_2).$$

For more details, we refer you to Exercise 89.

EXERCISES 2.5

In Exercises 1–16, solve each equation by moving the constant term to the right side and taking square roots.

1. $x^2 - 4 = 0$
2. $x^2 - 25 = 0$
3. $2x^2 - 32 = 0$
4. $3x^2 - 27 = 0$
5. $4x^2 - 18 = 0$
6. $5x^2 - 60 = 0$
7. $(x + 1)^2 - 2 = 0$
8. $(x - 2)^2 - 4 = 0$
9. $(5x - 2)^2 - 11 = 0$
10. $x^2 + 8 = 0$
11. $x^2 + 15 = 0$
12. $2x^2 + 18 = 0$

13. $(x + 3)^2 + 10 = 0$ **14.** $(x - 5)^2 + 24 = 0$
15. $(3x - 2)^2 + 20 = 0$ **16.** $(5x + 1)^2 + 32 = 0$

In Exercises 17–30, solve each equation by factoring.

17. $x^2 - 6x + 8 = 0$ **18.** $x^2 + 5x - 14 = 0$
19. $u^2 + 8u + 15 = 0$ **20.** $y^2 + 6y + 8 = 0$
21. $x^2 - 6x + 5 = 0$ **22.** $x^2 - 14x + 24 = 0$
23. $v^2 - 3v = 0$ **24.** $w^2 + 5w = 0$
25. $4x^2 - 5x - 6 = 0$ **26.** $5x^2 - 6x + 1 = 0$
27. $12x^2 + 5x = 3$ **28.** $20u^2 + 7u = 6$
29. $2y(4y - 3) = 9$ **30.** $2x(6x + 13) = 10$

In Exercises 31–52, complete the square and solve each equation.

31. $x^2 - 4x = 0$ **32.** $x^2 + 6x = 0$
33. $x^2 + 5x = 0$ **34.** $x^2 - 7x = 0$
35. $x^2 - 2x - 8 = 0$ **36.** $x^2 + 2x - 8 = 0$
37. $m^2 + 8m + 15 = 0$ **38.** $p^2 - 8p + 15 = 0$
39. $x^2 + 10x + 16 = 0$ **40.** $y^2 - 10y + 21 = 0$
41. $x^2 - 6x + 8 = 0$ **42.** $x^2 + 2x + 8 = 0$
43. $2x^2 - 8x + 4 = 0$ **44.** $3x^2 + 6x - 6 = 0$
45. $y^2 - y + \dfrac{2}{9} = 0$ **46.** $x^2 - \dfrac{2}{3}x - \dfrac{1}{3} = 0$
47. $2x^2 + 3x - \dfrac{1}{2} = 0$ **48.** $3x^2 - \dfrac{1}{2}x - \dfrac{1}{4} = 0$
49. $2x^2 - 3x + 2 = 0$ **50.** $z^2 - 5z + 7 = 0$
51. $3x^2 + 2x + \dfrac{1}{2} = 0$ **52.** $2x^2 - \dfrac{1}{2}x + \dfrac{1}{4} = 0$

In Exercises 53–66, use the quadratic formula to solve each equation.

53. $2x^2 + 4x - 5 = 0$ **54.** $2x^2 + 4x + 5 = 0$
55. $u^2 = 2 - u$ **56.** $3v^2 - 3 = -2v$
57. $p + 2p^2 = 3$ **58.** $q = 2 - 3q^2$
59. $x^2 + x = \dfrac{3}{2}$ **60.** $12 - 7x - x^2 = 0$
61. $5m^2 - 2m + 4 = 0$ **62.** $6r^2 - 11r + 7 = 0$
63. $x - 2 = x^2$ **64.** $3x^2 = 2x - 3$
65. $\dfrac{1}{2}x^2 - \dfrac{1}{3}x + 2 = 0$ **66.** $4x^2 - \dfrac{1}{4}x + 1 = 0$

In Exercises 67–76, transform each equation into a quadratic equation and solve it by any method you wish.

67. $x - \dfrac{60}{x} = 11$ **68.** $x + \dfrac{105}{x} = 22$

69. $3u = \dfrac{3}{u} - 8$ **70.** $4x = 11 - \dfrac{6}{x}$

71. $\dfrac{5}{x - 4} + \dfrac{4}{x + 2} = 3$ **72.** $\dfrac{6}{x + 5} - \dfrac{5}{x - 3} = 3$

73. $\dfrac{x - 2}{x + 1} + \dfrac{9}{4} = \dfrac{x + 2}{x - 1}$ **74.** $\dfrac{x + 4}{x + 6} - \dfrac{3}{5} = \dfrac{x - 3}{x + 1}$

75. $\dfrac{3x - 1}{x + 1} = \dfrac{5x}{x + 4}$ **76.** $\dfrac{2x - 4}{x - 4} = \dfrac{3x - 3}{3x + 2}$

C In Exercises 77–80, use the quadratic formula and a calculator to solve each equation. Round off your answers to two decimal places.

77. $0.6x^2 - 0.3x + 0.1 = 0$
78. $0.12x^2 + 0.2x + 0.18 = 0$
79. $1.31x^2 - 3.2x + 5.12 = 0$
80. $0.08x^2 - 0.01x + 0.03 = 0$

In Exercises 81–86, solve each quadratic equation and use the solution to factor the corresponding quadratic polynomial.

81. $2x^2 - 18 = 0$ **82.** $3x^2 - 12 = 0$
83. $2x^2 + 3x - 2 = 0$ **84.** $3x^2 - x - 4 = 0$
85. $x^2 - 2x - 1 = 0$ **86.** $x^2 - 4x + 1 = 0$

In Exercises 87–90, x_1 and x_2 denote the roots (real or complex, equal or distinct) of the quadratic equation $ax^2 + bx + c = 0$.

87. Prove that $x_1 + x_2 = -(b/a)$. (*Hint:* Use formula (2.12) and observe that the two radicals have opposite signs.)

88. Prove that $x_1 \cdot x_2 = c/a$. (*Hint:* Use formula (2.12) and the special product $(A + B)(A - B) = A^2 - B^2$.)

89. Using the results of Exercises 87 and 88, prove that

$$ax^2 + bx + c = a(x - x_1)(x - x_2).$$

90. Suppose that the coefficients a, b, and c are real numbers and that the discriminant $\Delta = b^2 - 4ac$ is negative (case of complex conjugate roots). Let

$$h = -\frac{b}{2a} \quad \text{and} \quad k = \frac{\sqrt{-\Delta}}{2a}$$

be the real and imaginary parts of the roots. Prove that the quadratic polynomial can always be rewritten in the following form:

$$ax^2 + bx + c = a[(x - h)^2 + k^2].$$

2.6   EQUATIONS REDUCIBLE TO QUADRATIC FORM

Some equations that are not in quadratic form can be transformed into a quadratic equation. We can then solve the new equation by the methods described in the previous sections and we must check *all* solutions in the original equation. Sometimes, the new equation has solutions that are *not* solutions of the original one. Such solutions are called *extraneous solutions*; we discard them.

EXAMPLE 1   Solve $\sqrt{x + 5} = x - 1$.

Solution: By squaring both sides, we eliminate the radical and transform the given equation into a quadratic equation:

$$x + 5 = (x - 1)^2$$
$$= x^2 - 2x + 1,$$

or

$$x^2 - 3x - 4 = 0.$$

The last equation has two solutions, $x = -1$ and $x = 4$. We must check these solutions into the original equation. Substituting -1 for x, we obtain

$$\sqrt{4} = -2,$$

which is false. (Recall that $\sqrt{4}$ denotes the *principal* square root of 4, so $\sqrt{4} = +2$.) Thus -1 is an extraneous solution. Substituting 4 for x in the original equation, we get

$$\sqrt{9} = 3,$$

which is true. Hence, the solution of the original equation is $x = 4$.

Practice Exercise 1   Solve the equation $2x = \sqrt{10x - 4}$.

Answer: $\dfrac{1}{2}$, 2

EXAMPLE 2   Solve $\sqrt{3x - 2} - \sqrt{x + 3} = 1$.

Solution: As a first step, it is useful to rewrite the equation so that one radical appears alone on one side of the equation:

$$\sqrt{3x - 2} = \sqrt{x + 3} + 1.$$

This is called *isolating the radical*. By squaring both sides, one of the radicals is eliminated:

$$3x - 2 = \left(\sqrt{x + 3} + 1\right)^2$$
$$= x + 3 + 2\sqrt{x + 3} + 1$$
$$= x + 4 + 2\sqrt{x + 3}$$

Next, we isolate the remaining radical and obtain

$$2x - 6 = 2\sqrt{x + 3},$$

or

$$x - 3 = \sqrt{x + 3}.$$

Squaring both sides of this equation now gives us

$$(x - 3)^2 = x + 3$$
$$x^2 - 6x + 9 = x + 3$$
$$x^2 - 7x + 6 = 0.$$

Hence

$$(x - 1)(x - 6) = 0,$$

which has solutions $x = 1$ and $x = 6$. By substituting into the original equation, it is easy to check that 6 is a solution and that 1 is an extraneous solution. Thus, the only valid solution to the problem is $x = 6$.

Practice Exercise 2   Find all solutions of the equation $\sqrt{x + 5} - \sqrt{x - 3} = 2$.

Answer: $x = 4$.

EXAMPLE 3   Solve the equation $\dfrac{1}{x} = \dfrac{1}{x^2}$.

Solution: When $x \neq 0$, the given equation is equivalent to

$$x^2 = x \quad \text{or} \quad x^2 - x = 0.$$

The last equation has solutions $x = 0$ and $x = 1$. Since $x = 0$ cannot be a solution of the original equation, the only solution is $x = 1$.

Practice Exercise 3   Solve the equation $\dfrac{1}{x^2} + \dfrac{1}{3x} = 0$.

Answer: $x = -3$

EXAMPLE 4   Find all solutions of $x^4 + x^2 - 12 = 0$.

Solution: By making the change of variable $x^2 = u$, so that $x^4 = u^2$, we obtain a quadratic equation in u:

$$u^2 + u - 12 = 0$$

or

$$(u + 4)(u - 3) = 0.$$

The solutions of this equation are $u = -4$ and $u = 3$. Since $x^2 = u$, we have

$$x^2 = -4 \quad \text{and} \quad x^2 = 3,$$

so

$$x = \pm 2i \quad \text{and} \quad x = \pm\sqrt{3}.$$

Substituting $2i$ for x into the given equation gives us

$$(2i)^4 + (2i)^2 - 12 = 2^4 i^4 + 2^2 i^2 - 12$$
$$= 16 - 4 - 12 \qquad [i^4 = 1, \quad i^2 = -1]$$
$$= 0.$$

Thus, $2i$ is a solution. Substituting $\sqrt{3}$ for x, we get

$$(\sqrt{3})^4 + (\sqrt{3})^2 - 12 = 9 + 3 - 12 = 0.$$

Thus, $\sqrt{3}$ is another solution. We leave it to you to check that $-2i$ and $-\sqrt{3}$ are also solutions. Therefore, the given equation has four solutions: $-2i$, $2i$, $-\sqrt{3}$, and $\sqrt{3}$.

Practice Exercise 4   Solve the equation $x^4 - 4 = 3x^2$.

Answer: $\pm 2, \pm i$

EXAMPLE 5   Find all solutions of the equation $x^{2/3} - 5x^{1/3} + 6 = 0$.

Solution: If we set $u = x^{1/3}$, so that $u^2 = x^{2/3}$, then the given equation is transformed into the quadratic equation

$$u^2 - 5u + 6 = 0$$

or

$$(u - 2)(u - 3) = 0,$$

which has solutions $u = 2$ and $u = 3$. Since $u = x^{1/3}$, we have

$$x^{1/3} = 2 \quad \text{and} \quad x^{1/3} = 3.$$

Hence

$$(x^{1/3})^3 = 2^3 \quad \text{and} \quad (x^{1/3})^3 = 3^3,$$

so that

$$x = 8 \quad \text{and} \quad x = 27.$$

You can check that 8 and 27 are both solutions of the given equation.

Practice Exercise 5   Solve the equation $x + 2x^{1/2} - 8 = 0$.

Answer: $x = 4$

EXERCISES 2.6

In Exercises 1–20, solve each equation. Check for extraneous solutions.

1. $\sqrt{x} - 3 = 9$ **2.** $15 - \sqrt{x} = 4$

3. $\sqrt{1 - 2x} = 5$ **4.** $\sqrt{3x + 1} = 4$

5. $x = \sqrt{7x - 12}$ **6.** $x = \sqrt{3x + 10}$

7. $3x = \sqrt{9x - 2}$ **8.** $2x = \sqrt{4x + 3}$

9. $x + 1 = 2\sqrt{x}$ **10.** $2x + 1 = \sqrt{8x}$

11. $\sqrt{x - 2} = x - 8$ **12.** $\sqrt{x + 17} = x - 3$

13. $3\sqrt{x - 8} = x - 6$ **14.** $5\sqrt{2x + 1} = 4x - 1$

15. $\sqrt{3x + 7} - 1 = x$ **16.** $\sqrt{7x + 8} - x = 2$

17. $\sqrt{2x + 2} = \sqrt{x + 2} + 1$

18. $\sqrt{x - 3} + 3 = \sqrt{2x + 8}$

19. $\sqrt{2x - 3} + \sqrt{x + 2} = 3$

20. $\sqrt{2x + 14} - \sqrt{x + 3} = 2$

In Exercises 21–62, transform each equation into a quadratic equation before solving it. Always check for extraneous solutions.

21. $1 + \dfrac{2}{x} - \dfrac{3}{x^2} = 0$ **22.** $3 + \dfrac{1}{y} + \dfrac{2}{y^2} = 0$

23. $u = \dfrac{2}{u} - 1$ **24.** $3 = \dfrac{3}{v^2} - \dfrac{2}{v}$

25. $x^{-1} - 4x^{-2} = 0$ **26.** $3x^{-1} - 5x^{-2} = 0$

27. $x^{-1} - 2x^{-2} = 1$

28. $4 - 4x^{-1} + x^{-2} = 0$

29. $(x - 1)^2 - 3(x - 1) - 4 = 0$

30. $4(y + 2)^2 - 7(y + 2) - 2 = 0$

31. $\dfrac{3}{(x - 1)^2} - \dfrac{5}{x - 1} - 2 = 0$

32. $1 - \dfrac{5}{t - 1} + \dfrac{6}{(t - 1)^2} = 0$

33. $\dfrac{3}{(x^2 - 1)^2} - \dfrac{4}{x^2 - 1} + 1 = 0$

34. $2 + \dfrac{5}{t^2 - 1} - \dfrac{3}{(t^2 - 1)^2} = 0$

35. $\left(\dfrac{x}{x + 1}\right)^2 - \dfrac{x}{x + 1} - 2 = 0$

36. $2\left(\dfrac{t}{t - 1}\right)^2 - \dfrac{7t}{t - 1} + 6 = 0$

37. $u^4 - 6u^2 - 8 = 0$ **38.** $v^4 - 8v^2 + 15 = 0$

39. $4x^4 + 6x^2 - 1 = 0$ **40.** $12u^4 - 2u^2 - 1 = 0$

41. $6(m - 2)^4 - 13(m - 2)^2 + 2 = 0$

42. $5(z + 1)^4 - 16(z + 1)^2 + 3 = 0$

43. $2x^{-4} + 3x^{-2} - 2 = 0$ **44.** $3y^{-4} - 6y^{-2} - 1 = 0$

45. $x + 6\sqrt{x} - 16 = 0$ **46.** $y + y^{1/2} - 12 = 0$

47. $p - 15p^{1/2} + 26 = 0$ **48.** $m - 4\sqrt{m} = 0$

49. $y^{-1/2} - 5y^{-1/4} + 6 = 0$

50. $2x^{-1/2} - 11x^{-1/4} + 5 = 0$

51. $x^{2/3} - 6x^{1/3} + 8 = 0$ **52.** $y^{2/3} + 5y^{1/3} - 14 = 0$

53. $(x - 2)^{2/3} = x^{1/3}$ **54.** $(x - 3)^{2/3} = (4x)^{1/3}$

55. $(x + 1)^{2/3} = (1 - x)^{1/3}$ **56.** $(x - 2)^{2/3} = (x + 4)^{1/3}$

57. $x^6 + 8x^3 + 15 = 0$ **58.** $r^6 - 14r^3 + 24 = 0$

59. $(x - 2)^2 - 4(x - 2) = 0$

60. $(x + 1)^2 + 7(x + 1) = 0$

61. $(2x + 1)^2 + 15 = 8(2x + 1)$

62. $(3x + 2)^2 + 21 = 10(3x + 2)$

In Exercises 63–68, solve each equation for the indicated variable. Assume that all variables represent positive numbers.

63. $A = \pi r^2$, for r **64.** $K = \dfrac{1}{2}mv^2$, for v

65. $F = G\dfrac{mM}{d^2}$, for d **66.** $F = \dfrac{4\pi^2 Mr}{T^2}$, for T

67. $s = s_0 + \dfrac{1}{2}at^2$, for t

68. $x = x_0 + \dfrac{(v + v_0)(v - v_0)}{2a}$, for v

69. The lateral area A of a circular cone of radius r and height h is given by the formula

$$A = \pi r\sqrt{r^2 + h^2}.$$

Solve for h in terms of A and r.

70. The following formula is used in physics:

$$p = \dfrac{mv}{\sqrt{1 - (v/c)^2}},$$

where p denotes the momentum of a particle of mass m moving with speed v. The constant c represents the speed of light. Solve for v in terms of p, m, and c.

2.7 WORD PROBLEMS INVOLVING QUADRATIC EQUATIONS

In many applications, we encounter problems whose solutions are given by one or both solutions of a quadratic equation. To handle such problems, you should follow the guidelines explained at the beginning of Section 2.2. As we already said, there are no fixed rules that tell you how to translate a given problem into a mathematical equation. The only way to develop and sharpen your problem-solving skills is to work as many problems as possible.

EXAMPLE 1

Find two consecutive odd natural numbers whose product is 143.

Solution: Any odd natural number can be written $2n + 1$, where n is a suitable natural number. For example, $7 = 2 \cdot 3 + 1$, $11 = 2 \cdot 5 + 1$, and so on. Moreover, if $2n + 1$ is an odd number, then $2n + 3$ is the following odd number, so that $2n + 1$ and $2n + 3$ are two consecutive odd natural numbers. If their product is 143, we obtain the equation

$$(2n + 1)(2n + 3) = 143.$$

Solving for n, we have

$$4n^2 + 8n - 140 = 0$$
$$n^2 + 2n - 35 = 0$$
$$(n - 5)(n + 7) = 0,$$

so $n = 5$ and $n = -7$. Since n represents a natural number, the negative solution must be discarded. Thus the two consecutive odd numbers are $2 \cdot 5 + 1 = 11$ and $2 \cdot 5 + 3 = 13$.

    Check: $11 \times 13 = 143$.

Practice Exercise 1

Find two consecutive even natural numbers whose product is 2808.

Answer: 52 and 54

EXAMPLE 2

A rectangular plot of ground measuring 9 ft by 15 ft is to be used as a flower garden. An inside pavement surrounding the entire border of the plot is to be built so that an area of 72 sq ft is left for the flowers. How wide should the pavement be?

Solution: If x denotes the width of the pavement, then $9 - 2x$ and $15 - 2x$ are the dimensions of the rectangular plot left for the flowers (Figure 2.16), whose area must be 72 sq ft. Thus, we have the equation

$$(9 - 2x)(15 - 2x) = 72.$$

Solving this equation for x, we get

$$4x^2 - 48x + 135 = 72$$
$$4x^2 - 48x + 63 = 0$$
$$x = \frac{48 \pm \sqrt{1296}}{8} = \frac{48 \pm 36}{8}$$

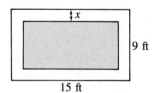

15 ft

Figure 2.16

Thus, $x = 10.5$ and $x = 1.5$ are the two values that might solve the problem. Since the width of the plot is only 9 ft, the value $x = 10.5$ must be discarded. Therefore, the pavement should be 1.5 ft wide.

Check: The dimensions of the inner rectangle are 6 ft and 12 ft, and its area is 72 sq ft.

Practice Exercise 2

Jane crochets a border of equal width around all sides of a rectangular tablecloth that measures 36 inches by 60 inches. How wide is the crocheted border if the area of the new rectangle is 2560 square inches?

Answer: 2 inches

EXAMPLE 3

On a river where the rate of the current is 2 mi/hr, a boat traveled 12 mi upstream and then returned to its departing point. If the round trip took 2.5 hr, find the speed of the boat in still water.

Solution: If x denotes the speed of the boat in still water, then $x - 2$ and $x + 2$ denote, respectively, the speeds of the boat traveling upstream and downstream. Since *distance = rate × time,* let us make the following table:

	Distance	*Rate*	*Time*
Upstream	12	$x - 2$	$12/(x - 2)$
Downstream	12	$x + 2$	$12/(x + 2)$

The total traveling time is the sum of the times to travel upstream and downstream, so

$$\frac{12}{x - 2} + \frac{12}{x + 2} = \frac{5}{2}.$$

Eliminating the denominators and solving for x, we obtain

$$24(x + 2) + 24(x - 2) = 5(x + 2)(x - 2)$$
$$24x + 48 + 24x - 48 = 5x^2 - 20$$
$$5x^2 - 48x - 20 = 0$$
$$(5x + 2)(x - 10) = 0,$$

so $x = -2/5$ and $x = 10$. Speed is always a positive quantity; thus the speed of the boat in still water is 10 mi/hr.

Practice Exercise 3

Fred can row a boat in still water at a rate of 6 miles per hour. On a certain stream, he rows 4 miles upstream and then 4 miles back downstream, in a total of 1 hour and 30 minutes. What is the rate of the current?

Answer: 2 miles per hour

EXAMPLE 4

A travel agent estimates that 20 people will take a sightseeing tour at a cost of $100 per person. She also believes that each $5 reduction of the price will bring two additional people. To receive a total of $2250, how many people should take the tour?

Solution: If n denotes the number of price reductions, then $20 + 2n$ is the number of people taking the tour at $100 - 5n$ dollars per person. Thus, the mathematical equation corresponding to this problem is

$$(20 + 2n)(100 - 5n) = 2250$$
$$2000 + 100n - 10n^2 = 2250$$
$$n^2 - 10n + 25 = 0$$
$$(n - 5)^2 = 0,$$

The number of reductions is then $n = 5$, and the number of people taking the tour is $20 + 2 \cdot 5 = 30$.

Practice Exercise 4

A farmer has 30 orange trees planted in his orchard, and they produce an average of 600 oranges per tree per year. He estimates that for each additional tree planted in the orchard, the annual yield of each tree decreases by 10 oranges. How many additional trees should he plant in order to produce a total average of 20,250 oranges per year?

Answer: 15 trees

Upward and Downward Motion of a Projectile

If air resistance is neglected, the quadratic equation

$$s = -16t^2 + v_0 t + s_0$$

gives the distance s (in feet) above the ground, at the time t seconds, of a projectile fired directly upward from a height of s_0 feet above the ground and with an initial velocity of v_0 feet per second.

EXAMPLE 5

Assume that a projectile is fired vertically from ground level with an initial velocity of 88 ft/s. If air resistance is neglected, find when the projectile is 96 ft above the ground.

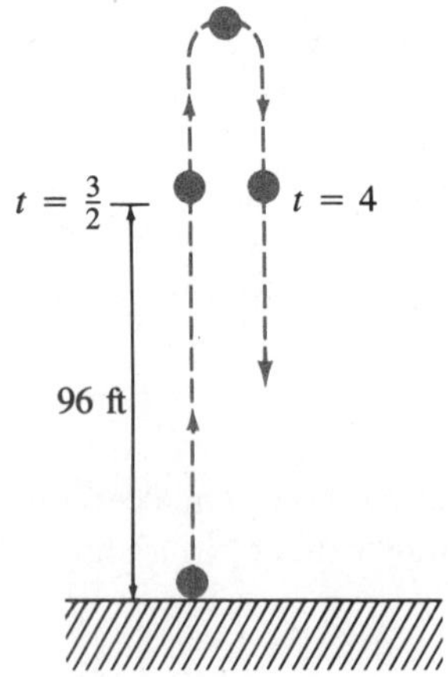

Solution: Since $s_0 = 0$ and $v_0 = 88$, the quadratic equation giving the height of the projectile is

$$s = -16t^2 + 88t.$$

Setting $s = 96$, we obtain

$$96 = -16t^2 + 88t.$$

Solving for t, we have

$$16t^2 - 88t + 96 = 0$$
$$2t^2 - 11t + 12 = 0$$
$$(2t - 3)(t - 4) = 0.$$

Thus, either $t = 3/2$ seconds or $t = 4$ seconds. In fact, both solutions are physically possible (Figure 2.17). Moving upward, the projectile takes $3/2$ seconds to

$t = \frac{3}{2}$ $t = 4$

96 ft

Figure 2.17

reach the altitude of 96 ft. Due to the earth's gravitational pull, the projectile's velocity decreases continuously from the initial value of 88 ft/s. When the velocity is zero, the projectile starts to fall back down. The time of 4 seconds corresponds to how long it takes for the projectile to reach its maximum altitude and then fall back to a height of 96 ft.

Practice Exercise 5   Assume that the projectile in Example 5 is fired upward from 50 ft above ground level. When will it be 162 ft above the ground?

Answer: 2 seconds and 3.5 seconds

EXERCISES 2.7

Solve the following problems.

1. The product of two consecutive natural numbers is 272. Find the numbers.
2. Find two consecutive positive integers whose product is 1260.
3. Find two consecutive even positive integers whose product is 624.
4. Find two consecutive odd natural numbers whose product is 675.
5. The sum of two numbers is 27 and their product is 162. Find the numbers.
6. Find two positive numbers whose difference is 7 and whose product is 120.
7. The sum of a number and its reciprocal is 26/5. Find the number.
8. The difference of a positive number and its reciprocal is 35/6. Find the number.
9. The sum of two integers is 16 and the sum of the squares of the integers is 136. Find the integers.
10. The sum of squares of two positive integers is 160. Find the integers if their difference is 8.
11. Find m so that one solution of the quadratic equation $x^2 - 16x - 4m = 0$ is 6. Determine the other solution.
12. One solution of the quadratic equation $2x^2 + kx - 40 = 0$ is -8. Find k and the other solution.
13. A rectangle has dimensions 3 in. by 6 in. If each side is increased by the same amount, the area of the new rectangle is three times the area of the old one. Find the dimensions of the new rectangle.
14. If the sides of an 8 m-by-12 m rectangle are decreased by the same amount, the area of the new rectangle is 5/8 the area of the original rectangle. What are the dimensions of the new rectangle?
15. The area of a rectangle is 96 sq ft and its length is 4 ft longer than its width. Find the dimensions of the rectangle.
16. What are the dimensions of a rectangle whose area is 65 m² and whose length is 3 m less than twice its width?
17. A rectangular field of area 28,800 sq yds is enclosed with 720 yds of fencing. Find the dimensions of the field.

18. A rectangular flower bed is bordered on three sides by 24 ft of ornamental fencing. The fourth and longer side, along a house, has no fencing. What are the dimensions of the plot if its area is 64 sq ft?
19. How long are the base and height of a triangle whose area is 45 cm² and whose height is 8 cm longer than twice the base?
20. The area of a triangle is 52 sq ft and its height is 5 ft longer than its base. Find the base and height of the triangle.
21. In a right triangle, the shorter leg is 2 ft shorter than the other leg and 4 ft shorter than the hypotenuse. Find the dimensions of the triangle.
22. The diagonal of a rectangle is 1 m longer than one of the sides and 8 m longer than the other side. Find the dimensions of the rectangle.
23. An open box is to be made from a square piece of cardboard by cutting out a 4-in. square from each corner and folding up the sides. If the volume of the box is to be 144 in.³, find the dimensions of the square piece of cardboard.

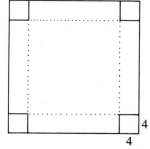

24. The length of a rectangular sheet of tin is twice its width. An open box is to be constructed by cutting out 2-in. squares from each corner and folding up the sides. What are the dimensions of the rectangular sheet of tin if the volume of the box is 192 in.³?
25. A pavement of uniform width is to be built around a rectangular plot 20 m by 30 m. If the area of the new rectangle is 704 m², find the width of the pavement.

26. The dimensions of a rectangular framed picture are 18 in. by 24 in. The picture itself is a rectangle whose area is 352 sq in. If the frame has the same width on all four sides, what is its width?

27. The radius of a circle is 8 ft. What change in the radius will increase the area by 36π ft?

28. The surface area of a sphere of radius r is given by $S = 4\pi r^2$. If $r = 4$ cm, what change in the radius will double the sphere's surface area?

29. If p_0 dollars are invested at an interest rate of $100r$ percent compounded annually, then the amount of principal p at the end of t years is given by

$$p = p_0(1 + r)^t.$$

If $p_0 = \$1200$, $p = \$1323$, and $t = 2$ years, find the interest rate r.

30. At what annual rate of interest, compounded annually, should \$5000 be invested so that at the end of two years it will increase to \$5832?

31. A group of people is chartering a flight for \$14,400. If they can find 12 more passengers, then each one will pay \$40 less for the flight. How many people are there?

32. A Princeton–Atlantic City bus was chartered for \$960. At the last moment, eight passengers had to cancel their trip, so that each remaining passenger had to pay an additional \$4 for the trip. How many passengers had originally chartered the bus?

33. To print 100 posters, a printer charges 20 cents per poster. For each increase of 50 posters, he decreases the price to print each poster by 2 cents. How many posters will he print for \$36?

34. A landlady has 100 apartments for rent. She estimates that at \$250 per month, all apartments will be rented, and that for each successive increase of \$10 in monthly rent of each apartment, two apartments will remain vacant. If she wants to receive \$27,000 per month from rentals, for how much should she rent each apartment?

35. The speed of a boat in still water is 14 mph. Knowing that the boat takes one hour longer to travel 48 miles up a river than it takes to return, find the rate of the current.

36. On a stream where the rate of the current is 3 mph, Janice rowed upstream for 5 miles and then returned to the starting point. She took $1\frac{3}{4}$ hours to complete the trip. How fast can Janice row in still water?

37. Working together, Peter and Mary can paint a garage in $1\frac{1}{5}$ days. Working alone, Mary can paint the garage in one day less than Peter. How long does it take each of them to paint the garage working alone?

38. A drain can empty a tank in 40 minutes more time than it takes a pipe to fill it. When both the drain and the pipe are open, it takes 1/2 hour to fill the tank. If the drain is closed, how long will it take the pipe to fill the tank?

39. One bus travels 10 mph faster than another bus. If the faster bus takes 1/2 hour less time to travel 150 miles, what are the rates of the two buses?

40. A plane doing meteorological observation flies from point A to point B and then back to A in 2.5 hours. The two points are 360 miles apart and the wind is blowing from A to B at a rate of 60 miles per hour. What is the speed of the plane in still air?

41. When Mary decreased her running speed by 25 m/min, it took her 4 min more to run 3000 m. What was her original running speed?

42. If Peter increases his running speed by 60 yd/min, then he can complete a 2400-yd running course in 2 min less than his previous time. Find Peter's original running speed.

43. A projectile is fired vertically from ground level with an initial velocity of 960 ft/s.
    (a) When will it be 8000 ft above ground?
    (b) When will it hit the ground?

44. A projectile is fired straight upward from a height of 32 ft above the ground and with an initial velocity of 96 ft/s.
    (a) When will the projectile be 112 ft above the ground?
    (b) When will it hit the ground?

45. A piece of wire of length l inches is cut into two pieces, one of which is 16 in. long. Each piece is bent into the shape of a square. If the sum of the areas of the squares is 272 in.^{2}, find the length l of the original piece of wire.

46. A wire 28 cm long is cut into two pieces. One piece is bent into the shape of a rectangle whose length is three times its width. The other piece is bent into the shape of a square. If the areas of the rectangle and square add up to 21 cm^2, what are the dimensions of the rectangle?

47. Suppose that in a certain town the daily demand equation for 6-volt batteries is $q = 4500/p$, where p is the price in dollars of each battery, and q is the quantity of batteries demanded in a day. Suppose also that the supply equation is $Q = 3000p - 1500$, where Q is the quantity of batteries that a manufacturer is willing to supply at p dollars per battery. At what price will $Q = q$; that is, at what price will supply equal demand?

48. The cost equation for a manufacturer of digital watches is $C = x^2 - 18x + 125$, where x is the number of units produced daily. If the company sells each watch for \$12, its daily revenue equation is $R = 12x$. Find the break-even points for the manufacturer.

49. The earth's mean radius is 6.371×10^6 m. How far is the horizon of the Earth from a balloon 2.056×10^3 m high?

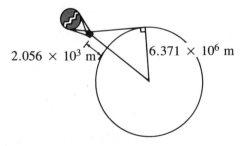

50. The area of an equilateral triangle is 15.85 sq ft. Find the dimension of one of its sides.

2.8 QUADRATIC AND RATIONAL INEQUALITIES

Now we turn our attention to the problem of solving *quadratic inequalities*, such as $ax^2 + bx + c \geq 0$, $ax^2 + bx + c > 0$, $ax^2 + bx + c \leq 0$, or $ax^2 + bx + c < 0$.

EXAMPLE 1

Solve the inequality $2x^2 - 5x - 3 \leq 0$ and sketch its solution set.

Solution: By factoring the quadratic polynomial, we see that the given inequality is equivalent to $(2x + 1)(x - 3) \leq 0$. Before attempting to solve this inequality, we must find the roots of the quadratic equation $2x^2 - 5x - 3 = 0$ or $(2x + 1)(x - 3) = 0$. The roots are $x = -1/2$ and $x = 3$, and they divide the real line into the three intervals: $(-\infty, -1/2)$, $(-1/2, 3)$, and $(3, +\infty)$. As can be seen, the factor $2x + 1$ is negative whenever $x < -1/2$ and positive whenever $x > -1/2$. Similarly, $x - 3$ is negative for all values of $x < 3$ and positive for all values of $x > 3$. Thus, we have to combine the signs of these two factors in each of the three intervals, to determine the sign of the product $(2x + 1)(x - 3) = 2x^2 - 5x - 3$. This can be done by arranging our work according to the following table.

	For x in each of the intervals			
Sign of	$-\infty$	$-1/2$	3	$-\infty$
$2x + 1$	$-$	0 $+$	$+$	
$x - 3$	$-$	$-$	0 $+$	
$(2x + 1)(x - 3)$	$+$	0 $-$	0 $+$	

The table shows the real line divided by the points $-1/2$ and 3 into three intervals. The signs of the factors $2x + 1$ and $x - 3$ are indicated in each of the intervals. For example, in the interval $(-1/2, 3)$, the sign of $2x + 1$ is *positive* while the sign of $x - 3$ is *negative*, so it follows that the sign of the product $(2x + 1)(x - 3) = 2x^2 - 5x - 3$ is *negative*. At all points in the other two intervals, the product is positive, so the inequality is not satisfied there. At $x = -1/2$ and at $x = 3$, the quadratic polynomial is zero. Thus, the solution set is the closed interval $[-1/2, 3]$.

$$\begin{array}{c} \underset{-1/2 \qquad\quad 3}{\overset{\hspace{1cm}[\quad\quad\quad]\hspace{1cm}\longrightarrow}{}} \end{array}$$

Practice Exercise 1

Find and sketch the solution set of the inequality $2x^2 - 5x - 3 > 0$.

Answer: $\left\{ x \in \mathbb{R} : x < -\dfrac{1}{2} \text{ or } x > 3 \right\}$

EXAMPLE 2

Solve the quadratic inequality $x^2 - 4x + 1 > 0$. Sketch the solution set.

Solution: First we must find the solutions of the quadratic equation

$x^2 - 4x + 1 = 0$. By using the quadratic formula, we get the solutions $x_1 = 2 - \sqrt{3}$ and $x_2 = 2 + \sqrt{3}$. According to our discussion about factoring of quadratic polynomials (at the end of Section 2.5), we can now factor the quadratic polynomial as follows:

$$x^2 - 4x + 1 = \left[x - \left(2 - \sqrt{3}\right)\right]\left[x - \left(2 + \sqrt{3}\right)\right].$$

Now, to solve the given inequality, we simply determine the signs of each linear factor appearing in the right-hand side of the last expression. Proceeding as in Example 1, we organize our work in tabular form as follows.

For x in each of the intervals

Sign of	$-\infty$	$2 - \sqrt{3}$		$2 + \sqrt{3}$	$+\infty$
$x - \left(2 - \sqrt{3}\right)$	$-$	0	$+$		$+$
$x - \left(2 + \sqrt{3}\right)$	$-$		$-$	0	$+$
$x^2 - 4x + 1$	$+$	0	$-$	0	$+$

The table shows that the solution set of the given inequality is the union of the intervals $\left(-\infty, 2 - \sqrt{3}\right)$ and $\left(2 + \sqrt{3}, +\infty\right)$. Notice that since we are solving a strict inequality, the points $2 - \sqrt{3}$ and $2 + \sqrt{3}$ are not included in the solution set.

Practice Exercise 2   Solve the inequality $x^2 - 4x + 1 \le 0$, and sketch the solution set.

Answer:

Rational Inequalities

In the examples that follow, we will show how we can solve certain inequalities involving rational expressions by the method just described.

EXAMPLE 3   Solve the inequality $\dfrac{x - 7}{x - 5} \le 2$.

Solution: The given inequality is equivalent to $\left[(x - 7)/(x - 5)\right] - 2 \le 0$. Next, we write the left-hand side of this inequality as a single fraction:

$$\frac{x - 7}{x - 5} - \frac{2(x - 5)}{x - 5} \le 0$$

$$\frac{x - 7 - 2x + 10}{x - 5} \le 0$$

$$\frac{-x + 3}{x - 5} \le 0.$$

We multiply both sides of the inequality by -1 so that the coefficient of x in the numerator becomes 1, and we obtain $(x - 3)/(x - 5) \geq 0$, an inequality equivalent to the original one. (Note the reversal of the sense of the inequality.) Now we can organize our work in tabular form.

	For x in each of the intervals			
Sign of	$-\infty$	3	5	$+\infty$
$x - 3$	$-$	0 $+$	$+$	
$x - 5$	$-$	$-$	0 $+$	
$(x - 3)/(x - 5)$	$+$	0 $-$	⁄⁄ $+$	

The dashes ⁄⁄ indicate that the rational expression $(x - 3)/(x - 5)$ has no meaning when $x = 5$. (Do you know why?) The table shows that $(x - 3)/(x - 5)$ is nonnegative in each of the intervals $(-\infty, 3]$ and $(5, +\infty)$. Thus, the solution set of the original inequality is the union of these two intervals: $(-\infty, 3] \cup (5, +\infty)$. Note that 3 is included in the solution set, while 5 is not.

Practice Exercise 3   Find the solution set of the inequality $(x - 7)/(x - 5) \geq 2$.

Answer: $[3, 5)$

EXAMPLE 4   Solve the inequality $(x^2 - x - 2)/(x - 3) > 0$.

Solution: When we factor the numerator, the given inequality becomes

$$\frac{(x + 1)(x - 2)}{x - 3} > 0,$$

where the left-hand side is now both a product and a quotient of linear terms. Next, notice that $x + 1 = 0$ when $x = -1$, and $x - 2 = 0$ when $x = 2$, and $x - 3 = 0$ when $x = 3$. The points -1, 2, and 3 divide the real line into four intervals as shown in the following table. In each one of the intervals we must determine the sign of each linear term. The sign of the rational expression $(x + 1)(x + 2)/(x - 3)$ in each one of the intervals is then obtained by the *rule of signs* for the product and/or quotient of signed numbers: an *even* number of negative signs indicates that the expression is *positive* for all values of x in that interval; an *odd* number of negative signs indicates that the expression is *negative* in that interval.

	For x in each of the intervals				
Sign of	$-\infty$	-1	2	3	$+\infty$
$x + 1$	$-$	0 $+$	$+$	$+$	
$x - 2$	$-$	$-$	0 $+$	$+$	
$x - 3$	$-$	$-$	$-$	0 $+$	
$(x + 1)(x - 2)/(x - 3)$	$-$	0 $+$	0 $-$	⁄⁄ $+$	

Thus, the solution set of the given inequality is the union of the intervals $(-1, 2)$ and $(3, +\infty)$. Why are the points -1, 2, and 3 excluded from the solution set?

Practice Exercise 4   Find the solution set of the inequality $(x^2 - x - 2)/(x - 3) \leq 0$.

Answer: $(-\infty, -1] \cup [2, 3)$

Applications

EXAMPLE **5**   Assume that a projectile is fired vertically upward from ground level with an initial velocity of 88 ft/s. If air resistance is neglected, during what time interval will the projectile be above 96 ft?

Solution: In Example 5 of Section 2.7, we explained that the equation $s = -16t^2 + 88t$ gives the height above ground of the projectile at time t. To find the time interval for which $s > 96$, we must solve the inequality

$$-16t^2 + 88t > 96,$$

or the equivalent one,

$$16t^2 - 88t + 96 < 0.$$

This inequality can be simplified by dividing both sides by 8:

$$2t^2 - 11t + 12 < 0.$$

Factoring the quadratic polynomial, we obtain the equivalent inequality

$$(2t - 3)(t - 4) < 0.$$

Now, we organize our work in tabular form.

	For t in each of the intervals			
Sign of	$-\infty$	$3/2$	4	$+\infty$
$2t - 3$	$-$	0 $+$	$+$	
$t - 4$	$-$	$-$	0 $+$	
$(2t - 3)(t - 4)$	$+$	0 $-$	0 $+$	

Thus, for all t such that $3/2 < t < 4$, the projectile will be above 96 ft.

Note that when $t = 3/2$ and $t = 4$, the projectile will be exactly at the height of 96 ft (see Example 5 of Section 2.7).

Practice Exercise 5   A travel agent estimates that 20 people will take a sightseeing tour at a cost of $100 per person. She also believes that each $5 reduction of the price will bring two additional people. How many reductions does she have to make to receive at least $2250? (*Hint:* See Example 4 of Section 2.7.)

Answer: Five reductions

> ### Guidelines for Solving Quadratic and Rational Inequalities
>
> 1. If possible, write the given expression as a product and/or quotient of linear factors.
> 2. Mark the points on the real line where each one of the factors is equal to zero. The points divide the line into disjoint intervals.
> 3. Make a sign table (as shown in the examples), displaying the sign of each linear factor in each one of the intervals.
> 4. Determine the sign of the given expression in each of the intervals, by counting the *number of negative signs* in each corresponding column. (An *even* number of negative signs indicates that the expression is *positive* in that interval, and an *odd* number of negative signs that the expression is *negative*.

EXERCISES 2.8

In Exercises 1–20, solve the inequalities and sketch the solution set.

1. $(x + 2)(x - 5) > 0$
2. $(x - 3)(x + 6) < 0$
3. $(x - 2)(x + 2) \geq 0$
4. $(x - 3)(x + 3) \leq 0$
5. $(2x + 1)(x - 1) \geq 0$
6. $(3x - 1)(2x + 1) > 0$
7. $x(x - 5) \leq 0$
8. $(x + 3)x > 0$
9. $4 - x^2 < 0$
10. $9 - x^2 \geq 0$
11. $3x^2 + 1 > 0$
12. $-4x^2 - 5 < 0$
13. $3x^2 + 5 < 0$
14. $2x^2 + 4 \leq 0$
15. $x^2 + 3x \geq 0$
16. $x^2 + 2x < 0$
17. $x^2 + 3x - 10 > 0$
18. $x^2 + x - 12 \geq 0$
19. $x^2 + 10 > 7x$
20. $x^2 + 2x < 15$

In Exercises 21–32, solve each inequality. You may first have to solve the corresponding equation and then factor the quadratic polynomial. (See Example 2 of this section.)

21. $2x^2 - 3 < 0$
22. $3x^2 - 5 \leq 0$
23. $x^2 - 4x + 1 \geq 0$
24. $x^2 - 6x + 7 < 0$
25. $2x^2 + 2x < 7$
26. $x^2 + x \geq 1$
27. $x^2 - 2x - 15 < 0$
28. $2x^2 - x + 1 > 0$
29. $4x^2 - 4x - 1 \geq 0$
30. $5x^2 + 5x - 1 < 0$
31. $x^2 - 2x + 3 > 0$
32. $3x^2 + 2x + 1 > 0$

In Exercises 33–52, find the solution set of each inequality.

33. $\dfrac{x - 1}{x + 3} \leq 0$
34. $\dfrac{x + 4}{x - 5} > 0$
35. $\dfrac{x(x - 2)}{x - 4} \geq 0$
36. $\dfrac{x + 5}{x(x - 3)} < 0$
37. $\dfrac{8}{x + 3} < 1$
38. $\dfrac{2x}{x + 2} \geq 1$
39. $\dfrac{3x + 2}{x + 5} \geq 1$
40. $\dfrac{2x + 3}{x - 1} < 1$
41. $\dfrac{2}{x - 2} > \dfrac{1}{x + 1}$
42. $\dfrac{2}{x + 3} \geq \dfrac{3}{x + 4}$
43. $\dfrac{3}{x^2} > 1$
44. $\dfrac{2}{x^2} \leq 5$
45. $\dfrac{x^2 - 3x + 2}{x + 3} \leq 0$
46. $\dfrac{x - 2}{x^2 - 2x - 15} > 0$
47. $\dfrac{x^2 - 3x - 4}{x^2 + 2x} \leq 0$
48. $\dfrac{3x^2 - x}{x^2 - 2x - 8} < 0$
49. $\dfrac{2x^2 + 5x - 3}{x^2 - 3x - 4} > 0$
50. $\dfrac{2x^2 - 5x + 3}{x^2 - 9} \geq 0$
51. $\dfrac{x}{2x - 1} < \dfrac{1}{x + 2}$
52. $\dfrac{2x}{2x - 1} > \dfrac{3}{x + 2}$

In Exercises 53–70, solve the word problems.

53. A projectile is fired vertically upward from ground level with an initial velocity of 960 ft/s. During what time interval will the projectile be **(a)** above 4400 ft? **(b)** above 7200 ft? **(c)** below 5184 ft?

54. A projectile is fired straight upward from a height of 32 ft above the ground with an initial velocity of 96 ft/s. During what time interval will the projectile be above 160 ft? Below 112 ft?

55. A manufacturer's weekly revenue is $R = 600p - 4p^2$, where p is the price in dollars of each unit of the product being manufactured. When is $R > 0$?

56. Suppose that when the price of a product is p dollars each, consumers will demand $225 - p^2$ units of the product. For what price range is the demand positive?

57. In a bacterial culture, the number of bacteria is given by $N = 1000 + 50t - 5t^2$, where t is the time in hours. When will the number of bacteria be smaller than 1000?

58. In a genetics lab, the number N of fruit flies at time t is given by $N = 500 + 50t - t^2$. When will the number of fruit flies be greater than 900?

59. A company's profit, in ten thousands of dollars, is given by $P = -4x^2 + 24x - 27$, where x is the amount, in thousands of dollars, spent on advertising. For what values of x does the company make a profit?

60. Let $P = 30 + 28x - 4x^2$ be the profit a printer makes by printing x hundreds of posters. For what values of x is $P \geq 75$?

61. The length of a rectangle is 4 ft greater than the width. For what values of the width will the area of the rectangle be greater than 45 sq ft?

62. The width of a rectangle is 6 m smaller than the length. Find the values of the length for which the area will be greater than 72 sq m.

63. Find all the values of the parameter k for which the equation $x^2 + kx + 4 = 0$ has two distinct real solutions. (*Hint:* The discriminant must be positive.)

64. For what values of the parameter k does the equation $-2x^2 + kx - 8 = 0$ have no real solutions?

65. In the study of environmental pollution, the equation

$$C = \frac{8x}{100 - x}$$

determines the cost, in ten thousands of dollars, to clean a polluted lake, where x is the percentage of pollutant removed from the lake. For what values of x will the cost be no smaller than \$20,000 and no greater than \$80,000?

66. Suppose that the cost C, in ten thousands of dollars, of cleaning a polluted pond is

$$C = \frac{5x}{1 - x},$$

where x is the amount of pollutant per unit of volume. Find all values of x for which C is no greater than \$150,000 and no smaller than \$30,000.

67. If a and b are real numbers with $a < b$, then show that $x^2 - (a + b)x + ab \leq 0$ in the interval $[a, b]$.

68. Show that $x^2 - 2ax + a^2 \geq 0$, for all $x \in \mathbb{R}$.

69. If a, b, and c are rational numbers, prove that the equation $ax^2 + bx + c = 0$ cannot have both a rational root and an irrational root.

70. What can you say about the relation between the solutions of the equation $ax^2 + bx + c = 0$ and the solutions of the equation $ax^2 - bx + c = 0$?

CHAPTER SUMMARY

In this chapter we discussed methods to solve *linear or quadratic equations* and *inequalities*. The underlying idea is to perform a sequence of admissible operations that replace the given equation or inequality by another *equivalent* equation or inequality whose solution is evident.

    A quadratic equation

$$ax^2 + bx + c = 0$$

can be solved either by *factoring* or by the method of *completing the square*. The second method allows us to derive the *quadratic formula*:

$$x = \frac{-b \pm \sqrt{b^2 - 4ac}}{2a}$$

In this formula the quantity $b^2 - 4ac$ is called the *discriminant* of the quadratic equation. It is the discriminant that determines the nature of the *roots* (or *solutions*) of a quadratic equation.

Discriminant	Roots
$b^2 - 4ac > 0$	two distinct real roots
$b^2 - 4ac = 0$	one real root
$b^2 - 4ac < 0$	no real roots

When solving a *linear inequality* such as

$$ax + b > 0,$$

where a and b are real numbers with $a \neq 0$, we first solve the equation

$$\boxed{ax + b = 0.}$$

Its unique solution $x = -b/a$ divides the line into two parts. In each of the half-lines that remain when the point $-b/a$ is removed, the sign of $ax + b$ stays the same: it is always *positive* on one of the half-lines and *negative* on the other.

When solving a *quadratic inequality* such as

$$\boxed{ax^2 + bx + c > 0,}$$

we first solve the corresponding quadratic equation. If the equation has real solutions, they divide the real line into disjoint intervals. We then study the sign of each factor of the quadratic polynomial in each of these intervals. For that purpose, a *sign table* (see Section 2.8) greatly simplifies our work. Certain *rational inequalities* can also be solved by the same method.

A good part of this chapter has also been devoted to the discussion of *word problems*. There are no fixed rules or formulas that enable you to solve a word problem; each new problem may present a new challenge. Simply do as many problems as you can. The more problems you solve, the more skillful you become. Always follow the *Guidelines for solving word problems* at the beginning of Section 2.2.

CHAPTER TEST

1. Solve for y the following equation:

$$\frac{3y - 3}{2} + \frac{1}{6} = y + 1 + \frac{y}{6}.$$

2. A businessman invested \$10,000 in two money market funds, yielding 6.0% and 8.5% per year. How much money was invested in each fund if, at the end of the year, he received \$750 in interest?

3. Find the value (or values) of x that satisfy the absolute value equation $|x + 5| = |-3|$.

4. What is the solution set of the inequality $2 \leq \dfrac{3x - 5}{4} \leq 4$?

5. Factor and solve the quadratic equation $15x^2 + 18x + 3 = 0$.

6. Solve for y the equation $y^{2/3} - 3y^{1/3} + 2 = 0$.

7. The area of a triangle is 210 sq m. If the height is 1 m greater than the base, find the lengths of the base and height.

8. Solve in inequality $(2x + 3)(3x - 3) < 0$.

9. Write in interval notation the solution set of the inequality $\dfrac{x - 2}{x - 5} < 2$.

10. Solve the equation $2y^2 - 4y - 5 = 0$ by using the quadratic formula.

REVIEW EXERCISES

In Exercises 1–36, solve each equation

1. $2x - 3(x - 1) = 4$

2. $4x - 6(x - 2) = 1$

3. $(2x - 1)(x + 3) = 2x(x - 4) + 1$

4. $3x(x - 1) - 8 = (x + 1)(3x - 2)$

5. $\dfrac{4}{x - 1} = \dfrac{2}{x - 3}$

6. $\dfrac{6}{x + 3} = \dfrac{2}{x - 4}$

7. $\dfrac{3}{2(x + 2)} = \dfrac{1}{4(x + 3)}$

8. $\dfrac{1}{5(x - 3)} = \dfrac{1}{2(x - 6)}$

9. $\dfrac{2}{x} - \dfrac{1}{3} = \dfrac{1}{2x} - \dfrac{1}{4}$

10. $\dfrac{1}{x} - \dfrac{2}{3} = \dfrac{1}{4} - \dfrac{1}{2x}$

11. $\dfrac{x - 7}{x^2 - 4x + 4} = \dfrac{6}{x - 2}$

12. $\dfrac{x + 10}{x^2 - x - 12} = \dfrac{5}{x + 3}$

13. $x^2 - 7x + 12 = 0$

14. $x^2 + x - 30 = 0$

15. $6x^2 - x - 2 = 0$

16. $8x^2 + 2x - 15 = 0$

17. $3x + 1 = \sqrt{12x}$

18. $3x + 2 = 2\sqrt{6x}$

19. $\sqrt{3x - 5} + 3 = x$

20. $\sqrt{1 - 2x} = x + 7$

21. $\dfrac{x-2}{3x+4} = \dfrac{x+1}{2x-5}$ **22.** $\dfrac{x+2}{x-3} = \dfrac{4x-1}{2x-3}$

23. $3 - \dfrac{6}{x} + \dfrac{2}{x^2} = 0$ **24.** $9 + \dfrac{6}{x} + \dfrac{1}{x^2} = 0$

25. $\dfrac{4}{x} = \dfrac{3}{x^2}$ **26.** $\dfrac{2}{x} = -\dfrac{5}{x^2}$

27. $\dfrac{4}{x-5} + \dfrac{5}{2} = \dfrac{9}{x+5}$ **28.** $\dfrac{5}{y-1} - \dfrac{3}{y} = \dfrac{3}{2}$

29. $2x = \dfrac{2}{x} - 3$ **30.** $r = \dfrac{10}{x-3}$

31. $x^{1/2} - 5x^{1/4} + 6 = 0$ **32.** $2y^{1/2} - 5y^{1/4} + 2 = 0$

33. $\dfrac{2}{x-1} + \dfrac{5}{x-7} = \dfrac{2}{4-x}$

34. $\dfrac{7}{x+1} + \dfrac{2}{1-x} = \dfrac{1}{x+5}$

35. $3\sqrt{x^2 - 9} = 4(x-2)$

36. $4\sqrt{y^2 - 13} = 3(y+1)$

In Exercises 37–44, we list several formulas that occur in mathematics and the applied sciences. Solve each of them for the indicated variable.

37. The area of a trapezoid of bases b_1 and b_2 and height h is given by $A = \dfrac{1}{2}(b_1 + b_2)h$. Solve it for b_1.

38. The equation $v = at + v_0$ expresses the linear velocity of an object at time t with initial velocity v_0 and constant acceleration a. Solve it for t.

39. The length of a rod at temperature t is given by $l = l_0(1 + \alpha t)$, where l_0 is the length at the temperature $t = 0°$ and α is a physical constant. Solve it for t.

40. Solve for h: $V = (1/3)\pi r^2 h$.

41. The formula $s = (a/2)t^2 + v_0 t$ gives the distance s traveled by an object at a time t, where v_0 is the initial velocity and a is the acceleration. Solve it for t.

42. Solve for r: $F = G(m_1 m_2)/r^2$.

43. Solve for r: $S = 2\pi r(r + h)$.

44. Solve for r: $V = \pi r^2 h$.

In Exercises 45–56, solve each inequality.

45. $\dfrac{x}{5} - 2 > \dfrac{x-3}{4} + 2$ **46.** $5 - \dfrac{x}{3} \le \dfrac{4-x}{4} - 2$

47. $\dfrac{x-5}{9} - \dfrac{3}{4} \le \dfrac{7x}{12} + \dfrac{x-1}{6}$

48. $\dfrac{x}{3} - \dfrac{2x-1}{2} > \dfrac{3x}{4} - \dfrac{x+1}{6}$

49. $\dfrac{x-1}{x+3} < 1$ **50.** $\dfrac{x+2}{x-4} \ge 2$

51. $\dfrac{x}{x-5} \ge 3$ **52.** $\dfrac{2x-1}{x-2} < 5$

53. $\dfrac{(2x-1)(x-3)}{3x-5} > 0$ **54.** $\dfrac{(x+3)(2x-7)}{5x-1} < 0$

55. $\dfrac{3x^2 + 5x - 2}{x^2 - 10x + 21} < 0$ **56.** $\dfrac{6x^2 - 13x - 5}{4x^2 + 7x - 2} > 0$

57. Find four consecutive integers so that twice the sum of the first three is the sum of 102 and three times the last number.

58. A wire 84 inches long is bent into the shape of a rectangle. Find the dimensions of the rectangle if its length is 8 in. longer than its width.

59. In a school board election, the winning candidate was elected with 990 votes. If 90 voters abstained and 45% of all voters voted against him, find how many people took part in the election.

60. Peter bought a record player at a 15% discount. A few months later, he sold it to a friend at a 20% discount from the already discounted price. If Peter received $163.20, what was the original price of the record player?

61. If the total box office receipts of a theater were $2537, and 70 more $6 tickets were sold than $8.50 tickets, how many $6 tickets were sold?

62. A tank contains 1200 liters of a 3% saline solution. Water evaporation increases the salt concentration of the solution. How many liters of water will have to evaporate to yield a 3.2% saline solution?

63. How many gallons of a 90% antifreeze solution should be added to 4 gallons of water to make a 30% antifreeze solution?

64. Two pipes fill a tank in 40 minutes. If one of the pipes takes 2 hours to fill the tank, how long does it take the other pipe to fill the tank?

65. Eight people will share equally the rental cost of a deep-sea fishing boat. If two more people decide to come along, then each share of the original eight will be reduced by $10. How much does it cost to rent the boat?

66. A company that makes digital watches has fixed costs of $2600 per month. The manufacturing cost of each watch is $25, and they can be sold for $35 each. How many watches must be manufactured per month so that the company has a profit of $2580?

67. If the area of a rectangle is 44 sq m and its length is 3 m less than twice its width, find the dimensions of the rectangle.

68. Find the values of m so that one solution of the equation $2x^2 + mx - m^2 = 0$ is $x = 3/2$.

69. Find the length of one side of an equilateral triangle whose area is $16\sqrt{3}$ sq m.

70. The speed of a faster runner is 2 mph more than the speed of a slower runner. If it takes the faster runner 5 minutes longer to run 6 mi than it takes the slower runner to run 4 mi, find the speed of each runner.

71. A shopowner purchased a shipment of glasses for $2400. His employees broke 10 glasses in the process of unpacking them. He sold the remaining glasses at a profit of $1.50 each and realized a total profit of $1765. How many glasses did he buy?

72. A bus company organizes a group excursion under the following conditions: If 150 people or fewer go, the ticket price is $50 per person, and for each person in excess of 150, the ticket price will be reduced by 12 cents. If the total received by the bus company was $9225, find how many people took the excursion.

73. In a right triangle, the hypotenuse and one leg are, respectively, 8 and 7 units longer than the other leg. Find the dimensions of the triangle.

74. Pipe B takes 20 minutes longer than pipe A to fill a tank. If both pipes can fill the tank in 24 minutes, how long does it take for pipe A to fill the tank?

75. The formula $I = c/d^2$ gives the intensity of illumination in foot-candles on a surface d feet from a light source of c candlepower. How far should a 90-candlepower light be placed from a surface to give the same intensity of illumination as a 40-candlepower light at 20 feet?

76. The radius of a circle is 3.125 cm. What change in the radius will decrease the area by 8.76 cm^2?

77. A sphere has radius 3.015 cm. What change in the radius will triple its surface area?

78. Find the approximate radius of a circle whose area equals that of a square of side 4.25 m.

79. If a search light has an intensity of 1.5×10^6 foot-candles at 55 ft, find its intensity at a distance of 1 mi (1 mi = 5280 ft).

80. How many kiloliters (kl) each of a 20.5% acid solution and a 41.2% acid solution should be mixed to prepare 5 kl of a 35.6% acid solution?

CHAPTER 3

■

COORDINATE GEOMETRY

René Descartes (1596–1650), a French philosopher and mathematician, is considered the founder of analytic geometry, in which algebra and geometry are blended into a single discipline. Analytic geometry is based upon the notion of coordinate systems on the line, on the plane, and in three-dimensional space. Geometric objects are represented by equations, and equations are given geometric meanings. In the first section of this chapter, we introduce the Cartesian coordinate system in a plane and we describe the midpoint and distance formulas. This is followed by a study of the straight line and linear relations. Next, we discuss graphs of simple equations and analyze different types of graph symmetry. Finally, the last part of this chapter is devoted to *conic sections* (circle, ellipse, parabola, and hyperbola), curves that are defined as intersections of a plane and a right circular cone.

3.1 CARTESIAN COORDINATE SYSTEM

In Section 1.3, we indicated how points on a line can be put into a one-to-one correspondence with real numbers. In this section, we briefly describe how points in a plane can be put into a one-to-one correspondence with ordered pairs of real numbers.

An *ordered pair* of real numbers is a pair of numbers, one of which is designated as the *first* number and the other as the *second* number of the pair. If x is the first number and y is the second number, then the ordered pair is denoted (x, y). It is important to notice that the ordered pair $(1, 2)$ is different from the ordered pair $(2, 1)$, and both are different from the *set* $\{1, 2\}$ (which is the same as the set $\{2, 1\}$, because the order of elements in sets is not distinguished). In general, the pair (a, b) is different from the pair (b, a) unless $a = b$ (and it is different from the set $\{a, b\}$ in all cases).

The correspondence between points in a plane and ordered pairs of real numbers is established as follows. In a plane, draw two perpendicular lines through a point O, called the *origin*, and define a coordinate system on each line. It is customary to choose the same unit of length on both lines, to take one of the lines horizontal with positive direction to the right, and to take the other line vertical with positive direction upward. The horizontal line is usually called the *x-axis* and the vertical line the *y-axis*.

If P is a point in the plane, then the two lines that pass through P and are parallel to the axes will intersect the axes at the points A and B with coordinates a and b, respectively (Figure 3.1). We assign to the point P the ordered pair of real numbers (a, b), called the *coordinates* of P. The number a is the *first coordinate*, *x-coordinate*, or *abscissa* of P, and the number b is the *second coordinate*, *y-coordinate*, or *ordinate* of P.

Conversely, every ordered pair (a, b) determines a unique point in a plane with coordinate axes: To the number a there corresponds a unique point A on the x-axis, and to the number b there corresponds a unique point B on the y-axis. The line through A parallel to the vertical axis intersects the line through B parallel to the horizontal axis at P, the point that corresponds to the pair (a, b).

A system of axes like the one we have described is called a *system of Cartesian coordinates* in the plane. The two axes divide the plane into four regions (Figure 3.1), each of which is called a *quadrant*. Points in the *first quadrant* (I) have both coordinates positive; points in the *third quadrant* (III) have both coordinates negative. A point in the *second quadrant* (II) has negative abscissa and positive ordinate, while a point in the *fourth quadrant* (IV) has positive abscissa and negative ordinate. Finally, points on the *x*-axis have ordinates equal to zero, and points on the *y*-axis have abscissas equal to zero. Figure 3.2 shows several points plotted in a coordinate plane.

We now obtain a formula that gives the coordinates of the midpoint of a line segment in terms of the coordinates of the endpoints of the segment.

Midpoint Formula

Let $P_1(x_1, y_1)$ and $P_2(x_2, y_2)$ be two points in a coordinate plane. Let M be the midpoint of the segment P_1P_2. The coordinates (x, y) of M are given by

$$x = \frac{x_1 + x_2}{2} \quad \text{and} \quad y = \frac{y_1 + y_2}{2}. \tag{3.1}$$

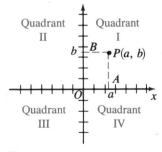

Figure 3.1

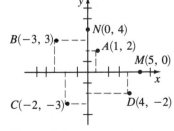

Figure 3.2

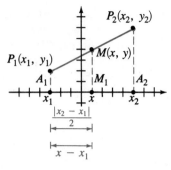

Figure 3.3

To obtain this formula, observe that the lines through P_1 and P_2 parallel to the y-axis intersect the x-axis at the points $A_1(x_1, 0)$ and $A_2(x_2, 0)$, respectively (Figure 3.3). Assuming $x_2 > x_1$, we have

$$d(A_1, A_2) = |x_2 - x_1| = x_2 - x_1.$$

From plane geometry, it follows that the line through M parallel to the y-axis intersects the x-axis at the midpoint M_1 of the segment $A_1 A_2$. Thus,

$$d(M_1, A_1) = \frac{x_2 - x_1}{2}.$$

On the other hand, if x denotes the abscissa of M_1, then

$$d(M_1, A_1) = x - x_1.$$

Equating the two expressions for $d(M_1, A_1)$ gives

$$x - x_1 = \frac{x_2 - x_1}{2},$$

so

$$x = x_1 + \frac{x_2 - x_1}{2}$$

$$= \frac{x_1 + x_2}{2}.$$

In the same way, the y-coordinate of M is

$$y = \frac{y_1 + y_2}{2}.$$

EXAMPLE 1   Find the midpoint M of the segment joining the points $A(2, 3)$ and $B(-4, 5)$ and plot the points.

Solution: According to the midpoint formula, we get

$$x = \frac{2 + (-4)}{2} = \frac{2 - 4}{2} = \frac{-2}{2} = -1,$$

$$y = \frac{3 + 5}{2} = \frac{8}{2} = 4.$$

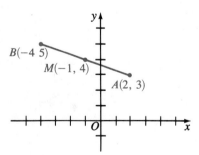

Figure 3.4

Practice Exercise 1   Find the midpoint M of the segment with endpoints $(-3, 5)$ and $(7, -2)$.

Answer: $(2, 3/2)$

Next, we derive a formula that gives the distance between two points in terms of the coordinates of those points.

Distance Formula

If $P_1(x_1, y_1)$ and $P_2(x_2, y_2)$ are two points in a coordinate plane, then the *distance* between P_1 and P_2 is given by

$$d(P_1, P_2) = \sqrt{(x_1 - x_2)^2 + (y_1 - y_2)^2}. \qquad (3.2)$$

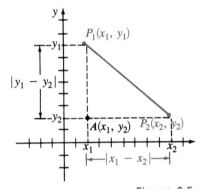

Figure 3.5

Assume that P_1 and P_2 are not on the same horizontal or vertical line. The triangle $P_1 A P_2$ (Figure 3.5) is a right triangle whose hypotenuse $P_1 P_2$ has length $d(P_1, P_2)$ and whose sides AP_2 and AP_1 have lengths $|x_1 - x_2|$ and $|y_1 - y_2|$, respectively. By the Pythagorean theorem,

$$[d(P_1, P_2)]^2 = |x_1 - x_2|^2 + |y_1 - y_2|^2 = (x_1 - x_2)^2 + (y_1 - y_2)^2.$$

Taking square roots, we obtain

$$d(P_1, P_2) = \sqrt{(x_1 - x_2)^2 + (y_1 - y_2)^2},$$

which is formula (3.2).

If P_1 and P_2 lie on the same horizontal line, then $y_1 = y_2$, and formula (3.2) becomes

$$d(P_1, P_2) = \sqrt{(x_1 - x_2)^2} = |x_1 - x_2|.$$

Similarly, if P_1 and P_2 lie on the same vertical line, then $x_1 = x_2$ and

$$d(P_1, P_2) = |y_1 - y_2|.$$

EXAMPLE 2   Find the distances between the points **(a)** $P(-2, 3)$, $Q(-1, -4)$, and **(b)** $A(2, 0)$, $B(-3, 0)$.

Solution: We use the distance formula for both pairs of points.

(a) $\begin{aligned} d(P, Q) &= \sqrt{[-2 - (-1)]^2 + [3 - (-4)]^2} \\ &= \sqrt{(-2 + 1)^2 + (3 + 4)^2} \\ &= \sqrt{(-1)^2 + 7^2} \\ &= \sqrt{1 + 49} \\ &= \sqrt{50} \\ &= 5\sqrt{2} \end{aligned}$

(b) $\begin{aligned} d(A, B) &= \sqrt{[2 - (-3)]^2 + (0 - 0)^2} \\ &= \sqrt{(2 + 3)^2 + 0^2} \\ &= \sqrt{25} \\ &= 5 \end{aligned}$

Since A and B are points on the x-axis with x-coordinates 2 and -3, respectively, their distance could also be obtained by using the formula for the distance between two points on an axis (Section 2.3):

$$d(A, B) = |2 - (-3)| = |2 + 3| = 5.$$

Practice Exercise 2   Find the distances between the following pairs of points: **(a)** $P(3, 4)$, $Q(-1, 6)$ **(b)** $M(0, -3)$, $N(0, 8)$.

Answer: **(a)** $2\sqrt{5}$ **(b)** 11

EXAMPLE 3   Plot the points $A(1, 3)$, $B(2, 1)$, and $C(6, 3)$ and show that the triangle ABC is a right triangle.

Solution: By the distance formula, we have

$$d(A, B) = \sqrt{(1 - 2)^2 + (3 - 1)^2} = \sqrt{1 + 4} = \sqrt{5},$$
$$d(B, C) = \sqrt{(2 - 6)^2 + (1 - 3)^2} = \sqrt{16 + 4} = \sqrt{20},$$
$$d(A, C) = \sqrt{(1 - 6)^2 + (3 - 3)^2} = \sqrt{25 + 0} = \sqrt{25}.$$

Now, to show that ABC is a right triangle, we use the *converse* of the Pythagorean theorem: if the lengths a, b, and c of the sides of a triangle satisfy the relation $c^2 = a^2 + b^2$, then the triangle is a right triangle. Since

$$(\sqrt{25})^2 = (\sqrt{20})^2 + (\sqrt{5})^2,$$

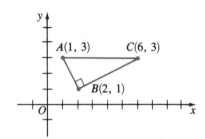

Figure 3.6   then ABC is a right triangle with hypotenuse AC.

Practice Exercise 3   Do the points $X(4, -1)$, $Y(1, 2)$, and $Z(8, 3)$ determine a right triangle?

Answer: Yes. (Obtain $d(X, Y) = 3\sqrt{2}$, $d(X, Z) = 4\sqrt{2}$, $d(Y, Z) = 5\sqrt{2}$ and use the converse of the Pythagorean theorem.)

EXERCISES 3.1

1. Plot the following points on a Cartesian coordinate system: $A(-1, 4)$, $B(-2, -5)$, $C(-1, -3)$, $M(2, 1)$, $N(-3, -2)$.
2. Plot the points $A(2, -4)$, $B(-4, 2)$, $C(1, 3)$, $D(3, 1)$ and draw the segments AB, AD, BC, and CD.

For the given points P and Q in Exercises 3–8, find the midpoint of the segment PQ.

3. $P(-1, 4)$, $Q(-2, 5)$ **4.** $P(2, 2)$, $Q(-3, -5)$
5. $P(-1, -3/2)$, $Q(1/4, 3)$
6. $P(2/3, -5)$, $Q(-2, 5/4)$
7. $P(2, 5)$, $Q(a, -5)$
8. $P(3, 4)$, $Q(-2, b)$

In Exercises 9–20, find the distance $d(P, Q)$.

9. $P(1, 1)$, $Q(5, 5)$
10. $P(-2, -2)$, $Q(-4, 4)$
11. $P(6, -2)$, $Q(5, -6)$
12. $P(-1, -2)$, $Q(2, 3)$
13. $P(0, -2/3)$, $Q(0, 5/3)$
14. $P(1, 1/2)$, $Q(-1/4, 1)$
15. $P(2, 2)$, $Q(3, -3)$
16. $P(0, 3)$, $Q(-5, 0)$
17. $P(1, b)$, $Q(3, -b)$
18. $P(b, -1)$, $Q(-1, b)$

19. $P(t, t + 2)$, $Q(t + 3, t - 2)$

20. $P(t - 1, s)$, $Q(t, s - 1)$

☐ In Exercises 21 and 22, use a calculator to find the distance between the two given points. Round off your answers to four significant digits.

21. $A(21.05, 1.314)$, $B(-3.114, 21.26)$

22. $M(\sqrt{2}, 3/8)$, $N(121/4, \sqrt{3})$

Solve Exercises 23–46.

23. Plot the points $A(7, 2)$, $B(4, 6)$, and $C(-4, 0)$ and show that the triangle ABC is a right triangle.

24. Show that the triangle with vertices $P(0, 0)$, $Q(2, 0)$, and $R(2, 6)$ is a right triangle.

25. Find the area of the triangle ABC in Exercise 23.

26. Find the area of the triangle PQR in Exercise 24.

27. Show that the triangle with vertices $A(1, -2)$, $B(-4, 2)$, and $C(1, 6)$ is an isosceles triangle.

28. Show that the triangle with vertices $P(1, 3)$, $Q(3, 1)$, and $R(4, 4)$ is an isosceles triangle.

29. Plot the points $A(-1, -2)$, $B(4, 3)$, $C(-1, 8)$, and $D(-6, 3)$ and show that they are the vertices of a square.

30. Show that the points $P(-1, -5)$, $Q(1, -2)$, $R(-1, 4)$, and $S(-3, 1)$ are the vertices of a parallelogram.

31. Determine b so that the triangle with vertices $O(0, 0)$, $A(2, 3)$, and $B(3, b)$ is a right triangle with right angle at O.

32. If $P(1, 2)$, $Q(4, 6)$, and $R(x, 3)$ are vertices of a right triangle with right angle at Q, find the value of x.

33. If $M(4, -7)$ is the midpoint of a segment whose endpoints are $P(x, y)$ and $Q(-5, 8)$, find x and y.

34. Given the points $A(2, 7)$ and $M(-3, -5)$, find the coordinates of the point B such that M is the midpoint of AB.

35. Find numbers a and b so that $M(a, 5)$ is the midpoint of a segment AB with $A(2, -6)$ and $B(-3, b)$.

36. Find m and n so that $M(-8, 4)$ is the midpoint of a segment whose endpoints are $P(m, -5)$ and $Q(3, n)$.

37. Given $A(3, 1)$ and $B(7, 5)$, find the coordinates of the point M on the segment AB such that $d(A, M) = (1/4)d(A, B)$. [*Hint:* If x is the abscissa of M, then $x - 3 = (7 - 3)/4$.]

38. Given $A(1, 2)$ and $B(-3, 3)$, find the coordinates of the point on the segment AB that is $2/3$ of the way from A to B.

39. Show that the point $P(4, 2)$ is on the perpendicular bisector of AB, where $A(-1, 3)$ and $B(3, -3)$. [*Hint:* If P is on the perpendicular bisector of AB, then $d(A, P) = d(B, P)$.]

40. Is $Q(-5, -4)$ also on the perpendicular bisector of the segment AB in Exercise 39?

41. Find the formula that expresses the fact that $P(x, y)$ is on the perpendicular bisector of AB, where $A(-1, 3)$ and $B(3, -3)$.

42. Let $A(3, -2)$ and $B(-3, 1)$. Write an equation for all points $P(x, y)$ such that $(d(A, P))^2 = (d(B, P))^2 + 1$.

43. What points (x, y) with $x = y$ are $\sqrt{5}$ units from $(2, 3)$?

44. Find all points $P(x, y)$ that satisfy the equation $x + y = 0$ and that are $\sqrt{10}$ units from the point $(-3, 5)$.

45. Find an equation for all points $P(x, y)$ that are 3 units from the point $(2, -4)$.

46. Find an equation for all points $P(x, y)$ that are 5 units from the point $(-3, 8)$.

In Exercises 47–50 check the relation $d(P, Q) = d(P, R) + d(R, Q)$ to determine whether the point R lies on the line segment PQ. It can be shown that R lies on the segment PQ exactly when $d(P, Q) = d(P, R) + d(R, Q)$.

47. $P(-2, -3)$, $Q(4, 3)$, $R(2, 1)$

48. $P(0, 2/5)$, $Q(2, 2)$, $R(-3, -2)$

49. $P(-3, 1/2)$, $Q(1/3, 2)$, $R(1, 1)$

50. $P(2/3, 0)$, $Q(-1, -2)$, $R(-2, 1)$

3.2   THE STRAIGHT LINE

A line is a geometric object. It is known from Euclidean geometry that two distinct points determine a unique line. If the two points lie on a Cartesian coordinate plane, the line also lies in the same plane and it can be described by an algebraic equation that relates the coordinates x and y of an arbitrary point on the line to the coordinates of the two given points. The line itself is called the *graph of the equation*.

In order to derive the equation of a line, we introduce the notion of *slope*, a number that measures the steepness of a line relative to the x-axis.

The Slope of a Line

Let $P_1(x_1, y_1)$ and $P_2(x_2, y_2)$ be two distinct points on a coordinate plane such that $x_1 \neq x_2$. The *slope m of the line l determined by these two points* is defined by

$$m = \frac{y_2 - y_1}{x_2 - x_1}. \qquad (3.3)$$

If $x_1 = x_2$, the line l is *vertical* (Figure 3.7a) and, in this case, the slope is *not defined*. Referring to Figure 3.7b, we call the differences $y_2 - y_1$ and $x_2 - x_1$ the *rise* and *run*, respectively, along l from P_1 to P_2, and we write

$$\text{slope} = \frac{\text{rise}}{\text{run}}.$$

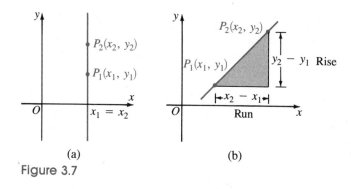

(a)   (b)

Figure 3.7

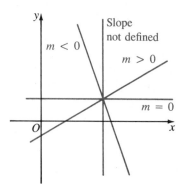

Figure 3.8

If the labeling of the points is reversed, the slope remains the same, since

$$m = \frac{y_2 - y_1}{x_2 - x_1} = \frac{y_1 - y_2}{x_1 - x_2}.$$

The slope of a line can be *any* real number; *positive, negative,* or *zero*. This corresponds to the fact that infinitely many lines can pass through a fixed point (Figure 3.8). If $m = 0$, then the line is parallel to the x-axis, so it is a *horizontal* line.

EXAMPLE **1**

When defined, find the slopes of the lines determined by the following pairs of points.   **(a)** $A(-1, 2)$, $B(2, 5)$ **(b)** $C(-1, -3)$, $D(-1, 2)$

Solution: **(a)** According to (3.3), we have

$$m = \frac{5 - 2}{2 - (-1)} = \frac{3}{3} = 1.$$

(b) Since C and D have the same x-coordinate (-1), the slope is undefined and the line through C and D is vertical.

Practice Exercise 1   Find the slopes of the lines through the following pairs of points. **(a)** $M(-3, 2)$, $N(4, 2)$ **(b)** $P(-3, 4)$, $Q(2, 1)$

Answer: **(a)** $m = 0$ **(b)** $m = -\dfrac{3}{5}$

Equations of a Line

Let l be the line determined by two distinct points $P_1(x_1, y_1)$ and $P_2(x_2, y_2)$, with $x_1 \neq x_2$, and let $P(x, y)$ be an arbitrary point on l (Figure 3.9). We want to find how the coordinates x and y of P are related to the coordinates of P_1 and P_2. By doing so we shall obtain an equation in x and y, called the *equation of the line l*.

Point–Slope Form

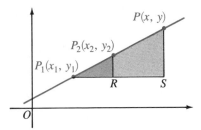

Figure 3.9

Since the triangles $P_1 R P_2$ and $P_1 S P$ in Figure 3.9 are similar, the slope of the line determined by P_1 and P_2 is equal to the slope of the line determined by P_1 and P, which we express as m:

$$\frac{y_2 - y_1}{x_2 - x_1} = \frac{y - y_1}{x - x_1} = m$$

Hence, we obtain

$$y - y_1 = m(x - x_1), \tag{3.4}$$

which is the *point–slope form* of the equation of the line l.

Notice that in this equation, x_1, y_1, and m are *known quantities* while x and y are *variables*. Actually, the ordered pair (x, y) represents the coordinates of an arbitrary point P on l. We can say that l is the set of all points $P(x, y)$ in the plane that satisfy the equation (3.4) and we can write

$$l = \{P(x, y): y - y_1 = m(x - x_1)\}.$$

Such a set is also called the *graph* of the equation (3.4). (Graphs of equations will be discussed in detail in the next section.)

EXAMPLE 2   Find an equation of the line with slope $-1/3$ that passes through the point $A(1, -2)$.

Solution: Using the point–slope form with $m = -1/3$, $x_1 = 1$, and $y_1 = -2$, we get

$$y - (-2) = -\frac{1}{3}(x - 1)$$

or

$$y + 2 = -\frac{1}{3}(x - 1).$$

We can write this equation in two equivalent forms. First, we can solve for y and simplify, to obtain

$$y = -\frac{1}{3}x - \frac{5}{3}.$$

Next, we can multiply by 3 and rearrange the terms to get

$$x + 3y + 5 = 0.$$

Practice Exercise 2

Find the point–slope equation of the line through $P(-3, 7)$ and with slope 5/3.

Answer: $y - 7 = \frac{5}{3}(x + 3)$

EXAMPLE   **3**

Find an equation of the line through the points $A(-1, 2)$ and $B(3, 5)$.

Solution: First, we must determine the slope of the line:

$$m = \frac{5 - 2}{3 - (-1)} = \frac{3}{4}.$$

Next, we use equation (3.4) with $m = 3/4$, $x_1 = -1$, and $y_1 = 2$ to get

$$y - 2 = \frac{3}{4}(x + 1).$$

This is the point–slope equation of the line with slope 3/4 passing through $A(-1, 2)$. The last equation can also be written in the following equivalent forms:

$$y = \frac{3}{4}x + \frac{11}{4}$$

and

$$3x - 4y + 11 = 0.$$

As an exercise, you can check that the last two equations can be obtained by writing the point–slope equation of the line through $B(3, 5)$ with slope $3/4$.

Practice Exercise 3

Find an equation of the line through $P(1, -2)$ and $Q(-5, 3)$.

Answer: $y + 2 = -\frac{5}{6}(x - 1)$ or $5x + 6y + 7 = 0$

Equation of a Vertical Line

If $x_1 = x_2$, then the line l passing through $P_1(x_1, y_1)$ and $P_2(x_2, y_2)$ is vertical. A point $P(x, y)$ lies on l if and only if

$$x = x_1. \tag{3.5}$$

This equation represents a vertical line passing through the point $P_1(x_1, y_1)$. Such a line can be viewed as the set of points in a Cartesian plane whose x-coordinates are equal to x_1 and whose y-coordinates are arbitrary numbers. We may write

$$l = \{P(x, y): x = x_1\}.$$

Slope–Intercept Form

If we solve equation (3.4) for y, we obtain

$$y = mx + (y_1 - mx_1).$$

Next, setting $b = y_1 - mx_1$ (a known quantity) gives us

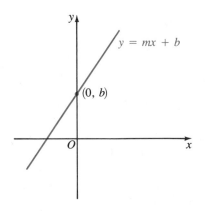

$$y = mx + b. \tag{3.6}$$

Figure 3.10

This is the *slope–intercept form* of the equation of the line l. Notice that in this form, the slope of the line is the *coefficient* of x. Moreover, if 0 is substituted for x in the equation (3.6), then $y = b$. This tells us that the point $(0, b)$ lies on the line and also on the y-axis. Thus, the line crosses the y-axis at the point $(0, b)$ (Figure 3.10). The number b is called the *y-intercept* of the line.

EXAMPLE 4

Find the slope–intercept equation of the line determined by the points $A(-1, 2)$ and $B(3, 5)$.

Solution: We saw, in Example 3, that a point–slope equation of this line is

$$y - 2 = \frac{3}{4}(x + 1).$$

Solving for y and simplifying gave us

$$y = \frac{3}{4}x + \frac{11}{4},$$

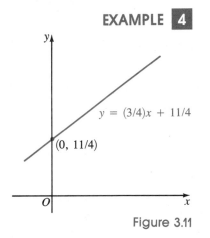

Figure 3.11

the slope–intercept equation of the line (Figure 3.11). The slope is 3/4, the coefficient of x, and the y-intercept is 11/4.

Practice Exercise 4

Find the slope–intercept equation of the line through $P(1, -2)$ and $Q(-5, 3)$.

Answer: $y = -\dfrac{5}{6}x - \dfrac{7}{6}$

It is important to observe that every nonvertical line has a *unique* slope–intercept equation. From the slope–intercept equation of a line, we can read off sufficient geometric data to characterize the line: the coefficient of x measures the steepness of the line relative to the x-axis, and the constant term is the y-coordinate of the point of intersection of the line and the y-axis.

Linear Equations

An equation of the form

$$Ax + By + C = 0, \tag{3.7}$$

with A and B not both zero, is called a *linear equation in x and y*.

Every equation of a straight line can be put into the form (3.7). For example, we saw that the equation

$$y = \frac{3}{4}x + \frac{11}{4}$$

can be rewritten $3x - 4y + 11 = 0$, which is of the form (3.7) with $A = 3$, $B = -4$, and $C = 11$. Similarly, the equation

$$x = \frac{2}{3}$$

representing the vertical line through the point $(2/3, 0)$ can be rewritten $3x - 2 = 0$. This is also an equation of the form (3.7) with $A = 3$, $B = 0$, and $C = -2$.

Conversely, every equation of the form (3.7) with A and B not both zero represents a line. We say that (3.7) is an equation of a line in *standard form*.

EXAMPLE 5   Write the equation $6x - 4y + 3 = 0$ in slope–intercept form. Find its slope and the x- and y-intercepts of the corresponding line, and sketch it.

Solution: The slope–intercept form is obtained by solving the given equation for y:

$$-4y = -6x - 3$$
$$y = \frac{3}{2}x + \frac{3}{4}.$$

Thus, the line has slope $3/2$ and y-intercept $3/4$.

The x-intercept is obtained by finding the point where the line crosses the x-axis. At this point, the y-coordinate is equal to zero. Thus, setting $y = 0$ in the equation $6x - 4y + 3 = 0$ and solving for x, we get

$$6x - 4 \cdot 0 + 3 = 0, \quad \text{or} \quad x = -\frac{3}{6} = -\frac{1}{2}.$$

Thus, $-1/2$ is the x-intercept.

To graph the line (Figure 3.12), we simply plot the x- and y-intercepts and join them by a straight line.

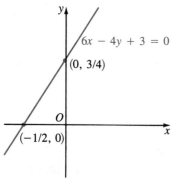

$6x - 4y + 3 = 0$

$(0, 3/4)$

$(-1/2, 0)$

Figure 3.12

Practice Exercise 5   Given the line $2x + 5y - 3 = 0$, find its slope and its x- and y-intercepts. Sketch the line.

Answer:
Slope: $-2/5$
x-intercept: $3/2$
y-intercept: $3/5$

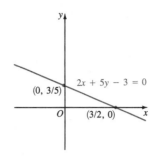

EXAMPLE 6   What lines have equations   **(a)** $6x - 5 = 0$?   **(b)** $5y + 7 = 0$?

Solution: **(a)** This is an equation of the form (3.7), with $A = 6$, $B = 0$, and $C = -5$. Solving for x, we get

$$x = \frac{5}{6}.$$

Thus, the given equation represents a vertical line through the point $(5/6, 0)$ (Figure 3.13).
(b) Here $A = 0$, $B = 5$, and $C = 7$. Solving for y, we obtain

$$y = -\frac{7}{5}.$$

In this case, the given equation represents a horizontal line (i.e., one that is parallel to the x-axis) through $(0, -7/5)$ (Figure 3.14).

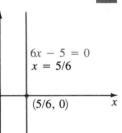

Figure 3.13

Practice Exercise 6   Find the slopes of the following lines.   **(a)** $3x - 12 = 0$ **(b)** $7y - 10 = 0$

Solution: **(a)** Slope undefined, vertical line through $(4, 0)$ **(b)** $m = 0$, horizontal line through $(0, 10/7)$.

Terminology

From now on, when no confusion is possible, we shall say *the line* $Ax + By + C = 0$, meaning *the line with equation* $Ax + By + C = 0$.

Parallel Lines

Two lines are said to be *parallel* if they lie in the same plane and do not intersect each other. Since the slope of a line is a number that measures how steep the line is in relation to the x-axis, it seems clear that parallel lines have the same slope. This is the *algebraic condition for parallel lines*, which we now state.

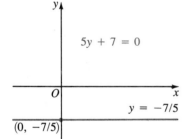

Figure 3.14

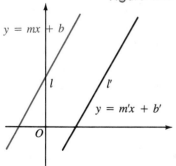

Figure 3.15

Condition for Parallel Lines

Let l and l' be two lines with slopes m and m', respectively (Figure 3.15). The lines are parallel if and only if

$$m = m'. \qquad (3.8)$$

EXAMPLE 7

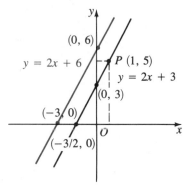

Figure 3.16

Find an equation of the line parallel to $2x - y + 6 = 0$ and passing through the point $P(1, 5)$. Sketch it.

Solution: First, we solve the given equation for y and find its slope:

$$y = 2x + 6.$$

Hence, the slope of the given line is $m = 2$. Since the line through $P(1, 5)$ is parallel to the given line, its slope is also 2. Thus, the point–slope form of the line through $P(1, 5)$ and parallel to $2x - y + 6 = 0$ is

$$y - 5 = 2(x - 1).$$

From this equation we can derive the point–intercept equation,

$$y = 2x + 3,$$

which is graphed in Figure 3.16.

Note that in the slope–intercept equation of two parallel lines, only the *constant* terms differ from each other. The linear terms (i.e., those containing x) are the same since the lines have the same slope.

Practice Exercise 7

Find an equation of the line passing through the point $P(-2, 4)$ and parallel to the line determined by the points $A(2, 0)$ and $B(3, 5)$.

Answer: $y - 4 = 5(x + 2)$

Perpendicular Lines

Two lines are *perpendicular* to each other if they intersect at a right angle. The *algebraic condition for perpendicular lines* is not as obvious as the condition for parallel lines; it can be stated as follows.

Condition for Perpendicular Lines

Let l and l' be two lines with slopes m and m' *both different from zero*. The two lines are perpendicular if and only if

$$m' = -\frac{1}{m}. \qquad (3.9)$$

Thus two lines (*neither of which is vertical*) are perpendicular if and only if *the slope of one line is the negative of the reciprocal of the slope of the other.* We will derive this relationship after an example.

EXAMPLE 8

Find an equation of the line perpendicular to the line $2x - 3y - 6 = 0$ and passing through $P(-1, 3)$. Sketch it.

Solution: First we find the slope of the given line:

$$2x - 3y - 6 = 0$$
$$-3y = -2x + 6$$
$$y = \frac{2}{3}x - 2,$$

so
$$m = \frac{2}{3}.$$

According to (3.9), every line perpendicular to the line $2x - 3y - 6 = 0$ has slope
$$m' = -\frac{1}{2/3} = -\frac{3}{2}.$$

Next we write the point–slope equation of the line through $P(-1, 3)$ with slope $-3/2$:
$$y - 3 = -\frac{3}{2}(x + 1).$$

This equation can also be written
$$y = -\frac{3}{2}x + \frac{3}{2} \quad \text{slope-intercept form}$$

or
$$3x + 2y - 3 = 0. \quad \text{standard form}$$

The line is graphed in Figure 3.17.

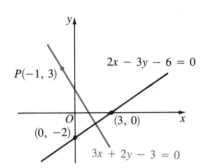

$2x - 3y - 6 = 0$

$P(-1, 3)$

$(0, -2)$

$(3, 0)$

$3x + 2y - 3 = 0$

Figure 3.17

Practice Exercise 8

Find an equation of the line perpendicular to the line $3x + 2y + 7 = 0$ and passing through $(-1, -2)$.

Answer: $y + 2 = \dfrac{2}{3}(x + 1)$

Now we show how to derive the perpendicularity condition (3.9).

The Perpendicularity Condition

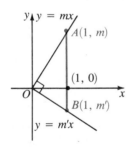

$y = mx$

$A(1, m)$

$(1, 0)$

$B(1, m')$

$y = m'x$

Figure 3.18

Let l and l' be two lines (neither of which is vertical) through the origin, with equations $y = mx$ and $y = m'x$ (Figure 3.18). Let $x = 1$ be the vertical line through the point $(1, 0)$. This line intersects l and l' at the points $A(1, m)$ and $B(1, m')$. (Can you see why?) Now, the lines l and l' are perpendicular if and only if the triangle AOB is a right triangle, with right angle at the vertex O. By the Pythagorean theorem,
$$d(A, B)^2 = d(O, A)^2 + d(O, B)^2.$$

Applying the distance formula (3.2) to each of these distances, we have
$$(m - m')^2 = (1 + m^2) + (1 + m'^2).$$

Squaring and simplifying give us
$$m^2 - 2mm' + m'^2 = 1 + m^2 + 1 + m'^2$$
$$-2mm' = 2$$
$$mm' = -1.$$

Hence
$$m' = -\frac{1}{m},$$

which is condition (3.9).

If the lines l and l' intersect at a point other than the origin, we can apply the same reasoning to the two lines that go through the origin and are parallel to l and l'.

Linear Relations

Two variables x and y that satisfy a linear equation $Ax + By + C = 0$ are said to be *linearly related*.

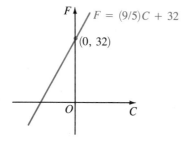

Figure 3.19

For example, the relationship between degrees Fahrenheit (F) and degrees Celsius (C) is given by the equation

$$F = \frac{9}{5}C + 32.$$

This equation can be rewritten $9C - 5F + 160 = 0$, which is an equation of the form (3.7) with $x = C, y = F, A = 9, B = -5$, and $C = 160$. Thus the variables C and F are linearly related.

Notice that in the (C, F) coordinate plane (Figure 3.19), the given equation represents a line whose slope is $9/5$ and whose y-intercept is 32.

EXAMPLE 9

Suppose that a variable s (space) is linearly related to a variable t (time). Find an equation relating s and t, knowing that $s = 6$ when $t = 3$, and that $s = 8$ when $t = 6$. Also, find s when $t = 10$.

Solution: First, we make a table of values as follows:

t	s
3	6
6	8

Next, we observe that if t and s are linearly related, they must satisfy a linear equation in t and s. This means that the equation represents the line, in the (t, s) coordinate plane, that passes through the points $(3, 6)$ and $(6, 8)$. To find the equation of this line, we compute its slope:

$$m = \frac{8 - 6}{6 - 3} = \frac{2}{3}.$$

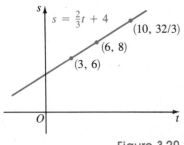

Figure 3.20

Then we write the point–slope equation:

$$s - 6 = \frac{2}{3}(t - 3)$$

$$s = \frac{2}{3}t + 4,$$

or, in standard form,

$$2t - 3s + 12 = 0.$$

This is a linear equation relating t and s. Setting $t = 10$, we obtain

$$s = \frac{2}{3} \cdot 10 + 4 = \frac{32}{3}.$$

Practice Exercise 9   Find a linear equation relating x and y if you know that $y = 2$ when $x = -1$, and that $y = -3$ when $x = 2$. Also, find y when $x = 0$.

Answer: $5x + 3y - 1 = 0$, $y = \dfrac{1}{3}$

EXERCISES 3.2

In Exercises 1–4, find the slope of the line passing through the given points.

1. $A(2, 2)$, $B(-1/2, -1/2)$ **2.** $A(1, 3)$, $B(-2/3, -2)$
3. $P(-2, -1/4)$, $Q(3, -2/3)$
4. $U(5, -1/4)$, $V(-5, 1/4)$

In Exercises 5–8, use slopes to determine whether or not the points P, Q, and R lie on a straight line.

5. $P(1, 2)$, $Q(3, 5)$, $R(5, 8)$
6. $P(-1, 3)$, $Q(2, -2)$, $R(0, 1)$
7. $P(2, 2)$, $Q(-2, 4)$, $R(0, 3)$
8. $P(0, 1)$, $Q(1/2, 0)$, $R(2, -3)$

In Exercises 9–20, write an equation of the line having the given properties.

9. Vertical and passing through $(-2, 5)$
10. Horizontal and passing through $(-1, -3)$
11. Having slope 3 and y-intercept -2
12. Having slope $-1/3$ and y-intercept 3
13. Passing through the origin and having slope $-2/5$
14. Having slope $-7/9$ and passing through $(-2, -5)$
15. The slope is -4 and the x-intercept is 2.
16. Having slope $-2/3$ and x-intercept 3
17. Passing through $(-6, -2)$ and having slope 0
18. Passing through $(-2, 0)$ and having slope $3/4$
19. The x-intercept is -3 and the y-intercept is 2.
20. The x-intercept is 5 and the y-intercept is -1.

In Exercises 21–28, write each of the lines in slope–intercept form. Find the slope, y-intercept, and x-intercept. Sketch the line.

21. $3x - 4y + 5 = 0$
22. $2x - 3y - 9 = 0$
23. $x = \dfrac{3}{5}y + 3$
24. $\dfrac{2}{3}x + \dfrac{2}{5}y - 4 = 0$
25. $2y + 3 = 0$
26. $4x - 3y = 2$
27. $\dfrac{x}{2} + \dfrac{y}{3} = 1$
28. $\dfrac{x}{4} - \dfrac{y}{2} = 1$

In Exercises 29–32, determine whether the given lines are parallel, perpendicular, or neither. Sketch the lines.

29. $2x + 7y - 2 = 0$, $7x - 2y + 1 = 0$
30. $y = \dfrac{3}{5}x - \dfrac{1}{4}$, $y = -\dfrac{5}{3}x + 2$
31. $x + y + 1 = 0$, $x - y = 0$
32. $\dfrac{x}{2} + \dfrac{y}{3} - 2 = 0$, $\dfrac{2}{3}x + y - 1 = 0$

In Exercises 33–42, find an equation of the line satisfying each of the given conditions. Write the equation in standard form. Sketch it.

33. Parallel to the line $y = 3x - 2$ and passing through $(0, 0)$
34. Parallel to the line $2x - 5y - 3 = 0$ and passing through $(-2, 3)$
35. Perpendicular to the line $y = 2x - 5$ and passing through $(-1, -2)$
36. Perpendicular to the line $6x - 2y + 5 = 0$ and passing through $(-1, -2)$
37. Perpendicular to the main diagonal (the line $y = x$) and passing through $(1, -4)$
38. Perpendicular to the diagonal $y = -x$ and passing through $(-2, 5)$

39. Parallel to the main diagonal and passing through $(-3, 7)$

40. Parallel to the main diagonal and passing through $(-5, -7)$

41. Parallel to $(1/2)x + (2/3)y - 3/4 = 0$ and passing through $P(3, -3)$

42. Perpendicular to $(3/5)x - (2/3)y + 1/2 = 0$ and passing through $Q(-1/3, 4/5)$

Work the following problems in Exercises 43–70.

43. Find a so that the line $ax + 3y - 5 = 0$ is parallel to the line $x - 4y + 1 = 0$.

44. Find a so that the line $ax - 4y + 3 = 0$ is perpendicular to the line $3x - 2y + 5 = 0$.

45. Find b so that the line $5x - by - 3 = 0$ is perpendicular to the line $2x + 2y - 5 = 0$.

46. Find b so that the line $3x + by = 5$ is parallel to the line $2x - 5y - 3 = 0$.

47. Find an equation of the line through $P(-2, -5)$ and perpendicular to the line determined by the points $Q(-2, 9)$ and $R(3, -10)$.

48. Find an equation of the line through $A(-1, 3)$ and parallel to the line determined by the points $B(2, 5)$ and $C(-3, -4)$.

49. Let $A(2, 5)$, $B(4, 9)$, and $C(6, 8)$ be the vertices of a triangle. Compute the slope of each side and show that the triangle is a right triangle.

50. By computing the slope of each side, show that the triangle with vertices $A(1, 3)$, $B(2, 1)$, and $C(8, 4)$ is a right triangle.

51. By writing the slope of each side, determine b so that the triangle with vertices $O(0, 0)$, $A(2, 3)$, and $B(3, b)$ is a right triangle with right angle at A.

52. Find x so that $P(1, 2)$, $Q(4, 6)$, and $R(x, 3)$ are vertices of a right triangle with right angle at Q.

53. Write an equation for the perpendicular bisector of the segment PQ with $P(2, -3)$ and $Q(6, -1)$.

54. Write an equation for the perpendicular bisector of the segment AB with $A(-1, 2)$ and $B(3, 1)$.

55. The variables u and v are linearly related. When $u = 5$ we have $v = 10$, and when $u = 8$ we have $v = 15$. Find an equation that relates u and v. Also, find v when $u = 2$.

56. Suppose that X and Y are linearly related in such a way that if $X = 2$ then $Y = 10$, and if $X = 5$ then $Y = 25$. Find **(a)** a linear equation relating X and Y, and **(b)** X when $Y = 0$.

57. Find a such that the point $A(1, 2)$ lies on the line $ax - 3y + 4 = 0$.

58. Find b such that the line $3x + by + 2 = 0$ has y-intercept $1/2$.

59. A w-lb weight suspended on a spring causes it to stretch s inches. Hooke's law states that the weight w and the stretch s are linearly related. If a 5-lb weight stretches the spring

1 in. and no weight causes no stretch, find **(a)** a linear equation relating s and w and **(b)** the stretch of the spring for a 2-lb weight.

60. A 4-lb weight stretches a spring 1 in. and an 8-lb weight stretches it 2 in. Assuming that Hooke's law (Exercise 59) holds, find a linear equation relating a w-lb weight to an s-in. stretch.

61. One of the depreciation methods used for tax purposes is the *straight-line method*, which apportions the deductions evenly in each year. Assume that a typewriter purchased for $1,000 is being depreciated linearly for a period of 5 years. (This means that after 5 years the value of the typewriter is $0.) Find **(a)** the linear equation giving the typewriter's value in t years; **(b)** its value after 3 years.

62. If the total cost y to manufacture a certain product is linearly related to the number x of units manufactured, find an equation relating x and y from the information that it costs $90 to manufacture 12 units and $105 to manufacture 15 units.

63. When air resistance is neglected, the velocity of a projectile fired upward with initial velocity v_0 (in ft/s) is given by the linear equation $v = -32t + v_0$, where t is time in seconds. Assuming that $v_0 = 192$ ft/s, find the velocity of the projectile after 3 s. When will the velocity be zero?

64. Suppose that in Exercise 63 the initial velocity is $v_0 = 240$ ft/s. Find the velocity after 5 s. When will the velocity be 120 ft/s?

65. Show that the lines $ax + by + c = 0$ and $bx - ay + d = 0$ are perpendicular to each other.

66. Show that the line passing through the point (x_0, y_0) and parallel to the line $ax + by + c = 0$ has equation $a(x - x_0) + b(y - y_0) = 0$. Write an equation of the line that passes through the point $P(7, -3)$ and parallel to the line $3x - 8y + 10 = 0$.

67. Show that an equation of the line through the point (x_0, y_0) and perpendicular to the line $ax + by + c = 0$ is $b(x - x_0) - a(y - y_0) = 0$.

68. If the x- and y-intercepts of a straight line are p and q, respectively, show that an equation of the line is $x/p + y/q = 1$.

69. If both the points $P_1(x_1, y_1)$ and $P_2(x_2, y_2)$, with $x_1 \neq x_2$, lie on the graph of the line $y = mx + b$, prove that $m = (y_2 - y_1)/(x_2 - x_1)$ and $b = (y_1 x_2 - y_2 x_1)/(x_2 - x_1)$.

70. Show that if three distinct points (x_1, y_1), (x_2, y_2), and (x_3, y_3) lie on the same line and if $x_1 \neq x_2$, then $x_1 \neq x_3$, $x_2 \neq x_3$, and

$$\frac{y_1 - y_2}{x_1 - x_2} = \frac{y_1 - y_3}{x_1 - x_3} = \frac{y_2 - y_3}{x_2 - x_3}.$$

3.3 GRAPHS OF EQUATIONS

To every set S of ordered pairs of real numbers (x, y), there is a corresponding set G of points $P(x, y)$ in a coordinate plane. The set G is called the *graph of S*.

In this section, we discuss graphs of simple equations in two variables, such as $x + y = 2$, $y = x^2$, $x = y^2$, $y = x^3$, and so on. Given an equation in two variables x and y, the *solution set* consists of all ordered pairs (x, y) that satisfy the equation. The corresponding set of points $P(x, y)$ in a coordinate plane is the *graph of the equation*.

Following an intuitive approach, we are going to sketch graphs of equations by *plotting points*. This is a rather crude and imprecise method that requires a certain amount of guessing and intuition. Nevertheless, the method provides good practice and we encourage you to graph a large number of equations. As we advance, we will explain better graphing techniques. As we shall see, a careful analysis of a given equation may unveil certain important features that minimize the plotting of points.

Graph Sketching

To sketch the graph of an equation, proceed as follows:

1. Examine the equation and, by reasoning, try to determine the general nature of the curve.
2. Construct a table of numerical values of x and y that satisfy the equation.
3. In a coordinate plane, plot the points (x, y) from the table.
4. Join the plotted points with a smooth curve.

EXAMPLE 1   Sketch the graph of the set $S = \{(x, y): x + y = 2\}$.

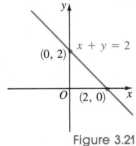

Figure 3.21

Solution: The set S is the solution set of the linear equation $x + y = 2$. As we already know, this equation describes a line. To sketch the line, we have to plot only two points, such as $(2, 0)$ and $(0, 2)$, the x- and y-intercepts of the line. In this case, a table of values is not really necessary. Figure 3.21 illustrates the graph of the equation $x + y = 2$.

Practice Exercise 1   Graph the equation $3x - 2y = 1$.

Answer:

EXAMPLE 2   Sketch the graph of the equation $y = x^2$.

Solution: We choose several values of x and find the corresponding values of y, making a table of values.

x	y
-3	9
-2	4
-1	1
0	0
1	1
2	4
3	9

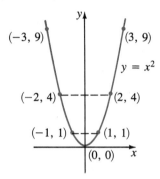

Figure 3.22

We plot the ordered pairs obtained from this table and connect them with a smooth curve, as illustrated in Figure 3.22. Since $y = x^2 \geq 0$ for all values of x, the graph is located in the upper half-plane. The curve is called a *parabola* and the point $(0, 0)$ is its *vertex*.

The parabola opens *upward*, and the vertex is its lowest point. Notice that if a pair (x, y) satisfies the equation $y = x^2$, then the pair $(-x, y)$ also satisfies the equation. This means that the parabola is *symmetric with respect to the y-axis,* and the y-axis is called the *axis of symmetry.* If the coordinate plane were folded along the axis of symmetry, the point $(1, 1)$ would coincide with $(-1, 1)$, the point $(2, 4)$ with $(-2, 4)$, and so on.

Practice Exercise 2   Sketch the graph of the equation $y = -x^2$.

Answer:

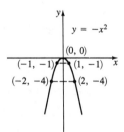

EXAMPLE 3   Graph the equation $x = y^2$.

Solution: In this case it is easier to assign values to y and find the corresponding values of x. Thus we make the following table:

x	y
9	−3
4	−2
1	−1
0	0
1	1
4	2
9	3

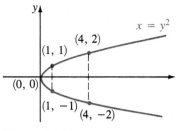

Figure 3.23

Plot the ordered pairs (x, y) obtained from this table and join them by a smooth curve as shown in Figure 3.23.

The graph is a parabola *opening to the right* and with vertex at the origin. Notice that if a point (x, y) lies on the graph, the point $(x, -y)$ also lies on the graph. In this case we have *symmetry with respect to the x-axis*. If the coordinate plane were folded along the x-axis, the point $(1, 1)$ would coincide with $(1, -1)$, the point $(4, 2)$ with $(4, -2)$, and so on.

Practice Exercise 3   Graph the equation $x = -y^2$.

Answer:

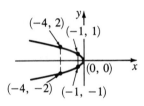

A parabola is a particular *conic section*, that is, a curve defined by the intersection of a plane and a right circular cone. We will discuss conic sections in detail in Section 3.4.

EXAMPLE 4   Graph the equation $y = x^3$.

Solution: We may assign any value to x; the corresponding y is then the cube of x. So, we obtain a table of values:

x	y
−2	−8
−1	−1
−(1/2)	−(1/8)
0	0
1/2	1/8
1	1
2	8

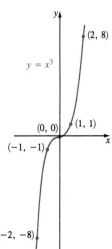

Figure 3.24

By plotting the corresponding pairs and connecting them with a smooth curve, we obtain the curve illustrated in Figure 3.24 on page 141. Now, observe that the equation $y = x^3$ is equivalent to the equation $(-y) = (-x)^3$. (Why?) Thus, if (x, y) belongs to the graph, then $(-x, -y)$ also belongs to the graph. This means that the graph is *symmetric with respect to the origin*. If we rotate the graph $180°$ about the origin, the point $(1, 1)$ will coincide with $(-1, -1)$, the point $(2, 8)$ with $(-2, -8)$, etc., so that the graph will coincide with itself.

Practice Exercise 4   Graph $y = -x^3$.

Answer:

The Equation of a Circle

As an application of the distance formula (3.2), we now show that each circle in the plane is the graph of an equation, called an *equation of the circle*. We shall also show, conversely, that equations of a certain form have graphs that are circles. Let C be a point in the plane and let r be a nonnegative real number. The *circle with center C and radius r* is defined, geometrically, to be the set of all points P in the plane whose distances from C are equal to r, that is, all points P that satisfy

$$d(P, C) = r. \tag{3.10}$$

If C has coordinates (h, k) and P has coordinates (x, y) (Figure 3.25), then from (3.2) we have

$$d(P, C) = \sqrt{(x - h)^2 + (y - k)^2},$$

and (3.10) becomes

$$\sqrt{(x - h)^2 + (y - k)^2} = r.$$

Squaring both sides yields

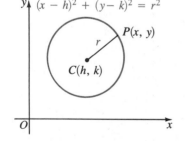

Figure 3.25

$$(x - h)^2 + (y - k)^2 = r^2. \tag{3.11}$$

This is called the *standard form* of the equation of a circle; it displays the coordinates (h, k) of the center and the radius r. When the center C coincides with the origin, equation (3.11) becomes

$$x^2 + y^2 = r^2. \tag{3.12}$$

EXAMPLE 5 Find an equation of the circle with center $C(1, 2)$ and radius 4.

Solution: By substituting $h = 1$, $k = 2$, and $r = 4$ into formula (3.11), we obtain

$$(x - 1)^2 + (y - 2)^2 = 16,$$

which is the desired equation.

Squaring and simplifying the last equation gives us

$$x^2 - 2x + 1 + y^2 - 4y + 4 = 16$$
$$x^2 + y^2 - 2x - 4y - 11 = 0,$$

which is another form of the equation of the circle centered at $C(1, 2)$ with radius 4.

Practice Exercise 5 Find an equation of the circle with center $C(-2, 3)$ and radius 2.

Answer: $(x + 2)^2 + (y - 3)^2 = 4$ or $x^2 + y^2 + 4x - 6y + 9 = 0$

If we expand the squares in (3.11), we obtain

$$x^2 - 2hx + h^2 + y^2 - 2ky + k^2 = r^2,$$

or

$$x^2 + y^2 - 2hx - 2ky + h^2 + k^2 - r^2 = 0,$$

which is an equation of the form

$$x^2 + y^2 + ax + by + c = 0, \qquad \textbf{(3.13)}$$

where $a = -2h$, $b = -2k$, and $c = h^2 + k^2 - r^2$.

EXAMPLE 6 Find the center and radius of the circle with the equation

$$x^2 + y^2 + 2x + 6y - 6 = 0.$$

Graph this equation.

Solution: Our aim is to replace the given equation with an equivalent one in standard form. This is achieved by using the method of *completing the square*. (See Section 2.5.) First, we group together the terms containing x and the terms containing y, arranging our work as follows:

$$(x^2 + 2x \quad) + (y^2 + 6y \quad) = 6.$$

Next, we complete the squares by adding 1 to the terms inside the first parentheses and 9 to the terms inside the second parentheses:

$$(x^2 + 2x + 1) + (y^2 + 6y + 9) = 6 + 1 + 9.$$

Notice that, in order to balance the equation, it is necessary to add 1 and 9 to the right-hand side of it. The last equation can be rewritten

$$(x + 1)^2 + (y + 3)^2 = 16 = 4^2.$$

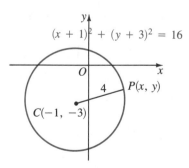

Figure 3.26

Comparing this equation with (3.11), we see that it represents a circle with center $C(-1, -3)$ and radius 4 (Figure 3.26 on page 143).

Practice Exercise 6

Write the equation $x^2 + y^2 - 8x + 2y + 15 = 0$ in standard form.

Answer: $(x - 4)^2 + (y + 1)^2 = 2$.

EXAMPLE   **7**

Determine whether or not the following equations represent circles: **(a)** $x^2 + y^2 - 2x - 6y + 10 = 0$,   **(b)** $x^2 + y^2 + 4x - 2y + 8 = 0$.

Solution: **(a)** Completing the square in the x and y terms, we obtain

$$(x^2 - 2x + 1) + (y^2 - 6y + 9) = -10 + 1 + 9$$
$$(x - 1)^2 + (y - 3)^2 = 0,$$

which is an equation of the form (3.11) with $h = 1$, $k = 3$, and $r = 0$. Since $x = 1$ and $y = 3$ are the only values that satisfy the last equation, its graph reduces to the *single point* $C(1, 3)$, so it is not a circle.
(b) Completing the squares, we have

$$(x^2 + 4x + 4) + (y^2 - 2y + 1) = -8 + 4 + 1$$
$$(x + 2)^2 + (y - 1)^2 = -3.$$

Since the right-hand side is negative while the left-hand side is always non-negative, there are *no* real numbers that can be substituted for x and y to satisfy the last equation. Thus the given equation does not represent a circle; its graph is the *empty set*.

Practice Exercise 7

Do the following equations represent circles?

(a) $x^2 + y^2 + 2x + 3 = 0$
(b) $x^2 + y^2 - 4x + 6y + 13 = 0$

Answer: **(a)** No   **(b)** No, the point $(2, -3)$ is its only solution.

As we have seen in Examples 6 and 7, an equation of the form (3.13) may represent a *circle*, a *point*, or the *empty set*.

EXAMPLE   **8**

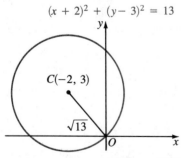

$(x + 2)^2 + (y - 3)^2 = 13$

Figure 3.27

Find an equation of a circle passing through the origin and having center $C(-2, 3)$. Graph the equation.

Solution: First we determine the radius of the circle. Since the origin is on the circle, the distance from the origin to the center $C(-2, 3)$ of the circle must equal the radius of the circle:

$$r = d(C, O) = \sqrt{(-2 - 0)^2 + (3 - 0)^2} = \sqrt{13}.$$

According to (3.11), we obtain the equation

$$(x + 2)^2 + (y - 3)^2 = 13,$$

whose graph is illustrated in Figure 3.27.

Practice Exercise 8   Find an equation in standard form for the circle with center at $(3, -1)$ and passing through $(1, 0)$.

Answer: $(x - 3)^2 + (y + 1)^2 = 5$

Circles are also conic sections of a particular kind. There will be more information about conic sections in Section 3.4.

Symmetry

In Examples 2, 3, and 4, we encountered graphs of equations with different types of symmetry. Now, we discuss the notion of symmetry in general.

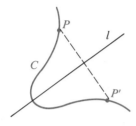

Figure 3.28

Symmetry with Respect to a Line

A curve C is said to be *symmetric with respect to a line* l if to each point P in C there corresponds a point P' in C, such that the line l is the perpendicular bisector of the segment PP' (Figure 3.28).

We say that l is a *line of symmetry* (or *axis of symmetry*) of the curve C and the points P and P' are *reflections* of each other with respect to the line l.

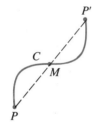

Figure 3.29

Symmetry with Respect to a Point

A curve C is *symmetric with respect to a point* M if to each point P in C there corresponds a point P' in C, such that M is the midpoint of the segment PP' (Figure 3.29).

The point M is called the *center of symmetry* of the curve C. If we rotate the curve $180°$ about the point M, then the point P will coincide with P'.

Symmetry in Terms of Coordinates

In a coordinate plane, the various notions of symmetry we just defined can be translated in terms of coordinates.

A curve C is symmetric with respect to the y-axis if, whenever $(x, y) \in C$, then $(-x, y) \in C$.

For example, the graph of $y = x^2$ (Figure 3.22) is symmetric with respect to the y-axis. Other examples are illustrated in Figure 3.30.

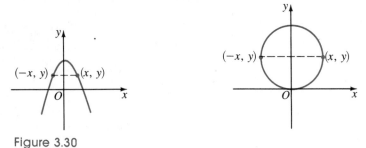

Figure 3.30

> A curve C is symmetric with respect to the x-axis if, whenever $(x, y) \in C$, then $(x, -y) \in C$.

Recall that the curve of equation $x = y^2$ (Figure 3.23) is symmetric with respect to the x-axis. Figure 3.31 shows two other examples of symmetry with respect to the x-axis.

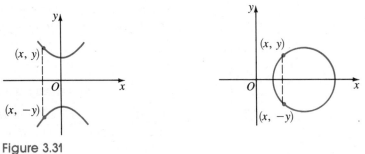

Figure 3.31

> A curve is symmetric with respect to the origin if, whenever $(x, y) \in C$, then $(-x, -y) \in C$.

The graph of $y = x^3$ (Figure 3.24) is symmetric with respect to the origin. More examples of symmetry about the origin are shown in Figure 3.32.

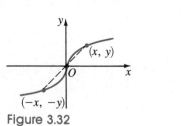

Figure 3.32

Symmetry with Respect to the Line *y* = *x*

Another important type of symmetry is *symmetry with respect to the line y = x.*

A curve C is symmetric with respect to the line $y = x$ if, whenever $(a, b) \in C$, then $(b, a) \in C$.

For example, the curves illustrated in Figure 3.33 are symmetric with respect to the line $y = x$. It is easy to see that the reflection of point (a, b) with respect to the line $y = x$ is the point (b, a), obtained by interchanging the coordinates.

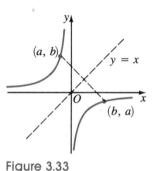

Figure 3.33

EXERCISES 3.3

In Exercises 1–8, plot the given points. Also, plot the points that are symmetric to each point with respect to (**a**) the *x*-axis, (**b**) the *y*-axis, (**c**) the origin, and (**d**) the line $y = x$.

1. $(2, -5)$ **2.** $(-3, 1)$

3. $(4, 0)$ **4.** $(-2, -2)$

5. $(2, 6)$ **6.** $(3, -8)$

7. $\left(\dfrac{3}{2}, 4\right)$ **8.** $\left(-6, \dfrac{5}{2}\right)$

In Exercises 9–22, graph each equation and check for symmetry.

9. $2x + y = 3$ **10.** $2x + 2y = 1$

11. $y = 1 - x^2$ **12.** $y = 4 - x^2$

13. $2y - 3x^2 = 0$ **14.** $3y + 4x^2 = 0$

15. $x - y^2 - 4 = 0$ **16.** $x + y^2 + 3 = 0$

17. $y + x^3 = 0$ **18.** $y = -5x^3$

19. $y = 2 - x^3$ **20.** $y = 2 + x^3$

21. $x + y^3 = 0$ **22.** $x = y^3$

In Exercises 23–34, write an equation for each of the following circles. Graph it.

23. Center $(1, 1)$; radius $\sqrt{5}$

24. Center $(-1, -3)$; radius 5

25. Center at the origin and passing through $(2, 2)$

26. Center at the origin and passing through $(-1, 4)$

27. Center $C(2, -1)$ and passing through the origin

28. Center $C(-1, -3)$ and passing through the origin

29. Center $C(3, 5)$ and tangent to the x-axis
30. Center $C(-2, -6)$ and tangent to the y-axis.
31. The points $P(-1, 3)$ and $Q(2, 5)$ lie on the circle and the segment PQ is a diameter.
32. The points $A(1, 0)$ and $B(0, 5)$ are endpoints of a diameter.
33. Center in the first quadrant, tangent to the coordinate axes, and has radius 3.
34. Center in the fourth quadrant, tangent to the coordinate axes, and has radius 5.

In Exercises 35–50, determine whether each equation represents a circle. If possible, write each equation in the form $(x - h)^2 + (y - k)^2 = r^2$ and state the center and radius.

35. $x^2 + y^2 - 2y - 8 = 0$ **36.** $x^2 + 2x + y^2 - 3 = 0$

37. $2x^2 + 2y^2 - 18 = 0$ **38.** $3x^2 + 3y^2 - 75 = 0$
39. $x^2 + y^2 - 6x - 4y + 9 = 0$
40. $x^2 + y^2 - 2x + 10y + 18 = 0$
41. $x^2 + y^2 - 4y - 5 = 0$
42. $x^2 + y^2 - x + 3y - 2 = 0$
43. $x^2 + y^2 + 6x + 4y + 13 = 0$
44. $x^2 + y^2 - 10x + 8y + 41 = 0$
45. $x^2 + y^2 + x - 3y = 0$
46. $x^2 + y^2 - 3x - 5y - 1 = 0$
47. $x^2 + y^2 - 2x + 4y + 6 = 0$
48. $x^2 + y^2 + 6x - 8y + 35 = 0$
49. $3x^2 + 3y^2 + 9x + 12y - 6 = 0$
50. $2x^2 + 2y^2 + 4x - 8y - 6 = 0$

3.4   CONIC SECTIONS (Optional)

Conic sections are curves defined by the intersections of a plane and a right circular cone. They were studied by Apollonius of Perga (262–190 BC), a Greek mathematician. His treatise "Conics" is still considered one of the most important works on the subject. The study of conic sections is of contemporary importance: the orbits of planets, satellites, and comets are elliptical; parabolic mirrors are used in telescopes; and certain navigational systems are based on properties of hyperbolas.

Since conic sections are plane curves, it follows that they can be defined by using only two-dimensional concepts. Moreover, in a Cartesian coordinate system, the conic sections can be described by quadratic equations in two variables. The study of such equations and their graphs is the main objective of this section.

Figure 3.34 illustrates four conic sections (parabola, ellipse, circle, and hyperbola) obtained as intersections of a plane and a cone.

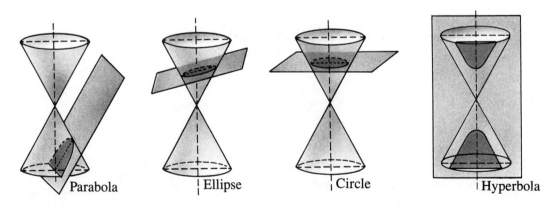

Parabola   Ellipse   Circle   Hyperbola

Figure 3.34

Parabolas

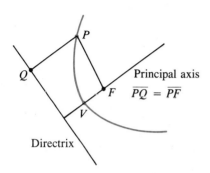

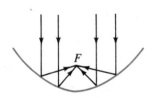

Figure 3.35

> ### Parabola
>
> A *parabola* is the set of points in a plane equidistant from a fixed point and a fixed line. The fixed point is called the *focus* of the parabola, and the fixed line is called the *directrix*.

The line through the focus F and perpendicular to the directrix is called the *principal axis* of the parabola (Figure 3.35). The point of intersection V of the curve and the principal axis is called the *vertex* of the parabola. It follows from the definition that the vertex V is midway between the focus and the directrix. The parabola is symmetric with respect to the principal axis, which, for this reason, is also called the *axis of symmetry*.

Parabolic forms are frequently encountered in the physical world, as well as in art and architectural design. If air resistance is neglected, the path described by a projectile under the force of gravity is a parabola. When a parabola is rotated about its principal axis, it generates a surface called *a paraboloid of revolution*. A parabolic mirror (a paraboloid of revolution) has a very important physical property; when a light ray parallel to the principal axis reaches the mirror, it is reflected to the focus (Figure 3.36). For this reason, reflector telescopes use parabolic mirrors. Also, the antenna of a radio telescope has the shape of a parabolic dish. Rays of light emitted from the focus are reflected off the surface of a parabolic mirror in rays parallel to the axis, as in automobile headlights.

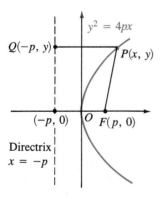

Figure 3.36

The Equation of a Parabola

In a Cartesian system of coordinates, a parabola is in *standard position* if its vertex coincides with the origin and its axis coincides with one of the coordinate axes.

To derive the equation of a parabola in standard position, consider the parabola illustrated in Figure 3.37, whose principal axis is the x-axis and whose focus F has coordinates $(p, 0)$ with $p > 0$. The directrix is the vertical line $x = -p$. If $P(x, y)$ lies on the parabola, then by definition,

$$d(P, F) = d(P, Q),$$

where $Q(-p, y)$ is the foot of the perpendicular from P to the directrix.

Figure 3.37

Since

$$d(P, Q) = |x + p|$$

and

$$d(P, F) = \sqrt{(x - p)^2 + y^2},$$

we have

$$\sqrt{(x - p)^2 + y^2} = |x + p|.$$

Squaring and simplifying, we obtain

$$(x - p)^2 + y^2 = (x + p)^2$$
$$\cancel{x^2} - 2px + \cancel{p^2} + y^2 = \cancel{x^2} + 2px + \cancel{p^2}.$$

Hence

$$y^2 = 4px. \tag{3.14}$$

Thus the coordinates (x, y) of a point P on the parabola satisfy the equation (3.14). Since all the algebraic steps are reversible, any point whose coordinates satisfy (3.14) must lie on the parabola.

Equation (3.14) is the *standard equation* of the parabola with focus $F(p, 0)$, $p > 0$, and directrix $x = -p$. Observe that the curve opens to the right.

If the focus is the point $F(-p, 0)$, with $p > 0$, and the directrix is the vertical line $x = p$, then by reasoning in the same manner as we did to derive (3.14), we get the standard equation:

$$y^2 = -4px. \tag{3.15}$$

This parabola, illustrated in Figure 3.38, opens to the left.

If the principal axis of the parabola coincides with the y-axis, then we obtain two other equations, as illustrated in Figure 3.39.

$$x^2 = 4py \qquad x^2 = -4py \tag{3.16}$$

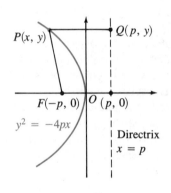

Figure 3.38

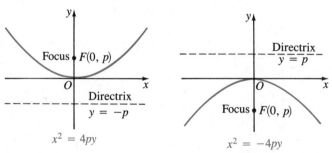

Figure 3.39

In Example 2 of Section 3.3, we indicated by plotting points that the graph of $y = x^2$ is a parabola. This equation can be rewritten $x^2 = 4(1/4)y$, which is an equation of the form (3.16). Its graph is then a parabola with vertex at the origin and directrix the horizontal line $y = -1/4$. A similar observation applies to the equation $y^2 = x$ discussed in Example 3 of Section 3.3.

EXAMPLE 1

$y^2 = 8x$

$F(2, 0)$

$x = -2$

Figure 3.40

Determine the vertex, focus, principal axis, and directrix of the parabola defined by $y^2 = 8x$. Graph this equation.

Solution: This is an equation of the form $y^2 = 4px$. Setting $4p = 8$, we get $p = 2$. Thus, the focus is $F(2, 0)$ and the directrix is the vertical line $x = -2$ (Figure 3.40). Also, the parabola has vertex at the origin and the principal axis is the x-axis.

Practice Exercise 1

Find the vertex, focus, principal axis, and directrix of the parabola $y^2 = -6x$. Graph this equation.

Answer:
Vertex $(0, 0)$
Focus $(-3/2, 0)$
Principal axis $y = 0$
Directrix $x = 3/2$

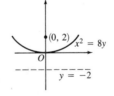

$y^2 = -6x$

$(-3/2, 0)$

$x = 3/2$

EXAMPLE 2

Sketch the parabola $y = -(1/6)x^2$. Locate the vertex, focus, principal axis, and directrix.

Solution: We first rewrite the given equation $x^2 = -6y$. This is now an equation of the form $x^2 = -4py$, with $p = 3/2$. It follows that the parabola has vertex at the origin and focus at $(0, -(3/2))$. The principal axis is the y-axis and the directrix is the horizontal line $y = 3/2$. We have graphed the equation in Figure 3.41.

Practice Exercise 2

Sketch the parabola $y = (1/8)x^2$, locating the vertex, focus, principal axis, and directrix.

Answer:
Vertex $(0, 0)$
Focus $(0, 2)$
Principal axis $x = 0$
Directrix $y = -2$

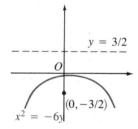

$y = 3/2$

$(0, -3/2)$

$x^2 = -6y$

Figure 3.41

$(0, 2)$

$x^2 = 8y$

$y = -2$

Parabola with Vertex at (*h, k*)

Four new equations can be derived for parabolas whose vertices are located at a point $V(h, k)$ and whose principal axes are parallel to one of the coordinate axes. For the parabola illustrated in Figure 3.42, the equation is

$$(y - k)^2 = 4p(x - h).$$

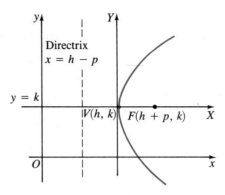

Figure 3.42

The other equations of parabolas are as follows.

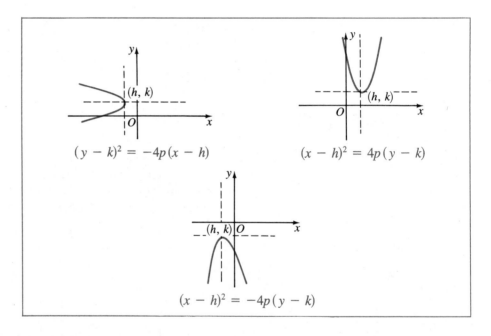

$(y - k)^2 = -4p(x - h)$ $(x - h)^2 = 4p(y - k)$

$(x - h)^2 = -4p(y - k)$

EXAMPLE 3   Find an equation of the parabola with vertex $V(-2, -1)$ and directrix $y = 1$.

Solution: On a coordinate plane, plot the vertex $V(-2, -1)$ and the directrix $y = 1$ (Figure 3.43). The axis of symmetry is the vertical line through the vertex, or $x = -2$. Since the vertex is midway between the directrix and the focus, it follows that the coordinates of the focus are $(-2, -3)$. Thus, $p = 2$ and the equation is

$$(x - (-2))^2 = -4 \cdot 2(y - (-1))$$

or

$$(x + 2)^2 = -8(y + 1).$$

Squaring and simplifying give us the equivalent equation

$$x^2 + 4x + 8y + 12 = 0.$$

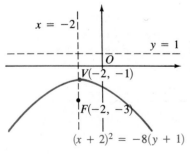

Figure 3.43

Practice Exercise 3

Find an equation of the parabola with vertex $V(2, 0)$ and focus $(5, 0)$.

Answer: $y^2 = 12(x - 2)$ or $y^2 - 12x + 24 = 0$

EXAMPLE 4

Show that $y = x^2 - 2x + 3$ is an equation of a parabola. Find its vertex, focus, principal axis, and directrix. Graph it.

Solution: Writing the equation

$$y - 3 = x^2 - 2x$$

and completing the square relative to x, we get

$$y - 3 + 1 = x^2 - 2x + 1$$

or

$$y - 2 = (x - 1)^2.$$

Now we rewrite the last equation as follows:

$$(x - 1)^2 = 4\left(\frac{1}{4}\right)(y - 2),$$

which is an equation of the form $(x - h)^2 = 4p(y - k)$ with $h = 1$, $k = 2$, and $p = 1/4$. This equation is equivalent to the original one, and it is the equation of a parabola with vertex $(1, 2)$ and principal axis $x = 1$ (Figure 3.44). The curve opens upward, since $y - 2$ is equal to a square and is thus always nonnegative. The focus, located on the principal axis, is $1/4$ unit above the vertex; thus it is $F(1, 9/4)$. The directrix, parallel to the x-axis, is placed $1/4$ unit below the vertex; its equation is $y = 7/4$.

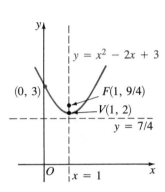

Figure 3.44

Practice Exercise 4

Graph the parabola $y^2 + 4x + 2y - 11 = 0$. Locate the vertex, focus, principal axis, and directrix.

Answer: Vertex $(3, -1)$, focus $(2, -1)$, principal axis $y = -1$, directrix $x = 4$

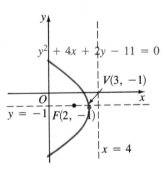

HISTORICAL NOTE

"No one, in Greek times, supposed that conic sections had any utility; at last, in the seventeenth century, Galileo discovered that projectiles move in parabolas, and Kepler discovered that planets move in ellipses. Suddenly, the work that the Greeks had done from pure love of theory became the key to warfare and astronomy."
From *A History of Western Philosophy*, by Bertrand Russell.

Ellipses

Ellipse

An *ellipse* is the set of points in a plane the sum of whose distances from two fixed points is constant. The two fixed points are called *foci*.

If we denote the foci by F_1 and F_2, the constant by $2a$, and a point on the ellipse by P, then by definition,

$$d(P, F_1) + d(P, F_2) = 2a.$$

The line passing through the foci is called the *principal axis* of the ellipse. The curve is *symmetric* with respect to the principal axis. The points V_1 and V_2 (Figure 3.45), where the ellipse intersects the principal axis, are called *vertices*. The *center* of the ellipse is the midpoint of the segment $V_1 V_2$.

A very simple method of drawing an ellipse is derived directly from the definition (Figure 3.46). Hold the two ends of a string of length $2a$ fixed at the foci F_1 and F_2. Trace the curve with a pencil, holding it taut against the string.

Ellipses are frequently encountered in the physical world. The orbits of planets, some comets, and satellites are elliptical. Some gears and cams and the domes of some buildings have elliptical forms.

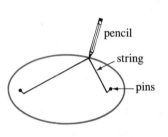

Figure 3.45

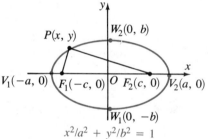

Figure 3.46

Equations for Ellipses

In a Cartesian coordinate plane, an ellipse is in *standard position* when its principal axis coincides with one of the coordinate axes and its center coincides with the origin.

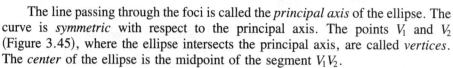

Figure 3.47

Consider an ellipse in standard position, such that its principal axis coincides with the x-axis (Figure 3.47). The foci are then symmetrically located with respect to the origin, and we may assume that they have coordinates $F_1(-c, 0)$ and $F_2(c, 0)$. If $P(x, y)$ belongs to the ellipse, then

$$d(P, F_1) + d(P, F_2) = 2a$$

or, according to the distance formula,

$$\sqrt{(x + c)^2 + y^2} + \sqrt{(x - c)^2 + y^2} = 2a.$$

Transposing one of the radicals, squaring and simplifying, we obtain

$$a\sqrt{(x - c)^2 + y^2} = a^2 - cx.$$

Squaring and simplifying again, we have

$$(a^2 - c^2)x^2 + a^2y^2 = a^2(a^2 - c^2),$$

so

$$\frac{x^2}{a^2} + \frac{y^2}{a^2 - c^2} = 1.$$

If we consider the triangle PF_1F_2 of Figure 3.47 and remember that the sum of the lengths of any two sides of a triangle is greater than the third side, we see that

$$d(P, F_1) + d(P, F_2) = 2a > 2c = d(F_1, F_2),$$

which means that $a > c$, so $a^2 - c^2 > 0$. This lets us define the following quantity.

For any ellipse, there is a number $b > 0$ such that
$$b^2 = a^2 - c^2.$$

Replacing $a^2 - c^2$ in the equation $(x^2/a^2) + [y^2/(a^2 - c^2)] = 1$, we obtain the equation of the ellipse in *standard form*:

$$\frac{x^2}{a^2} + \frac{y^2}{b^2} = 1 \qquad (a > b). \tag{3.17}$$

Setting $y = 0$ gives us the x-intercepts of the ellipse:

$$\frac{x^2}{a^2} + \frac{0}{b^2} = 1$$

$$\frac{x^2}{a^2} = 1$$

$$x^2 = a^2,$$

hence

$$x = \pm a.$$

Thus, the vertices V_1 and V_2 (Figure 3.47) have coordinates $(a, 0)$ and $(-a, 0)$. Similarly, if we set $x = 0$, we obtain the y-intercepts $\pm b$.

The segment V_1V_2 of length $2a$ is called the *major axis* of the ellipse. The segment W_1W_2 is the *minor axis*; it has length $2b$. Since $a > b$, the major axis is always greater than the minor axis. The numbers a and b are also called, respectively, the *semimajor* and the *semiminor* axes of the ellipse.

Until now we have assumed that the principal axis coincided with the x-axis. If the principal axis is the y-axis, then the roles of the x- and y-coordinates are interchanged. Proceeding as before, we obtain the equation in standard form:

$$\frac{x^2}{b^2} + \frac{y^2}{a^2} = 1 \qquad (a > b). \tag{3.18}$$

EXAMPLE 5

Graph the equation $16x^2 + 25y^2 = 400$.

Solution: We divide both sides by 400 and get $x^2/25 + y^2/16 = 1$, the standard equation of an ellipse. Setting $y = 0$, we obtain the x-intercepts, ± 5. Setting $x = 0$, we obtain the y-intercepts, ± 4. Thus the principal axis coincides with the x-axis (Figure 3.48). Since $c^2 = a^2 - b^2 = 25 - 16 = 9$, it follows that $c = 3$. Thus, the foci are $F_1(-3, 0)$ and $F_2(3, 0)$.

Practice Exercise 5

Graph the equation $4x^2 + y^2 = 16$. Determine the vertices, foci, and principal axis.

Answer:
$V(0, \pm 4)$
$F\left(0, \pm 2\sqrt{3}\right)$
Principal axis vertical

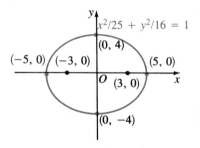

Figure 3.48

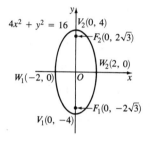

Ellipse with Center at (h, k)

For ellipses centered at a point $C(h, k) \neq O(0, 0)$ and major and minor axes parallel to the coordinate axes, the equations are as follows.

$$\frac{(x - h)^2}{a^2} + \frac{(y - k)^2}{b^2} = 1 \quad (a > b) \quad \text{major axis parallel to the } x\text{-axis}$$

$$\frac{(x - h)^2}{b^2} + \frac{(y - k)^2}{a^2} = 1 \quad (a > b) \quad \text{major axis parallel to the } y\text{-axis}$$

EXAMPLE 6

Find an equation of the ellipse with center $C(2, 1)$, focus $F(0, 1)$, and semimajor axis of length $a = 3$.

Solution: The center and focus lie on the horizontal line $y = 1$, as show in Figure 3.49. The other focus is then $F_2(4, 1)$ and the vertices are $V_1(-1, 1)$ and $V_2(5, 1)$. (Why?) Since $a = 3$ and $c = 2$, we have

$$
\begin{aligned}
b^2 &= a^2 - c^2 \\
&= 3^2 - 2^2 \\
&= 9 - 4 \\
&= 5.
\end{aligned}
$$

Hence, the semiminor axis has length $b = \sqrt{5}$. Thus, the equation of the ellipse in standard form is

$$\frac{(x - 2)^2}{9} + \frac{(y - 1)^2}{5} = 1.$$

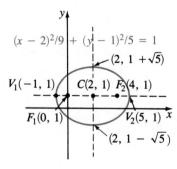

Figure 3.49

By eliminating the denominators, squaring, and simplifying, we get

$$5(x - 2)^2 + 9(y - 1)^2 = 45$$
$$5(x^2 - 4x + 4) + 9(y^2 - 2y + 1) = 45$$
$$5x^2 + 9y^2 - 20x - 18y - 16 = 0.$$

The latter is another equation for the ellipse.

Practice Exercise 6   Find an equation of the ellipse with center at $(0, 2)$, vertex at $(0, 6)$, and semiminor axis of length $b = 3$. Graph it.

Answer: $\dfrac{x^2}{9} + \dfrac{(y - 2)^2}{16} = 1$

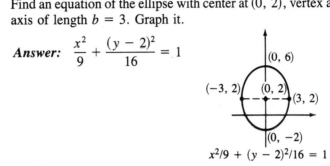

$x^2/9 + (y - 2)^2/16 = 1$

EXAMPLE 7   Show that the equation $4x^2 + 9y^2 - 8x + 36y + 4 = 0$ represents an ellipse. Find its center and its major and minor axes.

Solution: The method of solution is the same as the one we used to find centers and radii of circles. First, we complete the square on x and y, arranging our work as follows:

$$4(x^2 - 2x \quad) + 9(y^2 + 4y \quad) = -4$$
$$4(x^2 - 2x + 1) + 9(y^2 + 4y + 4) = -4 + 4 + 36.$$

Observe that we have added 1 inside the first parentheses and 4 inside the second parentheses. This has the effect of adding 4 and 36 to the left-hand side of the equation. To balance the equation, we have added 4 and 36 to the right-hand side. Next, we obtain

$$4(x - 1)^2 + 9(y + 2)^2 = 36.$$

Dividing both sides by 36, we get

$$\frac{(x - 1)^2}{9} + \frac{(y + 2)^2}{4} = 1.$$

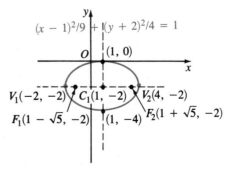

Figure 3.50

This is a standard equation of an ellipse with center at $(1, -2)$, semimajor axis $a = 3$, and semiminor axis $b = 2$. The graph is shown in Figure 3.50 (page 157).

Practice Exercise 7   Find the center and the major and minor axes of the ellipse

$$4x^2 + y^2 + 16x - 2y + 13 = 0.$$

Answer: $C(-2, 1)$, $a = 2$, $b = 1$, principal axis parallel to the y-axis

Eccentricity of an Ellipse

Eccentricity

The ratio

$$e = \frac{c}{a}$$

is called the *eccentricity* of an ellipse.

Since $0 < c < a$ for an ellipse, we always have $0 < e < 1$. That is, the eccentricity of an ellipse is a positive number less than 1.

EXAMPLE 8   Find the eccentricity of the ellipse $\dfrac{x^2}{25} + \dfrac{y^2}{16} = 1$.

Solution: As we have seen in Example 5, $a = 5$, $b = 4$, and $c = 3$. Thus,

$$e = \frac{3}{5} = 0.6.$$

Practice Exercise 8   What is the eccentricity of the ellipse $\dfrac{x^2}{1} + \dfrac{y^2}{4} = 1$?

Answer: $e = \dfrac{\sqrt{3}}{2}$

The eccentricity measures how much an ellipse differs from a circle. If a is fixed and c varies from 0 to a, then the corresponding ellipses vary in shape as illustrated in Figure 3.51. If c is equal to zero, the eccentricity e is equal to 0, and the ellipse is a circle. On the other hand, if c is equal to a, in which case the eccentricity is equal to 1, the ellipse reduces to a line segment.

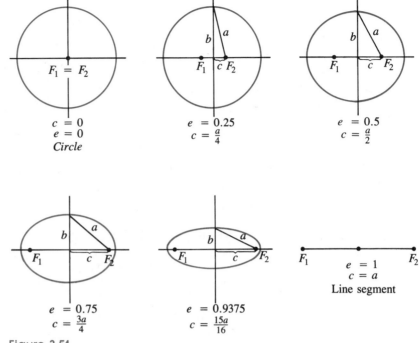

Figure 3.51

Eccentricities of Orbits

The eccentricity of the orbit of the Earth is approximately 0.017, a small number. Thus the orbit of the Earth is almost a circle. On the other hand, Halley's comet (which paid us a disappointing visit at the end of 1985 and the beginning of 1986) has a very elongated orbit. Its eccentricity is approximately 0.98, a number very close to 1!

ECCENTRICITIES OF ORBITS OF MAJOR PLANETS	
Planet	*Eccentricity*
Mercury	0.206
Venus	0.007
Earth	0.017
Mars	0.093
Jupiter	0.048
Saturn	0.056
Uranus	0.047
Neptune	0.009
Pluto	0.249

Hyperbolas

Hyperbola

A *hyperbola* is the set of points in a plane the difference of whose distances from two fixed points is constant. The two fixed points are called *foci*.

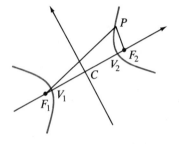

If F_1 and F_2 are the foci, P is a point on the hyperbola such that $d(P, F_1) > d(P, F_2)$, and $2a$ is the constant, then by definition we have

$$d(P, F_1) - d(P, F_2) = 2a.$$

The line through the foci is the *principal axis* of the hyperbola (Figure 3.52). The points of intersection of the curve and the principal axis, V_1 and V_2, are called *vertices* of the hyperbola. The midpont of the segment $V_1 V_2$ is the *center* of the hyperbola. It can be shown that the curve is symmetric with respect to the principal axis.

As with other conic sections, hyperbolas are encountered in the physical world. Under the action of an electric field, the path described by certain atomic particles is a branch of a hyperbola. Some comets are believed to move along hyperbolic orbits. Hyperbolic forms are found in optics and in some modern architectural structures.

Figure 3.52

Equations of Hyperbolas

In a rectangular coordinate system, a hyperbola is said to be in *standard position* if its center is the origin and its principal axis coincides with one of the coordinate axes.

We will now derive the equation of a hyperbola in standard position, assuming that the principal axis coincides with the x-axis and that the foci are $F_1(-c, 0)$ and $F_2(c, 0)$, with $c > 0$ (Figure 3.53).

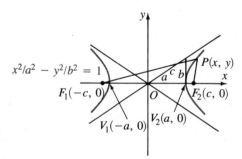

Figure 3.53

Let $P(x, y)$ be a point on the curve and assume that $d(P, F_1) > d(P, F_2)$. (The same reasoning applies if $d(P, F_2) > d(P, F_1)$.) We have, by definition,

$$d(P, F_1) - d(P, F_2) = 2a,$$

which, in terms of coordinates, can be written

$$\sqrt{(x + c)^2 + y^2} - \sqrt{(x - c)^2 + y^2} = 2a.$$

Transposing the second radical to the right-hand side, squaring, and simplifying, we get

$$cx - a^2 = a\sqrt{(x - c)^2 + y^2}.$$

Squaring again and simplifying give us

$$(c^2 - a^2)x^2 - a^2 y^2 = a^2(c^2 - a^2);$$

so

$$\frac{x^2}{a^2} - \frac{y^2}{c^2 - a^2} = 1.$$

In the triangle $PF_1 F_2$ (Figure 3.53), the difference between the lengths of the sides PF_1 and PF_2 is smaller than the length of the side $F_1 F_2$. Thus,

$$d(P, F_1) - d(P, F_2) = 2a < 2c = d(F_1, F_2),$$

which means that $c > a$, so $c^2 - a^2 > 0$. This lets us define the following quantity.

For any hyperbola, there is a number $b > 0$ such that

$$b^2 = c^2 - a^2.$$

Replacing $c^2 - a^2$ in the equation $(x^2/a^2) - [y^2/(c^2 - a^2)] = 1$, we get the equation of a hyperbola in *standard form*:

$$\frac{x^2}{a^2} - \frac{y^2}{b^2} = 1. \qquad\qquad \textbf{(3.19)}$$

If the principal axis of the hyperbola coincides with the y-axis and the foci are $F_1(0, -c)$ and $F_2(0, c)$, then with the same reasoning as we just used, we obtain the equation

---

$$\frac{y^2}{a^2} - \frac{x^2}{b^2} = 1. \qquad\qquad \textbf{(3.20)}$$

EXAMPLE 9

Find an equation of the hyperbola with foci $F_1(-5, 0)$ and $F_2(5, 0)$, and $a = 3$.

Solution: Since $a = 3$ and $c = 5$, then

$$b^2 = c^2 - a^2$$
$$= 5^2 - 3^2$$
$$= 16.$$

Hence, $b = 4$, and the equation of the hyperbola in standard form is

$$\frac{x^2}{9} - \frac{y^2}{16} = 1.$$

If we multiply both sides by $9 \cdot 16 = 144$, we obtain the equivalent equation

$$16x^2 - 9y^2 = 144.$$

Practice Exercise 9

Find an equation of the hyperbola with foci $F_1(0, -\sqrt{5})$ and $F_2(0, \sqrt{5})$, and $a = 2$.

Answer: $\dfrac{y^2}{4} - \dfrac{x^2}{1} = 1$

Ellipses versus Hyperbolas

You have probably noticed the minus sign that differentiates the standard equation of a hyperbola from that of an ellipse. More important, the relations among the numbers a, b, and c are different. For an ellipse $a > c$ and $b^2 = a^2 - c^2$. For a hyperbola $c > a$ and $b^2 = c^2 - a^2$.

Transverse Axis

If we set $y = 0$ in the equation

$$\frac{x^2}{a^2} - \frac{y^2}{b^2} = 1,$$

we get $x = \pm a$, so the coordinates of the vertices are $V_1(-a, 0)$ and $V_2(a, 0)$ (Figure 3.53). The segment $V_1 V_2$ joining the vertices is called the *transverse axis* of the hyperbola; its length is $2a$.

Conjugate Axis

If we now set $x = 0$ in the equation

$$\frac{x^2}{a^2} - \frac{y^2}{b^2} = 1,$$

we obtain $y^2 = -b^2$, which has no real solution. Therefore, the curve does not intersect the y-axis. Nevertheless, we may consider the points $W_1(0, -b)$ and $W_2(0, b)$ on the y-axis. We call the segment $W_1 W_2$ the *conjugate axis* of the hyperbola; its length is $2b$ and we refer to W_1 and W_2 as the *vertices* on the conjugate axis.

The Fundamental Rectangle

When sketching the graph of a hyperbola, it is very useful to consider the rectangle with sides parallel to the coordinate axes and with vertices (a, b), $(-a, b)$, $(-a, -b)$, and $(a, -b)$. This rectangle is called the *fundamental rectangle* (Figure 3.54). To locate the foci of the hyperbola, draw a circle with center at the origin and radius c (the *semidiagonal* of the fundamental rectangle, since $c^2 = a^2 + b^2$). The points where the circle intersects the principal axis are the foci (which are located at $(-c, 0)$ and $(c, 0)$).

Asymptotes

The two lines $y = \pm(b/a)x$, the extensions of the diagonals of the fundamental rectangle, are called *asymptotes* of the hyperbola. They have a very important geometric meaning. If $P(x, y)$ lies on the hyperbola and $P_1(x, y_1)$ is the point on the asymptote with the same x-coordinate as P, as shown in Figure 3.54, then the vertical distance $|y_1 - y|$ between P and P_1 becomes smaller and smaller as x becomes larger and larger.

The fundamental rectangle and the asymptotes are very helpful when graphing a hyperbola.

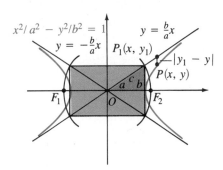

Figure 3.54

EXAMPLE 10

Graph the hyperbola $\dfrac{x^2}{16} - \dfrac{y^2}{9} = 1$.

Solution: In this case, $a = 4$ and $b = 3$. It follows that the x-intercepts (vertices) are at the points $(-4, 0)$ and $(4, 0)$. Also the vertices on the conjugate axis are $(0, -3)$ and $(0, 3)$. By plotting these points, we can draw the fundamental rectangle, as shown in Figure 3.55. Also, from the relation $b^2 = c^2 - a^2$, we get

$$c^2 = a^2 + b^2$$
$$= 4^2 + 3^2$$
$$= 16 + 9$$
$$= 25.$$

Thus, $c = 5$ is the length of the semidiagonal of the fundamental rectangle. It follows that the foci are $F_1(-5, 0)$ and $F_2(5, 0)$. By extending the diagonals of the fundamental rectangle, we have the two asymptotes and we can now draw a rough sketch of the graph. Additional points may be plotted for precision.

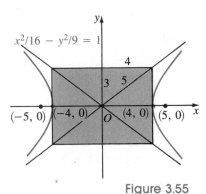

Figure 3.55

Practice Exercise 10   Sketch the graph of the hyperbola $\dfrac{y^2}{16} - \dfrac{x^2}{9} = 1$.

Answer:

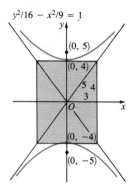

$y^2/16 - x^2/9 = 1$

Hyperbola Centered at (h, k)

If the center of a hyperbola is $C(h, k) \neq O(0, 0)$ and the principal axis is parallel to one of the coordinate axes, the equations are of the following form.

$$\frac{(x - h)^2}{a^2} - \frac{(y - k)^2}{b^2} = 1 \qquad \text{principal axis parallel to the } x\text{-axis}$$

$$\frac{(y - k)^2}{a^2} - \frac{(x - h)^2}{b^2} = 1 \qquad \text{principal axis parallel to the } y\text{-axis}$$

EXAMPLE 11   Find an equation of the hyperbola with center at $(1, 2)$, a vertex at $(1, 6)$, and $b = 3$.

Solution: The center and the vertex lie on the vertical line $x = 1$, four units away from each other. Thus $a = 4$. Since $b = 3$, we may write

$$\frac{(y - 2)^2}{16} - \frac{(x - 1)^2}{9} = 1.$$

The other vertex is at $(1, -2)$. The foci are $F_1(1, -3)$ and $F_2(1, 7)$. (Why?)

Practice Exercise 11   Find an equation of the hyperbola with center at $(-2, 1)$, the vertex at $(-7, 1)$, and conjugate axis of length 8.

Answer: $\dfrac{(x + 2)^2}{25} - \dfrac{(y - 1)^2}{16} = 1$

EXAMPLE 12   Show that the equation $x^2 - 4y^2 - 2x - 8y - 7 = 0$ represents a hyperbola.

Solution: We write the equation as follows:

$$(x^2 - 2x \qquad) - 4(y^2 + 2y \qquad) = 7,$$

and complete the square on x and y:

$$(x^2 - 2x + 1) - 4(y^2 + 2y + 1) = 7 + 1 - 4$$
$$(x - 1)^2 - 4(y + 1)^2 = 4.$$

Dividing both sides by 4, we get

$$\frac{(x - 1)^2}{4} - \frac{(y + 1)^2}{1} = 1.$$

This is an equation of a hyperbola with center $C(1, -1)$, $a = 2$, and $b = 1$.

Practice Exercise 12   Does the equation $y^2 - 3x^2 - 2y - 12x - 14 = 0$ represent a hyperbola?

Answer: Yes: $\dfrac{(y - 1)^2}{3} - \dfrac{(x + 2)^2}{1} = 1;$ $C(-2, 1)$, $a = \sqrt{3}$, $b = 1$

EXERCISES 3.4

In Exercises 1–8, determine the vertex, focus, principal axis, and directrix of each parabola. Graph each one.

1. $y = 4x^2$ **2.** $y = -2x^2$
3. $y = 3x^2 + 1$ **4.** $x = 4y^2 - 1$
5. $x^2 - 4y + 8 = 0$ **6.** $y^2 - 3x - 6 = 0$
7. $(x - 1)^2 = y + 1$ **8.** $(y + 2)^2 = 2(x - 1)$

In Exercises 9–18, find an equation for the parabola corresponding to the given information. Sketch the graph.

9. $V(0, 0)$, $F(3/2, 0)$ **10.** $V(0, 0)$, $F(0, -1/4)$
11. $V(1, 2)$, $F(2, 2)$ **12.** $V(-2, 1)$, $F(-2, 3)$
13. $V(3, 0)$, directrix $x = 1$
14. $V(0, -2)$, directrix $y = -3$
15. $V(-1, -1)$, directrix $y = -1/2$
16. $V(2, 0)$, directrix $x = 1/3$
17. $V(1, 2)$, principal axis $x = 1$, contains the point $(2, 6)$
18. $V(2, 3)$, principal axis $y = 3$, contains the point $(4, 5)$

In Exercises 19–22, analyze the properties of the curve each equation represents.

19. $2x^2 - 8y + 6 = 0$ **20.** $y^2 + x + 1 = 0$
21. $y^2 - 2y - x = 0$ **22.** $x^2 - 4x + 4y = 0$

In Exercises 23–26, find the required quantities.

23. Find the values of c for which the parabola $y = x^2 + 4x + c$ has two x-intercepts.
24. For which values of b does the parabola $x = y^2 + by + 9$ have two y-intercepts?

25. Find the points of intersection of the line $y = 2x + 1$ and the parabola $y = 3x^2 + 5x - 5$. (*Hint:* Set $2x + 1 = 3x^2 + 5x - 5$ and solve for x.)
26. Find the points of intersection of the parabolas $y = x^2 - 20$ and $y = 12 - x^2$.

In Exercises 27–32, determine the center, vertices, foci, and the major and minor axes of each ellipse. Graph each one.

27. $\dfrac{x^2}{16} + \dfrac{y^2}{9} = 1$ **28.** $\dfrac{x^2}{36} + \dfrac{y^2}{64} = 1$
29. $4x^2 + 9y^2 = 36$ **30.** $25x^2 + 16y^2 = 400$
31. $\dfrac{(x - 1)^2}{25} + \dfrac{(y + 1)^2}{4} = 1$
32. $\dfrac{(x + 2)^2}{9} + \dfrac{(y - 3)^2}{16} = 1$

In each of Exercises 33–44, find an equation of the ellipse corresponding to the given data.

33. Center at $(0, 0)$, a focus at $(-2, 0)$, length of the major axis 8
34. Center at $(0, 0)$, a focus at $(3, 0)$, length of the minor axis 8
35. Center at $(1, 2)$, one vertex at $(6, 2)$, and one endpoint of the minor axis at $(1, 6)$
36. Center at $(-3, -1)$, one vertex at $(-3, 4)$, and one endpoint of the minor axis at $(-1, -1)$
37. Foci at $(0, 0)$ and $(6, 0)$ and vertices at $(-2, 0)$ and $(8, 0)$
38. Foci at $(0, 1)$ and $(0, 9)$ and vertices at $(0, 0)$ and $(0, 10)$
39. A focus at $(-5, 0)$, center at $(0, 0)$, and eccentricity $1/3$

40. A focus at $(0, 6)$, center at $(0, 0)$, and eccentricity $3/4$
41. A focus at $(2, 0)$, center at $(0, 0)$, and a vertex at $(3, 0)$
42. A focus at $(0, -3)$, center at $(0, 0)$, and a vertex at $(0, 4)$
43. Vertices at $(-5, 0)$ and $(5, 0)$ and the ellipse contains the point $(4, 12/5)$.
44. Vertices at $(0, -13)$ and $(0, 13)$ and the ellipse contains the point $(60/13, 5)$.

Show that each equation in Exercises 45–48 represents an ellipse. Determine the center and the major and minor axes.

45. $9x^2 + 4y^2 - 18x + 8y - 23 = 0$
46. $x^2 + 4y^2 + 2y = 0$
47. $4x^2 + 9y^2 - 16x + 18y - 11 = 0$
48. $9x^2 + 25y^2 - 6x + 10y - 223 = 0$

According to Kepler's first law, the orbit of each planet is an ellipse with the sun at one focus. For each orbit, the vertex closest to the sun is called *perihelion* and the vertex farthest from the sun is called *aphelion*.

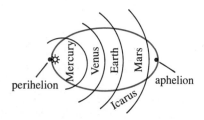

G49. The eccentricity of the orbit of the earth is 1.670×10^{-2} and the major axis has length 1.858×10^8 miles. How close is the earth to the sun at perihelion? How far is it at aphelion?
G50. The asteroid Icarus has perihelion of 1.7×10^7 miles and

aphelion of 1.83×10^8 miles. What is the eccentricity of its elliptical orbit?

In Exercises 51–54, determine the center, vertices, and foci of the hyperbola.

51. $\dfrac{x^2}{64} - \dfrac{y^2}{36} = 1$ **52.** $\dfrac{x^2}{25} - \dfrac{y^2}{144} = 1$

53. $\dfrac{(x - 2)^2}{81} - \dfrac{(y + 3)^2}{144} = 1$

54. $\dfrac{(x + 5)^2}{64} - \dfrac{(y + 7)^2}{36} = 1$

In Exercises 55–64, find the equation of the hyperbola corresponding to the given data.

55. Center at $(0, 0)$, focus at $(4, 0)$, $a = 2$
56. Center at $(2, 0)$, focus at $(4, 0)$, $a = 1$
57. Vertices at $(-2, 1)$ and $(6, 1)$, foci at $(-3, 1)$ and $(7, 1)$
58. Vertices at $(1, \pm 3)$ and foci at $(1, \pm 5)$
59. Vertices at $(\pm 1, 0)$ and asymptotes $y = \pm 2x$
60. Vertices at $(\pm 3, 0)$ and asymptotes $y = \pm(4/3)x$
61. Foci at $(-4, 1)$ and $(6, 1)$, $a = 4$
62. Foci at $(-6, -1)$ and $(10, -1)$, $a = 6$
63. Vertices at $(0, -4)$ and $(0, 4)$, contains the point $(9/4, 5)$
64. Vertices $(-8, 0)$ and $(8, 0)$, contains the point $(10, 9/2)$

Show that the equations in Exercises 65–70 represent hyperbolas. Determine the center and vertices.

65. $2y^2 + 4y - x^2 = 0$ **66.** $3x^2 - 6x - y^2 = 0$
67. $x^2 - y^2 + 4x + 2y + 2 = 0$
68. $y^2 - x^2 + 6y - 4x + 1 = 0$
69. $4x^2 - 9y^2 - 16x - 18y - 29 = 0$
70. $2x^2 + 12x - y^2 + 10y + 1 = 0$

CHAPTER SUMMARY

In a *Cartesian coordinate system*, there is a correspondence between points in a plane and ordered pairs of real numbers. To each point P there corresponds an ordered pair (x, y) of real numbers called the *coordinates* of P. The number x is the *abscissa* or *x-coordinate* of P and the number y is the *ordinate* or *y-coordinate* of P. Conversely, every ordered pair (x, y) of real numbers determines a unique point in the plane.

If $P_1(x_1, y_1)$ and $P_2(x_2, y_2)$ are the points in a plane, the coordinates of the *midpoint M* of the segment $P_1 P_2$ are

$$x = \frac{x_1 + x_2}{2}, \qquad y = \frac{y_1 + y_2}{2} \quad \text{midpoint formula}$$

The *distance* between P_1 and P_2 is

$$d = \sqrt{(x_1 - x_2)^2 + (y_1 - y_2)^2} \quad \text{distance formula}$$

Once we establish a coordinate system, we can represent geometric objects by equations and we can give equations a geometric meaning. Two distinct points determine a unique *straight line*. Every *nonvertical line* has a *slope m* defined by

$$m = \frac{y_2 - y_1}{x_2 - x_1},$$

where (x_1, y_1) and (x_2, y_2) are coordinates of two distinct points on the line. The slope of a vertical line is *not defined*. *Any* real number can be the slope of a line, and this corresponds to the fact that there are infinitely many lines passing through a point. If the slope is equal to zero, then the line is *horizontal*.

Among the equations that represent a given line, two are particularly important

$$y - y_1 = m(x - x_1) \qquad \text{point–slope form}$$

and

$$y = mx + b \qquad \text{slope–intercept form}$$

Both equations can be written

$$Ax + By + C = 0 \qquad \text{standard form}$$

with A and B not simultaneously zero. If two variables x and y satisfy a *linear equation* such as $Ax + By + C = 0$, then we say they are *linearly related*.

Two lines of slopes m and m' are *parallel* if $m = m'$. Two lines of slopes m and m' both different from zero are *perpendicular* if $m = -1/m'$.

Graphs of equations can be sketched by *plotting points*. Since this is a rather crude and tedious way to graph an equation, you should always analyze the equation and look for certain features that reduce the plotting of points to a minimum. One such feature is *symmetry*. In terms of coordinates, we can summarize the different symmetries of a curve C by the following table.

Symmetry of C with respect to	Properties of coordinates
x-axis	$(a, b) \in C$, and $(a, -b) \in C$
y-axis	$(a, b) \in C$, and $(-a, b) \in C$
origin	$(a, b) \in C$, and $(-a, -b) \in C$
line $y = x$	$(a, b) \in C$, and $(b, a) \in C$

Conic sections are curves defined by the intersections of a plane and a right circular cone. In a Cartesian coordinate system, we describe conic sections by quadratic equations in two variables.

Of all conic sections, the *circle* is the simplest one. An equation of a circle with center $C(h, k)$ and radius r is

$$(x - h)^2 + (y - k)^2 = r^2.$$

A *parabola* is the set of points in a plane equidistant from a fixed point, the *focus*, and a fixed line, the *directrix*. Equations of parabolas in standard position are as follows.

$$\begin{aligned} y^2 &= 4px \qquad &\text{axis of symmetry: } x\text{-axis} \\ x^2 &= 4py \qquad &\text{axis of symmetry: } y\text{-axis} \end{aligned}$$

An *ellipse* is the set of all points in a plane the sum of whose distances from two fixed points, the *foci*, is constant. Equations of ellipses in standard position are as follows.

$$\begin{aligned} \frac{x^2}{a^2} + \frac{y^2}{b^2} &= 1 \quad (a > b) \qquad &\text{major axis along } x\text{-axis} \\ \frac{x^2}{b^2} + \frac{y^2}{a^2} &= 1 \quad (a > b) \qquad &\text{major axis along } y\text{-axis} \end{aligned}$$

The *eccentricity*

$$e = \frac{c}{a} \qquad \left(c = \sqrt{a^2 - b^2}\right)$$

of an ellipse (a number between 0 and 1) measures by how much the ellipse deviates from a circle. Ellipses with eccentricity near zero approach circles; ellipses with eccentricity near 1 are very elongated curves.

A *hyperbola* is a set of points in a plane the difference of whose distances from two fixed points, the *foci*, is constant. Equations of hyperbolas in standard position are as follows.

$$\frac{x^2}{a^2} - \frac{y^2}{b^2} = 1 \qquad \textit{transverse axis} \text{ along } x\text{-axis}$$

$$\frac{y^2}{a^2} - \frac{x^2}{b^2} = 1 \qquad \textit{transverse axis} \text{ along } y\text{-axis}$$

Conic sections are frequently encountered in the physical sciences and engineering, as well as in art and architectural designs. The path of a projectile is a parabola; planets and comets move around the sun in elliptical orbits; satellites have been placed into circular orbits around the earth; and, under certain conditions, atomic particles move along one of the branches of a hyperbola.

CHAPTER TEST

1. Given the points $P(4, 4)$ and $Q(-4, -1)$, find **(a)** the midpoint of the segment PG, and **(b)** the distance between P and Q.

2. Given points $A(3, 3)$, $B(-1, 0)$ and $C(-2, y)$, find y so that ABC is a right triangle with the right angle at A.

3. What is the slope of the line passing through the points $P(7, -7)$ and $Q(-4, 6)$?

4. Write the slope–intercept equation of the line that passes through the point $P(2, 2)$ and has slope -1.

5. Determine whether the lines $2x + 4y - 2 = 0$ and $4x - 2y - 6 = 0$ are parallel, perpendicular, or neither.

6. Write an equation of the circle with center $C(1, -2)$ and that contains the point $P(5, 1)$.

7. If the equation $x^2 + y^2 - 6x - 8y - 2 = 0$ represents a circle, find its center and radius.

8. For what values of c does the parabola $y = 3x^2 - 4x + c$ have no x-intercepts?

9. An ellipse has center at $(-4, 2)$, a focus at $(-1, 2)$, and the semimajor axis is 6 units long. Find its equation in standard form.

10. Find the standard equation of the hyperbola centered at $(0, 0)$ with a focus at $(-2, 0)$ and such that $a = 1$.

REVIEW EXERCISES

In Exercises 1–4, the coordinates of P and Q are given. Find the coordinates of the midpoint of the segment PQ.

1. $P(2, 8)$, $Q(3, -5)$

2. $P(-3, -7)$, $Q(-2, -5)$

3. $P(1/3, -1/2)$, $Q(3/4, 1/5)$

4. $P(2/9, 1/4)$, $Q(3/5, 2/7)$

In Exercises 5–14, solve the given problem.

5. If $A(2, -3)$ and $M(4, -1)$, find the coordinates of B so that M is the midpoint of the segment AB.

6. If $B(-1, -3)$ and $M(-6, 2)$, find the coordinates of A so that M is the midpoint of the segment AB.

7. Find the points on the x-axis that are at a distance of 6 units from the point $P(2, -5)$.

8. Find the points on the main diagonal (i.e., the line $y = x$) that are at a distance of 5 units from the point $A(7, 6)$.

9. Show that the triangle with vertices $A(-4, 0)$, $B(1, 10)$, and $C(4, 6)$ is a right triangle. Find its area.

10. Plot the points $A(5, -6)$, $B(8, 5)$, and $C(1, -2)$. Show that ABC is a right triangle and find its area.

11. Show that the points $A(-3, -4)$, $B(1, -8)$, $C(7, -2)$, and $D(3, 2)$ are the vertices of a parallelogram.

12. Plot the points $A(-2, 1)$, $B(3, 7)$, $C(9, 2)$, and $D(4, -4)$ and show that they are the vertices of a square.

13. Find b so that the triangle with vertices $O(0, 0)$, $A(2, 3)$, and $B(3, b)$ is a right triangle with right angle at A.

14. Given $Q(0, -2)$ and $R(0, 3)$, find a point P on the positive x-axis so that QPR is a right triangle with right angle at P.

In Exercises 15–18, graph each equation and check for symmetry.

15. $3x = -5y^3$

16. $2y + 7x^3 = 0$

17. $y = x^2 - 4x$

18. $x = y^2 - 6y$

In Exercises 19–22, find the slope of the line containing the given points.

19. $A(-3, -5)$, $B(4, 7)$

20. $P(2, 1/3)$, $Q(1/4, 3)$

c **21.** $M(2.157, -3.019)$, $N(0.026, -3.107)$

c **22.** $R(\sqrt{3}, 1.414)$, $Q(-\sqrt{5}, 3.141)$

In Exercises 23–32, find an equation for each of the following lines. Write your answer in the form $Ax + By + C = 0$.

23. Passing through $(-2, -6)$ with slope $-1/3$
24. Passing through $(3, -8)$ with slope $3/11$
25. With y-intercept $3/2$ and slope $5/3$
26. With x-intercept -5 and slope $-1/2$
27. With x-intercept $3/5$ and y-intercept 6
28. With x-intercept -3 and y-intercept -8

c **29.** Passing through $(2.35, -7.21)$ and $(1.07, -2.15)$

c **30.** Passing through $(-1.18, 3.26)$ and $(4.03, -1.38)$

31. Parallel to $x/2 + y/3 = 1$ and passing through $(1, 6)$
32. Perpendicular to $x/3 - y/3 = 1$ and passing through $(1, 3)$

In Exercises 33–42, solve the given problem.

33. Find b so that the line $y = 5x + b$ contains the point $(-1, 3)$.
34. Find m so that the line $y = mx + 7$ contains the point $(-2, -5)$.

c **35.** Find the slope and the y-intercept of the line $5.17x - 3.21y + 1.72 = 0$.

c **36.** Find the slope and the x-intercept of the line $-3.371x + 1.016y - 2.371 = 0$.

37. A line passes through the point $(3, 5)$ and has equal x- and y-intercepts. Find its equation.
38. The y-intercept of a line is the opposite of its x-intercept. If the line passes through $(1, -4)$, find its equation.
39. The variable v (velocity) is linearly related to the variable t (time). Assume that $v = 15$ when $t = 0$, and $v = 45$ when $t = 1.5$. Find an equation relating v and t. Also, find t when $v = 85$.
40. The variables C (degrees Celsuis) and F (degrees Fahrenheit) are linearly related. Let $F = 32$ when $C = 0$, and $F = 212$ when $C = 100$. Find an equation relating both variables.
41. The value of a stamp collection is appraised at \$20,000. If it appreciates linearly at a rate of 6% per year, find its value V in t years.
42. A farmer buys a piece of equipment for \$36,000 and estimates that in 15 years its scrap value will be \$3,600. Assuming that the depreciation is linear, find a formula for the value V of the equipment in t years.

In Exercises 43–48, find an equation of the circle satisfying the stated conditions.

43. Center at the origin and passing through $(-1, 3)$
44. Center at $(3, 5)$ and passing through the origin
45. Center at $(-1, 1)$ and passing through $(2, -3)$
46. Center in the second quadrant, tangent to both axes, and has radius 2

47. Center in the third quadrant, tangent to both axes, and has radius $\sqrt{10}$
48. Diameter AB with $A(1, -3)$ and $B(-2, 5)$

In Exercises 49–54, find the center and radius of each circle.

49. $\dfrac{x^2}{2} + \dfrac{y^2}{2} = 8$ **50.** $\dfrac{x^2}{3} + \dfrac{y^2}{3} = 125$

51. $x^2 + y^2 - 3y - 1 = 0$ **52.** $x^2 + y^2 - x - 3 = 0$

53. $x^2 + y^2 + 3x + 5y - \dfrac{1}{2} = 0$

54. $x^2 + y^2 - 6x + 8y = 0$

In Exercises 55–60, find an equation of the parabola corresponding to the given data.

55. Focus at $(0, 3)$ and directrix $y = -3$
56. Focus at $(2, 0)$ and directrix $x = -2$
57. Vertex at $(0, 0)$, passes through the point $(2, 4)$, principal axis horizontal
58. Vertex at the origin, passes through $(-1, -2)$, principal axis vertical
59. Principal axis vertical, passes through $(0, 4)$, vertex at $(2, 1)$
60. Principal axis horizontal, passes through $(2, -2)$, vertex at $(-1, 2)$

Solve Exercises 61 and 62.

61. Find the points of intersection of the parabola $y = x^2$ with the line $3x - 2y - 1 = 0$.
62. Find the points of intersection of the parabola $y = (x - 1)^2$ and the line $x + y = 3$.

In Exercises 63–68, find an equation of the ellipse corresponding to the given data.

63. Center at $(-1, 0)$, focus at $(-1, 2)$, $a = 3$
64. Foci at $(\pm 5, 0)$ and $b = 2$
65. Foci at $(-2, 3)$ and $(4, 3)$, semimajor axis of length 5
66. Foci at $(0, \pm 4)$ and eccentricity $1/10$
67. Vertices at $V_1(-5/3, 0)$ and $V_2(5/3, 0)$, and passes through the point $(1, 1)$
68. The point $(\sqrt{3}, 2)$ lies on the curve, and the vertices are $V_1(-3, 0)$ and $V_2(3, 0)$.

In Exercises 69–74, find an equation of the hyperbola corresponding to the given data.

69. Foci at $(0, \pm 5)$ and vertices at $(0, \pm 3)$
70. Foci at $(0, \pm 6)$ and conjugate axis of length 2
71. Focus at $(-5, 0)$, vertex at $(-2, 0)$, and center at the origin
72. Vertices at $(\pm 4, 0)$ and asymptotes $y = \pm \dfrac{1}{2}x$
73. A focus at $(10, 0)$, center at the origin, and eccentricity $5/4$

74. Conjugate axis of length 6, vertices at $(\pm 5, 0)$.

Discuss each of the equations in Exercises 75–82.

75. $x^2 + y^2 - 4x - 6y + 4 = 0$

76. $x^2 + y^2 - 6x - 7 = 0$

77. $9x^2 + 4y^2 + 18x + 8y - 12 = 0$

78. $x^2 + 3y^2 + 6y - 6 = 0$

79. $y^2 - 6y - 3x = 0$

80. $2x^2 - 2x + y - 2 = 0$

81. $x^2 - 4y^2 - 2x - 8y - 12 = 0$

82. $x^2 - 4y^2 + 2y = 0$

In Exercises 83–90, solve the given problem.

83. For which values of b does the parabola $y = -2x^2 + bx - 5$ have only one x-intercept?

84. Find a so that the parabola $y = ax^2 - 6x + 1$ has only one x-intercept.

85. Find the points of intersection of the parabolas $y = 4 - (x - 1)^2$ and $y = x^2 - 2x - 3$.

86. Find the points of intersection of the parabolas $y = x^2 + 2x - 15$ and $y = 9 - (x + 2)^2$.

87. Find an equation of the circle passing through the points $A(1, 4)$ and $B(-5, 2)$ and whose center is the midpoint of the segment AB.

88. Find an equation of the circle passing through the points $(0, 0)$, $(8, 0)$, and $(0, -6)$.

C 89. The major axis of the orbit of Mars has length 2.834×10^8 miles, and the eccentricity of the orbit is 9.3×10^{-2}. How close is Mars to the sun at perihelion and how distant is it at aphelion?

C 90. In April of 1983, the astronauts aboard the space shuttle *Challenger* launched into orbit the first of three global communications satellites. The satellite was to be placed in a circular orbit over the equator, at a distance of approximately 22,300 miles above the earth. In such an orbit, the satellite would be synchronized with the rotation of the earth so that, seen from the earth, it would remain at the same point. Due to equipment failure, the satellite was placed instead in an elliptical orbit whose closest point to the earth's surface was 13,540 miles and whose farthest point was 21,950 miles. Find the eccentricity of that elliptical orbit. (radius of the earth = 3,959 miles)

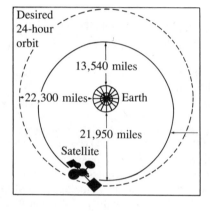

CHAPTER 4

■

FUNCTIONS

In this chapter we introduce the concept of functions, one of the most important and useful concepts in mathematics. First, we discuss the notions of domain and range, dependent and independent variables, and function values. Next, we define the graph of a function and describe the graphs of some simple functions. In Section 4.2, we discuss algebraic properties of functions and some examples of functions derived from the applied sciences. In Section 4.3, we describe several techniques to graph functions and define even and odd functions and increasing and decreasing functions. Finally, in the last section we discuss the notion of variation of one variable with respect to others.

4.1   FUNCTIONS

The notion of *function*, one of the most important concepts in mathematics, is linked to the notion of *related variables,* which we frequently encounter in the applied sciences and everyday life.

Here are some examples of related variables. At the end of the previous section, we saw that the relationship between degrees Fahrenheit, *F,* and degrees Celsius, *C,* is given by the linear equation

$$F = \frac{9}{5} C + 32.$$

The variables *F* and *C* are linearly related as defined in Chapter 3, page 136. To each value assigned to *C* there corresponds a unique value of *F,* and vice versa.

If an object moves along a line at a *constant rate* (or *speed*) *r,* then the *distance d* covered by the object is related to the *time t* by the equation

$$d = rt.$$

To each assigned value of the time *t* there corresponds a unique value of the distance *d.*

The formula

$$A = l^2$$

gives the area of a square in terms of the length *l* of the side of the square. It shows that to each positive value for *l* there corresponds a unique value of the area *A.*

According to Boyle's law, the pressure *P* of a gas enclosed in a container and kept at a constant temperature is related to the volume of the gas by the formula

$$PV = C,$$

where *C* is a constant. This formula can be written

$$P = \frac{C}{V},$$

showing that to each value of the volume *V* there corresponds a unique value of the pressure *P.* This formula indicates that pressure and volume are in *inverse proportion*. That is, a decrease in volume corresponds to an increase in pressure, and an increase in volume corresponds to a decrease in pressure.

There is a relationship between the cost of producing a unit of a certain item and the number of units produced. Within certain limits, the more a company produces, the less it costs to produce it.

These are examples of an important kind of relationship between two variables; this relationship characterizes the notion of *function*.

Function

A function *f* is a rule that assigns to each element *x* of a set *X,* called the *domain* of the function, a unique element *y* of a set *Y.* The element *y* is called the *image* of x under f, and the set of all images of elements of X is the *range* of the function.

Functions are sometimes illustrated by diagrams like Figure 4.1. The arrow indicates that the element y is the image of x under f.

As an example of how general the function concept can be, suppose that X is the set of all counties in the state of New Jersey and that Y denotes the set of all natural numbers. If to each county x we assign the number y of registered voters in the county, a function f from X into Y is defined. The domain of this function is the set of all counties in New Jersey and the range is a subset of the set of natural numbers.

As another example, suppose that to every individual we assign his or her age in years and months, with the number of months converted into a fraction of the year. This defines a function from the set of all individuals into the set of rational numbers.

It is important to observe that, in the definition of function, to each element x in X there corresponds a *unique* element in Y. However, the same y may be assigned to different elements in X. For instance, two different counties may have the same number of registered voters, and several individuals may have the same age.

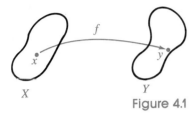

Figure 4.1

Notations

The domain of a function f is denoted by D_f or, when no confusion is possible, by D. If f is a function, x an element in the domain of f, and y the image of x under f, both notations

$$f\colon x \longrightarrow y \qquad \text{or} \qquad y = f(x)$$

indicate that the function f assigns y to the element x. The second notation, $y = f(x)$, which we read, "y is equal to f of x," is due to Euler.

Constant Function

A function that assigns to *every* element in its domain a *single element* is said to be a *constant function*. It follows that the range of a constant function consists of only *one* element.

For example, a teacher assigns as homework to each of his/her students the reading of the *same* section of this book.

Real-Valued Functions

A *real-valued function* is a rule that assigns to each element of the domain a unique real number of the range.

For example, the two functions described above (number of voters and age of individuals) are real-valued functions.

Function Values

For the most part, throughout this book we shall consider only *real-valued functions* whose domains are *sets of real numbers*.

For example, the correspondence that assigns to every *real* number x its square, x^2, defines a real-valued function denoted by

$$f: x \longrightarrow x^2 \qquad \text{or} \qquad f(x) = x^2.$$

Also, the correspondence that assigns to every *nonnegative number* x its principal square root, $\sqrt{x}$, defines the *square root* function, denoted by

$$g: x \longrightarrow \sqrt{x} \qquad \text{or} \qquad g(x) = \sqrt{x}.$$

If f is a real-valued function and x is an element in the domain of f, then the number $f(x)$ is called the *function value* of f at x.

Domains

Throughout this book, unless otherwise stated, domains of functions are assumed to be the largest possible sets for which the defining rules assign real numbers as function values. They are sometimes called *maximal* or *natural domains*.

EXAMPLE 1

Find the domain and range of the function $f(x) = x^2$. Also, find the function values $f(-5), f(3.5), f(12.1)$, and $f(2/3)$.

Solution: For any given number x, we can always compute x^2, so the domain of the function $f(x) = x^2$ is the set of all real numbers. Since every nonnegative number has a square root, it follows that the range of f is the set of all nonnegative numbers. Next, we have

$$f(-5) = (-5)^2 = 25,$$
$$f(3.5) = (3.5)^2 = 12.25,$$
$$f(12.1) = (12.1)^2 = 146.41,$$
$$f(2/3) = \left(\frac{2}{3}\right)^2 = \frac{4}{9}.$$

Practice Exercise 1

Find the domain and range of the function $F(x) = x^2 + 3$. Compute the function values $F(-1.2), F(0), F(10), F(1/4)$.

Answer: Domain: all real numbers; range: all numbers ≥ 3; $F(-1.2) = 4.44$, $F(0) = 3$, $F(10) = 103$, $F(1/4) = 49/16$

EXAMPLE 2

Find the domain and range of the function $g(x) = \sqrt{x}$. Is it possible to find the function values $g(0), g(16)$, and $g(-5)$?

Solution: The principal square root $\sqrt{x}$ is defined only when x is a *nonnegative number*. So, the domain of g is the set of all nonnegative numbers. On the other hand, let y be a nonnegative number. If we set $x = y^2$, then x is nonnegative and the principal root of x is y, that is, $y = \sqrt{x}$. Thus, the range of g is also the set of all nonnegative numbers. Next, we have

$$g(0) = \sqrt{0} = 0 \qquad \text{and} \qquad g(16) = \sqrt{16} = 4.$$

Since -5 does not belong to the domain of $g(x) = \sqrt{x}$, the expression $g(-5)$ is *not* defined.

Practice Exercise 2 Find the domain and range of $G(x) = \sqrt{x} + 2$. Can you find the function values $G(3)$, $G(9)$, and $G(-1)$?

Answer: Domain: $x \geq 0$; range: all numbers ≥ 2; $G(3) = \sqrt{3} + 2$, $G(9) = 5$, $G(-1)$ is undefined.

Independent and Dependent Variables

In the definition of a function f, the roles of the two variables are conceptually different: x denotes a freely chosen element in the domain of f, while y is the number that corresponds to x by f. For this reason, x is called the *independent variable* and y the *dependent variable*.

For example, consider the square root function g defined by $y = g(x) = \sqrt{x}$. If we take

$$x = \frac{4}{9}, \qquad x = 6, \qquad \text{and} \qquad x = 25$$

as our choices for the dependent variable, then the corresponding values of the dependent variable are

$$y = \frac{2}{3}, \qquad y = \sqrt{6}, \qquad \text{and} \qquad y = 5.$$

The independent variable is also called the *input* and the dependent variable the *output* of a function. For the square root function, if the input is 100, the output is 10.

Graphs of Functions

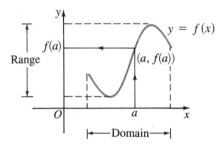

Figure 4.2

Let $y = f(x)$ be a real-valued function defined on D, a subset of the real numbers. The *graph* of f is the set of points in a Cartesian plane whose coordinates are $(x, f(x))$, for all $x \in D$.

By the definition of function, to each element a in the domain D there corresponds a unique element $f(a)$ in the range of f. Thus *every vertical line $x = a$, with $a \in D$, intersects the graph in a single point*. Given a function $y = f(x)$, to find the function value of f at $x = a$, start at the point a and move vertically until the graph is reached, and then move horizontally to the y-axis (Figure 4.2).

The Vertical Line Test

In a coordinate plane, a given set of points is the graph of a function when every vertical line intersects the set in *at most* one point.

For example, consider the parabola of the equation $y = x^2$ (Figure 4.3a). Since every vertical line intersects the curve in exactly one point, the curve is the

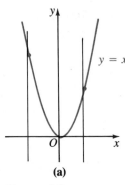 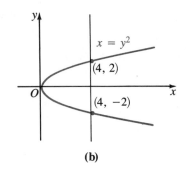

(a)   (b)

Figure 4.3

graph of a function that we may denote by $f(x) = x^2$. This function is also said to be *defined by the equation* $y = x^2$.

By contrast, there are vertical lines that intersect the graph of the equation $x = y^2$ in more than one point (Figure 4.3b). For example, the vertical line $x = 4$ intersects the graph at the points $(4, 2)$ and $(4, -2)$. In fact, for each nonnegative number x, there are two numbers y and $-y$ so that $y^2 = (-y)^2 = x$. Thus, the parabola illustrated in Figure 4.3b *cannot* be the graph of a function.

The Graph of $g(x) = \sqrt{x}$

The equation $x = y^2$ *does not* define y as a function of x because to each nonnegative number x there correspond *two* numbers: y, the principal square root of x, and $-y$, the negative square root of x. The equation itself does not tell us which of the two numbers to assign to x.

However, as we have already said, the function $g(x) = \sqrt{x}$ assigns to each x the principal square root of x. If we set $y = g(x) = \sqrt{x}$, then $y \geq 0$ and $y^2 = x$. It follows that the function $g(x) = \sqrt{x}$ is defined by the equation $y^2 = x$ with $y \geq 0$. Thus the graph of g (Figure 4.4) consists of all points on the graph of $x = y^2$ that lie above or on the x-axis.

x	$\sqrt{x}$
0	0
1	1
2	$\sqrt{2}$
3	$\sqrt{3}$
4	2

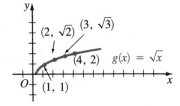

Figure 4.4

Graph Sketching

Graphs of functions can be sketched by plotting points, in the same way as graphs of equations (Section 3.3). The method is not foolproof and, in many cases, requires guesswork and intuition. Nevertheless, in simple situations, we can obtain a lot of relevant information concerning the graph of a function through simple analysis of the function, without having to rely on more advanced techniques. By using all the available information, the plotting of points can be reduced to a minimum.

EXAMPLE 3

Graph the function $f(x) = 2x + 3$.

Solution: If we set $y = f(x)$, then the graph of the given function is the same as the graph of the linear equation $y = 2x + 3$, which represents a straight line with slope 2 and y-intercept 3. Recall that to sketch the graph we have to plot only two points, such as the x- and y-intercepts, and join them by a straight line (Figure 4.5).

x	y
0	3
$-3/2$	0

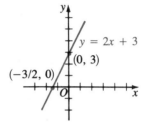

Figure 4.5

Practice Exercise 3

Sketch the graph of the function $f(x) = -3x + 6$.

Answer:

x	y
0	6
2	0

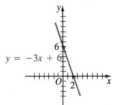

EXAMPLE 4

Sketch the graph of the function $f(x) = 4 - x^2$.

Solution: Setting $y = f(x)$, we see that the given function is defined by the equation $y = 4 - x^2$. The graph, illustrated in Figure 4.6, is a parabola obtained by adding 4 to the ordinate of each point on the graph of $y = -x^2$. (See Practice Exercise 2 of Section 3.3.) Or, if we write $y - 4 = -x^2$ and use the results of Section 3.4, we see that the graph is a parabola opening downward, with vertex at $(0, 4)$ and principal axis the y-axis. To double-check our work, a table of values accompanies the graph of $f(x) = 4 - x^2$.

x	y
-2	0
-1	3
0	4
1	3
2	0

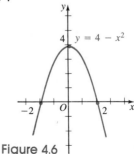

Figure 4.6

Practice Exercise 4   Sketch the graph of the function $f(x) = x^2 - 1$.

Answer:

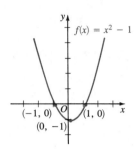

EXAMPLE 5   Find the domain of $f(x) = \sqrt{x + 4}$ and sketch the graph of the function.

Solution: The function f is defined whenever $x + 4 \geq 0$, that is, for all $x \geq -4$. Thus, the domain of f is the unbounded interval $[-4, +\infty)$. We can make a table of values and plot a few points, as shown in Figure 4.7, to sketch the graph of f.

However, it is better to proceed in a different manner. If we write $y = \sqrt{x + 4}$, then we can see that the graph is the part of the parabola $x + 4 = y^2$ that lies above the x-axis. (Why?)

x	y
-4	0
-3	1
0	2
5	3

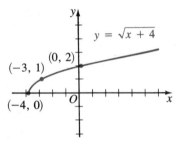

Figure 4.7

Practice Exercise 5   What is the domain of $f(x) = \sqrt{x - 2}$? Graph this function.

Answer: Domain: $[2, +\infty)$

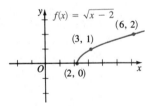

EXERCISES 4.1

In Exercises 1–14, find the domain and range of each of the following functions.

1. $f(x) = 3x - 4$

2. $g(x) = 4x - 1$

3. $F(x) = \dfrac{2x}{3} + \dfrac{1}{5}$

4. $G(x) = \dfrac{x}{5} - \dfrac{3}{4}$

5. $H(x) = 4 - x^2$

6. $h(x) = 1 - x^2$

7. $r(x) = 3x^2 - 1$

8. $s(x) = 4x^2 + 5$

9. $u(x) = \sqrt{x} - 3$

10. $v(x) = \sqrt{x} + 4$

11. $l(x) = 2\sqrt{x} + 1$

12. $m(x) = 3\sqrt{x} - 2$

13. $p(x) = \sqrt{x - \dfrac{1}{2}}$

14. $g(x) = \sqrt{x + \dfrac{2}{3}}$

For each function in Exercises 15–22, find the indicated function values. When necessary, use your calculator.

15. $f(x) = 3x - 4; f(-2), f(0), f(2/3),$ and $f(\sqrt{1.5})$

16. $g(x) = 4x - 1; g(-3/8), g(0), g(4),$ and $g(\sqrt{2})$

17. $F(x) = 4 - x^2; F(-3), F(-2), F(1),$ and $F(\sqrt{3})$

18. $G(x) = 1 - x^2; G(-\sqrt{5}), G(-3), G(1),$ and $G(5)$

19. $h(x) = 3 - \sqrt{x}; h(2), h(16), h(3.5), h(-8), h(1/4)$

20. $H(x) = \sqrt{x} + 5; H(0), H(-1), H(4), H(5.8), H(9/16)$

21. $k(x) = \sqrt{x} - 4; k(0), k(5), k(12), k(7.25), k(17/4)$

22. $K(x) = \sqrt{x + 1}; K(0), K(5), K(-12), K(5.5), K(-7/16)$

In Exercises 23–40, find the domain of each function. Sketch the graph.

23. $f(x) = 4x - 3$

24. $f(x) = 5x - 4$

25. $f(x) = 2 - \dfrac{3}{5}x$

26. $f(x) = 1 - \dfrac{3}{7}x$

27. $f(x) = x^2 - 4$

28. $f(x) = x^2 - 6$

29. $f(x) = 3 - x^2$

30. $f(x) = 9 - x^2$

31. $f(x) = \sqrt{x - 1}$

32. $f(x) = \sqrt{x + 3}$

33. $f(x) = \sqrt{4 - x}$

34. $f(x) = \sqrt{2 - x}$

35. $f(x) = \sqrt{x - 2}$

36. $f(x) = \sqrt{x + 4}$

37. $f(x) = \sqrt{x} - 3$

38. $f(x) = \sqrt{x} + 4$

39. $f(x) = 1 - \sqrt{x}$

40. $f(x) = 3 - \sqrt{x}$

4.2   ALGEBRA OF FUNCTIONS

Functions can be added, subtracted, multiplied, and divided to generate new functions.

Function Operations

Let f and g be two functions with domains D_f and D_g, respectively. The *sum* $f + g$, *difference* $f - g$, *product* $f \cdot g$, and *quotient* f/g are functions defined by

$$(f + g)(x) = f(x) + g(x),$$
$$(f - g)(x) = f(x) - g(x),$$
$$(f \cdot g)(x) = f(x) \cdot g(x),$$
$$(f/g)(x) = f(x)/g(x), \qquad \text{provided that } g(x) \neq 0.$$

In these definitions, the domain of $f + g, f - g,$ and $f \cdot g$ is the set of all points that belong to both domains D_f *and* D_g; that is, the domain of the three functions is the *intersection* of the domains D_f and D_g. The domain of f/g is the intersection of the domains *except* for the points at which $g(x) = 0$.

EXAMPLE **1**   Find the sum, difference, product, and quotient of the functions $f(x) = 3x - 2$ and $g(x) = x^2 - 4$, and determine the domains of definition.

Solution: Since f and g are defined for all real numbers, the functions $f + g$, $f - g$, and $f \cdot g$ are also defined for all real numbers. We have

$$(f + g)(x) = f(x) + g(x) = (3x - 2) + (x^2 - 4) = x^2 + 3x - 6,$$
$$(f - g)(x) = f(x) - g(x) = (3x - 2) - (x^2 - 4) = -x^2 + 3x + 2,$$

and

$$(f \cdot g)(x) = f(x) \cdot g(x) = (3x - 2)(x^2 - 4) = 3x^3 - 2x^2 - 12x + 8.$$

Since the polynomial $x^2 - 4$ is zero when $x = -2$ and $x = 2$, the domain of the quotient

$$(f/g)(x) = f(x)/g(x) = \frac{3x - 2}{x^2 - 4}$$

is the set $\mathbb{R}$ of all real numbers except -2 and 2. Thus, the domain is the union of the open intervals $(-\infty, -2)$, $(-2, 2)$, and $(2, +\infty)$.

Practice Exercise 1 Let $F(x) = 2x - 5$ and $G(x) = 1 - x^2$. Find $F + G$, $F - G$, $F \cdot G$, and F/G. Determine the domains of these functions.

Answer: $(F + G)(x) = -x^2 + 2x - 4$, defined for all real numbers
$(F - G)(x) = x^2 + 2x - 6$, for all real numbers
$(F \cdot G)(x) = -2x^3 + 5x^2 + 2x - 5$, for all real numbers
$(F/G)(x) = \dfrac{2x - 5}{1 - x^2}$, for all real numbers except -1 and 1

EXAMPLE 2 If $F(x) = 2x - 8$ and $G(x) = \sqrt{x + 3}$, find $(F/G)(x)$ and its domain of definition.

Solution: The function F is defined for all real numbers, while G is defined only for all real numbers x such that $x \geq -3$. Since $G(-3) = \sqrt{-3 + 3} = \sqrt{0} = 0$, the quotient

$$(F/G)(x) = \frac{F(x)}{G(x)} = \frac{2x - 8}{\sqrt{x + 3}}$$

is defined for all numbers x such that $x > -3$. That is, the domain of F/G is the unbounded interval $(-3, +\infty)$.

Practice Exercise 2 Given $f(x) = \sqrt{x - 4}$ and $g(x) = 3x - 8$, find f/g and its domain.

Answer: $(f/g)(x) = \dfrac{f(x)}{g(x)} = \dfrac{\sqrt{x - 4}}{3x - 8}$, defined for $x \geq 4$

Polynomial Functions

Among the simplest functions that we encounter are the functions defined by polynomials.

If n is a nonnegative integer, a *polynomial function of degree n* is a function defined by an expression of the form

$$f(x) = a_n x^n + a_{n-1} x^{n-1} + \cdots + a_1 x + a_0,$$

where $a_0, a_1, \ldots, a_{n-1}, a_n$ are real numbers and $a_n \neq 0$. The domain of definition of a polynomial function is the set $\mathbb{R}$ of all real numbers.

Constant Functions

If $n = 0$, then $f(x) = a_0$, and f is called a *constant function*.

Linear Functions

A polynomial function of degree 1 is said to be a *linear function*.

EXAMPLE 3

Let $f(x) = 2x + 1$ be a linear function. Find the function values $f(-3)$, $f(0)$, and $f(1)$. Also, if a and b are real numbers, find $f(a) + f(b)$ and $f(a + b)$.

Solution: We have

$$f(-3) = 2(-3) + 1 = -6 + 1 = -5,$$
$$f(0) = 2 \cdot 0 + 1 = 1,$$

and

$$f(1) = 2 \cdot 1 + 1 = 3.$$

Next, if a and b are real numbers, $f(a)$ and $f(b)$ are the values of the linear polynomial $2x + 1$ at $x = a$ and $x = b$. So,

$$f(a) = 2a + 1 \quad \text{and} \quad f(b) = 2b + 1.$$

It follows that

$$f(a) + f(b) = (2a + 1) + (2b + 1)$$
$$= 2a + 2b + 2.$$

Now, $f(a + b)$ is the function value of f at $x = a + b$, that is,

$$f(a + b) = 2(a + b) + 1$$
$$= 2a + 2b + 1.$$

Compare $f(a + b)$ and $f(a) + f(b)$, and notice that $f(a + b) \neq f(a) + f(b)$.

Practice Exercise 3

For $g(x) = 3x + 5$, find $g(-2)$, $g(4/3)$, and $g(5)$. Also find $g(a - b)$ and $g(a) - g(b)$.

Answer: $g(-2) = -1$, $g(4/3) = 9$, $g(5) = 20$, $g(a - b) = 3a - 3b + 5$, $g(a) - g(b) = 3a - 3b$

Quadratic Functions

A *quadratic function* is a polynomial function of degree 2.

EXAMPLE 4

Let $F(x) = x^2 - 5x + 6$ be a quadratic function. Find the function values $F(-2)$, $F(0)$, and $F(2)$. Also, find $2F(a)$ and $F(2a)$.

Solution: We have

$$F(-2) = (-2)^2 - 5(-2) + 6 = 20,$$
$$F(0) = 0^2 - 5 \cdot 0 + 6 = 6,$$
$$F(2) = 2^2 - 5 \cdot 2 + 6 = 0.$$

Since $F(a) = a^2 - 5a + 6$, then

$$2F(a) = 2(a^2 - 5a + 6)$$
$$= 2a^2 - 10a + 12.$$

On the other hand,

$$F(2a) = (2a)^2 - 5(2a) + 6$$
$$= 4a^2 - 10a + 6.$$

Are there numbers a for which the relation $F(2a) = 2F(a)$ holds?

Practice Exercise 4   If $G(x) = 3x^2 - 2x - 1$, find $G(-1/3)$, $G(0.5)$, and $G(4)$. Also, find $G(b/3)$ and $G(b)/3$.

Answer: $G(-1/3) = 0$, $G(0.5) = -1.25$, $G(4) = 39$,
$G(b/3) = b^2/3 - 2b/3 - 1$, $G(b)/3 = b^2 - 2b/3 - 1/3$

EXAMPLE 5   If $F(x) = x^2 - 5x + 6$ and a and h are real numbers, with $h \neq 0$, find

$$\frac{F(a + h) - F(a)}{h}.$$

Solution: We have

$$F(a + h) = (a + h)^2 - 5(a + h) + 6$$

and

$$F(a) = a^2 - 5a + 6.$$

Thus

$$\frac{F(a + h) - F(a)}{h} = \frac{[(a + h)^2 - 5(a + h) + 6] - [a^2 - 5a + 6]}{h}.$$

Simplifying the right-hand side of the last expression, we obtain

$$\frac{F(a + h) - F(a)}{h} = \frac{(a^2 + 2ah + h^2) - 5(a + h) + 6 - a^2 + 5a - 6}{h}$$

$$= \frac{\cancel{a^2} + 2ah + h^2 - \cancel{5a} - 5h + \cancel{6} - \cancel{a^2} + \cancel{5a} - \cancel{6}}{h}$$

$$= \frac{h(2a + h - 5)}{h}$$

$$= 2a + h - 5.$$

Practice Exercise 5   Let $G(x) = 3x^2 - 2x - 1$ and let b and t be real numbers, with $t \neq 0$. Find

$$\frac{G(b + t) - G(b)}{t}.$$

Answer: $6b - 2 + 3t$

An expression of the form

$$\frac{F(a + h) - F(a)}{h}$$

is sometimes called a *difference quotient*. Computations of the kind discussed in Example 5 are important. They form the foundation on which differential calculus is built.

Rational Functions

The quotient of two polynomials $f(x)$ and $g(x)$ defines a *rational function*,

$$r(x) = \frac{f(x)}{g(x)},$$

whose domain is the set of all real numbers x for which the denominator $g(x)$ is different from zero.

EXAMPLE 6   Find the domain of the function

$$f(x) = \frac{x + 2}{x + 5}.$$

If a, b, and $a - b$ are in the domain of f, find $f(a) - f(b)$ and $f(a - b)$.

Solution: Since f is the quotient of the polynomials $x + 2$ and $x + 5$, the domain D of f is the set of all real numbers for which $x + 5 \neq 0$, that is, $x \neq -5$. We can also say that D is the union of the intervals $(\infty, -5)$ and $(-5, +\infty)$. If $a, b \in D$, then

$$f(a) = \frac{a + 2}{a + 5}, \quad f(b) = \frac{b + 2}{b + 5},$$

and we have

$$\begin{aligned}
f(a) - f(b) &= \frac{a + 2}{a + 5} - \frac{b + 2}{b + 5} \\
&= \frac{(a + 2)(b + 5) - (b + 2)(a + 5)}{(a + 5)(b + 5)} \\
&= \frac{ab + 5a + 2b + 10 - ab - 5b - 2a - 10}{(a + 5)(b + 5)} \\
&= \frac{3a - 3b}{(a + 5)(b + 5)}.
\end{aligned}$$

If $a - b \in D$, then

$$f(a - b) = \frac{(a - b) + 2}{(a - b) + 5}.$$

In fact, in most cases, $f(a - b) \neq f(a) - f(b)$. (What happens if $a = -2$ and $b = 1$?)

Practice Exercise 6   Let

$$g(x) = \frac{3x - 5}{4x + 1}.$$

What is the domain of g? Find $g(a)$ and $g(1/a)$, assuming that these two function. values are defined.

Answer: Domain: All real numbers $\neq -1/4$; $g(a) = \dfrac{3a - 5}{4a + 1}$, $g\left(\dfrac{1}{a}\right) = \dfrac{3 - 5a}{4 + a}$

The operations of sum, difference, product, and quotient of functions can be combined with root extraction to produce new functions.

EXAMPLE 7   What is the domain of the function $G(x) = \sqrt{2x + 1}$? Find the function values $G(-1/2)$, $G(4)$, and $G(8)$. Assuming that a, b, and $ab \in D_G$, find $G(a) \cdot G(b)$ and $G(ab)$.

Solution: In order for G to be defined, the linear expression $2x + 1$ must be nonnegative. Thus, the domain of G consists of all numbers x satisfying the inequality

$$2x + 1 \geq 0,$$

that is, all numbers x so that $x \geq -1/2$.

Next, we have

$$G\left(-\frac{1}{2}\right) = \sqrt{2\left(-\frac{1}{2}\right) + 1} = \sqrt{-1 + 1} = \sqrt{0} = 0,$$
$$G(4) = \sqrt{2 \cdot 4 + 1} = \sqrt{9} = 3,$$

and

$$G(8) = \sqrt{2 \cdot 8 + 1} = \sqrt{17}.$$

Assuming that a and b belong to D_G, we have

$$G(a) = \sqrt{2a + 1} \quad \text{and} \quad G(b) = \sqrt{2b + 1},$$

so

$$\begin{aligned} G(a) \cdot G(b) &= \sqrt{2a + 1} \cdot \sqrt{2b + 1} \\ &= \sqrt{(2a + 1)(2b + 1)} \\ &= \sqrt{4ab + 2a + 2b + 1}. \end{aligned}$$

If $ab \in D_G$, then

$$G(ab) = \sqrt{2ab + 1}.$$

Notice that, in general, $G(ab) \neq G(a) \cdot G(b)$. Can you find a pair of numbers a, b in D_G for which $G(ab) = G(a) \cdot G(b)$?

Practice Exercise 7   Find the domain of $g(x) = \sqrt{2 - 3x}$. Find the function values $g(-1)$, $g(-2/3)$, and $g(1)$. Also, find $g(a)/g(b)$ and $g(a/b)$, assuming that these function values are defined.

Answer: Domain: $x \le 2/3$; $g(-1) = \sqrt{5}$, $g(-2/3) = 2$, $g(1)$ not defined;

$$\frac{g(a)}{g(b)} = \sqrt{\frac{2 - 3a}{2 - 3b}}; \quad g\left(\frac{a}{b}\right) = \sqrt{\frac{2b - 3a}{b}}.$$

Functions in Applications

Many formulas used in mathematics and the applied sciences provide examples of functions.

One of the greatest scientific discoveries of the Renaissance was Galileo's law of falling bodies. The law states that the distance s covered by an object falling freely from rest is proportional to the square of time; that is, the formula

$$s(t) = ct^2$$

expresses s as a function of t, where c is a constant.

HISTORICAL NOTE

GALILEO GALILEI (1564–1642)

Galileo was an Italian astronomer, mathematician, and physicist. He discovered the law of falling bodies: when a body is falling freely and the resistance of the air is neglected, acceleration is constant; furthermore, the acceleration is the same for all bodies. Galileo constructed telescopes with which he made important astronomical discoveries: he observed the phases of Venus, found out that the Milky Way consists of countless stars, and discovered the satellites of Jupiter. He ardently adopted Copernicus's theory that the earth and the planets move around the sun. Brought to trial by the Inquisition, he was forced to renounce all his beliefs and writings holding that the sun was the center of the planetary system. Galileo's persistent investigations of natural laws laid the foundations for modern experimental science.

EXAMPLE 8

When the distance s is measured in *feet* and the time t in *seconds*, the constant c in the preceding formula is approximately equal to 16 ft/s^2. How many feet does an object fall in **(a)** 2 seconds, **(b)** 3 seconds, **(c)** 6.5 seconds?

Solution: We have $s(t) = 16t^2$. So,

$$s(2) = 16 \cdot 2^2 = 64 \text{ ft,}$$
$$s(3) = 16 \cdot 3^2 = 144 \text{ ft,}$$
$$s(6.5) = 16 \cdot (6.5)^2 = 676 \text{ ft.}$$

Practice Exercise 8

The area A of a circle of radius r is equal to π times the square of the radius. Write a formula expressing A as a function of r. Also, find the area if $r = \sqrt{2}$, 4, and 6.2 units of length.

Answer: $A(r) = \pi r^2$, $A(\sqrt{2}) = 2\pi$, $A(4) = 16\pi$, $A(6.2) = 38.44\pi$

EXAMPLE 9 A manufacturer can produce a maximum of 60 car radios per day, at a cost of $35 per radio. If his daily operating expenses are $575, write a formula for the total cost of producing x radios per day. What is the domain of definition of this function?

Solution: If $C(x)$ denotes the total cost to produce x radios per day, then

$$C(x) = 575 + 35x,$$

with x a nonnegative integer less than or equal to 60. Thus, the domain of C is the set $\{x: 0 \le x \le 60, x \text{ an integer}\}$.

Practice Exercise 9 It costs $25 per day plus 25¢ per mile to rent a car. Find the cost per day, $C(x)$, as a function of x miles.

Answer: $C(x) = 25 + 0.25x$

EXERCISES 4.2

In Exercises 1–10, find the domain of each function.

1. $f(x) = \dfrac{3}{4 - 2x}$

2. $f(x) = \dfrac{-2}{3x - 9}$

3. $f(x) = \sqrt{4 - 3x}$

4. $f(x) = \sqrt{2x + 6}$

5. $f(x) = \dfrac{x}{(x - 1)^2}$

6. $f(x) = \dfrac{2x - 1}{(x + 2)^2}$

7. $f(x) = \dfrac{2x + 1}{x^2 - 3x + 2}$

8. $f(x) = \dfrac{5x - 3}{x^2 - 3x - 4}$

9. $f(x) = \sqrt{1 - x^2}$

10. $f(x) = \sqrt{x^2 + 2x}$

For each function in Exercises 11–18, find the indicated function values. When necessary, use your calculator.

11. $f(x) = 3x + 6; f(1), f(0), f(2/3), f(\sqrt{1.5})$

12. $g(x) = 3 - 5x; g(-2), g(0), g(4/5), g(\sqrt{6})$

13. $F(x) = -x^2 - x - 1; F(-1), F(0), F(1), F(\sqrt{2})$

14. $G(x) = x^2 - 3x + 4; G(-4), G(-1/2), G(0), G(3)$

15. $h(x) = \sqrt{2x - 4}; h(2), h(10), h(20/9), h(\sqrt{3})$

16. $H(x) = \sqrt{2x + 2}; H(-2/3), H(5), H(-34/74), H(14/3)$

17. $k(x) = \dfrac{x}{x + 3}; k(-4), k(-2), k(0), k(3.512)$

18. $K(x) = \dfrac{2x}{4 - x}; K(-1), K(0), K(3), K(-6.418)$

In Exercises 19–24, for each of the given functions find **(a)** $f(a)$, **(b)** $f(-a)$, **(c)** $-f(a)$, **(d)** $f(a) + f(b)$, and **(e)** $f(a + b)$.

19. $f(x) = 5 - 3x$

20. $f(x) = 2 - 7x$

21. $f(x) = x^2 - 4$

22. $f(x) = x^2 + 1$

23. $f(x) = \dfrac{x - 1}{x + 1}$

24. $f(x) = \dfrac{2 - x}{x + 2}$

In Exercises 25–30, for each of the given functions, find **(a)** $h(1/a)$, **(b)** $1/h(a)$, **(c)** $h(a) + h(1/a)$, and **(d)** $h\left(a + \dfrac{1}{a}\right)$.

25. $h(x) = \dfrac{1}{x - 5}$

26. $h(x) = \dfrac{1}{x + 3}$

27. $h(x) = \dfrac{1}{2x - 1}$

28. $h(x) = \dfrac{3x}{x - 1}$

29. $h(x) = 1 - x^2$

30. $h(x) = x^2 + 2$

In Exercises 31–36, find **(a)** $(f + g)(x)$, **(b)** $(f - g)(x)$, **(c)** $(f \cdot g)(x)$, and **(d)** $(f/g)(x)$. State the domain of each function.

31. $f(x) = 3x + 5, \quad g(x) = x^2 - 3$

32. $f(x) = x^2, \quad g(x) = 2x - 5$

33. $f(x) = \dfrac{3x}{2x - 1}, \quad g(x) = \dfrac{x + 1}{5x + 3}$

34. $f(x) = \dfrac{1}{x + 7}, \quad g(x) = x + 7$

35. $f(x) = 1 - \dfrac{3}{x}, \quad g(x) = 1 + \dfrac{3}{x}$

36. $f(x) = x + \dfrac{2}{x}, \quad g(x) = x - \dfrac{2}{x}$

In Exercises 37–46, find $\dfrac{f(a + h) - f(a)}{h}$ and simplify your answer.

37. $f(x) = 8$

38. $f(x) = -6$

39. $f(x) = 3x + 5$

40. $f(x) = 5 - 2x$

41. $f(x) = x^2 + 4$

42. $f(x) = x^2 - 1$

43. $f(x) = 3x^2 - x$

44. $f(x) = -3x^2 + x$

45. $f(x) = 2x^2 - 3x - 1$

46. $f(x) = x^2 - x + 1$

In Exercises 47–70, solve the given problem.

47. The length of a rectangle is 4 cm longer than twice its width. Write a formula for the area of the rectangle as a function of the width.

48. The height of a triangle is 5 ft longer than three times its base. Express the area as a function of the base.

49. Express the area of a square as a function of its perimeter.

50. Write a formula for the area of an equilateral triangle as a function of the length of a side.

51. The length of a rectangular box is four times the width and the height is half the width. What is the volume of the box as a function of the width?

52. The width of a rectangular box is twice the height and the length is twice the width. Write the lateral area of the box (top excluded) as a function of the height.

53. The height of a cylinder is four times the radius. Write the volume of the cylinder as a function of the height. (Recall that $V = \pi r^2 h$.)

54. The height of a cylinder is five times the radius. What is the volume of the cylinder as a function of the radius?

55. The volume of a cylindrical tin can is 16π cubic centimeters. Write a formula for $L(h)$, the lateral surface area of the can, as a function of the height h. (Recall that $L = 2\pi rh$.)

56. For the tin can in Exercise 55, write a formula for $S(r)$, the total surface area, as a function of the radius. (Recall that $S = 2\pi r^2 + 2\pi rh$.)

57. If \$1,000 is invested at a simple interest rate of 6% per year, express the principal p as a function of t years.

58. Suppose that if you invest p_0 dollars today, at a simple interest rate, you'll receive \$5,000 in 10 years. Write a formula giving p_0 as a function of the interest rate.

59. Suppose that in the simple lens equation

$$\frac{1}{f} = \frac{1}{d_1} + \frac{1}{d_2},$$

the object distance d_1 is five times the image distance d_2. Express the focal length f as a function of d_2.

60. The formula

$$\frac{1}{R} = \frac{1}{R_1} + \frac{1}{R_2}$$

gives the resistance R of an electric circuit when two resistances R_1 and R_2 are connected in parallel. If resistance R_2 is half of resistance R_1, find R as a function of R_1.

61. A landlady estimates that at \$250 per month she can rent all of her 100 apartments. However, for each \$10 increase in monthly rent, two apartments will remain vacant. What is her monthly revenue as a function of the number x of rental increases?

62. A bus company can organize an excursion at a cost of \$35 per person if 80 people decide to take the tour. Moreover, the company will reduce the cost at a rate of 12¢ per person in excess of 80. If x denotes the number of people taking the tour, write the total cost, C, as a function of x.

63. Margie has a total of A dollars in nickels, dimes, and quarters. If she has five fewer dimes than quarters and ten more dimes than nickels, write A as a function of the number x of nickels.

64. Tim has ten more one-dollar bills than five-dollar bills and six fewer ten-dollar bills than five-dollar bills. Express Tim's amount of money as a function of the number n of five-dollar bills.

65. The length of a rectangular sheet of tin is twice its width. An open box is to be constructed by cutting 2-inch squares from each corner and folding up the sides. Write the volume of the box as a function of the width of the sheet of tin.

66. An open box is to be made from a rectangular piece of cardboard having dimensions 24 inches by 36 inches, by cutting a square of side x inches from each corner and turning up the sides. Express the volume $V(x)$ of the box in terms of x. What is the domain of this function?

67. The height h of a cylinder is 3 times the radius.
  (a) Write a formula expressing $L(h)$, the lateral surface area of the cylinder, as a function of h.
  (b) Find the function values $L(0.25)$, $L(1)$, $L(1.25)$, and $L(2.012)$. Round off your answers to two decimal places.

68. The radius of a cylinder is $3/4$ the length of the height.
  (a) Write a formula expressing the lateral surface area $L(r)$ as a function of the radius r. (See Exercise 55.)
  (b) Find the function values $L(0.5)$, $L(2)$ and $L(3.12)$ and round off your answers to two decimal places.

69. The formula

$$T(l) = 2\pi\sqrt{\frac{l}{10}}$$

expresses the period T, in seconds, of a simple pendulum as a function of its length l, in meters. Find the function values $T(1/2)$, $T(2)$, and $T(5)$. Round off your answers to two decimal places.

70. For a simple pendulum, the period T, in seconds, is a function of the length l, in feet, given by

$$T(l) = 2\pi\sqrt{\frac{l}{32}}.$$

Find the function values $T(3/4)$, $T(1.5)$, and $T(2.25)$. Round off your answers to three decimal places.

4.3   GRAPH SKETCHING

In this section, we discuss several graphing techniques, such as the *stretching* and *shrinking* of graphs, and *horizontal* and *vertical translations* of graphs. Combined with the plotting of points, these techniques considerably simplify the task of graphing a function.

We start by discussing the stretching and shrinking techniques.

EXAMPLE 1   Graph the functions

(a) $f(x) = 3x^2$ (b) $f(x) = \dfrac{1}{4}x^2$.

Solution: (a) The graph of $f(x) = 3x^2$ may be obtained multiplying by 3 the ordinates of all points on the graph of the parabola $y = x^2$. Figure 4.8 illustrates the graphs of $f(x) = 3x^2$ and $y = x^2$ (dashed line). The graph of $f(x) = 3x^2$ is also a parabola with vertex at the origin and axis of symmetry the y-axis.

(b) If we multiply by $1/4$ the ordinates of points on the graph of $y = x^2$, we obtain the graph of $f(x) = (1/4)x^2$, shown in Figure 4.9.

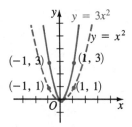

Figure 4.8

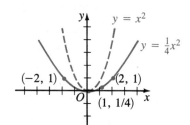

Figure 4.9

Practice Exercise 1   Graph the functions

(a) $f(x) = 2\sqrt{x}$,   (b) $f(x) = \dfrac{1}{2}\sqrt{x}$.

Answer:

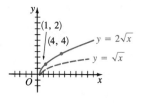

(a)

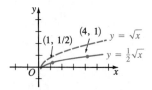

(b)

Example 1 and Practice Exercise 1 illustrate the following general fact.

> ### Stretching and Shrinking of Graphs
>
> To obtain the graph of $y = af(x)$, multiply by a the ordinates of points on the graph of $y = f(x)$. If $a > 1$, the graph of $y = f(x)$ is *stretched* vertically by a factor of a. If $0 < a < 1$, the graph is *shrunk* (or *squeezed*) vertically. If $a = -1$, the graph is reflected across the x-axis. For example, compare the graphs of $y = x^2$ and $y = -x^2$.

Now we discuss vertical translations of graphs.

EXAMPLE **2**

Graph the functions
(a) $f(x) = \sqrt{x} + 1$ **(b)** $f(x) = \sqrt{x} - 2$.

Solution: The graphs of both functions together with the graph of $y = \sqrt{x}$ (dashed lines) are illustrated in Figure 4.10.

To obtain the graph of $y = \sqrt{x} + 1$, we increase by 1 the ordinate of each point on the graph of $y = \sqrt{x}$. In other words, the graph of $y = \sqrt{x} + 1$ is obtained by shifting the graph of $y = \sqrt{x}$ one unit up. If we shift the graph of $y = \sqrt{x}$ two units down, the result will be the graph of $y = \sqrt{x} - 2$.

Notice that the x- and y-intercepts were changed as a result of this shifting. For example, the origin is both the x- and y-intercepts of the graph of $y = \sqrt{x}$. However, the x-intercept of the graph of $y = \sqrt{x} - 2$ is 4, and the y-intercept is -2.

Figure 4.10

Practice Exercise 2

Graph the functions
(a) $f(x) = x^2 + 1$ **(b)** $f(x) = x^2 - 1$.

Answer:

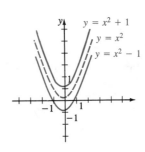

The technique of shifting that we considered in Example 2, called *vertical translation,* applies to any function and can be summarized as follows.

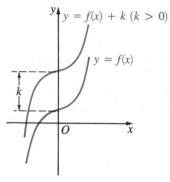

Figure 4.11

Vertical Translations of Graphs

The graph of

$$y = f(x) + k$$

may be obtained by shifting the graph of $y = f(x)$ as shown in Figure 4.11: k units up if $k > 0$ or $-k$ units down if $k < 0$.

Finally, we discuss the notion of horizontal translation of graphs.

EXAMPLE 3

Graph the functions
(a) $f(x) = (x - 1)^2$ **(b)** $f(x) = (x + 3)^2$.

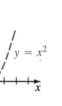

Figure 4.12

Solution: **(a)** Let $g(x) = x^2$. Since

$$f(a) = (a - 1)^2 = g(a - 1),$$

we see that the point with abscissa a on the graph of f has the same ordinate as the point with abscissa $a - 1$ on the graph of g. The graph of $f(x) = (x - 1)^2$ is then obtained by shifting the graph of $g(x) = x^2$ one unit to the *right* (Figure 4.12). The graph of $f(x) = (x - 1)^2$ is thus a parabola whose vertex is the point $(1, 0)$ and whose axis of symmetry is the vertical line $x = 1$.

(b) Again let $g(x) = x^2$. The relation

$$f(a) = (a + 3)^2 = g(a + 3)$$

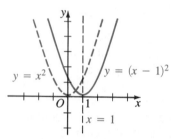

Figure 4.13

shows that the point of abscissa a on the graph of f has the same ordinate as the point of abscissa $a + 3$ on the graph of g. Thus the graph of $f(x) = (x + 3)^2$ is a parabola obtained by shifting the graph of $g(x) = x^2$ three units to the *left* (Figure 4.13). The point $(-3, 0)$ is the vertex of the parabola, and the line $x = -3$ is the axis of symmetry.

Practice Exercise 3

Graph the functions
(a) $f(x) = \sqrt{x - 2}$ **(b)** $f(x) = \sqrt{x + 4}$.

Answer:

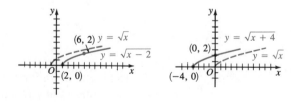

The technique described in Example 3, called *horizontal translation,* applies to any function. It can be summarized as follows.

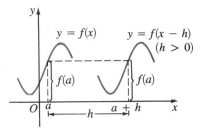

Figure 4.14

Horizontal Translation of Graphs

The graph of

$$y = f(x - h)$$

may be obtained from the graph of $y = f(x)$ by a *horizontal translation,* as shown in Figure 4.14. If $h > 0$, the graph of f is shifted h units to the right; if $h < 0$, the graph of f is shifted $-h$ units to the left.

Special Functions

Sometimes we encounter functions defined in special ways, such as in the following example.

EXAMPLE 4   Graph the function

$$f(x) = \begin{cases} x + 2 & \text{if} & x \le 0, \\ 4 - x^2 & \text{if} & 0 < x \le 2, \\ 3 & \text{if} & x > 2. \end{cases}$$

Solution: The given function is defined for all real values of x. The graph, illustrated in Figure 4.15, consists of three distinct parts. When $x \le 0$, the function is $f(x) = x + 2$, thus the graph is part of the line $y = x + 2$. On the interval $(0, 2]$, the graph is part of the parabola $y = 4 - x^2$. Finally, when $x > 2$, the function $f(x)$ is constant and equal to 3, and the graph is part of the horizontal line $y = 3$.

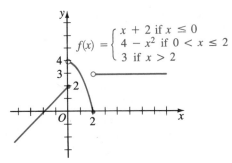

Figure 4.15

Practice Exercise 4   Graph the function defined as follows:

$$f(x) = \begin{cases} -x - 1 & \text{if} & x < 0, \\ x^2 - 4 & \text{if} & 0 \le x \le 2, \\ 2 & \text{if} & x > 2. \end{cases}$$

Answer:

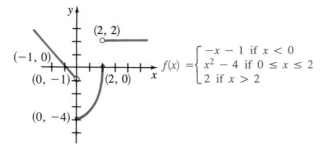

$$f(x) = \begin{cases} -x - 1 & \text{if } x < 0 \\ x^2 - 4 & \text{if } 0 \leq x \leq 2 \\ 2 & \text{if } x > 2 \end{cases}$$

The Absolute Value Function

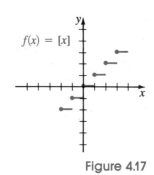

Figure 4.16

Consider the function denoted

$$f(x) = |x| = \begin{cases} x & \text{if} & x \geq 0 \\ -x & \text{if} & x < 0 \end{cases}$$

and called the *absolute value* function. Its domain of definition is the set of all real numbers and its range consists of the set of all nonnegative numbers. If $x > 0$, then $f(x) = |x| = x$, so the graph of the absolute value function coincides with the graph of $y = x$. If $x < 0$, then $f(x) = |x| = -x$, and the graph of f coincides with the graph of $y = -x$.

Notice that the graph is symmetric with respect to the y-axis.

The Greatest-Integer Function

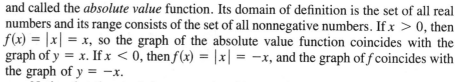

Figure 4.17

If x is a real number, denote by $[x]$ the *greatest integer that is less than or equal to x*. For example, the greatest integer less than or equal to $1/2$ is 0, the greatest integer less than or equal to 15 is 15, and the greatest integer less than or equal to $\sqrt{2}$ is 1. Thus we write

$$\left[\frac{1}{2}\right] = 0, \qquad [15] = 15, \qquad \text{and} \qquad [\sqrt{2}] = 1.$$

We also have

$$\left[-\frac{2}{3}\right] = -1, \qquad \left[\frac{25}{3}\right] = 8, \qquad \text{and} \qquad [\pi] = 3.$$

The rule that assigns to every real number x the integer $[x]$ defines a function called the *greatest-integer function*, denoted by $f(x) = [x]$. The domain of definition of f is the set of all real numbers and the range is the set of all integers.

Even and Odd Functions

A function f with domain D is said to be *even* if $f(x) = f(-x)$ for all x in D.
A function f is said to be *odd* if $f(-x) = -f(x)$, for all $x \in D$.

The definition says that the value of an even function does not change when

$-x$ is substituted for x. However, the function value of an odd function changes sign if $-x$ is substituted for x.

The two simplest examples of odd and even functions are, respectively, $f(x) = x$ and $f(x) = x^2$. The reason for the names *even* and *odd functions* comes from the fact that *even powers* of x define even functions, while *odd powers* of x define odd functions.

In Figure 4.18, we illustrate the graphs of the even functions $f(x) = x^2$ and $f(x) = x^4$. Observe that both graphs are symmetric with respect to the y-axis.

x	x^2	x^4
0	0	0
1/2	1/4	1/16
1/3	1/9	1/81
1	1	1
3/2	9/4	81/16
2	4	16

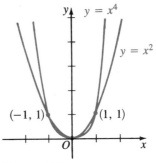

Figure 4.18

Figure 4.19 illustrates the graphs of $f(x) = x$ and $f(x) = x^3$, two odd functions. Note that both are symmetric with respect to the origin.

x	x^3
0	0
1/2	1/8
1	1
2	8

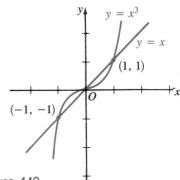

Figure 4.19

As an exercise, you can check that the graph of any even function is symmetric with respect to the y-axis, while the graph of any odd function is symmetric with respect to the origin.

Increasing and Decreasing Functions

Let f be a function and let I be a subset of the domain of f. If x_1 and x_2 are in I, then:

f is *increasing* on I if $f(x_1) < f(x_2)$ whenever $x_1 < x_2$, and
f is *decreasing* on I if $f(x_1) > f(x_2)$ whenever $x_1 < x_2$.

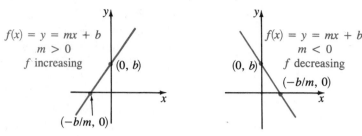

Figure 4.20

For example, we already know that the graph of a linear function

$$f(x) = mx + b$$

is a line with slope m. If $m > 0$, then f is an increasing function on $\mathbb{R}$, and if $m < 0$, then f is a decreasing function on $\mathbb{R}$ (Figure 4.20).

As another example, the function $f(x) = x^2$ is decreasing on the interval $(-\infty, 0]$ and increasing on the interval $[0, +\infty)$, as shown in Figure 4.21.

A constant function $f(x) = b$ is neither increasing nor decreasing (Figure 4.22).

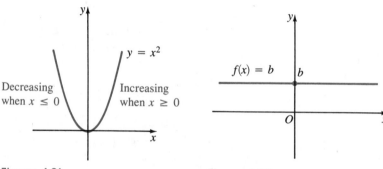

Figure 4.21   Figure 4.22

EXERCISES 4.3

In Exercises 1–22, graph each function. (Make use of stretching, shrinking, horizontal and vertical shifts.)

1. $f(x) = \dfrac{x^2}{2}$

2. $f(x) = 4x^2$

3. $f(x) = -3x^2$

4. $f(x) = -\dfrac{x^2}{4}$

5. $f(x) = 3\sqrt{x}$

6. $f(x) = \dfrac{\sqrt{x}}{2}$

7. $f(x) = -2\sqrt{x}$

8. $f(x) = \dfrac{-3\sqrt{x}}{4}$

9. $f(x) = x^2 - 4$

10. $f(x) = 1 - x^2$

11. $f(x) = \sqrt{x} + 3$

12. $f(x) = \sqrt{x} - 1$

13. $f(x) = 2 - \sqrt{x}$

14. $f(x) = 3 - \sqrt{x}$

15. $f(x) = (x - 3)^2$

16. $f(x) = (x + 4)^2$

17. $f(x) = -(x + 1)^2$

18. $f(x) = -(x - 2)^2$

19. $f(x) = \sqrt{x + 5}$

20. $f(x) = \sqrt{x - 1}$

21. $f(x) = (x - 1)^2 + 4$

22. $f(x) = (x + 4)^2 - 1$

In Exercises 23–28, graph each function.

23. $f(x) = \begin{cases} -1 & \text{if } x \leq 0 \\ 1 & \text{if } x > 0 \end{cases}$

24. $f(x) = \begin{cases} 2 & \text{if } x < 0 \\ -3 & \text{if } x \geq 0 \end{cases}$

25. $f(x) = \begin{cases} 2 & \text{if } x < -2 \\ -x & \text{if } -2 \leq x \leq 2 \\ -2 & \text{if } x > 2 \end{cases}$

26. $f(x) = \begin{cases} -1 & \text{if } x \leq -1 \\ x^3 & \text{if } -1 < x < 1 \\ 1 & \text{if } x \geq 1 \end{cases}$

27. $f(x) = \begin{cases} -2x - 1 & \text{if } -2 \leq x < -1 \\ x^2 & \text{if } 1 \leq x < 1 \\ \dfrac{x+2}{3} & \text{if } 1 < x \leq 2 \end{cases}$

28. $f(x) = \begin{cases} 3x + 4 & \text{if } x \leq 1 \\ x^2 - 1 & \text{if } x > 1 \end{cases}$

In Exercises 29–38, combine the definition of the absolute value function and the various graphing techniques to graph each function.

29. $f(x) = 3|x|$

30. $f(x) = \dfrac{|x|}{4}$

31. $f(x) = |x| - 2$

32. $f(x) = |x| + 3$

33. $f(x) = 4 - |x|$

34. $f(x) = 1 - |x|$

35. $f(x) = |x - 1|$

36. $f(x) = |x + 2|$

37. $f(x) = |x + 1| - 2$

38. $f(x) = |x - 3| + 1$

Graph each of the functions in Exercises 39–44. Use the definition of the greatest-integer function and the various graphing techniques discussed in Section 4.3.

39. $f(x) = 2[x]$

40. $f(x) = \dfrac{[x]}{4}$

41. $f(x) = [x] + 1$

42. $f(x) = [x] - 2$

43. $f(x) = [x - 1]$

44. $f(x) = [x - 2]$

In Exercises 45–60, determine whether each function is even, odd, or neither.

45. $f(x) = 3x^4 + x^2$

46. $f(x) = x^2 + 4x$

47. $F(x) = 2x^3 - 6x$

48. $F(x) = -x^4 - 3x^2$

49. $g(x) = x^6 - 2x^2 - 4$

50. $g(x) = x^5 - 1$

51. $G(x) = |x| + 2$

52. $G(x) = 4 - |x|$

53. $h(x) = |x|^3$

54. $h(x) = -|x|^5$

55. $H(x) = 5/x$

56. $H(x) = -3/x^2$

57. $k(x) = \dfrac{1}{x^2 - 1}$

58. $k(x) = \dfrac{2}{x - 1}$

59. $p(x) = \dfrac{x^3 + 5x}{x^2 + 1}$

60. $q(x) = \dfrac{2x^4 - 5x^2 - 8}{x^3 + x}$

Find intervals (each as large as you can find) in which each of the functions in Exercises 61–68 is increasing or decreasing. First, graph each function.

61. $f(x) = -5x + 3$

62. $f(x) = 4x - 2$

63. $f(x) = x^2 - 4$

64. $f(x) = 9 - x^2$

65. $f(x) = -|x - 3|$

66. $f(x) = |x + 5|$

67. $f(x) = x - |x|$

68. $f(x) = x + |x|$

In Exercises 69–80, solve each problem.

69. A coin collection is worth \$16,000 now and its value appreciates linearly at a rate of 8% per year. Find the function $V(x)$ giving the value of the coin collection in x years. Graph this function.

70. A piece of equipment purchased for \$48,000 is expected to have a scrap value of \$6,000 after 15 years. Assuming that the depreciation is linear, what will be the value $V(t)$ of the equipment after t years? Graph this function.

71. To mail a first-class letter costs 22¢ for the first ounce or fraction, plus 17¢ for each additional ounce or fraction thereof. Graph the function $f(x)$ that gives the cost of mailing an x-ounce letter.

72. A car rental company charges \$15 per day plus \$1.50 for each 10 miles or fraction of 10 miles. Graph the function $C(x)$ that expresses the daily cost of renting and driving a car for x miles.

73. Let $D(p) = 125 - \dfrac{p^2}{2}$ be the demand function for a certain item, where p is the price of the item. Graph this function.

74. Suppose that the cost to produce x units of a certain item is

$$C(x) = \frac{x^2}{4} + 25, \qquad 0 \leq x \leq 10.$$

Sketch the graph.

☐ **75.** Prove that the sum or difference of two even functions is an even function.

☐ **76.** Prove that the sum or difference of two odd functions is an odd function.

☐ **77.** Prove that the product or quotient of two odd functions is an even function.

☐ **78.** Prove that the product or quotient of two even functions is an even function.

☐ **79.** Show that for any function $f(x)$, the function

$$F(x) = \frac{1}{2}\big(f(x) + f(-x)\big)$$

is even, and that the function

$$G(x) = \frac{1}{2}\big(f(x) - f(-x)\big)$$

is odd.

☐ **80.** Prove that any function $f(x)$ can always be written as the sum of an even function and an odd function. (*Hint:* Use Exercise 79.)

4.4 VARIATION

In the beginning of Section 4.1, we described two simple but important examples of related variables. In the first one,

$$d = r \times t \qquad (\text{distance} = \text{rate} \times \text{time}),$$

if the rate r remains constant, the distance d *varies directly* as t; that is, d increases (or decreases) as t increases (or decreases). In the second one,

$$P = \frac{C}{V} \qquad (\text{Boyle's law}),$$

the pressure P of a gas enclosed in a container and kept at a constant temperature increases if the volume decreases, and decreases if the volume increases. This is an example of *inverse variation*.

In mathematics and the applied sciences, we often have to deal with relations involving a dependent variable and one or more independent variables. This leads us to the notion of variation.

Direct Variation

A variable *y varies directly as x,* or *y* is *directly proportional to x,* if there is a real number *k* such that

$$y = kx.$$

The number *k* is called the *constant of proportionality*.

For example, if a car is traveling at 55 mi/h, the distance d (in miles) traveled by the car in t hours is given by $d = 55t$. Thus, d is directly proportional to t, and 55 is the constant of proportionality.

Observe that in direct variation, the dependent variable increases (or decreases) numerically as the independent variable increases (or decreases).

Sometimes, a variable is directly proportional to a power of another variable. For example, in the formula $s = 16t^2$ giving the distance in feet that an object falls in t seconds, s *varies directly as the square* of the time t. The constant of proportionality is $k = 16$.

Next, we describe *inverse variation*.

Inverse Variation

A variable *y varies inversely as x,* or *y* is said to be *inversely proportional to x,* if

$$y = \frac{k}{x}$$

for some fixed number k, called the *constant of proportionality*.

If y varies inversely as x, then the product xy is constant and equal to the constant of proportionality k.

EXAMPLE 1   When the volume of air inside a bicycle pump is 54 in^3, the air pressure is 15 lb/in^2. Find the air pressure when the volume is equal to 18 in^3.

Solution: According to Boyle's law, the pressure P is inversely proportional to the volume V and we may write $PV = C$. If the volume changes to V', the corresponding pressure P' is such that $P'V' = C$. Hence

$$P'V' = PV.$$

Setting $V = 54$, $P = 15$, and $V' = 18$ in the last equation and solving it for P', we obtain

$$P' \times 18 = 15 \times 54,$$

or

$$P' = \frac{15 \times 54}{18} = 45 \text{ lb/in}^2,$$

which is the air pressure when the volume is 18 in^3.

In inverse variation, the dependent variable increases (or decreases) numerically as the independent variable decreases (or increases).

Sometimes y varies inversely as a power of x. For example, the intensity of illumination, I, produced by a light source varies inversely as the square of the distance, d, from the source; that is, $I = c/d^2$, where c is the constant of proportionality.

Practice Exercise 1   A light source at 20 feet produces an illumination of 60 foot-candles. What is the intensity of illumination produced at 10 feet?

Answer: 240 foot-candles

Several Independent Variables

In certain cases, the dependent variable may depend on more than one independent variable.

Joint Variation

A variable y is said to *vary jointly* as x and z if there is a number k (constant of proportionality) such that

$$y = kxz.$$

For example, in the formula $d = rt$, the distance d varies jointly as the speed r and the time t. A box of dimensions x, y, and z has volume $V = xyz$. Thus, the volume of a box varies jointly as the dimensions of its sides. In both examples, the constant of proportionality is 1.

In certain cases, a variable may vary jointly as powers of other variables. For instance, the volume V of a right cylinder of radius r and height h is given by the formula

$$V = \pi r^2 h,$$

which shows that V varies jointly as the square of the radius and the height of the cylinder. The constant of proportionality is π.

We can also have *combined variation*, such as in the formula

$$F = G\frac{mm'}{r^2}$$

(Example 5 of Section 1.3), which gives the magnitude of the force of attraction between two masses m and m' placed at a distance r from each other; G is a physical constant.

As a side remark, we mention that joint variation provides examples of *functions of several variables,* a topic studied in calculus. For instance, we may say that the volume V of a cylinder is a function of two variables: the radius r and the height h. The force F is a function of three variables: the masses m and m', and the distance r separating these masses.

EXAMPLE **2**

Suppose that w varies directly as u and inversely as the square of v. If $w = 2$ when $u = 16$ and $v = 2$, find an explicit formula for w. Also, find w when $u = 8$ and $v = 2$.

Solution: The first statement translates into

$$w = k\frac{u}{v^2}.$$

To evaluate the constant of proportionality, we substitute the given values for u, v, and w in the last equation and solve it for k:

$$2 = k\frac{16}{2^2} = 4k,$$

$$k = \frac{2}{4} = \frac{1}{2}.$$

Therefore, the explicit formula for w is

$$w = \frac{1}{2}\frac{u}{v^2}.$$

Now, substituting 8 for u and 2 for v, we obtain a numerical value for w:

$$w = \frac{1}{2} \cdot \frac{8}{2^2} = \frac{1}{2} \cdot \frac{8}{4} = 1.$$

Practice Exercise 2

A variable m varies directly as the square of n and inversely as p. Find **(a)** the constant of proportionality if $m = 3$ when $n = 3$ and $p = 2$, and **(b)** m when $n = 4$ and $p = 4$.

Answer: **(a)** $k = \dfrac{2}{3}$ **(b)** $m = \dfrac{8}{3}$

ⓒ EXAMPLE 3   Empirical observation indicates that the shoe size of a normal man varies directly as the 3/2 power of his height. On the average, a man 6 feet tall wears a size 11 shoe. What is the approximate shoe size of a basketball player 7 feet 3 inches tall?

Solution: We let s denote the shoe size and let h denote the height. Since s varies directly as the 3/2 power of h, we have

$$s = ch^{3/2}.$$

To determine the constant of proportionality c, we substitute 6 for h and 11 for s, obtaining

$$11 = c6^{3/2}.$$

Using a calculator we have

$$11 \simeq c(14.7),$$

hence

$$c \simeq 0.75.$$

Thus, the equation

$$s = 0.75h^{3/2}$$

relates the shoe size s of a normal man to his height h. Since $7'3'' = 7.25$ ft, substituting 7.25 for h in the last equation yields

$$s = 0.75(7.25)^{3/2}.$$

Now, by performing the following keystrokes on our calculator,

$$0.75 \;\boxed{\times}\; 7.25 \;\boxed{y^x}\; 1.5 \boxed{=}\; ,$$

we obtain

$$s \simeq 14.64 \simeq 15.$$

Alternate solution. We can also find the desired value of s without finding the constant of proportionality. We start with the equation $s = ch^{3/2}$. When $h = 6$, then $s = 11$, so that

$$11 = c6^{3/2}.$$

When $h = 7.25$, we have

$$s = c(7.25)^{3/2}.$$

Dividing this equation by the preceding one, we obtain

$$\frac{s}{11} = \frac{(7.25)^{3/2}}{6^{3/2}}$$

$$s = \frac{11 \times (7.25)^{3/2}}{6^{3/2}}$$

$$\simeq \frac{11 \times 19.52}{14.70}$$

$$\simeq 14.61 \simeq 15.$$

Practice Exercise 3   The frequency of vibration of a guitar string of constant length is proportional to the square root of the tension. If a tension of 40 units produces a frequency of 330 hertz (Hz), what is the frequency when the tension is doubled?

Answer: 467 Hz

EXERCISES 4.4

In Exercises 1–6, find an explicit formula for the dependent variable.

1. u varies directly as v, and $u = 18$ when $v = 6$.
2. r varies inversely as s, and $r = 48$ when $s = 1/16$.
3. y varies inversely as the square root of x, and $y = 1/6$ as $x = 4$.
4. p varies directly as the square root of q, and $p = 2$ when $q = 2$.
5. y varies inversely as x^2, and $y = 27$ when $x = 1/9$.
6. f varies jointly as p and q and inversely as r^2, and $f = 1$ when $p = 2$, $q = 3$, and $r = 3$.

Solve each of Exercises 7–30.

7. If w varies jointly as x and y, and $w = 6$ when $x = 3$ and $y = 4$, find w when $x = 1$ and $y = 6$.
8. If a varies directly as b and inversely as c, and $a = \dfrac{15}{2}$ when $b = 5$ and $c = 2$, find a when $b = 2$ and $c = 7$.
9. Let p be directly proportional to the square of q and inversely proportional to r. If $p = 9/4$ when $q = 3$ and $r = 2$, find p when $q = 2$ and $r = 3$.
10. Suppose that y is directly proportional to m and inversely proportional to n and r^2. If $y = 1/4$ when $m = 3$, $n = 1$, and $r = 2$, find y when $m = 2$, $n = 3$, and $r = 4$.
11. Hooke's law states that the force F required to stretch a spring a length l beyond its original length is directly proportional to the additional length. If a force of 6 pounds stretches a certain spring 2 inches, how much will a force of 28 pounds stretch the spring?
12. The pressure exerted by a liquid at a point is proportional to the depth of the point below the surface of the liquid. If the pressure at 18 meters below the surface of the liquid is 60 kg/cm², what is the pressure at 26 meters below the surface?
13. The intensity of illumination at a given point is inversely proportional to the square of the distance from the point to

the source. If a light source at 20 feet produces an illumination of 40 foot-candles, what is the illumination 8 feet from the light source?

14. In free-fall, the distance traveled by an object is proportional to the square of the time. If the distance traveled in 2 s is 20 m, find the distance traveled in 8 s.
15. Stefan–Boltzmann's law states that the energy E emitted by an object is proportional to the area A of the object and the fourth power of its Kelvin temperature T. Write a formula expressing E in terms of A and T.
16. In an artery, the resistance r to blood flow is directly proportional to the length l of the artery and inversely proportional to the fourth power of its diameter d. Write a formula expressing r in terms of l and d. What happens if l is increased by a factor of 4 and d is doubled?
17. The length of the skid marks caused by braking a car varies directly as the square of the initial velocity. Assume that the skid marks are 80 feet long when the initial velocity is 40 mi/h. Find the length of the skid marks when the velocity is 60 mi/h.
18. The force of attraction between two unit masses varies inversely as the square of their distance. If at a distance of 8 units the force is 3/128, what is the force at a distance of 4 units?
19. The frequency of vibration, f, of a guitar string is proportional to the square root of the tension T and inversely proportional to the length l of the string.
    (a) Write a formula for f in terms of T and l.
    (b) What happens to the frequency if the length is increased by a factor of 2 and the tension is increased by a factor of 4?
20. In Exercise 19, what happens to the frequency if both the tension and length increase by a factor of 4?
21. The volume of a sphere is directly proportional to the cube of its radius. If a sphere of radius 3 in. has a volume of 36π in³, find the formula for the volume of a sphere.

[C] Use this formula to find the volume of a sphere of radius $r = 2.25$ in.

22. The surface area of a sphere varies with the square of its radius.

    (a) Find the surface area formula knowing that for a sphere of radius 5 m, the surface area is $100\pi\, m^2$.

    [C] (b) Use the formula found in part a to approximate the surface area of a sphere of radius 2.5 m.

[C] 23. The weight of an object above the surface of the earth varies inversely with the square of its distance from the center of the earth. If an astronaut weighs 175 lb at the surface of the earth, what will be the weight of the astronaut 150 miles above the surface of the earth? (The radius of the earth is 3959 mi.)

[C] 24. What will be the weight of an astronaut 240 km above the surface of the earth if the astronaut's weight on the surface of the earth is 80 kg? (*Hint:* Use Exercise 23 and the fact that the radius of the earth is 6,300 km.)

[C] 25. The kinetic energy K of a particle varies directly as its mass m and the square of its velocity v. A particle of mass 4 g moving at 10 cm/sec has a kinetic energy of 200 ergs.

    (a) Find the formula for kinetic energy.

    [C] (b) Find the kinetic energy of a particle of mass 6.80 g moving at 18.5 cm/sec.

[C] 26. The electrical resistance of a wire varies inversely as the square of the diameter of the wire. If a wire of diameter

2×10^{-2} cm has a resistance of 10^{-1} ohm, what is the resistance of a wire of diameter 1.25×10^{-2} cm?

☐ 27. The period of a pendulum varies directly as the square root of its length and inversely as the square root of the acceleration due to gravity.

    (a) Find an explicit formula for the period, knowing that at a place where the acceleration due to gravity is g, a pendulum of length g/π^2 has period 2.

    [C] (b) Find the period of a pendulum 50 cm long at a place where $g = 9.8$ m/sec^2.

☐ 28. The formula found in Exercise 27a can also be used to find
[C] the period of an earth satellite whose orbit is a circle. In this case the length must be replaced by the radius r of the circle. Find the period (in minutes) of a satellite if the radius of its orbit is 6400 km.

☐ 29. According to Kepler's third law, the time required for a
[C] planet to make a complete revolution around the sun varies directly with the 3/2 power of its average (mean) distance from the sun. The earth–sun mean distance is 9.29×10^7 mi. Find how long it takes for Jupiter to complete one revolution around the sun if the Jupiter–sun mean distance is 4.83×10^8 mi.

☐ 30. Mars orbits the sun in 687 days. Use Exercise 29 to find the
[C] mean distance from Mars to the sun.

CHAPTER SUMMARY

A *function f* is a rule that assigns to each element x of a set a *unique* element y of another set. The element x is the *independent variable* or *input* of f, and the element y is the *dependent variable* or *output* of f. The *domain* of f is the set where f is defined, and the *range* is the set of *images* under f of all domain elements.

If f is a *real-valued function* and x is a number in the domain of f, the number $f(x)$ is the *function value* of f at x.

The *graph* of a function f with domain D is the set of points in a Cartesian coordinate plane with coordinates $(x, f(x))$ for all $x \in D$. For any $a \in D$, the vertical line through $(a, 0)$ intersects the graph of f at a single point. Conversely, a set of points in a coordinate plane is the graph of a function if every vertical line intersects the set in *at most* one point.

If f and g are two functions, you can add, subtract, multiply, and divide them to generate new functions.

$(f + g) = f(x) + g(x)$	sum
$(f - g)(x) = f(x) - g(x)$	difference
$(f \cdot g)(x) = f(x) \cdot g(x)$	product
$(f/g)(x) = f(x)/g(x) \quad (g(x) \neq 0)$	quotient

Polynomial functions are the simplest functions in mathematics. A *constant function*

$$f(x) = a, \qquad a \in \mathbb{R},$$

is a polynomial of degree 0. A *linear function*

$$f(x) = ax + b, \qquad a \neq 0,$$

is a polynomial of degree 1. A *quadratic function*

$$f(x) = ax^2 + bx + c, \qquad a \neq 0,$$

is a polynomial of degree 2.

If $f(x)$ and $g(x)$ are polynomials, the quotient

$$r(x) = f(x)/g(x)$$

is a *rational function.*

Graph sketching can be considerably simplified if the *plotting of points* is combined with the techniques of *stretching, shrinking,* and *translating* graphs, as explained in Section 4.3.

A function f with domain D is said to be *even,* if

$$f(-x) = f(x) \qquad \text{for all } x \in D$$

It is said to be *odd,* if

$$f(-x) = -f(x) \qquad \text{for all } x \in D$$

Let I be a subset of the domain of f and let $x_1, x_2 \in I$. Then f is *increasing* on I

$$\text{if } x_1 < x_2 \quad \text{implies} \quad f(x_1) < f(x_2)$$

The function f is *decreasing* on I

$$\text{if } x_1 < x_2 \quad \text{implies} \quad f(x_1) > f(x_2)$$

The notion of *variation* establishes how a given variable depends on a single variable or on several other variables. A variable y is said to be *directly proportional* to x if there is a number k, called a *constant of proportionality*, such that

$$y = kx.$$

A variable y is said to be *inversely proportional* to x if

$$y = k/x.$$

In both cases, we can say that y is a function of x. We may also have direct or inverse variation of y as a power of x or as several other independent variables.

CHAPTER TEST

1. What is the domain of the function $F(u) = \sqrt{3u - 4}$?
2. Given $f(u) = 2u^2 + 4$, find $f(3)$, $f(-3)$, and $f(\sqrt{5})$.
3. Find the domain of the function $\dfrac{2x + 5}{x^2 + 7x + 12}$.
4. Find $\dfrac{f(-1 + h) - f(-1)}{h}$ if $f(x) = 3x^2 - 3$.
5. Paula has a total of A dollars in nickels, dimes, and quarters. If she has eight fewer nickels than dimes, and six more quarters than dimes, write A as a function of the number x of dimes.
6. Use the techniques of horizontal and vertical translations to graph $f(x) = (x - 3)^2 - 3$.
7. Is the function $f(x) = 2x^6 - 4x^5$ odd, even, or neither?
8. Graph the function $f(x) = \sqrt{-3x + 7}$.
9. If y varies directly as the cube of x and inversely as z, and if $y = 1$ when $x = 2$ and $z = 3$, find y when $x = 2$ and $z = 4$.
10. Write an equation for the following statement: "p varies jointly as the square of x and the cube root of y, and inversely as the square of z."

REVIEW EXERCISES

In Exercises 1–14, find the domain of definition of each of the following functions.

1. $f(x) = \dfrac{3x + 1}{5 - 4x}$
2. $f(x) = \dfrac{-2x - 3}{3x - 7}$
3. $f(x) = \sqrt{6 - 9x}$
4. $f(x) = \sqrt{4x + 10}$
5. $f(x) = \dfrac{x + 1}{(x - 2)(3x + 5)}$
6. $f(x) = \dfrac{2x - 3}{(x + 4)(2x - 1)}$
7. $f(x) = \dfrac{3x - 8}{x^2 - x - 2}$
8. $f(x) = \dfrac{x - 4}{2x^2 - 5x + 3}$
9. $f(x) = \sqrt{2x^2 - 18}$
10. $f(x) = \sqrt{48 - 3x^2}$
11. $f(x) = \sqrt{\dfrac{x - 1}{x + 2}}$
12. $f(x) = \sqrt{\dfrac{2x + 3}{x - 4}}$
13. $f(x) = \sqrt{\dfrac{-x}{x^2 + 1}}$
14. $f(x) = \sqrt{\dfrac{x^2}{1 - x^2}}$

In Exercises 15–24, find the indicated function values. When necessary, use your calculator.

15. $f(x) = 5x - 1$, $f(-7/10)$, $f(2/5)$, $f(3)$, $f(3.12)$
16. $f(x) = 2 - 3x$, $f(-4/9)$, $f(5/3)$, $f(-8)$, $f(1.45)$

17. $f(x) = 3 - 2x^2, f(-5), f(0), f(5), f(\sqrt{10})$
18. $f(x) = 4 - 3x^2, f(-2), f(1), f(3), f(\sqrt{8})$
19. $f(x) = \sqrt{4 - 3x}, f(-7), f(-15), f(11/12), f(2.5)$
20. $f(x) = \sqrt{5 - 2x}, f(-10), f(-31/2), f(2/9), f(4.1)$

21. $f(x) = \dfrac{-2x}{x^2 + 1}, f(-3), f(0), f(3), f(4.25)$

22. $f(x) = \dfrac{x - 1}{x^2 - 2}, f(-2), f(10), f(2), f(0.05)$

23. $f(x) = \sqrt{\dfrac{x - 1}{x - 3}}, f(1), f(-10), f(7/2), f(9.5)$

24. $f(x) = \sqrt{\dfrac{2 - x}{x - 5}}, f(2), f(3/5), f(4), f(2.75)$

In Exercises 25–32, find $\dfrac{f(x + h) - f(x)}{h}$ and simplify your answer.

25. $f(x) = 3 - 7x$ **26.** $f(x) = 6x + 2$
27. $f(x) = 5x^2 - x$ **28.** $f(x) = 3x^2 - 2x + 4$

29. $f(x) = \dfrac{2}{x}$ **30.** $f(x) = \dfrac{1}{x - 1}$

31. $f(x) = \dfrac{x - 1}{x + 1}$ **32.** $f(x) = \dfrac{1 - x}{2x + 1}$

In Exercises 33–40, solve each problem.

33. If $f(x) = \dfrac{x + a}{x - a}$, find $f(a + b)$, $f\!\left(\dfrac{1}{a}\right) + f\!\left(\dfrac{1}{b}\right)$ and $f\!\left(\dfrac{1}{a} + \dfrac{1}{b}\right)$.

34. If $g(x) = x + \dfrac{1}{x}$, find $g(a) + \dfrac{1}{g(a)}$ and $g\!\left(a + \dfrac{1}{a}\right)$.

☐ **35.** If $f(x) = x$, show that $(f(a + b))^2 - (f(a - b))^2 = 4f(a)f(b)$.

☐ **36.** If $f(x) = \dfrac{1}{x}$, show that $f(a + 1) - f(a - 1) = -2f(a + 1)f(a - 1)$.

☐ **37.** Express the area of a triangle as a function of its height h, when its base is one-third of h.

☐ **38.** Let $S = 2ab + 2ac + 2bc$ be the lateral surface area of a box with sides a, b, and c. Express S as a function of a, when b is twice a, and c is one-half of a.

☐ **39.** The height of a right circular cylinder is five times its radius. Express the lateral surface area of the cylinder as a function of the radius.

☐ **40.** Express the total surface area of the cylinder in Exercise 39 as a function of the radius.

In Exercises 41–54, use the techniques of stretching, shrinking, and shifting to graph each function.

41. $f(x) = 8 - 2x^2$ **42.** $f(x) = 3x^2 - 27$

43. $f(x) = (x - \tfrac{1}{2})^2$ **44.** $f(x) = -(x + 4)^2$
45. $f(x) = -(2x + 3)^2$ **46.** $f(x) = (3x - 4)^2$
47. $f(x) = \sqrt{x - \tfrac{3}{5}}$ **48.** $f(x) = \sqrt{x + 5}$
49. $f(x) = \sqrt{4x - 2}$ **50.** $f(x) = \sqrt{9x + 6}$

51. $f(x) = \dfrac{|x - 2|}{3}$ **52.** $f(x) = 3|x + 5|$

53. $f(x) = 1 - |x + 3|$ **54.** $f(x) = 2 - |x + 1|$

Determine whether each function in Exercises 55–62 is even, odd, or neither.

55. $f(x) = x^3 - 6x$ **56.** $f(x) = x^2 - 20$
57. $f(x) = x(x^2 + 2)$ **58.** $f(x) = x(x^3 + 2x)$

59. $f(x) = \dfrac{x^2}{2} + |x|^3$ **60.** $f(x) = -3x^2 + 2|x|^5$

61. $f(x) = \dfrac{|x|}{x^2 - 1}$ **62.** $f(x) = \dfrac{|x|}{x^3 - 1}$

Find the largest possible intervals in which each of the functions in Exercises 63–70 is increasing or decreasing.

63. $f(x) = \dfrac{1}{2}x - 3$ **64.** $f(x) = 2 - \dfrac{5}{7}x$

65. $f(x) = 2x^2 - 5$ **66.** $f(x) = 3 - 4x^2$
67. $f(x) = -|x + 3|$ **68.** $f(x) = 2|x - 4|$
69. $f(x) = \sqrt{x - 4}$ **70.** $f(x) = -\sqrt{x + 2}$

Express each statement in Exercises 71–74 as an equation. Use k for the constant of proportionality.

71. The pressure in the ears of an underwater swimmer is directly proportional to the depth at which she is swimming.
72. At sea level, the distance that a person can see to the horizon is directly proportional to the square root of the height of the person's eyes above sea level.
73. The weight of a solid cylinder varies jointly as the height and the square of its radius.
74. The electric current I in a wire is directly proportional to the electromotive force E and inversely proportional to the resistance R.

Solve each of the following problems in Exercises 75–80.

75. If y varies directly as the square root of x, and $y = 3$ when $x = 4$, find y when $x = 12$.
76. If a varies directly as b and inversely as the square root of c, and $a = 2$ when $b = 2$ and $c = 16$, find a when $b = 4$ and $c = 8$.
☐ **77.** The striking force F of a moving car is proportional to its weight w and the square of its velocity v. By how much would the striking force increase if the velocity were doubled?
☐ **78.** A manufacturer estimates that the number of units u of a certain item sold per week is inversely proportional to the

quantity $p + 35$, where p is the price per unit in dollars. Suppose that at a price of $15 per unit, 300 units are sold per week. How many units per week would be sold if the price were $25 per unit?

79. At a constant speed, the number of gallons of gasoline used by a car is directly proportional to the length of time it travels. If the car uses 4 gallons in 1 hour and 20 minutes of traveling, how many gallons will it use in 4.5 hours?

80. The maximum safe load L that can be supported by a wooden beam of rectangular cross section varies jointly as the width w and the square of the height h of the cross section, and inversely as the length l of the beam. A beam 6 ft long with cross section of width 2 in. and height 4 in. will safely support 450 lb. How much weight can be safely supported by a beam of the same material which is 12 ft long and has a cross section of width 4 in. and height 8 in.?

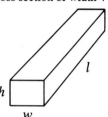

■

CHAPTER 5

POLYNOMIAL AND RATIONAL FUNCTIONS; COMPOSITE AND INVERSE FUNCTIONS

Polynomial and rational functions deserve our special attention, in view of their many applications. As we have already mentioned, polynomial functions are among the simplest functions. In Chapter 4 we discussed polynomial functions of degrees 0 and 1. In this chapter we shall study polynomial functions of degrees 2 and higher. First we discuss quadratic functions, whose graphs are parabolas. As we shall see, all relevant properties of the graph are derived from algebraic properties of the quadratic polynomial. To a certain extent, this is also true for graphs of other polynomials, particularly those that can be factored as a product of linear functions. Rational functions are more difficult to graph. In simple cases, though, a lot of information about the graph of a rational function (such as locations of vertical and horizontal asymptotes) can be derived from algebra. In the last section of this chapter we discuss composition of functions and study the important concept of inverse functions.

5.1   QUADRATIC FUNCTIONS

Let f be a quadratic function given by

$$f(x) = ax^2 + bx + c, \qquad (5.1)$$

where a, b, and c are real numbers and $a \neq 0$. The method of completing the square (Section 2.5 of Chapter 2) allows us to rewrite f as follows:

$$f(x) = a(x - h)^2 + k, \qquad (5.2)$$

where h and k are real numbers.

205

From this new expression and using horizontal and vertical translations of graphs as explained in Section 4.3 of Chapter 4, we can easily obtain the graph of f.

First, we shift the graph of $y = ax^2$ horizontally $|h|$ units, to the right if $h > 0$ or to the left if $h < 0$. The result is the graph of $y = a(x - h)^2$. Next, we shift the graph of $y = a(x - h)^2$ vertically $|k|$ units, upward if $k > 0$ or downward if $k < 0$.

The following examples show how powerful this simple method is.

EXAMPLE 1

Graph $f(x) = 2x^2 - 12x + 10$.

Solution: Our aim is to write f in the form (5.2). This can be done by arranging our work as follows:

$$f(x) = 2(x^2 - 6x \qquad) + 10.$$

Next we convert $x^2 - 6x$ into a perfect square by adding 9 inside the parentheses and obtaining $x^2 - 6x + 9 = (x - 3)^2$. Since 9 was added inside the parentheses, $18 = 2 \cdot 9$ must be subtracted outside the parentheses so that f remains unchanged:

$$f(x) = 2(x^2 - 6x + 9) + 10 - 18$$
$$= 2(x - 3)^2 - 8.$$

The last expression is of the form (5.2) with $a = 2$, $h = 3$, and $k = -8$. The graph, shown in Figure 5.1, is a parabola obtained by shifting the graph of $y = 2x^2$ three units to the right and eight units down. The parabola opens upward, and its vertex has coordinates $(3, -8)$. The minimum value of the function f is -8, corresponding to the y-coordinate of the vertex. Note that the minimum value is obtained by substituting 3 for x in the new expression for f:

$$f(3) = 2(3 - 3)^2 - 8 = -8.$$

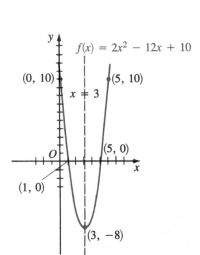

Figure 5.1

Also observe that the same result is obtained if 3 is substituted for x in the original form of f. Thus, comparing the results of the two substitutions is a good check to determine whether the arithmetic was done correctly.

The axis of symmetry of the curve is the line $x = 3$. Additional points on the graph are found from the intercepts. Setting $x = 0$, we get $f(0) = 2 \cdot 0^2 - 12 \cdot 0 + 10 = 10$. Thus the graph crosses the y-axis at the point $(0, 10)$. To obtain the x-intercepts, we solve the quadratic equation $2x^2 - 12x + 10 = 0$. This equation is equivalent to $(2x - 2)(x - 5) = 0$, so the solutions are $x = 1$ and $x = 5$. Thus the x-intercepts are 1 and 5.

Practice Exercise 1

Graph the function $f(x) = 2x^2 + 4x - 6$.

Answer:

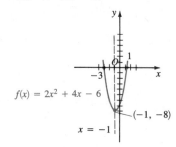

EXAMPLE **2**   Graph the function $f(x) = -x^2 - x + 2$.

Solution: We begin by completing the square as follows:

$$f(x) = -(x^2 + x \qquad) + 2$$

$$= -\left(x^2 + 2\left(\frac{1}{2}\right)x \qquad\right) + 2$$

$$= -\left(x^2 + 2\left(\frac{1}{2}\right)x + \frac{1}{4}\right) + 2 + \frac{1}{4}$$

$$= -\left(x + \frac{1}{2}\right)^2 + \frac{9}{4}.$$

The function f is now of the form (5.2) with $a = -1$, $h = -1/2$, and $k = 9/4$. Since $a < 0$, the graph opens downward. The vertex $(-1/2, 9/4)$ is the highest point of the graph, and the maximum value of f is $f(-1/2) = 9/4$. The line $x = -1/2$ is the axis of symmetry. The graph is illustrated in Figure 5.2. The y-intercept 2 is obtained by setting $x = 0$ in the equation of f: $f(0) = 2$. The x-intercepts -2 and 1 are the solutions of the quadratic equation $-x^2 - x + 2 = 0$.

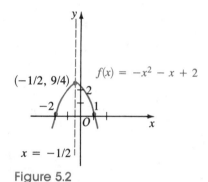

Figure 5.2

Practice Exercise 2   Graph the function $f(x) = -x^2 + 2x + 8$.

Answer:

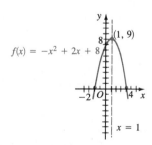

The results of Examples 1 and 2 can be summarized as follows:

The Graph of $y = ax^2 + bx + c$

To graph a quadratic function

$$f(x) = ax^2 + bx + c,$$

complete the square and write f in the form

$$f(x) = a(x - h)^2 + k.$$

    The graph is a parabola with vertex at (h, k) and axis of symmetry $x = h$.

    If $a > 0$, the parabola opens upward and the vertex is its lowest point. The function f has a *minimum*, obtained by evaluating f at $x = h$, that is, $f(h) = k$.

    If $a < 0$, the parabola opens downward and the vertex is its highest point. The function f has a *maximum* equal to $f(h) = k$.

    To find the y-intercept, evaluate $f(x)$ at $x = 0$.

    To find the x-intercepts (if any), solve the quadratic equation $ax^2 + bx + c = 0$.

The *x*-Intercepts of the Parabola $y = ax^2 + bx + c$

When graphing a quadratic function

$$f(x) = ax^2 + bx + c,$$

it is helpful to find the x-intercepts (if any) of the graph. They are the solutions of the quadratic equation

$$ax^2 + bx + c = 0.$$

As we said in Section 2.5 of Chapter 2, this equation has two real roots if the discriminant $\Delta = b^2 - 4ac$ is positive; it has only one root if $\Delta = 0$; and it has no real roots if Δ is negative.

    Examples 1 and 2 illustrate the case $\Delta > 0$. We consider the other two cases in the examples that follow.

EXAMPLE **3**   Graph the function $f(x) = 4x^2 - 4x + 1$.

Solution: The quadratic polynomial is a perfect square: $4x^2 - 4x + 1 = (2x - 1)^2$. Thus

$$f(x) = (2x - 1)^2.$$

Since $2x - 1 = 2(x - 1/2)$, we can write f as follows:

$$f(x) = 4\left(x - \frac{1}{2}\right)^2.$$

The graph of f, illustrated in Figure 5.3, can be obtained by translating the parabola $y = 4x^2$ horizontally $1/2$ unit to the right. Note that the graph just touches the x-axis at the point $(1/2, 0)$; that is the vertex of the parabola. When this happens, we say that the parabola is *tangent* to the x-axis.

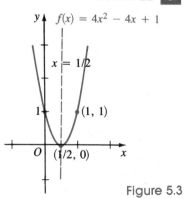

Figure 5.3

In this example the discriminant of the quadratic polynomial is

$$\Delta = b^2 - 4ac = 4^2 - 4 \cdot 4 \cdot 1 = 0.$$

The only solution of the equation $4x^2 - 4x + 1 = 0$ is $x = 1/2$. (Thus the x-intercept is $1/2$.) The y-intercept is 1 and the axis of symmetry is $x = 1/2$.

Practice Exercise 3   Graph $f(x) = -2x^2 - 4x - 2$.

Answer:

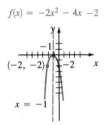

EXAMPLE 4   Graph the function $f(x) = -2x + 4x - 5$.

Solution: Here $a = -2$, $b = 4$, $c = -5$, and $\Delta = 4^2 - 4(-2)(-5) = 16 - 40 < 0$. The quadratic equation has no real roots, so the graph of f has no x-intercept. Since $a = -2 < 0$, the parabola opens downward and the vertex is its highest point. To find the coordinates of the vertex, as well as the axis of symmetry, we write f in the form (5.2):

$$\begin{aligned}
f(x) &= -2x^2 + 4x - 5 \\
&= -2(x^2 - 2x \quad\ ) - 5 \\
&= -2(x^2 - 2x + 1) - 5 + 2 \\
&= -2(x - 1)^2 - 3.
\end{aligned}$$

Thus the vertex has coordinates $(1, -3)$ and the axis of symmetry is $x = 1$. The graph is shown in Figure 5.4.

Figure 5.4

Practice Exercise 4   Graph $f(x) = x^2 + 2x + 5$.

Answer:

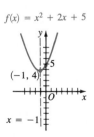

Tips on Graphing Quadratic Polynomials

We can now summarize our results.

Let $f(x) = ax^2 + bx + c$ be a quadratic function and let $\Delta = b^2 - 4ac$ be the discriminant of the quadratic polynomial.

1. If $\Delta > 0$, the graph of f has two x-intercepts.
2. If $\Delta = 0$, the graph of f has one x-intercept, and the graph is said to be *tangent* to the x-axis.
3. If $\Delta < 0$, the graph has no x-intercepts.

EXAMPLE 5

Mrs. Robinson owns a pizza parlor. By analyzing her costs, she estimates that the daily cost, $C(x)$, of operating her business is given by the formula

$$C(x) = 2x^2 - 240x + 7380,$$

where x is the number of pizzas she sells per day. Find the number of pizzas that she must sell to minimize her costs. What is the minimum cost?

Solution: This problem is equivalent to that of finding the lowest point on the graph of $C(x)$. By completing the square, we have

$$C(x) = 2(x^2 - 120x \qquad) + 7380$$
$$= 2(x^2 - 2 \cdot 60x + 3600) + 7380 - 7200$$
$$= 2(x - 60)^2 + 180.$$

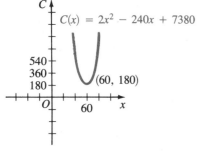

Figure 5.5

Thus the minimum point on the graph of $C(x)$ is $(60, 180)$, as shown in Figure 5.5.

Mrs. Robinson must sell 60 pizzas in order to minimize her costs, and the minimum daily cost is $C(60) = \$180$.

HISTORICAL NOTE

NICOLE ORESME

Nicole Oresme (1320–1382) was a chaplain to King Charles V of France. He later became the Bishop of Liseux in Normandy. He is interesting to us in his role as a mathematician because, in a work written about 1360, he graphed a function (of course, he did not use the modern notation). He called the vertical coordinate *latitude* and the horizontal coordinate *longitude* in analogy with the use of those terms in map-making, which had been practiced since ancient times. This work was reprinted as late as 1515. Thus while it was Descartes who first completely saw that coordinatization could turn geometry into algebra and conversely, the method of graphical representation that underlies analytic geometry was no less than 130 years older than Descartes.

Practice Exercise 5

The profit function for a manufacturer is

$$P(x) = 300x - 2x^2,$$

where x is the number of units produced. How many units must be produced in order to maximize profit? What is the maximum profit?

Answer: 75 units, $11,250

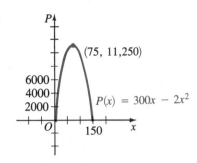

The following example and practice exercise are related to Example 4 and Practice Exercise 4 of Section 2.7 of Chapter 2.

EXAMPLE 6   A travel agent estimates that 20 people will take a sightseeing tour at a cost of $100 per person. She also believes that each $5 reduction of the price will bring two additional people. How many price reductions would maximize her revenue? What is the maximum revenue?

Solution: If x denotes the number of price reductions, then $20 + 2x$ is the number of people taking the tour at $100 - 5x$ per person. Thus the revenue is

$$R(x) = (20 + 2x)(100 - 5x)$$
$$= 2000 + 100x - 10x^2.$$

By completing the square, we can rewrite the quadratic function R as follows:

$$R(x) = -10(x - 5)^2 + 2250.$$

We see that the corresponding parabola has a maximum point with coordinates $(5, 2250)$. Thus $x = 5$ is the number of price reductions that maximizes the revenue. The maximum revenue is $R(5) = \$2250$.

Practice Exercise 6   A farmer has 30 orange trees in his orchard, which produces an average of 600 oranges per tree per year. He estimates that for each additional tree planted in the orchard, the annual yield of each tree decreases by ten oranges. How many additional trees should he plant to maximize the total yield of oranges? What is the maximum yield?

Answer: 15 trees, 20,250 oranges

EXERCISES 5.1

1. Graph $f(x) = ax^2$ if

   (a) $a = 3$ **(b)** $a = -\dfrac{1}{2}$ **(c)** $a = -4$

   (d) $a = -2$

2. Given $f(x) = ax^2 + c$, graph f if

   (a) $a = -3, c = 1$ **(b)** $a = 2, c = -1$

   (c) $a = \dfrac{1}{4}, c = -2$ **(d)** $a = -\dfrac{1}{2}, c = \dfrac{3}{5}$

3. Given $f(x) = (x - h)^2$, graph f if
     (a) $h = 2$ (b) $h = -3$ (c) $h = -5$
     (d) $h = 4$
4. Graph $f(x) = a(x - h)^2 + k$, if
     (a) $a = 2$, $h = 1$, $k = 3$
     (b) $a = -2$, $h = -3$, $k = 1$
     (c) $a = -1$, $h = 2$, $k = -1$
     (d) $a = 3$, $h = 0$, $k = -3$

Calculate the discriminant $\Delta = b^2 - 4ac$ of each quadratic function in Exercises 5–14. What can you say about the number of x-intercepts of the graph? Does the parabola open upward or downward?

5. $f(x) = 3x^2 + 1$ 6. $f(x) = -4x^2 - 3$
7. $f(x) = x^2 - 8x + 20$ 8. $f(x) = -x^2 + 6x - 2$
9. $f(x) = -2x^2 + 7x + 4$ 10. $f(x) = 3x^2 - 5x + 1$
11. $f(x) = 9x^2 - 12x + 4$
12. $f(x) = -4x^2 + 20x - 25$
13. $f(x) = -x^2 - 2x - 3$ 14. $f(x) = x^2 - x + 6$

In Exercises 15–28, graph each of the given quadratic functions. Give the vertex and the axis of each parabola.

15. $f(x) = x^2 - 9$ 16. $f(x) = 16 - 4x^2$
17. $f(x) = x^2 - 6x + 8$ 18. $f(x) = x^2 - 8x + 12$
19. $f(x) = x^2 - 6x + 9$ 20. $f(x) = x^2 + 8x + 16$
21. $f(x) = 2x^2 + 1$ 22. $f(x) = -3 - x^2$
23. $f(x) = 2x^2 - x + 1$ 24. $f(x) = 3x^2 + x + 1$
25. $f(x) = -x^2 + 2x$ 26. $f(x) = x^2 + 4x$
27. $f(x) = -2x^2 + 8x - 11$ 28. $f(x) = 3x^2 - 6x + 4$

In Exercises 29–50, solve each problem.

29. For what value of c is the graph of $f(x) = x^2 + 8x + c$ tangent to the x-axis?
30. For what value of k is the graph of $f(x) = x^2 + kx + k$ tangent to the x-axis?
31. Find all values of b for which the graph of $f(x) = 2x^2 + bx + 2$ does not cross the x-axis.
32. Find all values of c for which the graph of $f(x) = -x^2 + 3x - c$ does not intersect the x-axis.
33. Find two positive numbers whose sum is 36 and whose product is maximum.
34. Find two numbers whose difference is 48 and whose product is minimum.
35. What is the area of the largest rectangular field that can be enclosed with 1260 m of fencing?
36. What is the area of the largest rectangular plot that can be enclosed and divided into two equal rectangular parts with 240 m of fencing?
37. Mary is planting a vegetable garden on a rectangular plot bordered on one side by a wall. She wants to use 90 ft of fencing to fence off the other three sides and to divide the plot into two equal rectangular parts. Find the dimensions

of the plot that has maximum area. What is the maximum area?

38. A woman wants to make a vegetable garden on a rectangular plot bordered on one side by her house. She has 36 ft of fencing to fence off the other three sides of the plot. What should be the dimensions of the plot if the enclosed area is to be a maximum?
39. Peter operates a hot dog stand. If his profit $P(x)$ is given by the function

$$P(x) = 0.8x - 0.002x^2,$$

where x is the number of hot dogs, how many hot dogs must Peter sell to maximize his profit? What is the maximum profit?

40. The cost function for manufacturing electric shavers is given by

$$C(x) = 2x^2 - 60x + 825,$$

where x is the number (in hundreds) of units manufactured per week and C is the cost in hundreds of dollars. What weekly output corresponds to minimum cost? What is the minimum cost?

41. The demand function for a certain type of candy bar is given by

$$p = 80 - x,$$

where p is the price (in cents) when x thousand units are demanded. The revenue function is defined by

$$R(x) = (80 - x)x.$$

Find the maximum revenue and the corresponding price.

42. The demand function for a manufacturer's product is $p = 300 - 2x$, where p is the price (in dollars) per unit when x units are demanded. The revenue function $R(x)$ is defined by

$$R(x) = (300 - 2x)x.$$

Find the maximum revenue and the price corresponding to the maximum revenue.

43. If an object is thrown upward from the ground level with an initial velocity of 88 ft/s, then its height s (in feet) above the ground after t seconds is given by

$$s = -16t^2 + 88t.$$

Find the maximum height reached by the projectile. After how many seconds will the object hit the ground?

44. The height s (in meters) above the ground after t seconds of a stone that has been thrown upward from ground level is given by

$$s = -5t^2 + 35t.$$

(a) After how many seconds will the stone reach maximum height?

(b) Find the maximum height.

45. Wastewater containing biodegradable organic material is being dumped into a pond. If the amount A of organic material is given by

$$A(t) = 800 - 160t + 16t^2,$$

where t represents days, find after how many days the amount of organic material will be minimum. What is this minimum?

46. After t hours the number of bacteria in a certain culture is given by

$$N = 1000 + 50t - 5t^2.$$

When will the number of bacteria be a maximum? What is the maximum number?

☐ 47. To print 100 posters, a printer charges 20¢ per poster. For each increase of 50 posters, he decreases the price of printing by 2¢ per poster. How many posters should he print to maximize the revenue?

☐ 48. A merchant sells 80 pounds of coffee daily at $3.00 per pound. Each 10¢ increase in price decreases sales by five pounds. If his costs are $2.00 per pound, how much should

he charge per pound of coffee in order to maximize profit?

☐ 49. Show that every quadratic function

$$f(x) = ax^2 + bx + c$$

can be written

$$f(x) = a(x - h)^2 + k,$$

where

$$h = -\frac{b}{2a} \quad \text{and} \quad k = -\frac{\Delta}{4a} = -\frac{b^2 - 4ac}{4a}.$$

(*Hint:* Complete the square, as shown in Example 1 of this section.)

☐ 50. Show that the graph of

$$f(x) = ax^2 + bx + c$$

is a parabola with vertex at the point

$$\left(-\frac{b}{2a}, -\frac{\Delta}{4a} \right)$$

and symmetry axis $x = -\frac{b}{2a}$.

5.2 POLYNOMIAL FUNCTIONS

We already mentioned that a function of the form

$$f(x) = a_n x^n + a_{n-1} x^{n-1} + \cdots + a_1 x + a_0,$$

where $a_0, a_1, \ldots, a_n$ are real numbers and $a_n \neq 0$, is called a *polynomial function of degree n.*

We know from Chapter 4 that the graph of a linear polynomial $f(x) = ax + b$ is a straight line. In Section 5.1 we saw that the graph of a quadratic polynomial $f(x) = ax^2 + bx + c$ is always a parabola.

The graphs of most polynomial functions of degree greater than 2 are difficult to describe without using certain techniques studied in calculus. Of course, it is always possible to obtain an approximate graph of a polynomial by plotting points; the more points we plot, the more accurate the graph. However, our aim is to describe several graphing techniques that reduce the plotting of points to a minimum and still enable us to draw a reasonable sketch of the graph of a polynomial. First we discuss some simple cases.

EXAMPLE 1 Graph the function $f(x) = \frac{1}{2}x^3$.

Solution: Using the technique of shrinking (see Section 4.3), we see that the graph can be obtained by multiplying by $1/2$ the ordinates of all points on the graph of $y = x^3$. (Recall that the graph of $y = x^3$ was discussed in Example 4 of Section 3.3.) Figure 5.6 illustrates the graphs of $y = x^3$ (dashed line) and $y = (1/2)x^3$ (solid line) for comparison purposes.

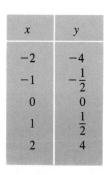

x	y
-2	-4
-1	$-\dfrac{1}{2}$
0	0
1	$\dfrac{1}{2}$
2	4

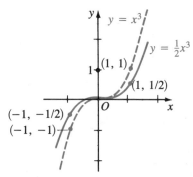

Figure 5.6

Note that f is an odd function; thus the graph is symmetric with respect to the origin. Also, since $y = (1/2)x^3 > 0$ when $x > 0$, that part of the graph lies above the x-axis. It lies below the x-axis when $x < 0$.

Practice Exercise 1   Graph the function $f(x) = (x - 1)^3$.

Answer:

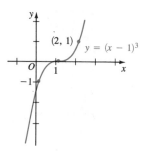

EXAMPLE **2**   Graph $f(x) = x^4 - 1$.

Solution: Using the technique of vertical translation (Section 4.3), it follows that the graph is obtained by vertically shifting the graph of $y = x^4$ one unit down.

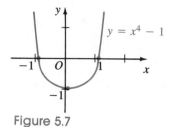

Figure 5.7

Practice Exercise 2   Graph $f(x) = (x + 1)^4$.

Answer:

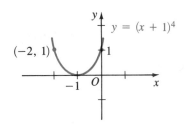

Polynomials in Factored Form

The task of graphing a polynomial is considerably simplified when the polynomial is written as a product of linear factors.

First, it can be shown in advanced courses that the graph of any polynomial is always a curve that contains no gaps, corners, jumps, or missing points.

Second, the graph is unbounded as $|x|$ increases without bound, that is, the graph cannot level off as $|x|$ becomes very large.

Third, for a polynomial $f(x)$, the x-intercepts are the real solutions of the equation $f(x) = 0$. These are easily obtained when the polynomial is written as a product of linear factors.

Fourth, the x-intercepts divide the real line into a finite number of intervals, in each of which the sign of the polynomial function remains constant. To determine the sign of the polynomial in any particular such interval, we may use the method for solving inequalities explained in Section 2.8. Alternatively, we can determine the sign by simply evaluating the polynomial at one conveniently chosen test point in the interval.

EXAMPLE **3**   Graph the function $f(x) = x^3 + x^2 - 2x$.

Solution: First we factor the polynomial as follows:

$$f(x) = x(x^2 + x - 2) = x(x - 1)(x + 2).$$

Next we observe that at the points $x = -2$, $x = 0$, and $x = 1$, the polynomial function is equal to zero. Thus, -2, 0, and 1 are the x-intercepts of the graph. They divide the line into four intervals: $(-\infty, -2)$, $(-2, 0)$, $(0, 1)$, and $(1, +\infty)$. By using the method described in Section 2.8, we can determine the sign changes of $f(x)$ as x varies on the real line.

Sign of	*For x in each of the intervals*								
	$-\infty$		-2		0		1		$+\infty$
x		$-$		$-$	0	$+$		$+$	
$x - 1$		$-$		$-$		$-$	0	$+$	
$x + 2$		$-$	0	$+$		$+$		$+$	
$f(x) = x(x - 1)(x + 2)$		$-$	0	$+$	0	$-$	0	$+$	

The table shows that f is negative in the intervals $(-\infty, -2)$ and $(0, 1)$, so the graph of f lies below the x-axis there. In the intervals $(-2, 0)$ and $(1, +\infty)$, we see that

f is positive, so the graph lies above the x-axis in those intervals. Using this information and plotting a few points, we obtain the curve illustrated in Figure 5.8.

It is often simpler to determine the sign changes of the polynomial function f by evaluating it at some arbitrarily chosen point inside each of the intervals between x-intercepts. As shown in Figure 5.8, $f(-1) = 2 > 0$ and $f(1/2) = -5/8 < 0$. As an exercise, check that $f(-3) = -12 < 0$ and $f(2) = 8 > 0$.

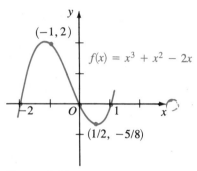

Figure 5.8

Practice Exercise 3   Graph $f(x) = -x^3 + 2x^2 + 3x$.

Answer:

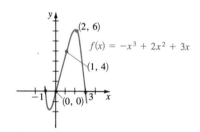

High and Low Points of Polynomial Functions

As we already know, the graph of a quadratic function always has a maximum (high point) or a minimum (low point). These are *turning points* of the graph. At a maximum, the function changes from increasing to decreasing; at a minimum it changes from decreasing to increasing.

Polynomials of degree greater than 2 may also have high and low points. It can be proved that a polynomial of degree n has *at most $n - 1$* turning points. For example, the graph of $f(x) = x^3 + x^2 - 2x$ (Figure 5.8) indicates that the function $f(x)$ attains a maximum at some point in the open interval $(-2, 0)$, while similarly, $f(x)$ attains a minimum in the open interval $(0,1)$. However, the problem of finding the maximum and minimum points of a polynomial function of degree ≥ 3 is difficult; there is no analogue of the simple method of completing the square that we used with quadratic functions. Historically, differential calculus was developed in order to solve this problem. The appropriate methods to handle the general problem thus lie beyond the scope of this book.

The examples and exercises considered in this section are fairly simple. If you determine the x-intercepts and plot a reasonable number of points, you will be able to determine the approximate location of the maxima and minima of a polynomial function.

EXAMPLE 4   Graph the function $f(x) = -x^3 - x^2 + 4x + 4$.

Solution: Factoring the polynomial, we get

$$f(x) = -x^2(x + 1) + 4(x + 1)$$
$$= -(x^2 - 4)(x + 1)$$
$$= -(x + 2)(x - 2)(x + 1).$$

It follows that the polynomial function $f(x)$ is zero at the points $x = -2$, $x = -1$, and $x = 2$. These points divide the real line into four intervals: $(-\infty, -2)$ $(-2, -1)$, $(-1, 2)$, and $(2, +\infty)$. We could now determine the sign changes of f by making a table like the one in Example 3. Instead, we are going to select a point in each interval and substitute it into the function to determine whether the function is positive or negative in that interval. Here is a selection of points.

Interval	Selected point	Function value
$(-\infty, -2)$	-3	$f(-3) = 10 > 0$
$(-2, -1)$	$-\dfrac{3}{2}$	$f\left(-\dfrac{3}{2}\right) = -\dfrac{7}{8} < 0$
$(-1, 2)$	0	$f(0) = 4 > 0$
$(2, +\infty)$	3	$f(3) = -20 < 0$

Using this information and plotting the points, we obtain the graph in Figure 5.9. Note that 4 is the y-intercept of the graph because $f(0) = 4$. Also, f has a minimum in the interval $(-2, -1)$ and a maximum in the interval $(-1, 2)$.

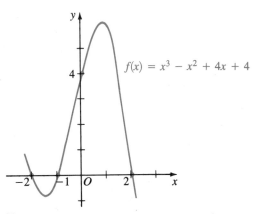

$f(x) = x^3 - x^2 + 4x + 4$

Figure 5.9

Practice Exercise 4   Graph the function $f(x) = x^3 - 2x^2 - x + 2$.

Answer: $f(x) = x^3 - 2x^2 - x + 2$

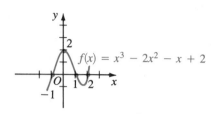

Guidelines for Graphing Polynomial Functions

1. If possible, completely factor the polynomial.
2. Find the x- and y-intercepts.
3. Determine the sign changes of the polynomial function as x varies along the x-axis.
4. Make a table of function values and plot the corresponding points.
5. Sketch the graph.

EXERCISES 5.2

Graph the following functions in Exercises 1–20.

1. $f(x) = (x - 2)^3$
2. $f(x) = (2 - x)^3$
3. $f(x) = (x + 3)^4$
4. $f(x) = (x - 2)^4$
5. $f(x) = x(x + 3)^2$
6. $f(x) = 2x(x - 1)^2$
7. $f(x) = x^2(x - 1)$
8. $f(x) = -x^2(x - 4)$
9. $f(x) = x^3 - 4x$
10. $f(x) = 2x^3 - 2x$
11. $f(x) = x(x - 2)(x - 4)$
12. $f(x) = (x - 1)(x + 1)(x + 2)$
13. $f(x) = x^2(x - 1)(x + 2)$
14. $f(x) = x^2(x + 1)(x + 3)$
15. $f(x) = x^4 - 2x^2 - 8$
16. $f(x) = x^4 - 8x^2 - 9$
17. $f(x) = x^3 + x^2 - 4x - 4$
18. $f(x) = x^3 + 3x^2 - x - 3$
19. $f(x) = 2x^3 + 3x^2 - 2x - 3$
20. $f(x) = x^3 + 2x^2 - 9x - 18$

In Exercises 21–34, the graph of each polynomial function is symmetric about a line or a point. Graph each function and name the line or point of symmetry.

21. $f(x) = 2x^3 + 4$
22. $f(x) = (2x - 1)^3$
23. $f(x) = 3x^4$
24. $f(x) = \frac{1}{2} x^5$
25. $f(x) = 2x^3 - \frac{1}{3}$
26. $f(x) = \frac{1}{2} x^6 + 1$
27. $f(x) = x^3 - 4$
28. $f(x) = 2 - x^4$
29. $f(x) = \frac{1}{2}(x + 1)^3$
30. $f(x) = -2(x - 3)^4$
31. $f(x) = (x - 2)^4 + 1$
32. $f(x) = (x + 1)^3 + 4$
33. $f(x) = (2x - 1)^3 + 4$
34. $f(x) = (3x + 1)^3 - 5$

In Exercises 35–40, use a calculator (if necessary) to determine several points on the graph of each polynomial. Sketch the graph.

35. $f(x) = x^3 + x$
36. $f(x) = -x^3 - 2x$
37. $f(x) = x^3 + x - 1$
38. $f(x) = x^3 - x^2 + 3x - 3$
39. $f(x) = 1 - x - x^3$
40. $f(x) = x^2 - x^3$

5.3 RATIONAL FUNCTIONS

Graphing a rational function

$$f(x) = \frac{P(x)}{Q(x)},$$

where $P(x) = a_n x^n + \cdots + a_1 x + a_0$ and $Q(x) = b_m x^m + \cdots + b_1 x + b_0$ are polynomial functions, is not a simple matter, particularly if the degrees of the polynomials arc large. Nevertheless, in simple cases, by combining the plotting of points with certain important properties of rational functions, it is possible to sketch the graph of a rational function fairly accurately.

As we already know, a rational function is defined for all real numbers x, except the numbers for which the denominator is equal to zero.

Two Initial Steps

When dealing with a rational function $f(x)$, we have to analyze, among other things, two important features.

1. The function values of f at x near the numbers for which the denominator is equal to zero.
2. The function values of f when $|x|$ becomes very large.

These ideas can be illustrated by studying the graph of

$$f(x) = \frac{1}{x},$$

which is the simplest nontrivial rational function. First, f is defined for all real numbers $x \neq 0$. Next, note that the graph of f is the same as the graph of the equation

$$y = \frac{1}{x}.$$

This equation is equivalent to the equation $x = 1/y$ or the equation $xy = 1$. We can now derive the following information about the coordinates (x, y) of a point on the graph of $y = 1/x$:

(a) x and y are multiplicative inverses of each other;

(b) neither x nor y can be zero, so the graph does not have x- or y-intercepts;

(c) x and y are both positive or both negative. This tells us that the graph lies on the first and third quadrants.

Now we make a table of values.

x	y
-3	$-\dfrac{1}{3}$
-2	$-\dfrac{1}{2}$
-1	-1
$-\dfrac{1}{2}$	-2
$\dfrac{1}{2}$	2
1	1
2	$\dfrac{1}{2}$
3	$\dfrac{1}{3}$

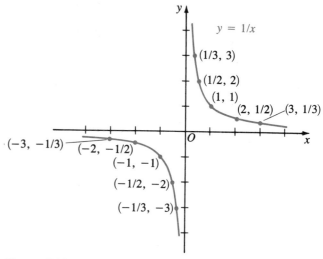

Figure 5.10

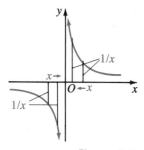

Figure 5.11

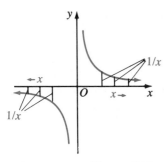

Figure 5.12

Plotting the points and connecting them with a smooth curve, we obtain the graph illustrated in Figure 5.10. You can check that the graph is symmetric with respect to the origin and with respect to the main diagonal.

Now we analyze what happens to the function $f(x) = 1/x$ as x approaches 0. If x is a small positive number, then $f(x) = 1/x$ is a large positive number. For instance, if $x = 0.1, 0.01$, and 0.001, then $y = 10, 100$, and 1000. On the other hand, if x is a negative number with small absolute value, then y is a negative number with large absolute value. For instance, if $x = -0.1, -0.01$, and -0.001, then $y = -10, -100$, and -1000. In both cases we can say that as x *approaches* 0 (*from the left or right*), *the absolute value of* $y = 1/x$ *becomes large and increases without bound* (Figure 5.11). For this reason, the vertical line $x = 0$ (i.e., the y-axis) is said to be a *vertical asymptote* of the graph of $f(x) = 1/x$.

Next let us see what happens to f as $|x|$ becomes very large. If x is a large positive number, then $y = 1/x$ is a small positive number. Moreover, as x increases without bound, y decreases and approaches 0. For instance, if $x = 10^4$, 10^9, and 10^{12}, then $y = 10^{-4}$, 10^{-9}, and 10^{-12}. We leave it to you to check that if x is negative with a large absolute value, then y is negative with a small absolute value. Summarizing these results, we can say that as $|x|$ increases without bound, the values of $y = 1/x$ approach 0 (Figure 5.12). Thus we say that the horizontal line $y = 0$ (i.e., the x-axis) is a *horizontal asymptote* of the graph of $y = 1/x$.

This discussion motivates the following definitions.

Vertical and Horizontal Asymptotes

A line $x = a$ is a *vertical asymptote* of the graph of a function $y = f(x)$ if $|f(x)|$ increases without bound as x approaches a from the left or right.

A line $y = b$ is a *horizontal asymptote* of the graph of a function $y = f(x)$ if the values $f(x)$ approach b as $|x|$ increases without bound.

EXAMPLE 1   Sketch the graph of $f(x) = \dfrac{2}{x - 5}$.

Solution: Using horizontal translation and stretching (see Section 4.3), we can obtain the graph of f by drawing the graph of $y = 1/x$, shifting it 5 units to the right, and then multiplying each ordinate by 2 (Figure 5.13). Thus the graph is merely a translation of the graph of $y = 1/x$ followed by a stretch. Note that when $x = 0$, we have $f(0) = -2/5$; so $-2/5$ is the y-intercept of the graph.

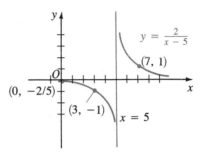

Figure 5.13

The graph indicates that the line $x = 5$ is a vertical asymptote and the line $y = 0$ is a horizontal asymptote.

To see that $x = 5$ is a vertical asymptote, note that f is not defined at 5, because at this number the denominator is zero. As x approaches 5 (from the left or right), the difference $x - 5$ approaches 0 so that $|f(x)| = 2/|x - 5|$ becomes large and increases without bound.

To see that the line $y = 0$ is a horizontal asymptote, we have to proceed in a different manner. If we divide the numerator and denominator of f by x, we can write

$$f(x) = \frac{\dfrac{2}{x}}{1 - \dfrac{5}{x}}.$$

Now we study the behavior of f as $|x|$ becomes very large. We already know that as $|x|$ increases without bound, then $1/x$ approaches 0. It follows that the two terms $2/x = 2 \cdot 1/x$ and $5/x = 5 \cdot 1/x$ also approach 0. Thus the numerator of f approaches 0 while the denominator approaches 1. Therefore f approaches 0 as $|x|$ increases without bound, and the line $y = 0$ is a horizontal asymptote.

Practice Exercise 1   Sketch the graph of $f(x) = \dfrac{3}{x + 1}$.

Answer:

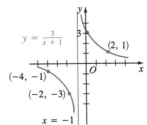

Vertical Asymptotes of the Graph of a Rational Function

The following is a general result concerning rational functions. Let

$$f(x) = \frac{P(x)}{Q(x)}$$

be a rational function. If a is a number such that $Q(a) = 0$ but $P(a) \neq 0$, then the line $x = a$ is a *vertical asymptote* of the graph of f.

EXAMPLE **2**   Discuss and sketch the graph of the function

$$f(x) = \frac{2x + 1}{x - 4}.$$

Solution: The domain of definition of this function is the set of all real numbers $x \neq 4$. When $x = 4$, the denominator is equal to zero, but the numerator is different from zero. Thus, the line $x = 4$ is a vertical asymptote of the graph.

In order to determine a horizontal asymptote, we divide (as in Example 1) both numerator and denominator of f by x:

$$f(x) = \frac{2x + 1}{x - 4} = \frac{2 + \dfrac{1}{x}}{1 - \dfrac{4}{x}}.$$

What happens to f for large values of $|x|$? As $|x|$ increases without bound, both $1/x$ and $4/x = 4 \cdot 1/x$ approach zero. It follows that the numerator of f approaches 2, while the denominator approaches 1. Thus f approaches $2/1 = 2$ as $|x|$ increases without bound and, therefore, the line $y = 2$ is a horizontal asymptote.

Now we determine the x- and y-intercepts. If $x = 0$, then $f(0) = -1/4$, so $-1/4$ is the y-intercept of the graph. To find the x-intercepts we must solve $f(x) = 0$, so we set

$$\frac{2x + 1}{x - 4} = 0.$$

For a fraction to be zero, the numerator must be zero but the denominator different from zero. Thus $2x + 1 = 0$, which gives us $x = -1/2$. Therefore $-1/2$ is the x-intercept of the graph.

Using all the preceding information and plotting a few points, we obtain the graph shown in Figure 5.14.

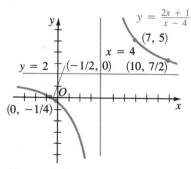

Figure 5.14

Practice Exercise 2   Discuss and graph the function $f(x) = \dfrac{x-1}{2x+4}$.

Answer:

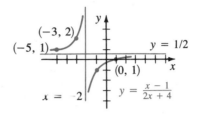

Reciprocals of Powers of *x*

When considering rational functions, it is important to study the reciprocals of powers of x. Here, we discuss the reciprocal of the square function. Consider the function

$$y = f(x) = \frac{1}{x^2},$$

whose domain is the set of all numbers $x \neq 0$. The line $x = 0$ is a vertical asymptote of the graph of $f(x) = 1/x^2$, because the denominator of f is zero when $x = 0$, while the numerator is always nonzero. The function f is even, thus the graph is symmetric with respect to the y-axis. Moreover, since $f(x)$ is always positive, the graph is located above the x-axis.

Next, we discuss what happens to $y = 1/x^2$ as $|x|$ increases without bound. If $|x|$ is a large number, then $y = 1/x^2$ is a small number. For instance, if $x = \pm 10, \pm 10^2$, and $\pm 10^6$, then $y = 10^{-2}, 10^{-4}$, and 10^{-12}, which are very small numbers.

At this point, we could compare the values of $y = 1/x^2$ with the values of $y = 1/x$. If $|x|$ is a large number, then $1/x$ is a small number and $1/x^2$ is even smaller. For example, if $x = 10^4$, then $1/x = 10^{-4}$ and $1/x^2 = 10^{-8}$, which is smaller. Thus we can say that as $|x|$ increases without bound, $1/x^2$ approaches 0 *faster* than $1/x$. These properties are all illustrated in the graph of f shown in Figure 5.15.

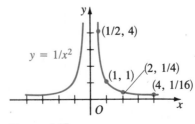

Figure 5.15

Let us make a final remark about the graph of a rational function of the form

$$f(x) = \frac{1}{x^n}.$$

If n is an *odd* number, the graph of f resembles that of $y = 1/x$. If n is an *even* number, the graph of f resembles the graph of $y = 1/x^2$.

In the next example we discuss the case of a rational function whose numerator and denominator are two simple quadratic polynomials.

EXAMPLE 3

Discuss and sketch the graph of $f(x) = \dfrac{x^2}{x^2 - 1}$.

Solution: The function f is defined for all real numbers except $x = -1$ and $x = 1$. Since the denominator is zero at $x = -1$ and $x = 1$, but the numerator is different from zero, the lines $x = -1$ and $x = 1$ are vertical asymptotes.

To find the horizontal asymptotes, divide both numerator and denominator by x^2 (the highest power of x in the expression defining f) and write

$$f(x) = \frac{x^2}{x^2 - 1} = \frac{1}{1 - \dfrac{1}{x^2}}.$$

As $|x|$ becomes large, the term $1/x^2$ approaches 0 and $f(x)$ approaches 1. Thus the line $y = 1$ is a horizontal asymptote.

Next check that 0 is both the x- and y-intercept of the graph of f.

The following table gives us a few points on the graph. We have selected only positive values of x because f is an even function ($f(x) = f(-x)$) and, as a consequence, the graph is symmetric with respect to the y-axis.

You should plot some additional points and check the graph illustrated in Figure 5.16.

x	$f(x)$
1/2	−1/3
2/3	−4/5
3/2	9/5
2	4/3

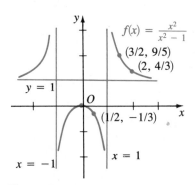

Figure 5.16

Practice Exercise 3   Discuss and sketch the graph of $f(x) = \dfrac{x^2 - 1}{x^2}$.

Answer:

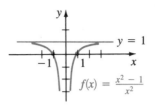

$$f(x) = \frac{x^2 - 1}{x^2}$$

Horizontal Asymptotes of the Graph of a Rational Function

We have seen in Example 2 that the horizontal asymptote of the graph of the function

$$f(x) = \frac{2x + 1}{x - 4}$$

is the line $y = 2$. Note that 2 is the quotient of 2 and 1, the leading coefficients in the numerator and denominator. Analogously, the horizontal asymptote of the graph of

$$f(x) = \frac{x^2}{x^2 - 1}$$

(Example 3) is $y = 1$. Also note that 1 is the quotient of the leading coefficients in the numerator and denominator.

These are no chance events; they are consequences of a general result about rational functions, which we now explain.

Method of Determining the Horizontal Asymptote

Let

$$f(x) = \frac{a_n x^n + \cdots + a_1 x + a_0}{b_m x^m + \cdots + b_1 x + b_0}$$

be a rational function. In order to find the horizontal asymptote of f, we have to consider three cases.

1. $n = m$. If the numerator and denominator have the *same* degree, then the line $y = a_n/b_m$ is a horizontal asymptote. Observe that a_n/b_m is the quotient of the leading coefficients of the two polynomials.
2. $n < m$. If the degree of the numerator is *smaller* than the degree of the denominator, then the line $y = 0$ is a horizontal asymptote. For example, $y = 0$ is the horizontal asymptote of the graph of $f(x) = 2/(x - 5)$.
3. $n > m$. If the degree of the numerator is *greater* than the degree of the denominator, then the graph of f has no horizontal asymptote.

These properties tell us that to find horizontal asymptotes of rational functions, we just have to compare the degrees of the numerator and denominator and, if they are equal, we look at their leading coefficients.

EXAMPLE 4   Determine the horizontal asymptote of each function.

(a) $f(x) = \dfrac{2x^2 - 5}{3x^2 - 2x - 1}$ **(b)** $f(x) = \dfrac{6x^2 - 1}{x^3 - 4x - 2}$

Solution: **(a)** Both polynomials have the same degree so that, according to case 1, the line $y = 2/3$ is the horizontal asymptote of the graph.
(b) In this case, the degree of the numerator is smaller than the degree of the denominator. According to case 2, the line $y = 0$ is the horizontal asymptote.

Practice Exercise 4   Find the horizontal asymptotes of each function.

(a) $f(x) = \dfrac{2x^3 - 1}{x^2 - x + 4}$ **(b)** $f(x) = \dfrac{6x^2 - 7}{3x^2 - 4x - 2}$

Answer: **(a)** no horizontal asymptote   **(b)** $y = 2$

EXAMPLE 5   Discuss and sketch the graph of

$$f(x) = \frac{x^2}{x^2 + 2x - 8}.$$

Solution: Since $x^2 + 2x - 8 = (x - 2)(x + 4)$, the function f is defined for all real numbers except $x = 2$ and $x = -4$. Thus the two lines $x = 2$ and $x = -4$ are vertical asymptotes. Since the numerator and denominator have the same degree, the line $y = 1/1 = 1$ is a horizontal asymptote.

Next, we determine the x- and y-intercepts. If $x = 0$ then $f(0) = 0$, so 0 is the y-intercept. Setting $f(x) = 0$ and solving for x we obtain $x = 0$, so 0 is also the x-intercept.

Now it is helpful to have an idea of the location of the graph of f. We can do this by investigating the sign of $f(x)$ as x varies in the domain of definition. In the intervals where $f(x) < 0$, the graph lies below the x-axis. To determine the sign changes of f, we can use the method of solving inequalities explained in Section 2.8. The first step is to factor the denominator of f and write

$$f(x) = \frac{x^2}{(x - 2)(x + 4)}.$$

We must now keep track of the signs of the various factors. For this purpose, we organize our work in tabular form as follows.

	For x in each of the intervals		
Sign of	$-\infty$ -4	2	$+\infty$
x^2	$+$ $+$		$+$
$x-2$	$-$ $-$	0	$+$
$x+4$	$-$ 0 $+$		$+$
$x^2/(x-2)(x+4)$	$+$ $-$		$+$

The table shows that the graph of f lies above the x-axis for all values of x in the intervals $(-\infty, -4)$ and $(2, +\infty)$, and below the x-axis for all values of x in the interval $(-4, 2)$.

Using all the available information and plotting a few points, we can sketch the graph of f as shown in Figure 5.17.

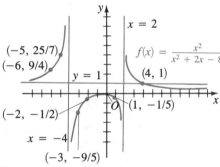

Figure 5.17

The graph in Figure 5.17 illustrates a new important feature of certain rational functions, which we now describe. If f is a rational function, its graph *never crosses a vertical asymptote*. (Why?) However, the graph *may cross a horizontal asymptote*. To find the points where the graph crosses the horizontal asymptote, set $f(x)$ equal to the number corresponding to the asymptote and solve the equation for x. For example, $y = 1$ is the horizontal asymptote of

$$f(x) = \frac{x^2}{x^2 + 2x - 8}.$$

If we set $f(x) = 1$, that is,

$$\frac{x^2}{x^2 + 2x - 8} = 1,$$

and solve for x, we obtain $x = 4$. Thus $f(4) = 1$, so $(4, 1)$ is a point that lies both on the graph of f and on the line $y = 1$.

Practice Exercise 5   Discuss and sketch the graph of

$$f(x) = \frac{x^2}{x^2 - 2x - 3}.$$

Answer:

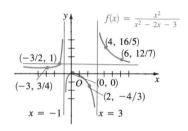

We end this section with a summary of all of the results concerning the graphing of rational functions.

Steps in Sketching the Graph of a Rational Function

To sketch the graph of a rational function $f(x)$, proceed as follows.

1. Determine the domain of the function.
2. Find any vertical and horizontal asymptotes.
3. Find all x- and y-intercepts of the graph.
4. Study the sign variation of f as x varies in the domain of definition.
5. Look for points where the graph may cross a horizontal asymptote.
6. Make a table of function values and plot the corresponding points.
7. Join the points with a smooth curve.

EXERCISES 5.3

In Exercises 1–10, find vertical and horizontal asymptotes of each function.

1. $f(x) = \dfrac{1}{2x + 1}$

2. $f(x) = \dfrac{-2}{3x - 2}$

3. $f(x) = \dfrac{2x - 4}{3x + 2}$

4. $f(x) = \dfrac{x}{4x - 5}$

5. $f(x) = \dfrac{2x^2 + 3}{x^2 - 4x + 1}$

6. $f(x) = \dfrac{3x^2 - 2x + 1}{x^2 + 4}$

7. $f(x) = \dfrac{2x + 3}{x^2 - 3x + 5}$

8. $f(x) = \dfrac{5x - 3}{3x^2 - 4x - 1}$

9. $f(x) = \dfrac{3x^2 - 4x + 1}{5x - 3}$

10. $f(x) = \dfrac{4x^3 - 2}{x^2 - 4x + 3}$

In Exercises 11–20, use the techniques of stretching, shrinking, reflecting, and/or shifting to sketch the graphs of the given functions.

11. $f(x) = \dfrac{5}{x}$

12. $f(x) = -\dfrac{3}{x}$

13. $f(x) = \dfrac{-3}{x + 5}$

14. $f(x) = \dfrac{2}{4 - x}$

15. $f(x) = 2 - \dfrac{3}{x}$

16. $f(x) = \dfrac{4}{x} + 3$

17. $f(x) = \dfrac{1}{x^2} + 3$

18. $f(x) = 2 - \dfrac{1}{x^2}$

19. $f(x) = \dfrac{2}{(x - 1)^2} + 3$

20. $f(x) = 4 - \dfrac{1}{(x + 1)^2}$

In Exercises 21–40, sketch the graph of each rational function. Find vertical and horizontal asymptotes and x- and y-intercepts. In order to improve the accuracy of your graph, you may use a calculator to determine more points on the graph.

21. $f(x) = \dfrac{-3}{x - 5}$

22. $f(x) = \dfrac{5}{1 - x}$

23. $f(x) = \dfrac{3}{1 - 2x}$

24. $f(x) = \dfrac{4}{3x - 2}$

25. $f(x) = \dfrac{x - 4}{2x + 1}$

26. $f(x) = \dfrac{2x - 1}{x + 3}$

27. $f(x) = \dfrac{x^2}{x^2 - 4}$

28. $f(x) = \dfrac{2x^2}{x^2 - 1}$

29. $f(x) = \dfrac{x^2}{x^2 + 1}$

30. $f(x) = \dfrac{-x^2}{x^2 + 1}$

31. $f(x) = \dfrac{3x}{x^2 - 4}$

32. $f(x) = \dfrac{4x}{x^2 - 1}$

33. $f(x) = \dfrac{3x - 1}{x^2 - 4}$

34. $f(x) = \dfrac{x + 3}{x^2 - 1}$

35. $f(x) = \dfrac{2x + 1}{(x - 1)(x + 3)}$

36. $f(x) = \dfrac{3x - 2}{(x - 2)(x + 1)}$

37. $f(x) = \dfrac{2x^2}{x^2 - x - 6}$

38. $f(x) = \dfrac{3x^2}{x^2 - x + 2}$

39. $f(x) = \dfrac{x^2}{x^2 - 2x + 1}$

40. $f(x) = \dfrac{x}{x^2 - 2x + 1}$

Solve Exercises 41 and 42.

☐ **41.** Sketch the graph of $F(x) = 1/x^3$ and compare it with the graph of $f(x) = 1/x$.

☐ **42.** Sketch the graph of $G(x) = 1/x^4$ and compare it with the graph of $g(x) = 1/x^2$.

Find the horizontal asymptote of the graph of each function in Exercises 43–48.

43. $f(x) = \dfrac{3x^2 - 1}{5x^2 + 3x + 2}$

44. $f(x) = \dfrac{4x^2 + x - 4}{3x^2 + 5}$

45. $f(x) = \dfrac{x^2 + 1}{3x^4 - 5}$

46. $f(x) = \dfrac{x^3 - 1}{3x^5 - 4x^2 + 1}$

47. $f(x) = \dfrac{3x^{99} - 1}{x^{100} + x^{50}}$

48. $f(x) = \dfrac{5x^{200} - 4x^{100} + 3}{6x^{200} + x^2}$

Solve Exercises 49 and 50.

☐ **49.** Let

$$f(x) = \dfrac{x^n}{3x^6 - 5x + 1}.$$

Find the horizontal asymptote of f if
(a) $n = 6$, **(b)** $n < 6$.

50. Let n be a positive integer and let

$$f(x) = \dfrac{2x^n + 5}{5x^{2n} + 6x^n + 1}.$$

Find the horizontal asymptote of f.

5.4   COMPOSITE AND INVERSE FUNCTIONS

In Chapter 4 we discussed ways of combining functions, and we defined the sum, difference, product, and quotient of functions. There is another important way of combining functions, called *composition of functions,* which we now explain. Composition of functions leads us to *inverse functions,* a concept of great significance in mathematics and its applications.

Composition of Functions

Consider the function

$$h(x) = \sqrt{3x - 2},$$

defined for all $x \geq 2/3$. The function h can be obtained from the functions $f(x) = 3x - 2$ and $g(x) = \sqrt{x}$ in the following manner. First, the function f assigns to each number x the value $3x - 2$. Next, whenever $3x - 2$ is nonnegative, we can take its square root and obtain $g(3x - 2)$, which is the value of h at x. Symbolically, we may write

$$x \xrightarrow{\;f\;} 3x - 2 \xrightarrow{\;g\;} \sqrt{3x - 2}.$$

We say that the function $h(x)$ is the *composition* of the functions $f(x)$ and $g(x)$.

With the help of a calculator we can illustrate the composition of these two functions. For example, to find the function value $h(4)$, we perform the following keystrokes:

$$4 \;\boxed{\times}\; 3 \;\boxed{-}\; 2 \;\boxed{=}\;\boxed{\sqrt{}}$$

and get the final display, 3.162277. You can see that in the sequence of keystrokes $4\,\boxed{\times}\,3\,\boxed{-}\,2\,\boxed{=}$, we have entered the number 4, multiplied it by 3, and subtracted 2 to obtain 10, which is the function value of $f(x)$ at $x = 4$. The last keystroke $\boxed{\sqrt{}}$ gives us the square root of 10, that is, $\sqrt{10} \simeq 3.162277$. Note that the keystrokes correspond exactly to the preceding diagram, which indicates the composition of f and g.

A word of caution is in order about the domain of the *composite function h.* The function f is defined for all real numbers, while g is defined only for nonnegative numbers. Thus the function h is defined for all real numbers x satisfying the inequality

$$3x - 2 \geq 0,$$

that is, for all $x \geq 2/3$. Thus the composition of f and g *is defined only for those values of x such that $f(x)$ lies in the domain of g.*

Composite Functions

Given two functions f and g, the *composite function $g \circ f$* is the function defined by

$$(g \circ f)(x) = g(f(x)),$$

for all x in the domain of f such that $f(x)$ lies in the domain of g (Figure 5.18).

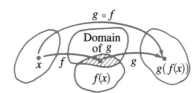

Domain of f Range of f Range of g
Figure 5.18

When we define a composite function, the order in which the functions are combined may affect the final result. For example, if $f(x) = 3x - 2$ and $g(x) = \sqrt{x}$, then the composite function $k = f \circ g$ is

$$k(x) = (f \circ g)(x) = f(g(x)) = 3\sqrt{x} - 2,$$

which is defined for all $x \geq 0$. The two composite functions,

$$h(x) = \sqrt{3x - 2} \quad \text{and} \quad k(x) = 3\sqrt{x} - 2,$$

are clearly distinct. For example, $k(4) = 4 \neq h(4) = \sqrt{10}$.

EXAMPLE 1   **(a)** If $f(x) = x^2 - 4$ and $g(x) = 1/x$, find $h(x) = (g \circ f)(x)$ and $k(x) = (f \circ g)(x)$, and determine the domain of definition of each function.
(b) Evaluate $(g \circ f)(-3)$ and $(f \circ g)(2)$.

Solution: **(a)** We have

$$h(x) = g(f(x)) = g(x^2 - 4) = \frac{1}{x^2 - 4}$$

and the domain of definition of h is the set $\{x \in \mathbb{R} \colon x \neq -2, \ x \neq 2\}$.
   On the other hand,

$$k(x) = f(g(x)) = f\left(\frac{1}{x}\right) = \left(\frac{1}{x}\right)^2 - 4 = \frac{1}{x^2} - 4,$$

defined for all numbers $x \neq 0$.
(b) Since $(g \circ f)(x) = h(x) = 1/(x^2 - 4)$, we see that

$$(g \circ f)(-3) = \frac{1}{(-3)^2 - 4} = \frac{1}{9 - 4} = \frac{1}{5}.$$

Since $(f \circ g)(x) = k(x) = (1/x^2) - 4$, it follows that

$$(f \circ g)(2) = \frac{1}{2^2} - 4 = \frac{1}{4} - 4 = -\frac{15}{4}.$$

Practice Exercise 1   **(a)** Let $f(x) = 2x - 4$ and $g(x) = \sqrt{x + 3}$. Find $(g \circ f)(x)$ and $(f \circ g)(x)$, and determine their domains of definition.
(b) Evaluate $(g \circ f)(5)$ and $(f \circ g)(6)$.

Answer: **(a)** $(g \circ f)(x) = \sqrt{2x - 1}$ for $x \geq 1/2$
and $(f \circ g)(x) = 2\sqrt{x + 3} - 4$ for $x \geq -3$
(b) 3, 2

One-to-One Functions

We have seen that a function f assigns to each x in its domain a unique value $y = f(x)$, the image of x under f. However, different elements in the domain of f may have the same image. For example, if $f(x) = x^2$, then $x = -2$ and $x = 2$ have the same image: $f(-2) = f(2) = 4$.

For some functions, however, each element in the range is the image of *exactly one* element in the domain. These are said to be *one-to-one functions*.

One-to-One Function

A function $y = f(x)$ is one-to-one if whenever $f(x_1) = f(x_2)$, then $x_1 = x_2$.
In other words, $f(x)$ is one-to-one if $x_1 \neq x_2$ implies $f(x_1) \neq f(x_2)$.

For example, the function $f(x) = x^3$ is one-to-one on $\mathbb{R}$, while the function $f(x) = x^2$ is not. This can be seen by looking at the graphs of both functions shown in Figure 5.19. We have already remarked that every vertical line through a point in the domain of a function intersects the graph at a single point. However, a horizontal line may intersect the graph in more than one point. Figure 5.19a shows a horizontal line intersecting the graph of $f(x) = x^2$ in two points. The x-coordinates of these points, x_1 and x_2, are such that $x_1^2 = x_2^2$; thus $f(x) = x^2$ is not a one-to-one function. Figure 5.19b indicates that every horizontal line intersects the graph of $f(x) = x^3$ in a single point. Thus $f(x) = x^3$ is a one-to-one function.

We now give the *horizontal line test*, which tells us whether or not a function is one-to-one.

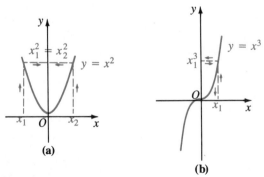

(a)

(b)

Figure 5.19

> ### Horizontal Line Test
>
> A function is one-to-one if each horizontal line intersects its graph in at most one point (Figure 5.20).

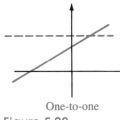
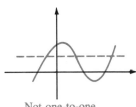

One-to-one   Not one-to-one

Figure 5.20

It is easy to see that every *increasing* or *decreasing* function is one-to-one.

EXAMPLE 2   Determine whether each of the following functions is one-to-one.
(a) $f(x) = 4x - 1$ **(b)** $g(x) = 5 - 3x^2$

Solution: **(a)** We may apply the definition of a one-to-one function. If $x_1 \neq x_2$, then $4x_1 \neq 4x_2$ and $4x_1 - 1 \neq 4x_2 - 1$. Thus $x_1 \neq x_2$ implies $f(x_1) \neq f(x_2)$, so that f is one-to-one. Note that $f(x) = 4x - 1$ is an increasing function.
(b) For $x = -1$ and $x = 1$, we have

$$g(-1) = 5 - 3 \cdot (-1)^2 = 2 = 5 - 3 \cdot (1)^2 = g(1);$$

that is, the function values of g at $x = -1$ and $x = 1$ are the same. Thus g is not one-to-one.

We could also apply the horizontal line test to show that f is one-to-one but g is not.

Practice Exercise 2   Which of the following functions is one-to-one?
(a) $f(x) = 3 - 4x$ **(b)** $g(x) = 3x^2 + 2$.

Answer: **(a)** f is one-to-one (decreasing)   **(b)** g is not one-to-one

Inverse Functions

Let f be a function with domain X and range Y. If f is a one-to-one function, then every element y in Y is the image under f of only one element x in X. Thus we can define a new function g by assigning to each y in Y the unique x in X such that $y = f(x)$, as shown by Figure 5.21. This new function g with *domain Y* and *range X* is called the *inverse* of f.

According to this definition, we can say that

$$y = f(x) \qquad \text{if and only if} \qquad x = g(y).$$

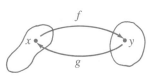

Figure 5.21

In other words, if the function f assigns y in Y to an element x in X, then the function

g assigns x to this element y. Conversely, if y is an element in Y and x its image under g, then the image of x under f is y.

We can summarize these results as follows.

Two Properties Concerning a Function and Its Inverse

If f is a one-to-one function with domain X and range Y and g is the inverse of f, we have

$$(g \circ f)(x) = x \qquad \text{for every } x \in X,$$
$$(f \circ g)(y) = y \qquad \text{for every } y \in Y.$$

These two relations say that if two functions are inverses of each other, then each function "undoes" what the other "does."

Notation

If g is the inverse of f, we often denote g by f^{-1} and read "f-inverse." With this notation, we may rewrite the two properties above as follows.

$$(f^{-1} \circ f)(x) = x \qquad \text{for every } x \in X,$$
$$(f \circ f^{-1})(y) = y \qquad \text{for every } y \in Y.$$

In certain cases, we can find the inverse of a function by solving an equation associated with the function.

EXAMPLE 3   Find the inverse of the function $f(x) = 4x - 1$.

Solution: As we already know (Example 2), f is a one-to-one function, so it has an inverse function g. To find the inverse, set $y = f(x)$, obtaining the equation

$$y = 4x - 1.$$

Solve this equation for x:

$$4x - 1 = y$$
$$4x = y + 1$$
$$x = \frac{y + 1}{4}.$$

The function

$$g(y) = \frac{y + 1}{4}$$

is the inverse of f. This can be verified by checking the composition of f and g. We have

$$(g \circ f)(x) = g(f(x)) = \frac{f(x) + 1}{4} = \frac{4x - 1 + 1}{4} = x$$

and

$$(f \circ g)(y) = f(g(y)) = 4g(y) - 1 = 4\left(\frac{y + 1}{4}\right) - 1 = y.$$

Since it is customary to use x as the independent variable of a function, we write g as follows:

$$g(x) = \frac{x + 1}{4}.$$

Practice Exercise 3   Find the inverse of the function $f(x) = \dfrac{3x + 1}{5}$.

Answer: $g(x) = \dfrac{5x - 1}{3}$

We now summarize the method described in Example 3.

Algebraic Method for Finding the Inverse

Let $f(x)$ be a one-to-one function. To find the inverse g of f, proceed as follows.

1. Set $y = f(x)$.
2. Solve this equation for x and get $x = g(y)$.
3. Interchange x and y so that the last equation becomes $y = g(x)$.
4. Check that $(g \circ f)(x) = x$ and $(f \circ g)(x) = x$.

EXAMPLE 4   Show that the functions $f(x) = 3 - 4x$ and $g(x) = (3 - x)/4$ are inverses of each other and sketch their graphs on the same coordinate plane.

Solution: To show that f and g are inverses of each other, we must show that

$$(g \circ f)(x) = x \qquad \text{and} \qquad (f \circ g)(x) = x$$

for all x. To verify this, we have

$$(g \circ f)(x) = g(f(x)) = \frac{3 - f(x)}{4} = \frac{3 - (3 - 4x)}{4} = x$$

and

$$(f \circ g)(x) = f(g(x)) = 3 - 4g(x) = 3 - 4\left(\frac{3 - x}{4}\right) = x.$$

Thus f and g are inverses of each other. Now we sketch the graph of both functions in Figure 5.22.

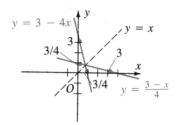

Figure 5.22

Practice Exercise 4   Show that $f(x) = 2x - 4$ and $g(x) = (x/2) + 2$ are inverses of each other. Graph these two functions on the same coordinate plane.

Answer:

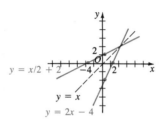

EXAMPLE 5   Show that the functions $f(x) = x^2$ and $g(x) = \sqrt{x}$, for $x \geq 0$, are inverses of each other. Graph these functions on the same coordinate plane.

Solution: We have

$$(g \circ f)(x) = g(f(x)) = \sqrt{x^2} = x, \quad \text{for all } x \geq 0,$$

and

$$(f \circ g)(x) = f(g(x)) = (\sqrt{x})^2 = x, \quad \text{for all } x \geq 0.$$

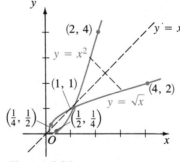

Figure 5.23

Practice Exercise 5   Show that the functions $f(x) = \sqrt{x - 1}$, for $x \geq 1$, and $g(x) = x^2 + 1$, for

$x \geq 0$, are inverses of each other. Graph both functions on the same coordinate plane.

Answer:

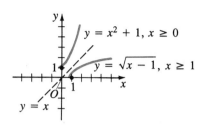

The Graph of an Inverse Function

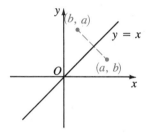

Figure 5.24

Consider the two functions $f(x) = 3 - 4x$ and $g(x) = (3 - x)/4$ of Example 4. Their graphs appear to be symmetric to each other with respect to the line $y = x$ (Figure 5.22). The same can be said of the graphs of $f(x) = x^2$ and $g(x) = \sqrt{x}$, for $x \geq 0$, shown in Figure 5.23.

The two examples illustrate a general fact about the graphs of two functions that are inverses of each other. Suppose that f^{-1} is the inverse of f; let (a, b) be an arbitrary point belonging to the graph of f. This means that $b = f(a)$. Since f^{-1} is the inverse of f, it follows that $a = f^{-1}(b)$, and the point (b, a) belongs to the graph of f^{-1}. Now, as shown in Figure 5.24, the points (a, b) and (b, a) are symmetric to each other with respect to the line $y = x$. This implies that the graph of f^{-1} can be obtained by *reflecting the graph of f about the line $y = x$.*

We may state the following result.

Symmetry about the Line $y = x$

The graphs of f and f^{-1} are symmetric to each other about the line $y = x$.

EXAMPLE 6

Find the inverse f^{-1} of the function $f(x) = 4 - x^2$ for $x \geq 0$. Graph f and f^{-1} on the same coordinate plane.

Solution: We set $y = f(x)$. The function f is defined for $x \geq 0$ and its range is the set of all numbers $y \leq 4$. (Can you see why?) Next we solve the equation for x,

$$y = 4 - x^2$$
$$x^2 = 4 - y$$
$$x = \sqrt{4 - y},$$

where $\sqrt{}$ denotes the principal square root. Thus

$$f^{-1}(y) = \sqrt{4 - y} \qquad (y \leq 4)$$

is the inverse of f. Changing y into x, we write the inverse as

$$y = f^{-1}(x) = \sqrt{4 - x} \qquad (x \leq 4).$$

Now we, graph the two functions. We start with $f(x) = 4 - x^2, x \geq 0$, whose graph is part of a parabola (Figure 5.25). We obtain the graph of f^{-1} by reflecting the graph of f about the line $y = x$.

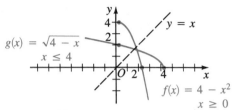

Figure 5.25

Practice Exercise 6   Find the inverse $f^{-1}(x)$ of the function $f(x) = \sqrt{4 + x}$, for $x \geq -4$. Graph both functions on the same coordinate plane.

Answer:

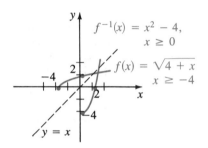

EXERCISES 5.4

In Exercises 1–8, find $(f \circ g)(x)$, $(g \circ f)(x)$, and $(f \circ f)(x)$.

1. $f(x) = 2x + 3, g(x) = 3x - 4$
2. $f(x) = 4x^2 - 3x, g(x) = 5x - 4$
3. $f(x) = 3x + 1, g(x) = \sqrt{x + 5}$
4. $f(x) = x^2 - 9, g(x) = \dfrac{1}{x + 1}$
5. $f(x) = 2x - 5, g(x) = \dfrac{1}{x^2 + 2}$
6. $f(x) = 2x^2 - 1, g(x) = \sqrt{2x + 1}$
7. $f(x) = x^3 + 1, g(x) = \sqrt[3]{x - 1}$
8. $f(x) = 2x - 3, g(x) = \dfrac{3x}{x - 4}$

In Exercises 9–20, let $f(x) = \sqrt{x + 2}$ and $g(x) = x^2 - 4$. Find the indicated function values.

9. $(f \circ g)(\sqrt{6})$
10. $(f \circ g)(2\sqrt{3})$
11. $(f \circ g)(1.8)$
12. $(f \circ g)(3.5)$

13. $(g \circ f)(\sqrt{3.25})$
14. $(g \circ f)(\sqrt{4.5})$
15. $(g \circ f)(34)$
16. $(g \circ f)(100)$
17. $(f \circ f)(23)$
18. $(f \circ f)(-7/4)$
19. $(g \circ g)(-\sqrt{2})$
20. $(g \circ g)(2)$

In Exercises 21–26, express each function as a composition of two functions chosen from $f(x) = \sqrt{x}$, $g(x) = x^2 + 1$, and $h(x) = 5x - 2$.

21. $F(x) = \sqrt{5x - 2}$
22. $G(x) = x + 1$
23. $H(x) = 25x^2 - 20x + 5$
24. $K(x) = \sqrt{x^2 + 1}$
25. $U(x) = x^4 + 2x^2 + 2$
26. $V(x) = 5\sqrt{x} - 2$

In Exercises 27–34, find the inverse of f and determine its domain of definition.

27. $f(x) = \dfrac{1}{2 - x}, \quad x \neq 2$

28. $f(x) = \dfrac{-2}{3x + 1}, \quad x \neq -\dfrac{1}{3}$

29. $f(x) = x^2 - 4, \quad x \geq 0$ **30.** $f(x) = x^2 - 4, \quad x \leq 0$

31. $f(x) = \sqrt{4x - 1}, \quad x \geq \dfrac{1}{4}$

32. $f(x) = \sqrt{2x - 3}, \quad x \geq \dfrac{3}{2}$

33. $f(x) = \sqrt[3]{2x - 4}$ **34.** $f(x) = x^3 - 5$

In Exercises 35–38, show that each of the following functions is its own inverse. What can you say about their graphs?

35. $f(x) = \dfrac{x}{x - 1}$ **36.** $g(x) = \dfrac{2x + 1}{x - 2}$

37. $F(x) = \dfrac{x + 1}{2x - 1}$ **38.** $G(x) = \dfrac{1 - 2x}{3x + 2}$

In Exercises 39–44, show that the given functions are inverses of each other and sketch their graphs on the same coordinate plane.

39. $f(x) = 4x + 5; \; g(x) = \dfrac{x - 5}{4}$

40. $f(x) = 2 - 3x; \; g(x) = \dfrac{2 - x}{3}$

41. $f(x) = \dfrac{1}{x - 1}, \, x \neq 1; \; g(x) = \dfrac{x + 1}{x}, \, x \neq 0$

42. $f(x) = \dfrac{1}{2 - x}, \, x \neq 2; \; g(x) = \dfrac{2x - 1}{x}, \, x \neq 0$

43. $f(x) = \sqrt{3x - 1}, \, x \geq \dfrac{1}{3}; \; g(x) = \dfrac{x^2 + 1}{3}, \, x \geq 0$

44. $f(x) = x^2 + 3, \, x \leq 0; \; g(x) = -\sqrt{x - 3}, \, x \geq 3$

In Exercises 45–50, solve each problem.

45. The demand D for a certain brand of toaster is given by $D = 300 - 0.002p^2$, where p is the price (in dollars) of each toaster. If the price p, as a function of the cost c (in dollars), is given by $p = 3c - 20$, find the demand in terms of the cost.

46. Suppose that the total number of units Q produced daily by a manufacturer is given by the following function of the number n of employees:

$$Q(n) = 6n - \dfrac{n^2}{10}.$$

Suppose also that the total revenue R received for selling Q units of the product is given by $R = 30Q$. Find $(R \circ Q)(n)$. What does this function represent?

47. Let f and f^{-1} be inverses of each other. Show that (a, b) belongs to the graph of f if and only if (b, a) belongs to the graph of f^{-1}.

48. Under what conditions on m and b does the linear function $f(x) = mx + b$ have an inverse?

49. Let $f(x) = mx + b$ and $g(x) = nx + d$. Under what conditions on m, b, n, and d do we have $f \circ g = g \circ f$?

50. Find a formula for the inverse of the function

$$f(x) = \dfrac{ax + b}{cx + d},$$

where a, b, c, and d are constants such that $ad - bc \neq 0$. Why is this restriction necessary?

CHAPTER SUMMARY

Polynomial functions are the simplest functions that can be defined by using algebraic operations. As you already know, the graph of a *linear function*

$$f(x) = ax + b$$

is a *straight line*. The graph of a quadratic function

$$f(x) = ax^2 + bx + c$$

is a *parabola*. All the information needed to sketch the parabola can be derived directly from the quadratic polynomial. For example:

$a > 0$	the parabola *opens upward*,
$a < 0$	the parabola *opens downward*.

The *discriminant*

$$\Delta = b^2 - 4ac$$

tells you how many x-intercepts the parabola may have.

$\Delta > 0$	two x-intercepts
$\Delta = 0$	parabola tangent to the x-axis
$\Delta < 0$	no x-intercepts

By completing the square, we can always write the quadratic function in the form

$$f(x) = a(x - h)^2 + k,$$

where $h = -b/2a$ and $k = -\Delta/4a^2$. From this we can extract additional information.

> The *vertex* has coordinates (h, k).
> The equation of the *axis of symmetry* is $x = h$.

If the parabola opens downward, the vertex is its highest point (*maximum*); if it opens upward, the vertex is its lowest point (*minimum*).

The task of graphing a polynomial of degree greater than 2 is considerably simplified if we can express the polynomial as a product of linear factors. Refer to the Guidelines for Graphing Polynomial Functions at the end of Section 5.2.

To graph a *rational function* you should first find the domain of definition and then determine *vertical* and *horizontal asymptotes*. Follow the Steps in Sketching the Graph of a Rational Function listed at the end of Section 5.3.

If f and g are two functions, the *composite function*

$$(g \circ f)(x) = g(f(x))$$

is defined for all x in the domain of f for which $f(x)$ lies in the domain of g.

Two functions f and g are *inverses* of each other if and only if

> $(g \circ f)(x) = x$ for all x in the domain of f
>   and
> $(f \circ g)(x) = x$ for all x in the domain of g.

The inverse of a function f is denoted by f^{-1}. Only *one-to-one functions* have inverses. In particular, *increasing* or *decreasing* functions have inverses. If the graph of a function is known, you may use the *horizontal line test* to check whether or not the function has an inverse. In simple cases, you may be able to find the inverse of a function just by solving an equation associated with the function. This is the *algebraic method for finding the inverse*. If two functions f and g are inverses of each other, then their graphs are *symmetric to each other about the line $y = x$*.

CHAPTER TEST

1. A parabola has an equation $y = 2(x + 4)^2 + 5$. What are the coordinates of its vertex and the equation of its axis of symmetry?
2. Graph the parabola with the equation $f(x) = x^2 + 4x + 3$, and give the coordinates of the vertex.
3. For what values of c does the parabola $y = 2x^2 + 3x + c$ have no x-intercepts?
4. Graph the polynomial $f(x) = (x - 1)(x + 1)(x + 2)$.
5. Graph the rational function $f(x) = \dfrac{3x + 1}{x - 3}$.
6. Find the vertical and horizontal asymptotes for the function

$$f(x) = \frac{4x^2 - 1}{x^2 - 7x + 12}.$$

7. If $f(x) = \sqrt{-x - 3}$ and $g(x) = 3x + 3$, find $(f \circ g)(x)$.
8. What is the function value $(f \circ g)(1)$ if $f(t) = 5t - 4$ and $g(t) = 1 - t^2$?

9. Find the inverse of the function $f(x) = \dfrac{4x - 4}{x + 2}$ and determine the domain of the inverse.
10. Let $f(x) = -6 - 5x^2$ be defined for $x \geq 0$. Find $f^{-1}(x)$ and determine its domain.

In Exercises 1–6, graph each of the following quadratic functions. Find x- and y-intercepts, and determine the axis of symmetry and the vertex.

1. $f(x) = 8 - 2x^2$ **2.** $f(x) = 27 - 3x^2$

3. $f(x) = 3x^2 + x - 4$ **4.** $f(x) = 4x^2 + 11x - 3$

5. $f(x) = -x^2 + 2x - 5$ **6.** $f(x) = -3x^2 + 2x - 4$

In Exercises 7–16, graph each polynomial function.

7. $f(x) = 3x^2(x + 2)$ **8.** $f(x) = 3x(x + 4)^2$

9. $f(x) = (x - 2)(x - 3)(x + 2)$

10. $f(x) = (x - 4)(x - 1)(x + 3)$

11. $f(x) = 2x^3 - x^2 - 8x + 4$

12. $f(x) = 3x^3 - 2x^2 - 3x + 2$

13. $f(x) = x(x + 1)^3$ **14.** $f(x) = x^3(x - 2)$

15. $f(x) = 16x^2 - x^4$ **16.** $f(x) = x^4 - 4x^2$

C In Exercises 17 and 18, sketch the graph of each polynomial function. If necessary, use a calculator to determine several points on each graph.

17. $f(x) = x^3 + 2x + 1$

18. $f(x) = 2x^3 - x^2 + 2x - 1$

In Exercises 19–32, graph each rational function. Find horizontal and vertical asymptotes and x- and y-intercepts.

C In order to improve the accuracy of your graph, use a calculator to plot more points of the graph.

19. $f(x) = \dfrac{5}{x + 4}$ **20.** $f(x) = \dfrac{-2}{x + 4}$

21. $f(x) = \dfrac{x + 1}{x - 4}$ **22.** $f(x) = \dfrac{x - 3}{x + 5}$

23. $f(x) = \dfrac{2x + 3}{x - 2}$ **24.** $f(x) = \dfrac{x - 1}{3x + 4}$

25. $f(x) = \dfrac{x^2}{3x^2 + 1}$ **26.** $f(x) = \dfrac{4x^2}{x^2 + 1}$

27. $f(x) = \dfrac{2x}{x^2 - 9}$ **28.** $f(x) = \dfrac{3x}{x^2 - 16}$

29. $f(x) = \dfrac{x - 1}{x^2 - 16}$ **30.** $f(x) = \dfrac{x + 2}{x^2 - 9}$

31. $f(x) = \dfrac{x^2}{x^2 + 5x + 4}$ **32.** $f(x) = \dfrac{x^2}{x^2 - 5x + 4}$

In Exercises 33–38, find the horizontal asymptotes of the following functions.

33. $f(x) = \dfrac{3x - 4}{2x + 5}$ **34.** $f(x) = \dfrac{5x - 2}{3x + 4}$

35. $f(x) = \dfrac{4x^2 - 5}{3x^2 - 6x - 1}$ **36.** $f(x) = \dfrac{5x^2 - 3}{4x^2 - 3x + 1}$

37. $f(x) = \dfrac{3x + 7}{x^3 - 6x + 1}$ **38.** $f(x) = \dfrac{8x^2 - 4}{5x^3 - 4x - 2}$

In Exercises 39–42, find $(f \circ g)(x)$, $(g \circ f)(x)$, and $(f \circ f)(x)$.

39. $f(x) = \dfrac{1}{3x + 1}$, $g(x) = \sqrt{2x - 3}$

40. $f(x) = \dfrac{4x + 1}{x - 4}$, $g(x) = \dfrac{1}{x^2}$

41. $f(x) = \dfrac{x + 1}{x - 3}$, $g(x) = \dfrac{3x + 1}{x - 1}$

42. $f(x) = \dfrac{2x - 1}{x + 2}$, $g(x) = \dfrac{2x + 1}{2 - x}$

In Exercises 43–46, show that the given functions are inverses of each other. Graph the functions on the same coordinate plane.

43. $f(x) = \dfrac{2x + 1}{x}$, $g(x) = \dfrac{1}{x - 2}$

44. $u(x) = \dfrac{1}{1 - 3x}$, $v(x) = \dfrac{x - 1}{3x}$

45. $F(x) = \sqrt{3 - x}$, $x \le 3$, $G(x) = 3 - x^2$, $x \ge 0$

46. $f(x) = \sqrt{3x - 2}$, $x \ge \dfrac{2}{3}$, $g(x) = \dfrac{x^2 + 2}{3}$, $x \ge 0$

For each function in Exercises 47–52, find the inverse and determine its domain.

47. $f(x) = \dfrac{2}{3x - 5}$ **48.** $f(x) = \dfrac{2x + 1}{x - 3}$

49. $f(x) = \sqrt{3x - 4}$, $x \ge \dfrac{4}{3}$

50. $f(x) = 4 - x^2$, $0 \le x \le 2$

51. $f(x) = 9 - x^2$, $-3 \le x \le 0$

52. $f(x) = \sqrt[3]{2x - 1}$

In Exercises 53–60, solve the given problem.

53. Show that if f and g are odd functions, then $f \circ g$ is an odd function.

54. Show that if f is an even function and g is an odd function, then $f \circ g$ is an even function.

55. Find the horizontal asymptote of the graph of

$$F(x) = \frac{x^n + 1}{1 - x^n} \text{ (} n \text{ a positive integer)}.$$

56. What is the horizontal asymptote of the graph of

$$G(x) = \frac{3x^n}{2x^{100} + 51}$$

when: **(a)** $n < 100$; **(b)** $n = 100$? What happens when $n > 100$?

57. The price P (in dollars) of a certain product is given by the equation

$$P = 100 - 5D^2,$$

where D is the demand (in hundreds of units). Express P as a function of the number t of years, if the demand has risen according to the equation

$$D = 3 + 2\sqrt{t}.$$

58. Show that if the linear function $f(x) = mx + b$ has an inverse, then the inverse is also a linear function.

59. Let $f(x) = mx + b$ and $g(x) = nx + d$. Under what conditions on m, n, b, and d are f and g inverses of each other?

60. Let

$$f(x) = \frac{ax + b}{cx + d},$$

where a, b, c, and d are constants such that $ad - bc \neq 0$. For what values of these constants is $f(x)$ its own inverse?

CHAPTER 6

EXPONENTIAL

AND

LOGARITHMIC

FUNCTIONS

Up to now we have considered only functions that are formed by a finite number of algebraic operations—sum, difference, product, quotient, and root extraction—on the constant and identity functions. Such functions are called *algebraic functions* and include among others polynomial and rational functions. A function that is not algebraic is called a *transcendental function*. In this chapter we study two of them: the exponential and logarithmic functions. Both play fundamental roles in mathematics and the applied sciences. Exponential functions describe certain growth and decay phenomena studied in biology, economics, and physics. Logarithmic functions are defined as inverses of exponential functions. Among them, the common logarithm (base 10 logarithm) was an important aid in numerical computations before the availability of calculators. The natural logarithm (base *e* logarithm) is of fundamental importance in both theoretical and applied mathematics.

 ## 6.1 EXPONENTIAL FUNCTIONS

It has been observed that, under favorable conditions, the number of bacteria in a culture doubles after a fixed period of time, independent of the size of the culture with which the experiment was started, and independent of the time when the first observation was made.

Suppose, for example, that the number of bacteria in a culture doubles every hour. If initially there are n_0 bacteria present, then the number of bacteria after one hour will be $n_0 2$. Since the number doubles again in another hour, two hours after starting the observation, the number of bacteria will be $n_0 2^2$. It follows similarly

243

that for each integer t, after t hours the number of bacteria in the culture will be given by the expression

$$n(t) = n_0 2^t.$$

Thus we obtain a function defined for *positive integral values* of t. But, according to the definition of rational exponents (Section 1.8), this function is also defined for *rational values* of t. For instance, we know that $2^{1/2} = \sqrt{2}$ and $2^{5/2} = \sqrt{2^5} = 4\sqrt{2}$, so it is reasonable to expect that the numbers

$$n\left(\frac{1}{2}\right) = n_0\sqrt{2} \quad \text{and} \quad n\left(\frac{5}{2}\right) = n_0 4\sqrt{2}$$

represent the number of bacteria in the culture after $1/2$ hour and $5/2$ hours, respectively.

Can the function we defined for rational values of t be defined for *all* real values of t? In other words, can we give a meaning to expressions like $2^{\sqrt{2}}$ or 2^{π}? The answer, which is affirmative, depends on concepts and properties of the real number system that can be fully explained only in more advanced courses. Nevertheless, we describe in intuitive terms what is involved in defining the number $2^{\sqrt{2}}$.

First, recall that $\sqrt{2}$ is an irrational number and, as such, has a nonterminating decimal representation $1.4142136\ldots$. Next, consider the rational numbers

$$1 < 1.4 < 1.41 < 1.414 < 1.4142 < 1.41421 < \cdots.$$

They are *decimal approximations* of $\sqrt{2}$ and form an *increasing sequence* of rational numbers. (Sequences are studied in Chapter 12.)

Given a rational number, such as 1.41, it follows from the definition of rational exponents that

$$2^{1.41} = 2^{141/100} = \sqrt[100]{2^{141}}.$$

An approximation to this power of 2 can be obtained with the help of calculator by performing the following keystrokes: $2\;\boxed{y^x}\;1.41\;\boxed{=}\;2.6573716$. Table 6.1 was obtained by using a calculator and rounding off the results to five decimal places.

TABLE 6.1

t	2^t
1	2
1.4	2.63902
1.41	2.65737
1.414	2.66475
1.4142	2.66512
1.41421	2.66514
⋮	⋮

The table indicates that the sequence of powers of 2

$$2 < 2^{1.4} < 2^{1.41} < 2^{1.414} < 2^{1.4142} < 2^{1.41421} < \cdots$$

increases as the exponents get closer to $\sqrt{2}$. Moreover, it can be proved that this sequence of powers of 2 approximates, to any desired degree of accuracy, a real number that we denote by $2^{\sqrt{2}}$. As an exercise, you should compare the approximations in the right column of the table above with the approximation

$$2^{\sqrt{2}} \simeq 2.6651441,$$

which is obtained by using the square root and power keys of a calculator.

Similarly, it is possible to give a meaning to the number 2^{π} by using decimal approximations of π. Generally, any real number t can be approximated to any degree of accuracy by an increasing sequence $r_1 < r_2 < \cdots < r_n < \cdots$ of rational numbers. (This is a fundamental property of the real number system.) The number 2^t can then be approximated to any degree of accuracy by the sequence of powers

$$2^{r_1} < 2^{r_2} < \cdots < 2^{r_n} < \cdots.$$

The previous discussion shows that

$$f(x) = 2^x$$

is a function defined for *all* real numbers x not merely for rational numbers x. The function defined in this way is called *the exponential function with base 2*. The graph of this function is illustrated in Figure 6.1. In the table accompanying the graph, we show function values corresponding to some integral values of x. This has been done in order to simplify the plotting of points and because they suffice to give us an idea of the curve.

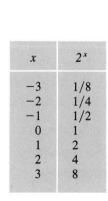

x	2^x
-3	$1/8$
-2	$1/4$
-1	$1/2$
0	1
1	2
2	4
3	8

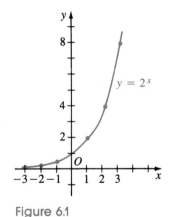

Figure 6.1

Since the function values of $f(x) = 2^x$ are always positive, the graph lies above the x-axis. Increasing values of x correspond to increasing values of 2^x; that is, if $x_1 < x_2$, then $2^{x_1} < 2^{x_2}$. Thus $f(x) = 2^x$ is an *increasing function* of x. Moreover, as x increases in the positive direction, the values of 2^x increase without bound. On the other hand, as x decreases in the negative direction the values of 2^x approach 0.

For future reference, we also sketch the graph of the function

$$f(x) = 2^{-x}$$

(Figure 6.2), which you should compare to the graph of $f(x) = 2^x$. Note that $2^{-x} = \dfrac{1}{2^x} = \left(\dfrac{1}{2}\right)^x$.

x	2^{-x}
-3	8
-2	4
-1	2
0	1
1	$\dfrac{1}{2}$
2	$\dfrac{1}{4}$
3	$\dfrac{1}{8}$

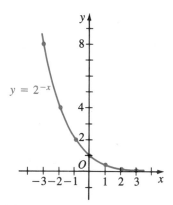

Figure 6.2

We see that the graph of $f(x) = 2^{-x}$ is the reflection about the y-axis of the graph of $f(x) = 2^x$. Also, observe that increasing values of x correspond to decreasing values of 2^{-x}. Thus, $f(x) = 2^{-x}$ is a *decreasing function*. As x increases in the positive direction, the values of 2^{-x} decrease and approach 0. As x moves leftward along the x-axis, the values of $f(x) = 2^{-x}$ increase without bound.

Exponential Functions

Let $a > 0$ and $a \neq 1$. The exponential function with *base a* is defined by

$$f(x) = a^x,$$

where x is a real number.

If $x = n$ is a natural number, then

$$f(n) = a^n = a \cdot a \cdots a \quad (n \text{ times}).$$

If $x = p/q$ is a positive rational number, then

$$a^x = a^{p/q} = \sqrt[q]{a^p}$$

and

$$a^{-x} = a^{-p/q} = \frac{1}{\sqrt[q]{a^p}}.$$

If x is an irrational number approximated by a sequence $r_1, r_2, \ldots$ of rational numbers, then the powers $a^{r_1}, a^{r_2}, \ldots$ are defined and give approximations of a^x.

Remark

In the definition of exponential functions, we have assumed that the base a is positive and not equal to 1, for the following reasons.

If $a = 1$, then $1^x = 1$ for each value of x, and the exponential function becomes trivial: it coincides with the constant function $f(x) = 1$.

If $a < 0$, then the exponential function is not defined for most values of x. For example, if $a = -3$ and $x = 1/2$, then $(-3)^{1/2} = \sqrt{-3}$ is a (nonreal) complex number, but we are considering only real-valued functions.

Properties of Exponential Functions

The exponential function $f(x) = a^x$, for $a > 0$, $a \neq 1$, satisfies the following properties.

   I. $a^x a^y = a^{x+y}$
  II. $(a^x)^y = a^{xy}$
 III. $a^{-x} = 1/a^x$
 IV. If $a > 1$ and $x < y$, then $a^x < a^y$.
  V. If $0 < a < 1$ and $x < y$, then $a^x > a^y$.
 VI. $a^x = a^y$ if and only if $x = y$.

We already know that properties I, II, and III are true when the exponents x and y are rational numbers (Section 1.8). Now, we are saying that these properties extend to the case of real exponents.

Property IV states that if $a > 1$, then the exponential function $f(x) = a^x$ is an *increasing function*. On the other hand, if $0 < a < 1$, property V states that $f(x) = a^x$ is a *decreasing function*.

Property VI says that every exponential function $f(x) = a^x$, $a > 0$, $a \neq 1$, is a one-to-one function. Proofs of properties I through VI will not be given here, because they require more advanced mathematics that fall beyond the scope of this book.

The graphs of exponential functions corresponding to both the case $a > 0$ and the case $0 < a < 1$ are illustrated in Figure 6.3. Since a^0 is always equal to 1 for all $a \neq 0$, we see that both graphs cross the y-axis at the point $(0, 1)$.

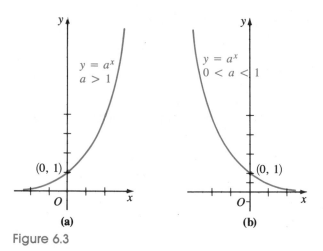

Figure 6.3

When $a > 1$, as the values of x increase without bound, the corresponding values a^x also increase without bound. As the values of x decrease without bound, the corresponding values of a^x decrease and approach 0 (Figure 6.3a).

Analogous remarks can be made in the case $0 < a < 1$ (Figure 6.3b). In both cases, the x-axis is a horizontal asymptote.

EXAMPLE **1**   Sketch the graphs of the exponential functions $f(x) = 2^x$ and $g(x) = (4/3)^x$ on the same coordinate plane and compare them.

Solution: Consider the table of values for the two functions. By plotting the corresponding points and joining them, we obtain the graphs illustrated in Figure 6.4.

x	2^x	$\left(\dfrac{4}{3}\right)^x$
-2	$\dfrac{1}{4}$	$\dfrac{9}{16}$
-1	$\dfrac{1}{2}$	$\dfrac{3}{4}$
0	1	1
1	2	$\dfrac{4}{3}$
2	4	$\dfrac{16}{9}$

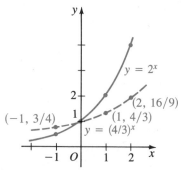

Figure 6.4

Notice that the two graphs cross at the point $(0, 1)$. When $x > 0$, the graph of $f(x) = 2^x$ lies above the graph of $f(x) = (4/3)^x$. When $x < 0$, the graph of $f(x) = 2^x$ lies below the graph of $f(x) = (4/3)^x$.

Example 1 illustrates the following properties of exponential functions.

If $b > a > 1$, then $b^x > a^x$ holds for all $x > 0$, but $b^x < a^x$ holds for all $x < 0$. For $x = 0$, it is always the case that $b^0 = a^0 = 1$.

Practice Exercise 1   Sketch and compare the graphs of the functions $f(x) = 2^x$ and $g(x) = (3/2)^x$.

Answer:

x	$y = 2^x$	$y = \left(\dfrac{3}{2}\right)^x$
-1	$\dfrac{1}{2}$	$\dfrac{2}{3}$
0	1	1
1	2	$\dfrac{3}{2}$
2	4	$\dfrac{9}{4}$

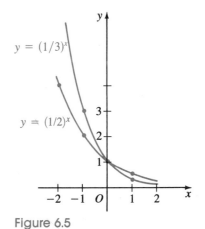

EXAMPLE **2**   Sketch and compare the graphs of the functions $f(x) = (1/2)^x$ and $g(x) = (1/3)^x$.

Solution: Consider the following table of values:

x	$\left(\dfrac{1}{2}\right)^x$	$\left(\dfrac{1}{3}\right)^x$
-2	4	9
-1	2	3
0	1	1
1	$\dfrac{1}{2}$	$\dfrac{1}{3}$
2	$\dfrac{1}{4}$	$\dfrac{1}{9}$

Figure 6.5

The two graphs illustrated in Figure 6.5 cross at the point $(0, 1)$. For negative values of x, the graph of $g(x) = (1/3)^x$ is located above the graph of $f(x) = (1/2)^x$, while for positive values of x, the graph of $g(x)$ is below the graph of $f(x)$.

Example 2 illustrates the following properties of exponential functions.

> If $0 < b < a < 1$, then $b^x < a^x$ holds for all $x > 0$, but $b^x > a^x$ holds for all $x < 0$. For $x = 0$, it is always the case that $b^0 = a^0 = 1$.

Practice Exercise 2

Sketch and compare the graphs of $f(x) = (2/3)^x$ and $g(x) = (1/2)^x$.

Answer:

x	$(2/3)^x$	$(1/2)^x$
-2	$9/4$	4
-1	$3/2$	2
0	1	1
1	$2/3$	$1/2$
2	$4/9$	$1/4$

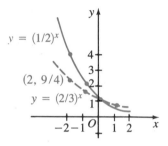

HISTORICAL NOTE

THE NUMBER e

It was Euler who introduced the letter e to denote the irrational number $2.71828\ldots$, which is the base of the natural logarithms.

The Number e

Among the examples of irrational numbers discussed in Section 1.2, we mentioned the number e, which is of fundamental importance in mathematics and the applied sciences. The following is an approximation of e to 20 decimal places:

$$e \simeq 2.71828\ 18284\ 59045\ 23536.$$

Since e is an irrational number, it can be approximated in many ways by increasing sequences of rational numbers. To obtain one of these sequences, we may proceed as follows. With each natural number n, we associate the rational number

$$\left(1 + \frac{1}{n}\right)^n = \left(\frac{n + 1}{n}\right)^n.$$

TABLE 6.2

n	$\left(1 + \dfrac{1}{n}\right)^n$
1	2
2	2.25000
3	2.37037
4	2.44141
5	2.48832
10	2.59374
10^2	2.70481
10^3	2.71692
10^4	2.71815
10^5	2.71827
$\vdots$	$\vdots$

Evaluating this expression for $n = 1, 2, 3, \ldots$, we obtain a sequence of rational numbers. As n increases without bound, it can be shown that this sequence increases and approaches a real number, denoted by e. Using a terminology of calculus, we can say that e is the *limit* of the sequence $\left\{ \left(1 + \dfrac{1}{n} \right)^n \right\}$. Table 6.2 on the preceding page shows several approximations for e that were obtained with the help of a calculator. For simplicity, the numbers in the right column were rounded off to five decimal places.

The Natural Exponential Function

The exponential function

$$f(x) = e^x,$$

whose base is the number e, is called the *natural exponential function*.

It is worth mentioning that since the number e can be approximated by $[1 + (1/n)]^n$, the function e^x can also be approximated by

$$\left[\left(1 + \frac{1}{n} \right)^n \right]^x \quad \text{for } n = 1, 2, 3, \ldots.$$

Because of the importance of the exponential functions with bases e and $1/e$, mathematicians have produced tables giving decimal approximations of the numerical values of e^x and e^{-x} for moderate-sized values of x. A short table of these values is given in the Appendix (Table 1). Nowadays, the use of such tables has been overshadowed by the advent of scientific calculators, which compute numerical values of these functions (correct to the number of decimal places the calculator can show) at the press of a key. Students who own scientific calculators are encouraged to use them to speed up their calculations as they work from this book. Students working without such calculators will find tables like the one in the Appendix indispensable.

Using Your Calculator

As we have already mentioned, the key $\boxed{y^x}$ on your calculator can be used to find functional values of exponential functions.

Certain scientific calculators have the key $\boxed{10^x}$ for the computation of powers of 10. For example, if you enter 3 and press the $\boxed{10^x}$ key, the display will show 1000.

Also, some calculators have the key $\boxed{e^x}$ or $\boxed{\text{EXP}}$ to compute the natural exponential function. If you own such a calculator, then keying in

$$1 \;\boxed{e^x}$$

gives you the number e to as many places as your calculator will display. Keying in

$$0.8 \;\boxed{e^x}$$

gives you an approximation for $e^{0.8}$.

However, there are calculators that do not have the key $\boxed{e^x}$. In order to display the number e, you have to enter 1 and press the keys $\boxed{\text{INV}}\boxed{\ln x}$. To obtain an approximation for $e^{0.8}$, enter 0.8 and press the keys $\boxed{\text{INV}}\boxed{\ln x}$. This will give you

$$e^{0.8} = 2.2255409.$$

The reason for this, as we shall explain in Section 6 of this chapter, is that the natural exponential function e^x is the inverse of the natural logarithmic function $\ln x$.

EXERCISES 6.1

In Exercises 1–16, sketch the graph of each function. Whenever applicable, use the techniques of shifting, stretching, and reflecting graphs.

1. $f(x) = 3^x$

2. $f(x) = \left(\dfrac{1}{4}\right)^x$

3. $f(x) = \left(\dfrac{3}{2}\right)^x$

4. $f(x) = \left(\dfrac{4}{3}\right)^x$

5. $f(x) = 2^x + 3$

6. $f(x) = 2^x - 1$

7. $f(x) = 3^x - 1$

8. $f(x) = 3^x + 2$

9. $f(x) = 2^{x-1}$

10. $f(x) = 2^{x+3}$

11. $f(x) = 3^{x+2}$

12. $f(x) = 3^{x-4}$

13. $f(x) = 3^{2-x}$

14. $f(x) = 2^{3-x}$

15. $f(x) = 1 - 2^x$

16. $f(x) = 3 - 2^{-x}$

Without using a table or a calculator, find the indicated function values in Exercises 17–22.

17. $f(x) = 4^x$; $f(2), f(-2), f(1/2)$

18. $f(x) = 3(2^x)$; $f(3), f(-3), f(0)$

19. $f(x) = 5(9^x)$; $f(1/2), f(-1/2), f(1)$

20. $f(x) = 2(16^{-x})$; $f(1/4), f(-1/2), f(1/2)$

21. $f(x) = 3 + 2(1/8)^{x+1}$; $f(-2), f(0), f(-2/3)$

22. $f(x) = 2 - 3(4)^{x-2}$; $f(2), f(0), f(5/2)$.

Using Table 1 in the Appendix, approximate the following values in Exercises 23–30.

23. e^2

24. e^{-2}

25. $e^{1/2}$

26. $e^{0.25}$

27. $e^{-1.2}$

28. $e^{-2.25}$

29. $e^{1/20}$

30. $e^{4/25}$

In Exercises 31–36, sketch the graph of the given function. Using Table 1 in the Appendix or a calculator, plot points corresponding to integral values of x and join them by a smooth curve.

31. $f(x) = e^x$

32. $f(x) = e^{-x}$

33. $f(x) = e^{x^2}$

34. $f(x) = e^{-x^2}$

35. $f(x) = \dfrac{e^x + e^{-x}}{2}$

36. $f(x) = \dfrac{e^x - e^{-x}}{2}$

c In Exercises 37–46, use a calculator to approximate the given numbers.

37. $3^{\sqrt{2}}$

38. $5^{-\sqrt{3}}$

39. $4^{-\sqrt{5}}$

40. 3^{π}

41. $\left(\sqrt{2}\right)^{\sqrt{2}}$

42. $\pi^2 - 2^{\pi}$

43. π^{π}

44. $3(2.015)^{3.271}$

45. $7\left(\dfrac{1}{2}\right)^{-0.15}$

46. $5(3)^{1.02}$

c Use a calculator to solve the problems in Exercises 47–50.

47. If $f(x) = x(2^x)$, find $f(-0.51)$, $f(0.01)$, and $f(0.51)$.

48. If $g(x) = 2xe^{-x}$, find $g(-0.5)$, $g(0.5)$, and $g(1.01)$.

49. Let $f(x) = [1 + (1/x)]^x$. Approximate the following functional values to five decimal places: $f(4)$, $f(10)$, $f(10^3)$, and $f(10^5)$. Compare your answers with Table 6.2.

50. Let $g(x) = (1 + x)^{1/x}$. Approximate the following functional values to five decimal places: $g(1/3)$, $g(1/5)$, $g(10^{-2})$, and $g(10^{-4})$. Compare your answers with Table 6.2. What conclusion can you draw from this?

**APPLICATIONS OF
EXPONENTIAL FUNCTIONS**

Exponential functions appear in many problems in mathematics and the applied sciences.

Compound Interest

Assume that a sum of money p_0 (the *initial principal*) is invested at an interest rate of $100r\%$ per year *compounded annually*. Then, at the end of one year, the amount of principal $p(1)$ is equal to the initial principal p_0 plus the earned interest $p_0 r$. That is,

$$p(1) = p_0 + p_0 r$$
$$= p_0(1 + r).$$

At the end of two years, the new principal $p(2)$ will be equal to the sum of $p_0(1 + r)$, the principal at the end of the first year, plus the earned interest $p_0(1 + r)r$. That is,

$$p(2) = p_0(1 + r) + p_0(1 + r)r$$
$$= p_0(1 + r)^2.$$

By repeating this reasoning, we see that the principal $p(t)$ at the end of t years is given by formula (6.1).

Interest Compounded Annually

$$p(t) = p_0(1 + r)^t, \qquad\qquad\text{(6.1)}$$

The function $p(t)$ is an exponential function of t. If we set $t = 0$, then $p(0) = p_0$ is the initial principal.

When money is deposited in a savings account, the interest is compounded *quarterly* (four times a year). There are certain types of investment accounts that compound interest on a monthly or even a daily basis. In such cases, formula (6.1) has to be changed.

Suppose that an interest rate of $100r\%$ per year is being *compounded N times a year*. Then, the interest rate will be $(100r/N)\%$ *per period* and, in t years, the *number of periods* will be Nt. Thus, the amount of principal $p(t)$ at the end of t years compounded N times a year is given by formula (6.2).

Interest Compounded N Times a Year

$$p(t) = p_0\left(1 + \frac{r}{N}\right)^{Nt}, \qquad\qquad\text{(6.2)}$$

This is also an exponential function.

EXAMPLE 1   A sum of $1000 is invested at an annual rate of 8% compounded quarterly. Find the principal at the end of   **(a)** one year,   **(b)** four years.

Solution: Since the interest is 8% per year, compounded quarterly, the variable r in formula (6.2) is equal to 0.08. The period is $N = 4$, and the rate per period is $r/N = 0.08/4 = 0.02$.

(a) By using (6.2) with $t = 1$, we get

$$p(1) = 1000(1 + 0.02)^4$$
$$= 1000(1.02)^4$$
$$\simeq 1082.43.$$

(b) By using (6.2) with $t = 4$, we obtain

$$p(4) = 1000(1.02)^{16}$$
$$\simeq \$1372.79.$$

In this example, the value of $(1.02)^{16}$ was obtained with the help of a calculator by performing the following keystrokes:

$$1000 \;\boxed{\times}\; 1.02 \;\boxed{x^y}\; 16 \;\boxed{=}\; 1372.7857$$

The final amount was rounded off to the nearest cent. If a calculator is unavailable, a compound interest table, found in most libraries, would have to be used.

Practice Exercise 1   If $2000 is invested at an annual rate of 10% compounded semiannually, find the principal   **(a)** in two years,   **(b)** in five years.

Answer: **(a)** $2431.01   **(b)** $3257.79

Returning to formula (6.2), we may ask what happens if the period N increases without bound. This corresponds to compounding the interest daily, hourly, every minute, and so on. First, write (6.2) as follows:

$$p(t) = p_0\left(1 + \frac{r}{N}\right)^{Nt}$$
$$= p_0\left(1 + \frac{1}{N/r}\right)^{Nt}.$$

Next, set $N/r = n$ or $N = nr$ and obtain

$$p(t) = p_0\left(1 + \frac{1}{n}\right)^{nrt}$$
$$= p_0\left[\left(1 + \frac{1}{n}\right)^n\right]^{rt}.$$

If N increases without bound, so does n. But it was remarked in the previous section that as n increases without bound, $[1 + (1/n)]^n$ approaches the number e. Therefore, as the number of periods N increases without bound, formula (6.2) becomes

---

Interest Compounded Continuously

$$p(t) = p_0 e^{rt}. \qquad (6.3)$$

This represents the amount of principal after t years if the interest is *compounded continuously*. We can say that formula (6.3) is obtained as a *limit* of formula (6.2) as the number N of periods increases without bound.

Illustration Table

Principal after one year from an initial amount of $10,000 deposited at 12% annual interest for various frequencies of compounding

	N	$p(1)$
Annually	1	$11,200.00
Quarterly	4	11,255.09
Weekly	52	11,273.41
Hourly	8760	11,274.95
Continuously (formula (6.3))		11,274.97

Compare the last two numbers. Do you realize that you earn only 2¢ more by compounding continuously instead of hourly?

EXAMPLE 2

A sum of $1000 is invested at an annual rate of 8%. Find the principal at the end of 10 years if the interest is   **(a)** compounded quarterly,   **(b)** compounded continuously.

Solution: **(a)** Applying (6.2) with $r = 0.08$, $N = 4$, and $t = 10$, we get

$$p(10) = 1000(1.02)^{40}$$
$$\simeq 1000 \times 2.20804$$
$$\simeq \$2208.04$$

The approximation of $(1.02)^{40}$ was obtained with the help of a calculator.
(b) By using (6.3) with $r = 0.08$ and $t = 10$, we obtain

$$p(10) = 1000e^{0.8}$$
$$\simeq 1000 \times 2.2255409$$
$$\simeq \$2225.54,$$

where the approximation for $e^{0.8}$ was obtained using a calculator. Notice that the two answers differ by only $17.50.

Practice Exercise 2

If \$4000 is invested at an annual rate of 12%, find the principal at the end of 8 years if the interest is compounded   **(a)** quarterly,   **(b)** continuously.

Answer: **(a)** \$10,300.33   **(b)** \$10,446.79

Population Growth

Under stable conditions, the population of a given species *increases* at a rate that is proportional to the population. If $r > 0$ denotes the *rate of increase per individual per unit of time* and P_0 denotes the size of the population at time $t = 0$, it can be shown that the formula

$$P(t) = P_0 e^{rt} \qquad\qquad (6.4)$$

gives the population at time t.

Notice that this last formula is analogous to formula (6.3).

EXAMPLE **3**

In 1960, the earth's human population was estimated at 3×10^9 and was increasing at a rate of 2% per year. Based on that information, what was the earth's estimated population in 1970?

Solution: If 1960 represents the time $t = 0$, then $P_0 = 3 \times 10^9$ in formula (6.4). Since the rate of increase is 2% per year, we have $r = 0.02$. Thus, we can write

$$P(t) = (3 \times 10^9)e^{0.02t}.$$

Setting $t = 10$, we obtain the estimated population in 1970:

$$
\begin{aligned}
P(10) &= (3 \times 10^9)e^{0.02 \times 10} \\
&= (3 \times 10^9)e^{0.2} \\
&= (3 \times 10^9) \times 1.221 \qquad \text{[From Table 1]} \\
&= 3.663 \times 10^9
\end{aligned}
$$

If you use a calculator, the following keystrokes

$$3 \;\boxed{\text{EE}}\; 9 \;\boxed{\times}\; 0.2 \;\boxed{e^x}\;\boxed{=}$$

display the answer 3.6642×10^9.

Practice Exercise 3

Using the data of Example 3, find the earth's estimated population in 1980.

Answer: 4.4755×10^9

The graph in Figure 6.6 illustrates the earth's estimated population based on the information of Example 3 and Practice Exercise 3.

Doubling Time

Every exponential growth problem can be formulated using the concept of *doubling*

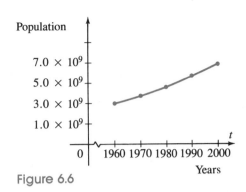

Figure 6.6

time. This is the time interval required for an exponentially growing quantity to double in size.

Assume that T is the doubling time of a certain population and that P_0 is the size of the initial population. Then formula (6.4) can be rewritten with an exponential in base 2 in the following way. If it takes a time interval T for the initial population P_0 to double, then, according to formula (6.4), we have

$$2P_0 = P(T) = P_0 e^{rT}$$

or

$$e^{rT} = 2,$$

hence

$$e^r = 2^{1/T}.$$

Substituting this expression for e^r in formula (6.4), we obtain

$$P(t) = P_0 2^{t/T}, \qquad (6.5)$$

a formula that gives the population at time t in terms of the doubling time T.

EXAMPLE **4**   A culture of bacteria growing exponentially doubles every 4 days. If the initial number of bacteria is 1500, find **(a)** the number of bacteria after 8 days,   **(b)** the number of bacteria after 10 days.

Solution: **(a)** Since the doubling time is 4 days and 8 is a multiple of 4, we can make the following table in order to solve the first part of the example.

t *days*	$P(t)$	
0	1500	initial number
4	3000	first doubling time
8	6000	second doubling time

Thus, after 8 days the number of bacteria is 6000.

(b) For the second part, we use formula (6.5) with $P_0 = 1500$ and $T = 4$:

$$P(t) = 1500 \cdot 2^{t/4}.$$

Thus,

$$
\begin{aligned}
P(10) &= 1500 \cdot 2^{10/4} \\
&= 1500 \cdot 2^{5/2} \\
&= 1500 \cdot \sqrt{2^5} \\
&= 1500 \cdot 2^2 \cdot \sqrt{2} \\
&= 6000\sqrt{2} \\
&\approx 8485.
\end{aligned}
$$

Practice Exercise 4　In a certain bacterial culture, the number of bacteria is doubling every day. If the original number of bacteria was 1200, find the number of bacteria after 3 days. What is the number after $4\frac{1}{2}$ days?

Answer:　9600;　27,153

Radioactive Decay

It has been observed that a radioactive material *decays exponentially* at a rate proportional to the amount of material present at a given time. If Q_0 is the initial amount of material and $r < 0$ is the *rate of decay per unit of time*, then the formula

$$Q(t) = Q_0 e^{rt} \qquad (r < 0) \tag{6.6}$$

gives the amount of material remaining at time t.

HISTORICAL NOTE

RADIOACTIVITY

In 1896, A. H. Becquerel (1852–1908), a French physicist, discovered radioactivity in uranium. He noted that if salts of uranium were brought close to an unexposed photographic plate that was protected from light, the plate became exposed in spite of its protective wrapping. Becquerel's work on radioactivity was extended by Marie and Pierre Curie, who discovered the highly radioactive element radium. The Curies shared with Becquerel the 1903 Nobel Prize in Physics.

---

c EXAMPLE 5　Carbon-14 decays at a rate of 0.012% per year. How much of a 60-g sample will be left in 1000 years?

Solution: Since the rate of decay is 0.012% per year, then $r = -0.012/100 = -0.00012$. According to formula (6.6), we write

$$Q(1000) = 60 \cdot e^{-0.00012 \times 1000}$$
$$= 60 \cdot e^{-0.12}$$
$$= 60 \times 0.8869204$$
$$= 53.215226$$
$$\simeq 53 \ g.$$

◧ Practice Exercise 5   A laboratory worker places in a container 100 mg of thorium-234, a radioactive element whose rate of decay is 2.88% per day. Find the amount left after 30 days.

Answer: 42 mg

Half-Life

The rate of decay of a radioactive element is often measured in terms of the *half-life* of the element. This is the time interval required for any initial amount of the element to be reduced by half.

If the half-life T of a radioactive element is known, then formula (6.6) can be rewritten as an exponential with base $1/2$, in the following way. If it takes a time interval T for an initial amount Q_0 of material to be reduced by half, then from formula (6.6) we have

$$\frac{Q_0}{2} = Q_0 e^{rT}$$

or

$$e^{rT} = \frac{1}{2},$$

so

$$e^r = \left(\frac{1}{2}\right)^{1/T}.$$

Substituting in (6.6), we obtain

$$Q(t) = Q_0 \left(\frac{1}{2}\right)^{t/T}, \tag{6.7}$$

a formula giving the amount at time t of a radioactive element whose half-life is T.

When the half-life of an element is given, the amount of the element can be easily computed, without the help of formulas (6.6) and (6.7), at periods of time that are multiples of the half-life.

EXAMPLE **6**

Radioactive iodine has a half-life of 25 minutes. If the initial amount of iodine was 100 mg, find **(a)** the amount of iodine left after 50 minutes, **(b)** the amount of iodine left after 62.5 minutes.

Solution: **(a)** Since half of the amount of iodine decays in 25 minutes, and half of the remaining amount decays in another 25 minutes, the following table suffices to solve the first part of this example. (See Figure 6.7.)

t min	$Q(t)$ mg
0	100
25	50
50	25

Thus, the amount of iodine remaining after 50 minutes is 25 mg.
(b) Here, we use formula (6.7) with $T = 25$ and $Q_0 = 100$. The equation

$$Q(t) = 100\left(\frac{1}{2}\right)^{t/25}$$

represents the amount of iodine at t minutes. Setting $t = 62.5$, we obtain

$$
\begin{aligned}
Q(t) &= 100\left(\frac{1}{2}\right)^{62.5/25} \\
&= 100\left(\frac{1}{2}\right)^{5/2} \\
&= 100\left(\frac{1}{\sqrt{2^5}}\right) \\
&= 100\left(\frac{1}{4\sqrt{2}}\right) \\
&= \frac{25}{\sqrt{2}} \\
&\approx 17.67767 \\
&\approx 18 \text{ mg.}
\end{aligned}
$$

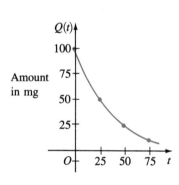

Figure 6.7

Practice Exercise 6

The half-life of polonium-128 is 3 minutes. What amount of a 120-g sample of polonium-128 will be left after 9 minutes? After 4.5 minutes?

Answer: 15 g, 42 g

Carbon Dating

Radioactive decay is used in archeology to estimate the age of fossils. When cosmic rays strike nitrogen-14 in the upper atmosphere, a radioactive form of carbon is produced. This is carbon-14 (^{14}C), whose half-life is 5750 years. Carbon-14 has two more neutrons in its nucleus than ordinary carbon-12 (^{12}C); in radioactive decay, ^{14}C loses those two neutrons and becomes ^{12}C. All living creatures absorb and metabolize both ordinary and radioactive carbon at a constant ratio: roughly one atom of carbon-14 to 10^{12} atoms of carbon-12. As long as a plant or an animal is alive, the level of carbon-14 in its tissue remains constant. When death occurs, the absorption process stops and the level of carbon-14 diminishes with time. By measuring the amount of carbon-14 in an organic sample of a fossil, it is possible to estimate the date of its death. This method is called *carbon-14 dating*.

EXAMPLE 7   Tests show that an animal skeleton contains 1/8 of the original amount of ^{14}C. Approximately how old is the skeleton?

Solution: Since the half-life of ^{14}C is 5750 years, formula (6.7) yields

$$Q(t) = Q_0 \left(\frac{1}{2}\right)^{t/5750}.$$

Now, we must find out how long it takes for an initial amount Q_0 of carbon-14 to be reduced to $Q_0/8$. We have

$$\frac{Q_0}{8} = Q_0 \left(\frac{1}{2}\right)^{t/5750}$$

or, after simplification,

$$\left(\frac{1}{2}\right)^3 = \left(\frac{1}{2}\right)^{t/5750}.$$

Since exponential functions are one-to-one functions (Property VI), it follows that

$$3 = \frac{t}{5750}$$

or

$$t = 5750 \times 3$$
$$= 17{,}250 \text{ years}.$$

Could you have solved Example 7 without using formulas (6.6) or (6.7)? Explain.

Practice Exercise 7   An animal bone found in an archeological site contains 25% of the original amount of carbon-14. Approximately how long ago did the animal die?

Answer: 11,500 years

Illustration Table

Half-life of certain radioactive elements

Element	Half-life
Bismuth-214	19.7 minutes
Polonium-218	3 minutes
Rubidium-87	6×10^{10} years
Thorium-234	24 days
Uranium-238	4.5×10^9 years

EXERCISES 6.2

In Exercises 1–6, exponential growth or decay is assumed. Solve each exercise by making a table as shown in Examples 4 and 6 of this section.

1. In a lab culture, bacteria is doubling every 12 hours. If the initial number of bacteria is 1800, what is the number of bacteria after one day? After 36 hours?

2. Money invested at an annual rate of 8% compounded quarterly doubles every 9 years. For a principal of $6000 invested today, what is the principal after 18 years?

3. The population of a country has been doubling every 25 years. If in 1985 the population is 120 million people, what was the population in 1910?

4. In a laboratory, fruit flies are doubling every 18 hours. If after 3 days the number of fruit flies is 3,600, what was the original number of fruit flies?

5. The half-life of polonium-128 is 3 minutes. Find the amount remaining from a 60-g sample after 6 minutes and after 9 minutes.

6. The half-life of thorium-234 is 24 days. How many days will it take for a sample of thorium to contain 12.5% of the original amount?

In Exercises 7–30, exponential behavior is assumed. Use a table or a calculator when necessary to solve each problem.

7. The sum of $10,000 is invested at a rate of 7.5% per year compounded quarterly. Find the amount of principal at the end of (a) one year, (b) two years. If the interest is compounded continuously, find the principal at the end of the same periods of time.

8. If $1200 is invested at an annual rate of 6%, find (a) the amount of money at the end of 5 years if the interest is compounded monthly; (b) the amount of money at the end of 5 years if the interest is compounded continuously.

9. How much would have to be invested at 6% today, compounded yearly, to have $8000 in 4 years?

10. How much would have to be invested at a rate of 7.5% per year, compounded quarterly, to have $1500 in 3.5 years?

11. In 1790 the population of the United States was 4 million. From 1790 to 1960 the average rate of growth of the population was 2.24% per year. Find the population of the United States in 1960.

12. In 1960 the population of a certain country was 100 million. If the rate of growth is 3% per year, what will the country's population be in the year 2000?

13. It is estimated that the population of a certain city is given by

$$P(t) = 20,000(1.02)^t,$$

where t is the number of years after 1980. How large was the population in 1983?

14. The population of a city is 1 million and is increasing at an annual rate of 3%. Find the population after (a) 2 years; (b) 5 years.

15. In a certain bacterial culture, the number of bacteria doubles every hour. If the initial number was 15,000, write a formula giving the number of bacteria after t hours. What is the number of bacteria after 3 hours?

16. In a bacterial culture, the number of bacteria quadruples in 6 hours. Find the number of bacteria after 9 hours if the initial number of bacteria was 1500.

17. At sea level, the atmospheric pressure is $p_0 = 760$ millimeters of mercury. At an altitude of x kilometers above

sea level, the atmospheric pressure in millimeters of mercury is given by the formula

$$p(x) = p_0 e^{-0.11445x}.$$

Find the atmospheric pressure at an altitude of **(a)** 2 km; **(b)** 5 km.

18. At an altitude of h miles above sea level, the atmospheric pressure $p(h)$ in pounds per square inch is

$$p(h) = 13.7 e^{-0.21h}.$$

Find the atmospheric pressure at **(a)** 1 mile; **(b)** 2.5 miles.

19. The half-life of radium is 1620 years. If the initial amount is 100 mg, write an equation giving the amount $Q(t)$ of radium after t years. Find the amount of radium after **(a)** 630 years; **(b)** 3240 years.

20. The half-life of krypton-91 is 10 seconds. If an initial amount of 500 mg is placed in a container, find the amount left after **(a)** 15 sec; **(b)** 20 sec.

21. A vertical beam of light with intensity I_0 when hitting the ocean surface has intensity $I = I_0 e^{-1.4d}$ at a depth of d meters. If $I_0 = 180$ lumens, what is the intensity of the light at a depth of 4 m?

22. At a depth of 2.5 meters below the ocean surface, the intensity of a vertical beam of light is 8 lumens. What is the intensity of the beam at the ocean surface? (See Exercise 21.)

23. A sample of charcoal from an archeological site contains 6.25% of the original amount of ^{14}C. How old is the charcoal?

24. Carbon-14 decomposes at a rate of 0.012%. How much of a 120-g sample decomposes in 2000 years?

25. Assume that it takes 15 seconds for half of a certain amount of sugar to dissolve in water and that the formula

$$A(t) = C\left(\frac{1}{2}\right)^{t/15}$$

gives the amount of sugar undissolved after t seconds. If 40 g of sugar are placed in water, how long will it take for all but 10 g of sugar to dissolve? How long will it take for all but 5 g of sugar to dissolve?

26. Referring to Exercise 25, assume that it takes T seconds for half of a certain amount of sugar to dissolve in water. If it is found that after 24 seconds, 1/8 of the original amount of sugar remains undissolved, find T.

27. Domestic wastewater is a mixture of biodegradable organic materials. In the presence of oxygen, such organic material decays as it is consumed by different kinds of bacteria. Under certain conditions, the decay is exponential and the amount L of organic material in wastewater is given by

$$L = L_0 e^{-kt},$$

where the constant k, called the *reaction rate*, is about 0.2 per day for domestic water. What percentage of the initial amount of organic material will have decayed in 3.5 days?

28. Suppose that the amount of organic material in wastewater after t days is given by the formula $L = L_0 e^{-kt}$. If after 3 days only half of the initial material remains, find what fraction of the initial material will have decayed after 9 days.

29. A chunk of sugar crystal is placed in a container holding a fixed volume of water. The solution is stirred and, from time to time, the concentration of the solution is measured. It is found that the concentration after t seconds is given by

$$C = C_e(1 - e^{-kt}),$$

where C_e is the equilibrium concentration and k is a constant. If after 10 seconds the concentration is half that of equilibrium, what is the concentration after 20 seconds?

30. Under the assumptions of Exercise 29, how long will it take for the concentration to be equal to $7C_e/8$?

6.3   LOGARITHMIC FUNCTIONS

We have seen in Section 6.2 that exponential functions are one-to-one functions (Property VI of Exponential Functions). Thus, we can define the inverse of an exponential function.

The Logarithmic Function

Let $a > 0$ be a real number, $a \neq 1$. The *logarithmic function with base a* is defined by its value at x:

$$y = \log_a x \quad \text{if and only if} \quad x = a^y. \tag{6.8}$$

The domain of definition is the set of all positive real numbers, and the range is the set of all real numbers.

From (6.8) we derive the following identities:

$$a^{\log_a x} = x$$

and

$$y = \log_a(a^y),$$

which express the fact that the two functions $y = a^x$ and $y = \log_a x$ are inverses of each other.

Also, (6.8) tells us that logarithmic equations can be converted into exponential equations and vice versa. For example:

Logarithmic form	Exponential form
$2 = \log_2 4$	$4 = 2^2$
$\log_{10} 1000 = 3$	$1000 = 10^3$
$\log_5\left(\dfrac{1}{25}\right) = -2$	$\dfrac{1}{25} = 5^{-2}$
$\log_e x = 4$	$x = e^4$

EXAMPLE 1

Rewrite each equation in logarithmic form: **(a)** $2^4 = 16$, **(b)** $5^{-3} = 1/125$.

Solution: **(a)** According to (6.8),

$$2^4 = 16 \quad \text{if and only if} \quad \log_2 16 = 4.$$

(b) Analogously,

$$5^{-3} = \frac{1}{125} \quad \text{if and only if} \quad \log_5\left(\frac{1}{125}\right) = -3.$$

Practice Exercise 1

Write as logarithms the following exponential equations:
(a) $3^5 = 243$, **(b)** $4^{-3} = 1/64$.

Answer: **(a)** $\log_3 243 = 5$ **(b)** $\log_4\left(\dfrac{1}{64}\right) = -3$

EXAMPLE 2   Find each of the following values: **(a)** $\log_2 8$,   **(b)** $\log_{10}(0.001)$.

Solution: **(a)** Set $y = \log_2 8$. According to (6.8), $y = \log_2 8$ is equivalent to $8 = 2^y$. Since $8 = 2^3$, we can rewrite $8 = 2^y$ as $2^3 = 2^y$, from which it follows that $y = 3$. Thus, $\log_2 8 = 3$.
(b) We have

$$\log_{10}(0.001) = y \quad \text{exactly when} \quad 0.001 = 10^y$$

Since $0.001 = 10^{-3}$, the last expression can be rewritten as $10^{-3} = 10^y$. Thus, $y = -3$ and $\log_{10}(0.001) = -3$.

Practice Exercise 2   Find the following numbers: **(a)** $\log_4 2$,   **(b)** $\log_3(1/27)$.

Answer: **(a)** $\frac{1}{2}$ **(b)** -3

Graphing Logarithmic Functions

As we explained in Section 5.4, given a function $f(x)$, the graph of $f^{-1}(x)$ is the reflection of the graph of f about the line $y = x$. Thus, the graph of $y = \log_a x$ is the reflection of the graph of $y = a^x$ about the line $y = x$. We have two cases to consider, depending on whether $a > 1$ or $0 < a < 1$.

The graph of $y = \log_a x$ when $a > 1$ is illustrated in Figure 6.8. In the same coordinate plane, we have sketched the graph of a^x (light color) and the graph of $\log_a x$ (bright color).

The graph shows the following important properties of the function $f(x) = \log_a x$, when $a > 1$.

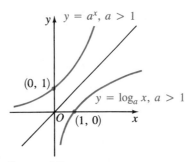

Figure 6.8

1. The logarithmic function—defined only for positive values of x—is an increasing function of x; that is if $x < y$, then $\log_a x < \log_a y$.
2. The x-intercept of the graph is 1. In other words, $\log_a 1 = 0$ because $a^0 = 1$.
3. If $0 < x < 1$, then $\log_a x < 0$; and, as x approaches 0, the quantity $\log_a x$ decreases without bound.
4. As x increases without bound, $\log_a x$ increases without bound.

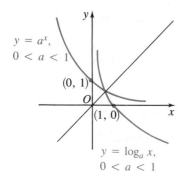

$y = a^x,$
$0 < a < 1$

(0, 1)

O

(1, 0) x

$y = \log_a x,$
$0 < a < 1$

Figure 6.9

We rarely use $\log_a x$ when $0 < a < 1$. But it is useful to see the graph of $f(x) = \log_a x$ for $0 < a < 1$. It is illustrated in Figure 6.9, where the graphs of $g(x) = a^x$, $0 < a < 1$, and its inverse (the logarithmic function) are shown.

When $0 < a < 1$, the following properties of the function $f(x) = \log_a x$ hold.

1. The logarithmic function—defined only for positive values of x—is a decreasing function of x; that is, if $x < y$, then $\log_a x > \log_a y$.
2. The x-intercept of the graph is 1, because $\log_a 1 = 0$.
3. If $0 < x < 1$, then $\log_a x > 0$. Moreover, as x approaches 0, the logarithm function increases without bound.
4. As x increases without bound, the function decreases without bound.

Notice that since exponential functions are one-to-one, logarithmic functions are also one-to-one.

Common Logarithms

When the base is the number 10, we write $\log x$ instead of $\log_{10} x$. Logarithms to the base 10 are called *common logarithms*. Before the development of electronic calculators and computers, common logarithms were extensively employed in numerical computations, due to the fact that we use the decimal system for writing numerals. Nowadays, common logarithms are rarely used in numerical computations. However, they are still found in many applications, such as measuring the acidity or basicity of chemical solutions, the loudness of sound, the magnitude of earthquakes, and the magnitude of stars.

Natural Logarithms

When the base is the number e, we write $\ln x$ (read "natural log of x") instead of $\log_e x$. Logarithms to the base e are called *natural* or *Napierian logarithms,* after John Napier (1550–1617), a Scottish mathematician who invented logarithms. Natural logarithms play an important role in theoretical work in mathematics and the applied sciences.

EXAMPLE 3

Evaluate x in each of the following equations.
(a) $\log 1000 = x$ (b) $\ln e^5 = x$
(c) $\log_2 x = 3$ (d) $\log_x 64 = 3$

Solution: By using (6.8), we can transform each logarithmic equation into an equivalent one involving exponents.
(a) $\log 1000 = x$ if and only if $1000 = 10^x$, thus $x = 3$.
(b) $\ln e^5 = x$ if and only if $e^5 = e^x$, so $x = 5$.
(c) $\log_2 x = 3$ if and only if $x = 2^3 = 8$.
(d) $\log_x 64 = 3$ if and only if $64 = x^3$, so $x = 4$.

Practice Exercise 3   Find x such that the following equations hold.
(a) $\log_2 128 = x$ **(b)** $\ln x = 2$
(c) $\log x = 4$ **(d)** $\log_x 125 = 3$

Answer: **(a)** $x = 7$ **(b)** $x = e^2$ **(c)** $x = 10,000$ **(d)** $x = 5$

Using Your Calculator

If your calculator has the keys $\boxed{\log}$ and $\boxed{\ln}$, it is very easy to obtain numerical values for the functions $\log x$ and $\ln x$. You simply enter x and press the desired key. For example, if $x = 5.92$, then entering 5.92 and pressing the $\boxed{\log}$ key will give you

$$\log 5.92 \simeq 0.7723217.$$

If you enter 5.92 and press the $\boxed{\ln}$ key, you obtain

$$\ln 5.92 \simeq 1.7783364.$$

Some calculators have keys for each of the functions e^x and $\ln x$. Others do not have a key for e^x. If this is the case, you must calculate e^x as the *inverse* of $\ln x$, using the keys $\boxed{\text{INV}}\,\boxed{\ln}$. For example, to find $e^{2.81}$, enter 2.81 and press the keys $\boxed{\text{INV}}\,\boxed{\ln}$, obtaining $e^{2.81} \simeq 16.609918$.

Properties of Logarithms

The properties of the exponential functions listed in Section 6.1 imply properties of the logarithmic functions that are of fundamental importance in both practical and theoretical work with logarithms.

I. The logarithm of a product is the sum of the logarithms:

$$\log_a(mn) = \log_a m + \log_a n. \qquad (6.9)$$

To prove this, set

$$u = \log_a m \quad \text{and} \quad v = \log_a n.$$

According to (6.8),

$$m = a^u \quad \text{and} \quad n = a^v.$$

Multiplying m and n and using property I of exponential functions, we get

$$mn = a^u a^v = a^{u+v}.$$

From (6.8) it follows that

$$u + v = \log_a(mn).$$

Substituting $\log_a m$ for u and $\log_a n$ for v, we obtain (6.9).

II. The logarithm of the reciprocal of a number is the additive inverse of the logarithm of the number:

$$\log_a \frac{1}{n} = -\log_a n. \qquad (6.10)$$

Indeed, if $u = \log_a n$, then $n = a^u$ and vice versa. By property II of exponential functions,

$$a^{-u} = \frac{1}{a^u} = \frac{1}{n}.$$

It follows from (6.8) that

$$-u = \log_a \frac{1}{n}.$$

Substituting $\log_a n$ for u, we obtain

$$-\log_a n = \log_a \frac{1}{n},$$

which is relation (6.10).

Combining properties I and II, we obtain the following property.

III. The logarithm of a quotient is the difference of the logarithms:

$$\log_a \frac{m}{n} = log_a m - \log_a n. \qquad (6.11)$$

The proof is left as an exercise.

IV. The logarithm of a power of a number is the exponent of that power multiplied by the logarithm of the number:

$$\log_a (m^r) = r \log_a m. \qquad (6.12)$$

This proof is also left as an exercise.

Since a radical is a fractional power, from property IV of logarithms we derive the following property concerning the logarithm of a radical.

$$\log_a \sqrt[n]{m} = \frac{1}{n}\log_a m. \qquad\qquad (6.13)$$

EXAMPLE 4

If log 2 = 0.3010 and log 3 = 0.4771, find the following numbers.

(a) log 6   **(b)** $\log\frac{2}{3}$ **(c)** log 81   **(d)** $\log \sqrt[5]{3}$

Solution: **(a)** Since $6 = 2\cdot 3$, we write

$$\begin{aligned} \log 6 &= \log(2\cdot 3) \\ &= \log 2 + \log 3 \qquad \text{[Property I]} \\ &= 0.3010 + 0.4771 \\ &= 0.7781. \end{aligned}$$

(b) By property III,

$$\begin{aligned} \log\frac{2}{3} &= \log 2 - \log 3 \\ &= 0.3010 - 0.4771 \\ &= -0.1761. \end{aligned}$$

(c) Since $81 = 3^4$, then according to property IV,

$$\begin{aligned} \log 81 &= \log 3^4 \\ &= 4\log 3 \\ &= 4 \times 0.4771 \\ &= 1.9084. \end{aligned}$$

(d) Using (6.13), we obtain

$$\begin{aligned} \log \sqrt[5]{3} &= \frac{1}{5}\log 3 \\ &= \frac{1}{5}(0.4771) \\ &= 0.0954. \end{aligned}$$

■ A word of caution   Notice that

$$\log(2\cdot 3) \ne (\log 2)(\log 3).$$

From part a of Example 4, $\log(2\cdot 3) = 0.7781$. However, $(\log 2)(\log 3) = (0.3010)(0.4771) = 0.1436071$.

Similarly, $\log(2/3) \ne \log 2/\log 3$, because $\log(2/3) = -0.1761$ and $\log 2/\log 3 = 0.3010/0.4771 = 0.6308949$.

Practice Exercise 4

Given log 4 = 0.6021 and log 5 = 0.6990, find the following numbers.

(a) $\log 20$ **(b)** $\log \dfrac{5}{4}$ **(c)** $\log 4^5$ **(d)** $\log \sqrt[5]{4}$

Answer: **(a)** 1.3011   **(b)** 0.0969   **(c)** 3.0105   **(d)** 0.12042

EXAMPLE 5

(a) Express $\log_a(x^5y^2/\sqrt[3]{z})$ in terms of $\log_a x$, $\log_a y$, and $\log_a z$.

(b) Write $\log \sqrt[4]{x^2(x-1)(x-2)^3}$ in terms of $\log x$, $\log(x-1)$, and $\log(x-2)$.

Solution: **(a)** Using the properties of logarithms (6.9)–(6.13), we get

$$\log_a(x^5y^2/\sqrt[3]{z}) = \log_a(x^5y^2) - \log_a \sqrt[3]{z}$$
$$= \log_a x^5 + \log_a y^2 - \log_a \sqrt[3]{z}$$
$$= 5\log_a x + 2\log_a y - \frac{1}{3}\log_a z.$$

(b) Analogously, we have

$$\log \sqrt[4]{x^2(x-1)/(x-2)^3} = \frac{1}{4}\log x^2(x-1)/(x-2)^3$$

$$= \frac{1}{4}[\log x^2(x-1) - \log(x-2)^3]$$

$$= \frac{1}{4}[\log x^2 + \log(x-1) - \log(x-2)^3]$$

$$= \frac{1}{4}[2\log x + \log(x-1) - 3\log(x-2)]$$

$$= \frac{1}{2}\log x + \frac{1}{4}\log(x-1) - \frac{3}{4}\log(x-2).$$

Practice Exercise 5

Use the properties of logarithms to write $\log m^3(\sqrt{(x-1)/(y+2)^4})$ in terms of $\log m$, $\log(x-1)$, and $\log(y+2)$.

Answer: $3\log m + \dfrac{1}{2}\log(x-1) - 2\log(y+2)$

EXAMPLE 6

Write each of the following expressions as a single logarithm.

(a) $2\log 5 - \log 7 + \log 4 - \log 3$ **(b)** $\ln(x+1) - 3\ln x + \dfrac{1}{2}\ln(x+2)$

Solution: **(a)** We have

$$2\log 5 - \log 7 + \log 4 - \log 3 = \log 5^2 + \log 4 - (\log 7 + \log 3)$$
$$= \log(5^2 \cdot 4) - \log(7 \cdot 3)$$
$$= \log 100 - \log 21$$
$$= \log \frac{100}{21}.$$

(b) Similarly,

$$\ln(x + 1) - 3 \ln x + \frac{1}{2}\ln(x + 2) = \ln(x + 1) - \ln x^3 + \ln\sqrt{x + 2}$$
$$= \ln[(x + 1)\sqrt{x + 2}] - \ln x^3$$
$$= \ln\frac{(x + 1)\sqrt{x + 2}}{x^3}.$$

Practice Exercise 6   Write the following expression as a single logarithm.

$$\ln 15 + 5 \ln x - 2 \ln(x + 1) + \frac{1}{2}\ln(x - 1)$$

Answer: $\ln\dfrac{15x^5 \sqrt{x - 1}}{(x + 1)^2}$

EXERCISES 6.3

In Exercises 1–10, use the definition of logarithms to find each of the following values.

1. $\log_2 16$

2. $\log_3 27$

3. $\log_2\dfrac{1}{8}$

4. $\log_5 125$

5. $\log 10^4$

6. $\ln e^{1/2}$

7. $\log(0.1)^2$

8. $\log\dfrac{1}{1000}$

9. $\ln\left(\dfrac{1}{e}\right)^3$

10. $\log_4\dfrac{1}{64}$

In Exercises 11–18, rewrite each expression in logarithmic form.

11. $2^5 = 32$

12. $3^4 = 81$

13. $4^{-2} = \dfrac{1}{16}$

14. $5^{-4} = \dfrac{1}{625}$

15. $\left(\dfrac{1}{3}\right)^4 = \dfrac{1}{81}$

16. $\left(\dfrac{1}{2}\right)^{-6} = 64$

17. $a^b = c$

18. $u^{-v} = w$

In Exercises 19–26, rewrite each expression in exponential form.

19. $\log_2 128 = 7$

20. $\log_3 243 = 5$

21. $\log_2\dfrac{1}{1024} = -10$

22. $\log_5\dfrac{1}{3125} = -5$

23. $\log 100 = 2$

24. $\log 1000 = 3$

25. $\log q = r$

26. $\log_n m = -p$

In Exercises 27–40, given $\log 2 \approx 0.30$, $\log 3 \approx 0.48$, $\log 4 \approx 0.60$, and $\log 5 \approx 0.70$, find approximations of the following numbers.

27. $\log 10$

28. $\log 12$

29. $\log 18$

30. $\log 15$

31. $\log 20$

32. $\log 60$

33. $\log\dfrac{3}{5}$

34. $\log\dfrac{5}{2}$

35. $\log\dfrac{10}{3}$

36. $\log\dfrac{4}{15}$

37. $\log 8$

38. $\log 16$

39. $\log\sqrt{10}$

40. $\log\sqrt[3]{15}$

Use the definition and properties of logarithms to find x in Exercises 41–54.

41. $\log_6 x = 3$

42. $\log_4 x = 0$

43. $\log_{3/2} x = 4$

44. $\log_{2/3} x = 3$

45. $\log_x 4 = 5$

46. $\log_x 216 = 3$

47. $\log_{2x} 8 = 3$

48. $\log_{3x} 225 = 2$

49. $\log_7 49 = x$

50. $\log_5 125 = x$

51. $\log_3(2x + 1) = 2$

52. $\log_2(3x - 2) = 2$

53. $\log_2(1 - 2x) = -3$

54. $\log_3(2 - 4x) = -2$

In Exercises 55–66, write each expression as a single logarithm.

55. $\log 8 + \log 6 - \log 12$

56. $\ln 12 - \ln 4 + \ln 5$

57. $3 \log_5 2 + 2 \log_5 3 - \dfrac{1}{2}\log_5 4$

58. $2 \log 8 - 5 \log 2 - \dfrac{1}{3} \log 27$

59. $2 \ln x - \dfrac{1}{4} \ln y + 4 \ln z$

60. $\dfrac{1}{2} \ln x + 3 \ln y - 2 \ln x$

61. $2 \log x + \dfrac{1}{3} \log y - 4 \log x$

62. $\dfrac{1}{4} \log x - \dfrac{1}{2} \log y + \dfrac{1}{4} \log z$

63. $\log(x + 1) - \log(x - 2) + 2 \log x$

64. $3 \log x - \dfrac{1}{2} \log(x + 1) - \log(2x - 1)$

65. $3 \log(x + 2) + 2 \log x - 4 \log(x + 1)$

66. $4 \log(x + 1) + 2 \log(x - 1) - \dfrac{1}{2} \log(x + 3)$

In Exercises 67–72, write each expression in terms of $\log x$ and $\log(x - 2)$.

67. $\log \dfrac{x^2}{(x - 2)^3}$

68. $\log \dfrac{\sqrt{x}}{(x - 2)^2}$

69. $\log[x^2(x - 2)^4]$

70. $\log \sqrt{\dfrac{x^3}{x - 2}}$

71. $\log \sqrt{(x - 2)^3 x^4}$

72. $\log\left(\dfrac{x}{x - 2}\right)^4$

In Exercises 73–76, write each expression in terms of $\log x$, $\log(x + 1)$ and $\log(x - 2)$.

73. $\log \dfrac{x^2(x + 1)^3}{(x - 2)^5}$

74. $\log \dfrac{x(x - 2)^3}{\sqrt{x + 1}}$

75. $\log \left(\dfrac{1}{x}\right) \sqrt[3]{\dfrac{x + 1}{(x - 2)^2}}$

76. $\log \sqrt{\dfrac{x^2(x + 1)^3}{(x - 2)^5}}$

Use properties of logarithms to prove each of the following identities in Exercises 77–80.

77. $\log \dfrac{x + \sqrt{x^2 - 1}}{x - \sqrt{x^2 - 1}} = 2 \log(x + \sqrt{x^2 - 1})$

78. $\log \dfrac{x + \sqrt{x^2 - 1}}{x - \sqrt{x^2 - 1}} = -2 \log (x - \sqrt{x^2 - 1})$

79. $\dfrac{\log(x + h) - \log x}{h} = \log\left(1 + \dfrac{h}{x}\right)^{\frac{1}{h}}$

80. $\log \dfrac{y}{a + \sqrt{a^2 + y^2}} = \log \dfrac{\sqrt{a^2 + y^2} - a}{y}$

6.4   COMPUTING COMMON LOGARITHMS; APPLICATIONS OF COMMON LOGARITHMS

In the past, common logarithms were an important computational tool. They were widely used in astronomy, engineering, and statistics to reduce the computational work for certain numerical expressions. For this purpose, the use of tables, such as Table 2 in the Appendix, was absolutely necessary.

Today the importance of tables has considerably declined. Inexpensive calculators can perform power and root-extraction operations, and they compute exponential, logarithmic, and trigonometric functions with great accuracy and speed.

Despite this fact, it is useful for students to learn how to use tables. Situations sometimes arise where no calculator is available, or sometimes teachers feel that, at this stage, calculators should not be used.

Figure 6.10 reproduces a portion of Table 2 found in the Appendix. The table gives approximations of common logarithms to four decimal places. Suppose that we want to find the common logarithm of 592. First, we write the number in scientific notation

$$592 = 5.92 \times 10^2.$$

Next, we use properties of logarithms to obtain

$$\log 592 = \log(5.92 \times 10^2)$$
$$= \log 5.92 + \log 10^2 \qquad \text{Property I of logarithms}$$
$$= \log 5.92 + 2 \qquad \log 10^2 = 2$$

n	0	1	2	3	4	5	6	7	8	9
5.5	.7404	.7412	.7419	.7427	.7435	.7443	.7451	.7459	.7466	.7474
5.6	.7482	.7490	.7497	.7505	.7513	.7520	.7528	.7536	.7543	.7551
5.7	.7559	.7566	.7574	.7582	.7589	.7597	.7604	.7612	.7619	.7627
5.8	.7634	.7642	.7649	.7657	.7664	.7672	.7679	.7686	.7694	.7701
5.9	.7709	.7716	.7723	.7731	.7738	.7745	.7752	.7760	.7767	.7774
6.0	.7782	.7789	.7796	.7803	.7810	.7818	.7825	.7832	.7839	.7846
6.1	.7853	.7860	.7868	.7875	.7882	.7889	.7896	.7903	.7910	.7917
6.2	.7924	.7931	.7938	.7945	.7952	.7959	.7966	.7973	.7980	.7987
6.3	.7993	.8000	.8007	.8014	.8021	.8028	.8035	.8041	.8048	.8055
6.4	.8062	.8069	.8075	.8082	.8089	.8096	.8102	.8019	.8116	.8122

Figure 6.10

Now, we have to find the value of log 5.92. Referring to the table, we first locate the number 5.9 in the left column and then move across until we reach the column headed by 2, which is the last digit of 5.92, to obtain

$$\log 5.92 \simeq 0.7723.$$

Therefore,

$$\log 592 = \log 5.92 + 2$$
$$\simeq 0.7723 + 2$$
$$= 2.7723.$$

Observe that in our calculations, we have obtained an *integral part*, 2, called the *characteristic* of the logarithm, and 0.7723, a *decimal number between 0 and 1*, called the *mantissa* of the logarithm.

In general, if a number n is written in scientific notation as

$$n = m \times 10^k,$$

where m is a number such that $1 \le m < 10$ and k is an integer, then the common logarithm of n is

$$\log n = \log(m \times 10^k)$$
$$= \log m + \log 10^k$$
$$= \log m + k.$$

The integer k is the *characteristic* and the number $\log m$ is the *mantissa* of $\log n$. Since $\log 1 = 0$, $\log 10 = 1$, and the common logarithm is an increasing function, we have

$$0 \le \log m < 1.$$

Thus the mantissa of a logarithm is always a nonnegative number less than 1.

EXAMPLE 1   Find each of the following: **(a)** log 57,200, **(b)** log 0.0583. Determine the characteristic and mantissa of each logarithm.

Solution: **(a)** We have

$$\log 57,200 = \log(5.72 \times 10^4)$$
$$= \log 5.72 + \log 10^4 \qquad \text{Property I of logarithms}$$
$$\simeq 0.7574 + 4 \qquad\qquad \text{Table 2}$$
$$= 4.7574.$$

The characteristic of log 57,200 is 4 and the mantissa is 0.7574.

(b)
$$\log 0.0583 = \log(5.83 \times 10^{-2})$$
$$= \log 5.83 + \log 10^{-2}$$
$$\simeq 0.7657 - 2$$
$$= -1.2343$$

The characteristic of log 0.0583 is −2 and the mantissa is 0.7657.

If you are using a calculator, it is not necessary to convert the given numbers into scientific notation. The sequence of keystrokes

$$57,200 \quad \boxed{\log} \boxed{=}$$

will give you log 57,200 ≃ 4.757396. Also, using your calculator, you can check that

$$\log 0.0583 \simeq -1.2343314.$$

Practice Exercise 1   Approximate the following numbers: **(a)** log 6240, **(b)** log 0.00635.

Answer: **(a)** log 6240 ≃ 3.7952 **(b)** log 0.00635 ≃ −2.1972

Antilogarithms

Now we will consider the opposite problem: suppose we are given the logarithm of a number and want to find the number or an approximation of it.

EXAMPLE **2**   Find x such that log $x = 5.7664$. The number x is called the *antilogarithm* of 5.7664.

Solution: First, we write

$$\log x = 5.7664$$
$$= 0.7664 + 5,$$

in order to determine the characteristic, 5, and the mantissa, 0.7664. Next, according to the table (Figure 6.10), log 5.84 ≃ 0.7664, so

$$\log x = 0.7664 + 5$$
$$\simeq \log 5.84 + \log 10^5 \qquad 5 = \log 10^5$$
$$= \log(5.84 \times 10^5) \qquad \text{Property I of logarithms}$$

Hence

$$x \simeq 5.84 \times 10^5.$$

Practice Exercise 2   If $\log x = 2.7832$, then find an approximation for x.

Answer: $x \simeq 607$

Before using the table, make sure that the mantissa of the logarithm is a number between 0 and 1, and that the characteristic is an integer.

EXAMPLE 3   Find x if $\log x = -1.2396$.

Solution: As before, we write

$$\log x = -1.2396$$
$$= -0.2396 - 1.$$

Since -0.2396 is negative, it *cannot* be the mantissa of $\log x$, since the mantissa of a logarithm is always positive and less than 1. Thus we have to express $\log x$ as a sum of a number between 0 and 1 and an integer. This can be achieved by adding and subtracting 2 to the right-hand side of that equality:

$$\log x = (2 - 1.2396) - 2$$
$$= 0.7604 - 2.$$

The number 0.7604 is the mantissa of $\log x$ and -2 is its characteristic. Referring to the table (Figure 6.10), we see that $\log 5.76 \simeq 0.7604$. Thus,

$$\log x = 0.7604 - 2$$
$$\simeq \log 5.76 + \log 10^{-2}$$
$$= \log(5.76 \times 10^{-2});$$

hence

$$x \simeq 5.76 \times 10^{-2}.$$

Practice Exercise 3   Determine x if $\log x = -2.1952$.

Answer: $x \simeq 6.38 \times 10^{-3}$

If you are using a calculator, it is very simple to compute the antilogarithm of a number. For example, to find x in Example 2, enter 5.7664 and press the keys $\boxed{\text{INV}}\ \boxed{\text{log}}$ to obtain

$$x \simeq 583982.72$$

or, in scientific notation,

$$x \simeq 5.8398 \times 10^{5}.$$

To solve Example 3 with a calculator, enter 1.2396, and press the keys $\boxed{+/-}$, $\boxed{\text{INV}}$, and $\boxed{\text{log}}$. You will get

$$x \simeq 0.057597,$$

or

$$x \simeq 5.7597 \times 10^{-2}.$$

All the numbers considered in Examples 1, 2, and 3 were found in the table of logarithms. If a number is not found in the table, then the method of *linear interpolation,* discussed in the Appendix, can be used to approximate the logarithm or antilogarithm of a number. When speed and more accuracy are desired, a calculator should be used.

Applications of Common Logarithms

Logarithmic functions are used in chemistry, physics, psychology, and other applied sciences.

Measuring pH

For example, the acidity or basicity of a chemical solution, denoted by pH, is defined by the formula

$$pH = -\log[H^+],$$

where $[H^+]$ is the hydrogen ion concentration in moles per liter. Acid solutions have pH < 7 and basic solutions have pH > 7.

EXAMPLE 4 (a) Find the pH of eggs, knowing that $[H^+] = 1.6 \times 10^{-8}$ moles per liter.
☐ (b) Find the hydrogen ion concentration, in moles per liter, of tomatoes whose pH is 4.2.

Solution: (a) We have

$$
\begin{aligned}
pH &= -\log(1.6 \times 10^{-8}) \\
&= -\log 1.6 - \log 10^{-8} \\
&= -\log 1.6 + 8 \\
&\simeq -0.2 + 8 \\
&= 7.8.
\end{aligned}
$$

(b) If pH = 4.2, then $\log[H^+] = -4.2$. Hence, using a calculator,

$$[H^+] \simeq 6.31 \times 10^{-5} \text{ moles per liter.}$$

Practice Exercise 4 (a) Apricots have a pH of 3.8. Find the hydrogen ion concentration in moles per liter.
(b) Find the pH of apples if $[H^+] \simeq 8 \times 10^{-4}$ moles per liter.

Answer: (a) 1.6×10^{-4} moles per liter   (b) 3.1

Measuring Loudness of Sound

The human ear is sensitive to sound over an enormous range of intensity, from 10^{-16} to 10^{-4} W/cm^2. For this reason, the loudness of sound is measured on a logarithmic scale. The sound level L, in decibels (dB), of a sound intensity I is given by

$$L = 10 \log\frac{I}{I_0},$$

where $I_0 = 10^{-16}$ W/cm^2 is the minimum detectable intensity.

EXAMPLE **5**   (a) Find the sound level, in decibels, of ordinary conversation, where $I = 10^{-10}$ W/cm^2.

G (b) The clatter of a chain saw is at 105 dB. What is the corresponding sound intensity?

Solution: (a) Substitute 10^{-10} for I and 10^{-16} for I_0 into the given equation, obtaining

$$L = 10 \log\frac{10^{-10}}{10^{-16}}$$

$$= 10 \log 10^6$$

$$= 60 \text{ dB}.$$

(b) We have

$$105 = 10 \log\frac{I}{10^{-16}}$$

or

$$\log(10^{16}I) = 10.5.$$

Hence

$$10^{16}I = 10^{10.5} \simeq 3.1623 \times 10^{10},$$

or

$$I \simeq 3.1623 \times 10^{-6} \text{ W/cm}^2.$$

Practice Exercise 5   (a) The sound level inside a car moving at 55 mph is 72 dB. Find the sound intensity.

(b) What is the sound level of a whisper whose sound intensity is 10^{-14} W/cm^2?

Answer: (a) 1.6×10^{-9} W/cm^2 (b) 20 dB

EXERCISES 6.4

In Exercises 1–10, use Table 2 in the Appendix or a calculator to approximate each of the following to four decimal places.

1. log 3.72

2. log 5.38

3. log 3250

4. log 1230

5. log 71300

6. log 10500

7. log 0.0225

8. log 0.00518

9. log 51.2

10. log 5120

In Exercises 11–20, use Table 2 in the Appendix or a calculator to find an approximation of x.

11. log $x = 3.5391$

12. log $x = 5.6513$

13. log $x = 4.9253$

14. log $x = 2.9713$

15. log $x = -2.2636$

16. log $x = -0.4609$

17. $\log x = 1.3181$

18. $\log x = -2.0141$

19. $\log x = 5.3243 - 8$

20. $\log x = 2.1004 - 5$

In Exercises 21–40, solve the given problem.

21. Given $[H^+]$, find the pH of each substance: **(a)** carrots: $[H^+] = 1.1 \times 10^{-5}$ moles per liter; **(b)** soil: $[H^+] = 3.1 \times 10^{-7}$ moles per liter.

22. Compare the acidity of the following substances: **(a)** milk: $[H^+] = 4 \times 10^{-7}$ moles per liter; **(b)** vinegar: $[H^+] = 6.3 \times 10^{-3}$ moles per liter.

23. Find the hydrogen ion concentration $[H^+]$ in each of the following substances: **(a)** sea water: pH $\simeq 7.8$; **(b)** soil: pH $\simeq 5$.

24. Gastric juice has pH $\simeq 1.7$ and pancreatic juice has pH $\simeq 7.8$. Determine their hydrogen ion concentrations.

25. To cultivate roses, the soil pH must be kept between 6.0 and 8.0. Find the corresponding range for the hydrogen ion concentration.

26. If $[H^+]$ varies from 1.1×10^{-5} to 5.2×10^{-5} moles per liter, find the pH range.

27. Sound levels above 90 dB are considered dangerous to the human ear. The sound intensity of a jet airliner during takeoff is approximately 10^{-4} W/cm^2. Is the decibel level dangerous?

28. Determine the sound intensity corresponding to **(a)** 90 dB **(b)** 110 dB

29. On the Richter scale, the magnitude R of an earthquake of intensity I is given by the formula

$$R = \log \frac{I}{I_0},$$

where I_0 is a minimum intensity used for comparison. Find the magnitude on the Richter scale of an earthquake of intensity $10^{6.5} I_0$.

30. If the intensity of an earthquake is $10^{7.9}$ times the minimum intensity I_0, what is its magnitude on the Richter scale?

31. Measured on the Richter scale, the magnitude of a quake is 5.6. How many times more intense is it than the minimum intensity?

32. If on the Richter scale the magnitude of an earthquake is 8.2, compare its intensity to the minimum intensity, I_0.

33. Two earthquakes registered magnitudes of 7.5 and 6.5, respectively, on the Richter scale. Compare their intensities.

34. The San Francisco quake of 1906 is believed to have had a magnitude of 8.2. The Seattle quake of 1965 measured 7 on the Richter scale. How many times more intense was the San Francisco quake than the Seattle quake?

35. The magnitude M of a star is defined by

$$M = -2.5 \log kI,$$

where k is a constant and I is the intensity of the light from the star. Sirius, the brightest star in the sky, has magnitude -1.42, and Canopus, the second brightest star, has magnitude -0.72. What is the ratio of the intensities of light from Sirius and Canopus?

36. Vega, in the Lyra constellation, has magnitude 0.04, while the magnitude of Canopus is -0.72. Find the ratio between the light intensities emanating from Canopus and Vega.

37. If the light intensity of a star is $1/50$ that of Canopus (see Exercise 35), find the magnitude of the star.

38. Find the magnitude of a star whose light intensity is $1/100$ that of Sirius (see Exercise 35).

39. Students in a precalculus course took a final exam. Each month thereafter, they were given an equivalent form of the exam, in order to test how much they still remembered. Suppose (according to a mathematical model from psychology) that the function

$$F(t) = 65 - 15 \log(t + 1)$$

gives the average test score after t months. (The function $F(t)$ is called the *forgetting function*.)
(a) Find the average test score of the final exam ($t = 0$).
(b) Find the average test scores after 2 and 6 months.

40. In Exercise 39, find the average test scores after 8 and 12 months. How long will it take for the average to drop below 50?

EXPONENTIAL AND LOGARITHMIC EQUATIONS; CHANGE OF BASE

In certain equations the unknown may appear as an exponent or a logarithm. To solve such equations, we use the properties of the exponential and logarithmic functions.

EXAMPLE 1   Solve the equation $4^x = 2^{3x-5}$.

Solution: Since $4 = 2^2$, we rewrite the given equation so that both sides are powers of 2:

$$2^{2x} = 2^{3x-5}.$$

Since exponential functions are one-to-one functions, it follows that the two exponents must be equal, that is,

$$2x = 3x - 5.$$

Solving for x, we obtain $x = 5$.

Practice Exercise 1   Solve the exponential equation $4^x = 8^{x+1}$ for x.

Answer: $x = -3$

Often the two sides of an exponential equation are not easily written with a common base. When this happens, we take the logarithm of both sides of the equation, as shown in the following example.

EXAMPLE 2   Solve the equation $3^{2x-1} = 11$.

Solution: Taking common logarithms of both sides, we have

$$\log 3^{2x-1} = \log 11$$
$$(2x - 1)\log 3 = \log 11$$
$$2x - 1 = \frac{\log 11}{\log 3},$$

so

$$x = \frac{1}{2}\left(1 + \frac{\log 11}{\log 3}\right)$$

This is the exact solution of the given equation. If an approximation is desired, then according to Table 2 in the Appendix,

$$x \simeq \frac{1}{2}\left(1 + \frac{1.0414}{0.4771}\right)$$
$$\simeq 1.5913.$$

If a calculator is used, we perform the following keystrokes

$$1 \boxed{\div} 2 \boxed{\times} \boxed{(} 1 \boxed{+} 11 \boxed{\log} \boxed{\div} 3 \boxed{\log} \boxed{)} \boxed{=}$$

and obtain

$$x \simeq 1.5913292.$$

Practice Exercise 2   (a) Find the exact solution of the equation $4^{3x+1} = 30$.
© (b) Approximate the solution to four decimal places.

Answer: (a) $x = \frac{1}{3}\left[\left(\frac{\log 30}{\log 4}\right) - 1\right]$ (b) $x \simeq 0.4845$

EXAMPLE **3**   Solve the equation

$$\log(3x - 1) = 1 + \log(1 - x).$$

Solution: The given equation can be written as follows:

$$\log(3x - 1) - \log(1 - x) = 1 = \log 10$$

or, using property III of logarithms,

$$\log \frac{3x - 1}{1 - x} = \log 10;$$

which is equivalent to

$$\frac{3x - 1}{1 - x} = 10.$$

Solving for x, assuming $x \neq 1$, we obtain

$$3x - 1 = 10(1 - x)$$
$$13x = 11$$
$$x = \frac{11}{13}.$$

You can check that this is the desired solution by substituting $11/13$ for x into the original equation.

Practice Exercise 3   Find the solution of the logarithmic equation $\log(2x - 1) - 1 = \log(x - 1)$.

Answer: $x = \dfrac{9}{8}$

EXAMPLE **4**   Solve $\log(2x - 1) + \log(x - 1) = 1$.

Solution: We use property I of logarithms to write the given equation in an equivalent form:

$$\log(2x - 1)(x - 1) = 1.$$

From definition (6.8), we obtain

$$(2x - 1)(x - 1) = 10$$
$$2x^2 - 3x + 1 = 10$$
$$2x^2 - 3x - 9 = 0$$
$$(2x + 3)(x - 3) = 0.$$

Hence

$$x = 3 \quad \text{or} \quad x = -\frac{3}{2}.$$

■ **Note carefully!** If $x = -3/2$, then $2x - 1$ and $x - 1$ are negative and so $\log(2x - 1)$ and $\log(x - 1)$ are not defined. The number $-3/2$ is called an *extraneous solution,* and it must be discarded. **The only solution of the given equation is then $x = 3$.**

Practice Exercise 4   Solve the equation

$$\log(x + 1) = 1 - \log(2x + 1).$$

Answer: $x = \dfrac{3}{2}$

EXAMPLE 5   Solve the equation $(e^x - e^{-x})/2 = 1$.

Solution: We write the given equation as

$$e^x - e^{-x} - 2.$$

Since $e^{-x} = 1/e^x$, it follows that if we denote e^x by u, then the last equation becomes

$$u - \frac{1}{u} = 2.$$

Multiplying both sides by u (which cannot be zero), we obtain the quadratic equation

$$u^2 - 1 = 2u$$

or

$$u^2 - 2u - 1 = 0.$$

Solving it for u, we obtain the solutions

$$u = 1 \pm \sqrt{2}.$$

Since $u = e^x$ is always positive, the negative solution $1 - \sqrt{2}$ must be discarded. The positive solution $1 + \sqrt{2}$ yields the exponential equation

$$e^x = 1 + \sqrt{2},$$

which can be solved by taking natural logarithms of both sides:

$$x \ln e = \ln\left(1 + \sqrt{2}\right),$$

or

$$x = \ln\left(1 + \sqrt{2}\right).$$

This is the exact solution of the given equation. If an approximation is desired, we obtain

$$x \simeq \ln 2.4142$$
$$\simeq 0.8814.$$

Practice Exercise 5   What is the exact solution of the exponential equation $(10^x - 10^{-x})/2 = 1$? Approximate the solution to four decimal places.

Answer: $x = \log\left(1 + \sqrt{2}\right) \simeq 0.3828$

Doubling Time and Rate of Increase of a Population

In Section 6.2, we dealt with two formulas that describe population growth:

$$P(t) = P_0 e^{rt} \quad \text{and} \quad P(t) = P_0 2^{t/T}.$$

In the first formula, the positive number r denotes the *rate of increase per individual per unit of time;* in the second formula, T denotes the *doubling time.*

Sometimes it is useful to derive a relation between these two quantities, r and T. Assuming that both formulas represent the growth of the same population, it follows that

$$P_0 e^{rt} = P_0 2^{t/T}$$
$$e^{rt} = 2^{t/T}$$
$$[e^r]^t = [2^{1/T}]^t, \text{ for all } t.$$

If we set $t = 1$, then we obtain

$$e^r = 2^{1/T}.$$

By taking natural logarithms of both sides, an expression for the rate of increase in terms of the doubling time is obtained:

$$r = \frac{\ln 2}{T}.$$

This shows that the rate of increase is *inversely proportional* to the doubling time, and the *constant of proportionality* is $\ln 2$.

EXAMPLE 6

In a culture of bacteria growing exponentially, the doubling time is 4 days. What is the daily rate of increase?

Solution: According to the above formula,

$$r = \frac{\ln 2}{4}$$
$$= \frac{0.6931471}{4}$$
$$= 0.1732868$$
$$\approx 0.17.$$

Thus, the rate of increase is approximately 17% per day.

Practice Exercise 6

What is the doubling time for a population increasing at a rate of 2% per year?

Answer: 34.657 years $\approx$ 35 years

MATHEMATICAL VIGNETTE

GROWTH OF THE WORLD'S POPULATION

The following table shows the growth of the earth's population. Figures for 1999, 2010, and 2025 are projections on the basis of present growth.

Year	Population (in billions)
1830	1
1930	2
1960	3
1975	4
1986	5
1999	6
2010	7
2025	8

Note that it took 100 years (from 1830 to 1930) for the population to double. It doubled again in 1975 after 45 years. It is estimated that it will double again in about 50 years.

Half-Life and Rate of Decrease of a Radioactive Element

Two formulas were used in Section 6.2 to describe the decay of a radioactive element:

$$Q(t) = Q_0 e^{rt} \quad \text{and} \quad Q(t) = Q_0 \left(\frac{1}{2}\right)^{t/T},$$

where r (a negative number) represents the *rate of decay per unit of time* and T is the *half-life* of the radioactive element.

Proceeding in exactly the same manner as for the rate of increase of a population and the doubling time, you can check that the following relation holds:

$$r = -\frac{\ln 2}{T}.$$

That is, the rate of decay of a radioactive element varies *inversely* as the half-life of the element, and the *constant of proportionality* is $-\ln 2$.

EXAMPLE **7**

What is the half-life of radioactive iodine if 2.77% of the iodine decays every minute?

Solution: The rate of decay is $r = -2.77/100 = -0.0277$. Since

$$r = -\frac{\ln 2}{T},$$

then

$$T = -\frac{\ln 2}{r}.$$

Substituting -0.0277 for r, we get

$$T = \frac{\ln 2}{0.0277}$$
$$= 25.023364$$
$$\approx 25 \text{ minutes.}$$

Practice Exercise 7   The half-life of polonium-128 is 3 minutes. Determine its rate of decrease.

Answer: 23.1% per minute

Change of Base

When dealing with logarithms, it is sometimes necessary to change the base. For example, suppose that we want to determine the value of $\log_2 7$. We would need a table of logarithms to the base 2, but such a table is unavailable. On the other hand, we cannot use a calculator directly, since the only keys available are for common and natural logarithms. However, we can change the base and express $\log_2 7$ as the quotient of the common logarithms of 7 and 2. This is done as follows. If we set $x = \log_2 7$, then $2^x = 7$. Taking common logarithms of both sides gives us

$$x \log 2 = \log 7,$$

or

$$x = \frac{\log 7}{\log 2}.$$

We conclude that

$$\log_2 7 = \frac{\log 7}{\log 2}.$$

The method described in this example applies to logarithms to any base. Suppose that we want to express $\log_a x$ in terms of $\log_b x$. If we set

$$y = \log_a x,$$

then

$$a^y = x.$$

Taking logarithms to the base b of both sides gives us

$$y \log_b a = \log_b x,$$

or

$$y = \frac{\log_b x}{\log_b a}.$$

Substituting for y, we obtain the *change of base formula:*

$$\log_a x = \frac{\log_b x}{\log_b a}.$$

If we set $x = b$ in the last formula and use the fact that $\log_b b = 1$, we obtain

$$\log_a b = \frac{1}{\log_b a}.$$

EXAMPLE 8   Find an approximation for $\log_6 8$ using natural logarithms.

Solution: Apply the change of base formula with $x = 8$, $a = 6$, and $b = e$, to get

$$\log_6 8 = \frac{\log_e 8}{\log_e 6} = \frac{\ln 8}{\ln 6}.$$

Using Table 3 in the Appendix or a calculator, you can obtain the approximation

$$\log_6 8 \simeq \frac{2.0794}{1.7918}$$
$$= 1.1605.$$

■ Note   If you are using a calculator, it is very easy to compute the number $(\ln 8)/(\ln 6)$. Here are the keystrokes:

$$8\ \boxed{\ln}\ \boxed{\div}\ 6\ \boxed{\ln}\boxed{=}.$$

Practice Exercise 8   Approximate the value $\log_2 7$.

Answer: $\log_2 7 \simeq 2.807$

EXERCISES 6.5

In Exercises 1–8, use only properties of exponents to solve each of the given equations.

1. $2^{x-1} = 16$

2. $3^{x-4} = 9$

3. $3^{2x-6} = 27^{3-x}$

4. $5^{x-5} = 125^{3-2x}$

5. $\left(\frac{1}{4}\right)^{x+4} = 8^{2x+4}$

6. $27^{2x-1} = \left(\frac{1}{9}\right)^{3-x}$

7. $\left(\sqrt{2}\right)^{6x+1} = 16^{x+2}$

8. $25^{2-x} = \left(\sqrt{5}\right)^{3x-1}$

In Exercises 9–30, solve each of the given equations.

9. $7^x = 2$ **10.** $10^x = 5$

11. $3^{-x} = 8$ **12.** $2^{-x} = 10$

13. $10^{5x+1} = 21$ **14.** $12^{3x-1} = 24$

15. $2^{3x-1} = 3^{1-2x}$ **16.** $5^{2-x} = 6^{x+1}$

17. $\log x - \log 2 = 1$ **18.** $\log x + \log 7 = 2$

19. $\log(2x + 4) = 1 + \log(x - 2)$

20. $\log x - \log(x + 1) = 2$

21. $\log(x + 1) = 1 - \log(x + 2)$

22. $\log(x^2 + 1) - \log(x - 1) = 1 + \log(x + 1)$

23. $(\log x)^2 + \log x^2 = 0$ **24.** $\log x^3 = (\log x)^2$

25. $\log(\log x) = 1$ **26.** $\log(\log x^2) = 1$

27. $\dfrac{e^x + e^{-x}}{2} = 1$ **28.** $\dfrac{4^x - 4^{-x}}{2} = 1$

29. $\dfrac{e^x - e^{-x}}{e^x + e^{-x}} = \dfrac{1}{2}$ **30.** $\dfrac{e^x + e^{-x}}{e^x - e^{-x}} = 4$

Given $\ln 2 \simeq 0.7$, $\ln 3 \simeq 1.1$, and $\ln 5 \simeq 1.6$, evaluate each of the logarithms in Exercises 31–38 without the use of tables or calculators.

31. $\log_5 4$ **32.** $\log_3 25$

33. $\log 18$ **34.** $\log 15$

35. $\log_6 30$ **36.** $\log_4 12$

37. $\log_{15} 24$ **38.** $\log_9 45$

Given $\log 2 \simeq 0.30$, $\log 3 \simeq 0.50$ and $\log 5 \simeq 0.70$, evaluate each of the logarithms in Exercises 39–46 without using tables or calculators.

39. $\log_2 9$ **40.** $\log_3 8$

41. $\log_3 12$ **42.** $\log_4 20$

43. $\log_6 24$ **44.** $\log_{12} 10$

45. $\log_4 \dfrac{1}{6}$ **46.** $\log_5 \dfrac{4}{9}$

In Exercises 47–54, using the change of base formula together with Table 2, Table 3, or a calculator, find each of the following numbers to four decimal places.

47. $\log_7 8$ **48.** $\log_4 15$

49. $\log_2 5.4$ **50.** $\log_5 3.1$

51. $\log_3 1.23$ **52.** $\log_4 2.01$

53. $\log_4(2.3 \times 10^{-2})$ **54.** $\log_3(7 \times 10^{-3})$

In Exercises 55–70, solve each of the given problems.

55. If $\log_4 11 \simeq 1.7$ and $\log_4 2 \simeq 0.5$, find $\log_2 11$.

56. If $\log_3 12 \simeq 2.2619$, find $\log_{12} 3$.

57. Assume that the formula $n(t) = n_0 2^{2t}$ gives the number of bacteria in a certain culture at time t. If $n_0 = 10,000$, how long will it take for the number of bacteria to equal **a)** 320,000? **b)** 2,560,000?

58. The formula

$$Q(t) = 100\left(\frac{1}{2}\right)^{t/24}$$

gives the amount, in grams, of thorium-234 after t days. How many days will it take for the amount to be 25 grams? 6.25 grams?

59. Iodine-128 decays according to the formula

$$Q(t) = Q_0\left(\frac{1}{2}\right)^{t/25},$$

where Q_0 is the initial amount and $Q(t)$ is the amount of iodine after t minutes. How long will it take for the remaining amount to be $1/10$ of the initial amount?

60. The population of a city of 20,000 inhabitants is growing at a rate of 3% per year. After t years the population will be $P(t) = 20,000e^{0.03t}$. How long will it take for the population to double?

61. The half-life of rubidium-87 is 6×10^{11} years. Find its rate of decay.

62. The equation $Q(t) = Q_0 e^{-0.00012t}$ represents the amount remaining, after t years, of an original amount Q_0 of carbon-14. Find the half-life of carbon-14.

63. The formula $p(h) = 14.7e^{-0.12h}$ gives the atmospheric pressure in pounds per square inch at an altitude of h miles above sea level. At what altitude will the atmospheric pressure be one half of the sea level pressure? One third of the sea level pressure?

64. The intensity I of a vertical beam of light at a depth of x feet below the surface of Crystal Lake in Wisconsin is given by the formula

$$I = I_0 e^{-0.0485x}.$$

At what depth is the light intensity equal to $1/10$ of its value at the surface of the lake?

65. The formula

$$PV = Ae^{-rt}$$

gives the dollar amount (called *present value*) that should be invested now at an annual interest rate of $100r\%$ compounded continuously to achieve A dollars in t years.

(a) What is the present value at 8% interest of a gift of $5000 to be made 5 years from now?

(b) What is the annual interest rate of a savings account if the present value of $8000 in 10 years is $4282?

66. The present value (see Exercise 65) at 7.5% of a gift to be made in 6 years is $6376.29. What will be the value of the gift?

67. According to Newton's law of cooling, the temperature T of a body after time t is given by

$$T = A + Ce^{-kt},$$

where A is the temperature of the surrounding medium and C and k are constants. A bowl containing water at 100°C is placed in a freezer at a temperature of −5°C. If after 4 min the temperature of the water is 16°C, find its temperature after 6 min.

68. A cup of coffee at a temperature of 150°F is placed in a room at a temperature of 75°F. After 3/4 hr it has cooled to 100°F. Find its temperature after 1/2 hr.

69. If $5000 is invested at an annual rate of 12% compounded monthly, how long will it take for it to double?

70. A sum of $1200 is invested at an annual rate of 7.5% compounded quarterly. How long will it take for it to double? When will the principal exceed $5000?

CHAPTER SUMMARY

If a is any real number such that $a > 0$ and $a \neq 1$, the *exponential function with base a* is defined by

$$f(x) = a^x$$

for all real numbers x. Exponential functions play a fundamental role in mathematics and the applied sciences. If $a > 1$, such functions describe exponential growth; if $0 < a < 1$, they describe exponential decay. Of special interest is the *natural exponential function:*

$$f(x) = e^x, \qquad e \simeq 2.71828 \ldots .$$

If an *initial principal p_0* is invested for t years at an *annual interest rate r compounded N times a year,* then the *principal* $p(t)$ is given by

$$p(t) = p_0\left(1 + \frac{r}{N}\right)^{Nt}$$

When interest is compounded continuously, the principal $p(t)$ after t years is

$$p(t) = p_0 e^{rt}.$$

If $r > 0$ denotes the *rate of increase per individual per unit of time* and P_0 denotes the amount of population *at time $t = 0$,* the formula

$$P(t) = P_0 e^{rt}$$

gives the population at time t.

In radioactive decay, the equation

$$Q(t) = Q_0 e^{rt} \qquad (r < 0)$$

represents the amount remaining, after t years, of an initial sample Q_0 of a radioactive material whose *rate of decay per unit of time* is r.

A *logarithmic function*

$$f(x) = \log_a x \qquad (a > 0, a \neq 1)$$

is defined as the inverse of an exponential function:

$$y = \log_a x \qquad \text{if and only if} \qquad x = a^y.$$

Of great importance are the *common logarithm* (base 10), denoted by $\log x$, and the *natural logarithm* (base e), denoted by $\ln x$. Common logarithms are still used in chemistry, engineering, geology, and physics. Natural logarithms are used in theoretical and applied mathematics.

The pH of a chemical solution is defined by the formula

$$\text{pH} = -\log[\text{H}^+],$$

where $[\text{H}^+]$ is the hydrogen ion concentration. Acid solutions have pH < 7 and basic solutions have pH > 7.

The *sound level L*, measured in *decibels* (dB), is given by

$$L = 10 \log(I/I_0),$$

where I is the sound *intensity* (in W/cm^2), and $I_0 = 10^{-16}$ W/cm^2 is the minimum detectable intensity.

When dealing with exponential growth or decay, the concepts of *doubling time* and *half-life* are of particular importance. The doubling time T is related to the rate of increase r as follows:

$$r = \frac{\ln 2}{T},$$

where $r > 0$ is the rate of increase and T is the doubling time. Similarly, the relationship between the rate of decay r and the half-life of a radioactive element is

$$r = \frac{-\ln 2}{T},$$

where $r < 0$ is the rate of decay and T is the half-life.

In view of the practical importance of logarithmic and exponential functions, tables have been constructed as an aid to numerical computations. However, tables are now becoming obsolete due to the availability of inexpensive calculators that can compute logarithms and exponents with great accuracy and speed.

CHAPTER TEST

1. Graph $g(x) = 1/2^{-x-1}$.
2. Let $F(x) = xe^{3x}$. Approximate to three decimal places the function values $F(-0.4)$, $F(0.4)$, and $F(1.25)$.
3. Suppose that the initial amount of a radioactive material is 140 milligrams and that its half-life is 33 minutes. Make a table showing the amount of material left after 33, 66, and 132 minutes.
4. The population of a town of 26,000 people is increasing exponentially at an annual rate of 3%. Estimate the population after 6 years.
5. Use properties of logarithms to write the number $\log_2 (50/80)$ as a sum or difference of multiples of $\log_2 2 = 1$, $\log_2 3$, and $\log_2 5$.

6. Write $4 \log x - \dfrac{2}{3}\log(x + 1) + \log(3x + 1)$ as one logarithm.
7. Find the sound intensity I corresponding to 55 dB. [*Hint:* Use the formula for the sound level in decibels.]
8. Let $A(t) = A_0 r^t$ where A_0 and r are positive constants. Find A_0 and r if $A(0) = 256$ and $A(2) = 243$.
9. On a Richter scale, the magnitudes of two earthquakes of intensities I_1 and I_2 are 6.0 and 6.6, respectively. What is the ratio I_1/I_2 of the two intensities? [*Hint:* Use the equation $R = \log(I/I_0)$.]
10. Solve the logarithmic equation $\log(6x + 1) = 1 + \log(x - 5)$ for x.

REVIEW EXERCISES

In Exercises 1–10, solve each problem as described.

1. Find each of the following values.
   (a) $5^{\log_5 8}$ (b) $3^{\log_3 7}$ (c) $e^{\ln 25}$ (d) $10^{\log 2}$
2. Determine the numbers.
   (a) $\log_{1/2} 4$ (b) $\log_{0.1} 100$ (c) $\log_{1/4} 64$
   (d) $\log_{1/5} 625$
3. Find each of the functional values.
   (a) $f(x) = \left(\dfrac{1}{4}\right)^x$; $f(3), f(2), f(-2)$
   (b) $g(x) = x^2(2^{-x})$; $g(-0.5), g(0.01), g(0.5)$

4. Find each of the functional values.
   (a) $F(x) = 5\left(\dfrac{1}{2}\right)^x$; $F(-16), F(-4), F(2)$
   (b) $G(x) = \sqrt{x}(3^{x-2})$; $G(0.2), G(0.4), G(2.5)$
5. Let $q(t) = ab^t$, where a and b are constants. If $q(0) = 10$ and $q(1) = 20$, find a and b. Next, find $q(3)$.
6. Let $p(t) = p_0 r^t$, where p_0 and r positive are constants. Determine p_0 and r, knowing that $p(0) = 128$ and that $p(2) = 32$. Find $p(5)$.
7. The function $S(t)$ is defined by $S(t) = S_0\left(1 + \dfrac{r}{2}\right)^t$, where

S_0 and r are constants.
(a) If $S(0) = 16$ and $S(3) = 250$, find S_0 and r.
(b) Find $S(2)$.

8. Let $A(t) = A_0\left(1 + \dfrac{r}{2}\right)^t$. If $A(0) = 16$ and $A(3) = 54$, find the constants A_0 and r. Also, find $A(2)$.

9. Graph the following functions.
(a) $f(x) = 2^x - 2^{-x}$ (b) $f(x) = \dfrac{2}{e^x - e^{-x}}$

10. Sketch the graph of each function.
(a) $f(x) = 3^x + \left(\dfrac{1}{3}\right)^x$ (b) $f(x) = \dfrac{2}{e^x + e^{-x}}$

In Exercises 11–32, solve the equations.

11. $\log_4(x + 2) = 2$

12. $\log_2(3x + 1) = 4$

13. $\log_8(2x + 1) = \dfrac{2}{3}$

14. $\log_{16}(3x + 2) = \dfrac{3}{4}$

15. $\log(2x + 4) = \log 10 - \log 6$

16. $\log(3 - 2x) = \log 4 - \log 9$

17. $2 \log_3 x = 3 \log_3 4$

18. $2 \log(x + 1) = 4 \log 5$

19. $\log x - \log(x - 1) = 2 \log 4$

20. $\log_6(2x - 1) - \log_6(x + 1) = \log_6 5 - \log_6 4$

21. $4^{2x-1} = 16$

22. $3^{x+1} = 9$

23. $4^{2x+1} = 8^{x-3}$

24. $9^{4x-3} = 27^{x-7}$

25. $\left(\dfrac{1}{3}\right)^{1-2x} = 9^{3x-2}$

26. $5^{3x-4} = \left(\dfrac{1}{25}\right)^{3-4x}$

27. $(\sqrt[3]{2})^{x+3} = 4^{x-2}$

28. $(\sqrt[4]{3})^{3x+1} = 9^{1-2x}$

29. $3^{2x+1} = 4^{3x-2}$

30. $5^{3-x} = 2^{3x+2}$

31. $\dfrac{3^x + 3^{-x}}{2} = 1$

32. $\dfrac{2^x - 2^{-x}}{2} = 1$

In Exercises 33–40, write each of the following expressions as a single logarithm.

33. $\ln 7 + \ln 12 - \ln 5$

34. $\log 9 + \log 15 - \log 21$

35. $3 \ln 4 - 2 \ln 5 + 3 \ln 8$

36. $2 \log 8 - 3 \log 4 + \dfrac{1}{5} \log 12$

37. $7 \log a - \dfrac{1}{2} \log b + \dfrac{3}{4} \log c$

38. $3 \ln u - \dfrac{2}{3} \ln v + 4 \ln w$

39. $2 \log(x + 1) - 3 \log(x + 2) + 4 \log x$

40. $3 \ln(x + 2) - \dfrac{1}{3} \ln(x + 1) + 2 \ln(x + 3)$

In Exercises 41–44, write each expression in terms of $\ln x$, $\ln(x - 1)$, and $\ln(2x + 1)$.

41. $\ln \dfrac{x(x - 1)^2}{(2x + 1)^3}$

42. $\ln \dfrac{x^3\sqrt{x - 1}}{\sqrt[3]{2x + 1}}$

43. $\ln \sqrt{\dfrac{x(x - 1)^3}{(2x + 1)^4}}$

44. $\ln \dfrac{x^2 \sqrt[3]{(x - 1)^2}}{(2x + 1)^3}$

In Exercises 45–64, solve each of the given problems.

45. Approximate each of the following logarithms.
(a) $\log(32.14 \times 10^{-3})$ (b) $\log(257.1 \times 10^{-5})$
(c) $\log 0.006254$ (d) $\log 4516000$

46. Approximate x in each of the following.
(a) $\log x = 5.3313$ (b) $\log x = 1.4449$
(c) $\log x = -4.4409$ (d) $\log x = -2.3079$

47. Find the pH of each substance. Round off your answer to two significant digits.
(a) Lemon juice: $[H^+] = 5.1 \times 10^{-3}$ moles per liter
(b) Beer: $[H^+] = 5 \times 10^{-5}$ moles per liter

48. Find the hydrogen ion concentration $[H^+]$ in moles per liter of each substance. Round off your answers to two significant digits.
(a) Blood: pH $= 7.4$
(b) Intestinal juice: pH $= 7.7$

49. Find the sound level, in decibels, of heavy traffic, where the sound intensity is $I = 10^{-8}$ W/cm^2.

50. The sound level of a pneumatic drill is 110 dB. Find the corresponding sound intensity.

51. Compare the intensities of two earthquakes of magnitudes 5.5 and 7 measured on the Richter scale.

52. Compare the intensity of an earthquake of magnitude 6.5 on the Richter scale with the minimum intensity, I_0.

53. Vega, a star in the constellation Lyra, has magnitude 0.04. Find the magnitude of a star with light intensity $1/50$ that of Vega. (The formula for the magnitude of a star is given in Exercise 35 of Section 6.4.)

54. How many times more intense is the light emanating from Canopus, a star of magnitude -0.72, than the light emanating from Alpha Centauri, a star of magnitude -0.27? (See Exercise 35 of Section 6.4.)

55. The population of a country is increasing at a rate of 2.1% a year. Find the doubling time.

56. The half-life of thorium-230 is 80,000 years. Find its yearly rate of decay.

57. Potassium-40 has a half-life of 1.3×10^9 years. Given an initial amount of 1.25×10^{-3} g of potassium-40, find how much is left after 5×10^9 years.

58. According to a recently completed census, the population of China reached 1 billion in 1982 and is growing at an annual rate of 1.45%. Estimate how large the population of China will be at the turn of the century.

C **59.** How much would have to be invested today at an annual rate of 7.5% compounded continuously to have $12,500 in 8 years?

C **60.** A sum of $25,120 is invested at an annual rate of 8.25% compounded continuously. How long will it take for the sum to double?

C **61.** The 200% *declining balance method* was one of the depreciation methods allowed by the IRS. According to this method, the original cost C of an item depreciates over N years at a rate of $(200/N\%)$ per year. Thus the value V of the item after t years is

$$V = C\left(1 - \frac{2}{N}\right)^t.$$

If a typewriter costing $1200 is depreciated by this method over 6 years, what is its value after 4 years? 6 years?

C **62.** The value of a truck costing $15,000 is $556 after 3 years. Using the formula of Exercise 61, find the number of years over which the truck was depreciated.

C **63.** An apple pie baked at 325°F is removed from the oven and placed in a room at a constant temperature of 70°F. One minute later the temperature of the pie is 250°F. What is its temperature after 5 minutes? (See Exercise 67 of Section 6.5)

C **64.** A thermometer registering 50°C is placed in a freezer that is kept at a constant temperature of -10°C. What temperature will the thermometer register after 5 minutes if it registers 5°C after 2 minutes?

In Exercises 65–70, use the definitions of exponential and logarithmic functions to prove the following relationships.

65. $\log_a(m/n) = \log_a m - \log_a n$

66. $\log_a \sqrt[n]{m} = (1/n)\log_a m$

67. $\log_{ab} x = \dfrac{\log_a x}{1 + \log_a b}$

68. $\log_{a/b} x = \dfrac{\log_a x}{1 - \log_a b}$

69. $\log_a x = 2\log_{a^2} x$

70. $(\log_a x)(\log_b y) = (\log_b x)(\log_a y)$

■

CHAPTER 7

RIGHT TRIANGLE
TRIGONOMETRY

The branch of mathematics that studies the relationships between the angles and sides of a triangle is trigonometry. Literally, the word means triangle measuring. It has, however, acquired a broader meaning so as to include the study of circular functions and their graphs, as well as functions that describe periodic motions. The chapter begins with a discussion of angles, their measurement in degrees and radians, and conversion formulas. Trigonometric functions are defined as ratios between the lengths of the sides of a right triangle. This simple and concrete approach is directly related to the method of solving right triangles. The values of the trigonometric functions for special angles (multiples of 30° and 45°) are constructed in this chapter, and several examples and applications involving the use of trigonometric tables and/or calculators are also discussed.

7.1 ANGLES

Let l_1 and l_2 be two distinct half-lines lying on a plane and having the same origin O (Figure 7.1). An angle can be generated when one of the half-lines is rotated around O until it coincides with the other half-line.

For example, we can say that the angle α is obtained when the half-line l_1 is rotated *counterclockwise* around O until it coincides with the half-line l_2. Similarly, the angle β can be obtained by rotating the half-line l_1 clockwise around O until it coincides with the half-line l_2. In both cases, l_1 is called the *initial side*, l_2 the *terminal side*, and O the *vertex* of each angle. Also, if A and B are points on l_1 and l_2, respectively, we may refer to the angle α as the angle AOB and denote the angle by $\angle \alpha$ or $\angle AOB$.

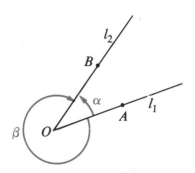

Figure 7.1

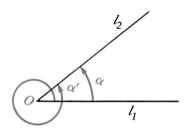

Figure 7.2

Two angles are said to be *coterminal* if they have the same initial and terminal sides. The angles α and β in Figure 7.1 are coterminal.

An angle is said to be *positive* when its initial side is rotated in a counterclockwise direction to its terminal side. It is *negative* if its initial side is rotated in a clockwise direction. In Figure 7.1, the angle α is positive and the angle β is negative.

In trigonometry, the amount or direction of rotation is not restricted in any way. We shall consider angles whose initial sides make several revolutions about O, in a clockwise or counterclockwise direction, before stopping at their terminal sides. The angles α and α' shown in Figure 7.2 are positive, coterminal, but distinct. The initial side l_1 of α' makes a complete revolution about O before coinciding with the terminal side l_2.

Angles in Standard Position

In a rectangular coordinate system, an angle is said to be in *standard position* when *its vertex coincides with the origin and its initial side coincides with the positive x-axis*. If the terminal side of an angle in standard position lies in the first quadrant, then the angle is said to be a *first-quadrant angle*. Analogously, there are *second-, third-,* and *fourth-quadrant* angles. When the terminal side lies on a coordinate axis, the angle is called *quadrantal angle*. In Figure 7.3, the angles α, β, γ, and δ are, respectively, first-, second-, third-, and fourth-quadrant angles.

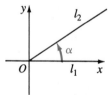

First-quadrant angle

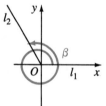

Second-quadrant angle

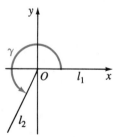

Third-quadrant angle

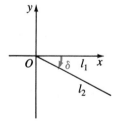

Fourth-quadrant angle

Figure 7.3

Measures of Angles

As you already know, there are several units of length: centimeters, meters, inches, feet, miles; and several units of area: square meters, square feet, square miles, acres. Similarly, there are several units for measuring angles, of which *degrees* and *radians* are the two most commonly used. Before defining them, we introduce the notion of a *central angle*.

Central Angle

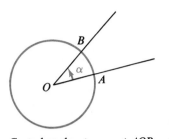

Central angle ∠α = ∠AOB

Figure 7.4

A *central angle* of a circle is an angle whose vertex coincides with the center of the circle.

Let α be a central angle (Figure 7.4). If A and B are, respectively, the intersections of the initial and terminal sides of α with the circle, they determine the arc *AB*. We say that the angle α is *subtended by the arc AB*, or that the arc *AB* *subtends* the angle α.

Degrees

If a circle is divided into 360 equal parts, each central angle subtended by any one of these parts is said to have a measure of *one degree*, denoted by 1°.

In standard position, an angle of 1 degree is obtained by rotating the positive *x*-axis counterclockwise 1/360 of a complete revolution.

Degrees are subdivided into *minutes* and *seconds*. Each degree has 60 minutes (denoted by ′) and each minute has 60 seconds (denoted by ″). Thus

$$1° = 60' \text{ and } 1' = 60''.$$

The use of degrees for angle measurement goes back to the Babylonians, who used a sexagesimal (base 60) system of numeration. The convenience of working with degrees comes from the fact that the integer 360 has a large number of factors. With the growing use of computers and calculators, angles are also measured in *decimal degrees*; for example 15.24° denotes 15 degrees and 24 hundredths of a degree.

An angle of measure 90° is called a *right angle* (Figure 7.5). An angle whose measure is greater than 0° but less than 90° is called an *acute angle*. An angle whose measure is greater than 90° but smaller than 180° is an *obtuse angle*.

For simplicity and when no confusion is possible, we say "α is a 45° angle" or "α = 45°" instead of saying "α is an angle measuring 45°."

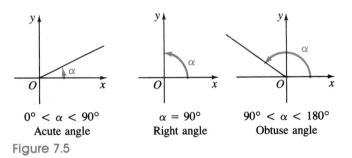

| 0° < α < 90° | α = 90° | 90° < α < 180° |
| Acute angle | Right angle | Obtuse angle |

Figure 7.5

EXAMPLE 1   Find a positive angle and a negative angle that are coterminal with a 45° angle in standard position.

Solution: There are many angles that are coterminal with a 45° angle (see Figure 7.6). For example,

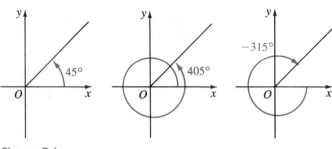

Figure 7.6

$$45° + 360° = 405°$$

is a positive angle coterminal with a 45° angle. A negative angle coterminal with a 45° angle is

$$45° - 360° = -315°.$$

(Can you find other angles?)

Practice Exercise 1   Find a positive angle and a negative angle that are coterminal with a 30° angle in standard position.

Answer: For example, 750° and −690°.

   Note that every angle is coterminal with some angle measuring between 0° and 360°.

EXAMPLE 2   Find the angle θ that is coterminal with an angle measuring 780° and such that $0° \leq \theta \leq 360°$.

Solution: We divide 780 by 360. The remainder, being positive and less than 360, gives us the desired angle. We have

$$780 = 360 \times 2 + 60,$$

so

$$\theta = 60°;$$

see Figure 7.7.

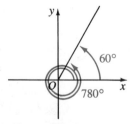

Figure 7.7

Practice Exercise 2   Find α, knowing that $0 \leq \alpha < 360°$ and α is coterminal with an angle measuring 500°.

Answer: $\alpha = 140°$

Radians

Let C be a circle with center at the origin and radius r units of length (Figure 7.8). Let $\angle BOQ$ be an angle in standard position, where B and Q are the points of intersection of C with the initial and terminal sides of the angle. As the initial side rotates toward the terminal side, the point B describes the arc BQ on the circle C, whose measure is assumed to be s units of length. This measure is called the *arc length* of the arc BQ. The arc length is positive for a counterclockwise rotation and negative for a clockwise rotation.

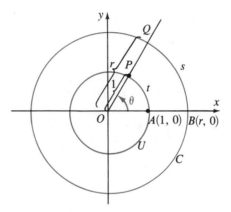

Figure 7.8

Definition
The ratio
$$\theta = \frac{s}{r} \qquad (7.1)$$
is the measure of the angle $\angle BOQ$ in *radians*.

It follows that *one radian* is the measure of *a central angle that is subtended by an arc of length r units on a circle of radius r units.*

EXAMPLE 3   Find the radian measure θ of a central angle subtended by an arc 25 cm long on a circle of radius 8 cm.

Solution: According to (7.1), we have

$$\theta = \frac{25}{8} = 3.125 \text{ radians.}$$

Practice Exercise 3

On a circle of radius 5 in., a central angle θ is subtended by an arc 12 in. long. Find the radian measure of θ.

Answer: $\theta = 2.4$ radians

Figure 7.8 also illustrates the unit circle U with center at the origin. The points A and P are the intersections of the initial and terminal sides of the angle BOQ with U. Assume that the length of the arc AP is t units. From plane geometry, it follows that the ratio of the arc lengths of AP and BQ is the same as the ratio of the lengths of the radii, that is,

$$\frac{t}{1} = \frac{s}{r},$$

or

$$t = \frac{s}{r} = \theta.$$

This shows that *the radian measure of an angle in standard position is numerically equal to the length of the arc on the unit circle that subtends the angle.*

Since the circumference of a unit circle has length equal to 2π units, the radian measure of an angle of one complete counterclockwise revolution is 2π radians. Figure 7.9 illustrates several angles, in standard position, measured in radians.

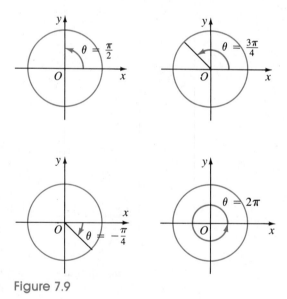

Figure 7.9

In what follows, when no unit of measure is given for an angle, it is assumed that the angle is measured in radians. Thus, we say "θ is the angle $\pi/2$" or "$\theta = \pi/2$," meaning that "θ is an angle whose measure is $\pi/2$ radians."

EXAMPLE **4**

An angle in standard position measures $5\pi/4$ radians. In which quadrant does the terminal side of the angle lie?

Solution: Figure 7.10 shows that the terminal side of the angle $\theta = 5\pi/4$ lies in the third quadrant.

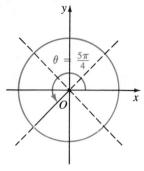

Figure 7.10

Practice Exercise 4

An angle in standard position measures 300°. In which quadrant does the terminal side of the angle lie?

Answer: Fourth quadrant

Conversion Formulas

We have seen that an angle of one complete counterclockwise revolution has measure 2π radians or 360°. Thus,

$$2\pi \text{ radians} = 360 \text{ degrees}$$

or

$$\pi \text{ radians} = 180 \text{ degrees}.$$

Hence, it follows that

$$1 \text{ radian} = \frac{180}{\pi} \text{ degrees}$$

or, approximately, 1 radian $\simeq 57.29578°$. Analogously, we see that

$$1 \text{ degree} = \frac{\pi}{180} \text{ radians}$$

or, approximately, $1° \simeq 0.0174533$.

EXAMPLE 5

Convert each of the following degree measures to radians.
(a) 60°   **(b)** 225°.

Solution: **(a)** Since $1° = \pi/180$ radians, we have

$$60° = 60\left(\frac{\pi}{180}\right) = \frac{60\pi}{180} = \frac{\pi}{3} \text{ radians.}$$

(b) Analogously,

$$225° = 225\left(\frac{\pi}{180}\right) = \frac{225\pi}{180} = \frac{5\pi}{4} \text{ radians.}$$

Practice Exercise 5   Express the following angles in radians. (a) 120° (b) 210°

Answer: (a) $\dfrac{2\pi}{3}$ (b) $\dfrac{7\pi}{6}$

EXAMPLE 6   Convert to degrees each of the following radian measures.

(a) $\dfrac{3\pi}{4}$ (b) $\dfrac{\pi}{12}$

Solution: (a) Since 1 radian = $180/\pi$ degrees, then

$$\frac{3\pi}{4} \text{ radians} = \frac{\overset{3}{\cancel{3\pi}}}{\cancel{4}}\left(\frac{\overset{45}{\cancel{180}}}{\cancel{\pi}}\right) \text{ degrees} = 135°.$$

(b) Analogously,

$$\frac{\pi}{12} \text{ radians} = \frac{\cancel{\pi}}{\cancel{12}}\left(\frac{\overset{15}{\cancel{180}}}{\cancel{\pi}}\right) \text{ degrees} = 15°.$$

Practice Exercise 6   Express in degrees the following radian measures. (a) $4\pi/3$ (b) $3\pi/2$

Answer: (a) 240° (b) 270°

⊙ EXAMPLE 7   (a) Find the approximate value in degrees of an angle $\theta = 4$ radians. Express the angle in degrees, minutes, and seconds.
(b) Convert 60°15′ to radians.

Solution: (a) Since 1 radian ≃ 57.29578°, it follows that 4 radians ≃ 4 × 57.29578° = 229.18312°. Now we must convert the decimal 0.18312° into minutes and seconds. There are 60′ in each degree, so 0.18312° corresponds to

$$0.18312 \times 60' = 10.9872'.$$

Since there are 60″ in each minute, the number of seconds in 0.9872′ is

$$0.9872 \times 60'' = 59.232''.$$

Thus,

$$4 \text{ radians} = 229°10'59''.$$

(b) First, we convert 60°15′ to decimal degrees. Since 15′ = (15/60)° = 0.25°, it follows that 60°15′ = 60.25°. Now, 1° ≃ 0.0174533 radians, so

$$60°15' \simeq 60.25 \times 0.0174533 \text{ radians} = 1.0515609 \text{ radians}.$$

Practice Exercise 7   (a) Convert 2.5 radians to degrees.   (b) Convert 52°18′ to radians.

Answer: (a) 143.23945° = 143°14′22″   (b) 0.9128072 radians

Using Your Calculator

Scientific calculators allow conversion of degrees into radians, and vice versa. In some calculators, such as the TI-30-II, degrees, minutes, and seconds must be first converted to decimal degrees before being entered into the calculator. When converting degrees and minutes to decimal degrees, proceed as in Example 7b. If seconds are also involved, remember that 1° = 60′ = 3600″.

In the TI-30-II, the key sequence ⎡INV⎤ ⎡DRG⎤ instructs the calculator to change the angular mode and, at the same time, converts a displayed value from one angular mode into another. For example, to convert 60°15′ into radians, set the calculator in degree mode, enter 60.25 and press the keys ⎡INV⎤ ⎡DRG⎤. The display

$$\boxed{1.0515609}$$

corresponds to 60°15′ in radians.

Some calculators, such as the Casio fx-910, are equipped with keys that allow direct conversion of degrees, minutes, and seconds into decimal degrees, and vice versa. If you own such a calculator, consult the user's manual for more details about angle conversion.

Finally, most scientific calculators operate also in the *grad mode*. One grad is the measure of a central angle that subtends an arc 1/400 of the circumference of a circle. For example, a right angle measures 100 grads, and 2π radians correspond to 400 grads. However, throughout this book we shall consider angle measurements only in degrees or radians.

Applications

Suppose that a wheel is rotating around its center at a constant rate (Figure 7.11). As the point A travels a distance AP along the rim, the radius OA sweeps the angle AOP.

We give the name *angular speed ω* of A (or the wheel) to the rate at which the measure θ of the angle AOP changes per unit of time, and we write

$$\omega = \frac{\theta}{t}. \tag{7.2}$$

Angular speeds are often measured in *radians per second* (rad/s). For instance, if a wheel is rotating at a rate of one revolution per second, then its angular speed is 2π rad/s.

If r = OA denotes the radius of the wheel, then the *linear speed v* of A is defined by

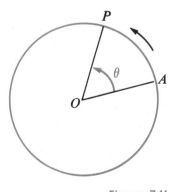

Figure 7.11

$$v = r\omega. \tag{7.3}$$

EXAMPLE 8

On a turntable, a 7-in. diameter record is rotating at a rate of 45 revolutions per minute (rpm). Find: **(a)** the angular speed of the record and the angle swept in 2 seconds; **(b)** the linear speed of a point on its rim and the distance traveled by the point in 2 seconds.

Solution: **(a)** The rate of 45 rpm corresponds to $45/60 = 3/4$ revolutions per second. Since the angular speed of one revolution per second is 2π rad/s, the angular speed of the record is

$$\omega = \frac{3}{4}(2\pi) = \frac{3\pi}{2}\text{rad/s}.$$

During one second, the angle swept by the radius is $3\pi/2$ radians. According to (7.2), the angle θ swept in 2 seconds is

$$\theta = 2 \times \frac{3\pi}{2} = 3\pi\,\text{rad}.$$

(b) Since the radius of the record is $r = 3.5$ in., the linear speed of a point on the rim of the record is given by (7.3):

$$v = (3.5) \times \frac{3\pi}{2} = 5.25\pi \text{ in./s} \simeq 16.5 \text{ in./s}.$$

Thus, the distance traveled by a point on the rim in 2 s is $2 \times (5.25\pi)$ in. = 10.5π in. $\simeq 33$ in. Observe that the same result could have been derived from formula (7.1). Since in 2 s the angle $\theta = 3\pi$ rad is swept by the radius $r = 3.5$ in., the distance traveled by a point on the rim is then

$$s = 3\pi(3.5) = 10.5\pi \text{ in.} \simeq 33 \text{ in.}$$

Practice Exercise 8

A wheel of radius 3 ft is rotating at the rate of 36 rpm. Find the angular speed of the wheel and the angle swept in 3 s.

Answer: $\omega = \dfrac{6\pi}{5}$ rad/s; $\theta = \dfrac{18\pi}{5}$ radians

EXERCISES 7.1

In Exercises 1–6, place each angle in standard position and find a positive angle and a negative angle that are coterminal with the given angle.

1. 150°

2. 210°

3. −60°

4. $\dfrac{5\pi}{6}$

5. $\dfrac{7\pi}{3}$

6. $-\dfrac{9\pi}{4}$

In Exercises 7–12, in which quadrant does the terminal side of each of the following angles lie? Sketch the angles in standard position.

7. 156°

8. −105°

9. −318°

10. $-\dfrac{3\pi}{4}$

11. $-\dfrac{5\pi}{4}$

12. $\dfrac{4\pi}{3}$

In Exercises 13–18, convert each degree measure to radians. Write your answers as rational multiples of π. Do not use a calculator.

13. 150°   **14.** 210°
15. −240°   **16.** −45°
17. −120°   **18.** 300°

In Exercises 19–24, convert each radian measure to degrees. Do not use a calculator.

19. $\dfrac{5\pi}{6}$ **20.** $\dfrac{3\pi}{4}$

21. $-\dfrac{\pi}{6}$ **22.** $\dfrac{7\pi}{6}$

23. $-\dfrac{9\pi}{4}$ **24.** $-\dfrac{3\pi}{2}$

In Exercises 25–28, a central angle θ is subtended by an arc of length s on a circle of radius r. Find the measure of θ in radians for each given arc length and radius.

25. $s = 20$ cm. $r = 4$ cm   **26.** $s = 38$ ft, $r = 8$ ft
27. $s = 6$ m, $r = 24$ cm   **28.** $s = 4$ ft, $r = 8$ in.

In Exercises 29–32, assume that the circumference of a circle is divided into n equal parts. Find the measure, in degrees and radians, of the central angle subtended by each arc for each given n.

29. $n = 6$ **30.** $n = 18$
31. $n = 12$ **32.** $n = 24$

In Exercises 33–38, find the required quantity.

33. Find the length of an arc that subtends a central angle of 5 radians on a circle of radius 12 cm.
34. Find the length of an arc that subtends a central angle of 6.2 radians on a circle of radius 6 ft.
35. If a central angle of 30° is subtended by a circular arc of length 12 m, find the radius of the circle.
36. A central angle of 45° is subtended by an arc 18 ft long. What is the radius of the circle?
37. A central angle of 4.05 radians is subtended by an arc of 9.72 cm. Find the radius of the circle.
38. If a central angle of 5.6 radians is subtended by an arc of 19.6 ft, find the radius of the circle.

ⓒ In Exercises 39–46, convert each angle to decimal degrees. Round off your answers to two decimal places.

39. $\alpha = 30°44'$ **40.** $\alpha = 45°16'$
41. $\alpha = 85°30'45''$ **42.** $\alpha = 36°18'24''$
43. $\theta = 4$ radians   **44.** $\theta = 2$ radians
45. $\theta = 1.5$ radians   **46.** $\theta = 3.2$ radians

ⓒ In Exercises 47–52, find the approximate measure of each angle in degrees, minutes, and seconds.

47. $\theta = 5$ radians   **48.** $\theta = 3$ radians
49. $\theta = 1.2$ radians   **50.** $\theta = 2.6$ radians
51. $\theta = 0.15$ radians   **52.** $\theta = 0.21$ radians

ⓒ In Exercises 53–58, convert each of the degree measures to radians. Round off your answers to four decimal places.

53. 45°   **54.** 60°
55. 38°   **56.** 52°
57. 126°   **58.** 218°

ⓒ In Exercises 59–64, express each angle in radians. Round off your answers to four decimal places.

59. 38°18'   **60.** 46°24'
61. 67°40'   **62.** 120°45'
63. 12°30'15''   **64.** 210°20'18''

In Exercises 65–80, solve each of the given problems.

65. Through how many radians does the hour hand of a clock turn in 1 hour? In $\frac{1}{2}$ hour? In 15 hours?
66. Through how many radians does the hour hand of a clock turn in 15 minutes? In two hours? In 210 minutes?
67. A stone attached to a string 1 m long and swung in a circle executes 32 revolutions per minute (rpm). Find **(a)** its angular and linear speed, and **(b)** the length of the arc described by the stone in 5 s.
68. Find the angular speed of a wheel rotating around its center at a rate of 1200 rpm. What is the linear speed of a point on the wheel located 6 in. from the center?
69. A record with a diameter of 12 in. is rotating at a rate of 78 rpm. Find **(a)** the angular velocity in radians per second, and **(b)** the linear speed, in inches per second, of a point on its rim.
ⓒ **70.** An earth satellite in a circular orbit 219 miles high makes a complete revolution around the earth in 115 minutes. What is its linear speed in miles per hour? (Earth's radius: 4000 miles)
ⓒ **71.** Assume that the earth's orbit around the sun is circular, with the sun located at its center. Determine, in radians, the angle swept out by a line from the sun to the earth in **(a)** 10 days, **(b)** one month. [*Hint:* The earth revolves around the sun in 365 days.]
72. Jupiter revolves around the sun in 12 years. Assuming that Jupiter's orbit is circular, with the sun at the center, find the radian measure of the angle swept out by a line from the sun to Jupiter in **(a)** 5 years, **(b)** 90 months.
73. When a rope is pulled down, it rotates (without slippage) a pulley whose diameter measures 30 cm. Through how many radians does the pulley turn when 120 cm of rope is pulled down?
74. A belt encircling a pulley with a diameter of 1 m rotates it without slippage. How many radians does the pulley turn when 3 m of the belt has gone around it?
ⓒ **75.** The radius of the earth is approximately 6.371×10^3 km. If the distance between two points P and Q on the surface of the earth is measured along the circumference of a *great*

circle, one whose center O is at the center of the earth, find the distance PQ if the angle POQ has measure **(a)** $\pi/6$; **(b)** $60°$; **(c)** 1 radian.

76. For the points P and Q in Exercise 75, find the distance in miles when the angle POQ measures **(a)** $30°$; **(b)** $\pi/12$ radians. (Earth's radius: 3.959×10^3 miles)

77. A *nautical mile* is defined as the length of an arc on a *great circle* (see Exercise 75) drawn on the surface of the earth and subtending a central angle of 1 minute. Find the length of a nautical mile in miles.

78. The radius of the earth is about 2.0903×10^7 ft. Find the length of a nautical mile in feet.

79. New York City, at $41°$ of latitude north, and Lima, Peru, at $12°$ of latitude south, are approximately on a north–south line. Find the distance in miles between these two cities. (Radius of the earth: 3.959×10^3 miles)

80. St. John's, Newfoundland, at $47.34°$ of latitude north, and Cachoeira do Sul, Brazil, at $30.02°$ of latitude south, are approximately on a north–south line. What is the distance in kilometers between these two cities? (Earth's radius: 6.371×10^3 km)

7.2 RIGHT TRIANGLE TRIGONOMETRY

Recall that a triangle is said to be a *right triangle* if one of its angles measures $90°$ (or $\pi/2$ radians). The other two angles are acute angles and their measures add to $90°$, since the sum of the measures of all angles in a triangle equals $180°$.

In its simplest form, trigonometry is the study of relationships between the angles and sides of a right triangle. Consider the right triangle ACB, shown in Figure 7.12, with sides of length a, b, hypotenuse of length c, and angles of measures α, β, and $\gamma = 90°$.

Throughout this chapter, whenever we refer to a right triangle ACB, we shall always denote by C the vertex of the right angle, and by α and β the measures of the acute angles with vertices A and B, respectively. The length of the hypotenuse will be denoted by c; the length of the side BC *opposite* to the angle α and *adjacent* to the angle β will be denoted by a; finally, the length of the side AC *opposite* to the angle β and *adjacent* to the angle α will be denoted by b.

There are six numbers that correspond to the angle α in the right triangle ACB: the *sine, cosine, tangent, cotangent, secant,* and *cosecant* of α, called the *trigonometric functions* of α and denoted by $\sin \alpha$, $\cos \alpha$, $\tan \alpha$, $\cot \alpha$, $\sec \alpha$, $\csc \alpha$, respectively. These numbers are defined as ratios between the lengths of the sides of the triangle in the following way.

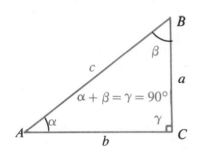

Figure 7.12

$$\sin \alpha = \frac{a}{c} \qquad \csc \alpha = \frac{c}{a}$$

$$\cos \alpha = \frac{b}{c} \qquad \sec \alpha = \frac{c}{b} \qquad (7.4)$$

$$\tan \alpha = \frac{a}{b} \qquad \cot \alpha = \frac{b}{a}$$

As this table summarizes, the sine of α is *the ratio between the length of the opposite side and the length of the hypotenuse*; the cosine of α is *the ratio between the length of the adjacent side and the length of the hypotenuse*; and the tangent of α is *the ratio between the length of the opposite side and the length of the adjacent side*. Similar statements can be made about the other three trigonometric functions of α: cot α, sec α, and csc α.

It is important to observe that sine, cosine, and tangent are the three basic trigonometric functions. The other three functions are obtained by taking reciprocals. As an exercise, you can check the following relations:

$$\csc \alpha = \frac{1}{\sin \alpha}, \qquad \sec \alpha = \frac{1}{\cos \alpha}, \qquad \cot \alpha = \frac{1}{\tan \alpha}.$$

This is why calculators have only the keys $\boxed{\sin}$, $\boxed{\cos}$, and $\boxed{\tan}$ for computing the trigonometric functions of an angle. To obtain the remaining functions, you have to use the reciprocal key $\boxed{1/x}$.

As an exercise, you can derive from (7.4) two important relations:

$$\tan \alpha = \frac{\sin \alpha}{\cos \alpha}, \qquad \cot \alpha = \frac{\cos \alpha}{\sin \alpha}.$$

Formulas similar to those in (7.4) can be obtained for the angle β in the triangle ACB (Figure 7.12). For each of the two acute angles of a right triangle, we have the following relations, where "opp" stands for opposite, "adj" for adjacent, and "hyp" for hypotenuse.

$$
\begin{aligned}
\sin &= \frac{\text{opp}}{\text{hyp}} & \csc &= \frac{\text{hyp}}{\text{opp}} \\[1em]
\cos &= \frac{\text{adj}}{\text{hyp}} & \sec &= \frac{\text{hyp}}{\text{adj}} \\[1em]
\tan &= \frac{\text{opp}}{\text{adj}} & \cot &= \frac{\text{adj}}{\text{opp}}
\end{aligned}
\qquad (7.5)
$$

These relations indicate that the sine of an acute angle in a right triangle is the ratio between the length of the side opposite the angle and the length of the hypotenuse; the cosine is the ratio between the length of the side adjacent to the angle and the hypotenuse; and so on.

The relations (7.5) are frequently used when working with right triangles; they should be memorized.

EXAMPLE $\boxed{1}$ Given the triangle in Figure 7.13, find the values of the trigonometric functions of α.

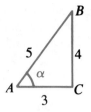

Figure 7.13

Solution: According to the formulas (7.4), we have

$$\sin \alpha = \frac{4}{5}, \qquad \csc \alpha = \frac{5}{4},$$

$$\cos \alpha = \frac{3}{5}, \qquad \sec \alpha = \frac{5}{3},$$

$$\tan \alpha = \frac{4}{3}, \qquad \cot \alpha = \frac{3}{4}.$$

Practice Exercise 1   Given the following right triangle, find the values of the trigonometric functions of α.

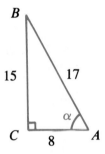

Answer: $\sin \alpha = \dfrac{15}{17}$, $\cos \alpha = \dfrac{8}{17}$, $\tan \alpha = \dfrac{15}{8}$, $\cot \alpha = \dfrac{8}{15}$, $\sec \alpha = \dfrac{17}{8}$,

$\csc \alpha = \dfrac{17}{15}$

EXAMPLE **2**   Given the triangle in Figure 7.14, find the values of the six trigonometric functions of the angle α.

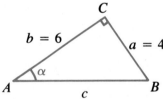

Figure 7.14

Solution: First, we have to find the length c of the hypotenuse. By the Pythagorean theorem,

$$c^2 = a^2 + b^2$$
$$= 4^2 + 6^2$$
$$= 16 + 36$$
$$= 52;$$

hence

$$c = \sqrt{52} = 2\sqrt{13}.$$

Next, according to (7.4), we obtain

$$\sin \alpha = \frac{4}{2\sqrt{13}} = \frac{2\sqrt{13}}{13}, \qquad \csc \alpha = \frac{2\sqrt{13}}{4} = \frac{\sqrt{13}}{2},$$

$$\cos \alpha = \frac{6}{2\sqrt{13}} = \frac{3\sqrt{13}}{13}, \qquad \sec \alpha = \frac{2\sqrt{13}}{6} = \frac{\sqrt{13}}{3},$$

$$\tan \alpha = \frac{4}{6} = \frac{2}{3}, \qquad \cot \alpha = \frac{6}{4} = \frac{3}{2}.$$

Notice that in writing trigonometric ratios, we usually rationalize the denominators.

Practice Exercise 2   Find the values of the six trigonometric functions of the angle β.

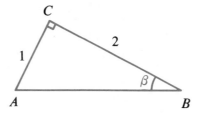

Answer: $\sin \beta = \dfrac{\sqrt{5}}{5}$, $\cos \beta = \dfrac{2\sqrt{5}}{5}$, $\tan \beta = \dfrac{1}{2}$, $\cot \beta = 2$, $\sec \beta = \dfrac{\sqrt{5}}{2}$, $\csc \beta = \sqrt{5}$

□ EXAMPLE 3   Find the values of the six trigonometric functions of the angle β in Figure 7.15. Round off your answers to four decimal places.

Solution: We have

$$\sin \beta = \frac{2.9087}{5.0712} = 0.5736, \qquad \csc \beta = \frac{5.0712}{2.9087} = 1.7435,$$

$$\cos \beta = \frac{4.1541}{5.0712} = 0.8192, \qquad \sec \beta = \frac{5.0712}{4.1541} = 1.2208,$$

$$\tan \beta = \frac{2.9087}{4.1541} = 0.7002, \qquad \cot \beta = \frac{4.1541}{2.9087} = 1.4282.$$

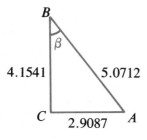

Figure 7.15

Practice Exercise 3   Given the following right triangle, find the values of the six trigonometric functions of α. Round off your answers to three decimal places.

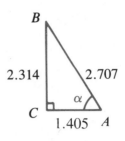

Answer: $\sin \alpha = 0.855$, $\cos \alpha = 0.519$, $\tan \alpha = 1.647$, $\cot \alpha = 0.607$,
$\sec \alpha = 1.927$, $\csc \alpha = 1.170$

EXAMPLE 4

Let α be an acute angle such that $\sin \alpha = 1/2$. Find the values of the remaining five trigonometric functions of α.

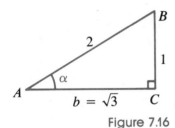

Figure 7.16

Solution: First, we draw a right triangle with one of its sides measuring 1 unit of length and the hypotenuse measuring 2 units of length. We label as α the angle facing the side of length 1 (Figure 7.16). Next, using the Pythagorean theorem, we can determine the length b of the other side:

$$b^2 = 2^2 - 1^2$$
$$= 4 - 1$$
$$= 3.$$

Hence

$$b = \sqrt{3}.$$

Now we have

$$\cos \alpha = \frac{\sqrt{3}}{2} \quad \text{and} \quad \tan \alpha = \frac{1}{\sqrt{3}} = \frac{\sqrt{3}}{3}.$$

We can obtain the remaining trigonometric values by taking reciprocals. If $\tan \alpha = \sqrt{3}/3$, then $\cot \alpha = 3/\sqrt{3} = \sqrt{3}$. If $\cos \alpha = \sqrt{3}/2$, then $\sec \alpha = 2/\sqrt{3} = 2\sqrt{3}/3$. Finally, if $\sin \alpha = 1/2$, then $\csc \alpha = 2/1 = 2$.

Practice Exercise 4

Let β be an acute angle such that $\cos \beta = 3/4$. Find the values of the remaining five trigonometric functions of β.

Answer: $\sin \beta = \dfrac{\sqrt{7}}{4}$, $\tan \beta = \dfrac{\sqrt{7}}{3}$, $\cot \beta = \dfrac{3\sqrt{7}}{7}$, $\sec \beta = \dfrac{4}{3}$,

$\csc \beta = \dfrac{4\sqrt{7}}{7}$

Similar Triangles

Let ACB and $A'C'B'$ be two right triangles (Figure 7.17) and suppose that the angles of vertices A and A' are *equal*.

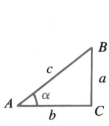
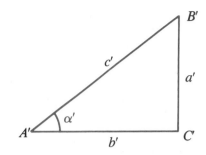

Figure 7.17

Then the trigonometric functions

$$\sin \alpha = \frac{a}{c} \quad \text{and} \quad \sin \alpha' = \frac{a'}{c'}$$

are also equal.

To see this, we recall that two triangles are said to be *similar* if the corresponding angles are equal. Since the angle α in the right triangle ACB is equal to the angle α' in the right triangle $A'C'B'$, it follows that *all* corresponding angles are equal and, therefore, the two triangles are similar.

In similar triangles, the lengths of corresponding sides (that is, facing equal angles) are proportional. For the triangles ACB and $A'C'B'$, this means that

$$\frac{a}{a'} = \frac{b}{b'} = \frac{c}{c'}. \tag{7.6}$$

Leaving aside the middle term of the relationship (7.6), we can write

$$\frac{a}{a'} = \frac{c}{c'}$$

or, equivalently,

$$\frac{a}{c} = \frac{a'}{c'}.$$

Thus

$$\sin \alpha = \sin \alpha'.$$

Notice that the triangles ACB and $A'C'B'$ are *not congruent* but similar, and we may say that the triangle $A'C'B'$ is a magnification of the triangle ACB. However, since $\alpha = \alpha'$, the sine of this angle can be computed using either one of the two triangles. The lengths a and a' are different, and c and c' are different; but the ratios a/c and a'/c' are equal.

Similarly, from (7.6) we also derive that

$$\cos \alpha = \frac{b}{c} = \frac{b'}{c'} \quad \text{and} \quad \tan \alpha = \frac{a}{b} = \frac{a'}{b'}.$$

EXAMPLE **5**   The two triangles shown in Figure 7.18 are similar. Find b'.

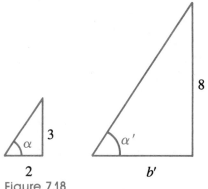

Figure 7.18

Solution: **First method.** Using the definition of the tangent function and the dimensions indicated in both triangles, we have:

$$\tan \alpha = \frac{3}{2} \quad \text{and} \quad \tan \alpha' = \frac{8}{b'}.$$

Thus,

$$\frac{3}{2} = \frac{8}{b'}$$

and, solving for b', we obtain

$$b' = \frac{2 \times 8}{3} = \frac{16}{3}.$$

Second method. Substituting 3 for a, 8 for a', and 2 for b in (7.6), we obtain

$$\frac{3}{8} = \frac{2}{b'}.$$

Hence

$$b' = \frac{2 \times 8}{3} = \frac{16}{3}.$$

Practice Exercise 5   The following two triangles are similar. Find a'.

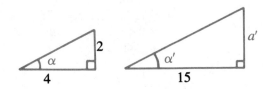

Answer: $a' = \dfrac{15}{2}$

◯ EXAMPLE 6   The two triangles in Figure 7.19 are similar. Find sin α and sin α'. Are they equal?

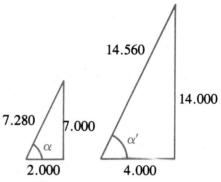

Figure 7.19

Solution: We have

$$\sin \alpha = \frac{7.000}{7.280} = 0.962$$

and

$$\sin \alpha' = \frac{14.000}{14.560} = 0.962.$$

Thus

$$\sin \alpha = \sin \alpha'.$$

Practice Exercise 6   The following two triangles are similar. Verify that cos α = cos α'.

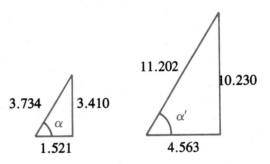

Answer: cos α = 0.407 = cos α'

Complementary Angles

Two angles are said to be *complementary* if the sum of their measures is 90° (or $\pi/2$ radians). As an example, the acute angles α and β of a right triangle (Figure 7.12) are complementary.

Since the side of length a is opposite the angle α and adjacent to the angle β, we have

$$\sin \alpha = \frac{a}{c} = \cos \beta.$$

Analogously,

$$\cos \alpha = \frac{b}{c} = \sin \beta, \qquad \tan \alpha = \frac{a}{b} = \cot \beta,$$

and so on.

For example, consider the right triangle ACB (Figure 7.20) with sides $a = 5$, $b = 12$, and $c = 13$.

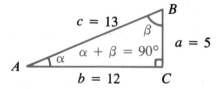

Figure 7.20

We have $\sin \alpha = 5/13$ and $\cos \beta = 5/13$. Thus, $\sin \alpha = \cos \beta$. Analogously,

$$\cos \alpha = \frac{12}{13} = \sin \beta,$$

and

$$\tan \alpha = \frac{5}{12} = \cot \beta.$$

Since $\beta = 90° - \alpha$, we obtain the following relations concerning complementary angles.

$$
\begin{array}{ll}
\sin \alpha = \cos(90° - \alpha) & \csc \alpha = \sec(90° - \alpha) \\
\cos \alpha = \sin(90° - \alpha) & \sec \alpha = \csc(90° - \alpha) \\
\tan \alpha = \cot(90° - \alpha) & \cot \alpha = \tan(90° - \alpha)
\end{array}
$$

The sine and cosine functions are said to be *cofunctions* of each other (the prefix *co-* refers to *complementary*). The same can be said of the tangent and cotangent functions, and the secant and cosecant functions. These relations show that any trigonometric function of an acute angle equals the cofunction of the complementary angle. For example,

$$\sin 30° = \cos 60°, \qquad \cos 40° = \sin 50°, \qquad \tan \frac{\pi}{3} = \cot \frac{\pi}{6}.$$

Special Angles

The angles of measure $30° = \pi/6$ radian, $45° = \pi/4$ radian, and $60° = \pi/3$ radians are called *special angles* because the exact value of their trigonometric functions can be obtained using the definitions and the Pythagorean theorem.

The Values of the Trigonometric Functions of a 45° Angle

In a right triangle, if one acute angle measures 45°, then the other acute angle also measures 45° and the triangle is isosceles. Suppose that in such a triangle the equal sides measure 1 unit of length. Then, by the Pythagorean theorem, the hypotenuse measures $\sqrt{2}$ units of length. From Figure 7.21, we can now read off the values of the trigonometric functions of $\alpha = 45°$.

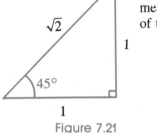

Figure 7.21

$$\sin 45° = \frac{1}{\sqrt{2}} = \frac{\sqrt{2}}{2} \qquad \csc 45° = \sqrt{2}$$

$$\cos 45° = \frac{1}{\sqrt{2}} = \frac{\sqrt{2}}{2} \qquad \sec 45° = \sqrt{2}$$

$$\tan 45° = 1 \qquad\qquad \cot 45° = 1$$

The Values of the Trigonometric Functions of a 60° Angle

A triangle is said to be *equilateral* if its three sides are equal. All the angles are also equal, measuring 60° (or $\pi/3$ radians) each. A line drawn from any vertex of an equilateral triangle, perpendicular to the opposite side, passes through the midpoint of the opposite side (Figure 7.22a).

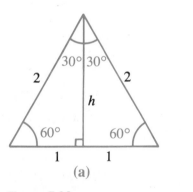
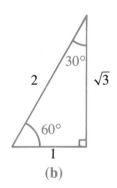

(a)   (b)

Figure 7.22

This divides the equilateral triangle into two right triangles with angles measuring 30°, 60°, and 90° (Figure 7.22a). Assume that each side of the equilateral triangle measures 2 units of length. It is easy to check, using the Pythagorean theorem, that the height h of the equilateral triangle is $\sqrt{3}$. From the right triangle in Figure 7.22b, we can read off the values of the trigonometric functions of the angle of measure 60° ($= \pi/3$ radians).

$$\sin 60° = \frac{\sqrt{3}}{2} \qquad \csc 60° = \frac{2}{\sqrt{3}} = \frac{2\sqrt{3}}{3}$$

$$\cos 60° = \frac{1}{2} \qquad \sec 60° = 2$$

$$\tan 60° = \sqrt{3} \qquad \cot 60° = \frac{1}{\sqrt{3}} = \frac{\sqrt{3}}{3}$$

The Values of the Trigonometric Functions of a 30° Angle

The values of the trigonometric functions of a 30° angle may now be obtained by using the values of the trigonometric functions of a 60° angle and the relations about complementary angles. Since $\alpha = 30°$ and $90° - \alpha = 60°$ are complementary angles, we have $\sin \alpha = \cos(90° - \alpha)$, so

$$\sin 30° = \cos 60° = \frac{1}{2}.$$

In a similar fashion,

$$\cos 30° = \sin 60° = \frac{\sqrt{3}}{2},$$

and

$$\tan 30° = \cot 60° = \frac{\sqrt{3}}{3}.$$

By taking reciprocals, we can obtain the values of the remaining trigonometric functions of a 30° angle:

$$\cot 30° = \frac{3}{\sqrt{3}} = \sqrt{3}, \quad \sec \ 30° = \frac{2}{\sqrt{3}} = \frac{2\sqrt{3}}{3}, \quad \text{and} \quad \csc 30° = 2.$$

For future reference, we list the values of the trigonometric functions of the special angles in table form.

SPECIAL ANGLES

α degrees	α radians	$\sin \alpha$	$\cos \alpha$	$\tan \alpha$	$\cot \alpha$	$\sec \alpha$	$\csc \alpha$
30°	$\frac{\pi}{6}$	$\frac{1}{2}$	$\frac{\sqrt{3}}{2}$	$\frac{\sqrt{3}}{3}$	$\sqrt{3}$	$\frac{2\sqrt{3}}{3}$	2
45°	$\frac{\pi}{4}$	$\frac{\sqrt{2}}{2}$	$\frac{\sqrt{2}}{2}$	1	1	$\sqrt{2}$	$\sqrt{2}$
60°	$\frac{\pi}{3}$	$\frac{\sqrt{3}}{2}$	$\frac{1}{2}$	$\sqrt{3}$	$\frac{\sqrt{3}}{3}$	2	$\frac{2\sqrt{3}}{3}$

Since special angles are frequently used in applications, it is a good idea either to memorize this table or to be able to find quickly the values of the trigonometric functions for the special angles. Notice that it is enough to memorize only the trigonometric values of sine for the three angles. The remaining values are obtained by reversing the entire sine column (for cosine) taking ratios (for tangent and cotangent), and taking reciprocals (for secant and cosecant). As for the sine column, observe the pattern: $1/2 = \sqrt{1}/2, \sqrt{2}/2, \sqrt{3}/2$.

Trigonometric Tables

To compute trigonometric functions of an acute angle other than the special angles we just considered, we need the help of a trigonometric table or a calculator.

In Figure 7.23 we reproduce a portion of Table 4, found in the Appendix. The table uses complementary angles measured in degrees and minutes or in radians. The trigonometric values have been rounded off to four decimal places. To determine the trigonometric function of an angle between 0° and 45°, locate the value of the angle in the left column; find the desired function at the top of the table and read down to the indicated row. For example, Figure 7.23 shows how to find

$$\cos 30°20' = 0.8631.$$

For an angle between 45° and 90°, locate the value of the angle in the *right* column; find the desired function at the *bottom* of the table, and read *up*. For example, the figure shows how to find

$$\tan 59°30' = 1.698.$$

Observe that

$$\cos 30°20' = 0.8631 = \sin 59°40',$$

because 30°20' and 59°40' are complementary angles.

If the given angle is not found in the table, then we can use the method of *linear interpolation,* discussed in the Appendix, to approximate the values of the trigonometric functions of the angle. When speed and more accuracy are desired, a scientific calculator should be used. Calculators usually operate in at least two

Degrees	Radians	Sin	Cos	Tan	Cot	Sec	Csc	
⋮	⋮	⋮	⋮	⋮	⋮	⋮	⋮	⋮
30°00'	.5236	.5000	.8660	.5774	1.732	1.155	2.000	1.0472	**60°00'**
10	265	025	646	812	720	157	1.990	443	**50**
→ 20	294	050	631	851	709	159	980	414	**40**
30	.5323	.5075	.8616	.5890	1.698	1.161	1.970	1.0385	**30** ←
40	352	100	601	930	686	163	961	356	**20**
50	381	125	587	969	675	165	951	327	**10**
31°00'	.5411	.5150	.8572	.6009	1.664	1.167	1.942	1.0297	**59°00'**
10	440	175	557	048	653	169	932	268	**50**
20	469	200	542	088	643	171	923	239	**40**
30	.5498	.5225	.8526	.6128	1.632	1.173	1.914	1.0210	**30**
40	527	250	511	168	621	175	905	181	**20**
50	556	275	496	208	611	177	896	152	**10**
⋮	⋮	⋮	⋮	⋮	⋮	⋮	⋮	⋮
	Cos	Sin	Cot	Tan	Csc	Sec	Radians	Degrees

Figure 7.23

modes: degree or radian. Some of them also operate in grad mode. Be sure to convert measurements in degrees, minutes, and seconds into decimal degrees if your calculator requires it.

⊙ EXAMPLE 7

Use a calculator to approximate the following numbers.
(a) sin 79°   (b) cot 79°

Solution: We must first make sure that our calculator is in degree mode.
(a) Enter 79 and press the key $\boxed{\sin}$ to obtain

$$\sin 79° = 0.9816272.$$

The value given by Table 4 in the Appendix is 0.9816.
(b) To evaluate cot 79°, we must use the relation cot $\alpha = 1/\tan \alpha$. Entering 79 and pressing the keys $\boxed{\tan}$ and $\boxed{1/x}$, we obtain

$$\cot 79° = 0.1943803.$$

The table value in the Appendix is 0.1944.

⊙ Practice Exercise 7

Approximate the following numbers.
(a) cos 36° (b) csc 36°

Answer: (a) 0.809017 (b) 1.7013016

⊙ EXAMPLE 8

Use a calculator to approximate the following numbers.
(a) cos 30°20′ (b) sec 30°20′

Solution: (a) We set the calculator in degree mode and convert degrees and minutes into decimal degrees:

$$30°20′ = 30° + (20/60)° = 30.333333°.$$

Next we press the key $\boxed{\cos}$, to obtain

$$\cos 30°20′ = 0.8631019.$$

(b) To evaluate sec 30°20′, we must use the relation sec $x = 1/\cos x$. With the calculator in degree mode, we enter 30.333333 and press the keys $\boxed{\cos}$ and $\boxed{1/x}$, obtaining

$$\sec 30°20′ = 1.1586118.$$

⊙ Practice Exercise 8

Use a calculator to find the following numbers.
(a) sin 42°30′ (b) sec 42°30′

Answer: (a) 0.6755902 (b) 1.3563417

⊙ EXAMPLE 9

Approximate the following numbers.

(a) $\tan \dfrac{2\pi}{11}$ (b) $\cot \dfrac{2\pi}{11}$

Solution: We make sure that our calculator is in radian mode.

(a) Performing the key strokes 2 $\boxed{\times}$ $\boxed{\pi}$ $\boxed{\div}$ 11 $\boxed{=}$ $\boxed{\tan}$, we obtain

$$\tan\frac{2\pi}{11} = 0.642661.$$

(b) The key strokes 2 $\boxed{\times}$ $\boxed{\pi}$ $\boxed{\div}$ 11 $\boxed{=}$ $\boxed{\tan}$ $\boxed{1/x}$ give us

$$\cot\frac{2\pi}{11} = 1.5560304.$$

◖ Practice Exercise 9   Use your calculator to approximate the following numbers.

(a) $\cos\dfrac{3\pi}{10}$ **(b)** $\sec\dfrac{3\pi}{10}$

Answer: **(a)** 0.5877853 **(b)** 1.7013016

EXERCISES 7.2

In Exercises 1–6, find the trigonometric functions for α.

1.

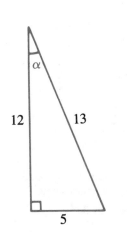

2.

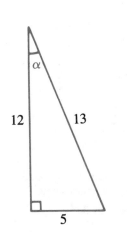

3.

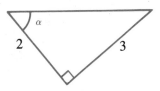

4.

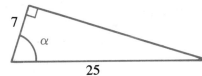

5.

6.

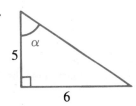

🅲 In Exercises 7–10, find the trigonometric functions of the angle α. Round off your answers to four decimal places.

7.

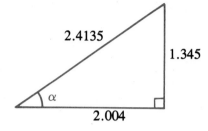

8.

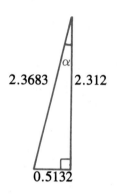

9.

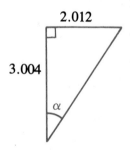

10.

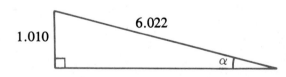

In Exercises 11–14, the given triangles are similar. Find the length of the indicated side.

11.

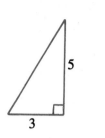
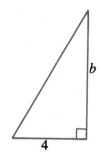

12.

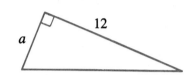

🅲 **13.**

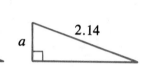
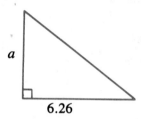

🅲 **14.**

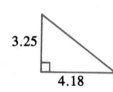

In Exercises 15–24, a trigonometric function of the angle α is given. Find the exact values of the remaining five trigonometric functions of α. Do not use your calculator.

15. $\sin \alpha = \dfrac{5}{8}$ **16.** $\sin \alpha = \dfrac{4}{7}$

17. $\cos \alpha = \dfrac{3}{4}$ **18.** $\cos \alpha = \dfrac{5}{11}$

19. $\tan \alpha = \dfrac{3}{5}$ **20.** $\cot \alpha = \dfrac{7}{2}$

21. $\cot \alpha = \sqrt{2}$ **22.** $\tan \alpha = 5$

23. $\csc \alpha = \dfrac{3}{2}$ **24.** $\sec \alpha = \sqrt{3}$

In Exercises 25–30, express each trigonometric value as the value of the cofunction of the complementary angle.

25. $\sin 48°$ **26.** $\cos 32°$

27. $\tan\dfrac{5\pi}{12}$

28. $\cot\dfrac{3\pi}{15}$

29. $\sec 78°$

30. $\csc 12°$

In Exercises 31–37, find the required trigonometric function.

31. Let $\cos(90° - \alpha) = 0.5299$. Find $\sin \alpha$.

32. If $\sin\left(\dfrac{\pi}{2} - \alpha\right) = 0.9063$, find $\cos \alpha$.

33. Find $\sin \alpha$, if $\sin\left(\dfrac{\pi}{2} - \alpha\right) = \dfrac{4}{5}$.

34. Find $\cos \alpha$, if $\cos\left(\dfrac{\pi}{2} - \alpha\right) = \dfrac{5}{13}$.

35. If $\sin\left(\dfrac{\pi}{2} - \alpha\right) = \dfrac{3}{4}$, find $\tan \alpha$.

36. Find $\tan \alpha$, if $\cos\left(\dfrac{\pi}{2} - \alpha\right) = \dfrac{5}{7}$.

37. Let $\tan(90° - \alpha) = \dfrac{7}{3}$. Find $\sin \alpha$.

38. Let $\tan\left(\dfrac{\pi}{2} - \alpha\right) = \dfrac{3}{5}$. Find $\cos \alpha$.

In Exercises 39–46, use Table 4 in the Appendix to approximate the given trigonometric values.

39. $\tan 15°30'$

40. $\sec 25°10'$

41. $\cos 56°40'$

42. $\sin 65°50'$

43. $\sin 0.1047$

44. $\sec 1.4573$

45. $\tan 0.3927$

46. $\cot 1.1257$

In Exercises 47–54, approximate your answers to four decimal places.

47. $\sin 22.6°$

48. $\cos 46.1°$

49. $\tan 12.05°$

50. $\sec 36.25°$

51. $\sin 23°15'$

52. $\cos 35°15'$

53. $\tan 48°12'$

54. $\sec 52°26'$

In Exercises 55–60, solve the given problems.

55. Use formulas (7.4) to show that $\tan \alpha = \dfrac{\sin \alpha}{\cos \alpha}$ and that

$$\sec \alpha = \dfrac{1}{\cos \alpha}.$$

56. Use formulas (7.4) to show that $\cot \alpha = \dfrac{\cos \alpha}{\sin \alpha}$ and that

$$\csc \alpha = \dfrac{1}{\sin \alpha}.$$

57. If α is an acute angle, prove that $(\sin \alpha)^2 + (\cos \alpha)^2 = 1$. [*Hint*: Use the Pythagorean theorem.]

58. Let α be an acute angle. Prove that

$$(\cot \alpha)^2 + 1 = (\csc \alpha)^2.$$

[*Hint*: Use Exercises 56 and 57.]

59. Let α be an acute angle. Prove that

$$(\tan \alpha)^2 + 1 = (\sec \alpha)^2.$$

[*Hint*: Use Exercises 55 and 57.]

60. If a straight line with equation $y = mx + b$, where $m > 0$, forms an acute angle x with the positive x-axis, prove that $m = \tan \alpha$.

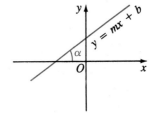

7.3   SOLVING RIGHT TRIANGLES

In the previous section, the trigonometric functions of an acute angle were defined as ratios between the lengths of the sides of a right triangle. Introduced in this manner, trigonometric functions are closely related to the solving of triangles, one of the most important applications of trigonometry. To *solve a triangle* means to determine all of its elements: *the measures of its angles* and *the lengths of its sides*.

In this section we consider only *right triangles*. The solving of general triangles will be discussed in Chapter 9. If we are given *one side and an acute angle* or *two sides* of a right triangle, then the other sides and angles can be determined by using formulas (7.4) and relying on Table 4 or a calculator.

Throughout this section, we use the same notations for the vertices, angles, and sides of a right triangle as the ones used in Figure 7.12 of the previous section. In particular, C will always denote the vertex of the right angle and c the length of the hypotenuse.

EXAMPLE 1

Solve the right triangle ACB if $c = 8.75$ ft and $\alpha = 26°$. (See Figure 7.24.)

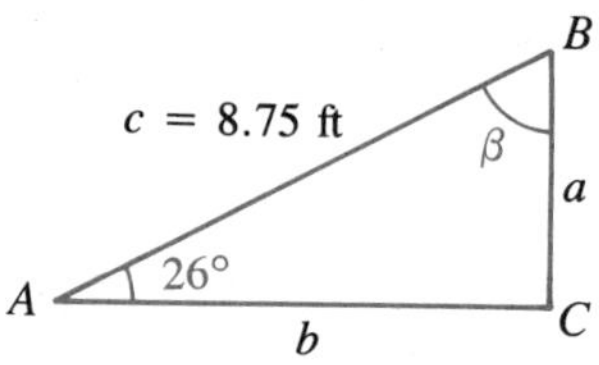

Solution: Since $\alpha + \beta = 90°$, we find that

$$\beta = 90° - 26° = 64°$$

According to formulas (7.4), we have

$$\sin 26° = \frac{a}{8.75}.$$

Hence

$$a = (8.75) \sin 26°$$
$$= (8.75)(0.4384) \quad \text{from table or calculator}$$
$$= 3.84 \text{ ft} \quad \text{rounding off to two decimal places}$$

Also,

$$\cos 26° = \frac{b}{8.75};$$

hence

$$b = (8.75) \cos 26°$$
$$= (8.75)(0.8988)$$
$$= 7.86 \text{ ft.}$$

Figure 7.24

Practice Exercise 1

Solve the right triangle ACB if $c = 5.68$ m and $\beta = 42°$.

Answer: $a = 4.22$ m, $b = 3.80$ m, $\alpha = 48°$

EXAMPLE 2

If $\alpha = 35°20'$ and $b = 18$ m in a right triangle ACB, find the other elements. (See Figure 7.25.)

Solution: First, determine the angle β:

$$\beta = 90° - 35°20' = 54°40'.$$

Since

$$\tan 35°20' = \frac{a}{18},$$

we obtain

$$a = 18 \tan 35°20'$$
$$= 18(0.7089)$$
$$\simeq 12.76 \text{ m}$$
$$\simeq 13 \text{ m.}$$

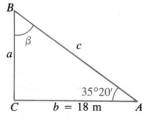

Figure 7.25

The length of the hypotenuse c can be found by using the cosine function:

$$\cos 35°20' = \frac{18}{c}.$$

Hence

$$c = \frac{18}{\cos 35°20'}$$

$$= \frac{18}{0.8158}$$

$$\simeq 22.06 \text{ m}$$

$$\simeq 22 \text{ m}.$$

If you are using a calculator, here are the keystrokes to find c:

$$18 \boxed{\div} \boxed{(} 35 \boxed{+} 20 \boxed{\div} 60 \boxed{)} \cos \boxed{=}$$

Practice Exercise 2   In a right triangle ACB, you are given $a = 12$ ft and $\beta = 28°40'$. Find the other elements of the triangle.

Answer: $b = 7$ ft, $c = 14$ ft, $\alpha = 61°20'$

EXAMPLE **3**   Solve the right triangle ACB if $a = 4$ ft and $c = 5$ ft. (See Figure 7.26.)

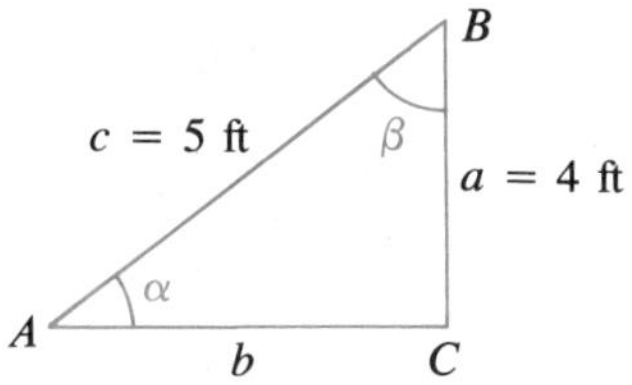

Solution: First, by applying the Pythagorean theorem, we obtain the length of the side b:

$$4^2 + b^2 = 5^2$$

$$b^2 = 25 - 16 = 9$$

$$b = 3 \text{ ft}.$$

Figure 7.26

Next, we have to find the measures of the angles α and β. Since

$$\sin \alpha = \frac{4}{5} = 0.8,$$

in order to find the degree measure of α we must use a table or a calculator.

Referring to Table 4 in the Appendix, observe that the value 0.8 is not found in the sine column of that table. At this point, we could use the method of linear interpolation as explained in the Appendix to get an approximation for α. Since 0.8 lies between the value 0.7986 corresponding to $\sin 53°$ and the value 0.8004 corresponding to $\sin 53°10'$, and since 0.8 is closer to 0.8004 than to 0.7986, we can take as an approximation for α the value $53°10'$. Using this approximation for α, we get the measure of β.

$$\beta = 90° - 53°10' \simeq 36°50'.$$

If a calculator is used, then a better approximation for α can be obtained. By setting the calculator in degree mode, entering the value 0.8, and pressing the keys $\boxed{\text{INV}}\,\boxed{\sin}$, we obtain the display 53.130102. Thus,

$$\alpha \simeq 53.130102°,$$

which, when converted to degrees, minutes, and seconds, is approximately equal to

$$\alpha \simeq 53°7'48''.$$

Next, we obtain the measure of β:

$$\beta \simeq 90° - 53°7'48''$$

$$= 36°52'12''.$$

Practice Exercise 3   Solve the right triangle ACB if $a = 5$ and $b = 12$.

Answer: $c = 13$, $\alpha \simeq 22.62° = 22°37'$, $\beta \simeq 67.38° = 67°23'$

EXAMPLE 4

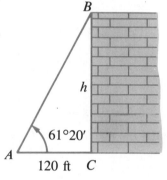

Figure 7.27

Find the height of a building if it is known that the *angle of elevation* (as shown by an arrow in Figure 7.27) from a point 120 ft from the base of the building is $61°20'$.

Solution: According to Figure 7.27, we have

$$\tan 61°20' = \frac{h}{120};$$

so

$$h = 120 \tan 61°20'$$

$$= 120(1.829)$$

$$= 219.48$$

$$\simeq 219 \text{ ft}.$$

Practice Exercise 4   From a point 120 ft from the base of a flagpole, the angle of elevation to the top is $30°$. Find the height of the flagpole.

Answer: 69 ft

EXAMPLE 5

From a promontory overlooking the ocean, a man sees a buoy. If the man is 55 m above sea level and the *angle of depression* of the man's line of sight (as shown in Figure 7.28) is $15°10'$, find the approximate distance from the buoy to the promontory.

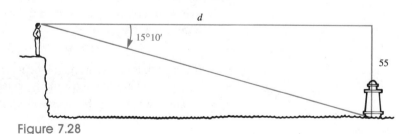

Figure 7.28

Solution: If d denotes the distance from the buoy to the promontory, then

$$\tan 15°10' = \frac{55}{d},$$

$$d = \frac{55}{\tan 15°10'} \quad \text{or} \quad d = 55 \cot 15°10'.$$

According to Table 4, we obtain

$$d = 55(3.689)$$

$$= 202.895$$

$$\doteq 203 \text{ m.}$$

On a calculator, the sequence of keystrokes

$$15 \boxed{+} \; 10 \; \boxed{÷} \; 60 \; \boxed{=} \; \boxed{\tan} \; \boxed{1/x} \; \boxed{×} \; 55 \; \boxed{=}$$

yields 202.9001, which we round off to 203.

Practice Exercise 5 A building is 250 m high and the angle of depression from the top of the building to a point on the ground is 62°30′. How far is the point on the ground from the top of the building?

Answer: 282 m

EXERCISES 7.3

In Exercises 1–10, two elements of a right triangle are given. Using Table 4 in the Appendix, approximate the other elements.

1. $\alpha = 60°$, $a = 10$
2. $\beta = 35°$, $b = 4$
3. $a = b = 3$
4. $a = 5$, $c = 10$
5. $\beta = 25°$, $a = 7$
6. $\alpha = 54°$, $c = 12$
7. $\alpha = 32°10'$, $a = 15$
8. $\beta = 15°30'$, $c = 11$
9. $\alpha = 75°20'$, $b = 12$
10. $\beta = 44°$, $a = 25$

[c] In Exercises 11–20, two elements of a right triangle are given. Use a calculator to find the other elements.

11. $a = 5$, $b = 12$
12. $a = 6$, $b = 8$
13. $\alpha = 30°20'$, $b = 10.5$
14. $\beta = 45°30'$, $c = 12.6$
15. $\beta = 72°31'$, $c = 18.5$
16. $\alpha = 24°16'$, $a = 25.4$
17. $\alpha = 32°26'$, $a = 125.6$
18. $\beta = 15°21'$, $c = 13.4$
19. $\alpha = 36°15'$, $b = 218.3$
20. $\beta = 55°32'$, $b = 131.6$.

In Exercises 21–40, solve the given problem.

21. A straight road climbs at a constant angle of 5°. How many vertical feet does a car climb after traveling 2 miles? (1 mi = 5,280 ft)
22. Find the length l of the shadow of a flagpole 30 ft high when the angle of elevation of the sun is 55°.

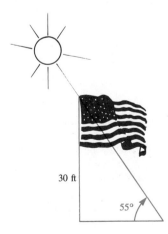

30 ft

55°

[C] 23. An antenna 50 ft tall casts a shadow 40 ft long on level ground. To the nearest minute, find the angle of elevation of the sun.

[C] 24. At a certain time, a pole 4 ft long casts a shadow 8 ft long. To the nearest second, what is the angle of elevation of the sun?

25. A wire 60 ft long is stretched taut from the top of an antenna to a bolt in the ground 45 ft from the base of the antenna. To the nearest minute, find the angle that the wire makes with the ground.

26. A ladder leaning against a vertical wall makes a 60° angle with the ground. If the foot of the ladder is 10 ft away from the wall, find **(a)** the length of the ladder; **(b)** the height of the top of the ladder from the ground.

27. A surveyor is on one bank of a river, directly across from a post on the opposite bank. He moves 30 ft along the river bank and finds that the angle between the river bank and his line of sight to the post is 75°. How wide is the river?

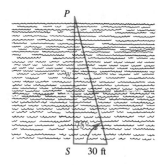

28. The angle of elevation of a kite is 58°20′. Assuming that the string is taut and measures 75 m, find the height of the kite.

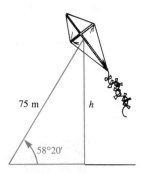

29. Find the area of a regular hexagon inscribed in a circle of radius 8 cm.

30. Find the area of a regular octagon inscribed in a circle of radius 10 m.

31. For an observer on level ground, the angle of elevation of the peak of a mountain is 45°. If the observer walks 1,500 m along a straight line toward the base of the mountain, he finds that the angle of elevation changes to 60°. How high is the mountain?

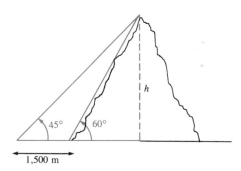

32. Find the approximate length of a ramp that rises to 5 ft above level ground and makes an angle of 8° with the horizontal.

[C] 33. A jet airliner flying at an altitude of 2 mi and with a constant speed of 400 mph begins to climb at an angle of 10°. How long will it take, to the nearest minute, for the airliner to reach an altitude of 4 mi?

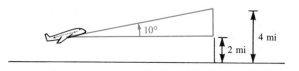

[C] 34. An airplane pilot flying at an altitude of 3500 ft starts to make his landing approach along a straight line and at an angle of 8°. How far does the plane travel until it reaches the runway?

35. From a point 300 ft directly across from a town hall building with a steeple, the angle of elevation to the bottom of the steeple is 15°. From the same point, the angle of elevation to the top of the steeple is 22°. How high is the steeple? What is the vertical distance from the ground to the top of the steeple?

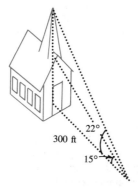

300 ft   22°

15°

36. A person located 2,000 m from the launching pad of a rocket observes that at a certain instant, the angle of elevation is 12°. A few seconds later the angle of elevation is 45°. How far did the rocket rise between the two observations?

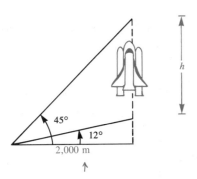

45°

12°

2,000 m

37. A point on level ground is 50 m from the base of a building. The angle of depression from the top of the building to the point on the ground is 40°. How high is the building?

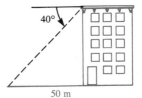

40°

50 m

38. The top of a lighthouse is 250 ft above sea level. The angle of depression from its top to a ship at sea is 8°30′. How far is the ship from the lighthouse?

39. In order to measure the height of a cloud, a spotlight is beamed vertically upward to produce a lighted spot on the cloud. An observer 350 m away from the spotlight finds that the angle of elevation of the lighted spot is 65°. How high is the cloud?

40. An airplane is flying horizontally at an altitude of 12,000 ft. If the angle of depression from the airplane to a control tower is 22°10′, how far must the plane fly to be directly above the tower?

CHAPTER SUMMARY

Consider two coinciding half-lines with origin O. When one of the half-lines is rotated *clockwise* or *counterclockwise* about O, it generates or sweeps an *angle*. The fixed half-line is said to be the *initial side* of the angle, while the half-line that was rotated is the *terminal side* of the angle. The point O is called the *vertex* of the angle. An angle is said to be in *standard position* when its vertex coincides with the origin of a Cartesian coordinate system and its initial side coincides with the positive x-axis.

Angles are often measured in *degrees* or *radians*. Degrees can be converted into radians and vice versa.

$$1 \text{ degree} = \frac{\pi}{180} \text{ radians}$$

$$1 \text{ radian} = \frac{180}{\pi} \text{ degrees}$$

When a record is rotating on a turntable at a constant rate, two different notions of speed are defined.

$$\omega = \frac{\theta}{t} \qquad angular\ speed$$

$$v = r\omega \qquad linear\ speed$$

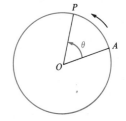

The *angular speed* is the rate of change per unit of time of the angle swept out by the radius of the record. It is often measured in *radians per second*. The *linear speed* is the rate of change per unit of time of the arc length described by a point at a fixed distance from the center. It is measured in *units of length per second*.

The six *trigonometric ratios* (or *functions*) of an acute angle in a right triangle are defined according to the following table.

$$\sin = \frac{\text{opp}}{\text{hyp}} \qquad \csc = \frac{\text{hyp}}{\text{opp}}$$

$$\cos = \frac{\text{adj}}{\text{hyp}} \qquad \sec = \frac{\text{hyp}}{\text{adj}}$$

$$\tan = \frac{\text{opp}}{\text{adj}} \qquad \cot = \frac{\text{adj}}{\text{opp}}$$

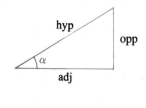

That is, $\sin \alpha$ is the *ratio* between *the length of the side opposite to α* and *the length of the hypotenuse*. Similar statements hold for the other ratios. Note that there are two *basic* trigonometric functions of an angle α: $\sin \alpha$ and $\cos \alpha$. The others are obtained from these two.

$$\tan \alpha = \frac{\sin \alpha}{\cos \alpha} \qquad \cot \alpha = \frac{\cos \alpha}{\sin \alpha}$$

$$\sec \alpha = \frac{1}{\cos \alpha} \qquad \csc \alpha = \frac{1}{\sin \alpha}$$

If two right triangles are *similar*, then the lengths of the sides facing equal angles are proportional to each other.

$$\frac{a}{a'} = \frac{b}{b'} = \frac{c}{c'}$$

The trigonometric values of each angle can be computed by referring to either one of the two triangles.

The acute angles α and β in a right triangle are *complementary* to each other, that is,

$$\alpha + \beta = 90°$$

The following are relations concerning complementary angles;

$$\sin \alpha = \cos(90° - \alpha) \qquad \cos \alpha = \sin(90° - \alpha).$$

Special angles are angles whose measures are $30° = \pi/6$, $45° = \pi/4$, or $60° = \pi/3$. Since special angles are frequently used in applications, it is a good idea to memorize the following table of values.

α Degrees	α Radians	$\sin \alpha$	$\cos \alpha$	$\tan \alpha$	$\cot \alpha$	$\sec \alpha$	$\csc \alpha$
30°	$\frac{\pi}{6}$	$\frac{1}{2}$	$\frac{\sqrt{3}}{2}$	$\frac{\sqrt{3}}{3}$	$\sqrt{3}$	$\frac{2\sqrt{3}}{3}$	2
45°	$\frac{\pi}{4}$	$\frac{\sqrt{2}}{2}$	$\frac{\sqrt{2}}{2}$	1	1	$\sqrt{2}$	$\sqrt{2}$
60°	$\frac{\pi}{3}$	$\frac{\sqrt{3}}{2}$	$\frac{1}{2}$	$\sqrt{3}$	$\frac{\sqrt{3}}{3}$	2	$\frac{2\sqrt{3}}{3}$

To *solve a triangle* means to determine the *measures of its angles* and the *lengths of its sides*. When solving a right triangle, use the Pythagorean theorem together with the trigonometric ratios of an angle.

CHAPTER TEST

1. Convert 336° into radians. Write your answer as a simplified multiple of π.
2. Convert $4\pi/15$ radians to degrees.
3. A central angle of 105° is subtended by an arc of 52 cm. What is the radius of the circle?
4. A wheel is rotating around its center at a rate of 60 revolutions per minute (rpm). What is the angular speed of the wheel? What is the linear speed of a point 4 in. from the center of the wheel?
5. Let α be an acute angle of a right triangle such that $\sin \alpha = 5/9$. Find the exact values of the five remaining trigonometric ratios.

6. If $\sin(90° - \alpha) = 3/8$, find $\sin \alpha$.
7. In a right triangle ACB, suppose $a = 8.4$ ft and $\beta = 30°30'$. Solve the triangle.
8. A pole 4 m long casts a shadow 12 m long. Find the angle of elevation of the sun to the nearest tenth of a degree.
9. If $\tan(90° - x) = 5/3$, find the exact values of $\sin x$ and $\cos x$.
10. What is the radian measure of a central angle whose arc is $4\pi/3$ units long if the radius of the circle is 1 unit long? If the radius is 2 units long?

REVIEW EXERCISES

1. Find the radian measure of the angles with degree measures 72°; 15°; −50°; 135°; −85°.
2. Convert to radians the following degree measures: 36°; 30°; −25°; 270°; −170°.
3. Convert to degrees the following radian measures: $2\pi/9$; $7\pi/5$; $3\pi/15$; $7\pi/18$; $\pi/12$.
4. Find the degree measure of the angles with radian measures $3\pi/5$; $7\pi/10$; $5\pi/9$; $11\pi/18$; $5\pi/12$.
5. Convert to radians the following degree measures: 25°15′; 120°36′; 75°10′; 68°10′25″; 210°5′12″.
6. Convert to radians the following degree measures: 35°10′; 75°25′; 185°21′; 5°10′20″; 15°30′45″.
7. Convert to decimal degrees the following radian measures: 4; 10; 1.5; 4.8; 2.25. Round off your answers to two decimal places.
8. Find to the nearest minute the measure of the angles whose radian measures are 5; 8; 2.5; 5.6; 1.12.
9. If a central angle θ subtends an arc 60 cm long on a circle of radius 8 cm, find the measure of θ in radians.
10. On a circle of radius 12 ft, a central angle α subtends an arc of 35 ft. What is the radian measure of α?
11. If a central angle of 6.4 radians is subtended by an arc 12.16 ft long, find the radius of the circle.
12. On a circle of radius 3.8 m, an angle of 2.6 radians is subtended by an arc s. What is the length of s in meters?
13. What is the length of an arc that subtends an angle of 7.5° on a circle of radius 12 cm?
14. On a circle of radius 3.2 ft, an angle of 11.25° is subtended by an arc s. Find the length of s.

15. A wheel of radius 2.5 m is rotating at a rate of 2400 rpm. Find (a) its angular speed; (b) the distance traveled by a point on the rim of the wheel in one second.
16. An automobile tire has a 36 in. diameter and it is rotating at a rate of 514 rpm. How fast is the car traveling in mph? (1 mile = 5280 ft)
17. The space shuttle, in a circular orbit 175 mi high, makes one complete revolution around the earth every 90 minutes. Find its linear speed. (The radius of the earth is about 4000 mi.)
18. A record with a diameter of 12 in. is rotating at the rate of $33\frac{1}{3}$ rpm. Find the linear velocity, in in./s, of a point on its rim.
19. The wheels of a bicycle have a diameter of 22 in. If they are rotating at a rate of 16 rpm, how far does the bike travel in one minute?
20. The radius of the earth is 6.371×10^6 m. Find the linear speed in km/hr of a point on the equator.

In Exercises 21–26, ACB is a right triangle. Find the trigonometric values of the indicated angle.

21.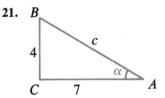

22.

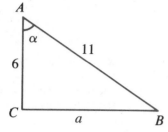

23.

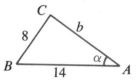

24.

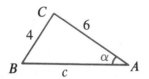

25.

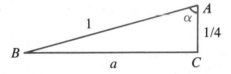

26.

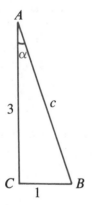

In each of Exercises 27–30, a trigonometric function of an angle x is given. Find the exact values of the remaining trigonometric functions of x. Do not use your calculator.

27. $\sin x = \dfrac{3}{8}$

28. $\cos x = \dfrac{1}{4}$

29. $\tan x = 3$

30. $\cot x = 5$

31. Let $\sin(90° - x) = \dfrac{2}{5}$. Find $\tan x$.

32. Let $\cos(90° - x) = \dfrac{3}{8}$. Find $\cot x$.

In Exercises 33–36, two elements of a right triangle are given. Use Table 4 in the Appendix to approximate the other elements.

33. $\alpha = 30°$, $a = 6$

34. $\beta = 34°$, $b = 8$

35. $\beta = 25°10'$, $b = 12$

36. $\alpha = 38°30'$, $c = 9$

 In Exercises 37–40, two elements of a right triangle are given. Use your calculator to approximate the other elements.

37. $a = 7$, $b = 24$

38. $a = 9$, $c = 41$

39. $\alpha = 45°20'$, $c = 10.4$

40. $\beta = 28°30'$, $a = 5.6$

41. If a road rises 5 m per 120 horizontal m, find the angle (to the nearest minute) that it makes with the horizontal.

42. A man 6 ft tall casts a shadow 9 ft 2 in. long. Find the angle of elevation of the sun.

43. A guywire to a pole makes an angle of $65°20'$ with level ground and is attached to a bolt in the ground 18 ft from the pole. How high is the attachment of the guywire to the pole?

44. Find the angle of elevation of a kite, when the kite is 150 ft. high and 210 ft of string is out.

45. A mountain climber on level ground observes that the angle of elevation to the top of the mountain is $30°$. After walking 1,200 meters along a straight line toward the mountain, he observes that the angle of elevation is $60°$. How high is the mountain?

46. An observer on level ground finds that the angle of elevation of a balloon is $72°$. If the horizontal distance from the observer to a point directly under the balloon is 240 ft, determine the height of the balloon.

47. In aerial navigation, angular directions are given clockwise from the north in degrees and minutes. Thus, north is $0°$, east is $90°$, south is $180°$, and west is $270°$. If an airplane is flying in a direction $150°$ from an airport, how far south of the airport will the plane be after traveling 300 miles? How far east will it be?

48. An airplane travels at a speed of 350 mph in a direction $225°$ from New York. After 3 hours, how far south of New York will the plane be?

49. The angle of depression from an airplane to a control tower is $30°16'$. If the plane is flying at an altitude of 10,500 ft, how far must it fly to be directly above the tower?

50. An airplane travels for 2 hours at a rate of 250 mph in a direction of $24°30'$ from Dallas. At the end of this time, how far north of Dallas is the plane? How far east?

■

CHAPTER 8

TRIGONOMETRIC
OR
CIRCULAR
FUNCTIONS

In this chapter, we redefine sine and cosine as coordinates of a point on the unit circle. We consider them as functions whose domain is the set of all real numbers. Next, we define the four remaining trigonometric functions as ratios and reciprocals of the sine and cosine functions, and we explain functional properties such as periodicity, oddness, and evenness of these functions. We derive the fundamental trigonometric identity and discuss several applications. When dealing with angles in standard position, it is useful to know alternate definitions of trigonometric functions, such as the ones we discuss in Section 8.2. In Section 8.3 we will show how trigonometric functions can be evaluated with the help of a table or a calculator and by making use of the notion of reference angles. In Section 8.4, we discuss graphs of trigonometric functions in detail. Oscillatory phenomena, including the motion of a pendulum and periodic changes of plant and animal population, can be described by trigonometric functions (Section 8.5). The last section of this chapter is devoted to the study of inverse trigonometric functions and their graphs.

8.1 CIRCULAR FUNCTIONS

In Chapter 7 we considered the trigonometric functions only for acute angles. This is enough if our only objective is to solve right triangles. However, if our aim is to solve a general triangle, it is necessary to consider trigonometric functions of obtuse angles. In fact, we don't want to restrict the amount and direction of rotation of an angle in standard position in any way, so that the measure of the angle can be any positive or negative number. Thus, it is necessary to extend the definitions of trigonometric functions given in the previous chapter so as to include angles of measure greater than 90° ($\pi/2$ radians) and also negative angles.

In order to do this, we start by redefining the sine and cosine functions as coordinates of a point on the unit circle. Let θ be the measure (in degrees, radians, or grads) of an angle in standard position. The initial side of the angle intersects the unit circle U at the point $A(1, 0)$, while the terminal side intersects the circle at the point $P(x, y)$ (Figure 8.1). As the point P moves along the unit circle, clockwise or counterclockwise, the angle $\angle AOP$ changes and its measure θ takes on all possible values.

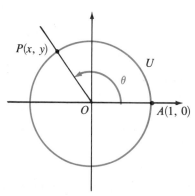

Figure 8.1

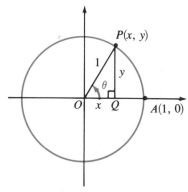

Figure 8.2

Suppose, for a moment, that θ is an acute angle; consider the right triangle OQP in Figure 8.2. The hypotenuse OP has length 1, the side OQ has length x, and PQ has length y. Thus, according to formulas (7.4) of Chapter 7,

$$\cos \theta = \frac{x}{1} = x \qquad \text{and} \qquad \sin \theta = \frac{y}{1} = y.$$

We conclude that for an acute angle θ in standard position, $\sin \theta$ is the *ordinate* of P and $\cos \theta$ is the *abscissa* of P. This justifies the following definition.

•

Definition

If θ denotes the measure of $\angle AOP$ in standard position and (x, y) denotes the coordinates of P, the point of intersection of the terminal side of the angle and the unit circle, then we define the cosine and sine of θ by

$$\cos \theta = x \qquad \text{and} \qquad \sin \theta = y. \tag{8.1}$$

Using this definition, sine and cosine are functions defined for all values (positive, negative, or zero) of the angle θ. Moreover, since the point P lies on the

unit circle, *the function values of the sine and cosine are always between* −1 *and*
1, that is,

$$-1 \leq \sin \theta \leq 1 \qquad \text{and} \qquad -1 \leq \cos \theta \leq 1. \qquad (8.2)$$

If θ is an acute angle, definitions (8.1) agree with the definitions of sine and
cosine given in (7.4). The advantage of the new definitions is that they apply to all
values of θ, while the definitions given in Chapter 7 are restricted to acute angles
of a right triangle.

From (8.1) we also derive the fact that the values of both the sine and cosine
of a first-quadrant angle are positive. For a second-quadrant angle, the sine is
positive, while the cosine is negative; for a third-quadrant angle, both the sine and
cosine are negative; finally, the sinc of a fourth-quadrant angle is negative and the
cosine is positive. The four cases are illustrated in Figure 8.3.

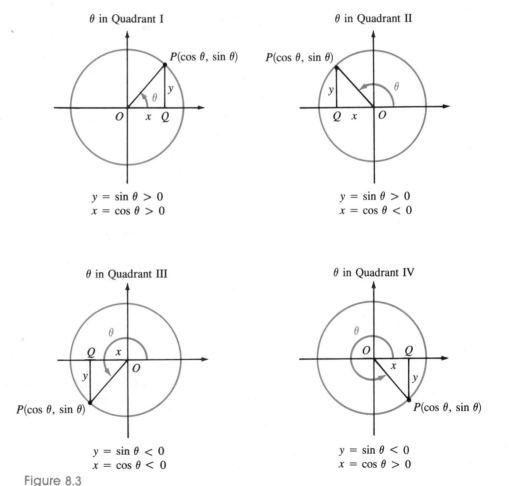

Figure 8.3

In certain simple cases, it is possible to compute the functional values of the
sine and cosine directly from the definitions (8.1).

EXAMPLE 1

Evaluate $\sin \theta$ and $\cos \theta$ when **(a)** $\theta = 135°$ and **(b)** $\theta = -\pi/2$.

Solution: We must find the coordinates of P, the point of intersection of the terminal side of each given angle with the unit circle.

(a) As shown in Figure 8.4, the point P lies on the second-quadrant diagonal and thus has coordinates $(-x, x)$. Since P also lies on the unit circle, we have

$$(-x)^2 + x^2 = 1,$$

or

$$2x^2 = 1.$$

Hence

$$x = \frac{1}{\sqrt{2}} = \frac{\sqrt{2}}{2}.$$

Therefore, the coordinates of P are $\left(-\sqrt{2}/2, \sqrt{2}/2\right)$ and, according to (8.1),

$$\sin 135° = \frac{\sqrt{2}}{2} \qquad \text{and} \qquad \cos 135° = -\frac{\sqrt{2}}{2}.$$

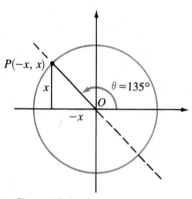

Figure 8.4

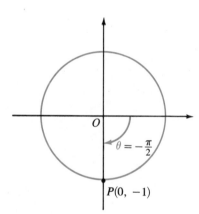

Figure 8.5

(b) The terminal side of the angle measuring $-\pi/2$ radians coincides with the negative y-axis (Figure 8.5), and the point P has coordinates $(0, -1)$. Thus,

$$\sin\left(-\frac{\pi}{2}\right) = -1 \qquad \text{and} \qquad \cos\left(-\frac{\pi}{2}\right) = 0.$$

Practice Exercise 1

Find $\sin \theta$ and $\cos \theta$ when **(a)** $\theta = \pi/4$ and **(b)** $\theta = 180°$.

Answer: **(a)** $\sin\dfrac{\pi}{4} = \cos\dfrac{\pi}{4} = \dfrac{\sqrt{2}}{2}$ **(b)** $\sin 180° = 0$, $\cos 180° = -1$

Periodicity of Sine and Cosine

A function $F(x)$ is said to be *periodic* if there is a positive number p such that $F(x + p) = F(x)$ for all x. The smallest such p, if it exists is called the *period* of $F(x)$.

If integral multiples of 2π are added to the radian measure θ of an angle, then the values of sine and cosine remain the same, because the terminal sides of the angles coincide. Moreover, it can be shown that 2π is the smallest of such numbers. Thus sine and cosine are periodic functions with period 2π.

$$\sin(\theta + 2\pi) = \sin \theta, \qquad \cos(\theta + 2\pi) = \cos \theta \qquad (8.3)$$

Similarly, if α is the degree measure of an angle, then

$$\sin(\alpha + 360°) = \sin \alpha, \qquad \cos(\alpha + 360°) = \cos \alpha.$$

Two Properties of Sine and Cosine

Figure 8.6 illustrates the following important properties of the sine and cosine.

$$\sin(-\theta) = -\sin \theta, \qquad \cos(-\theta) = \cos \theta \qquad (8.4)$$

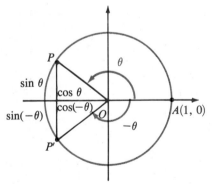

Figure 8.6

Relations (8.4) show that the sine is an *odd* function and the cosine is an *even* function.

In what follows, θ will always denote the radian measure of an angle.

Tangent and Cotangent Functions

As we already know for acute angles, the tangent and cotangent functions can be defined as quotients of the sine and cosine functions. Thus, for all angles, we set

$$\tan \theta = \frac{\sin \theta}{\cos \theta}, \quad \theta \neq \frac{\pi}{2} + k\pi, \text{ with } k \text{ any integer,} \tag{8.5}$$

and

$$\cot \theta = \frac{\cos \theta}{\sin \theta}, \quad \theta \neq k\pi, \text{ with } k \text{ any integer.} \tag{8.6}$$

Observe that the restrictions imposed on θ are necessary in order to prevent division by 0. When the angle θ is equal to $\pi/2$, $3\pi/2$, $-\pi/2$, or, more generally, when $\theta = \pi/2 + k\pi$, with k any integer, the corresponding point P on the unit circle coincides with the point $B(0, 1)$ or $B'(0, -1)$ (see Figure 8.7a). Hence

$$\cos\left(\frac{\pi}{2} + k\pi\right) = 0, \quad \text{for all integers } k,$$

and the tangent function is *not defined* for those angles.

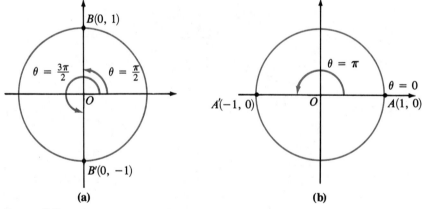

Figure 8.7

When $\theta = 0$, π, $-\pi$, or more generally, when $\theta = k\pi$, with k any integer, the corresponding point P coincides with the point $A(1, 0)$ or $A'(-1, 0)$ (Figure 8.7b). Hence

$$\sin k\theta = 0, \quad \text{for all integers } k,$$

and the cotangent function is *undefined* for those angles.

For use in future applications, we now describe geometrical interpretations for the tangent and cotangent functions. In Figure 8.8, the line l through the point

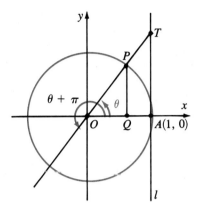

Figure 8.8

$A(1, 0)$ is parallel to the y-axis, and the terminal side of the angle θ has been extended so that it intersects l at the point T. In the two similar triangles OQP and OAT, let $\overline{AT}$ and $\overline{PQ}$ denote the lengths of the sides facing the angle θ, and $\overline{OQ}$ and $\overline{OA}$ denote the lengths of the sides facing the equal angles with vertices P and T, respectively. As we know from Chapter 7, in similar triangles the lengths of corresponding sides are proportional to each other; thus,

$$\frac{\overline{AT}}{\overline{PQ}} = \frac{\overline{OA}}{\overline{OQ}}.$$

But $\overline{PQ} = \sin\theta$, $\overline{OQ} = \cos\theta$, and $\overline{OA} = 1$. Hence

$$\frac{\overline{AT}}{\sin\theta} = \frac{1}{\cos\theta};$$

that is,

$$\overline{AT} = \frac{\sin\theta}{\cos\theta} = \tan\theta.$$

We conclude that the length of the segment AT on the line l represents the tangent of θ. For this reason, the line l is sometimes called the *tangent axis*. You can check that the tangent of a first- or third-quadrant angle is *positive*, while the tangent of a second- or fourth-quadrant angle is *negative*. Moreover, Figure 8.8 also illustrates the following relation:

$$\tan\theta = \tan(\theta + \pi), \tag{8.7}$$

which indicates that the tangent function is *periodic with period π*.

With a similar argument, it can be shown that the length of the segment BT in Figure 8.9 represents the cotangent of θ. Notice that BT lies on the line l' passing through $B(0, 1)$ and parallel to the x-axis. Figure 8.9 also illustrates the following relation:

$$\cot\theta = \cot(\theta + \pi), \tag{8.8}$$

showing that the cotangent function is also *periodic with period π*.

From formulas (8.4) and the definitions of tangent and cotangent [formulas (8.5) and (8.6)], it follows that

$$\tan(-\theta) = -\tan\theta \quad \text{and} \quad \cot(-\theta) = -\cot\theta. \tag{8.9}$$

That is, both tangent and cotangent are *odd* functions.

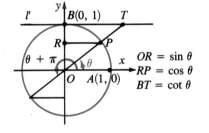

$OR = \sin\theta$
$RP = \cos\theta$
$BT = \cot\theta$

Figure 8.9

Secant and Cosecant Functions

These functions are defined as reciprocals of the sine and cosine functions:

$$\sec \theta = \frac{1}{\cos \theta}, \quad \theta \neq \frac{\pi}{2} + k\pi, \text{ with } k \text{ any integer,} \qquad (8.10)$$

and

$$\csc \theta = \frac{1}{\sin \theta}, \quad \theta \neq k\pi, \text{ with } k \text{ any integer.} \qquad (8.11)$$

Now we summarize the definitions of all six trigonometric functions of an angle in terms of the coordinates of the corresponding point on the unit circle.

Let θ be the (degree, radian, or grad) measure of $\angle AOP$ in standard position, and let $P(x, y)$ be the point of intersection of the terminal side of the angle and the unit circle. We have the following definitions.

$$\sin \theta = y \qquad\qquad \csc \theta = \frac{1}{y}, \quad \text{if } y \neq 0$$

$$\cos \theta = x \qquad\qquad \sec \theta = \frac{1}{x}, \quad \text{if } x \neq 0 \qquad (8.12)$$

$$\tan \theta = \frac{y}{x}, \quad \text{if } x \neq 0 \qquad \cot \theta = \frac{x}{y}, \quad \text{if } y \neq 0$$

EXAMPLE 2   Evaluate the six trigonometric functions by using (8.12) when **(a)** $\theta = \frac{\pi}{4}$ and **(b)** $\alpha = 270°$.

Solution: **(a)** As we already know,

$$\sin \frac{\pi}{4} = \cos \frac{\pi}{4} = \frac{\sqrt{2}}{2}.$$

Now, we use (8.12) to find the remaining values:

$$\tan \frac{\pi}{4} = \frac{\sin(\pi/4)}{\cos(\pi/4)} = \frac{\sqrt{2}/2}{\sqrt{2}/2} = 1,$$

$$\cot \frac{\pi}{4} = \frac{\cos(\pi/4)}{\sin(\pi/4)} = \frac{\sqrt{2}/2}{\sqrt{2}/2} = 1,$$

$$\sec \frac{\pi}{4} = \frac{1}{\cos(\pi/4)} = \frac{2}{\sqrt{2}} = \sqrt{2},$$

$$\csc \frac{\pi}{4} = \frac{1}{\sin(\pi/4)} = \frac{2}{\sqrt{2}} = \sqrt{2}.$$

(b) The terminal side of the angle $\alpha = 270°$ coincides with the negative y-axis, and P has coordinates $(0, -1)$. Thus,

$$\sin 270° = -1 \qquad \text{and} \qquad \cos 270° = 0.$$

It follows that

$$\cot 270° = \frac{\cos 270°}{\sin 270°} = \frac{0}{-1} = 0,$$

and

$$\csc 270° = \frac{1}{\sin 270°} = \frac{1}{-1} = -1.$$

The two remaining functions, tangent and secant, are undefined; do you see why?

Practice Exercise 2   Evaluate the six trigonometric functions by using (8.12) when **(a)** $\theta = 135°$ and **(b)** $\theta = -\pi/2$.

Answer: **(a)** $\sin \theta = \dfrac{\sqrt{2}}{2}, \quad \cos \theta = -\dfrac{\sqrt{2}}{2}, \quad \tan \theta = -1, \quad \cot \theta = -1,$
$\sec \theta = -\sqrt{2},\ \csc \theta = \sqrt{2}$
(b) $\sin \theta = -1, \quad \cos \theta = 0, \quad \tan \theta$ undefined, $\cot \theta = 0, \quad \sec \theta$ undefined, $\csc \theta = -1$

The Fundamental Pythagorean Identity

In Figure 8.2, the point $P(x, y)$ lies on the unit circle. Thus, we have

$$x^2 + y^2 = 1.$$

Since $x = \cos \theta$ and $y = \sin \theta$, we obtain the expression

$$(\cos \theta)^2 + (\sin \theta)^2 = 1.$$

According to standard notations for the square of trigonometric functions, we write

$$(\sin \theta)^2 = \sin \theta \cdot \sin \theta = \sin^2 \theta$$

and

$$(\cos \theta)^2 = \cos \theta \cdot \cos \theta = \cos^2 \theta.$$

Thus, the above expression can be written as

$$\sin^2 \theta + \cos^2 \theta = 1. \qquad (8.13)$$

This relation, called the *fundamental Pythagorean identity,* is satisfied for all values of the angle θ.

■ A word of caution   Do not confuse $\sin^2 \theta$ and $\sin \theta^2$. In general, $\sin^2 \theta \neq \sin \theta^2$. The first expression represents the *square* of $\sin \theta$, while the second one represents the *sine of the square* of θ.

HISTORICAL NOTE

HIPPARCHUS

The beginnings of trigonometry in classical antiquity are thought to have appeared in lost works of the astronomer Hipparchus, who flourished about 161–126 B.C. in Alexandria. He is reported by other authors to have constructed the equivalent of a table of sines, using the same 360° angle measure that we use today. It is also clear that he knew and used the equivalent of the Pythagorean identity $\sin^2\theta + \cos^2\theta = 1$. Although we cannot say that Hipparchus invented or discovered the sine function, he was working with it at this early date. Hipparchus used spherical trigonometry to find the number of degrees in the arc of the circle that a star appears to describe in the sky between its rising and its setting.

Other Identities

The identity (8.13) implies two other important trigonometric identities. If $\cos\theta \neq 0$, then dividing both sides of (8.13) by $\cos^2\theta$ gives

$$\frac{\sin^2\theta}{\cos^2\theta} + 1 = \frac{1}{\cos^2\theta}$$

or

$$\left(\frac{\sin\theta}{\cos\theta}\right)^2 + 1 = \left(\frac{1}{\cos\theta}\right)^2$$

Hence

$$\tan^2\theta + 1 = \sec^2\theta. \qquad (8.14)$$

Similarly, if $\sin\theta \neq 0$, it is easy to show that

$$\cot^2\theta + 1 = \csc^2\theta. \qquad (8.15)$$

Identities (8.13), (8.14), and (8.15) are very important and should be memorized. In the examples that follow, we use them to determine the values of all trigonometric functions of an angle θ, given the value of a trigonometric function of θ and an additional condition on θ.

EXAMPLE 3   If $\sin\theta = 1/3$ and θ is a second-quadrant angle, find the exact values of the remaining trigonometric functions.

Solution: From (8.13), we obtain

$$\cos^2\theta = 1 - \sin^2\theta.$$

Thus,

$$\cos^2\theta = 1 - \left(\frac{1}{3}\right)^2 = 1 - \frac{1}{9} = \frac{8}{9}.$$

Taking square roots, we obtain

$$\cos\theta = \pm\sqrt{\frac{8}{9}} = \pm\frac{2\sqrt{2}}{3}.$$

So we have determined the value of $\cos\theta$ except for sign, and we must now determine that. Here we use the additional information we have about θ. Since θ is a second-quadrant angle (Figure 8.10), its cosine is *negative*. Thus, $\cos\theta =$

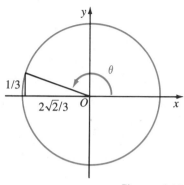

Figure 8.10

$-2\sqrt{2}/3$. The values of the four remaining trigonometric functions for the angle θ are obtained as follows:

$$\tan \theta = \frac{\sin \theta}{\cos \theta} = \frac{1/3}{-2\sqrt{2}/3} = -\frac{1}{2\sqrt{2}} = -\frac{\sqrt{2}}{4},$$

$$\cot \theta = \frac{\cos \theta}{\sin \theta} = \frac{-2\sqrt{2}/3}{1/3} = -2\sqrt{2},$$

$$\sec \theta = \frac{1}{\cos \theta} = \frac{1}{-2\sqrt{2}/3} = -\frac{3}{2\sqrt{2}} = -\frac{3\sqrt{2}}{4},$$

$$\csc \theta = \frac{1}{\sin \theta} = \frac{1}{1/3} = 3.$$

Practice Exercise 3

Find the exact values of the trigonometric functions of a third-quadrant angle θ, knowing that $\cos \theta = -1/3$.

Answer: $\sin \theta = -\dfrac{2\sqrt{2}}{3}$, $\tan \theta = 2\sqrt{2}$, $\cot \theta = \dfrac{\sqrt{2}}{4}$, $\sec \theta = -3$,

$\csc \theta = -\dfrac{3\sqrt{2}}{4}$

EXAMPLE 4

If $\tan \theta = 2$ and θ is a third-quadrant angle, find the exact values of the other trigonometric functions.

Solution: Since θ is a third-quadrant angle, $\sin \theta < 0$ and $\cos \theta < 0$ (Figure 8.11). From (8.14), we get

$$\sec^2 \theta = 1 + 2^2 = 5,$$

hence

$$\sec \theta = \pm\sqrt{5}.$$

Since $\cos \theta < 0$, it follows that $\sec \theta = 1/\cos \theta < 0$, so

$$\sec \theta = -\sqrt{5}.$$

Next, we have

$$\cos \theta = \frac{1}{\sec \theta} = -\frac{1}{\sqrt{5}} = -\frac{\sqrt{5}}{5}.$$

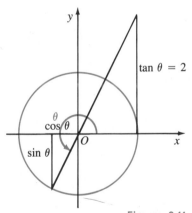

Figure 8.11

Since $\tan \theta = 2$ and $\cot \theta = 1/\tan \theta$, we get $\cot \theta = 1/2$. On the other hand, $\tan \theta = \sin \theta/\cos \theta$ with $\tan \theta = 2$ and $\cos \theta = -\sqrt{5}/5$; that is,

$$2 = \frac{\sin \theta}{-\sqrt{5}/5}.$$

Solving for $\sin \theta$, we obtain

$$\sin \theta = -\frac{2\sqrt{5}}{5}.$$

Finally,

$$\csc \theta = \frac{1}{\sin \theta} = -\frac{5}{2\sqrt{5}} = -\frac{\sqrt{5}}{2}.$$

Practice Exercise 4

If θ is a second-quadrant angle and $\cot \theta = -4$, find the values of the remaining trigonometric functions of θ.

Answer: $\sin \theta = \dfrac{\sqrt{17}}{17}$, $\cos \theta = -\dfrac{4\sqrt{17}}{17}$, $\tan \theta = -\dfrac{1}{4}$, $\sec \theta = -\dfrac{\sqrt{17}}{4}$, $\csc \theta = \sqrt{17}$

In the next example, computations were performed with the help of a calculator.

© EXAMPLE 5

If $\sin \alpha = 0.6052$ and $\sec \alpha < 0$, approximate the values of the five remaining trigonometric functions to four decimal places.

Solution: From (8.13), we get

$$\begin{aligned}
\cos \alpha &= \pm\sqrt{1 - \sin^2\alpha} \\
&= \pm\sqrt{1 - (0.6052)^2} \\
&= \pm\sqrt{0.6337} \\
&= \pm 0.7961.
\end{aligned}$$

Since $\sec \alpha = 1/\cos \alpha$ is negative, $\cos \alpha$ is also negative; therefore,

$$\cos \alpha = -0.7961.$$

Next, we obtain

$$\tan \alpha = \frac{0.6052}{-0.7961} = -0.7602,$$

$$\cot \alpha = -\frac{0.7961}{0.6052} = -1.3154,$$

$$\sec \alpha = \frac{1}{-0.7961} = -1.2561,$$

$$\csc \alpha = \frac{1}{0.6052} = 1.6523.$$

© Practice Exercise 5

If $\tan \beta = 1.204$ and $\sin \beta > 0$, approximate the values of the five remaining trigonometric functions to three decimal places.

Answer: $\sin \beta = 0.769$, $\cos \beta = 0.639$, $\cot \beta = 0.831$, $\sec \beta = 1.565$, $\csc \beta = 1.300$

EXERCISES 8.1

In Exercises 1–14, find the coordinates of the point on the unit circle corresponding to the given angle. Evaluate the six trigonometric functions of the angle. Do not use a calculator or table. Use your knowledge of special angles.

1. $\theta = \dfrac{5\pi}{4}$ **2.** $\theta = -\dfrac{3\pi}{2}$

3. $\alpha = 225°$ **4.** $\alpha = -45°$

5. $\alpha = \dfrac{5\pi}{2}$ **6.** $\alpha = \dfrac{7\pi}{4}$

7. $\theta = -135°$ **8.** $\theta = 315°$

9. $\theta = 3\pi$ **10.** $\theta = -6\pi$

11. $\theta = 100\pi$ **12.** $\theta = 205\pi$

13. $\theta = -180°$ **14.** $\theta = -225°$

In Exercises 15–22, the values of two trigonometric functions of θ are given. Usc the relations among trigonometric functions to find the other four trigonometric values.

15. $\sin \theta = \dfrac{2}{3}$, $\cos \theta = \dfrac{\sqrt{5}}{3}$

16. $\sin \theta = -\dfrac{3}{4}$, $\cos \theta = \dfrac{\sqrt{7}}{4}$

17. $\sin \theta = \dfrac{1}{4}$, $\sec \theta = -\dfrac{4\sqrt{15}}{15}$

18. $\cos \theta = -\dfrac{1}{2}$, $\csc \theta = \dfrac{2}{\sqrt{3}}$

19. $\tan \theta = \dfrac{\sqrt{2}}{4}$, $\sin \theta = -\dfrac{1}{3}$

20. $\cot \theta = -\dfrac{4}{3}$, $\sin \theta = \dfrac{3}{5}$

21. $\sin \theta = -\dfrac{3}{4}$, $\sec \theta = \dfrac{4\sqrt{7}}{7}$

22. $\cos \theta = -\dfrac{3}{5}$, $\tan \theta = \dfrac{4}{3}$

In Exercises 23–30, the value of a trigonometric function of θ and a condition on θ are given. Find the other five trigonometric functions of θ. Use the Pythagorean identities.

23. $\sin \theta = \dfrac{3}{4}$ and $\cos \theta < 0$

24. $\cos \theta = -\dfrac{1}{6}$ and $\sin \theta < 0$

25. $\tan \theta = 3$ and $\sin \theta > 0$

26. $\cot \theta = -2$ and $\sec \theta > 0$

27. $\cos \theta = \dfrac{1}{3}$ and θ is a fourth-quadrant angle.

28. $\sin \theta = -\dfrac{2}{5}$ and θ is a third-quadrant angle.

29. $\tan \theta = -1$ and θ is a second-quadrant angle.

30. $\sec \theta = -2$ and θ is a third-quadrant angle.

c In Exercises 31–34, given the approximate value of a trigonometric function of θ and a condition on θ, approximate the other five trigonometric functions to four decimal places.

31. $\sin \theta = 0.8192$ and $\cos \theta < 0$

32. $\cos \theta = 0.7431$ and $\sin \theta < 0$

33. $\tan \theta = 1.1925$ and $\cos \theta < 0$

34. $\sec \theta = 1.3091$ and $\csc \theta > 0$

In Exercises 35–38, prove the required relationships.

35. Show that $\sec(-\theta) = \sec \theta$ and $\csc(-\theta) = -\csc \theta$.

36. Show that the tangent and cotangent are odd functions.

37. Use thc definitions of tangent and cotangent to show that $\cot \theta = 1/\tan \theta$.

38. Derive the following identity from the fundamental Pythagorean identity:

$$\cot^2 \theta + 1 = \csc^2 \theta.$$

In Exercises 39 and 40, the circles are the unit circle.

39. Prove that $\overline{OM} = \sec \theta$.

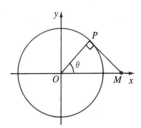

40. Prove that $\overline{ON} = \csc \theta$.

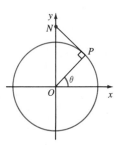

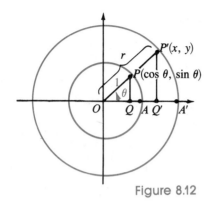

Figure 8.12

8.2 ALTERNATE DEFINITIONS OF TRIGONOMETRIC FUNCTIONS

In Section 8.1 we defined the trigonometric functions in terms of coordinates of a point on the unit circle [formulas (8.12)]. We can also define them in terms of the coordinates of a point on a concentric circle with arbitrary radius. Consider two concentric circles centered at the origin with radii 1 and r, respectively (Figure 8.12), and let θ be the measure of the equal angles: $\angle AOP = \angle A'OP'$. Since the triangles OQP and $OQ'P'$ are similar, the following relations hold:

$$\frac{\overline{OQ'}}{\overline{OQ}} = \frac{\overline{OP'}}{\overline{OP}} \quad \text{and} \quad \frac{\overline{Q'P'}}{\overline{QP}} = \frac{\overline{OP'}}{\overline{OP}}.$$

Since $\overline{OP} = 1$, $\overline{OQ} = \cos\theta$, $\overline{OQ'} = x$, $\overline{OP'} = r$, $\overline{QP} = \sin\theta$, and $\overline{Q'P'} = y$, the two relations above can be rewritten

$$\frac{x}{\cos\theta} = \frac{r}{1} \quad \text{and} \quad \frac{y}{\sin\theta} = \frac{r}{1}.$$

Hence

$$\cos\theta = \frac{x}{r} \quad \text{and} \quad \sin\theta = \frac{y}{r}. \qquad (8.16)$$

Using these two relations and the definitions of tangent, cotangent, secant, and cosecant, we can express the six trigonometric functions as the following ratios.

$$\sin\theta = \frac{y}{r} \qquad\qquad \csc\theta = \frac{r}{y}, \quad \text{if } y \neq 0$$

$$\cos\theta = \frac{x}{r} \qquad\qquad \sec\theta = \frac{r}{x}, \quad \text{if } x \neq 0 \qquad (8.17)$$

$$\tan\theta = \frac{y}{x}, \quad \text{if } x \neq 0 \qquad \cot\theta = \frac{x}{y}, \quad \text{if } y \neq 0$$

These formulas show that the trigonometric functions of an angle θ in standard position can be determined in terms of *the coordinates of a point on the terminal side of the angle and the distance from the point to the origin.*

EXAMPLE 1

Let θ be the measure of an angle in standard position and let $P(-3, 4)$ be a point on its terminal side. Find the exact values of the trigonometric functions of θ.

Solution: The point P is located in the second quadrant (Figure 8.13) and the

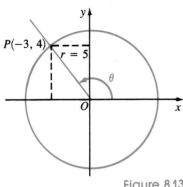

Figure 8.13

distance from P to the origin is

$$r = \sqrt{(-3)^2 + 4^2} = \sqrt{25} = 5.$$

According to formulas (8.17), we have

$$\sin \theta = \frac{4}{5}, \qquad \csc \theta = \frac{5}{4},$$

$$\cos \theta = -\frac{3}{5}, \qquad \sec \theta = -\frac{5}{3},$$

$$\tan \theta = -\frac{4}{3}, \qquad \cot \theta = -\frac{3}{4}.$$

Practice Exercise 1   The point $A(8, -6)$ lies on the terminal side OA of an angle α in standard position. Find the exact values of the trigonometric functions of α.

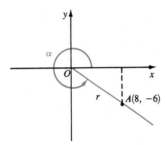

Answer: $\sin \alpha = -\dfrac{3}{5}$, $\cos \alpha = \dfrac{4}{5}$, $\tan \alpha = -\dfrac{3}{4}$, $\cot \alpha = -\dfrac{4}{3}$, $\sec \alpha = \dfrac{5}{4}$,

$\csc \alpha = -\dfrac{5}{3}$

EXAMPLE 2   The terminal side of a fourth-quadrant angle of measure θ lies on the line $y = -x/2$. Find the exact values of the trigonometric functions of θ.

Solution: First, we determine the coordinates of some point on the terminal side of θ. If, for example, we set $x = 2$ in the equation of the line $y = -x/2$, we obtain $y = -1$. Thus, the point $P(2, -1)$ in the fourth quadrant lies on the terminal side of the angle of measure θ (Figure 8.14). The distance from P to the origin is

$$r = \sqrt{2^2 + (-1)^2} = \sqrt{5}.$$

Thus, according to formulas (8.17), we have

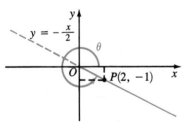

Figure 8.14

$$\sin \theta = -\frac{1}{\sqrt{5}} = -\frac{\sqrt{5}}{5}, \qquad \csc \theta = -\sqrt{5},$$

$$\cos \theta = \frac{2}{\sqrt{5}} = \frac{2\sqrt{5}}{5}, \qquad \sec \theta = \frac{\sqrt{5}}{2},$$

$$\tan \theta = -\frac{1}{2}, \qquad \cot \theta = -2.$$

Note that $\tan \theta = -1/2$, which is the slope of the line $y = -x/2$ where the terminal side of the angle θ lies.

■   **Remark**   Instead of substituting 2 for x in the equation $y = -x/2$, we could have substituted any other *positive* value for x, to obtain the coordinates of another point on the terminal side of the angle. The coordinates of points in the fourth quadrant and on the line $y = -x/2$ change, but the trigonometric values of θ remain the same. As an exercise, you can check this assertion by substituting other values for x, such as 1, 3, or 5.

Practice Exercise 2   The terminal side of a second-quadrant angle of measure β lies on the line $y = -x/2$. Find the exact values of the trigonometric functions of β.

Answer: $\sin \beta = \dfrac{\sqrt{5}}{5}$, $\cos \beta = -\dfrac{2\sqrt{5}}{5}$, $\tan \beta = -\dfrac{1}{2}$, $\cot \beta = -2$,

$\sec \beta = -\dfrac{\sqrt{5}}{2}$, $\csc \beta = \sqrt{5}$

EXAMPLE 3   If $\tan \theta = 3/5$ and θ is a third-quadrant angle, find the remaining values of the trigonometric functions of θ.

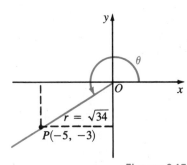

Figure 8.15

Solution: In the third quadrant, both coordinates of any given point are negative. If $P(x, y)$ lies on the terminal side of θ, then $\tan \theta = y/x$ [formulas (8.17)]. Setting $x = -5$ and $y = -3$ (Figure 8.15), we have

$$\tan \theta = \frac{y}{x} = \frac{-3}{-5} = \frac{3}{5}.$$

Thus, the point $P(-5, -3)$ in the third quadrant lies on the terminal side of the angle of measure θ. Taking reciprocals, we obtain

$$\cot \theta = \frac{x}{y} = \frac{-5}{-3} = \frac{5}{3}.$$

Since the distance from $P(-5, -3)$ to the origin is

$$r = \sqrt{(-5)^2 + (-3)^2} = \sqrt{34},$$

the other four trigonometric functions of θ are

$$\sin \theta = -\frac{3}{\sqrt{34}} = -\frac{3\sqrt{34}}{34}, \qquad \csc \theta = -\frac{\sqrt{34}}{3},$$

$$\cos \theta = -\frac{5}{\sqrt{34}} = -\frac{5\sqrt{34}}{34}, \qquad \sec \theta = -\frac{\sqrt{34}}{5}.$$

Practice Exercise 3   If $\cot \alpha = -1/2$ and α is a second-quadrant angle, find the values of all the trigonometric functions of α.

Answer: $\sin \alpha = \dfrac{2\sqrt{5}}{5}$, $\cos \alpha = -\dfrac{\sqrt{5}}{5}$, $\tan \alpha = -2$, $\cot \alpha = -\dfrac{1}{2}$,

$\sec \alpha = -\sqrt{5}$, $\csc \alpha = \dfrac{\sqrt{5}}{2}$

EXAMPLE 4

If $\sin \alpha = 12/13$ and α is a first-quadrant angle, find the remaining five trigonometric functions of α.

Solution: The point with y-coordinate 12 and distance from the origin $r = 13$ is on the terminal side of α (Figure 8.16). Using the Pythagorean theorem, we find that the x-coordinate of the point is 5. Thus,

$$\sin \alpha = \frac{12}{13}, \qquad \csc \alpha = \frac{13}{12},$$

$$\cos \alpha = \frac{5}{13}, \qquad \sec \alpha = \frac{13}{5},$$

$$\tan \alpha = \frac{12}{5}, \qquad \cot \alpha = \frac{5}{12}.$$

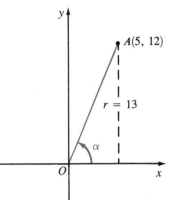

Figure 8.16

Practice Exercise 4

If $\cos \beta = -8/17$ and β is a second-quadrant angle, find $\sin \beta$, $\tan \beta$, $\cot \beta$, $\sec \beta$, and $\csc \beta$.

Answer: $\dfrac{15}{17}, \ -\dfrac{15}{8}, \ -\dfrac{8}{15}, \ -\dfrac{17}{8}, \dfrac{17}{15}$

EXERCISES 8.2

In Exercises 1–10, the given point is on the terminal side of θ, an angle in standard position. Find the values of the six trigonometric functions of θ. Give exact answers. Do not use a calculator.

1. $A(2, 5)$ **2.** $B(3, 6)$
3. $C(-2, 6)$ **4.** $D(-3, 5)$
5. $M(-3, -7)$ **6.** $N(-2, -3)$
7. $P(2, -4)$ **8.** $Q(3, -4)$
9. $X(4, -5)$ **10.** $Y(-4, 5)$

In Exercises 11–20, the terminal side of θ lies in the given quadrant and on the line with the given equation. Find exact values for the six trigonometric functions of θ.

11. first quadrant; $y = 2x$
12. second quadrant; $y = -4x$
13. second quadrant; $y = -3x$
14. first quadrant; $y = x/4$
15. third quadrant; $x - y = 0$
16. fourth quadrant; $x + y = 0$
17. first quadrant; $2x - 3y = 0$
18. second quadrant; $2x + 5y = 0$
19. third quadrant; $3x - 2y = 0$
20. fourth quadrant; $3x + y = 0$

In Exercises 21–28, the value of a trigonometric function of an angle and a condition on the angle are given. Find the remaining five trigonometric functions of the angle.

21. $\tan \theta = \dfrac{3}{5}$ and θ is a first-quadrant angle.

22. $\tan \theta = -4$ and θ is a second-quadrant angle.
23. $\cot \alpha = -2$ and $\sin \alpha > 0$.

24. $\cot \beta = \dfrac{1}{2}$ and $\cos \beta < 0$.

25. $\sin \theta = \dfrac{1}{3}$ and θ is a second-quadrant angle.

26. $\sin \theta = -\dfrac{3}{4}$ and $\cos \theta < 0$.

27. $\cos \theta = -\dfrac{2}{5}$ and θ is a third-quadrant angle.

28. $\cos \theta = \dfrac{1}{5}$ and $\sin \theta < 0$.

Solve Exercises 29 and 30.

29. Find the coordinates of the point where the half-line from the origin to the point $(4, -3)$ intersects the unit circle.
30. Find the coordinates of the point where the half-line from the origin to the point $(-7, -24)$ intersects the unit circle.

8.3 EVALUATING TRIGONOMETRIC FUNCTIONS

We can compute function values of trigonometric functions by using the special angles table, trigonometric tables, or a calculator. If the measure of a given angle is negative or greater than $90°$ ($\pi/2$ radians), then we can determine the function values of the angle by the *reference angle* of the given angle.

Reference Angle

The *reference angle* of a nonquadrantal angle θ in standard position is the *positive acute angle θ'* formed by the terminal side of θ and the *x*-axis.

For example, the reference angle of $\theta = 5\pi/6$ is $\theta' = \pi/6$, as shown in Figure 8.17a. The reference angle of $\theta = 225°$ is $\theta' = 45°$ (Figure 8.17b), and $\theta' = 60°$ is the reference angle of $\theta = -60°$ (Figure 8.17c).

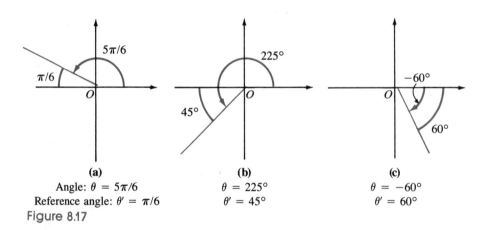

(a)	**(b)**	**(c)**
Angle: $\theta = 5\pi/6$	$\theta = 225°$	$\theta = -60°$
Reference angle: $\theta' = \pi/6$	$\theta' = 45°$	$\theta' = 60°$

Figure 8.17

If the reference angle of a given angle is a special angle, then the values of the trigonometric functions of the angle can be computed directly by using the special angle table of Section 7.2 (which by now you should have memorized) and by considering the geometry of the given angle in standard position. In any case, *it is extremely important that you make a sketch of the angle whose function values you are trying to evaluate.*

EXAMPLE **1**   Find the sine and cosine of the angles **(a)** $5\pi/6$ and **(b)** $225°$.

Solution: **(a)** The angle $\theta = 5\pi/6$ and its reference angle $\theta' = \pi/6$ are shown in Figure 8.18, where the circle represents the unit circle. Referring to the right triangle *OQP,* we see that

$$\sin \theta = \sin \theta' \quad \text{and} \quad \cos \theta = -\cos \theta'.$$

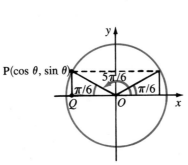

Figure 8.18   Figure 8.19

Thus, from the special angle table,

$$\sin \frac{5\pi}{6} = \sin \frac{\pi}{6} = \frac{1}{2}$$

and

$$\cos \frac{5\pi}{6} = -\cos \frac{\pi}{6} = -\frac{\sqrt{3}}{2}.$$

(b) Both the 225° angle and its reference angle measuring 45° are shown in Figure 8.19. It follows that

$$\sin 225° = -\sin 45° = -\frac{\sqrt{2}}{2}$$

and

$$\cos 225° = -\cos 45° = -\frac{\sqrt{2}}{2}.$$

Practice Exercise 1   Find **(a)** sin 135° and **(b)** $\cos \frac{4\pi}{3}$.

Answer: **(a)** $\dfrac{\sqrt{2}}{2}$ **(b)** $-\dfrac{1}{2}$

EXAMPLE 2   Find the values of the trigonometric functions of $\dfrac{5\pi}{3}$.

Solution: As shown in Figure 8.20, the quantity $5\pi/3$ is the radian measure of a fourth-quadrant angle whose reference angle measures $\pi/3$. Thus,

$$\sin \frac{5\pi}{3} = -\sin \frac{\pi}{3} = -\frac{\sqrt{3}}{2},$$

$$\cos \frac{5\pi}{3} = \cos \frac{\pi}{3} = \frac{1}{2},$$

$$\tan \frac{5\pi}{3} = -\tan \frac{\pi}{3} = -\sqrt{3},$$

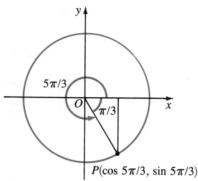

Figure 8.20

$$\cot \frac{5\pi}{3} = -\cot \frac{\pi}{3} = -\frac{\sqrt{3}}{3},$$

$$\sec \frac{5\pi}{3} = \sec \frac{\pi}{3} = 2,$$

$$\csc \frac{5\pi}{3} = -\csc \frac{\pi}{3} = -\frac{2\sqrt{3}}{3}.$$

Practice Exercise 2   Find the values of the trigonometric functions of $\frac{7\pi}{6}$.

Answer: $\sin \frac{7\pi}{6} = -\frac{1}{2}$, $\cos \frac{7\pi}{6} = -\frac{\sqrt{3}}{2}$, $\tan \frac{7\pi}{6} = \frac{\sqrt{3}}{3}$, $\cot \frac{7\pi}{6} = \sqrt{3}$,

$\sec \frac{7\pi}{6} = -\frac{2\sqrt{3}}{3}$, $\csc \frac{7\pi}{6} = -2$

EXAMPLE 3   Evaluate the trigonometric functions of $-210°$.

Solution: The given angle and its reference angle measuring $30°$ are illustrated in Figure 8.21. We have

$$\sin(-210°) = \sin 30° = \frac{1}{2},$$

$$\cos(-210°) = -\cos 30° = -\frac{\sqrt{3}}{2},$$

$$\tan(-210°) = -\tan 30° = -\frac{\sqrt{3}}{3},$$

$$\cot(-210°) = -\cot 30° = -\sqrt{3},$$

$$\sec(-210°) = -\sec 30° = -\frac{2\sqrt{3}}{3},$$

$$\csc(-210°) = \csc 30° = 2.$$

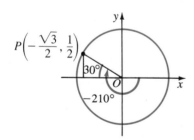

Figure 8.21

Practice Exercise 3   Find the values of the six trigonometric functions of $-60°$.

Answer: $\sin(-60°) = -\frac{\sqrt{3}}{2}$, $\cos(-60°) = \frac{1}{2}$, $\tan(-60°) = -\sqrt{3}$,

$\cot(-60°) = -\frac{\sqrt{3}}{3}$, $\sec(-60°) = 2$, $\csc(-60°) = -\frac{2\sqrt{3}}{3}$

In Examples 1, 2, and 3, all reference angles were special angles, so that the exact values of their trigonometric functions could be obtained from the special angle table. If the reference angle of a given angle is not a special angle, then we have to use a trigonometric table or a calculator to find the values of the trigonometric functions of the angle.

EXAMPLE 4   Evaluate (a) cot 239° and (b) sec 149°20′.

Solution: (a) The angle of measure 239° is a third-quadrant angle whose reference angle has measure $239° - 180° = 59°$. Using the trigonometric table displayed in Chapter 7 (Figure 7.23), and taking into account the fact that the cotangent function is positive in the third quadrant, we obtain

$$\cot 239° = \cot 59° = 0.6009.$$

If we use a calculator, then we don't need to determine the reference angle of the given angle. With the calculator in degree mode, we enter 239 and press the keys $\boxed{\tan}$ and $\boxed{1/x}$ to get

$$\cot 239° = 0.6008606.$$

(b) The reference angle has measure $180° - 149°20′ = 30°40′$ and the secant function is negative for a second-quadrant angle. According to the table in Figure 7.23,

$$\sec 149°20′ = -\sec 30°40′ = -1.163.$$

If we are using a calculator, we first set our calculator in degree mode and convert the given angle to decimal degrees:

$$149°20′ = 149° + (20/60)° = 149.33333°.$$

Next, we press the keys $\boxed{\cos}$ and $\boxed{1/x}$ to obtain
$$\sec 149°20′ = -1.1625891.$$

Practice Exercise 4   Using a trigonometric table, find the numbers (a) tan 210°30′ and (b) csc 149°.

Answer: (a) 0.5890 (b) 1.942

ᴳ EXAMPLE 5   Approximate the following trigonometric values: (a) $\tan(-5\pi/8)$; (b) $\csc(7\pi/5)$.

Solution: We set our calculator in radian mode.
(a) We perform the keystrokes

$$5 \boxed{\times} \boxed{\pi} \boxed{\div} 8 \boxed{=} \boxed{+/-} \boxed{\tan}$$

to obtain $\tan(-5\pi/8) = 2.4142136$.
(b) Since $\csc(7\pi/5) = 1/\sin(7\pi/5)$, the keystrokes

$$7 \boxed{\times} \boxed{\pi} \boxed{\div} 5 \boxed{=} \boxed{\sin} \boxed{1/x}$$

give us

$$\csc \frac{7\pi}{5} = -1.0514622.$$

c Practice Exercise 5   Find the approximate values of **(a)** $\cos(-3\pi/5)$ and **(b)** $\cot(4\pi/7)$.

   Answer: **(a)** -0.309017 **(b)** -0.2282435

EXERCISES 8.3

In Exercises 1–12, find the reference angle of the given angle.

1. $120°$

2. $135°$

3. $-320°$

4. $350°$

5. $\dfrac{2\pi}{3}$

6. $\dfrac{7\pi}{4}$

7. $-\dfrac{9\pi}{4}$

8. $\dfrac{11\pi}{3}$

9. $-400°$

10. $750°$

11. $-\dfrac{6\pi}{7}$

12. $\dfrac{5\pi}{7}$

In Exercises 13–24, find the exact value of each trigonometric function without using a trigonometric table or a calculator.

13. $\sin\dfrac{2\pi}{3}$

14. $\cos\dfrac{5\pi}{4}$

15. $\sec 210°$

16. $\csc 315°$

17. $\cot\left(-\dfrac{3\pi}{4}\right)$

18. $\sec\left(-\dfrac{\pi}{6}\right)$

19. $\cos\dfrac{11\pi}{6}$

20. $\cot\dfrac{5\pi}{3}$

21. $\sin(-390°)$

22. $\cos 405°$

23. $\cos(-300°)$

24. $\sin(-150°)$

In Exercises 25–36, use Table 4 in the Appendix to evaluate the given trigonometric functions.

25. $\tan 15°30'$

26. $\cos(-25°10')$

27. $\sec(-56°40')$

28. $\tan(-45°20')$

29. $\sin 105°20'$

30. $\csc 120°30'$

31. $\sin 0.1047$

32. $\sec 1.4573$

33. $\tan 0.3927$

34. $\cot 1.1257$

35. $\tan(-0.8639)$

36. $\cos(-0.6720)$

c In Exercises 37–50, use your calculator to approximate each number to four decimal places.

37. $\sin 22.6°$

38. $\tan 12.05°$

39. $\sin 23°15'$

40. $\cot(-105°26')$

41. $\csc(-26°32')$

42. $\tan 48°12'$

43. $\sin 42°30'16''$

44. $\cos 25°16'15''$

45. $\cot\left(-\dfrac{2\pi}{7}\right)$

46. $\csc\dfrac{12\pi}{5}$

47. $\cos\dfrac{\sqrt{2}\,\pi}{5}$

48. $\tan\dfrac{5\pi}{8}$

49. $\sec(-12.38)$

50. $\cot 2.318$

8.4	**GRAPHS OF TRIGONOMETRIC FUNCTIONS**

Up to now, we defined trigonometric functions as functions of θ, the degree or radian measure of an angle. However, in most applications in mathematics, trigonometric functions are regarded as *functions of a real number x,* with x representing the *radian measure* of an angle.

   In Figure 8.22, let U be the unit circle and assume that $\angle AOP$ has measure x radians. As we explained in Section 7.1, immediately after the definition of radian measure of an angle, the arc AP measures x units of length. Thus, to each real number x, a unique point P can be assigned on the unit circle so that the length of the arc AP or the radian measure of $\angle AOP$ is equal to x. Since the length of the unit circle is 2π, the point P also corresponds to the numbers $x - 2\pi$ and $x + 2\pi$.

More generally, P is the point on the unit circle that corresponds to all numbers of the form $x + 2k\pi$, with k any integer.

The Sine Function

The function

$$f(x) = \sin x$$

is defined by assigning to each real number x, the y-coordinate of P, the corresponding point on the unit circle (Figure 8.22). Notice that, except for the change of the name of the variable from θ to x, this is the definition of sine given in Section 8.1. The domain of definition of $f(x) = \sin x$ is the real line and the range is the closed interval $[-1, 1]$. Since the same point P on the unit circle corresponds to both of the numbers x and $x + 2\pi$, then

$$\sin x = \sin(x + 2\pi),$$

that is, the sine function has period 2π. This implies, as we shall see, that the graph of $f(x) = \sin x$ has a cyclical behavior. Also, in order to determine the graph, it suffices to consider only the function values of $\sin x$ for x in the interval $[0, 2\pi]$.

To sketch the graph of $y = \sin x$, we could form a table of values, plot the corresponding points, and join them by a smooth curve, as illustrated in Figure 8.23.

Figure 8.22

x	$\sin x$
0	0
$\dfrac{\pi}{6}$	$\dfrac{1}{2}$
$\dfrac{\pi}{2}$	1
$\dfrac{2\pi}{3}$	$\dfrac{\sqrt{3}}{2}$
π	0
$\dfrac{5\pi}{4}$	$-\dfrac{\sqrt{2}}{2}$
$\dfrac{3\pi}{2}$	-1
$\dfrac{5\pi}{6}$	$-\dfrac{1}{2}$
2π	0

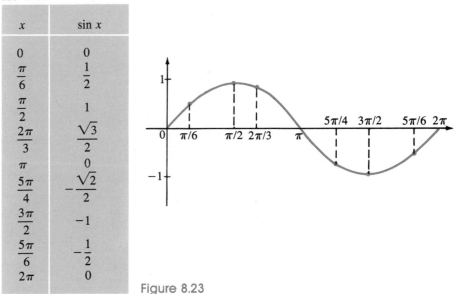

Figure 8.23

This method has a serious drawback, since it is difficult to plot points for which one or both coordinates are irrational numbers, such as $(\pi/6, 1/2) \simeq (0.5236, 0.5)$ or $(2\pi/3, \sqrt{3}/2) \simeq (2.094, 0.8660)$. For this reason, when studying the graph of $f(x) = \sin x$, it is preferable to examine the variation of $\sin x$ as x varies in the interval $[0, 2\pi]$ and the point P moves along the unit circle.

Figure 8.24 illustrates two identical coordinate systems. On the first one, P is a point on the unit circle such that the length of the arc AP or the radian measure

HISTORICAL NOTE

TRIGONOMETRIC FUNCTIONS

The notation and the names of the trigonometric functions as we use them today first appeared in the treatise *Introduction to the Analysis of Infinites* by the Swiss mathematician Leonhard Euler (1707–1783), published in 1748.

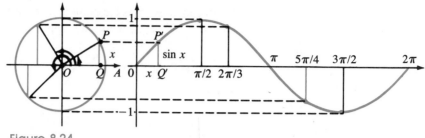

Figure 8.24

of $\angle AOP$ is x. The segment PQ represents $\sin x$. On the second coordinate system, the length of the segment OQ' is x and the segment $P'Q'$ is equal to the segment PQ. Thus, the point P' has coordinates $(x, \sin x)$. As the point P moves counterclockwise on the unit circle, the arc length x varies from 0 to 2π. When $x = 0$, the point P coincides with A, so that $\sin 0 = 0$. As x varies from 0 to $\pi/2$, the function values of $\sin x$ are positive and increase from 0 to 1; when $x = \pi/2$, $\sin \pi/2 = 1$. As x varies from $\pi/2$ to π, the values of $\sin x$ are still positive but decrease from 1 to $0 = \sin \pi$. As x varies from π to $3\pi/2$, the values of $\sin x$ are negative and decrease from 0 to -1; when $x = 3\pi/2$, $\sin 3\pi/2 = -1$. Finally, as x varies from $3\pi/2$ to 2π, the values of $\sin x$, still negative, increase from -1 to $0 = \sin 2\pi$. The behavior of the function $f(x) = \sin x$ can be summarized in the following table:

As x varies from	$y = \sin x$
0 to $\dfrac{\pi}{2}$	is positive and increases from 0 to 1
$\dfrac{\pi}{2}$ to π	is positive and decreases from 1 to 0
π to $\dfrac{3\pi}{2}$	is negative and decreases from 0 to -1
$\dfrac{3\pi}{2}$ to 2π	is negative and increases from -1 to 0

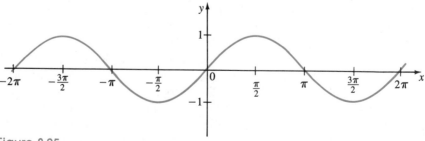

Figure 8.25

Because of the periodicity of the sine function, the same graph pattern repeats at intervals of length 2π along the x-axis. Thus, we obtain the curve illustrated in Figure 8.25. The x-intercepts occur when $y = \sin x = 0$, that is, when $x = k\pi$, with k any integer. If $x = 0$, then $\sin 0 = 0$, and the y-intercept is the origin.

The Cosine Function

The function

$$f(x) = \cos x$$

is defined by assigning to each real number x, the first coordinate of P, the corresponding point on the unit circle (Figure 8.22). The domain of definition of $f(x) = \cos x$ is the real line and the range is the closed interval $[-1, 1]$. Since $\cos(x + 2\pi) = \cos x$, the cosine function is periodic with period 2π, and its graph also exhibits a cyclical behavior, as in the case of the sine function. We can make the same observation concerning the plotting of points for the graph of $y = \cos x$ as we did about the graph of $y = \sin x$. Thus, to study the graph of the cosine function, it is better to examine the variation of $\cos x$ as x varies from 0 to 2π, plus the function values of $\cos x$ at some key points, such as $x = 0$, $\pi/2$, π, $3\pi/2$, and 2π.

Referring to the unit circle in Figure 8.24, the segment OQ represents $\cos x$. You can check that the behavior of $y = \cos x$ in the interval $[0, 2\pi]$ is summarized in the following table.

As x varies from	$y = \cos x$
0 to $\dfrac{\pi}{2}$	is positive and decreases from 1 to 0
$\dfrac{\pi}{2}$ to π	is negative and decreases from 0 to -1
π to $\dfrac{3\pi}{2}$	is negative and increases from -1 to 0
$\dfrac{3\pi}{2}$ to 2π	is positive and increases from 0 to 1

Proceeding in a similar manner as for $y = \sin x$ and taking into account the periodicity of the cosine function, we obtain the graph illustrated in Figure 8.26.

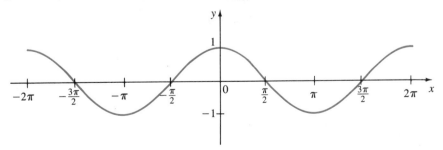

Figure 8.26

The x-intercepts occur when $x = \pm\pi/2, \pm3\pi/2, \ldots$, that is, when $x = (2k + 1)\pi/2$, with k any integer. When $x = 0$, $\cos 0 = 1$, so the graph crosses the y-axis at the point $(0, 1)$. Observe that the graph of $y = \cos x$ can be obtained by translating the graph of $y = \sin x$ to the left by $\pi/2$ units.

More General Functions

Now we want to use the graphs of $y = \sin x$ and $y = \cos x$ as models to graph more general trigonometric functions of the form $y = A \sin(Bx + C)$ and $y = A \cos(Bx + C)$. First we will discuss several cases, as a preparation for the discussion of the general case. In our first example, the function $\sin x$ is simply multiplied by a constant.

Amplitude

EXAMPLE **1**   Graph the function defined by $y = 3 \sin x$.

Solution: The given function has the same period and x-intercepts as those of the function $y = \sin x$. The graph of $y = 3 \sin x$ may be obtained by multiplying the ordinates of points on the graph of $y = \sin x$ by 3. Figure 8.27 illustrates the graphs of $y = \sin x$ (dashed line) and $y = 3 \sin x$ (solid line).

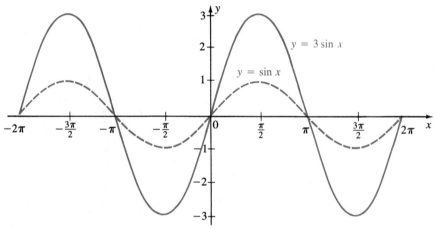

Figure 8.27

Practice Exercise 1   Sketch the graph of $f(x) = 2 \cos x$.

Answer:

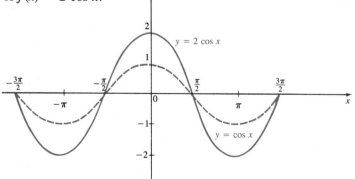

Example 1 indicates that the graph of $y = A \sin x$ can be obtained from the graph of $y = \sin x$ by multiplying each ordinate value by A. Both functions have the same period 2π and the same x-intercepts. However, the maximum value of $\sin x$ is 1, while the maximum value of $A \sin x$ is $|A|$. Also, the minimum value of $A \sin x$ is $-|A|$. Thus, the range of $f(x) = A \sin x$ is the closed interval $[-|A|, |A|]$. The same results apply to the function defined by $y = A \cos x$, and they can be summarized as follows.

Amplitude

The graphs of $y = A \sin x$ and $y = A \cos x$ have the same form as the graphs of $y = \sin x$ and $y = \cos x$, respectively. The range of both functions is the closed interval $[-|A|, |A|]$, and the number $|A|$ is called the *amplitude* of the function.

Changes in Period

If instead of $y = \sin x$ we want to graph $y = \sin Bx$, where B is a constant, then we must pay attention to a change in the period of the new function. Here is a simple example followed by a practice exercise.

EXAMPLE 2   Graph the function $f(x) = \sin 2x$.

Solution: The given function is defined for all real values of x and has amplitude 1. The period of $f(x) = \sin 2x$ is π, because

$$\sin[2(x + \pi)] = \sin(2x + 2\pi) = \sin 2x.$$

As illustrated in Figure 8.28, the graph of $f(x) = \sin 2x$ has the same form as the graph of $y = \sin x$, but the new curve has to be adjusted to fit the new period.

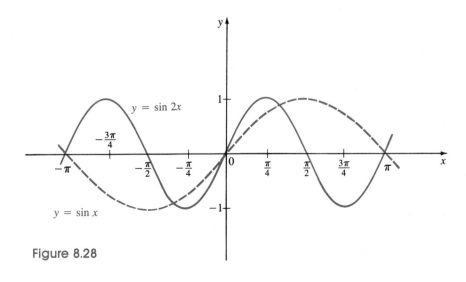

Figure 8.28

Notice that in addition to the x-intercepts at $x = k\pi$, which characterize $y = \sin x$, the function $f(x) = \sin 2x$ also has x-intercepts at $x = (2k + 1)\pi/2$.

Practice Exercise 2   What is the period of the function $f(x) = \cos(x/2)$? Graph this function.

Answer: Period $= 4\pi$

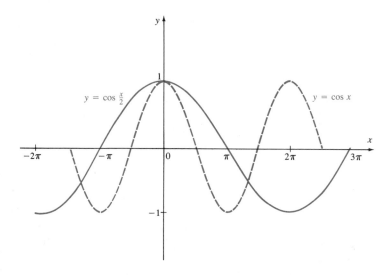

Example 2 and Practice Exercise 2 suggest the following generalization.

Period of $y = \sin Bx$ and $y = \cos Bx$, $B > 0$

The graphs of $y = \sin Bx$ and $y = \cos Bx$ are similar to the graphs of $y = \sin x$ and $y = \cos x$, respectively. If $B > 0$, the period of both functions is $2\pi/B$.

When graphing functions such as $y = A \sin Bx$ or $y = A \cos Bx$, we can always assume that $B > 0$. If $B < 0$, then by using the fact that the sine is an odd function and the cosine is an even function, we can rewrite the function so as to be back to the case $B > 0$. Here are an example and a practice exercise.

EXAMPLE 3   Determine the amplitude and period of the function defined by $y = 2 \sin(-2\pi x)$. Sketch its graph.

Solution: In this example, the coefficient of x is -2π, a negative number. Since sine is an odd function, we can write

$$y = 2 \sin(-2\pi x) = -2 \sin(2\pi x).$$

This is a function of the form $y = A \sin Bx$, with $A = -2$ and $B = 2\pi > 0$. Thus, the amplitude is $|A| = |-2| = 2$ and the period is $2\pi/2\pi = 1$. The graph is shown in Figure 8.29.

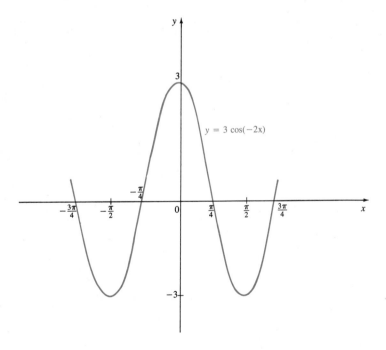

Figure 8.29

Practice Exercise 3   Graph the function $y = 3 \cos(-2x)$. Find the amplitude and period.

Answer: [Cosine is an even function; thus $y = 3 \cos(-2x) = 3 \cos 2x$.]
Amplitude = 3, period = π

Phase Shift

In the example and practice exercise that follow, we discuss cases where shifts in the sine or cosine waves may occur.

EXAMPLE **4**   Graph the function $f(x) = \sin\left(x - \dfrac{\pi}{2}\right)$.

Solution: The given function has amplitude 1. The period is 2π, because

$$\sin\left[(x + 2\pi) - \frac{\pi}{2}\right] = \sin\left[\left(x - \frac{\pi}{2}\right) + 2\pi\right] = \sin\left(x - \frac{\pi}{2}\right).$$

The graph of $y = \sin[x - (\pi/2)]$, illustrated in Figure 8.30, can be obtained by shifting the graph of $y = \sin x$ to the right by $\pi/2$ units. (See the subsection of Section 4.3 on horizontal translations of graphs.)

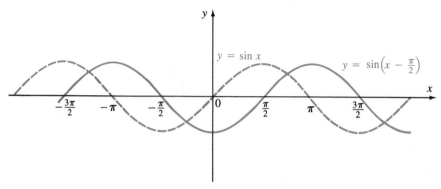

Figure 8.30

Practice Exercise 4   Sketch the graph of $y = \cos\left(x + \dfrac{\pi}{2}\right)$.

Answer:

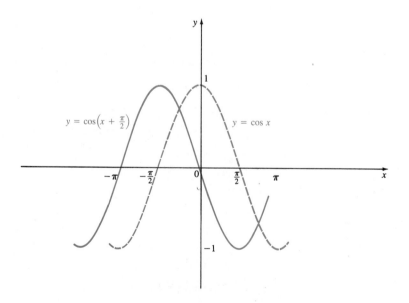

We can summarize the results of Exercise 4 and Practice Exercise 4 as follows.

> **Phase Shift**
>
> The graph of $y = \sin(x - D)$ can be obtained by translating the graph of $y = \sin x$ horizontally by $|D|$ units. The horizontal translation is to the right if $D > 0$ and to the left if $D < 0$. The number D is called the *phase shift*. The same results apply to the graph of $y = \cos(x - D)$.

The General Case

All functions discussed in the previous examples are particular cases of the function

$$y = A \sin(Bx + C), \quad B > 0,$$

which has amplitude $|A|$ and period $2\pi/B$. If we write

$$y = A \sin(Bx + C) = A \sin\left[B\left(x + \frac{C}{B}\right)\right]$$

and recall the discussion about horizontal translation of graphs (Section 4.3), then we see that the graph of $y = A \sin(Bx + C)$ is a horizontal translate of the graph of $y = A \sin Bx$. The graph is translated $|C/B|$ units to the right if $C < 0$ and to the left if $C > 0$. The number $-C/B$, which is the solution of the equation $Bx + C = 0$, is called the *phase shift*. Similar results hold for the graph of $y = A \cos(Bx + C)$.

EXAMPLE 5   Find the amplitude, period, and phase shift of the function $y = 3 \sin(2x + \pi)$. Sketch the graph.

Solution: In this example, $A = 3$, $B = 2$, and $C = \pi$. Thus, the amplitude is 3 and the period is $2\pi/2 = \pi$. To find the phase shift, we set $2x + \pi = 0$ and solve for x, obtaining $x = -\pi/2$. Thus the phase shift is $-C/B = -\pi/2$.

The graph of $y = 3 \sin(2x + \pi)$ is obtained by shifting the graph of $y = 3 \sin 2x$ horizontally $\pi/2$ units to the left. Both graphs are illustrated in Figure 8.31.

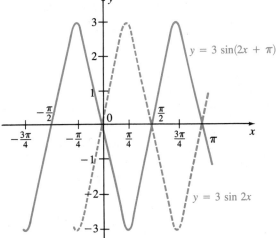

Figure 8.31

Practice Exercise 5   For the function $y = 4 \cos(2x - \pi)$, find the amplitude, period, and phase shift. Sketch the graph.

Answer: Amplitude = 4, period = π, phase shift = $\pi/2$

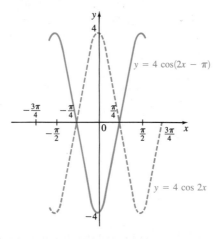

> ### Tips on Graphing Sine or Cosine Functions
>
> 1. Using the oddness of sine or the evenness of cosine, you can *always* write the function in the form
>
> $$y = A \sin(Bx + C) \quad \text{or} \quad y = A \cos(Bx + C), \quad \text{with} \quad B > 0.$$
>
> 2. Remember that for each of these functions, the *amplitude* is $|A|$, the *period* is $2\pi/B$, and the *phase shift* is $-C/B$.

In the rest of this section, we will briefly discuss the graphs of the remaining trigonometric functions.

The Tangent Function

The tangent function,

$$f(x) = \tan x,$$

is defined as the quotient $\sin x / \cos x$. Its domain of definition consists of all real numbers except those of the form $(2k + 1)\pi/2$, with k any integer, since at those values the cosine function is equal to zero. Since $\sin x \neq 0$ at these points, the lines $x = (2k + 1)\pi/2$, with k any integer, are vertical asymptotes of the graph of $y = \tan x$. The range of the tangent function is the whole real line, $(-\infty, +\infty)$. According to formula (8.7), the tangent function has period π. Thus, in order to sketch its graph, it suffices to analyze the behavior of $f(x) = \tan x$ within an interval of length π where the function is defined. We choose $(-\pi/2, \pi/2)$ as such an interval.

Figure 8.32 illustrates two identical coordinate systems. On the first one, AP is an arc of length x on the unit circle. As we explained in Section 8.1, the number tan x is represented by the length of the segment AT on the vertical line l through the point $A(1, 0)$. In the second graph, the length x of AP is transferred to the x-axis, so that M denotes a point with abscissa x. The lengths of the segments MN and AT are equal, so that the point N has coordinates $(x, \tan x)$ and lies on the graph of $y = \tan x$. Similarly, $M'N' = AT'$ and the point $N'(-x, -\tan x)$ also lies on the graph of $y = \tan x$.

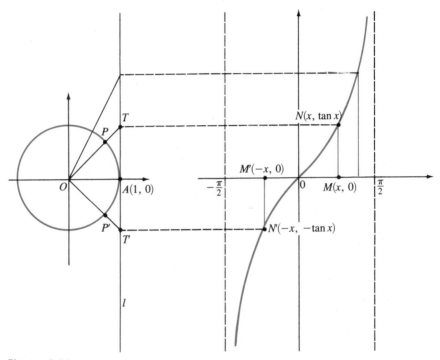

Figure 8.32

The behavior of $y = \tan x$, for x in the interval $(-\pi/2, \pi/2)$, can be summarized as follows.

As x varies from	$y = \tan x$
0 to $\dfrac{\pi}{2}$	is positive and increases without bound
0 to $-\dfrac{\pi}{2}$	is negative and decreases without bound

Because of the periodicity of $y = \tan x$, the same graph pattern is repeated in the intervals $(-3\pi/2, -\pi/2)$, $(\pi/2, 3\pi/2)$, and so on, as illustrated in Figure 8.33.

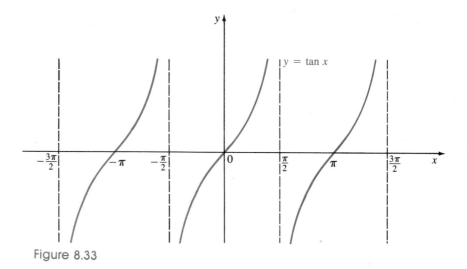

Figure 8.33

The Cotangent Function

The cotangent function

$$f(x) = \cot x,$$

defined as the quotient $\cos x / \sin x$, is the reciprocal of the tangent function. The domain of definition of the cotangent function is the set $\mathbb{R}$, except those numbers of the form $k\pi$, with k any integer, where $\sin x$ is equal to zero. The range is the whole line.

To determine the graph of $y = \cot x$ (Figure 8.34), notice that the lines $x = k\pi$, with k any integer, are vertical asymptotes, and that the x-intercepts occur at the points $x = (2k + 1)\,\pi/2$, with k any integer. Since $\cot x = 1/\tan x$, the graph of $y = \cot x$ can be obtained from the graph of $y = \tan x$ by taking recip-

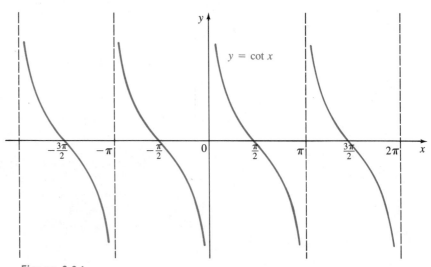

Figure 8.34

rocals of the ordinates. Notice that if x is positive and approaches 0, then tan x is positive and approaches 0, so cot x is positive and increases without bound. On the other hand, if x is negative and approaches 0, then tan x is negative and approaches 0, so cot x is negative and decreases without bound. A similar behavior occurs near each point $k\pi$, with k any integer.

Secant and Cosecant Functions

The secant function,

$$f(x) = \sec x,$$

is defined as the reciprocal of cos x, so it has for its domain of definition the set $\mathbb{R}$, except those numbers of the form $(2k + 1)\pi/2$, with k any integer. The graph of $y = \sec x$ can be obtained from the graph of $y = \cos x$ by taking reciprocals of the ordinates. In Figure 8.35, the graph of $y = \cos x$ is sketched in dashed lines and the graph of $y = \sec x$ is shown in solid lines. The vertical lines $x = (2k + 1)\pi/2$, with k any integer, are vertical asymptotes. The range of the secant function is the union of the intervals $(-\infty, -1]$ and $[1, +\infty)$.

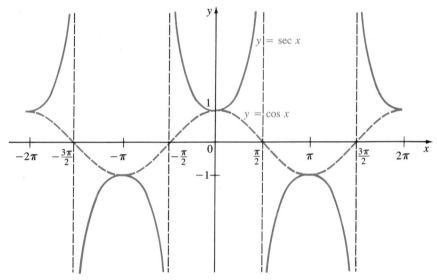

Figure 8.35

The cosecant function,

$$f(x) = \csc x,$$

is defined as the reciprocal of sin x, so it has for its domain of definition the set $\mathbb{R}$, except those numbers of the form $k\pi$, with k any integer. Its range is the union of the intervals $(-\infty, -1]$ and $[1, +\infty)$. Figure 8.36 shows the graphs of sin x (dashed lines) and sec x (solid lines). The vertical lines $x = k\pi$, with k any integer, are vertical asymptotes.

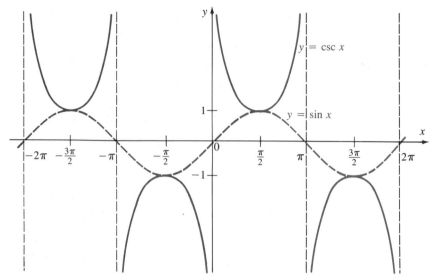

Figure 8.36

EXAMPLE 6   Graph the function $y = \tan 2x$.

Solution: First, we observe that the period of this function is $\pi/2$, because

$$\tan\left[2\left(x + \frac{\pi}{2}\right)\right] = \tan(2x + \pi) = \tan 2x.$$

As we see in Figure 8.37, the graph looks like the graph of $y = \tan x$, but the new curve must be adjusted to fit the new period.

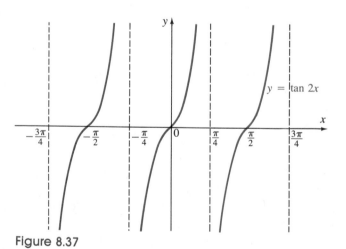

Figure 8.37

Practice Exercise 6   Find the period of $y = \cot 3x$. Graph this function.

Answer: Period $= \dfrac{\pi}{3}$

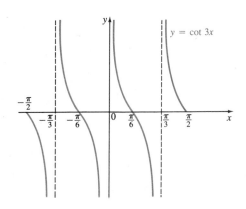

$y = \cot 3x$

EXERCISES 8.4

In Exercises 1–10, graph the functions.

1. $f(x) = \sin(-x)$ **2.** $f(x) = -\cos x$

3. $f(x) = 3 \cos x$ **4.** $f(x) = 2 \sin x$

5. $f(x) = \dfrac{1}{2} \sin x$ **6.** $f(x) = \dfrac{5}{4} \cos x$

7. $f(x) = -3 \sin x$ **8.** $f(x) = -4 \cos x$

9. $f(x) = -\dfrac{3}{4} \cos x$ **10.** $f(x) = -\dfrac{5}{2} \cos x$

In Exercises 11–30, find the amplitude, period, and phase shift of the given functions. Sketch their graphs.

11. $y = \cos 4x$ **12.** $y = \sin 3x$

13. $y = \sin \dfrac{x}{3}$ **14.** $y = \cos \dfrac{x}{4}$

15. $y = 3 \sin(-2x)$ **16.** $y = 4 \cos(-3x)$

17. $y = 3 \cos \dfrac{\pi x}{2}$ **18.** $y = 2 \sin \dfrac{\pi x}{3}$

19. $y = \sin\left(x + \dfrac{\pi}{3}\right)$ **20.** $y = \cos\left(x - \dfrac{\pi}{4}\right)$

21. $y = \sin \pi(x - 1)$ **22.** $y = \cos \pi(x + 1)$

23. $y = 2 \cos\left(x - \dfrac{\pi}{3}\right)$ **24.** $y = 4 \sin\left(x - \dfrac{\pi}{2}\right)$

25. $y = 3 \sin(2x - \pi)$ **26.** $y = 5 \cos(2x + \pi)$

27. $y = \tan 3x$ **28.** $y = \cot 2x$

29. $y = \tan\left(-\dfrac{x}{2}\right)$ **30.** $y = \cot\left(-\dfrac{x}{4}\right)$

In Exercises 31–36, use the technique of vertical translation to sketch the graphs of the following functions.

31. $f(x) = \cos x - 1$ **32.** $f(x) = 2 - \sin x$

33. $f(x) = 2 - \sin \pi x$ **34.** $f(x) = \cos 3x - 1$

35. $f(x) = \tan x + 2$ **36.** $f(x) = 1 - \cot x$

Are the functions in Exercises 37–40 even, odd, or neither?

37. $f(x) = \sin |x|$ **38.** $f(x) = |\cos x|$

39. $f(x) = \tan(-x)$ **40.** $f(x) = \cot x^3$

8.5   SIMPLE HARMONIC MOTION

The study of oscillatory phenomena such as the motion of a pendulum, the displacement of a mass–spring system, sound waves, alternating electric currents, climatic cycles, the human heartbeat, and periodic changes of plant and animal populations can be described in mathematical terms by the trigonometric equations

$$y = A \sin(Bx + C) \quad \text{and} \quad y = A \cos(Bx + C),$$

studied in the previous section. Each equation is said to represent a *simple harmonic motion* and its graph is called a *simple harmonic curve*.

As a first example of simple harmonic motion, consider a point moving around a circle with constant angular speed. When a point P (Figure 8.38) moves counterclockwise with constant angular speed ω around a circle of radius a, centered at the origin of a Cartesian coordinate system, its *projection* Q moves back and forth along the x-axis in oscillatory motion. Let us find an expression giving the position of Q at time t.

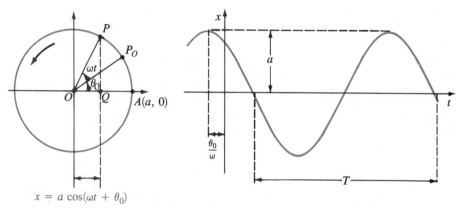

$$x = a \cos(\omega t + \theta_0)$$

Figure 8.38

Suppose that when $t = 0$ (that is, at the beginning of our observation), the point P is at the position P_0 and $\angle AOP_0$ measures θ_0 radians. Since the angular speed is ω, the measure θ of $\angle AOP$ at time t is $\theta = \omega t + \theta_0$. Thus, at time t, the *abscissa of P on the circle* or the *position of Q on the x-axis* is given by

$$x = a \cos(\omega t + \theta_0). \tag{8.18}$$

In this formula, a is the *amplitude of the oscillation* and θ_0 is called the *phase angle*. Setting $t = 0$ in (8.18) gives us

$$x = a \cos \theta_0;$$

this is the initial position of the point Q.

According to our discussion in Section 8.4, the *period T of the oscillation* is given by $T = 2\pi/\omega$, and it represents the time it takes for the moving point to complete *one oscillation* (or *full cycle*). The reciprocal of the period,

$$\nu = \frac{1}{T} = \frac{\omega}{2\pi},$$

is called the *frequency,* and it corresponds to the *number of oscillations* (or *cycles*) *per second*. When the period T is measured in seconds, the frequency ν is measured in *hertz*. One hertz (abbreviated Hz) corresponds to *one cycle per second*. (For example, the alternating current supplied by utility companies has a frequency of 60 Hz.)

The angular speed ω is sometimes referred to as the *angular frequency* of the oscillation; it is measured in radians per second (rad/s). Notice that it is the quantity ω that determines the period and frequency of the oscillation. Figure 8.38 illustrates the graph of equation (8.18) as well as the amplitude, period, and phase shift.

EXAMPLE 1

A wheel of radius 6 in. centered at the origin is rotating counterclockwise at a constant rate of 30 revolutions per minute (rpm). Describe the motion of the perpendicular projection along the x-axis of a point on the rim (Figure 8.38), knowing that at $t = 0$ the point on the rim has coordinates $\left(3\sqrt{2}, 3\sqrt{2}\right)$.

Solution: First we must determine the angular speed ω. The rate of 30 rpm corresponds to $30/60 = 1/2$ revolution per second. Since one revolution covers an angle of 2π radians, the angular speed is then

$$\omega = \frac{1}{2}(2\pi) = \pi \text{ rad/s}.$$

From this we obtain the period $T = 2\pi/\omega = 2\pi/\pi = 2$ seconds and the frequency $\nu = 1/T = 1/2$ Hz. Next, we determine the phase angle θ_0. If the initial position, P_0, of the point on the rim has coordinates $\left(3\sqrt{2}, 3\sqrt{2}\right)$, then the angle between the segment OP_0 and the x-axis is $\theta_0 = \pi/4$. (Why?) Since $a = 6$, $\omega = \pi$, and $\theta_0 = \pi/4$, if follows that

$$x = 6 \cos\left(\pi t + \frac{\pi}{4}\right).$$

Thus, the amplitude of the oscillation is 6 in., the period is 2 s, and the frequency $1/2$ Hz.

Practice Exercise 1

A point moves counterclockwise around a circle of radius 5 cm at a rate of 45 rpm. Describe the motion of the perpendicular projection of the point on a diameter, if the phase angle is $\pi/3$ radians. Find the amplitude and period of the motion.

Answer: $x = 5 \cos\left(\frac{3\pi}{2}t + \frac{\pi}{3}\right)$; amplitude = 5 cm; period = $4/3$ s

Mass–Spring System

In physics, a mass–spring system consists of a mass m attached to the lower end of a spring whose upper end is fastened to a rigid support. When the mass is pulled down and released, it will bob up and down in an oscillatory motion. The displacement of the mass is measured along a vertical coordinate axis with positive direction pointing downward and origin coinciding with the position of the mass at rest.

If air resistance and any other friction are neglected, it can be shown that the equation

$$y = a \cos \omega t$$

gives the position at time t of a mass m that was pulled down a units of length and released at time $t = 0$. In this equation, $\omega = \sqrt{k/m}$, where k is a constant measuring the *stiffness* of the spring, that is, the amount of force required to stretch the spring one unit of distance.

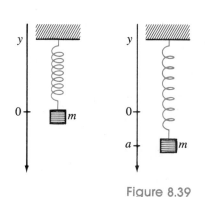

Figure 8.39

EXAMPLE 2

In a mass–spring system, a mass of 75 g is attached to a spring of stiffness $k = 1.2 \times 10^3$ dynes/cm. If the mass is pulled down 5 cm and then released, find (a) an equation describing the vertical displacements of the mass, and (b) the period and frequency of the motion.

Solution: (a) Since $m = 75$ and $k = 1.2 \times 10^3$, the angular frequency is

$$\omega = \sqrt{\frac{1.2 \times 10^3}{75}} = 4.$$

Thus, the equation describing the vertical displacement of the mass is

$$y = 5 \cos 4t.$$

(b) The period is

$$T = \frac{2\pi}{4} \simeq 1.57 \text{ s}$$

and the frequency is

$$\nu = \frac{4}{2\pi} \simeq 0.64.$$

Practice Exercise 2

The equation

$$y = 0.1 \cos 8.1t,$$

where y is measured in meters and t in seconds, describes the oscillation of a 0.5-kg mass attached to the end of a spring. Find the amplitude, period, and frequency of the oscillation. What is the stiffness constant of the spring?

Answer: Amplitude = 0.1   m;   period = 0.8   s;   frequency = 1.3   Hz; $k = 32.8$ N/m

The Simple Pendulum

A *simple pendulum* consists of a point mass m suspended by a weightless string of length l moving along a vertical circular arc (Figure 8.40). We assume that the string does not stretch.

If y denotes a small arc displacement from the vertical position (positive to the right and negative to the left), then y is given by the equation

$$y = a \cos \omega t,$$

where a is the initial displacement and $\omega = \sqrt{g/l}$, where g is the acceleration of gravity. The period of a simple pendulum is then $T = 2\pi/\omega = 2\pi\sqrt{l/g}$.

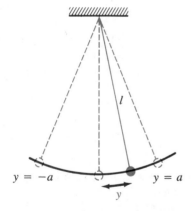

$y = -a$ $y = a$

y

Figure 8.40

⊙ EXAMPLE 3

In a location where the acceleration of gravity is 9.8 m/s², a pendulum of length 1.2 m is pulled to the right through an arc of length 0.2 m and then released. Write the equation describing the oscillation of the pendulum. What is the period of the pendulum?

Solution: First we determine the angular frequency of the oscillation:

$$\omega = \sqrt{\frac{g}{l}} = \sqrt{\frac{9.8}{1.2}} = 2.9.$$

Since $a = 0.2$ m, the equation

$$y = 0.2 \cos 2.9t$$

describes the oscillation of the pendulum. The period of the oscillation is

$$T = \frac{2\pi}{\omega} = \frac{2\pi}{2.9} \approx 2.2 \text{ seconds.}$$

ⓒ Practice Exercise 3   What is the length of a pendulum whose period is 1 s in a location where $g = 32$ ft/s²? Write the pendulum equation corresponding to an initial displacement of 0.22 ft.

Answer: $l = 0.81$ ft; $y = 0.22 \cos 6.28t$

EXERCISES 8.5

In Exercises 1–10, each equation describes a simple harmonic motion. Find **(a)** the amplitude, **(b)** the angular speed, **(c)** the period, **(d)** the frequency, and **(e)** the phase angle.

1. $x = 5 \cos\left(t - \dfrac{\pi}{4}\right)$

2. $x = 6 \cos\left(t + \dfrac{\pi}{2}\right)$

3. $x = 4 \cos(\pi t - 2)$

4. $x = 8 \cos(3\pi t + 1)$

5. $x = \dfrac{3}{4} \cos\left(2\pi t + \dfrac{\pi}{4}\right)$

6. $x = \dfrac{2}{5} \cos\left(\pi t - \dfrac{\pi}{8}\right)$

7. $x = 2 \cos\left(\dfrac{\pi}{2}t - \dfrac{\pi}{3}\right)$

8. $x = 3 \cos\left(\dfrac{\pi}{3}t + \dfrac{\pi}{4}\right)$

9. $x = 6 \cos\left(\pi t + \dfrac{\pi}{2}\right) + 1$

10. $x = 5 \cos(3\pi t - \pi) - 2$

In Exercises 11–30, solve each of the given problems.

11. A point moves counterclockwise at a rate of 30 cps around a circle with center at the origin and radius 4 units of length. If at $t = 0$ the coordinates of the point are $(0, 4)$, write the equation describing the oscillation of its projection on the x-axis.

12. Under the assumptions of Exercise 11, suppose that the radius of the circle measures 6 cm, the rate is 20 cpm, and the initial position of the point is $(3, 3\sqrt{3})$. What is the equation describing the motion of the projection of the point?

13. A 2-lb mass vibrates according to the equation

$$y = 0.25 \cos 2.5t,$$

where y is measured in feet and t in seconds. Determine **(a)** the amplitude of the motion, **(b)** the frequency, and **(c)** the stiffness constant.

14. A 2.0-kg mass attached to a spring is pulled down 0.24 m from equilibrium and then released. What is the equation describing the motion of the mass, if the stiffness constant is 92.5 N/m? What is the frequency of the oscillation?

15. The displacement of a particle is given by the equation

$$x = 6 \cos\left(20\pi t + \dfrac{\pi}{2}\right).$$

Find the amplitude, period, and frequency. What is the position of the particle after 1/120 s?

16. A mass–spring system is oscillating vertically with displacement given by

$$y = 24 \sin\left(3\pi t - \dfrac{\pi}{6}\right).$$

Find **(a)** the starting position of the mass (explain your answer), **(b)** the first instant after $t = 0$ that the mass reaches the starting position again, and **(c)** the first instant after $t = 0$ that the mass reaches its highest point.

17. What is the value of the angle θ_0 in formula (8.18) if, at $t = 0$, the point Q is at **(a)** $x = a$, **(b)** $x = 0$ and moving to the left, **(c)** $x = a/2$ and moving to the right?

18. Find the phase angle θ_0 in formula (8.18) if, at $t = 0$, the point Q is at **(a)** $x = -a$, **(b)** $x = 0$ and moving to the right, **(c)** $x = a\sqrt{2}/2$ and moving to the left.

19. The intensity I (in amperes) of an electrical current being produced by an alternating current generator is given by

$$I = 20 \sin(120\pi t),$$

where the time t is measured in seconds. Determine **(a)** the period and frequency of the current, **(b)** the maximum intensity of the current.

20. For an electrical circuit, the voltage E is given by

$$E = 2.4 \cos(50\pi t),$$

where t is the time in seconds. **(a)** Find the amplitude, period, and frequency. **(b)** Find E when $t = 0.03, 0.2, 1$, and 1.02 s.

21. If a pendulum of length 1.5 ft is pulled to the right through an arc of 0.05 ft and then released, find the equation describing the displacement of the pendulum at time t. What is the period of the pendulum? (acceleration of gravity: 32 ft/s²)

22. A pendulum 0.48 m long is released at an arc 0.06 m from the vertical. **(a)** Write the equation for the motion of the pendulum. **(b)** Find the period and frequency. (acceleration of gravity: 9.80 m/s²)

23. By what factor does the period of oscillation of a mass–spring system change if the mass attached at the end of the spring is quadrupled?

24. How does the period of a pendulum change if the length of the pendulum is cut in half?

25. Experimental observation shows that in certain predator–prey ecosystems, the numbers of predator and prey animals vary periodically through the years. Suppose that in a certain region where wolves are predators and caribous are prey, the caribou population, C, in t years is given by

$$C = 1500 - 250 \sin\left(\frac{\pi}{4}t\right).$$

What is the maximum population of caribous? After how many years is this maximum reached?

26. The deer population of a certain region is given by

$$D = 3600 + 400 \sin\left(\frac{\pi}{3}t\right).$$

where t is the time in years. What is the period of variation of this population? What are the maximum and minimum populations? After how many years is the minimum population reached?

MATHEMATICAL VIGNETTE

PREDATOR–PREY CYCLES

The feeding of one organism on another is called *predation*. A fox is a predator on rabbits (the prey). Ants are predators on grasshoppers. Predation is an environmental factor that may control and limit the size of a population. In a predator–prey relationship, each population continually determines the size of the other. The populations increase and decrease in cycles that can be represented graphically as follows.

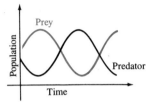

An increase in the predator population causes a decrease in the prey population. Lack of food eventually causes a decrease in the predator population, at which point the prey population starts to increase again.

27. On the surface of the moon, the acceleration of gravity is six times smaller than the acceleration of gravity on earth. Find the period of a pendulum 3 ft long on the surface of the moon.

28. Find the period of a simple pendulum on Mars, where the acceleration of gravity is about 0.4 times that on earth, knowing that the pendulum has a period of 1.2 s on earth.

29. When the temperature is 68°F, the period of the pendulum in a grandfather clock is 2 s. Due to an increase in temperature, the length of the pendulum increases by 10^{-4} of its original length. How many seconds will the clock gain or lose in one day?

30. A small insect of mass 0.15 g is caught in a spider web. Assuming that the web vibrates as a spring–mass system with a frequency of 20 Hz, find the stiffness of the web. At what frequency would the web vibrate if an insect of mass 0.30 g were caught in the web?

8.6   INVERSE TRIGONOMETRIC FUNCTIONS

In many practical situations, we are faced with the problem of determining the measure of an angle when we know the value of a trigonometric function of the angle. For example, suppose that we want to find the angle x in the following right triangle.

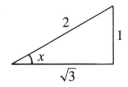

Since $\sin x = 1/2$ and x is an acute angle, then $x = 30°$ or $\pi/6$. Thus, in this case, x can be determined without ambiguity.

However, if we just ask for an angle x such that $\sin x = 1/2$, then x could be equal to $\pi/6$ ($= 30°$), or $5\pi/6$ ($= 150°$), or $-7\pi/6($ $= -210°$), as shown in Figure 8.41.

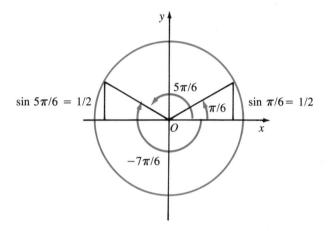

Figure 8.41

As a matter of fact, there are infinitely many values of x for which $\sin x = 1/2$, as Figure 8.42 illustrates. These are the abscissas of the points on the graph of $y = \sin x$ that are intercepted by the horizontal line $y = 1/2$.

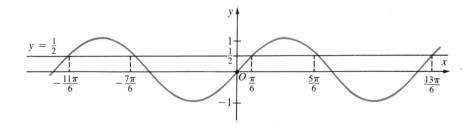

Figure 8.42

As we see, the function $f(x) = \sin x$ is *not a one-to-one function* but rather a *many-to-one* function, in the sense that many values of x have the same function value $\sin x$. The same observation applies to the other trigonometric functions.

Since trigonometric functions are not one-to-one, their inverses cannot be defined. However, if we restrict their domains of definition to suitable intervals, they become one-to-one and their inverses can be defined.

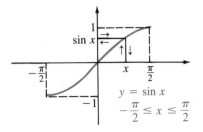

Figure 8.43

The Inverse Sine Function

Consider the function $y = \sin x$ defined only on the closed interval $[-\pi/2, \pi/2]$. A look at the graph (Figure 8.43) convinces us that this new function is one-to-one: every horizontal line through a point between -1 and 1 on the y-axis intercepts the graph at a single point. Thus, for the function $f(x) = \sin x$ restricted to the interval $[-\pi/2, \pi/2]$, we can define an inverse.

Definition

The *inverse sine function*, denoted by $y = \sin^{-1} x$, is defined by the following relation:

$$y = \sin^{-1} x, \quad -1 \leq x \leq 1, \quad \text{if and only if} \quad x = \sin y, \quad -\frac{\pi}{2} \leq y \leq \frac{\pi}{2}. \quad (8.19)$$

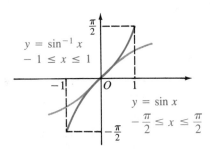

Figure 8.44

This definition indicates that the domain of the inverse sine function is the closed interval $[-1, 1]$ and the range is the closed interval $[-\pi/2, \pi/2]$. The following relations express the fact that $\sin x$ and $\sin^{-1} x$ are inverses of each other:

$$\sin(\sin^{-1} x) = x, \quad \text{for all } x \in [-1, 1],$$

$$\sin^{-1}(\sin x) = x, \quad \text{for all } x \in \left[-\frac{\pi}{2}, \frac{\pi}{2}\right].$$

Moreover, according to Section 5.4, the graph of $y = \sin^{-1} x$ (Figure 8.44) is the reflection about the line $y = x$ of the graph of $y = \sin x$, $-\pi/2 \leq x \leq \pi/2$.

The inverse sine function is also called the *arcsine function*, which is denoted by $y = \arcsin x$.

EXAMPLE 1   Find $\sin^{-1} \dfrac{1}{2}$.

Solution: If we set $y = \sin^{-1}(1/2)$, then according to (8.19),

$$y = \sin^{-1} \frac{1}{2} \quad \text{if and only if} \quad \frac{1}{2} = \sin y, \quad -\frac{\pi}{2} \leq y \leq \frac{\pi}{2}.$$

Thus, $y = \pi/6$, because $\pi/6$ is the only number between $-\pi/2$ and $\pi/2$ whose sine is equal to $1/2$.

Practice Exercise 1   Find $\sin^{-1} \dfrac{\sqrt{2}}{2}$.

Answer: $\dfrac{\pi}{4}$

⒞ EXAMPLE 2   In the example that follows, a table or calculator is required.
Evaluate arcsin 0.3338.

Solution: According to our definition, we must find a number y such that $-\pi/2 \leq y \leq \pi/2$ and sin $y = 0.3338$. Referring to Table 4 in the Appendix, we find that the value 0.3338 corresponds to the angle $y = 0.3403$ radians or 19°30′. Thus

$$\text{arcsin } 0.3338 = 0.3403 \text{ radians (or } 19°30′).$$

If we use a calculator, we can obtain a more accurate value. With the calculator in degree mode, we enter 0.3338 and press the keys $\boxed{\text{INV}}\,\boxed{\text{SIN}}$, obtaining

$$\sin^{-1}(0.3338) = 19.499583°$$
$$= 19°29′58\,''.$$

With the calculator in radian mode, we obtain

$$\sin^{-1}(0.3338) = 0.3403319.$$

⒞ **Practice Exercise 2**   Evaluate arcsin 0.8975.

Answer: 63.831375° ≃ 63°50′ (or 1.114068 radians)

EXAMPLE 3   Find $\sin^{-1}(\cos 30°)$ without using a table or calculator.

Solution: Since cos 30° = $\sqrt{3}/2$, we write $y = \sin^{-1}(\cos 30°) = \sin^{-1}(\sqrt{3}/2)$. Now, we must find a value of y between $-\pi/2$ and $\pi/2$ such that $\sqrt{3}/2 = \sin y$. Recalling the values of special angles, we obtain $y = \pi/3$ or $y = 60°$.

Practice Exercise 3   Without using a table or calculator, find $\sin^{-1}\left(\cos\dfrac{\pi}{3}\right)$.

Answer: $\dfrac{\pi}{6}$

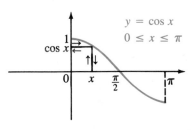

Figure 8.45

The Inverse Cosine Function

If the domain of $f(x) = \cos x$ is restricted to the interval $[0, \pi]$, we obtain a new function whose range is the interval $[-1, 1]$. As illustrated in Figure 8.45, the function $f(x) = \cos x$ defined on $[0, \pi]$ is one-to-one, so the inverse function can be defined.

Definition

The *inverse cosine function,* denoted by $y = \cos^{-1}x$, is defined by

$$y = \cos^{-1}x, \quad -1 \leq x \leq 1, \quad \text{if and only if} \quad x = \cos y, 0 \leq y \leq \pi. \quad (8.20)$$

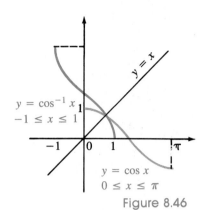

Figure 8.46

The domain of definition of the inverse cosine function is the interval $[-1, 1]$, and the range is the interval $[0, \pi]$. The following relations hold:

$$\cos(\cos^{-1}x) = x, \quad \text{for all } x \in [-1, 1],$$
$$\cos^{-1}(\cos x) = x, \quad \text{if all } x \in [0, \pi].$$

The graph of $y = \cos^{-1}x$, $-1 \le x \le 1$, is illustrated in Figure 8.46; it is the reflection about the line $y = x$ of the graph of $y = \cos x$, $0 \le x \le \pi$.

The inverse cosine function is also called the *arccosine function,* which is denoted by $y = \arccos x$.

EXAMPLE 4   Find $\cos^{-1}\left(-\dfrac{\sqrt{3}}{2}\right)$.

Solution: Let $y = \cos^{-1}(-\sqrt{3}/2)$ so that, by (8.20), we have $-\sqrt{3}/2 = \cos y$, $0 \le y \le \pi$. Thus, $y = 5\pi/6$ (or $150°$).

Practice Exercise 4   Find $\arccos\left(-\dfrac{1}{2}\right)$.

Answer: $\dfrac{2\pi}{3}$ (or $120°$)

EXAMPLE 5   Without using a table or calculator, find $\cos^{-1}\left(\tan\dfrac{\pi}{4}\right)$.

Solution: Since $\tan \pi/4 = 1$, we write

$$y = \cos^{-1}\left(\tan\frac{\pi}{4}\right) = \cos^{-1}(1),$$

so that $1 = \cos y$, $0 \le y \le \pi$. Thus $y = 0$.

Practice Exercise 5   Find $\cos^{-1}(\sin 45°)$ without using a table or a calculator.

Answer: $45°$ or $\dfrac{\pi}{4}$

Inverse Tangent Function

If the domain of the tangent function is restricted to the open interval $(-\pi/2, \pi/2)$ (Figure 8.33), we obtain a one-to-one function whose range is the interval $(-\infty, +\infty)$.

Definition

The *inverse tangent function,* denoted by $y = \tan^{-1}x$ or $y = \arctan x$, is defined by

$$y = \tan^{-1}x, \quad -\infty < x < +\infty, \quad \text{if and only if} \quad x = \tan y, \quad -\frac{\pi}{2} < y < \frac{\pi}{2}. \quad (8.21)$$

The domain of definition of $y = \tan^{-1}x$ is the entire real line $\mathbb{R}$, and the range is the open interval $(-\pi/2, \pi/2)$. The following relations hold:

$$\tan(\tan^{-1}x) = x, \quad \text{for all } x \in \mathbb{R},$$

$$\tan^{-1}(\tan x) = x, \quad \text{for all } x \in \left(-\frac{\pi}{2}, \frac{\pi}{2}\right).$$

Figure 8.47 illustrates the graph of $y = \tan^{-1}x$, obtained by reflecting the graph of $y = \tan x$, $-\pi/2 < x < \pi/2$, about the line $y = x$.

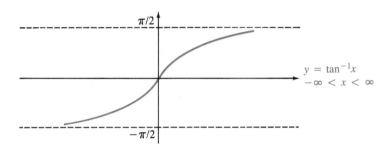

Figure 8.47

EXAMPLE 6

Find $\tan^{-1}(\sqrt{3})$.

Solution: Let $y = \tan^{-1}(\sqrt{3})$. By (8.21), we write $\sqrt{3} = \tan y$, $-\pi/2 < y < \pi/2$. Recalling the special angles table (Section 7.2), we obtain $y = \pi/3$.

Practice Exercise 6

Evaluate $\arctan 1$.

Answer: $\dfrac{\pi}{4}$

EXAMPLE 7

Find $\cos\left(\arctan \dfrac{\sqrt{3}}{3}\right)$.

Solution: **First method.** We let $y = \arctan \sqrt{3}/3$, so that $\tan y = \sqrt{3}/3$, with $-\pi/2 < y < \pi/2$. It follows from the special angles table that $y = \pi/6$; hence

$$\cos\left(\arctan \frac{\sqrt{3}}{3}\right) = \cos\left(\frac{\pi}{6}\right) = \frac{\sqrt{3}}{2}.$$

Second method. Again, we let $y = \arctan \sqrt{3}/3$, so that $\tan y = \sqrt{3}/3$. We consider now the right triangle with hypotenuse of length $2\sqrt{3}$ and sides of length 3 and $\sqrt{3}$ (see figure), such that y is the angle opposite the side of length $\sqrt{3}$. It follows that

$$\cos\left(\arctan \frac{\sqrt{3}}{3}\right) = \cos y = \frac{3}{2\sqrt{3}} = \frac{\sqrt{3}}{2}.$$

Practice Exercise 7   Find $\cos(\tan^{-1} 1)$.

Answer: $\dfrac{\sqrt{2}}{2}$

In the following example, the definitions of inverse trigonometric functions are combined with the trigonometric identities of Section 8.1

EXAMPLE 8   Evaluate $\sin\left(\cos^{-1}\dfrac{\sqrt{5}}{3}\right)$.

Solution: **First method.** If we set $y = \cos^{-1}(\sqrt{5}/3)$, then $\cos y = \sqrt{5}/3$, $0 \le y \le \pi$. Since $\sin^2 y + \cos^2 y = 1$, we obtain

$$\sin y = \sqrt{1 - \cos^2 y}$$
$$= \sqrt{1 - \frac{5}{9}}$$
$$= \sqrt{\frac{4}{9}}$$
$$= \frac{2}{3}.$$

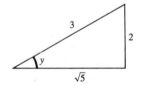

Here we have taken the positive square root because $\sin y$ is nonnegative when $0 \le y \le \pi$.

Second method. If $y = \cos^{-1}(\sqrt{5}/3)$, then $\cos y = \sqrt{5}/3$. Considering the right triangle with hypotenuse of length 3 and sides of length 2 and $\sqrt{5}$ (see figure), such that y is the angle opposite the side of length 2, we obtain that

$$\sin\left(\cos^{-1}\frac{\sqrt{5}}{3}\right) = \sin y = \frac{2}{3}.$$

Practice Exercise 8   Find $\cos\left(\tan^{-1}\dfrac{3}{4}\right)$.

Answer: $\dfrac{4}{5}$

EXAMPLE 9   Let $y = 3 \sin 2x$, $-\pi/4 \le x \le \pi/4$. Solve the given equation for x.

Solution: We first observe that since x lies in the interval $[-\pi/4, \pi/4]$, it follows that the function $f(x) = 3 \sin 2x$ is one-to-one, thus it has an inverse. Multiplying both sides of the given equation by $1/3$, we have

$$\frac{y}{3} = \sin 2x.$$

According to the definition of the inverse sine function, we obtain

$$2x = \sin^{-1}\left(\frac{y}{3}\right)$$

hence

$$x = \frac{1}{2} \sin^{-1}\left(\frac{y}{3}\right).$$

Practice Exercise 9   Solve the equation $y = 5 \cos 4x$, $0 \le x \le \pi/4$.

Answer: $x = \frac{1}{4} \arccos \frac{y}{5}$

EXERCISES 8.6

Using the special angle table (Section 7.2) for Exercises 1–12, find the following numbers.

1. $\arcsin \dfrac{\sqrt{3}}{2}$

2. $\cos^{-1}\left(-\dfrac{1}{2}\right)$

3. $\cos^{-1} 0$

4. $\sin^{-1}\left(-\dfrac{\sqrt{3}}{2}\right)$

5. $\tan^{-1}\left(-\dfrac{\sqrt{3}}{3}\right)$

6. $\tan^{-1}(-1)$

7. $\arccos \dfrac{\sqrt{2}}{2}$

8. $\arctan(-\sqrt{3})$

9. $\arcsin \dfrac{\sqrt{2}}{2}$

10. $\arcsin\left(-\dfrac{1}{2}\right)$

11. $\arctan \dfrac{\sqrt{3}}{3}$

12. $\arcsin 1$

In Exercises 13–20, use Table 4 in the Appendix or a calculator to find the corresponding angles to the nearest ten minutes.

13. $\arcsin 0.1994$

14. $\arccos 0.9147$

15. $\cos^{-1} 0.3907$

16. $\sin^{-1}(0.8936)$

17. $\arctan 0.9490$

18. $\tan^{-1}(1.228)$

19. $\tan^{-1}(0.9380)$

20. $\arcsin 0.7092$

In Exercises 21–40, find the values without the use of tables or calculators.

21. $\cos\left(\sin^{-1}\dfrac{\sqrt{2}}{2}\right)$

22. $\sin\left(\cos^{-1}\left(-\dfrac{1}{2}\right)\right)$

23. $\tan\left(\cos^{-1}\dfrac{\sqrt{3}}{2}\right)$

24. $\sec\left(\sin^{-1}\dfrac{1}{2}\right)$

25. $\cot\left(\sin^{-1}\dfrac{\sqrt{3}}{2}\right)$

26. $\tan\left(\sin^{-1}\left(-\dfrac{\sqrt{2}}{2}\right)\right)$

27. $\csc\left(\cos^{-1}\left(-\dfrac{1}{2}\right)\right)$

28. $\sin(\tan^{-1} 0)$

29. $\cos\left(\tan^{-1}\dfrac{\sqrt{3}}{3}\right)$

30. $\sin(\tan^{-1}(-\sqrt{3}))$

31. $\cos\left(\sin^{-1}\dfrac{\sqrt{2}}{3}\right)$

32. $\sin\left(\cos^{-1}\dfrac{\sqrt{3}}{4}\right)$

33. $\tan\left(\cos^{-1}\dfrac{\sqrt{3}}{5}\right)$

34. $\cot\left(\cos^{-1}\dfrac{2}{3}\right)$

35. $\cos(\tan^{-1}(-2))$

36. $\sec\left(\tan^{-1}\left(-\dfrac{\sqrt{11}}{2}\right)\right)$

37. $\sin(\tan^{-1}(\sqrt{5}))$

38. $\cot\left(\tan^{-1}\dfrac{\sqrt{6}}{4}\right)$

39. $\csc\left(\tan^{-1}\left(-\dfrac{2}{3}\right)\right)$

40. $\tan\left(\sin^{-1}\left(-\dfrac{\sqrt{2}}{5}\right)\right)$

In Exercises 41–50, find each of the following.

41. $\arcsin(-0.4563)$

42. $\cos^{-1}(0.5701)$

43. $\tan^{-1}(-5.3641)$

44. $\arctan(12.3516)$

45. $\sin(\cos^{-1}(-0.6324))$

46. $\cos(\sin^{-1}(0.8156))$

47. $\tan(\sin^{-1}(0.3535))$

48. $\tan(\cos^{-1}(0.7725))$

49. $\sin(\tan^{-1}(15.3214))$

50. $\cos(\tan^{-1}(-8.0023))$

In Exercises 51–60, solve each equation for x. Assume that in each case x varies in an interval where the trigonometric function is one-to-one.

51. $y = \sin 2x$

52. $y = \cos \dfrac{x}{2}$

53. $y = 2 \sin 3x$

54. $y = 3 \cos 5x$

55. $y = \dfrac{1}{3} \cos(x - 2)$

56. $y = \dfrac{1}{2} \sin(3x - 1)$

57. $y = 3 \cos^{-1} \dfrac{x}{2}$

58. $y = 2 \sin^{-1} \dfrac{x}{4}$

59. $y = \dfrac{2}{3} \arctan 2x$

60. $y = 3 \tan^{-1} \dfrac{x}{2}$

CHAPTER SUMMARY

To each real number x, a unique point P can be assigned on the unit circle so that the length of the arc AP or the radian measure of the central angle AOP is equal to x. The coordinates of P are then $\cos x$ and $\sin x$. In this way, the two basic trigonometric functions *cosine* and *sine* are defined as coordinates of a point on a unit circle. The domain of definition of both functions is the real line and their range the closed interval $[-1, 1]$. Moreover, they are *periodic* with *period* 2π:

$$\sin(x + 2\pi) = \sin x \quad \text{and} \quad \cos(x + 2\pi) = \cos x.$$

The remaining trigonometric functions, with their domains and periods, are as follows:

Function	Domain	Period
$\tan x = \dfrac{\sin x}{\cos x}$	$x \neq (2k+1)\pi/2$, k any integer	π
$\cot x = \dfrac{\cos x}{\sin x}$	$x \neq k\pi$, k any integer	π
$\sec x = \dfrac{1}{\cos x}$	$x \neq (2k+1)\pi/2$, k any integer	2π
$\csc x = \dfrac{1}{\sin x}$	$x \neq k\pi$, k any integer	2π

The relation

$$\sin^2 x + \cos^2 x = 1,$$

which is satisfied for all values of x, is called the *fundamental Pythagorean identity*. From this identity we derive two other identities:

$$\tan^2 x + 1 = \sec^2 x \quad \text{and} \quad \cot^2 x + 1 = \csc^2 x.$$

If $P(x, y)$ lies on the terminal side of an angle θ in standard position, then

$$\sin \theta = \frac{y}{r}, \qquad \csc \theta = \frac{r}{y}, \quad y \neq 0,$$

$$\cos \theta = \frac{x}{r}, \qquad \sec \theta = \frac{r}{x}, \quad x \neq 0,$$

$$\tan \theta = \frac{y}{x}, \quad x \neq 0, \qquad \cot \theta = \frac{x}{y}, \quad y \neq 0,$$

The *reference angle* of a nonquadrantal angle in standard position is the *positive acute* angle formed by the terminal side of the angle and the x-axis.

Certain oscillatory phenomena can be described by one of the trigonometric equations

$$y = A \sin(Bx + C) \quad \text{and} \quad y = A \cos(Bx + C),$$

where $B > 0$. Each equation represents a *simple harmonic motion* and has a graph called a *simple harmonic curve*. The oscillatory motion represented by either equation has amplitude $= A$, period $= 2\pi/B$, and phase shift $= -C/B$.

The inverse of a trigonometric function can be defined if the domain of the function is restricted in order to obtain a one-to-one function. For the sine, cosine, and tangent functions, inverses are defined as follows.

$$y = \sin^{-1} x, \quad -1 \leq x \leq 1, \quad \text{if and only if} \quad x = \sin y, \quad -\pi/2 \leq y \leq \pi/2$$
$$y = \cos^{-1} x, \quad -1 \leq x \leq 1 \quad \text{if and only if} \quad x = \cos y, \quad 0 \leq y \leq \pi$$
$$y = \tan^{-1} x, \quad -\infty < x < +\infty \quad \text{if and only if} \quad x = \tan y, \quad -\pi/2 < y < \pi/2$$

CHAPTER TEST

1. An angle in standard position measures 45°. What are the coordinates of the point of intersection of its terminal side and the unit circle?
2. Given $\sin \alpha = -2/3$ and $\cos \alpha = \sqrt{5}/3$, find the four remaining trigonometric function values of α.
3. What are the five remaining trigonometric function values of θ, given that $\tan \theta = 4/5$ and $\sin \theta > 0$?
4. The point $(-2, 4)$ is on the terminal side of an angle α in standard position. Find exact values for $\sin \alpha$ and $\cos \alpha$.
5. The terminal side of an angle β lies in the third quadrant and on the line $y = 3x/4$. Find $\sin \beta$, $\cos \beta$, and $\tan \beta$.
6. Given $x = -330°$, find **(a)** its reference angle and **(b)** the exact values of $\sin x$ and $\cos x$.

7. What are the amplitude, period, and phase shift of $f(x) = \sin(2x - 2\pi)$? Graph this function.
8. A 0.2-lb mass attached to the end of a spring bobs up and down according to the equation $y = 0.11 \cos(3.8t)$, where y is measured in feet, and t in seconds. Find **(a)** the amplitude and **(b)** the frequency of the motion.
9. Evaluate the number $\cos(\tan^{-1}(-1/4))$ without the use of tables or calculators.
10. Solve for x the equation $y = \sin 7x + 1$, with $-\pi/14 \le x \le \pi/14$.

REVIEW EXERCISES

In Exercises 1–4, find the coordinates of the point on the unit circle corresponding to the given angle. Evaluate the six trigonometric functions of the angle. Do not use a calculator or table. Use your knowledge of special angles.

1. $\alpha = \dfrac{3\pi}{4}$ 2. $\beta = \dfrac{5\pi}{6}$

3. $\gamma = -120°$ 4. $\delta = -210°$

In Exercises 5–8, the values of two trigonometric functions of an angle are given. Find the values of the four remaining trigonometric functions.

5. $\cos \alpha = \dfrac{2\sqrt{2}}{3}$, $\csc \alpha = 3$

6. $\sin \alpha = \dfrac{5}{6}$, $\sec \alpha = -\dfrac{6\sqrt{11}}{11}$

7. $\sin \beta = \dfrac{2}{5}$, $\tan \beta = -\dfrac{2\sqrt{21}}{21}$

8. $\cot \beta = -\dfrac{4}{3}$, $\csc \beta = -\dfrac{5}{3}$

In Exercises 9–14, the value of a trigonometric function of θ and a condition on θ are given. Find the values of the other five trigonometric functions of θ.

9. $\cos \theta = \dfrac{2}{5}$ and θ is a fourth-quadrant angle.

10. $\tan \theta = 2$ and θ is a third-quadrant angle.

11. $\cot \theta = -3$ and $\cos \theta < 0$

12. $\sin \theta = -\dfrac{2}{3}$ and $\tan \theta > 0$

13. $\sin \theta = 0.5948$ and $\cos \theta < 0$
14. $\cos \theta = -0.5062$ and $\sin \theta > 0$

In Exercises 15–20, you are given the coordinates of a point lying on the terminal side of an angle in standard position. Find the six trigonometric functions of the angle. Do not use tables or calculators.

15. $P(3, -5)$ 16. $P(-2, 8)$
17. $X(-3, 6)$ 18. $Y(-2, -7)$
19. $U(-3, -2)$ 20. $V(-4, 5)$

In Exercises 21–24, the terminal side of an angle lies in the given quadrant and on the given line. Find the exact values of the six trigonometric functions of the angle.

21. First quadrant and $y = 2x$
22. Second quadrant and $y = -5x$
23. Third quadrant and $3x - y = 0$
24. Fourth quadrant and $2x + 3y = 0$

In each of Exercises 25–28, provide the required quantities.

25. Find the reference angle of each of the following angles.

   (a) $130°$ **(b)** $220°$ **(c)** $-\dfrac{13\pi}{2}$ **(d)** $-\dfrac{5\pi}{6}$

   (e) $210°15'$ **(f)** $155°30'$

26. Using the special angle table, evaluate each of the following trigonometric functions.

   (a) $\cos \dfrac{4\pi}{3}$ **(b)** $\sin\left(-\dfrac{5\pi}{4}\right)$ **(c)** $\tan\left(-\dfrac{3\pi}{4}\right)$

   (d) $\cot \dfrac{2\pi}{3}$ **(e)** $\sin 405°$ **(f)** $\cos(-390°)$

27. Use Table 4 in the Appendix to find the following numbers.

    (a) $\sin 45°20'$ **(b)** $\cos 15°30'$ **(c)** $\sec 25°30'$

    (d) $\csc 56°20'$ **(e)** $\cos 0.1484$ **(f)** $\sin 0.4102$

28. Use a calculator to approximate the following numbers to four decimal places.

    (a) $\cos 32°18'$ **(b)** $\sin 42°12'$ **(c)** $\cot 35°26'$

    (d) $\tan 120°18'$ **(e)** $\cot 0.2449$ **(f)** $\sin 0.5337$

In Exercises 29–34, sketch the graphs of the given trigonometric functions.

29. $f(x) = 3 \sin x$ **30.** $f(x) = 2 \cos x$

31. $f(x) = 2 - \cos x$ **32.** $f(x) = \sin x + 3$

33. $f(x) = 2 - \sin\left(x - \dfrac{\pi}{2}\right)$ **34.** $f(x) = \cos(x - \pi) + 2$

In Exercises 35–40, find the amplitude, period, and phase shift of each function.

35. $y = 2 \cos \dfrac{x}{5}$ **36.** $y = -3 \sin \dfrac{x}{3}$

37. $y = \cos \pi(x - 2)$ **38.** $y = \sin \pi(x + 3)$

39. $y = -2 \sin(\pi x - 4)$ **40.** $y = 4 \cos(\pi x + 1)$

In Exercises 41–46, graph each function.

41. $y = 3 \sin \dfrac{\pi x}{2}$ **42.** $y = 2 \cos \dfrac{\pi x}{3}$

43. $y = \sin\left(x - \dfrac{\pi}{3}\right)$ **44.** $y = \cos\left(x + \dfrac{\pi}{3}\right)$

45. $y = \tan \dfrac{x}{2}$ **46.** $y = \cot 3x$

In Exercises 47–52, solve the given problems.

47. A spring vibrates with a frequency of 32 Hz when a 0.4-kg mass hangs from it. What will be the frequency when a 0.8-kg mass hangs from the spring?

48. A tuning fork vibrates with a frequency of 260 Hz. What is the period of the vibration? What is the angular frequency?

49. What is the equation describing the motion of a spring that, when stretched 20 cm from equilibrium and released, oscillates with a period of 1.5 s? What will be the displacement after 1.8 s?

50. A metal blade is tightly clamped to the edge of a table. When the tip of the blade is pulled 0.5 cm away from its resting position and released, the blade oscillates with a frequency of 150 Hz. Write the equation describing the motion of the tip of the blade. What are the period and angular frequency?

51. At a particular location on the earth, a simple pendulum whose length is 40 cm has a frequency of 0.787 Hz. What is the acceleration of gravity at this location? If the amplitude of the pendulum is 0.5 cm, write the equation giving the position of the pendulum at time t.

52. The displacement of a particle is given by the equation

$$x = 0.12 \cos\left(0.48t + \dfrac{\pi}{4}\right).$$

Find the amplitude, period, and frequency. What is the position of the particle after 0.5 s?

In Exercises 53–58, find the given values without using tables or calculators.

53. $\sin\left(\sin^{-1}\dfrac{\sqrt{2}}{2}\right)$ **54.** $\cos^{-1}\left(\cos\dfrac{3\pi}{4}\right)$

55. $\cos^{-1}\left(\sin\dfrac{4\pi}{3}\right)$ **56.** $\cot\left(\sin^{-1}\left(-\dfrac{1}{2}\right)\right)$

57. $\sec(\tan^{-1} 4)$ **58.** $\csc\left(\sin^{-1}\dfrac{2}{3}\right)$

In Exercises 59–64; find each of the following angles to the nearest minute.

59. $\arcsin 0.8312$ **60.** $\arccos 0.5012$

61. $\cos^{-1}(-0.4715)$ **62.** $\arctan(-3.013)$

63. $\sin^{-1}(0.7071)$ **64.** $\tan^{-1}(0.5774)$

In Exercises 65–70, approximate each of the following numbers to four decimal places.

65. $\cos(\sin^{-1}(-0.7413))$ **66.** $\sin(\cos^{-1}(0.9013))$

67. $\tan(\cos^{-1}(0.5561))$ **68.** $\cot(\sin^{-1}(0.3213))$

69. $\cos(\sin^{-1}(0.7071))$ **70.** $\sin(\cos^{-1}(0.7071))$

CHAPTER 9

TRIGONOMETRIC IDENTITIES AND APPLICATIONS OF TRIGONOMETRY

Several topics are discussed in this chapter. We start with trigonometric identities. Next, we study the addition and subtraction formulas, as well as the double-angle and half-angle formulas, which are extremely useful in many applications. We then investigate the problem of solving trigonometric equations. The solving of triangles, started in Chapter 7, is now completed with the study of the laws of sine and cosine. We study vectors in the plane and describe properties of the vector algebra. We also discuss two important applications of trigonometry to complex numbers: the representation of complex numbers in polar form, and De Moivre's formula. The chapter ends with a section on polar coordinates.

9.1 TRIGONOMETRIC IDENTITIES

In Section 8.1, we derived the Pythagorean identity:

$$\sin^2\theta + \cos^2\theta = 1, \qquad (9.1)$$

as well as the two identities

$$\tan^2\theta + 1 = \sec^2\theta \qquad (9.2)$$

and

$$\cot^2\theta + 1 = \csc^2\theta. \qquad (9.3)$$

These relationships are called *identities* because they are true for *all* allowable values of θ. For example, (9.1) is true for all values of θ without exception; (9.2) is true for all values of θ except the numbers $(\pi/2) + k\pi$, with k any integer; and (9.3) is true for all values of θ different from $k\pi$, with k any integer.

A *trigonometric identity* is a relation among trigonometric functions that is satisfied for all allowable values of the variable.

Identities (9.1), (9.2), and (9.3), called the *basic trigonometric identities*, are used in many applications. In Section 8.1, they were used to determine all the trigonometric values of an angle θ when we were given a trigonometric value of θ and a condition on θ. We now show how they can be used to verify other trigonometric identities.

EXAMPLE 1   Verify the identity

$$\frac{\sin^2\theta}{\cos\theta} + \cos\theta = \sec\theta.$$

Solution: The given expression is defined for all values of $\theta \neq (\pi/2) + k\pi$, with k any integer. Starting with the left-hand side, we find a common denominator and add, to obtain

$$\frac{\sin^2\theta}{\cos\theta} + \cos\theta = \frac{\sin^2\theta + \cos^2\theta}{\cos\theta}$$

$$= \frac{1}{\cos\theta} \quad \text{Use identity (9.1)}$$

$$= \sec\theta.$$

Hence, the given relationship is satisfied for all values of $\theta \neq (\pi/2) + k\pi$, with k any integer.

From now on, when we write a trigonometric identity, it is understood that we are considering only those values of the variable for which the various terms of the identity are defined.

Practice Exercise 1   Verify the identity

$$\frac{\cos^2\alpha}{\sin\alpha} = \csc\alpha - \sin\alpha.$$

EXAMPLE 2   Show that $\dfrac{1 - (\cos x - \sin x)^2}{\sin x} = 2\cos x.$

Solution: We have

$$\frac{1 - (\cos x - \sin x)^2}{\sin x}$$

$$= \frac{1 - \cos^2 x + 2 \sin x \cos x - \sin^2 x}{\sin x} \qquad \text{Square and remove parentheses}$$

$$= \frac{1 - (\cos^2 x + \sin^2 x) + 2 \sin x \cos x}{\sin x} \qquad \text{Use (9.1) and simplify}$$

$$= \frac{2 \sin x \cos x}{\sin x}$$

$$= 2 \cos x.$$

Practice Exercise 2   Derive the identity

$$\sec u - 2 \sin u = \frac{(\sin u - \cos u)^2}{\cos u}.$$

Hints for Verifying Trigonometric Identities

When verifying trigonometric identities, you should follow these guidelines.

1. Start with one side of the relationship—usually the more complicated one.
2. Perform allowable algebraic operations and use the basic trigonometric identities to obtain the other side of the relationship.
3. In some cases, it may be simpler to work with both sides of the relationship and try to reduce them to the same expression. If all steps on both sides can be reversed, then the given relationship is an identity.

EXAMPLE 3   Verify the identity $\dfrac{\tan^2 \theta - 1}{1 - \cot^2 \theta} = \tan^2 \theta$.

Solution: Working with the left-hand side yields

$$\frac{\tan^2 \theta - 1}{1 - \cot^2 \theta} = \frac{\dfrac{\sin^2 \theta}{\cos^2 \theta} - 1}{1 - \dfrac{\cos^2 \theta}{\sin^2 \theta}}$$

$$= \frac{\dfrac{\sin^2 \theta - \cos^2 \theta}{\cos^2 \theta}}{\dfrac{\sin^2 \theta - \cos^2 \theta}{\sin^2 \theta}}$$

$$= \frac{\sin^2 \theta - \cos^2 \theta}{\cos^2 \theta} \cdot \frac{\sin^2 \theta}{\sin^2 \theta - \cos^2 \theta}$$

$$= \frac{\sin^2 \theta}{\cos^2 \theta} = \tan^2 \theta.$$

Practice Exercise 3   Verify the identity $\dfrac{\cot^2 \alpha - 1}{1 - \tan^2 \alpha} = \cot^2 \alpha$.

EXAMPLE 4   Verify the identity $\dfrac{1}{\sec x + \tan x} = \dfrac{\cot x}{\csc x + 1}$.

Solution: We transform the left-hand side of the given expression as follows:

$$\frac{1}{\sec x + \tan x} = \frac{1}{\dfrac{1}{\cos x} + \dfrac{\sin x}{\cos x}} = \frac{1}{\dfrac{1 + \sin x}{\cos x}} = \frac{\cos x}{1 + \sin x}.$$

Next, we work with the right-hand side as follows:

$$\frac{\cot x}{\csc x + 1} = \frac{\dfrac{\cos x}{\sin x}}{\dfrac{1}{\sin x} + 1} = \frac{\dfrac{\cos x}{\sin x}}{\dfrac{1 + \sin x}{\sin x}}$$

$$= \frac{\cos x}{\sin x} \cdot \frac{\sin x}{1 + \sin x} = \frac{\cos x}{1 + \sin x}.$$

Since the two resulting expressions are the same and the steps we performed are reversible, the given relationship is an identity.

Practice Exercise 4   Verify the identity

$$\frac{\tan u}{\sec u + 1} = \frac{1}{\csc u + \cot u}.$$

EXERCISES 9.1

In Exercises 1–10, perform the indicated operations. Use the definitions of trigonometric functions and the basic trigonometric identities to simplify your result as much as possible.

1. $(\sin \theta)(\cot \theta + \csc \theta)$ **2.** $(\cos \theta)(\tan \theta - \sec \theta)$

3. $\dfrac{\sin x}{\csc x} + \dfrac{\cos x}{\sec x}$ **4.** $\dfrac{\sec x}{\cos x} - \dfrac{\tan x}{\cot x}$

5. $\dfrac{\tan \alpha}{\sec \alpha} + \dfrac{\cot \alpha}{\csc \alpha}$ **6.** $\dfrac{1}{\csc \alpha} - \dfrac{1}{\sec \alpha}$

7. $(\sin u + \cos u)^2$ **8.** $(\sin u - \cos u)^2$

9. $\dfrac{1}{1 - \sin \theta} + \dfrac{1}{1 + \sin \theta}$ **10.** $\dfrac{1}{1 - \cos \theta} - \dfrac{1}{1 + \cos \theta}$

In Exercises 11–60, verify each of the following trigonometric identities.

11. $\sin \theta \csc \theta = 1$ **12.** $\cos \theta \sec \theta = 1$

13. $\tan \theta \cos \theta = \sin \theta$ **14.** $\cot \theta \sin \theta = \cos \theta$

15. $\tan \theta = \sin \theta \sec \theta$ **16.** $\cot \theta = \cos \theta \csc \theta$

17. $\cos^2 \theta - \sin^2 \theta = 2 \cos^2 \theta - 1$

18. $\cos^2 \theta - \sin^2 \theta = 1 - 2 \sin^2 \theta$

19. $(\sin \alpha - \cos \alpha)^2 = 1 - 2 \sin \alpha \cos \alpha$

20. $(\sin \alpha + \cos \alpha)^2 = 1 + 2 \sin \alpha \cos \alpha$

21. $\tan \alpha + \cot \alpha = \sec \alpha \csc \alpha$

22. $\dfrac{\sin \alpha}{\csc \alpha} + \dfrac{\cos \alpha}{\sec \alpha} = 1$

23. $2 \cos^2 x - 1 = 1 - 2 \sin^2 x$

24. $\tan^2 x - \sin^2 x = \tan^2 x \sin^2 x$

25. $\cot^2 x - \cos^2 x = \cot^2 x \cos^2 x$

26. $(\sin^2 x)(\csc^2 x - 1) = \cos^2 x$

27. $\dfrac{1 + \sin u}{\cos u} + \dfrac{\cos u}{1 + \sin u} = 2 \sec u$

28. $\dfrac{\sin u}{1 - \cos u} + \dfrac{\sin u}{1 + \cos u} = 2 \csc u$

29. $\sec u = \dfrac{\cos u}{1 + \sin u} + \tan u$

30. $\csc u = \dfrac{\sin u}{1 + \cos u} + \cot u$

31. $\dfrac{1 - \sin u}{\cos u} = \dfrac{\cos u}{1 + \sin u}$

32. $\dfrac{1 - \cos u}{\sin u} = \dfrac{\sin u}{1 + \cos u}$

33. $\dfrac{\cot \theta}{\csc \theta + 1} = \dfrac{\csc \theta - 1}{\cot \theta}$

34. $\dfrac{1}{\sin \theta - 2} = \dfrac{\cot \theta}{\cos \theta - 2 \cot \theta}$

35. $\dfrac{\tan^2 \theta - 1}{1 - \cot^2 \theta} = \tan^2 \theta$ **36.** $\dfrac{1 - \sin^2 \theta}{\sin^2 \theta} = \cot^2 \theta$

37. $\dfrac{1 - \sin \alpha}{1 + \sin \alpha} = (\sec \alpha - \tan \alpha)^2$

38. $(\csc \alpha - \cot \alpha)^2 = \dfrac{1 - \cos \alpha}{1 + \cos \alpha}$

39. $\dfrac{1}{1 + \cos \alpha} + \dfrac{1}{1 - \cos \alpha} = 2 \csc^2 \alpha$

40. $\dfrac{1}{1 + \sin \alpha} + \dfrac{1}{1 - \sin \alpha} = 2 \sec^2 \alpha$

41. $\dfrac{\sec x + 1}{\sin x + \tan x} = \csc x$

42. $\dfrac{\csc x + 1}{\cos x + \cot x} = \sec x$

43. $\dfrac{\sec x}{1 + \sec x} - \dfrac{\sec x}{1 - \sec x} = 2 \csc^2 x$

44. $\dfrac{\tan x - \cot x}{\sin x \cos x} = \sec^2 x - \csc^2 x$

45. $\dfrac{\sin x}{1 + \cos x} = \csc x - \cot x$

46. $\dfrac{\cos x}{1 - \sin x} = \sec x + \tan x$

47. $\dfrac{\tan \beta + \sin \beta}{\tan \beta - \sin \beta} = \dfrac{\sec \beta + 1}{\sec \beta - 1}$

48. $\dfrac{\tan \beta + \sin \beta}{\sin \beta \tan \beta} = \dfrac{\sin \beta \tan \beta}{\tan \beta - \sin \beta}$

49. $\dfrac{\sin a \cos b + \cos a \sin b}{\cos a \cos b - \sin a \sin b} = \dfrac{\tan a + \tan b}{1 - \tan a \tan b}$

50. $\dfrac{\tan a + \tan b}{1 - \tan a \tan b} = \dfrac{\cot a + \cot b}{\cot a \cot b - 1}$

51. $\sin^4 x - \cos^4 x = 2 \sin^2 x - 1$

52. $\cos^4 x - \sin^4 x = 2 \cos^2 x - 1$

53. $\dfrac{\sec^4 x - 1}{\tan^2 x} = 2 + \tan^2 x$

54. $\dfrac{\tan^4 x - 1}{\sec^2 x} = \sec^2 x - 2$

55. $\dfrac{\sin^2 \theta + 3 \sin \theta + 2}{\cos^2 \theta} = \dfrac{\sin \theta + 2}{1 - \sin \theta}$

56. $\dfrac{\cos^2 \theta - \cos \theta - 2}{\sin^2 \theta} = \dfrac{\cos \theta - 2}{1 - \cos \theta}$

57. $\dfrac{\tan \theta}{\sin \theta + 3 \tan \theta} = \dfrac{1}{\cos \theta + 3}$

58. $\dfrac{\cos \theta - 2 \cot \theta}{\cot \theta} = \sin \theta - 2$

59. $\dfrac{\sin^3 \theta - \cos^3 \theta}{\sin \theta - \cos \theta} = 1 + \sin \theta \cos \theta$

60. $\sin^4 \theta + \cos^2 \theta = \sin^2 \theta + \cos^4 \theta$

9.2

THE ADDITION AND SUBTRACTION FORMULAS

In this section we derive identities that express the trigonometric functions of a sum or a difference of angles in terms of functions of the individual angles.

First, we establish the following identity.

> ### Cosine of the Difference of Two Angles
>
> $$\cos(\alpha - \beta) = \cos \alpha \cos \beta + \sin \alpha \sin \beta. \qquad (9.4)$$

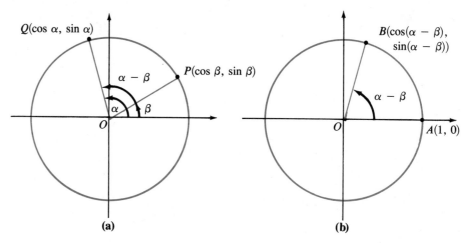

Figure 9.1

Let α, β, and $\alpha - \beta$ be the angles illustrated in Figure 9.1a. Then, the points of intersection of the terminal sides of β and α with the unit circle are $P(\cos \beta, \sin \beta)$ and $Q(\cos \alpha, \sin \alpha)$, respectively. If both points, P and Q, are rotated clockwise β angular units, then the point P coincides with the point $A(1, 0)$ and the point Q coincides with the point $B(\cos(\alpha - \beta), \sin(\alpha - \beta))$, as shown in Figure 9.1b. Since the distances $d(P, Q)$ and $d(A, B)$ are equal, it is also true that

$$d(P, Q)^2 = d(A, B)^2,$$

which, by the distance formula (3.2) of Chapter 3, we can write as

$$(\cos \alpha - \cos \beta)^2 + (\sin \alpha - \sin \beta)^2 = (\cos(\alpha - \beta) - 1)^2 + \sin^2(\alpha - \beta).$$

Expanding the three squares that have the form $(a - b)^2$, we obtain

$$\cos^2\alpha + \cos^2\beta - 2 \cos \alpha \cos \beta + \sin^2\alpha + \sin^2\beta - 2 \sin \alpha \sin \beta$$
$$= \cos^2(\alpha - \beta) + 1 - 2 \cos(\alpha - \beta) + \sin^2(\alpha - \beta).$$

Using the identity (9.1) three times, we derive

$$2 - 2(\cos \alpha \cos \beta + \sin \alpha \sin \beta) = 2 - 2 \cos(\alpha - \beta).$$

Hence

$$\cos(\alpha - \beta) = \cos \alpha \cos \beta + \sin \alpha \sin \beta,$$

which is the desired identity.

■ **Important note** Observe that in the proof of the identity (9.4), we used a new geometric argument: the clockwise rotation of Figure 9.1a by β angular units. Because of this, the identity (9.4) contains information that the Pythagorean identities do not. Consequently, this identity has a large number of applications; we can now solve problems we could not handle before. Here is a simple one.

EXAMPLE 1   Evaluate $\cos \pi/12$ without using tables or calculator.

Solution: Since

$$\frac{\pi}{12} = \frac{\pi}{3} - \frac{\pi}{4},$$

we may apply (9.4) with $\alpha = \pi/3$ and $\beta = \pi/4$:

$$\cos \frac{\pi}{12} = \cos\left(\frac{\pi}{3} - \frac{\pi}{4}\right)$$

$$= \cos \frac{\pi}{3} \cos \frac{\pi}{4} + \sin \frac{\pi}{3} \sin \frac{\pi}{4}$$

$$= \frac{1}{2} \cdot \frac{\sqrt{2}}{2} + \frac{\sqrt{3}}{2} \cdot \frac{\sqrt{2}}{2}$$

$$= \frac{\sqrt{2} + \sqrt{6}}{4}.$$

Practice Exercise 1   Without a table or calculator, compute $\cos 5\pi/12$. [*Hint*: $5\pi/12 = 2\pi/3 - \pi/4$]

Answer: $\dfrac{\sqrt{6} - \sqrt{2}}{4}$

Complementary Angle Formulas

As an application of the identity (9.4), we can derive the two *complementary angle formulas* (see Section 7.2):

$$\cos\left(\frac{\pi}{2} - \alpha\right) = \sin \alpha \qquad \text{and} \qquad \sin\left(\frac{\pi}{2} - \alpha\right) = \cos \alpha.$$

If we set $\alpha = \pi/2$ and $\beta = \alpha$ in formula (9.4), we obtain

$$\cos\left(\frac{\pi}{2} - \alpha\right) = \cos \frac{\pi}{2} \cos \alpha + \sin \frac{\pi}{2} \sin \alpha$$

$$= (0) \cos \alpha + (1) \sin \alpha$$

$$= \sin \alpha,$$

which is the first of the two formulas. Next, if we write $\alpha = \pi/2 - [(\pi/2) - \alpha]$, we have

$$\cos \alpha = \cos\left[\frac{\pi}{2} - \left(\frac{\pi}{2} - \alpha\right)\right]$$

$$= \sin\left(\frac{\pi}{2} - \alpha\right), \quad \text{Using the first formula}$$

which is the second of the two complementary angle formulas.

Now we derive the following identity.

Cosine of the Sum of Two Angles

$$\cos(\alpha + \beta) = \cos \alpha \cos \beta - \sin \alpha \sin \beta \qquad (9.5)$$

Writing $\alpha + \beta = \alpha - (-\beta)$ and applying (9.4), we have

$$\begin{aligned}
\cos(\alpha + \beta) &= \cos[\alpha - (-\beta)] \\
&= \cos \alpha \cos(-\beta) + \sin \alpha \sin(-\beta) \\
&= \cos \alpha \cos \beta - \sin \alpha \sin \beta,
\end{aligned}$$

where we have used the fact that $\cos \beta = \cos(-\beta)$ and $-\sin \beta = \sin(-\beta)$.

EXAMPLE **2**

Find the exact value of $\cos 7\pi/12$. Do not use a table or calculator.

Solution: Since

$$\frac{7\pi}{12} = \frac{\pi}{3} + \frac{\pi}{4},$$

we apply (9.5) to get

$$\begin{aligned}
\cos \frac{7\pi}{12} &= \cos\left(\frac{\pi}{3} + \frac{\pi}{4}\right) \\
&= \cos \frac{\pi}{3} \cos \frac{\pi}{4} - \sin \frac{\pi}{3} \sin \frac{\pi}{4} \\
&= \frac{1}{2} \cdot \frac{\sqrt{2}}{2} - \frac{\sqrt{3}}{2} \cdot \frac{\sqrt{2}}{2} = \frac{\sqrt{2} - \sqrt{6}}{4}.
\end{aligned}$$

Practice Exercise 2

Find the exact value of $\cos 11\pi/12$ without using a table or calculator. [*Hint:* $11\pi/12 = 2\pi/3 + \pi/4$]

Answer: $\dfrac{-\sqrt{2} - \sqrt{6}}{4}$

The sine of a sum or difference of angles can be expressed in terms of functions of individual angles, as follows.

Sine of the Sum and Difference of Two Angles

$$\sin(\alpha + \beta) = \sin \alpha \cos \beta + \cos \alpha \sin \beta \qquad (9.6)$$

$$\sin(\alpha - \beta) = \sin \alpha \cos \beta - \cos \alpha \sin \beta \qquad (9.7)$$

Let us prove identity (9.6), leaving the proof of (9.7) as an exercise. First, using the complementary angle formulas, we write

$$\sin(\alpha + \beta) = \cos\left[\frac{\pi}{2} - (\alpha + \beta)\right] = \cos\left[\left(\frac{\pi}{2} - \alpha\right) - \beta\right].$$

Next, using (9.4) and the complementary angle formulas again, we obtain

$$\begin{aligned}
\sin(\alpha + \beta) &= \cos\left[\left(\frac{\pi}{2} - \alpha\right) - \beta\right] \\
&= \cos\left(\frac{\pi}{2} - \alpha\right)\cos\beta + \sin\left(\frac{\pi}{2} - \alpha\right)\sin\beta \\
&= \sin\alpha\cos\beta + \cos\alpha\sin\beta,
\end{aligned}$$

which is the desired identity.

EXAMPLE 3

Find the exact value of $\sin 5\pi/12$. Do not use a table or calculator.

Solution: Since

$$\frac{5\pi}{12} = \frac{\pi}{4} + \frac{\pi}{6},$$

we have

$$\begin{aligned}
\sin\frac{5\pi}{12} &= \sin\left(\frac{\pi}{4} + \frac{\pi}{6}\right) \\
&= \sin\frac{\pi}{4}\cos\frac{\pi}{6} + \cos\frac{\pi}{4}\sin\frac{\pi}{6} \\
&= \frac{\sqrt{2}}{2}\cdot\frac{\sqrt{3}}{2} + \frac{\sqrt{2}}{2}\cdot\frac{1}{2} \\
&= \frac{\sqrt{6} + \sqrt{2}}{4}.
\end{aligned}$$

Practice Exercise 3

Without a table or calculator, compute the exact value of $\sin \pi/12$.

Answer: $\dfrac{\sqrt{6} - \sqrt{2}}{4}$

From identities (9.4), (9.5), (9.6), and (9.7), we have the following identities for the tangent of a sum or a difference of angles.

Tangent of the Sum and Difference of Two Angles

$$\tan(\alpha + \beta) = \frac{\tan\alpha + \tan\beta}{1 - \tan\alpha\tan\beta} \qquad (9.8)$$

$$\tan(\alpha - \beta) = \frac{\tan\alpha - \tan\beta}{1 + \tan\alpha\tan\beta} \qquad (9.9)$$

As an illustration, we derive identity (9.8).

$$\tan(\alpha + \beta) = \frac{\sin(\alpha + \beta)}{\cos(\alpha + \beta)}$$

$$= \frac{\sin \alpha \cos \beta + \cos \alpha \sin \beta}{\cos \alpha \cos \beta - \sin \alpha \sin \beta} \qquad \text{By (9.5) and (9.6)}$$

$$= \frac{\dfrac{\sin \alpha \cos \beta}{\cos \alpha \cos \beta} + \dfrac{\cos \alpha \sin \beta}{\cos \alpha \cos \beta}}{\dfrac{\cos \alpha \cos \beta}{\cos \alpha \cos \beta} - \dfrac{\sin \alpha \sin \beta}{\cos \alpha \cos \beta}} \qquad \begin{array}{l}\text{Divide the numerator and} \\ \text{denominator by } \cos \alpha \cos \beta\end{array}$$

$$= \frac{\tan \alpha + \tan \beta}{1 - \tan \alpha \tan \beta} \qquad \text{Simplify}$$

The verification of (9.9) is left as an exercise.

EXAMPLE 4   Evaluate tan 75°. Do not use a table or calculator.

Solution: Since 75° = 45° + 30°, we apply (9.8):

$$\tan 75° = \frac{\tan 45° + \tan 30°}{1 - \tan 45° \tan 30°}$$

$$= \frac{1 + \sqrt{3}/3}{1 - \sqrt{3}/3}$$

$$= \frac{3 + \sqrt{3}}{3 - \sqrt{3}}$$

$$= \frac{12 + 6\sqrt{3}}{6}$$

$$= 2 + \sqrt{3}.$$

Practice Exercise 4   Find the value tan 105°, without the help of a table or calculator.

Answer: $-\dfrac{\sqrt{3} + 1}{\sqrt{3} - 1} = -2 - \sqrt{3}$

EXAMPLE 5   If α is a first-quadrant angle such that sin $\alpha = 1/3$ and β is a second-quadrant angle such that cos $\beta = -1/2$, find $\sin(\alpha - \beta)$ and $\cos(\alpha - \beta)$. In which quadrant does $\alpha - \beta$ lie?

Solution: First, we must evaluate cos α and sin β. Since α is a first-quadrant angle, we have cos $\alpha > 0$; since β is a second-quadrant angle, we have sin $\beta > 0$. Their values are obtained by using the fundamental identity (9.1):

$$\cos \alpha = \sqrt{1 - \left(\frac{1}{3}\right)^2} = \frac{2\sqrt{2}}{3}$$

and

$$\sin \beta = \sqrt{1 - \left(-\frac{1}{2}\right)^2} = \frac{\sqrt{3}}{2}.$$

According to (9.7),

$$\sin(\alpha - \beta) = \sin \alpha \cos \beta - \cos \alpha \sin \beta$$
$$= \frac{1}{3} \cdot \left(-\frac{1}{2}\right) - \frac{2\sqrt{2}}{3} \cdot \frac{\sqrt{3}}{2}$$
$$= -\frac{1}{6} - \frac{\sqrt{6}}{3}$$
$$= -\frac{1 + 2\sqrt{6}}{6}$$

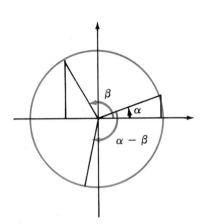

Figure 9.2

and, according to (9.4),

$$\cos(\alpha - \beta) = \cos \alpha \cos \beta + \sin \alpha \sin \beta$$
$$= \frac{2\sqrt{2}}{3} \cdot \left(-\frac{1}{2}\right) + \frac{1}{3} \cdot \frac{\sqrt{3}}{2}$$
$$= -\frac{\sqrt{2}}{3} + \frac{\sqrt{3}}{6}$$
$$= \frac{\sqrt{3} - 2\sqrt{2}}{6}.$$

Since $\left(\sqrt{3} - 2\sqrt{2}\right)/6 < 0$ (why?), both $\sin(\alpha - \beta)$ and $\cos(\alpha - \beta)$ are negative numbers. Thus $\alpha - \beta$ is a third-quadrant angle (Figure 9.2).

Practice Exercise 5   Let α and β be second-quadrant angles such that $\sin \alpha = 1/3$ and $\cos \beta = -1/2$. Find $\sin(\alpha - \beta)$ and $\cos(\alpha - \beta)$. In which quadrant does $\alpha - \beta$ lie?

Answer: $\dfrac{2\sqrt{6} - 1}{6}$, $\dfrac{2\sqrt{2} + \sqrt{3}}{6}$, first quadrant

Product Identities

We complete this section by deriving the *product identities*.

If identities (9.6) and (9.7) are added, the $\cos \alpha \sin \beta$ terms cancel out, and we obtain a new identity.

$$\sin(\alpha + \beta) + \sin(\alpha - \beta) = 2 \sin \alpha \cos \beta. \qquad (9.10)$$

If (9.7) is subtracted from (9.6), the result is

$$\sin(\alpha + \beta) - \sin(\alpha - \beta) = 2 \cos \alpha \sin \beta. \qquad (9.11)$$

Analogously, by adding and subtracting (9.4) and (9.5), we have

$$\cos(\alpha - \beta) + \cos(\alpha + \beta) = 2 \cos \alpha \cos \beta, \qquad (9.12)$$
$$\cos(\alpha - \beta) - \cos(\alpha + \beta) = 2 \sin \alpha \sin \beta. \qquad (9.13)$$

EXAMPLE 6   Express each of the following products as a sum or a difference.
(a) $2 \sin 3\theta \cos 2\theta$ **(b)** $\cos 7x \cos 2x$

Solution: **(a)** According to (9.10) with $\alpha = 3\theta$ and $\beta = 2\theta$, we have

$$2 \sin 3\theta \cos 2\theta = \sin(3\theta + 2\theta) + \sin(3\theta - 2\theta)$$
$$= \sin 5\theta + \sin \theta.$$

(b) Using (9.12) with $\alpha = 7x$ and $\beta = 2x$, we obtain

$$\cos 7x \cos 2x = \frac{1}{2}[\cos(7x - 2x) + \cos(7x + 2x)]$$

$$= \frac{1}{2}[\cos 5x + \cos 9x]$$

$$= \frac{\cos 5x + \cos 9x}{2}.$$

Practice Exercise 6   **(a)** Write $2 \cos 5x \sin 3x$ as a difference of sines.
(b) Express $2 \cos a \cos 3b$ as a sum of cosines.

Answer: **(a)** $\sin 8x - \sin 2x$ **(b)** $\cos(a - 3b) + \cos(a + 3b)$

EXERCISES 9.2

In Exercises 1–20, use the addition and subtraction formulas to find the exact function values. Do not use a table or calculator.

1. $\sin 15°$

2. $\cos 15°$

3. $\tan 15°$

4. $\sec 15°$

5. $\cos 75°$

6. $\sin 75°$

7. $\sin 105°$

8. $\cot 105°$

9. $\cos 210°$

10. $\sin 330°$

11. $\cos \dfrac{5\pi}{12}$

12. $\tan \dfrac{5\pi}{12}$

13. $\sec \dfrac{5\pi}{12}$

14. $\sin \dfrac{5\pi}{4}$

15. $\cos \dfrac{3\pi}{4}$

16. $\tan \dfrac{4\pi}{3}$

17. $\sin \dfrac{4\pi}{3}$

18. $\cos \dfrac{7\pi}{6}$

19. $\csc \dfrac{7\pi}{6}$

20. $\sin \dfrac{5\pi}{3}$

Using the addition and subtraction formulas, write each expression in Exercises 21–30 as one trigonometric function of one angle.

21. $\sin 32° \cos 28° + \cos 32° \sin 28°$

22. $\cos 18° \cos 46° - \sin 18° \sin 46°$

23. $\cos 76° \cos(-21°) + \sin 76° \sin(-21°)$

24. $\sin 25° \cos(-56°) - \cos 25° \sin(-56°)$

25. $\cos \dfrac{\pi}{5} \cos \dfrac{\pi}{4} - \sin \dfrac{\pi}{5} \sin \dfrac{\pi}{4}$

26. $\sin \dfrac{\pi}{3} \cos \dfrac{5\pi}{6} + \cos \dfrac{\pi}{3} \sin \dfrac{5\pi}{6}$

27. $\sin \dfrac{7\pi}{6} \cos\left(-\dfrac{\pi}{3}\right) - \cos \dfrac{7\pi}{6} \sin\left(-\dfrac{\pi}{3}\right)$

28. $\cos \dfrac{11\pi}{6} \cos\left(-\dfrac{7\pi}{4}\right) + \sin \dfrac{11\pi}{6} \sin\left(-\dfrac{7\pi}{4}\right)$

29. $\sin 5 \cos 3 - \cos 5 \sin 3$

30. $\cos 2 \cos 5 + \sin 2 \sin 5$

In Exercises 31–36, find the exact value of each expression.

31. $\sin\left[\cos^{-1}\left(\dfrac{4}{5}\right) + \sin^{-1}\left(\dfrac{4}{5}\right)\right]$

32. $\cos\left[\sin^{-1}\left(\dfrac{8}{17}\right) + \sin^{-1}\left(\dfrac{15}{17}\right)\right]$

33. $\sin\left[\tan^{-1}\left(\dfrac{3}{4}\right) - \cos^{-1}\left(\dfrac{5}{13}\right)\right]$

34. $\sin\left[\sin^{-1}\left(\dfrac{8}{10}\right) + \tan^{-1}\left(\dfrac{8}{15}\right)\right]$

35. $\cos\left[\tan^{-1}(3) - \sin^{-1}\left(\dfrac{\sqrt{5}}{3}\right)\right]$

36. $\cos\left[\tan^{-1}\left(\dfrac{2}{3}\right) + \cos^{-1}\left(\dfrac{2}{\sqrt{5}}\right)\right]$

In Exercises 37–46, solve the given problem.

37. If α and β are first-quadrant angles such that $\sin \alpha = 4/5$ and $\cos \beta = 5/13$, find $\cos(\alpha - \beta)$ and $\sin(\alpha + \beta)$.

38. If α and β are second-quadrant angles such that $\sin \alpha = 3/5$ and $\cos \beta = -8/17$, find $\cos(\alpha + \beta)$ and $\sin(\alpha - \beta)$.

39. Let α be an angle in quadrant I such that $\sin \alpha = 5/13$, and let β be an angle in quadrant III such that $\cos \beta = -3/5$. Find $\cos(\alpha + \beta)$ and $\cos(\alpha - \beta)$.

40. Let α be an angle in quadrant II such that $\sin \alpha = 12/13$, and let β be an angle in quadrant IV such that $\sin \beta = -7/25$. Find $\sin(\alpha + \beta)$ and $\sin(\alpha - \beta)$.

41. If α is a second-quadrant angle such that $\sin \alpha = 1/3$ and β is a first-quadrant angle such that $\cos \beta = 2/3$, find $\sin(\alpha + \beta)$, $\cos(\alpha + \beta)$, and the quadrant containing $\alpha + \beta$.

42. If α and β are second-quadrant angles with $\cos \alpha = -2/5$ and $\cos \beta = -4/5$, find $\sin(\alpha - \beta)$, $\cos(\alpha - \beta)$, and the quadrant containing $\alpha - \beta$.

43. If α and β are acute angles such that $\sin \alpha = 4/5$ and $\cos \beta = 2/5$, find $\sin(\alpha - \beta)$, $\tan(\alpha - \beta)$, and the quadrant containing $\alpha - \beta$.

44. If α is a second-quadrant angle such that $\sin \alpha = 2/3$ and β is a third-quadrant angle such that $\tan \beta = 2$, find $\sin(\alpha + \beta)$ and $\tan(\alpha + \beta)$. In which quadrant does $\alpha + \beta$ lie?

45. If α is a fourth-quadrant angle such that $\cos \alpha = 3/4$ and β is a second-quadrant angle such that $\cos \beta = -2/3$, find $\cos(\alpha - \beta)$ and $\tan(\alpha - \beta)$. In which quadrant does $\alpha - \beta$ lie?

46. If α and β are fourth-quadrant angles such that $\cos \alpha = 4/5$ and $\cos \beta = 1/5$, find $\sin(\alpha - \beta)$, $\cos(\alpha - \beta)$, and the quadrant containing $\alpha - \beta$.

In Exercises 47–58, express each product as a sum or a difference.

47. $2 \sin 6\theta \cos 2\theta$

48. $2 \cos 8\theta \sin 3\theta$

49. $\sin \theta \cos 3\theta$

50. $\cos 2\theta \sin 5\theta$

51. $2 \cos 5x \cos 2x$

52. $2 \sin 6x \sin 3x$

53. $\sin 3x \sin 8x$

54. $\cos 5x \sin 7x$

55. $4 \sin 8a \cos 3b$

56. $3 \cos 7a \sin 4b$

57. $3 \sin 5a \sin a$

58. $2 \cos 2a \cos 3a$

In Exercises 59–70, verify each identity.

59. $\sin(a - b) = \sin a \cos b - \cos a \sin b$

60. $\tan(a - b) = \dfrac{\tan a - \tan b}{1 + \tan a \tan b}$

61. $\tan a - \tan b = \dfrac{\sin(a - b)}{\cos a \cos b}$

62. $\cot b - \cot a = \dfrac{\sin(a - b)}{\sin a \sin b}$

63. $\cot a - \tan b = \dfrac{\cos(a + b)}{\sin a \cos b}$

64. $\cot a \cot b - 1 = \dfrac{\cos(a + b)}{\sin a \sin b}$

65. $\cot(a + b) = \dfrac{\cot a \cot b - 1}{\cot a + \cot b}$

66. $\cot(a - b) = \dfrac{\cot a \cot b + 1}{\cot b - \cot a}$

☐ 67. $\sin a + \sin b = 2 \sin \dfrac{a + b}{2} \cos \dfrac{a - b}{2}$

$\left[\textit{Hint}: \text{Set } \alpha = \dfrac{a + b}{2} \text{ and } \beta = \dfrac{a - b}{2} \text{ in formula (9.10).}\right]$

☐ 68. $\sin a - \sin b = 2 \cos \dfrac{a + b}{2} \sin \dfrac{a - b}{2}$

☐ 69. $\cos a + \cos b = 2 \cos \dfrac{a + b}{2} \cos \dfrac{a - b}{2}$

☐ 70. $\cos a - \cos b = -2 \sin \dfrac{a + b}{2} \sin \dfrac{a - b}{2}$

9.3   DOUBLE- AND HALF-ANGLE IDENTITIES

Setting $\alpha = \beta$ in identity (9.6) gives an expression for $\sin 2\alpha$:

$$\begin{aligned}
\sin 2\alpha &= \sin(\alpha + \alpha) \\
&= \sin \alpha \cos \alpha + \cos \alpha \sin \alpha
\end{aligned}$$

Thus we have a double-angle formula for the sine function.

Sine of a Double Angle

$$\sin 2\alpha = 2 \sin \alpha \cos \alpha \qquad \text{(9.14)}$$

Analogously, setting $\alpha = \beta$ in identity (9.5) yields

$$\begin{aligned}
\cos 2\alpha &= \cos(\alpha + \alpha) \\
&= \cos \alpha \cos \alpha - \sin \alpha \sin \alpha.
\end{aligned}$$

This gives us a double-angle formula for the cosine.

Cosine of a Double Angle

$$\cos 2\alpha = \cos^2 \alpha - \sin^2 \alpha \qquad \text{(9.15)}$$

Formulas (9.14) and (9.15), called *double-angle identities,* are important in many applications.

Using the identities $\cos^2 \alpha = 1 - \sin^2 \alpha$ and $\sin^2 \alpha = 1 - \cos^2 \alpha$ allows us to rewrite the identity (9.15) in two ways.

Cosine of a Double Angle

$$\cos 2\alpha = 1 - 2 \sin^2 \alpha \qquad \text{(9.16)}$$
$$\cos 2\alpha = 2 \cos^2 \alpha - 1 \qquad \text{(9.17)}$$

From (9.8), we also obtain an identity for $\tan 2\alpha$.

Tangent of a Double Angle

$$\tan 2\alpha = \frac{2 \tan \alpha}{1 - \tan^2\alpha}. \qquad (9.18)$$

EXAMPLE 1   If $\sin \alpha = 3/5$ and α is a second-quadrant angle, find $\sin 2\alpha$, $\cos 2\alpha$, and $\tan 2\alpha$.

Solution: First, let us find $\cos \alpha$.

$$\cos \alpha = -\sqrt{1 - \sin^2\alpha} = -\sqrt{1 - \frac{9}{25}}$$

$$= -\sqrt{\frac{16}{25}} = -\frac{4}{5}$$

The negative square root was taken since α is a second-quadrant angle. Using (9.14), we get

$$\sin 2\alpha = 2\left(\frac{3}{5}\right)\left(-\frac{4}{5}\right) = -\frac{24}{25}.$$

Identity (9.15) yields

$$\cos 2\alpha = \left(-\frac{4}{5}\right)^2 - \left(\frac{3}{5}\right)^2 = \frac{7}{25}.$$

Finally,

$$\tan 2\alpha = \frac{\sin 2\alpha}{\cos 2\alpha}$$

$$= \frac{-24/25}{7/25}$$

$$= -\frac{24}{7}.$$

Practice Exercise 1   If $\cos \alpha = 3/4$ and α is a first-quadrant angle, find $\sin 2\alpha$, $\cos 2\alpha$, and $\tan 2\alpha$.

Answer: $\dfrac{3\sqrt{7}}{8}, \dfrac{1}{8}, 3\sqrt{7}$

EXAMPLE 2   Verify the identity $\sin 2\alpha = \dfrac{2 \tan \alpha}{1 + \tan^2\alpha}$.

Solution: Starting with the right-hand side, we have

$$\frac{2 \tan \alpha}{1 + \tan^2\alpha} = \frac{2\dfrac{\sin \alpha}{\cos \alpha}}{1 + \dfrac{\sin^2\alpha}{\cos^2\alpha}} = \frac{2\dfrac{\sin \alpha}{\cos \alpha}}{\dfrac{\cos^2\alpha + \sin^2\alpha}{\cos^2\alpha}}$$

$$= \frac{2\dfrac{\sin \alpha}{\cos \alpha}}{\dfrac{1}{\cos^2\alpha}}$$

$$= 2\frac{\sin \alpha}{\cos \alpha} \cdot \cos^2\alpha$$

$$= 2 \sin \alpha \cos \alpha$$

$$= \sin 2\alpha.$$

Practice Exercise 2   Verify the identity $\cos 2\theta = \dfrac{\cot \theta - \tan \theta}{\cot \theta + \tan \theta}$.

Half-Angle Identities

Consider the two double-angle identities $\cos 2\alpha = 1 - 2 \sin^2\alpha$ and $\cos 2\alpha = 2 \cos^2\alpha - 1$. Solving the first one for $\sin^2\alpha$ and the second one for $\cos^2\alpha$, we obtain

$$\sin^2\alpha = \frac{1 - \cos 2\alpha}{2}$$

and

$$\cos^2\alpha = \frac{1 + \cos 2\alpha}{2}.$$

Next we set $\alpha = \theta/2$ and take square roots:

$$\sin \frac{\theta}{2} = \pm\sqrt{\frac{1 - \cos \theta}{2}} \qquad (9.19)$$

and

$$\cos \frac{\theta}{2} = \pm\sqrt{\frac{1 + \cos \theta}{2}}. \qquad (9.20)$$

These are the *half-angle identities*, which play an important role in calculus. In these two identities, the choice of the sign is determined by the quadrant in which the angle $\theta/2$ is located.

Note that from (9.19) and (9.20), we obtain the formula for the tangent of a half-angle:

$$\tan \frac{\theta}{2} = \pm \sqrt{\frac{1 - \cos \theta}{1 + \cos \theta}}.$$

It is not necessary to memorize this last formula: once you know $\sin \frac{\theta}{2}$ and $\cos \frac{\theta}{2}$, a simple division will give you the value of $\tan \frac{\theta}{2}$.

EXAMPLE **3**   Find $\sin(\theta/2)$ and $\cos(\theta/2)$ if $\cos \theta = 1/3$ and $3\pi/2 < \theta < 2\pi$.

Solution: If θ is such that $3\pi/2 < \theta < 2\pi$, then $3\pi/4 < \theta/2 < \pi$. That is, $\theta/2$ is a second-quadrant angle and, consequently, $\sin(\theta/2) > 0$ and $\cos(\theta/2) < 0$. According to (9.19), we have

$$\sin \frac{\theta}{2} = \sqrt{\frac{1 - \frac{1}{3}}{2}} = \sqrt{\frac{1}{3}} = \frac{\sqrt{3}}{3}.$$

From (9.20), it follows that

$$\cos \frac{\theta}{2} = -\sqrt{\frac{1 + \frac{1}{3}}{2}} = -\sqrt{\frac{2}{3}} = -\frac{\sqrt{6}}{3}.$$

Practice Exercise 3   Find $\sin(\alpha/2)$ and $\cos(\alpha/2)$ if $\cos \alpha = -1/2$ and α is such that $\pi/2 < \alpha < \pi$.

Answer: $\dfrac{\sqrt{3}}{2}, \dfrac{1}{2}$

EXAMPLE **4**   Determine $\sin(\theta/2)$ and $\cos(\theta/2)$ if $\tan \theta = 3/4$ and $\pi < \theta < 3\pi/2$.

Solution: To make use of formulas (9.19) and (9.20), we must first find $\cos \theta$. Since θ is a third-quadrant angle, $\cos \theta$ is negative. Using identity (9.2), we obtain

$$\sec \theta = -\sqrt{1 + \frac{9}{16}} = -\sqrt{\frac{25}{16}} = -\frac{5}{4}.$$

Hence,

$$\cos \theta = -\frac{4}{5}.$$

As an exercise, you can check that the value of $\cos \theta$ can also be obtained by drawing a right triangle, labeling one acute angle θ, the opposite side 3, and the adjacent side 4.

Since $\pi < \theta < 3\pi/2$, it follows that $\pi/2 < \theta/2 < 3\pi/4$. Thus, $\theta/2$ is a second-quadrant angle. Consequently, $\sin(\theta/2) > 0$ and $\cos(\theta/2) < 0$. From formulas (9.19) and (9.20), we obtain

$$\sin \frac{\theta}{2} = \sqrt{\frac{1 - \left(-\frac{4}{5}\right)}{2}} = \sqrt{\frac{1 + \frac{4}{5}}{2}}$$

$$= \sqrt{\frac{9}{10}} = \frac{3\sqrt{10}}{10}$$

and

$$\cos\frac{\theta}{2} = -\sqrt{\frac{1 + \left(-\frac{4}{5}\right)}{2}} = -\sqrt{\frac{1 - \frac{4}{5}}{2}}$$

$$= -\sqrt{\frac{1}{10}} = -\frac{\sqrt{10}}{10}.$$

Practice Exercise 4   Let θ be such that $\pi/2 < \theta < \pi$ and $\cot\theta = -5/12$. Find $\sin(\theta/2)$ and $\cos(\theta/2)$.

Answer: $\dfrac{3\sqrt{13}}{13}, \dfrac{2\sqrt{13}}{13}$

EXERCISES 9.3

Find the exact values of $\sin 2\alpha$, $\cos 2\alpha$, and $\tan 2\alpha$ from the information given in Exercises 1–8.

1. $\cos\alpha = \dfrac{3}{5}$, $0 < \alpha < \dfrac{\pi}{2}$ **2.** $\sin\alpha = \dfrac{1}{3}$, α acute

3. $\sin\alpha = -\dfrac{3}{4}$, $180° < \alpha < 270°$

4. $\cos\alpha = -\dfrac{2}{5}$, $\pi < \alpha < \dfrac{3\pi}{2}$

5. $\tan\alpha = \dfrac{4}{3}$, α acute

6. $\cot\alpha = -2$, $\dfrac{\pi}{2} < \alpha < \pi$

7. $\sec\alpha = \dfrac{5}{2}$, $\dfrac{3\pi}{2} < \alpha < 2\pi$

8. $\csc\alpha = \dfrac{4}{3}$, α acute

In Exercises 9–20, find the exact values of $\sin\alpha/2$ and $\tan\alpha/2$, subject to the given conditions.

9. $\sin\alpha = -\dfrac{1}{2}$, $\pi < \alpha < \dfrac{3\pi}{2}$

10. $\cos\alpha = \dfrac{4}{7}$, α acute

11. $\cos\alpha = \dfrac{1}{4}, \dfrac{3\pi}{2} < \alpha < 2\pi$

12. $\sin\alpha = -\dfrac{1}{3}$, $\dfrac{3\pi}{2} < \alpha < 2\pi$

13. $\tan\alpha = 3$, α acute

14. $\cot\alpha = -1$, $\dfrac{\pi}{2} < \alpha < \pi$

15. $\csc\alpha = -4$, $180° < \alpha < 270°$

16. $\sec\alpha = \dfrac{4}{3}$, $0° < \alpha < 90°$.

17. Let $\sin\alpha = 12/13$, with $\pi/2 < \alpha < \pi$. Find $\cos(\alpha/2)$.
18. Let $\tan\alpha = 4/3$, with $\pi < \alpha < 3\pi/2$. Find $\cos(\alpha/2)$.
19. Find $\cos\alpha$ if $\sin 2\alpha = 24/25$, with $\pi/4 < \alpha < \pi/2$.
20. Find $\cos\alpha$ if $\tan 2\alpha = -24/7$, with $3\pi/4 < \alpha < \pi$.

In Exercises 21–28, use the half-angle formulas to find the given trigonometric values. Do not use a table or calculator.

21. $\sin\dfrac{\pi}{8}$ 　　　　　　　　 **22.** $\tan\dfrac{\pi}{8}$

23. $\cos\dfrac{\pi}{12}$ 　　　　　　　 **24.** $\tan 15°$

25. $\sin 75°$ 　　　　　　　　　 **26.** $\cos\dfrac{5\pi}{12}$

27. $\cos 22°30'$ 　　　　　　　 **28.** $\sin 112°30'$

In Exercises 29–34, find the exact value of each expression without using a table or calculator.

29. $\sin\left[2\cos^{-1}\left(\dfrac{4}{5}\right)\right]$ 　　 **30.** $\cos\left[2\cos^{-1}\left(\dfrac{4}{5}\right)\right]$

31. $\cos\left[2\tan^{-1}\left(-\dfrac{5}{12}\right)\right]$ 　 **32.** $\sin\left[2\tan^{-1}\left(-\dfrac{5}{12}\right)\right]$

33. $\sin\left[\dfrac{1}{2}\sin^{-1}\left(\dfrac{3}{5}\right)\right]$ **34.** $\cos\left[\dfrac{1}{2}\sin^{-1}\left(-\dfrac{3}{5}\right)\right]$

In Exercises 35–46, verify the identities.

35. $(\sin\theta - \cos\theta)^2 = 1 - \sin 2\theta$

36. $1 + \cos 2\theta = \sin 2\theta \cot \theta$

37. $1 + \cot^2\theta = 2\cot\theta\csc 2\theta$

38. $\sin 2\theta = \tan\theta(1 + \cos 2\theta)$

39. $\cot 2\theta = \dfrac{\cot^2\theta - 1}{2\cot\theta}$ **40.** $\sec 2\theta - \dfrac{\sec^2\theta}{2 - \sec^2\theta}$

41. $\csc 2\theta = \dfrac{1}{2}(\tan\theta + \cot\theta)$

42. $\tan 2\theta = \dfrac{2\tan\theta}{2 - \sec^2\theta}$

43. $\tan\dfrac{\theta}{2} = \dfrac{1 - \cos\theta}{\sin\theta}$ **44.** $\cot\dfrac{\theta}{2} = \dfrac{1 + \cos\theta}{\sin\theta}$

45. $\cot\theta - \tan\theta = \dfrac{2\cos 2\theta}{\sin 2\theta}$

46. $\sin 2\theta = \dfrac{2\cot\theta}{1 + \cot^2\theta}$

🅒 Solve Exercises 47–50 using a calculator.

47. Approximate $\sin 2\alpha$ if $\cos\alpha = -0.5736$ and $\pi/2 < \alpha < \pi$.

48. Approximate $\cos 2\alpha$ if $\cos\alpha = 0.1736$ and α is an acute angle.

49. Approximate $\cos(\theta/2)$ if $\cos\theta = 0.8660$ and $3\pi/2 < \theta < 2\pi$.

50. Approximate $\sin(\theta/2)$ if $\cos\theta = -0.7660$ and $\pi < \theta < 3\pi/2$.

9.4 TRIGONOMETRIC EQUATIONS

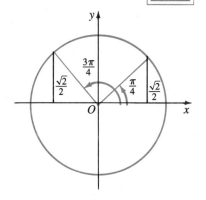

Figure 9.3

In previous sections of this book we discussed algebraic equations, as well as equations involving exponents and logarithms. We shall now consider equations involving trigonometric functions.

You should be aware of the distinction between a trigonometric identity and a trigonometric equation. A trigonometric identity is satisfied for *all* possible values of the variable, while a trigonometric equation is satisfied only for *some* values of the variable. The relationship $\sin^2\theta + \cos^2\theta = 1$ is true for all values of θ, so it is an identity. However, $\sin\theta = \cos\theta$ is a trigonometric equation; its solutions are the numbers $\dfrac{\pi}{4} + k\pi$, where k can be any integer.

There is no fixed set of rules leading us to the solution of a given trigonometric equation. To solve a trigonometric equation we have to combine algebraic methods, known properties of trigonometric functions, and trigonometric identities.

EXAMPLE 1   Find all solutions of the equation $2\sin\theta - \sqrt{2} = 0$.

Solution: The given equation is equivalent to

$$\sin\theta = \frac{\sqrt{2}}{2}.$$

According to the special angle table (Section 7.2), $\theta = \pi/4$ is a solution of the last equation. However, $\theta = 3\pi/4$ is also a solution (Figure 9.3). Since the sine function has period 2π, all solutions of the given equation are the numbers

$$\frac{\pi}{4} + 2k\pi \quad \text{and} \quad \frac{3\pi}{4} + 2k\pi, \quad \text{with } k \text{ any integer.}$$

Practice Exercise 1   Find all solutions of $2\cos\theta - \sqrt{3} = 0$.

Answer: $\pm\dfrac{\pi}{6} + 2k\pi$, k any integer

EXAMPLE 2   Find all solutions of $\sqrt{3}\,\tan\alpha - 3 = 0$. How many solutions are there in the interval $(-\pi/2, \pi/2)$?

Solution: Solving for $\tan\alpha$, we get

$$\tan\alpha = \frac{3}{\sqrt{3}} = \sqrt{3}.$$

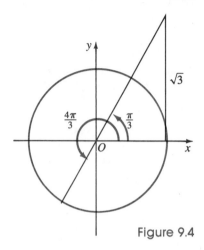

According to the special angle table, a solution is $\alpha = \pi/3$ (Figure 9.4). Since the tangent function has period π, all solutions of the given equation are the numbers

$$\frac{\pi}{3} + k\pi, \quad \text{with } k \text{ any integer.}$$

In the interval $(-\pi/2, \pi/2)$, the number $\pi/3$ is the unique solution of the given equation. (See the definition of arctan x, Section 8.6.)

Figure 9.4

Practice Exercise 2   Find all solutions of $\sqrt{3}\,\cot x - 1 = 0$ in the interval $(0, \pi)$.

Answer: $\dfrac{\pi}{3}$

EXAMPLE 3   Find the solutions of $\sin 2\alpha - \sqrt{3}\,\cos\alpha = 0$ in the interval $[0, 2\pi)$.

Solution: According to the trigonometric identity (9.14), the given equation is equivalent to

$$2\sin\alpha\cos\alpha - \sqrt{3}\,\cos\alpha = 0$$

or

$$\left(2\sin\alpha - \sqrt{3}\right)\cos\alpha = 0.$$

A product is zero if at least one of its factors is zero. Thus, we have to solve the following two equations in the interval $[0, 2\pi)$.

$$
\begin{array}{c|c}
\cos\alpha = 0 & 2\sin\alpha - \sqrt{3} = 0 \\[4pt]
 & \sin\alpha = \dfrac{\sqrt{3}}{2} \\[6pt]
\alpha = \dfrac{\pi}{2} \ \text{ and } \ \dfrac{3\pi}{2} & \alpha = \dfrac{\pi}{3} \ \text{ and } \ \dfrac{2\pi}{3}
\end{array}
$$

Hence the solutions of the given equation are $\pi/3$, $\pi/2$, $2\pi/3$, and $3\pi/2$.

Practice Exercise 3   Find all solutions of $\cos 2x - 2\sin^2 x = 0$ in the interval $[0, 2\pi)$.

Answer: $\dfrac{\pi}{6}, \dfrac{5\pi}{6}, \dfrac{7\pi}{6}, \dfrac{11\pi}{6}.$

EXAMPLE 4   Find all solutions of $2\sin^2\alpha + 1 = 3\sin\alpha$.

Solution: The given equation is equivalent to

$$2 \sin^2 \alpha - 3 \sin \alpha + 1 = 0,$$

which is a quadratic equation in $\sin \alpha$. Factoring gives us

$$(2 \sin \alpha - 1)(\sin \alpha - 1) = 0.$$

Next, we solve the two equations $2 \sin \alpha - 1 = 0$ and $\sin \alpha - 1 = 0$. In view of the periodicity of the sine function, it suffices to find the solutions of these equations in the interval $[0, 2\pi)$. Once they are found, we can obtain all solutions by adding multiples of 2π. In $[0, 2\pi)$ the only solution of $\sin \alpha = 1$ is $\pi/2$, while the solutions of $\sin \alpha = 1/2$ are $\pi/6$ and $5\pi/6$. Hence,

$$\frac{\pi}{2} + 2k\pi, \ \frac{\pi}{6} + 2k\pi, \quad \text{and} \quad \frac{5\pi}{6} + 2k\pi, \quad \text{with } k \text{ any integer,}$$

are all the solutions of the original equation.

Practice Exercise 4   Find all solutions of $2 \cos^2 \beta - 1 = 0$ in the interval $[0, 2\pi)$.

Answer: $\dfrac{\pi}{4}, \dfrac{3\pi}{4}, \dfrac{5\pi}{4}, \dfrac{7\pi}{4}$

ⓒ EXAMPLE 5   Approximate the solutions of $2 \cos^2 \alpha + 2 \sin \alpha - 1 = 0$ in the interval $(-\pi/2, \pi/2)$.

Solution: Replacing $\cos^2 \alpha$ by $1 - \sin^2 \alpha$ in the given equation gives us

$$2(1 - \sin^2 \alpha) + 2 \sin \alpha - 1 = 0,$$
$$2 - 2 \sin^2 \alpha + 2 \sin \alpha - 1 = 0,$$
$$2 \sin^2 \alpha - 2 \sin \alpha - 1 = 0,$$

which is a quadratic equation in $\sin \alpha$. Setting $u = \sin \alpha$ and solving $2u^2 - 2u - 1 = 0$ for u, we get the solutions

$$u = \frac{1 + \sqrt{3}}{2} \simeq 1.3660 \quad \text{and} \quad u = \frac{1 - \sqrt{3}}{2} \simeq -0.3660.$$

The first solution must be discarded, since $\sin \alpha$ can never be greater than 1. Thus, we must find an angle α in the interval $(-\pi/2, \pi/2)$ such that $\sin \alpha = -0.3660$. With the help of a calculator, we obtain

$$\alpha \simeq -0.3747 \text{ radians.}$$

ⓒ Practice Exercise 5   Approximate the solution of $4 \sin^2 x + 4 \cos x - 3 = 0$ in the interval $(0, \pi)$.

Answer: 1.7794 radians

EXAMPLE 6   Find the solutions of $\tan 2\alpha - \cot \alpha = 0$ in the interval $(-\pi/2, \pi/2)$.

Solution: Using the identity (9.18) and the definition of the cotangent, we rewrite the given equation as follows.

$$\frac{2 \tan \alpha}{1 - \tan^2 \alpha} - \frac{1}{\tan \alpha} = 0 \qquad \text{Find a common denominator and add}$$

$$\frac{2 \tan^2 \alpha - (1 - \tan^2 \alpha)}{(1 - \tan^2 \alpha)\tan \alpha} = 0 \qquad \text{Simplify}$$

$$\frac{3 \tan^2 \alpha - 1}{(1 - \tan^2 \alpha)\tan \alpha} = 0$$

Since a fraction is zero exactly when its numerator is zero and its denominator is different from 0, the solutions of the given equation are the solutions of

$$3 \tan^2 \alpha - 1 = 0$$

or

$$\tan \alpha = \pm \frac{1}{\sqrt{3}}.$$

In the interval $(-\pi/2, \pi/2)$, the last equation has two solutions: $\pi/6$ and $-\pi/6$.

Practice Exercise 6   Find all solutions of the equation $\tan 2x + 3 \cot x = 0$ in the interval $(0, \pi)$.

Answer: $\dfrac{\pi}{3}, \dfrac{2\pi}{3}$

Tips on Solving Trigonometric Equations

1. Always remember that an *identity* is satisfied for *all* values of the variable, while an *equation* is satisfied for *some* values of the variable. Also, there are equations with *no* solution.
2. When solving a trigonometric equation, you may have to combine algebraic methods, properties of trigonometric functions, and trigonometric identities in order to reduce the given equation to a simpler one.

EXERCISES 9.4

In Exercises 1–24, find the solutions of the given equations in the specified intervals.

1. $2 \sin \alpha - 1 = 0, \quad \left[-\dfrac{\pi}{2}, \dfrac{\pi}{2}\right]$

2. $2 \cos \alpha + \sqrt{3} = 0, \quad [0, \pi]$

3. $\sqrt{3} \tan \alpha - 1 = 0, \quad \left[-\dfrac{\pi}{2}, \dfrac{\pi}{2}\right]$

4. $\tan \alpha + \sqrt{3} = 0, \quad [0, 2\pi]$

5. $2 \cos x + 5 = 4, \quad [0°, 360°)$

6. $3 \csc x + 2 = 8, \quad [0°, 360°)$

7. $3 \csc x + 2 = 2 \csc x + 4, \quad [0°, 180°)$

8. $4 \tan x + 1 = 5 \tan x + 2, \quad [0°, 360°)$

9. $\sin \beta + \cos \beta = 0, \quad [0, 2\pi)$

10. $\sin \beta = \cos \beta, \quad [0, 2\pi)$

11. $(\sin x - 1)(\tan x - 1) = 0$, $[0°, 360°)$
12. $(\cos x + 1)(\cot x + 1) = 0$, $[0°, 360°)$
13. $(\sqrt{2} \sin x - 1)(\sec x + 2) = 0$, $[0, \pi)$
14. $(2 \cos x - 1)(\csc x - \sqrt{2}) = 0$, $[0, \pi)$
15. $(2 \sin x - \sqrt{3})(\tan x + \sqrt{3}) = 0$, $[0°, 180°)$
16. $(3 \cot x - \sqrt{3})(2 \cos x + 1) = 0$, $[0, 180°)$

17. $\sin \theta \tan \theta - \sin \theta = 0$, $\left[0, \dfrac{\pi}{2}\right]$

18. $\cos^2 \theta - \cos \theta = 0$, $[0, \pi]$
19. $2 \cos^2 x + 3 \cos x + 1 = 0$, $[0, 2\pi)$

20. $2 \sin^2 x - 3 \sin x + 1 = 0$, $\left[-\dfrac{\pi}{2}, \dfrac{\pi}{2}\right]$

21. $\dfrac{2 \sin x}{\sin^2 x - 3} = 1$, $[0°, 360°)$

22. $\dfrac{2 \cos^2 x - 5}{9 \cos x} = 1$, $[0°, 360°)$

23. $\sin \theta + \cos \theta = 1$, $\left[-\dfrac{\pi}{2}, \dfrac{\pi}{2}\right]$

24. $\tan \theta - \sec \theta = 1$, $[0, 2\pi)$

In Exercises 25–38, find all solutions for each equation in the given interval.

25. $\sin 2\alpha = 0$, $[0°, 360°)$
26. $\cos 2\alpha = 1$, $[0°, 360°)$
27. $\tan 2x = \sqrt{3}$, $[0°, 360°)$
28. $\sqrt{3} \cot 2x = -1$, $[0°, 360°)$

29. $2 \cos 2\beta - 1 = 0$, $\left[0, \dfrac{\pi}{2}\right]$

30. $2 \sin 2\beta - \sqrt{3} = 0$, $\left[-\dfrac{\pi}{4}, \dfrac{\pi}{4}\right]$

31. $\sin 2\theta - \sin \theta = 0$, $[0, \pi]$

32. $\cos 2\theta = 1 - \sin \theta$, $\left[-\dfrac{\pi}{2}, \dfrac{\pi}{2}\right]$

33. $\cos 2x = \sin 2x - 1$, $[0, \pi]$

34. $2 \sin^2 x - \sin 2x = 0$, $\left[-\dfrac{\pi}{2}, \dfrac{\pi}{2}\right]$

35. $\sin \dfrac{\theta}{2} = \cos \theta$, $[0°, 360°)$

36. $\cos \dfrac{\theta}{2} = \sin \theta$, $[0°, 360°)$

37. $\cos \theta = \cos \dfrac{\theta}{2} - 1$, $[0, 2\pi)$

38. $\cos \theta = \sin \dfrac{\theta}{2} + 1$, $[0, 2\pi)$

In Exercises 39–54, find all solutions of the given equations.

39. $2\sqrt{3} \cos \theta - 3 = 0$ **40.** $\sqrt{2} \sin \theta + 1 = 0$
41. $2 \sin \alpha - 1 = 0$ **42.** $2 \cos \alpha - \sqrt{2} = 0$
43. $\tan \alpha + 1 = 0$ **44.** $\cot \alpha - 1 = 0$
45. $\sec^2 u = 2$ **46.** $3 \csc^2 u = 4$
47. $(1 - \sin t)(1 + \cos t) = 0$
48. $(\tan t - 1)(\tan t - \sqrt{3}) = 0$
49. $\sin 2u + \cos u = 0$ **50.** $\cos 2u = \cos u - 1$
51. $\cos^2 u - \cos u - 6 = 0$
52. $2 \sin^2 \alpha - \sin \alpha - 6 = 0$
53. $\sec^2 x = \tan x + 1$ **54.** $2 \sin^2 \alpha = \cos \alpha + 2$

C In Exercises 55–60, approximate the solutions of the given equations in the given interval.

55. $2 \sin^2 x + 2 \cos x - 1 = 0$, $[0, \pi]$
56. $\cos^2 x - 2 \cos x - 1 = 0$, $[0, \pi]$

57. $2 \tan^2 x - 2 \tan x - 1 = 0$, $\left[-\dfrac{\pi}{2}, \dfrac{\pi}{2}\right]$

58. $2 \cot^2 \alpha - 7 \cot \alpha + 6 = 0$, $[0, \pi]$
59. $\cos^2 u - 2 \sin^2 u + 1 = 0$, $[0, \pi]$

60. $\tan^2 \beta + \sec^2 \beta = 4$, $\left[-\dfrac{\pi}{2}, \dfrac{\pi}{2}\right]$

9.5   THE LAW OF SINES

If none of the angles of a triangle is a right angle, then the triangle is called an *oblique triangle*. In this section and the following one, we discuss the problem of solving oblique triangles. In order to solve them, we shall have to prove and use the *law of sines* and the *law of cosines*.

The Law of Sines

If ABC is a triangle with angles of measure α, β, and γ opposite sides of lengths a, b, and c, respectively, then

$$\frac{a}{\sin \alpha} = \frac{b}{\sin \beta} = \frac{c}{\sin \gamma}.$$ **(9.21)**

In other words, the lengths of the sides are directly proportional to the sines of the opposite angles.

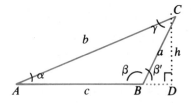

Figure 9.5

To prove the law of sines, consider the triangle ABC illustrated in Figure 9.5, where β is an obtuse angle and h is the length of the altitude from the vertex C to the line extending the opposite side, AB. Considering the right triangle ADC and remembering the definition of the sine as a ratio, we see that

$$\sin \alpha = \frac{h}{b} \quad \text{or} \quad h = b \sin \alpha.$$

Similarly, from the right triangle BDC we get

$$\sin \beta' = \frac{h}{a} \quad \text{or} \quad h = a \sin \beta'.$$

Thus,

$$a \sin \beta' = h = b \sin \alpha.$$

Since $\beta' = \pi - \beta$, we have

$$
\begin{aligned}
\sin \beta' &= \sin(\pi - \beta) \\
&= \sin \pi \cos \beta - \cos \pi \sin \beta \\
&= \sin \beta.
\end{aligned}
$$

Replacing $\sin \beta'$ in the previous equation, we obtain

$$a \sin \beta = b \sin \alpha,$$

or

$$\frac{a}{\sin \alpha} = \frac{b}{\sin \beta}.$$

This is half of the law of sines. To establish the rest of it, let us consider triangle ABC as illustrated in Figure 9.6, in which k is the length of the altitude from the vertex B to the opposite side, AC.

From the right triangles AEB and CEB, we derive the equations

$$\sin \alpha = \frac{k}{c} \quad \text{or} \quad k = c \sin \alpha$$

and

$$\sin \gamma = \frac{k}{a} \quad \text{or} \quad k = a \sin \gamma.$$

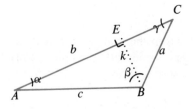

Figure 9.6

Thus,

$$a \sin \gamma = k = c \sin \alpha,$$

so

$$\frac{a}{\sin \alpha} = \frac{c}{\sin \gamma}.$$

Combining this equality with the previous one, we obtain the law of sines, (9.21).

As an exercise, you should prove the law of sines in the case where all the angles of the triangle *ABC* are acute angles.

HISTORICAL NOTE

THE LAWS OF SINES AND COSINES

It is believed that the law of sines for plane triangles, in its most general form, was known to the Persian mathematician al-Biruni (973–1048). The law of cosines in the plane was stated by François Viète (1540–1603).

Solving Oblique Triangles

We start with an example showing how to use the law of sines to solve a triangle when *two angles and the length of a side are given.*

EXAMPLE 1

Solve the triangle *ABC* with $\alpha = 40°$, $\beta = 120°$, and $a = 8.0$.

Solution: We must find the sides b and c and the angle γ (Figure 9.7). Since $\alpha + \beta + \gamma = 180°$, we can immediately calculate

$$\gamma = 180° - (40° + 120°) = 20°.$$

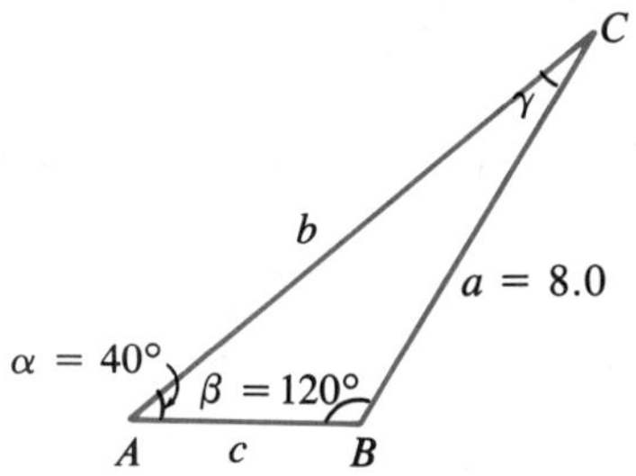

Figure 9.7

Next we apply the law of sines to obtain the sides b and c. First,

$$\frac{8}{\sin 40°} = \frac{b}{\sin 120°}.$$

Hence

$$b = \frac{8 \sin 120°}{\sin 40°} \simeq \frac{8 \times 0.8660}{0.6428} \simeq 10.8.$$

Similarly,

$$c = \frac{8 \sin 20°}{\sin 40°} \simeq \frac{8 \times 0.3420}{0.6428} \simeq 4.3.$$

Practice Exercise 1

In an oblique triangle *ABC*, $\alpha = 35°$, $\beta = 105°$, and $b = 12.3$. Find the other elements of the triangle.

Answer: $\gamma = 40°$, $a = 7.3$, $c = 8.2$

The Ambiguous Case

Whenever *two angles and the length of a side* of a triangle are given (such as in Example 1 and Practice Exercise 1), the problem of solving the triangle offers no conceptual difficulty: the third angle is the difference between 180° and the sum of the two angles, and direct application of the law of sines gives us the lengths of the other sides. However, the situation may be different if *only one angle and the lengths of two sides* are given.

First of all, if *two sides and the included angle* are given, then we cannot apply the law of sines. It will be necessary to start with the *law of cosines,* as we shall explain in the next section.

Second, if *two sides and one angle, not included between the two sides,* are given, then the problem may have *two solutions, a unique solution,* or *no solution.* In the next example and practice exercise, we discuss cases for which two solutions occur (the *ambiguous case*). The other cases (one or no solutions) are illustrated in Figure 9.9.

◻ EXAMPLE 2   Solve a triangle ABC with $\alpha = 45°$, $a = 4$, and $b = 5$. Approximate the angles to the nearest minute.

Solution: There are two triangles that meet the given conditions. To see this, draw an angle $\alpha = 45°$ in standard position and a line segment AC with length 5 units (Figure 9.8). Now, with a compass centered at C, strike off a circular arc of radius 4 units. The arc will intersect the horizontal axis at two points, B and B'. Thus, we obtain two triangles, ABC and $AB'C$, that satisfy the given conditions. Triangle $AB'C$ has an obtuse angle at B', while all the angles of ABC are acute.

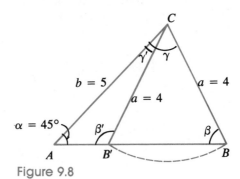

Figure 9.8

We can solve both triangles by using the law of sines. From (9.21), we have

$$\sin \beta = \frac{b \sin \alpha}{a}$$

$$= \frac{5 \sin 45°}{4}$$

$$\simeq \frac{5 \times 0.7071}{4}$$

$$\simeq 0.8839.$$

There are two angles, *one acute* and *one obtuse,* that satisfy the last relation. The acute one is

$$\beta = \sin^{-1}(0.8839) \simeq 62.11° \simeq 62°7',$$

and the obtuse one is

$$\beta' = 180° - 62°7' = 117°53'.$$

Corresponding to β and β', we have

$$\gamma \simeq 180° - (45° + 62°7') = 72°53'$$

and

$$\gamma' \simeq 180° - (45° + 117°53') = 17°7',$$

respectively. If c is the length of the side AB opposite angle γ, then

$$c = \frac{a \sin \gamma}{\sin \alpha}$$

$$= \frac{4 \sin(72°53')}{\sin 45°}$$

$$\simeq 5.$$

Similarly, if c' is the length of AB', then

$$c' = \frac{a \sin \gamma'}{\sin \alpha}$$

$$= \frac{4 \sin(17°7')}{\sin 45°}$$

$$\simeq 2.$$

C Practice Exercise 2   Solve a triangle ABC if $\alpha = 30°$, $a = 5.0$, and $b = 7.0$. Approximate the angles to the nearest tenth of a degree.

Answer: Two triangles: $\beta = 44.4°$, $\gamma = 105.6°$, $c = 9.6$; and $\beta' = 135.6°$, $\gamma' = 14.4°$, $c' = 2.5$

Notice that the data given in Example 2 were not enough to determine the triangle ABC in a *unique* way. Additional information, such as the type of the angle at the vertex B, is needed for the triangle ABC to be uniquely determined. In most applications one of the two triangles is specified, so ambiguity is avoided.

We illustrate in Figure 9.9 the four possible cases that may arise when an acute angle of measure α and the sides of length a and b are given.

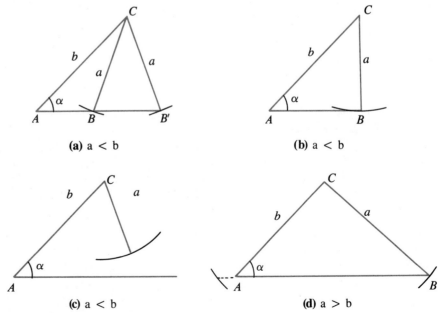

(a) a < b   **(b)** a < b

(c) a < b   **(d)** a > b

Figure 9.9

Note that $a < b$ in cases (a), (b), and (c), while $a > b$ in case (d). When $a < b$, the circular arc with center C and radius a may

1. intersect the horizontal axis in two points, determining two triangles (Figure 9.9a);
2. be tangent to the horizontal axis, determining a unique triangle (Figure 9.9b); or
3. not intersect the horizontal axis, in which case no triangle is formed (Figure 9.9c).

When $a > b$, only one triangle is determined (Figure 9.9d).

Application to Navigation

In navigation and surveying problems, *directions* or *bearings* are expressed by acute angles with initial sides along the north-south line and terminal sides in the indicated quadrant. For example, N35°E is the angle whose initial side is the northerly direction and whose measure is 35° eastward (Figure 9.10a). Similarly, S62°W is the angle whose initial side is the southerly direction and whose measure is 62° westward (Figure 9.10b).

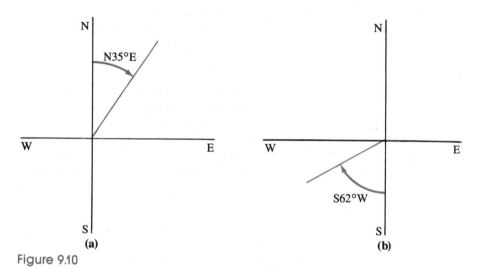

Figure 9.10

☉ EXAMPLE 3

A ship is sailing due south at a rate of 15 mph. At 12:00 noon it records the bearing from a lighthouse as S40°30′E. Three hours later the bearing from the same lighthouse is S72°15′E. How far is the ship from the lighthouse when the second bearing is recorded? Assuming that the ship follows the same course and speed, how close to the lighthouse will it pass?

Solution: Suppose that the lighthouse is located at C, as illustrated in Figure 9.11. At 12:00 noon the ship is at A, and at 3:00 P.M. the ship is at B. The distance from A to B is 45 miles and the angles α, β, and γ are 40°30′, 107°45′ ($= 180° - 72°15′$), and 31°45′, respectively. Using the law of sines, we get

$$\frac{a}{\sin 40°30′} = \frac{45}{\sin 31°45′},$$

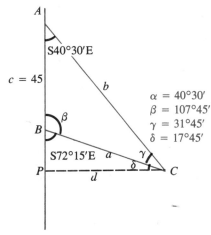

Figure 9.11

so

$$a = \frac{45 \sin 40.5°}{\sin 31.75°}.$$

With our calculator in degree mode, we perform the following keystrokes:

45 $\boxed{\times}$ 40.5 $\boxed{\sin}$ $\boxed{\div}$ 31.75 $\boxed{\sin}$ $\boxed{=}$

to obtain 55.538557. Rounding off gives us $a \simeq 56$ miles.

The closest distance d from the ship to the lighthouse is obtained by solving the right triangle BPC. Since the angle δ ($=\angle BCP$) measures $17°45' = 17.75°$ (why?), we have

$$d = a \cos 17.75°$$
$$= 55.538557 \times 0.9523958$$
$$= 52.894689$$
$$\simeq 53 \text{ miles}.$$

 Practice Exercise 3   An airplane is flying north at the speed of 350 km/h. At 9:00 A.M. it records the bearing of a radio tower as N10°20′W. One hour later, it records the bearing of the same radio tower as S45°30′W. How far is the airplane from the radio tower at 10:00 A.M.?

Answer: 76 km

EXERCISES 9.5

Solve each triangle in Exercises 1–12. Approximate each angle to the nearest ten minutes.

1. $\alpha = 38°$, $\beta = 75°$, $a = 12$

2. $\alpha = 45°$, $\beta = 75°$, $b = 15$

3. $\alpha = 105°$, $\beta = 15°$, $c = 9$

4. $\beta = 120°$, $\gamma = 35°$, $a = 10$

5. $\beta = 60°$, $\gamma = 45°$, $a = 8$
6. $\alpha = 30°$, $\gamma = 65°$, $c = 12$
7. $\beta = 42°10'$, $\gamma = 50°30'$, $a = 6$
8. $\alpha = 35°40'$, $\gamma = 18°20'$, $c = 18$
9. $\alpha = 30°$, $a = 4$, $b = 8$
10. $\beta = 60°$, $a = 4$, $b = 2\sqrt{3}$
11. $\beta = 60°$, $a = 4$, $b = 2$
12. $\alpha = 30°$, $a = 5$, $b = 12$

c In Exercises 13–24, use the law of sines to solve each triangle. Approximate the angles to the nearest minute and the side lengths to the nearest tenth.

13. $\alpha = 61°20'$, $\gamma = 38°30'$, $b = 5$
14. $\beta = 55°10'$, $\gamma = 47°20'$, $a = 14$
15. $\alpha = 35°$, $a = 6$, $b = 8$
16. $\beta = 65°$, $a = 16$, $b = 15$
17. $\alpha = 30°$, $a = 10$, $b = 8$
18. $\beta = 45°$, $a = 6$, $b = 12$
19. $\gamma = 65°$, $b = 9.2$, $c = 7.5$
20. $\alpha = 70°$, $a = 5.6$, $b = 8.4$
21. $\beta = 35°15'$, $\gamma = 42°30'$, $a = 5.2$
22. $\alpha = 62°30$, $\gamma = 42°15'$, $b = 4.5$
23. $\gamma = 40°20'$, $b = 10$, $c = 8$
24. $\alpha = 38°30'$, $a = 8$, $b = 10$

c In Exercises 25–34, solve each triangle. Approximate the angles to the nearest hundredth and the side lengths to two decimal places.

25. $\alpha = 42°$, $a = 12$, $b = 16$, $0° < \beta < 90°$
26. $\alpha = 75°$, $a = 44$, $b = 45$, $90° < \beta < 180°$
27. $\alpha = 28°40'$, $a = 18$, $b = 24$, $90° < \beta < 180°$
28. $\alpha = 31°30'$, $a = 25$, $b = 30$, $0° < \beta < 90°$
29. $\alpha = 30°$, $\beta = 86°$, $a = 5.75$
30. $\alpha = 45°$, $\beta = 60°$, $a = 10.25$
31. $\alpha = 35.5°$, $\beta = 62.25°$, $a = 13.5$
32. $\beta = 15.75°$, $\gamma = 81.4°$, $b = 21.6$
33. $\alpha = 12.5°$, $\gamma = 105.8°$, $c = 30.20$
34. $\beta = 40.6°$, $\alpha = 31.5°$, $a = 19.5$

In Exercises 35–38, find the area of each triangle.

35. $\beta = 30°$, $a = 2.5\,\text{m}$, $c = 12.8\,\text{m}$
36. $\alpha = 60°$, $b = 4\sqrt{3}\,\text{ft}$, $c = 5\,\text{ft}$
c 37. $\gamma = 38.5°$, $a = 40.6\,\text{cm}$, $b = 12.8\,\text{cm}$
c 38. $\beta = 42.1°$, $a = 21.5\,\text{in.}$, $c = 7.8\,\text{in.}$

In Exercises 39–50, solve the given problem.

39. A surveyor wants to determine the distance between two points A and C lying on opposite banks of a river. He chooses a point B situated 320 yards from A and measures the angles BAC and ABC to be 62° and 48°, respectively. Find the distance from A to C.

40. A pole 18 ft high leans away from the sun at an angle of 10° to the vertical. If the angle of elevation of the sun is 56°, find the length of the shadow cast by the pole on horizontal ground.

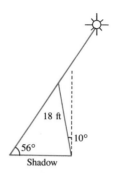

41. Two fire lookout stations A and B, which are 24 miles apart, spot a fire at C. If station A reports the fire at an angle $BAC = 60°20'$, and station B reports the fire at an angle $ABC = 35°40'$, how far is the fire from station A? From station B?

42. Two observers located 250 m apart at A and B on the bank of a river look at a point C on the opposite bank. If the angle CAB measures $46°30'$ and the angle ABC measures $53°20'$, find the distances from the two observers to the point C.

43. A vertical pole 45 ft long stands by the side of a road. The road makes a 12° angle with the horizontal. Find the length of the shadow cast by the pole directly downhill along the road when the angle of elevation of the sun is 36°.

44. Suppose that in problem 43 the angle of elevation of the sun is 60° and the length of the shadow is 24 ft. Find the length of the pole.

c 45. Two observation posts situated along the coast at points A and B, 8 km apart, detect a ship at a point C on the ocean. If the angle CAB measures $28°15'$ and the angle ABC measures $32°20'$, how far is the ship from observation post A? Assuming that the shore forms a straight line joining the two posts, how far is the ship from shore?

c 46. A cruiser sailing due east at a rate of 15 mph records the position of a radio beacon as N30°27′E. One hour later it records the position of the same radio beacon as N52°15′W. How far was the cruiser from the radio beacon at the first recording? How close to the radio beacon did the cruiser pass?

c 47. An airplane is flying due north at the speed of 150 mph. At 1 P.M. it records the bearing of a radio tower as N8°30′E. At 2 P.M. the bearing of the same radio tower is N70°15′E. How far is the airplane from the radio tower at 2 P.M.?

C 48. Two observers A and B on level ground and a balloon C in the sky are in the same vertical plane. The distance from A to B is 530 yards and the angles CAB and ABC are $26°18'$ and $48°15'$, respectively. Find **(a)** the distances from each observer to the balloon; and **(b)** the height of the balloon above the ground.

49. Prove the law of sines when all the angles of a triangle ABC are acute angles.

50. Prove that the area of triangle ABC in Figure 9.5 is given by $\dfrac{cb \sin \alpha}{2}$.

9.6   THE LAW OF COSINES

The Pythagorean theorem states that in a right triangle, the square of the length of the hypotenuse is equal to the sum of squares of the lengths of the sides. The *law of cosines,* which is valid in any triangle, generalizes that theorem as follows. (Refer to triangle ABC in Figure 9.12.)

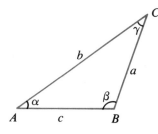

Figure 9.12

> ### Law of Cosines
>
> If ABC is a triangle with angles of measure α, β, and γ and sides of length a, b, and c opposite these angles, then
>
> $$a^2 = b^2 + c^2 - 2bc \cos \alpha,$$
> $$b^2 = a^2 + c^2 - 2ac \cos \beta,$$
> $$c^2 = a^2 + b^2 - 2ab \cos \gamma.$$
>
> In other words, the square of the length of any side is equal to the sum of squares of the lengths of the other two sides *minus* twice the product of the lengths of the other two sides and the cosine of the included angle.

In particular, if one of the angles of the triangle is $\pi/2$ radians or $90°$, we obtain the Pythagorean theorem. For example, if $\alpha = \pi/2$ (Figure 9.13), then $\cos \alpha = 0$ and the first of the three formulas becomes

$$a^2 = b^2 + c^2.$$

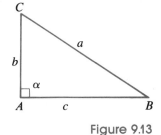

Figure 9.13

We now prove the first formula stated in the law of cosines. The proofs for the second and third formulas are left as exercises.

Consider a coordinate system as illustrated in Figure 9.14, so that the angle α of an arbitrary triangle ABC is in standard position. From formulas (8.16) of Section 8.2, it follows that the vertex C has coordinates $(b \cos \alpha, b \sin \alpha)$. Also, B has coordinates $(c, 0)$ and D has coordinates $(b \cos \alpha, 0)$. Since BDC is a right triangle, it follows by the Pythagorean theorem that

$$d(B, C)^2 = d(B, D)^2 + d(D, C)^2;$$

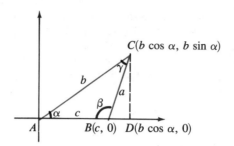

Figure 9.14

that is,

$$
\begin{aligned}
a^2 &= (b \cos \alpha - c)^2 + (b \sin \alpha)^2 \\
&= b^2 \cos^2\alpha - 2bc \cos \alpha + c^2 + b^2 \sin^2\alpha \\
&= b^2(\cos^2\alpha + \sin^2\alpha) + c^2 - 2bc \cos \alpha \\
&= b^2 + c^2 - 2bc \cos \alpha,
\end{aligned}
$$

which is the desired formula.

Solving Oblique Triangles

If we want to solve an oblique triangle, we would first try the law of sines, because it is simpler. However, in certain cases, we cannot use the law of sines directly. Such cases occur when we are given *two sides and the included angle* or *all three sides* of a triangle. First we have to use the law of cosines and *then* proceed with the law of sines, as shown in the following examples.

EXAMPLE 1   Solve the triangle ABC if $a = 4$, $b = 5$, and $c = 6$. Approximate the measures of the angles to the nearest minute.

Solution: Since the three sides are known, we have to use the law of cosines. It is advisable to find the largest angle first, in case it is obtuse. The largest angle is γ, facing the largest side $c = 6$. Thus

$$
\begin{aligned}
c^2 &= a^2 + b^2 - 2ab \cos \gamma \\
6^2 &= 4^2 + 5^2 - 2 \cdot 4 \cdot 5 \cos \gamma \\
36 &= 16 + 25 - 40 \cos \gamma \\
\cos \gamma &= \frac{1}{40}(16 + 25 - 36) \\
&= \frac{5}{40} = \frac{1}{8} = 0.125.
\end{aligned}
$$

Thus

$$
\gamma \simeq 82°49',
$$

At this point, we use the law of sines to find $\beta \simeq 55°46'$. Finally, we find the angle α: since $\alpha = 180° - (\beta + \gamma)$, it follows that $\alpha \simeq 41°25'$.

Practice Exercise 1 Find the angles of a triangle whose sides are $a = 3$, $b = 4$, and $c = 5$.

Answer: $\alpha \simeq 36.87° \simeq 36°52'$, $\beta \simeq 53.13° \simeq 53°8'$, $\gamma = 90°$

EXAMPLE 2 Solve the triangle ABC if $\alpha = 60°$, $b = 12$, and $c = 4$. (See Figure 9.15.)

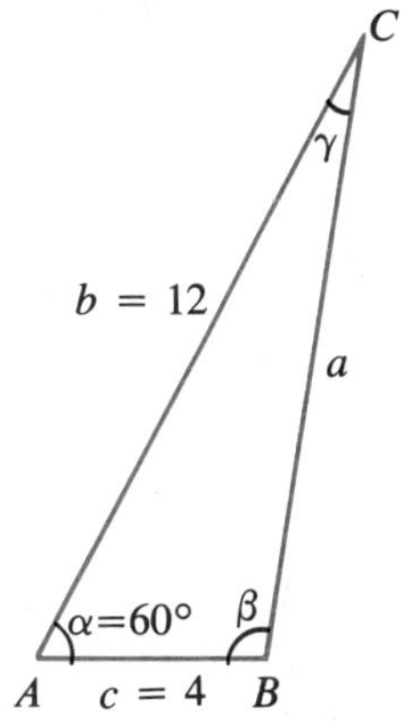

Figure 9.15

Solution: According to the law of cosines, we have

$$\begin{aligned} a^2 &= 12^2 + 4^2 - 2 \cdot 12 \cdot 4 \cos 60° \\ &= 144 + 16 - 48 \\ &= 112, \end{aligned}$$

so

$$a = \sqrt{112} = 4\sqrt{7} \simeq 11.$$

To determine the angles β and γ, we can use either the law of sines or the law of cosines. Let us use the law of sines, because the arithmetic computations are simpler. Also, it is better to find the angle γ opposite the shortest side $c = 4$, since this angle is acute and a table or calculator will give us the angle measurement directly. From (9.21), we have

$$\begin{aligned} \sin \gamma &= \frac{c \sin \alpha}{a} \\ &= \frac{4 \sin 60°}{4\sqrt{7}} \\ &= \frac{\sqrt{3}}{2\sqrt{7}} \simeq 0.3273, \end{aligned}$$

so

$$\gamma \simeq 19.11° \simeq 19°6'.$$

Finally, $\beta = 180° - (\alpha + \gamma)$, or

$$\begin{aligned} \beta &\simeq 180° - (60° + 19°6') \\ &\simeq 100°54'. \end{aligned}$$

Practice Exercise 2 Solve a triangle ABC if $a = 8$, $c = 4\sqrt{3}$, and $\beta = 30°$.

Answer: $b = 4$, $\alpha = 90°$, $\gamma = 60°$

C EXAMPLE 3 Let ABC be a triangle whose side lengths are $a = 9.48$ m, $b = 15.87$ m, and $c = 21.13$ m. Find γ to the nearest hundredth of a degree.

Solution: According to the law of cosines,

$$\begin{aligned} c^2 &= a^2 + b^2 - 2ab \cos \gamma \\ 2ab \cos \gamma &= a^2 + b^2 - c^2 \\ \cos \gamma &= \frac{a^2 + b^2 - c^2}{2ab}. \end{aligned}$$

Thus,

$$\cos \gamma = \frac{(9.48)^2 + (15.87)^2 - (21.13)^2}{2(9.48)(15.87)}.$$

The sequence of keystrokes

9.48 $\boxed{x^2}$ $\boxed{+}$ 15.87 $\boxed{x^2}$ $\boxed{-}$ 21.13 $\boxed{x^2}$ $\boxed{=}$ $\boxed{\div}$ $\boxed{(}$ 2 $\boxed{\times}$ 9.48 $\boxed{\times}$ 15.87 $\boxed{)}$ $\boxed{=}$

gives us

$$\cos \gamma = -0.3481265.$$

Next, we find γ. With the calculator in degree mode and displaying -0.3481265, we press the keys $\boxed{\text{INV}}$ $\boxed{\cos}$ to obtain

$$\gamma = 110.37277.$$

Rounding to the nearest hundredth, we get $\gamma \simeq 110.37°$.

Practice Exercise 3   In Example 3, find α to the nearest hundredth of a degree.

Answer: $\alpha \simeq 24.87°$

EXAMPLE **4**   The distance between two houses located near a pond cannot be measured across the water (Figure 9.16). A surveyor finds that the distances from a point A to each of the houses are 240 and 180 yards, and that the angle formed by the two lines from A to the houses is 30°. Determine the distance between the two houses.

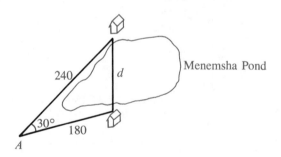

Figure 9.16

Solution: According to the law of cosines, if d denotes the distance between the two houses, then

$$d^2 = 240^2 + 180^2 - 2 \cdot 240 \cdot 180 \cos 30°$$
$$= 57600 + 32400 - 86400 \cdot \frac{\sqrt{3}}{2}$$
$$\simeq 57600 + 32400 - 74824.595$$
$$= 15175.405,$$

so

$$d \simeq \sqrt{15175.405}$$
$$\simeq 123 \text{ yards.}$$

Practice Exercise 4   Points A and B are on opposite sides of a lake. From a third point C, the distances AC and BC are 320 m and $150\sqrt{2}$ m, respectively. Find the distance AB if the angle between the lines of sight from C to A and to B is 45°.

Answer: 227 m

EXERCISES 9.6

In Exercises 1–8, solve each triangle. Approximate the angles to the nearest ten minutes.

1. $a = 5$, $b = 12$, $c = 13$
2. $a = 10$, $b = 6$, $c = 8$
3. $a = 5$, $b = 7$, $c = 9$
4. $a = 8$, $b = 10$, $c = 12$
5. $\beta = 45°$, $a = 3$, $c = 8$
6. $\gamma = 72°$, $a = 10$, $b = 15$
7. $\alpha = 110°$, $b = 12$, $c = 18$
8. $\beta = 150°$, $a = 8$, $c = 11$

In Exercises 9–16, solve each triangle. Approximate the angles to the nearest minute.

9. $a = 5.2$, $b = 8.3$, $c = 10.5$
10. $a = 2.5$, $b = 6$, $c = 6.5$
11. $a = 10.25$, $b = 5.60$, $c = 8.75$
12. $a = 5.15$, $b = 8.21$, $c = 7.25$
13. $\alpha = 56°18'$, $b = 5.62$, $c = 10.25$
14. $\beta = 32°30'$, $a = 12.05$, $c = 8.75$
15. $\beta = 105°25'$, $a = 4.75$, $c = 16.05$
16. $\gamma = 120°$, $a = 5.75$, $b = 10.25$

In Exercises 17–26, solve each triangle. Approximate the angles to the nearest hundredth of a degree and the side lengths to two decimal places.

17. $a = 12.5$, $b = 30$, $c = 32.5$
18. $a = 5.25$, $b = 12.05$, $c = 8.30$
19. $a = 4.25$, $b = 3.75$, $c = 5.18$
20. $a = 12.4$, $b = 5.15$, $c = 10.25$
21. $\beta = 42.5°$, $a = 3.10$, $c = 8.25$
22. $\alpha = 120.10°$, $b = 12.5$, $c = 7.18$
23. $\gamma = 75.12°$, $a = 10.5$, $b = 8.25$
24. $\beta = 80.5°$, $a = 5.8$, $c = 12.75$
25. $\alpha = 117.20°$, $b = 6.25$, $c = 8.45$
26. $\gamma = 100.25°$, $a = 6.25$, $b = 12.75$

In Exercises 27–46, solve the given problem.

27. The hour hand of a clock is 5 in. long and the minute hand is 8 in. long. Find the distance between the outer ends of the hour and minute hands at ten o'clock.

28. How far apart are the outer ends of the hour and minute hands of a clock at 4 P.M., if the hour hand is 3 cm long and the minute hand is 5 cm long?

29. An equilateral triangle is inscribed in a circle of radius 15 cm. How long is each side of the triangle?

30. A regular pentagon is inscribed in a circle of radius 10 m. How long is each side of the pentagon?

31. The sides of a triangle are $a = 4.75$ ft, $b = 10.75$ ft, and $c = 12.25$ ft. Find β to the nearest hundredth of a degree.

32. The sides of a triangle are $a = 10.50$ m, $b = 8.75$ m, and $c = 4.80$ m. Find γ to the nearest hundredth of a degree.

33. Find the distance between two points P and Q, knowing that a point R is 360 yards from P and 480 yards from Q, and the angle PRQ is 38°40'.

34. Two sides of a triangular plot are 150 m and 250 m long and form an angle of 42°30'. Determine the length of the third side. What is the area of the plot?

35. The sides of a parallelogram are 5 ft and 8 ft. One angle is 60° while another is 120°. What are the lengths of the diagonals?

36. Find the lengths of the diagonals of a parallelogram whose sides are 4 m and $2\sqrt{3}$ m and whose angles are 30° and 150°.

37. Two adjacent sides of a parallelogram meet at an angle of 58°30' and have lengths of 18 and 24 ft. Find the length of each diagonal.

38. The sides of a parallelogram are 120 in. and 156 in. long, and the largest angle measures 120°. Find the length of the shorter diagonal.

39. Two joggers starting from the same point run along two directions that make an angle of 24°. One of them runs at 8 mph and the other at 10 mph. How far apart will they be after 2 hours?

40. Two ships leave Boston at the same time on courses 36° apart. One ship travels at 24 mph and the other at 30 mph. How far apart are the ships after 1 hour?

41. Points A and B are 6 miles apart. If Mary runs for 2 miles along a straight line from A making an angle of 5.4° with AB, how far will she be from point B?

42. New York and Washington are approximately 200 miles apart. An airplane 105 miles from New York is 6°40' off course. How far is the airplane from Washington?

43. An airplane flies 95 miles from an airport in the direction 150° and then flies 185 miles in the direction 225°. How far is the plane from the airport?

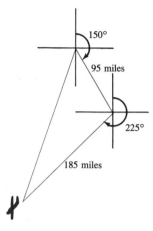

44. Two airplanes leave Kennedy Airport at 12:00 noon. One travels in the direction 65° at 420 mph and the other travels in the direction 220° at 500 mph. At 2:00 P.M., how far apart will the airplanes be?

45. Two boats leave a dock at the same time. One travels in the direction N35°20′E at 20 km/hr. The other travels in the direction S65°40′E at 35 km/hr. After two hours, how far apart are the two boats?

46. A surveyor stands at point *A* located 356 m from one end, *B*, of a pond and 475 meters from the other end, *C*, of the pond. The angle *BAC* is measured to be 110°15′. Find the distance from *B* to *C*.

Heron's Area Formula If *a*, *b*, and *c* are the lengths of the sides of a triangle, then the area *A* of the triangle is given by the formula

$$A = \sqrt{s(s-a)(s-b)(s-c)},$$

where

$$s = \frac{1}{2}(a+b+c)$$

is the *semiperimeter* of the triangle. This is called *Heron's area formula,* and it can be derived from the law of cosines.

In Exercises 47–54, use Heron's area formula to find the area of each triangle.

47. *a* = 9 m, b = 40 m, *c* = 41 m

48. *a* = 21 cm, *b* = 20 cm, *c* = 29 cm

49. *a* = 15 in., *b* = 9 in., *c* = 18 in.

50. *a* = 10 ft, *b* = 25 ft, *c* = 31 ft

51. *a* = 12.5 yards, *b* = 8.9 yards, *c* = 13.4 yards

52. *a* = 25.4 ft, *b* = 10.3 ft, *c* = 18.8 ft

53. What is the area of a triangular plot measuring 75 m by 40 m by 85 m?

54. The lengths of the sides of a triangular cornfield are 189 yards, 180 yards, and 261 yards. Find the area of the field.

In Exercises 55–60, the letters *a*, *b*, and *c* represent the lengths of the sides of a triangle, *α*, *β*, and *γ* the measures of the corresponding angles, and *s* the semiperimeter.

55. Prove that $1 + \cos \alpha = \dfrac{(b+c+a)(b+c-a)}{2bc}$.

[*Hint*: Use the law of cosines.]

56. Prove that $1 - \cos \alpha = \dfrac{(a+b-c)(a-b+c)}{2bc}$.

[*Hint*: Use the law of cosines.]

57. Prove that $\cos \dfrac{\alpha}{2} = \sqrt{\dfrac{s(s-a)}{bc}}$.

$\left[\textit{Hint: } \cos^2\left(\dfrac{\alpha}{2}\right) = \dfrac{1 + \cos \alpha}{2}\right]$

58. Prove that $\sin \dfrac{\alpha}{2} = \sqrt{\dfrac{(s-b)(s-c)}{bc}}$.

$\left[\textit{Hint: } \sin^2\left(\dfrac{\alpha}{2}\right) = \dfrac{1 - \cos \alpha}{2}\right]$

59. Let $A = \dfrac{1}{2}bc \sin \alpha$ be the area of the triangle. (See Exercise 50, Section 9.5.) Show that the area can be rewritten

$$A = \frac{bc}{2}\sqrt{(1 - \cos \alpha)(1 + \cos \alpha)}.$$

60. Use the formula $A = \dfrac{1}{2}bc \sin \alpha$ and Exercises 57, 58, and 59 to prove Heron's area formula:

$$A = \sqrt{s(s-a)(s-b)(s-c)}.$$

9.7   VECTORS IN THE PLANE

In the physical world we encounter two types of quantities: *scalar quantities* and *vector quantities*. A scalar quantity is completely specified by a single real number and an appropriate unit of measurement. As examples we mention mass, volume, temperature, the price of an item, the loudness of sound. Scalar quantities have magnitude only and obey the familiar rules of algebra.

On the other hand, vector quantities, each of which is represented by several numbers, have *direction* and *magnitude*. Also, they obey the rules of *vector algebra*, some of which will be described below. As examples of vector quantities, we mention the displacement of a body, velocity, force, and acceleration.

Two-Vectors

A *vector* **v** *in the plane* (or a *two-vector*) is an ordered pair $\langle a, b \rangle$ of real numbers. We write $\mathbf{v} = \langle a, b \rangle$ and call a and b the *components* of the vector **v**.

For example, $\mathbf{u} = \langle -3, 2 \rangle$, $\mathbf{v} = \langle 4, -1 \rangle$, $\mathbf{i} = \langle 1, 0 \rangle$ are vectors in the plane. In a similar manner, *vectors in space* (or *three-vectors*) are defined as ordered triples $\langle a, b, c \rangle$ of real numbers. More generally, an *n-vector* **v** is an ordered *n*-tuple $\langle v_1, v_2, \ldots, v_n \rangle$ of real numbers. The number v_i is called the *i*th *component* of the vector **v**. Throughout this book, we shall consider only two-vectors and their properties. Vectors in space and *n*-vectors are studied in more advanced courses.

Geometric Representation

Up to now, we have used ordered pairs of real numbers to label points in a Cartesian plane. Now vectors have also been defined as ordered pairs of real numbers (although with a different notation). Does that mean that vectors and points are the same mathematical concept? The answer is negative, and there is a subtle and important difference between them, which we will now explain. When we say that $\mathbf{v} = \langle a, b \rangle$ is a vector, we are not interpreting the ordered pair $\langle a, b \rangle$ as a point but as a *change in position*: $|a|$ units horizontally (to the right if $a > 0$, to the left if $a < 0$) and $|b|$ units vertically (up if $b > 0$, down if $b < 0$). For this reason, a two-vector $\mathbf{v} = \langle a, b \rangle$ can be represented by a *directed line segment* **PQ** (Figure 9.17), originating at *any* point P with coordinates (x, y) and proceeding to Q, a point with coordinates $(x + a, y + b)$. We call P the *initial point* and Q the *terminal point* of the vector **v**.

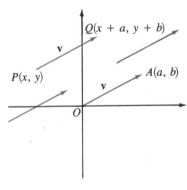

Figure 9.17

Figure 9.17 also illustrates several other directed line segments representing **v**. For simplicity and when no confusion is possible, we say "vector **PQ**" instead of saying "**PQ** is a directed line segment representing **v**." However, you should keep in mind that a vector is an abstract notion that specifies change in position and that there are infinitely many ways of representing a vector by a directed line segment.

In particular, the directed line segment **OA**, originating at the origin and ending at the point $A(a, b)$, is also a geometric representation of **v**. Note in this representation the close relationship between **v** and the terminal point A: the numbers a and b are both the *components* of **v** and the *coordinates* of A. When **v** is represented by **OA**, we say that **v** (or **OA**) is a *position vector*. The terminal point of a position vector $\mathbf{v} = \langle a, b \rangle$ always has coordinates (a, b).

EXAMPLE **1**

Given $P(-1, 3)$ and $Q(2, -4)$, find the vector represented by the directed line segment **PQ**.

Solution: If $\mathbf{v} = \langle a, b \rangle$ is the vector represented by **PQ**, then we have

$$2 = -1 + a \quad \text{and} \quad -4 = 3 + b.$$

Thus

$$a = 3 \quad \text{and} \quad b = -7.$$

Practice Exercise 1

Given $\mathbf{v} = \langle 3, -2 \rangle$ and $P(-1, 4)$, find the coordinates of a point Q so that **PQ** is a geometric representation of $\mathbf{v}$.

Answer: $Q(2, 2)$

Magnitude of a Vector

If $\mathbf{v} = \langle a, b \rangle$ is a two-vector, the *magnitude* of $\mathbf{v}$, denoted by $|\mathbf{v}|$, is defined to be the number

$$|\mathbf{v}| = \sqrt{a^2 + b^2}.$$

The magnitude of $\mathbf{v}$ is also called the *length* or the *norm* of $\mathbf{v}$.

If $\mathbf{v}$ is represented by the directed line segment **OA**, the magnitude of $\mathbf{v}$ corresponds to *the distance from A to the origin, O*. If $\mathbf{v}$ is represented by **PQ**, then the magnitude of $\mathbf{v}$ is *the length of the segment PQ* or *the distance from P to Q*.

The Zero Vector

The vector with both components equal to zero is called the *zero vector* and denoted by $\mathbf{O} = \langle 0, 0 \rangle$.

The zero vector may be represented by *any* point, and it has *no specific direction*. Besides, it is the only vector whose length is zero.

Equality of Two Vectors

If $\mathbf{u} = \langle u_1, u_2 \rangle$ and $\mathbf{v} = \langle v_1, v_2 \rangle$ are two vectors in the plane, we say $\mathbf{u} = \mathbf{v}$ if and only if $u_1 = v_1$ and $u_2 = v_2$.

In other words, two vectors are *equal* exactly when they have the *same corresponding components*.

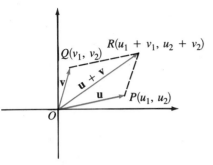

Figure 9.18

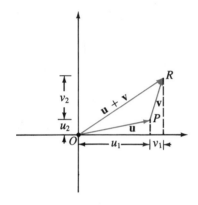

Figure 9.19

Addition of Vectors

If $\mathbf{u} = \langle u_1, u_2 \rangle$ and $\mathbf{v} = \langle v_1, v_2 \rangle$ are two vectors in the plane, we define the *sum* of $\mathbf{u}$ and $\mathbf{v}$, written $\mathbf{u} + \mathbf{v}$, as follows:

$$\mathbf{u} + \mathbf{v} = \langle u_1 + v_1, u_2 + v_2 \rangle.$$

Thus we add vectors by *adding the corresponding components*. For example, $\langle -3, 2 \rangle + \langle 4, -1 \rangle = \langle 1, 1 \rangle$ and $\langle -8, 3 \rangle + \langle 1, 0 \rangle = \langle -7, 3 \rangle$.

The sum $\mathbf{u} + \mathbf{v}$ is also called the *resultant* of the vectors $\mathbf{u}$ and $\mathbf{v}$. Each of the vectors is called a *component* of the sum.

Geometric Interpretation of Vector Addition

If $\mathbf{u} = \langle u_1, u_2 \rangle$ and $\mathbf{v} = \langle v_1, v_2 \rangle$ are represented by the directed line segments **OP** and **OQ**, respectively (Figure 9.18), then it can be shown that the sum $\mathbf{u} + \mathbf{v}$ can be represented by the directed line segment **OR**, the diagonal of the parallelogram *OPRQ*. This is called the *parallelogram law of vector addition*.

As we already know, the directed segments **OQ** and **PR** represent the same vector $\mathbf{v}$. Thus we may think of the sum of $\mathbf{u}$ and $\mathbf{v}$ as being performed by translating the directed line segment $\mathbf{v}$ parallel to itself, so that it originates at the endpoint of the directed segment representing $\mathbf{u}$. The resulting directed line segment represents $\mathbf{u} + \mathbf{v}$, as shown in Figure 9.19. This is called the *tail-to-tip method* of adding vectors. We may also write **OR** = **OP** + **PR**.

EXAMPLE 2   Let $A(-3, 1)$, $B(4, 2)$, and $C(-1, 5)$ be three points in the plane.
(a) Find the sum of the vectors represented by the directed line segments **AB** and **AC**.

(b) If **AD** represents the sum of the two vectors in part a, find the coordinates of D.

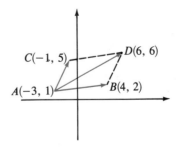

Figure 9.20

Solution: **(a)** If $\mathbf{v}$ and $\mathbf{w}$ denote the vectors represented by **AB** and **AC**, respectively, then $\mathbf{v} = \langle 7, 1 \rangle$ and $\mathbf{w} = \langle 2, 4 \rangle$. Thus, $\mathbf{v} + \mathbf{w} = \langle 9, 5 \rangle$.

(b) If x and y are the coordinates of D, then $x = -3 + 9 = 6$ and $y = 1 + 5 = 6$.

Practice Exercise 2

Let A, B, and C be the points considered in Example 2.
(a) Find the vector **BD** so that **BD** = **BA** + **BC**.
(b) What are the coordinates of D?

Answer: **(a)** **BD** = $\langle -12, 2 \rangle$　**(b)** $D = (-8, 4)$

EXAMPLE　3

Two forces $\mathbf{f}_1$ and $\mathbf{f}_2$ of magnitudes 5 and 6 N, respectively, act on a point and make an angle of 60°. Find the magnitude of the resultant of the two forces.

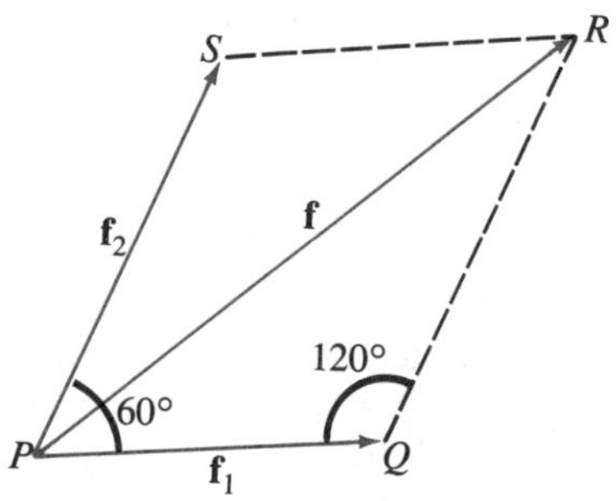

Solution: The resultant **f**, illustrated in Figure 9.21, is the sum of the two forces $\mathbf{f}_1$ and $\mathbf{f}_2$. The magnitude of **f** is the length of the diagonal PR of the parallelogram $PQRS$. Since, in the triangle PQR, the angle Q measures 120°, we can apply the law of cosines:

$$|\mathbf{f}|^2 = 5^2 + 6^2 - 2(5)(6)\cos 120°$$
$$= 25 + 36 + 30$$
$$= 91.$$

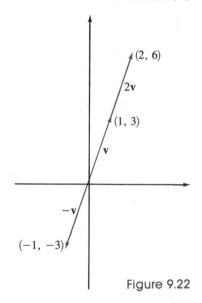

Figure 9.21

Thus the magnitude of the resultant is

$$|\mathbf{f}| = \sqrt{91} \approx 9.54.$$

Practice Exercise 3

Suppose that the two forces considered in Example 3 make a 120° angle. What is the magnitude of the resultant?

Answer: $\sqrt{31} \approx 5.57$

Scalar Multiplication

If $\mathbf{v} = \langle v_1, v_2 \rangle$ is a vector and α is a real number, then the *scalar product* (or *scalar multiple*) $\alpha\mathbf{v}$ is the vector

$$\alpha\mathbf{v} = \langle \alpha v_1, \alpha v_2 \rangle.$$

Thus to multiply a vector by a real number, we *multiply the components of the vector by the real number.* Notice that

$$|\alpha\mathbf{v}| = \sqrt{(\alpha v_1)^2 + (\alpha v_2)^2} = |\alpha|\sqrt{v_1^2 + v_2^2} = |\alpha|\,|\mathbf{v}|;$$

that is, *the magnitude of the scalar multiple $\alpha\mathbf{v}$ is $|\alpha|$ times the magnitude of* **v** .

Figure 9.22

EXAMPLE　4

If $\mathbf{v} = \langle 1, 3 \rangle$, find $2\mathbf{v}$ and $(-1)\mathbf{v}$. Represent these vectors geometrically and find their magnitudes.

Solution: If $\mathbf{v} = \langle 1, 3 \rangle$, then by the definition of the scalar product, we obtain

$$2\mathbf{v} = \langle 2, 6 \rangle \quad \text{and} \quad (-1)\mathbf{v} = \langle -1, -3 \rangle \quad \text{(See Figure 9.22.)}$$

The magnitudes of these vectors are

$$|\mathbf{v}| = \sqrt{1^2 + 3^2} = \sqrt{10},$$
$$|2\mathbf{v}| = \sqrt{2^2 + 6^2} = \sqrt{40} = 2\sqrt{10},$$
$$|-1\mathbf{v}| = \sqrt{1^2 + 3^2} = \sqrt{10}.$$

Practice Exercise 4   Let $\mathbf{v} = \langle 2, -1 \rangle$. Find $3\mathbf{v}$, $-2\mathbf{v}$, and the magnitudes of these vectors.

Answer: $3\mathbf{v} = \langle 6, -3 \rangle$, $-2\mathbf{v} = \langle -4, 2 \rangle$, $|3\mathbf{v}| = 3\sqrt{5}$, $|-2\mathbf{v}| = 2\sqrt{5}$

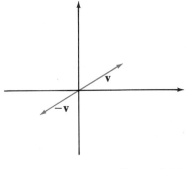

Figure 9.23

The Negative of a Vector

If $\mathbf{v} = \langle v_1, v_2 \rangle$ is a two-vector, we define $-\mathbf{v} = (-1)\mathbf{v} = \langle -v_1, -v_2 \rangle$ as the *negative* of the vector $\mathbf{v}$.

The vector $-\mathbf{v}$ has the *same magnitude* as $\mathbf{v}$ but *opposite direction*, as shown in Figure 9.23.

As we mentioned earlier, the scalar product $\alpha\mathbf{v}$ has magnitude $|\alpha|$ times the magnitude of $|\mathbf{v}|$. It has the *same* direction as $\mathbf{v}$ if α is *positive*, and the *opposite* direction if α is *negative*. (See Figure 9.24.)

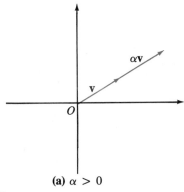

(a) $\alpha > 0$

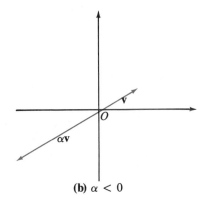

(b) $\alpha < 0$

Figure 9.24

Properties of Vector Addition

If $\mathbf{u}$, $\mathbf{v}$, and $\mathbf{w}$ are vectors, then the following properties hold.

$$\mathbf{u} + \mathbf{v} = \mathbf{v} + \mathbf{u}$$

$$(\mathbf{u} + \mathbf{v}) + \mathbf{w} = \mathbf{u} + (\mathbf{v} + \mathbf{w})$$

$$\mathbf{u} + \mathbf{O} = \mathbf{u}$$

$$\mathbf{u} + (-\mathbf{u}) = \mathbf{O}$$

As you see, addition of vectors satisfies the same properties as addition of real numbers. The first property expresses the *commutativity* of the sum. The second is the *associative property*. The third indicates that the zero vector, **O**, is the *additive identity* for addition. Finally, the fourth property tells us that $-\mathbf{u}$ is the *additive inverse* of **u**.

Each one of these properties follows immediately from the definition of vector sum and the properties of real numbers. To show the first one, if $\mathbf{u} = \langle u_1, u_2 \rangle$ and $\mathbf{v} = \langle v_1, v_2 \rangle$, then

$$
\begin{aligned}
\mathbf{u} + \mathbf{v} &= \langle u_1, u_2 \rangle + \langle v_1, v_2 \rangle \\
&= \langle u_1 + v_1, u_2 + v_2 \rangle \\
&= \langle v_1 + u_1, v_2 + u_2 \rangle \quad \text{Commutativity of real number addition} \\
&= \langle v_1, v_2 \rangle + \langle u_1, u_2 \rangle \\
&= \mathbf{v} + \mathbf{u}.
\end{aligned}
$$

The proofs of the other properties are left as exercises.

Difference of Vectors

The *difference* $\mathbf{u} - \mathbf{v}$ of the vectors $\mathbf{u} = \langle u_1, u_2 \rangle$ and $\mathbf{v} = \langle v_1, v_2 \rangle$ is defined by

$$
\mathbf{u} - \mathbf{v} = \mathbf{u} + (-\mathbf{v}) = \langle u_1 - v_1, u_2 - v_2 \rangle.
$$

That is, the difference $\mathbf{u} - \mathbf{v}$ is the *sum of* **u** *and the additive inverse* (*negative*) *of* **v**.

EXAMPLE 5   If $\mathbf{u} = \langle 1, -2 \rangle$ and $\mathbf{v} = \langle -3, 2 \rangle$, find $2\mathbf{u} - 3\mathbf{v}$.

Solution: We have

$$
\begin{aligned}
2\mathbf{u} - 3\mathbf{v} &= 2\langle 1, -2 \rangle - 3\langle -3, 2 \rangle \\
&= \langle 2, -4 \rangle - \langle -9, 6 \rangle \\
&= \langle 2 - (-9), -4 - 6 \rangle \\
&= \langle 11, -10 \rangle.
\end{aligned}
$$

Practice Exercise 5   Let $\mathbf{v} = \langle 2, -2 \rangle$ and $\mathbf{w} = \langle 3, 1 \rangle$. Find $3\mathbf{v} - 4\mathbf{w}$.

Answer: $\langle -6, -10 \rangle$

Properties of Scalar Multiplication

If α and β are real numbers and **u** and **v** are vectors, then the following properties hold.

$$
\begin{aligned}
\alpha(\mathbf{u} + \mathbf{v}) &= \alpha\mathbf{u} + \alpha\mathbf{v} \\
(\alpha + \beta)\mathbf{u} &= \alpha\mathbf{u} + \beta\mathbf{u} \\
\alpha(\beta\mathbf{u}) &= (\alpha\beta)\mathbf{u}
\end{aligned}
$$

To show the first property, let $\mathbf{u} = \langle u_1, u_2 \rangle$ and $\mathbf{v} = \langle v_1, v_2 \rangle$. We have

$$\alpha(\mathbf{u} + \mathbf{v}) = \alpha\langle u_1 + v_1, u_2 + v_2 \rangle$$
$$= \langle \alpha(u_1 + v_1), \alpha(u_2 + v_2) \rangle$$
$$= \langle \alpha u_1 + \alpha v_1, \alpha u_2 + \alpha v_2 \rangle$$
$$= \langle \alpha u_1, \alpha u_2 \rangle + \langle \alpha v_1, \alpha v_2 \rangle$$
$$= \alpha\mathbf{u} + \alpha\mathbf{v}.$$

The proofs of the other two properties are left as exercises.

Unit Vectors

A *unit vector* $\mathbf{u}$ is one with magnitude 1, that is, $|\mathbf{u}| = 1$.

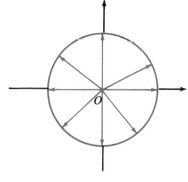

The endpoints of all position unit vectors in the plane lie on the unit circle.

Figure 9.25

Given any vector $\mathbf{v} \neq \mathbf{O}$, there is always a unit vector that has the same direction as $\mathbf{v}$. It is the vector $\mathbf{u} = \mathbf{v}/|\mathbf{v}|$. In fact, since $1/|\mathbf{v}|$ is a *positive* number (why?), it follows that $\mathbf{u} = \mathbf{v}/|\mathbf{v}|$ has same direction as $\mathbf{v}$. To see that $\mathbf{u}$ is a unit vector, we compute its length:

$$|\mathbf{u}| = \sqrt{\left(\frac{v_1}{|\mathbf{v}|}\right)^2 + \left(\frac{v_2}{|\mathbf{v}|}\right)^2} = \frac{\sqrt{v_1^2 + v_2^2}}{|\mathbf{v}|} = 1.$$

EXAMPLE 6

Find the unit vector in the direction of the vector $\mathbf{v} = \langle -3, 1 \rangle$.

Solution: First, we find the length of $\mathbf{v}$:

$$|\mathbf{v}| = \sqrt{(-3)^2 + 1^2} = \sqrt{10}.\bullet$$

The unit vector $\mathbf{u}$ in the direction of $\mathbf{v}$ is the scalar multiple $1/\sqrt{10}$ of $\mathbf{v}$:

$$\mathbf{u} = (1/\sqrt{10})\mathbf{v} = \langle -3/\sqrt{10}, 1/\sqrt{10} \rangle.$$

Practice Exercise 6

What is the unit vector in the direction of $\mathbf{w} = \langle 3, -4 \rangle$?

Answer: $\mathbf{u} = \langle 3/5, -4/5 \rangle$

The Unit Vectors i and j

Among all unit vectors in the plane, there are two that play a special and important role. They are the vectors $\mathbf{i} = \langle 1, 0 \rangle$, which points in the positive x-direction, and $\mathbf{j} = \langle 0, 1 \rangle$, which points in the positive y-direction (Figure 9.26).

Every vector $\mathbf{v}$ can be written as a sum of scalar multiples of $\mathbf{i}$ and $\mathbf{j}$. For if $\mathbf{v} = \langle v_1, v_2 \rangle$, then

$$\mathbf{v} = \langle v_1, 0 \rangle + \langle 0, v_2 \rangle$$
$$= v_1\langle 1, 0 \rangle + v_2\langle 0, 1 \rangle$$
$$= v_1\mathbf{i} + v_2\mathbf{j}.$$

We say that $\mathbf{v}$ is a *linear combination* of $\mathbf{i}$ and $\mathbf{j}$.

This notation is very useful in manipulations involving sums and scalar products of vectors. For example, if

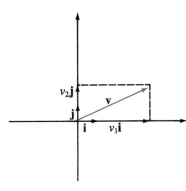

Figure 9.26

$$\mathbf{u} = u_1\mathbf{i} + u_2\mathbf{j} \qquad \text{and} \qquad \mathbf{v} = v_1\mathbf{i} + v_2\mathbf{j},$$

then

$$\begin{aligned}
\mathbf{u} + \mathbf{v} &= (u_1\mathbf{i} + u_2\mathbf{j}) + (v_1\mathbf{i} + v_2\mathbf{j}) \\
&= (u_1 + v_1)\mathbf{i} + (u_2 + v_2)\mathbf{j}
\end{aligned}$$

and

$$\begin{aligned}
\alpha\mathbf{u} &= \alpha(u_1\mathbf{i} + u_2\mathbf{j}) \\
&= (\alpha u_1)\mathbf{i} + (\alpha u_2)\mathbf{j}.
\end{aligned}$$

EXAMPLE 7

If $\mathbf{u} = 3\mathbf{i} + 2\mathbf{j}$ and $\mathbf{v} = 2\mathbf{i} - 4\mathbf{j}$, express $5\mathbf{u} - 2\mathbf{v}$ as a linear combination of $\mathbf{i}$ and $\mathbf{j}$.

Solution:

$$\begin{aligned}
5\mathbf{u} - 2\mathbf{v} &= 5(3\mathbf{i} + 2\mathbf{j}) - 2(2\mathbf{i} - 4\mathbf{j}) \\
&= 15\mathbf{i} + 10\mathbf{j} - 4\mathbf{i} + 8\mathbf{j} \\
&= (15 - 4)\mathbf{i} + (10 + 8)\mathbf{j} \\
&= 11\mathbf{i} + 18\mathbf{j}.
\end{aligned}$$

Practice Exercise 7

Let $\mathbf{a} = -2\mathbf{i} + 3\mathbf{j}$ and $\mathbf{b} = 5\mathbf{i} - 2\mathbf{j}$. Write $4\mathbf{a} + 2\mathbf{b}$ as a linear combination of $\mathbf{i}$ and $\mathbf{j}$.

Answer: $2\mathbf{i} + 8\mathbf{j}$

Horizontal and Vertical Components

When a vector $\mathbf{v} = \langle v_1, v_2 \rangle$ is written in the form

$$\mathbf{v} = v_1\mathbf{i} + v_2\mathbf{j},$$

we call $v_1\mathbf{i}$ the *horizontal component* and $v_2\mathbf{j}$ the *vertical component* of $\mathbf{v}$. We also say that the vector has been *decomposed* (or *resolved*) into its horizontal and vertical components. This is an important technique frequently used in the study and application of vectors.

Direction Angle

Let $\mathbf{v} = \langle v_1, v_2 \rangle$ be the position vector illustrated in Figure 9.27. The angle θ (measured in the counterclockwise direction), whose initial side is the positive

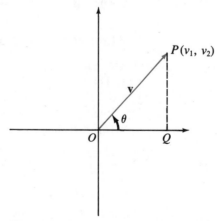

Figure 9.27

x-axis and terminal side the segment OP, is called the *direction angle* of **v**. From the triangle OQP, we obtain

$$\cos \theta = \frac{v_1}{|\mathbf{v}|} \quad \text{and} \quad \sin \theta = \frac{v_2}{|\mathbf{v}|},$$

or

$$v_1 = |\mathbf{v}| \cos \theta \quad \text{and} \quad v_2 = |\mathbf{v}| \sin \theta.$$

These two formulas express the components of a vector in terms of the *norm* and *direction angle* of the vector.

EXAMPLE 8

The length of a vector **v** is 5 units and its direction angle measures 30°. Write **v** as the sum of its horizontal and vertical components.

Solution: The horizontal and vertical components of **v** are obtained as follows:

$$v_1 = 5 \cos 30° = \frac{5\sqrt{3}}{2} \quad \text{and} \quad v_2 = 5 \sin 30° = \frac{5}{2}.$$

Thus

$$v = \left(\frac{5\sqrt{3}}{2}\right)\mathbf{i} + \left(\frac{5}{2}\right)\mathbf{j}.$$

Practice Exercise 8

If $\mathbf{u} = -2\mathbf{i} + 2\sqrt{3}\mathbf{j}$, find the direction angle of **u**.

Answer: 120° or $2\pi/3$ rad

The Inner Product

Let $\mathbf{u} = \langle u_1, u_2 \rangle$ and $\mathbf{v} = \langle v_1, v_2 \rangle$ be two vectors in the plane. The *inner product* (or *dot product* or *scalar product*) of **u** and **v** is defined by

$$\mathbf{u} \cdot \mathbf{v} = u_1 v_1 + u_2 v_2.$$

For example, if $\mathbf{u} = \langle 4, 1 \rangle$ and $\mathbf{v} = \langle 3, -5 \rangle$, then $\mathbf{u} \cdot \mathbf{v} = (4)(3) + (1)(-5) = 7$. Also, if $\mathbf{a} = -5\mathbf{i} + 2\mathbf{j}$ and $\mathbf{b} = 3\mathbf{i} - 2\mathbf{j}$, then $\mathbf{a} \cdot \mathbf{b} = (-5)(3) + (2)(-2) = -19$. Notice that the inner product of two vectors is a *scalar quantity*.

Properties of the Inner Product

If **u**, **v**, **w** are vectors and λ is a real number, then the following properties hold.

$$\mathbf{u} \cdot \mathbf{u} = |\mathbf{u}|^2$$
$$\mathbf{u} \cdot \mathbf{v} = \mathbf{v} \cdot \mathbf{u}$$
$$\mathbf{u} \cdot (\mathbf{v} + \mathbf{w}) = \mathbf{u} \cdot \mathbf{v} + \mathbf{u} \cdot \mathbf{w}$$
$$(\lambda \mathbf{u}) \cdot \mathbf{v} = \lambda (\mathbf{u} \cdot \mathbf{v})$$

To prove the first property, let $\mathbf{u} = \langle u_1, u_2 \rangle$. Then,

$$\mathbf{u} \cdot \mathbf{u} = u_1 u_1 + u_2 u_2 = u_1^2 + u_2^2 = |\mathbf{u}|^2,$$

by the definition of vector magnitude. This property of the inner product can be used to redefine the magnitude of a vector:

$$|\mathbf{u}| = \sqrt{\mathbf{u} \cdot \mathbf{u}}.$$

The proofs of the other properties are left as exercises.

Geometric Interpretation

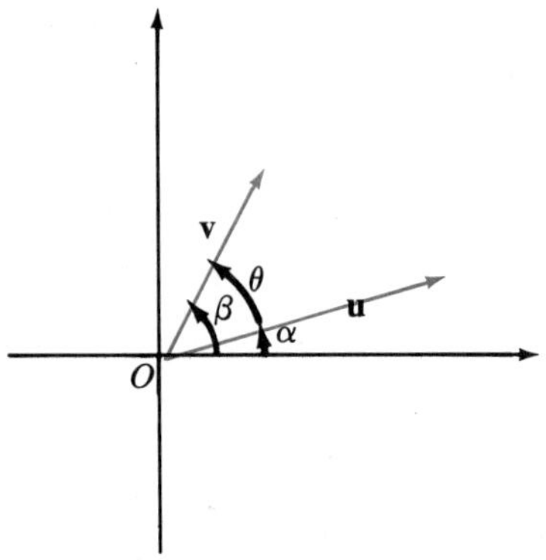

Let $\mathbf{u} = \langle u_1, u_2 \rangle$ and $\mathbf{v} = \langle v_1, v_2 \rangle$ be two vectors in the plane, and let θ be the angle between them, as illustrated in Figure 9.28. Let α and β be the direction angles of the vectors $\mathbf{u}$ and $\mathbf{v}$, respectively.

As we already know,

$$\cos \alpha = \frac{u_1}{|\mathbf{u}|}, \qquad \sin \alpha = \frac{u_2}{|\mathbf{u}|},$$

and

$$\cos \beta = \frac{v_1}{|\mathbf{v}|}, \qquad \sin \beta = \frac{v_2}{|\mathbf{v}|}.$$

Figure 9.28

Since $\theta = \beta - \alpha$, we use the formula for the cosine of a difference to obtain

$$
\begin{aligned}
\cos \theta = \cos(\beta - \alpha) &= \cos \beta \cos \alpha + \sin \beta \sin \alpha \\
&= \left(\frac{v_1}{|\mathbf{v}|}\right)\left(\frac{u_1}{|\mathbf{u}|}\right) + \left(\frac{v_2}{|\mathbf{v}|}\right)\left(\frac{u_2}{|\mathbf{u}|}\right) \\
&= \frac{u_1 v_1 + u_2 v_2}{|\mathbf{u}||\mathbf{v}|} \\
&= \frac{\mathbf{u} \cdot \mathbf{v}}{|\mathbf{u}||\mathbf{v}|}.
\end{aligned}
$$

Hence

$$\mathbf{u} \cdot \mathbf{v} = |\mathbf{u}||\mathbf{v}| \cos \theta,$$

that is, *the dot product of two vectors is the product of their lengths and the cosine of the angle between them.*

EXAMPLE 9 Find the angle between the vectors $\mathbf{u} = 2\mathbf{i} + \mathbf{j}$ and $\mathbf{v} = 3\mathbf{i} - 2\mathbf{j}$.

Solution: We have

$$
\begin{aligned}
\mathbf{u} \cdot \mathbf{v} &= (2)(3) + (1)(-2) = 4, \\
|\mathbf{u}| &= \sqrt{2^2 + 1^2} = \sqrt{5}, \\
|\mathbf{v}| &= \sqrt{3^2 + (-2)^2} = \sqrt{13}.
\end{aligned}
$$

Thus

$$\cos \theta = \frac{4}{\sqrt{5}\sqrt{13}} = \frac{4}{\sqrt{65}} \simeq 0.4961$$

and

$$\theta = \cos^{-1}\left(\frac{4}{\sqrt{65}}\right) \simeq 60.26°.$$

Practice Exercise 9   Let $\mathbf{a} = \langle -2, 2 \rangle$ and $\mathbf{b} = \langle 3, -1 \rangle$. Find **(a)** the cosine of the angle between these two vectors, and **(b)** the angle between them.

Answer: **(a)** $\cos \theta = -2/\sqrt{5}$ **(b)** $\theta \simeq 153.4°$

Orthogonality

From the formula $\mathbf{u} \cdot \mathbf{v} = |\mathbf{u}||\mathbf{v}| \cos \theta$, it follows that $\mathbf{u} \cdot \mathbf{v} = 0$ if and only if $\cos \theta = 0$.

Orthogonal Vectors

We say that two vectors $\mathbf{u}$ and $\mathbf{v}$ are *orthogonal* (or *perpendicular*) if and only if $\mathbf{u} \cdot \mathbf{v} = 0$.

EXAMPLE 10   Let $\mathbf{a} = \langle 3, 5 \rangle$, $\mathbf{b} = \langle -2, 3 \rangle$, and $\mathbf{c} = \langle 6, 4 \rangle$. Among these three vectors, find a pair of perpendicular vectors.

Solution: Computing inner products, we have

$$\mathbf{a} \cdot \mathbf{b} = (3)(-2) + (5)(3) = 9,$$
$$\mathbf{a} \cdot \mathbf{c} = (3)(6) + (5)(4) = 38,$$
$$\mathbf{b} \cdot \mathbf{c} = (-2)(6) + (3)(4) = 0.$$

Thus **b** and **c** are perpendicular vectors.

Practice Exercise 10   Is the vector $\mathbf{u} = \langle 5, 3 \rangle$ perpendicular to both of the vectors $\mathbf{v} = \langle -3, 5 \rangle$ and $\mathbf{w} = \langle 3, -5 \rangle$?

Answer: Yes

Applications

EXAMPLE 11   An airplane is flying due north at 120 mph in still air. Suddenly, an easterly wind of 50 mph begins to blow. What is the resulting speed of the plane? In which direction is the plane flying?

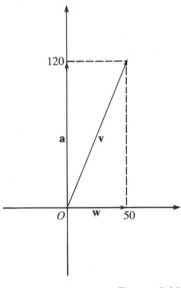

Figure 9.29

Solution: Two velocity vectors representing the speeds and directions of both the airplane and the wind are shown in Figure 9.29. The resulting velocity of the airplane is the sum of the two vectors. The magnitude of the resultant, obtained by using the Pythagorean theorem, is the new speed of the airplane:

$$|\mathbf{v}| = \sqrt{120^2 + 50^2} = 130 \text{ mph.}$$

To find the direction, we determine the angle θ that the vector $\mathbf{v}$ makes with the easterly direction:

$$\tan \theta = \frac{120}{50},$$

or

$$\theta = \arctan\left(\frac{12}{5}\right) \approx 67.38°.$$

Thus the plane is flying 67.38° north of east.

Practice Exercise 11

A man walks 4.5 km 30° east of north and then 6.0 km 60° west of north. Find **(a)** the man's final orientation relative to his starting point, and **(b)** the net distance walked by the man.

Answer: **(a)** 23.1° west of north   **(b)** 7.5 km

EXAMPLE 12

Neglecting friction, what is the force necessary to pull a 200-lb box up a ramp that makes a 30° angle with the ground?

Solution: The weight of the box is represented by the vector $\mathbf{w}$ in the diagram in Figure 9.30. We resolve the vector $\mathbf{w}$ into two components, $\mathbf{f}_p$ and $\mathbf{f}_n$. The component $\mathbf{f}_n$, perpendicular to the ramp, represents the force with which the box pushes against the ramp. It is balanced by the ramp (why?). The component $\mathbf{f}_p$, parallel to the ramp, represents the force necessary to pull the box up the ramp. To find the magnitude of $\mathbf{f}_p$, we use right-triangle trigonometry and obtain

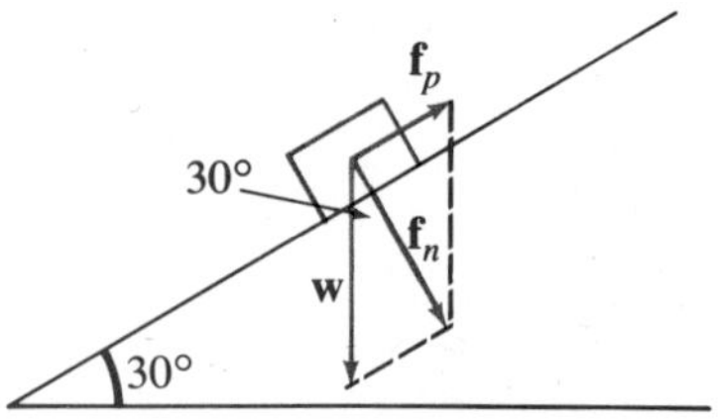

$$|\mathbf{f}_p| = |\mathbf{w}| \sin 30°$$
$$= 200\left(\frac{1}{2}\right) = 100 \text{ lb.}$$

Figure 9.30

Practice Exercise 12

In Example 12, find the magnitude of the component $\mathbf{f}_n$.

Answer: $100\sqrt{3} \approx 173.2$ lb

EXAMPLE 13

A father and his son are carrying a 150-lb crate supported by two ropes. The rope held by the father makes a 50° angle with the horizontal, and the rope held by the son makes a 30° angle with the horizontal. How much weight is each of them carrying?

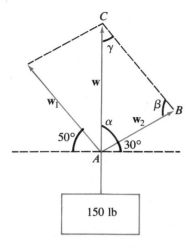

Figure 9.31

Solution: The weights $\mathbf{w}_1$ (father) and $\mathbf{w}_2$ (son) are distributed in such a way that their sum $\mathbf{w}$ (Figure 9.31) has magnitude 150 lb (the weight of the crate) and vertical direction. Referring to the triangle ABC in Figure 9.31, we apply the law of sines, where $\alpha = 60°$, $\beta = 80°$, and $\gamma = 40°$:

$$\frac{150}{\sin 80°} = \frac{|\mathbf{w}_1|}{\sin 60°} = \frac{|\mathbf{w}_2|}{\sin 40°}.$$

Thus we have

$$|\mathbf{w}_1| = \frac{150 \sin 60°}{\sin 80°} \simeq 132 \text{ lb}$$

and

$$|\mathbf{w}_2| = \frac{150 \sin 40°}{\sin 80°} \simeq 98 \text{ lb}.$$

Practice Exercise 13   A 500-lb weight is suspended from a ceiling by two ropes. One of the ropes makes an angle of 60° with the ceiling and the other makes a 45° angle. What is the tension on each rope?

Answer: 366 lb on the rope making a 60° angle, 259 lb on the rope making a 45° angle

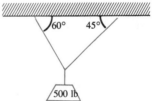

 EXERCISES 9.7

In Exercises 1–4, two points A and B are given. Find the vector $\mathbf{AB}$.

1. $A(2, 3)$, $B(4, -1)$ **2.** $A(-4, -2)$, $B(2, 3)$

3. $A(-1, -2)$, $B(3, 5)$ **4.** $A(-5, -3)$, $B(2, 1)$

In Exercises 5–8, the initial point of a vector and the vector itself are given. Find the coordinates of the terminal point.

5. $A(-1, 1)$, $\mathbf{v} = \langle -2, 5 \rangle$ **6.** $P(-2, 3)$, $\mathbf{w} = \langle 4, -2 \rangle$

7. $U(2, -6)$, $\mathbf{u} = \langle -3, 3 \rangle$ **8.** $A(-2, 2)$, $\mathbf{a} = \langle -4, 1 \rangle$

In Exercises 9–12, a vector and its terminal point are given. Find the coordinates of the initial point.

9. $\mathbf{a} = \langle -4, 1 \rangle$, $A(-2, 2)$ **10.** $\mathbf{u} = \langle -3, 3 \rangle$, $U(2, -6)$

11. $\mathbf{b} = \langle -3, 5 \rangle$, $O\langle 0, 0 \rangle$ **12.** $\mathbf{v} = \langle 5, -2 \rangle$, $Q(1, 1)$

In Exercises 13–16, find the required quantities.

13. Let $A(4, 2)$, $B(1, -3)$, and $C(-1, 5)$ be three points in the plane. Find the sum of the vectors $\mathbf{AB}$ and $\mathbf{AC}$.

14. If A, B, and C are the points considered in Exercise 13, find the sum of the vectors $\mathbf{CA}$ and $\mathbf{CB}$.

15. Vectors $\mathbf{u}$ and $\mathbf{v}$ make an angle of 150°. Find $|\mathbf{u} + \mathbf{v}|$ and $|\mathbf{u} - \mathbf{v}|$ if $|\mathbf{u}| = 4$ and $|\mathbf{v}| = 6$.

16. If $\mathbf{PQ}$ and $\mathbf{PR}$ make an angle of 45° where $|\mathbf{PQ}| = 3$ and $|\mathbf{PR}| = 5$, find $|\mathbf{PQ} + \mathbf{PR}|$ and $|\mathbf{PQ} - \mathbf{PR}|$.

In Exercises 17–20, find the magnitude of each of the following vectors.

17. OA is a position vector terminating at $A(-3, 5)$.

18. PQ is a vector from $P(-2, -3)$ to $Q(3, 2)$

19. $v = 3i - 5j$ **20.** $w = -2i + 6j$

In Exercises 21–24, you are given $a = \langle -2, 3 \rangle$ and $b = \langle 3, -1 \rangle$. Find each of the following vectors.

21. $3a - b$ **22.** $a + 4b$

23. $2a - 5b$ **24.** $-3a - 2b$

In Exercises 25–28, do the following problems on unit vectors.

25. What is the unit vector in the direction of the vector $v = \langle -5, 12 \rangle$? In the opposite direction?

26. Find the unit vector in the direction of the vector $w = \langle 6, -8 \rangle$. What is the unit vector in the opposite direction?

27. Let $u = -2i + 5j$ and $v = 3i - j$. Express $3u - 4v$ as a linear combination of i and j.

28. If $u = 3i - 7j$ and $v = -2i + 2j$, write $2u - 6v$ as a linear combination of i and j.

In Exercises 29–32, the magnitude and direction angle of a vector are given. Write each vector as the sum of its horizontal and vertical components.

29. 2, 60°   **30.** 4, 45°

31. 6, 150°   **32.** 5, 210°

Find the direction angle of each of the following vectors in Exercises 33–36.

33. $2i + 2j$ **34.** $2i - 2j$

35. $i - \sqrt{3}\, j$ **36.** $2\sqrt{3}\, i - 2j$

Find the dot product for each of the following pairs of vectors in Exercises 37–40.

37. $u = 4i - 3j, v = -5i + 2j$

38. $a = -5i - 2j, b = 3i + 4j$

39. $p = \langle 6, 0 \rangle, q = \langle -2, -4 \rangle$

40. $r = \langle -3, -7 \rangle, s = \langle -2, 2 \rangle$

Find the angle between each of the following pairs of vectors in Exercises 41–44.

41. $a = 4i - 3j, b = -2i + j$

42. $a = -2i - 5j, b = 4i + 3j$

43. $u = \langle 3, -7 \rangle, v = \langle 7, 3 \rangle$

44. $r = \langle 3, 8 \rangle, s = \langle -3, -8 \rangle$

In Exercises 45–70, solve the given problem.

45. Find a number x so that the vector $u = xi + 4j$ is perpendicular to the vector $v = -6i + 2j$.

46. Let $u = \langle a, -5 \rangle$ and $v = \langle -3, -4 \rangle$. Find the number a such that $u \cdot v = 0$.

47. Two forces of 12 and 18 N, making an angle of 135°, act on a point. Find the magnitude of the resultant.

48. Find the magnitude of the resultant of two forces of 80 and 120 lb that act on a point and make an angle of 30°.

49. If friction is neglected, what is the force necessary to push a 350-kg box up a ramp inclined 15° with the horizontal?

50. What force is required to keep a 3500-lb vehicle parked on a hill that makes a 10° angle with the horizontal?

51. The resultant of two forces has a magnitude of 180 lb. One of the forces has magnitude 70 lb and makes an angle of 30° with the resultant. Find the magnitude of the other force. What is the angle between the two forces?

52. Under the assumptions of Exercise 51, suppose that there is a 90° angle between the force with magnitude 70 lb and the resultant. What is the magnitude of the other force, and what is the angle between the two forces?

53. Two forces of equal magnitude act on a point and make an angle of 60°. If the resultant has magnitude $400\sqrt{3}$ lb, find the magnitude of each force.

54. Two forces act on a point and make an angle of 120°. One of the forces is twice the other. Find their magnitudes if the resultant has magnitude $150\sqrt{3}$ N.

55. An airplane is traveling 750 km/h in a direction 30° west of north. **(a)** Find the components of the velocity vector in the easterly and northerly directions. **(b)** How far north has the plane traveled after 2 hours?

56. A boat is sailing in a direction 45° east of south at 18 mph. If i and j denote the unit vectors in the easterly and northerly directions, write the velocity vector as a linear combination of i and j. How far south will the boat be after 3 hours?

57. A boat heads directly across a river that has a current of 1 m/s. If the boat can travel 3 m/s in still water, find the resulting velocity (magnitude and direction) of the boat relative to the shore.

58. The speed of a cruise boat relative to the earth is 12 km/h. If a vacationer walks 5 km/h directly across the boat, find the resulting velocity (magnitude and direction) of the vacationer relative to the earth.

59. Lisa can swim 1.50 m/s in still water. If she swims directly across a river that is 120 m wide and has a current of 0.8 m/s, how far downstream will she land?

60. A ferryboat whose speed in still water is 10.5 ft/s travels across a channel 500 ft wide. If the speed of the current is 2.1 ft/s, what is the upstream angle (with respect to a line perpendicular to the shore) with which the pilot must aim the ferryboat so that it reaches a point directly across from where it started?

61. A 1000-lb box is suspended by two cables. One of the cables makes a 30° angle and the other makes a 60° angle, both with respect to the horizontal. What is the tension on each cable?

62. Suppose that one of the cables in Exercise 61 makes a 45° angle with the horizontal, while the other cable remains in a horizontal position. Find the tension on each cable.

63. An airplane flies 450 mph. Starting at point A, it flies 30°

east of south for an hour, then it turns and flies 50° west of south for another hour. How far is the plane from *A*? What is its bearing relative to point *A*?

64. Starting from New York, a cruiser sails 26 miles on a bearing of N20°E and then 26 miles on a bearing of N80°E. How far is the cruiser from New York? What is its bearing?

65. A 60-lb sign supported by two bars is attached to a wall as shown in the figure.

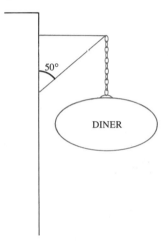

Find the magnitude of the forces acting on the supporting bars.

66. What are the magnitudes of the forces on the beams *AB* and *AC* that support a 500-lb weight as illustrated in the figure?

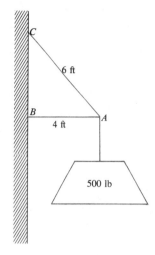

67. If **u**, **v**, **w** are vectors, prove that $(\mathbf{u} + \mathbf{v}) + \mathbf{w} = \mathbf{u} + (\mathbf{v} + \mathbf{w})$, $\mathbf{u} + \mathbf{O} = \mathbf{u}$, and $\mathbf{u} + (-\mathbf{u}) = \mathbf{O}$.

68. If α and β are real numbers and **u** is a vector, prove that $(\alpha + \beta)\mathbf{u} = \alpha\mathbf{u} + \beta\mathbf{u}$ and $\alpha(\beta\mathbf{u}) = (\alpha\beta)\mathbf{u}$.

69. Let $\mathbf{u} = \langle u_1, u_2 \rangle$, $\mathbf{v} = \langle v_1, v_2 \rangle$, and $\mathbf{w} = \langle w_1, w_2 \rangle$. Prove that $\mathbf{u} \cdot (\mathbf{v} + \mathbf{w}) = \mathbf{u} \cdot \mathbf{v} + \mathbf{u} \cdot \mathbf{w}$.

70. If $\mathbf{u} = \langle u_1, u_2 \rangle$ and $\mathbf{v} = \langle v_1, v_2 \rangle$ are vectors and λ is a real number, prove that $(\lambda\mathbf{u}) \cdot \mathbf{v} = \lambda(\mathbf{u} \cdot \mathbf{v})$.

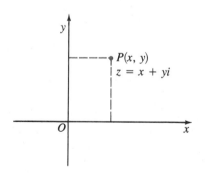

Figure 9.32

9.8   THE TRIGONOMETRIC FORM OF COMPLEX NUMBERS

Before reading this section, you should review the definition and basic facts about complex numbers discussed in Section 1.9.

Complex numbers can be represented by points in a plane. With every complex number $z = x + yi$ we associate the point $P(x, y)$ in a Cartesian coordinate system. Conversely, to every point in a coordinate plane there corresponds a unique complex number. The correspondence is one-to-one, so the set $\mathbb{C}$ of all complex numbers can be identified with the set of all points in a coordinate plane (Figure 9.32). Under such an identification, the plane is called the *complex plane,* the *x*-axis is called the *real axis,* and the *y*-axis is the *imaginary axis.*

EXAMPLE 1   Plot each of the following complex numbers in a complex plane.
(a) $z = 2 + 3i$ **(b)** $\bar{z} = 2 - 3i$ **(c)** $v = -5i$ **(d)** $w = -3$

Solution: The four corresponding points are shown in Figure 9.33. Notice that the points representing the conjugate numbers $2 + 3i$ and $2 - 3i$ are symmetric

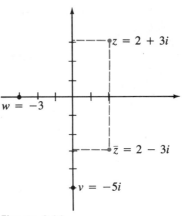

Figure 9.33

with respect to the *x*-axis. In general, the points in a complex plane representing two conjugate complex numbers z and $\bar{z}$ are symmetric with respect to the *x*-axis.

Practice Exercise 1   Plot each of the following complex numbers in a complex plane. **(a)** $-4 + 2i$ **(b)** $-2 - 3i$ **(c)** $-4i$ **(d)** $-4 - 2i$

The Absolute Value of a Complex Number

If $z = x + yi$ is a complex number, then its *absolute value* is defined by

$$|z| = \sqrt{z \cdot \bar{z}} = \sqrt{x^2 + y^2}. \tag{9.22}$$

Notice that if $y = 0$, that is, if $z = x$ is a real number, then

$$|z| = \sqrt{x^2} = |x|.$$

Thus, when z is a real number, the absolute values of z as a real number and as a complex number are the same.

The absolute value of a complex number z is also called the *modulus* of z.

EXAMPLE 2   Find the absolute values of the following complex numbers.
(a) $|2 + 4i|$ **(b)** $|8i|$

Solution: **(a)** According to the definition (9.22), we have

$$|2 + 4i| = \sqrt{2^2 + 4^2} = \sqrt{4 + 16} = \sqrt{20} = 2\sqrt{5}.$$

(b) In this case $x = 0$ and $y = 8$, so

$$|8i| = \sqrt{0^2 + 8^2} = \sqrt{64} = 8.$$

Practice Exercise 2   Compute the absolute values of the complex numbers $-5 + 2i$ and $-4i$.

Answer: $|-5 + 2i| = \sqrt{29}, |-4i| = 4$

Polar Form of Complex Numbers

If $z = x + yi$ is represented by the point $P(x, y)$ in the complex plane (Figure 9.34), then the absolute value $|z| = \sqrt{x^2 + y^2}$ of z represents the distance $d(O, P)$ from P to the origin.

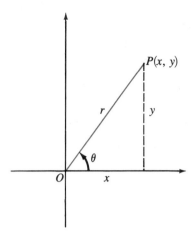

Figure 9.34

Setting $r = |z|$ and denoting by θ the angle in standard position whose terminal side is OP, the following relations are obtained.

$$\begin{cases} x = r \cos \theta \\ y = r \sin \theta \end{cases} \qquad (9.23)$$

Thus, we can write

$$z = x + yi = r \cos \theta + ir \sin \theta$$

or

$$z = r(\cos \theta + i \sin \theta). \qquad (9.24)$$

This is the *trigonometric form* or *polar form* of the complex number z. The real number $r \geq 0$ is the absolute value or the modulus of z. The angle θ is called the *argument* of z.

Notice that in the trigonometric representation of z, the angle θ is not unique: if we add any multiple of 2π to the argument θ, the representation changes while z remains the same. For this reason, we often restrict θ to the interval $[0, 2\pi)$.

EXAMPLE **3**

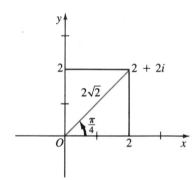

Figure 9.35

Write each of the following complex numbers in trigonometric form.
(a) $2 + 2i$ **(b)** $-\sqrt{3} - i$

Solution: **(a)** If we set $z = 2 + 2i$, then $|z| = 2\sqrt{2}$. (See Figure 9.35.) Next, we multiply and divide $2 + 2i$ by $2\sqrt{2}$, as follows:

$$z = 2 + 2i = 2\sqrt{2} \cdot \frac{2 + 2i}{2\sqrt{2}} = 2\sqrt{2}\left(\frac{1}{\sqrt{2}} + \frac{1}{\sqrt{2}}i\right).$$

Since $\cos(\pi/4) = \sin(\pi/4) = 1/\sqrt{2}$, we obtain

$$z = 2\sqrt{2}\left(\cos \frac{\pi}{4} + i \sin \frac{\pi}{4}\right).$$

(b) Let $w = -\sqrt{3} - i$. Then $|w| = 2$ and, proceeding as in part a, we write

$$w = -\sqrt{3} - i = 2\left(-\frac{\sqrt{3}}{2} - \frac{1}{2}i\right) = 2\left(\cos \frac{7\pi}{6} + i \sin \frac{7\pi}{6}\right)$$

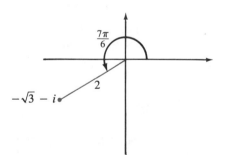

Figure 9.36

Practice Exercise 3

Write each of the following complex numbers in polar form:
(a) $3 - 3i$ **(b)** $1 + \sqrt{3}i$

Answer: **(a)** $3\sqrt{2}\left(\cos \frac{\pi}{4} - i \sin \frac{\pi}{4}\right)$ **(b)** $2\left(\cos \frac{\pi}{3} + i \sin \frac{\pi}{3}\right)$

The Product of Complex Numbers in Polar Form

Let $z_1 = r_1(\cos \theta_1 + i \sin \theta_1)$ and $z_2 = r_2(\cos \theta_2 + i \sin \theta_2)$ be two complex numbers in polar form. By multiplying them and using identities (9.5) and (9.6), we obtain the formula

$$z_1 z_2 = r_1 r_2 [\cos(\theta_1 + \theta_2) + i \sin(\theta_1 + \theta_2)]. \qquad \text{(9.25)}$$

To show this, we proceed as follows:

$$z_1 z_2 = r_1(\cos \theta_1 + i \sin \theta_1) \cdot r_2(\cos \theta_2 + i \sin \theta_2)$$
$$= r_1 r_2 [(\cos \theta_1 \cos \theta_2 - \sin \theta_1 \sin \theta_2) + i(\sin \theta_1 \cos \theta_2 + \cos \theta_1 \sin \theta_2)]$$
$$= r_1 r_2 [\cos(\theta_1 + \theta_2) + i \sin(\theta_1 + \theta_2)].$$

EXAMPLE 4   If $z_1 = 12(\cos 75° + i \sin 75°)$ and $z_2 = 3(\cos 35° + i \sin 35°)$, find $z_1 z_2$.

Solution: According to (9.25), we have

$$z_1 z_2 = 12(\cos 75° + i \sin 75°) \cdot 3(\cos 35° + i \sin 35°)$$
$$= 12 \cdot 3[\cos(75° + 35°) + i \sin(75° + 35°)]$$
$$= 36(\cos 110° + i \sin 110°).$$

Practice Exercise 4   Find the product of $u = 2\left(\cos \dfrac{\pi}{3} + i \sin \dfrac{\pi}{3}\right)$ and $v = 3\left(\cos \dfrac{\pi}{4} - i \sin \dfrac{\pi}{4}\right)$.

$$\left[\textit{Hint: } v = 3\left(\cos\left(-\frac{\pi}{4}\right) + i \sin\left(-\frac{\pi}{4}\right)\right) \text{ (Why?)}\right]$$

Answer: $uv = 6\left(\cos \dfrac{\pi}{12} + i \sin \dfrac{\pi}{12}\right)$

The Reciprocal of a Complex Number in Polar Form

Let $z = r(\cos \theta + i \sin \theta) \neq 0$ and let $z^{-1} = \rho(\cos \phi + i \sin \phi)$ be the reciprocal of z. We have, by definition,

$$zz^{-1} = 1.$$

or

$$r(\cos \theta + i \sin \theta) \cdot \rho(\cos \phi + i \sin \phi) = 1.$$

The number 1 can be represented in polar form as $1 = 1(\cos 0 + i \sin 0)$. On the other hand, by using (9.25), the last relation can be rewritten as follows:

$$r\rho[\cos(\theta + \phi) + i \sin(\theta + \phi)] = 1(\cos 0 + i \sin 0).$$

These two complex numbers are equal if and only if $r\rho = 1$, $\cos(\theta + \phi) = \cos 0 = 1$, and $\sin(\theta + \phi) = \sin 0 = 0$. The first relation gives us

$$\rho = r^{-1}.$$

Since the sine and cosine functions have period 2π, the last two relations imply that

$$\theta + \phi = 2\pi k, \quad \text{with } k \text{ any integer.}$$

Taking $k = 0$, we obtain $\phi = -\theta$. Thus, the following is a representation for the reciprocal z^{-1} of z:

$$z^{-1} = r^{-1}(\cos(-\theta) + i \sin(-\theta)). \qquad (9.26)$$

Since the sine is an odd function, while the cosine is an even function, (9.26) can be written

$$z^{-1} = r^{-1}(\cos \theta - i \sin \theta). \qquad (9.27)$$

EXAMPLE 5   If $z = 5\left(\cos \dfrac{\pi}{6} + i \sin \dfrac{\pi}{6}\right)$, find z^{-1}.

Solution: According to (9.26), we have

$$z^{-1} = \frac{1}{5}\left[\cos\left(-\frac{\pi}{6}\right) + i \sin\left(-\frac{\pi}{6}\right)\right]$$

or

$$z^{-1} = \frac{1}{5}\left(\cos \frac{\pi}{6} - i \sin \frac{\pi}{6}\right).$$

Practice Exercise 5   Find the reciprocal of $z = 4(\cos 75° + i \sin 75°)$.

Answer: $z^{-1} = \dfrac{1}{4}(\cos 75° - i \sin 75°)$

Formulas (9.26) and (9.27) give us representations of the reciprocal z^{-1} of a complex number z in terms of the modulus and argument of z.

The Quotient of Complex Numbers in Polar Form

Let $z_1 = r_1(\cos \theta_1 + i \sin \theta_1)$ and $z_2 = r_2(\cos \theta_2 + i \sin \theta_2) \neq 0$ be two complex numbers. Combining formulas (9.25) and (9.26), we obtain the trigonometric representation of the quotient z_1/z_2:

$$\frac{z_1}{z_2} = \frac{r_1}{r_2}[\cos(\theta_1 - \theta_2) + i \sin(\theta_1 - \theta_2)]. \qquad (9.28)$$

The proof of (9.28) is left as exercise.

EXAMPLE 6   If $z_1 = 2\left(\cos \dfrac{\pi}{4} + i \sin \dfrac{\pi}{4}\right)$ and $z_2 = 5\left(\cos \dfrac{\pi}{6} + i \sin \dfrac{\pi}{6}\right)$, find z_1/z_2.

Solution: According to (9.28), we have

$$\frac{z_1}{z_2} = \frac{2}{5}\left[\cos\left(\frac{\pi}{4} - \frac{\pi}{6}\right) + i \sin\left(\frac{\pi}{4} - \frac{\pi}{6}\right)\right]$$

$$= \frac{2}{5}\left(\cos \frac{\pi}{12} + i \sin \frac{\pi}{12}\right).$$

Practice Exercise 6   Find the quotient of $v = 3(\cos 30° + i \sin 30°)$ by $w = 6(\cos 180° + i \sin 180°)$.

Answer: $\dfrac{v}{w} = \dfrac{1}{2}(\cos 150° - i \sin 150°)$

EXERCISES 9.8

In Exercises 1–10, write each complex number in trigonometric form.

1. $3 - 3i$

2. $-2 - 2i$

3. $\sqrt{3} - i$

4. $1 - i\sqrt{3}$

5. $-2 - 2i\sqrt{3}$

6. $-1 + i$

7. $5i$

8. $-3i$

9. -15

10. $\sqrt{7}$

In Exercises 11–14, find $z_1 z_2$ and z_1/z_2 for each of the following pairs of complex numbers.

11. $z_1 = 5\left(\cos \dfrac{\pi}{4} + i \sin \dfrac{\pi}{4}\right)$, $z_2 = 8\left(\cos \dfrac{2\pi}{3} + i \sin \dfrac{2\pi}{3}\right)$

12. $z_1 = 4\left(\cos \dfrac{5\pi}{6} + i \sin \dfrac{5\pi}{6}\right)$,

$z_2 = 3\left(\cos \dfrac{4\pi}{3} + i \sin \dfrac{4\pi}{3}\right)$

13. $z_1 = 6(\cos 240° + i \sin 240°)$,
$z_2 = 9(\cos 105° + i \sin 105°)$

14. $z_1 = 3(\cos 112° + i \sin 112°)$,
$z_2 = 3(\cos 76° + i \sin 76°)$

In Exercises 15–18, find $z_1 z_2$ and z_1/z_2 as follows: **(a)** use the definitions of product and quotient described in Section 1.9, and **(b)** change the numbers to their trigonometric form and use (9.25) and (9.28).

15. $z_1 = \sqrt{3} + i$, $z_2 = 2 + 2i\sqrt{3}$

16. $z_1 = 1 - i\sqrt{3}$, $z_2 = -1 + i\sqrt{3}$

17. $z_1 = 5 + 5i$, $z_2 = -3i$

18. $z_1 = -1 - i$, $z_2 = 4\sqrt{3} + 4i$.

☐ **19.** If $z = r(\cos \theta + i \sin \theta)$, show that $z^2 = r^2(\cos 2\theta + i \sin 2\theta)$ and that $z^3 = r^3(\cos 3\theta + i \sin 3\theta)$. If n is a natural number, can you find z^n?

☐ **20.** Prove formula (9.28).

9.9   DE MOIVRE'S FORMULA; THE nTH ROOTS OF A COMPLEX NUMBER

If $z = r(\cos \theta + i \sin \theta)$ is a complex number in polar form, then

$$z^2 = [r(\cos \theta + i \sin \theta)][r(\cos \theta + i \sin \theta)]$$

$$= r^2[(\cos^2\theta - \sin^2\theta) + i(2\sin\theta\cos\theta)]$$
$$= r^2(\cos 2\theta + i\sin 2\theta). \quad [\text{By (9.14) and (9.15)}]$$

Similarly, it can be shown that

$$z^3 = r^3(\cos 3\theta + i\sin 3\theta).$$

More generally, we have

$$z^n = r^n(\cos n\theta + i\sin n\theta), \quad \text{for all natural numbers } n. \qquad (9.29)$$

This relation, called *De Moivre's formula*, can be proved using the *principle of induction* discussed in Chapter 12.

EXAMPLE 1   Find $(1 + i)^{11}$.

Solution: First we write $1 + i$ in polar form:

$$1 + i = \sqrt{2}\left(\cos\frac{\pi}{4} + i\sin\frac{\pi}{4}\right),$$

noticing that the argument $\pi/4$ is an angle between 0 and 2π. Next, using De Moivre's formula, we obtain

$$(1 + i)^{11} = (\sqrt{2})^{11}\left(\cos\frac{11\pi}{4} + i\sin\frac{11\pi}{4}\right).$$

Now the argument $11\pi/4$ is greater than 2π, so we reduce it to an angle between 0 and 2π. Since $11\pi/4 = 3\pi/4 + 2\pi$, we can write

$$(1 + i)^{11} = 2^5\sqrt{2}\left(\cos\frac{3\pi}{4} + i\sin\frac{3\pi}{4}\right).$$

In rectangular form the answer is

$$(1 + i)^{11} = 32\sqrt{2}\left(\cos\frac{3\pi}{4} + i\sin\frac{3\pi}{4}\right)$$
$$= 32\sqrt{2}\left(-\frac{\sqrt{2}}{2} + i\frac{\sqrt{2}}{2}\right) = -32 + 32i.$$

Practice Exercise 1   Find $(1 - i)^5$.

Answer: $(1 - i)^5 = 4\sqrt{2}\left(\cos\dfrac{5\pi}{4} - i\sin\dfrac{5\pi}{4}\right) = -4 + 4i.$

Combining (9.29) and (9.26), we obtain De Moivre's formula for *negative exponents:*

$$z^{-n} = r^{-n}(\cos(-n\theta) + i\sin(-n\theta)), \quad \text{for all natural numbers } n. \qquad (9.30)$$

EXAMPLE **2**   Find $(1 - i)^{-5}$.

Solution: We have

$$1 - i = \sqrt{2}\left[\cos\left(-\frac{\pi}{4}\right) + i\,\sin\left(-\frac{\pi}{4}\right)\right].$$

According to (9.30), we get

$$(1 - i)^{-5} = (\sqrt{2})^{-5}\left[\cos\frac{5\pi}{4} + i\,\sin\frac{5\pi}{4}\right]$$

$$= \frac{\sqrt{2}}{8}\left(\cos\frac{5\pi}{4} + i\,\sin\frac{5\pi}{4}\right).$$

In rectangular form, the answer is

$$(1 - i)^{-5} = \frac{\sqrt{2}}{8}\left(\cos\frac{5\pi}{4} + i\,\sin\frac{5\pi}{4}\right)$$

$$= \frac{\sqrt{2}}{8}\left(-\frac{\sqrt{2}}{2} - i\frac{\sqrt{2}}{2}\right) = -\frac{1}{8} - \frac{1}{8}i.$$

Practice Exercise 2   Find $(1 + i)^{-11}$.

Answer: $(1 + i)^{-11} = \dfrac{1}{32\sqrt{2}}\left(\cos\dfrac{3\pi}{4} - i\,\sin\dfrac{3\pi}{4}\right) = -\dfrac{1}{64} - \dfrac{i}{64}.$

Formulas (9.29) and (9.30) can be combined in one equation:

De Moivre's Formula

$z^n = r^n(\cos n\theta + i\,\sin n\theta),$ for all integers n.   **(9.31)**

The *n*th Roots of a Complex Number

Let z be a complex number and n a natural number. An *nth root* of z is a complex number w such that $w^n = z$. As we shall see, there are exactly n *distinct* nth roots of a complex number z. If z is given in polar form, De Moivre's formula provides a simple way of finding all the nth roots of z.

     Let $z = r(\cos\theta + i\,\sin\theta)$ and suppose that $w = \rho(\cos\phi + i\,\sin\phi)$ is a number such that $w^n = z$. By (9.29),

$$\rho^n(\cos n\phi + i\,\sin n\phi) = r(\cos\theta + i\,\sin\theta).$$

In order for these two complex numbers to be equal, we must have

$$\rho^n = r, \quad\quad \cos n\phi = \cos\theta, \quad\text{and}\quad \sin n\phi = \sin\theta.$$

The first relation implies that

$$\rho = r^{1/n},$$

HISTORICAL NOTE

POLAR FORM OF A COMPLEX NUMBER

It was Leonhard Euler who discovered the important connections between trigonometric and exponential functions (in the complex plane). Euler introduced the notation

$$re^{i\theta} = r(\cos\theta + i\,\sin\theta)$$

to represent a complex number with absolute value r and argument θ. De Moivre's theorem, relating trigonometry and root extraction in the complex plane,

$$e^{in\theta} = (\cos\theta + i\,\sin\theta)^n$$
$$= \cos n\theta + i\,\sin n\theta,$$

appeared first in Euler's *Introduction to the Analysis of Infinites*, in 1748.

while the other two relations imply that the arguments $n\phi$ and θ differ by an integral multiple of 2π, that is,

$$n\phi = \theta + 2k\pi,$$

or

$$\phi = \frac{\theta + 2k\pi}{n},$$

where k is an arbitrary integer. Thus, we can write

$$w = r^{1/n}\left[\cos\left(\frac{\theta + 2k\pi}{n}\right) + i \sin\left(\frac{\theta + 2k\pi}{n}\right)\right],$$

where k is any integer. It looks as if there are infinitely many nth roots of z. However, because of the periodicity of the sine and cosine functions, we can restrict the values of k to $k = 0, 1, \ldots, n - 1$ and obtain n *distinct* nth roots of z. No other value of k will yield a different root. Our results can be summarized as follows.

---

nth Root Formula

If $z = r(\cos \theta + i \sin \theta)$ is a complex number different from 0 and n is any natural number, then z has exactly n distinct nth roots, given by

$$w_k = r^{1/n}\left[\cos\left(\frac{\theta + 2k\pi}{n}\right) + i \sin\left(\frac{\theta + 2k\pi}{n}\right)\right], \qquad k = 0, 1, \ldots, n - 1. \qquad (9.32)$$

EXAMPLE 3   Find the four distinct fourth roots of $z = 1 + i\sqrt{3}$.

Solution: First we write z in trigonometric form:

$$z = 2\left(\cos \frac{\pi}{3} + i \sin \frac{\pi}{3}\right)$$

(Figure 9.37a). According to (9.32) with $n = 4$ and $\theta = \pi/3$, the four distinct roots are given by

$$w_k = 2^{1/4}\left[\cos\left(\frac{\pi/3 + 2k\pi}{4}\right) + i \sin\left(\frac{\pi/3 + 2k\pi}{4}\right)\right], \qquad k = 0, 1, 2, 3.$$

All the roots have modulus $\sqrt[4]{2}$, lie on the circle of radius $\sqrt[4]{2}$ centered at the origin, and are equally spaced, because the arguments of two successive roots differ by $\pi/2$ (Figure 9.37b). Substituting 0, 1, 2, and 3 for k, we obtain

$$w_0 = \sqrt[4]{2}\left(\cos \frac{\pi}{12} + i \sin \frac{\pi}{12}\right),$$

$$w_1 = \sqrt[4]{2}\left(\cos \frac{7\pi}{12} + i \sin \frac{7\pi}{12}\right),$$

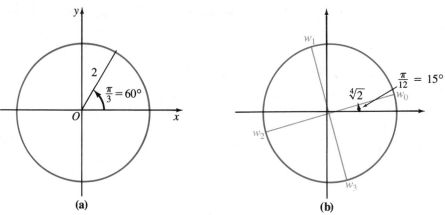

Figure 9.37

$$w_2 = \sqrt[4]{2}\left(\cos\frac{13\pi}{12} + i\sin\frac{13\pi}{12}\right),$$

$$w_3 = \sqrt[4]{2}\left(\cos\frac{19\pi}{12} + i\sin\frac{19\pi}{12}\right).$$

Practice Exercise 3   Find the two distinct second roots of $z = -\dfrac{1}{2} - \dfrac{\sqrt{3}}{2}i$.

Answer: $w_0 = \cos\dfrac{2\pi}{3} + i\sin\dfrac{2\pi}{3}, \quad w_1 = \cos\dfrac{5\pi}{3} + i\sin\dfrac{5\pi}{3}$

■ **A general remark** According to formula (9.32), all the nth roots of z have modulus $\sqrt[n]{r}$, thus they lie on the circle of radius $\sqrt[n]{r}$ centered at the origin. Moreover, since the arguments of two consecutive roots differ by $2\pi/n$, it follows that the roots are equally spaced along the circumference.

EXAMPLE **4**   Find all of the six distinct sixth roots of $z = -\dfrac{1}{2} - \dfrac{\sqrt{3}}{2}i$. Sketch the graph.

Solution: In polar form, the given number is $z = \cos 240° + i\sin 240°$. (See Figure 9.38.) According to (9.32) with $r = 1$, $\theta = 240°$, and $n = 6$, the six roots of z are given by

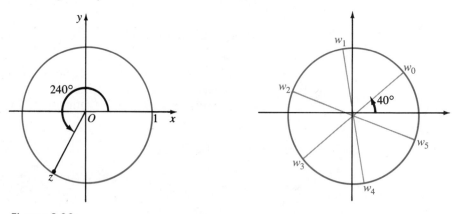

Figure 9.38

$$w_k = \cos\left(\frac{240° + 360k°}{6}\right) + i \sin\left(\frac{240° + 360k°}{6}\right), \qquad k = 0, 1, 2, 3, 4, 5.$$

Thus

$$w_0 = \cos 40° + i \sin 40°,$$
$$w_1 = \cos 100° + i \sin 100°,$$
$$w_2 = \cos 160° + i \sin 160°,$$
$$w_3 = \cos 220° + i \sin 220°,$$
$$w_4 = \cos 280° + i \sin 280°,$$
$$w_5 = \cos 340° + i \sin 340°.$$

Practice Exercise 4   Find the three distinct third roots of $z = 2(\cos 60° + i \sin 60°)$.

Answer: $w_0 = \sqrt[3]{2}(\cos 20° + i \sin 20°)$, $w_1 = \sqrt[3]{2}(\cos 140° + i \sin 140°)$,
$w_2 = \sqrt[3]{2}(\cos 260° + i \sin 260°)$.

The nth Roots of Unity

An important special case occurs when $z = 1$. The n distinct nth roots of $z = 1$ are then called the *nth roots of unity*. Since in this case $r = |z| = 1$ and $\theta = 0$, we derive from (9.32) the formula that gives the n distinct nth roots of unity:

$$u_k = \cos \frac{2k\pi}{n} + i \sin \frac{2k\pi}{n}, \qquad k = 0, 1, \ldots, n - 1. \quad \textbf{(9.33)}$$

EXAMPLE **5**   Find the fifth roots of unity.

Solution: According to (9.33) with $n = 5$,

$$u_k = \cos \frac{2k\pi}{5} + i \sin \frac{2k\pi}{5}, \qquad k = 0, 1, 2, 3, 4.$$

Thus

$$u_0 = \cos 0 + i \sin 0 = 1,$$
$$u_1 = \cos \frac{2\pi}{5} + i \sin \frac{2\pi}{5},$$
$$u_2 = \cos \frac{4\pi}{5} + i \sin \frac{4\pi}{5},$$
$$u_3 = \cos \frac{6\pi}{5} + i \sin \frac{6\pi}{5},$$
$$u_4 = \cos \frac{8\pi}{5} + i \sin \frac{8\pi}{5}.$$

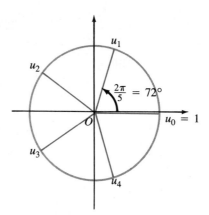

Figure 9.39

Geometrically, the five distinct roots are equally spaced points on the unit circle, forming the vertices of a regular pentagon.

Practice Exercise 5　　Find the cube roots of unity.

Answer:　$u_0 = 1,$　$u_1 = \cos \dfrac{2\pi}{3} + i \sin \dfrac{2\pi}{3} = -\dfrac{1}{2} + \dfrac{\sqrt{3}}{2}i,$

$u_2 = \cos \dfrac{4\pi}{3} + i \sin \dfrac{4\pi}{3} = -\dfrac{1}{2} - \dfrac{\sqrt{3}}{2}i.$

Regular Polygons

In general, the n distinct nth roots of unity are equally spaced points on the unit circle with one of them being the point $(1, 0)$. The n points form the vertices of a *regular n-gon*.

　　In Figure 9.40, we illustrate the square and cube roots of unity.

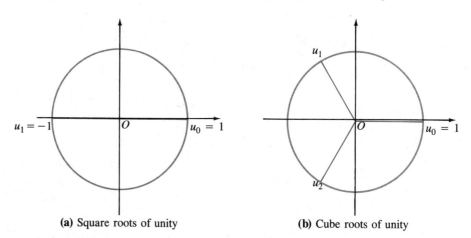

(a) Square roots of unity　　　　　(b) Cube roots of unity

Figure 9.40

Zeros of Polynomials

The techniques described in the preceding pages can be used to find all complex zeros of certain simple polynomials.

EXAMPLE **6**   Find all complex zeros of the following two polynomials.
(a) $P(x) = x^3 - 8$ **(b)** $Q(x) = x^3 + 1$

Solution: **(a)** We have to solve the equation

$$x^3 - 8 = 0.$$

or

$$x^3 = 8.$$

In other words, we must find all distinct cube roots of 8. Clearly, $x = 2$ is one of the roots. To find the others, write 8 in polar form: $8 = 8(\cos 0 + i \sin 0)$. The three distinct roots illustrated in Figure 9.41 are obtained from (9.32).

$$w_0 = 2(\cos 0 + i \sin 0) = 2$$

$$w_1 = 2\left(\cos \frac{2\pi}{3} + i \sin \frac{2\pi}{3}\right) = -1 + i\sqrt{3}$$

$$w_2 = 2\left(\cos \frac{4\pi}{3} + i \sin \frac{4\pi}{3}\right) = -1 - i\sqrt{3}$$

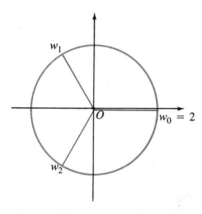

Figure 9.41

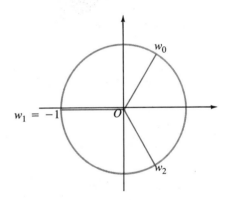

Figure 9.42

(b) In this case we have to solve the equation

$$x^3 + 1 = 0,$$

or

$$x^3 = -1.$$

The equation has a real root, $x = -1$, and two complex roots. The three roots are the cube roots of -1. To find them, we write -1 in polar form:

$$-1 = \cos \pi + i \sin \pi,$$

and use formula (9.32), obtaining

$$w_0 = \cos \frac{\pi}{3} + i \sin \frac{\pi}{3} = \frac{1}{2} + \frac{\sqrt{3}}{2}i,$$

$$w_1 = \cos \pi + i \sin \pi = -1,$$

$$w_2 = \cos \frac{5\pi}{3} + i \sin \frac{5\pi}{3} = \frac{1}{2} - \frac{\sqrt{3}}{2}i.$$

Practice Exercise 6   Find all complex roots of the following two polynomials.
(a) $A(x) = x^3 + 8$ (b) $B(x) = x^3 - 1$

Answer:

(a) $w_0 = 2\left(\cos \dfrac{\pi}{3} + i \sin \dfrac{\pi}{3}\right) = 1 + i\sqrt{3}$ (b) $w_0 = 1$

$w_1 = 2(\cos \pi + i \sin \pi) = -2$

$w_1 = \cos \dfrac{2\pi}{3} + i \sin \dfrac{2\pi}{3}$

$= -\dfrac{1}{2} + \dfrac{\sqrt{3}}{2} i$

$w_2 = 2\left(\cos \dfrac{5\pi}{3} + i \sin \dfrac{5\pi}{3}\right) = 1 - i\sqrt{3}$

$w_2 = \cos \dfrac{4\pi}{3} + i \sin \dfrac{4\pi}{3}$

$= -\dfrac{1}{2} - \dfrac{\sqrt{3}}{2} i$

EXERCISES 9.9

In Exercises 1–10, use De Moivre's formula to find each of the following numbers. Whenever possible, write your answers in rectangular form.

1. $(-1 + i)^5$ 2. $(-1 - i)^8$
3. $(-\sqrt{3} - i)^{11}$ 4. $(\sqrt{3} + i)^9$
5. $[2(\cos 60° + i \sin 60°)]^7$
6. $(\cos 30° + i \sin 30°)^8$
7. $\left[\sqrt{3}(\cos 120° - i \sin 120°)\right]^6$
8. $\left[\sqrt{2}(\cos 15° - i \sin 15°)\right]^8$
9. $(-1 + i\sqrt{3})^{-4}$ 10. $(\sqrt{3} + i)^{-5}$

In Exercises 11–20, find all the nth roots of z, for n and z as indicated. Whenever possible, write your answers in rectangular form.

11. $z = 1 + i\sqrt{3}, n = 2$ 12. $z = i, n = 2$
13. $z = -i, n = 3$ 14. $z = -1 - i\sqrt{3}, n = 4$

15. $z = -\sqrt{3} + i, n = 5$ 16. $z = -8i, n = 3$
17. $z = 64 - 64i\sqrt{3}, n = 6$
18. $z = -32 - 32i, n = 5$
19. $z = 16(\cos 120° + i \sin 120°), n = 4$
20. $z = 27(\cos 45° + i \sin 45°), n = 3$

Find the solutions of the given equations in Exercises 21–28.

21. $x^3 + 27 = 0$ 22. $x^3 - 64 = 0$
23. $x^4 - 16 = 0$ 24. $x^4 + 81 = 0$
25. $x^6 + 1 = 0$ 26. $x^5 + 32 = 0$
27. $x^3 - 8i = 0$ 28. $x^3 + 64i = 0$

☐ 29. Show that $\cos \dfrac{\pi}{4} - i \sin \dfrac{\pi}{4}$ is a solution of $x^4 - x^2 + 1 = i$.

☐ 30. Show that $\dfrac{\sqrt{2}}{2} + \dfrac{\sqrt{2}}{2} i$ is a solution of $x^6 + x^4 + x^2 + 1 = 0$.

9.10   POLAR COORDINATES

As you already know, in a Cartesian coordinate system, each point P is represented by an ordered pair (x, y) of numbers, where x and y are the distances from P to two fixed perpendicular axes. The same point P (see Figure 9.34) can be represented by the ordered pair (r, θ), where r is the *distance from P to the origin O* and θ is the *direction angle* of the vector OP.

The Polar Coordinate System

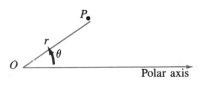

Figure 9.43

To define a polar coordinate system, we start with a fixed point O, called the *pole* or *origin,* and a fixed half-line, called the *polar axis,* that originates from O. In principle, this half-line can be any half-line, but it is customary to choose it as a horizontal line pointing to the right. In this manner the polar axis may be identified with the positive x-axis in a rectangular coordinate system. If P is any point other than the pole, then r, the distance from O to P, is a positive real number. Let θ be the angle determined by the polar axis and the half-line passing from O through P (Figure 9.43). The ordered pair (r, θ) is a pair of *polar coordinates* assigned to P.

EXAMPLE 1

Use polar coordinates to plot the following points: $(1, \pi/3)$, $(3, \pi/2)$, $(4, 2\pi/3)$, $(2, 7\pi/6)$, $(3, 5\pi/3)$.

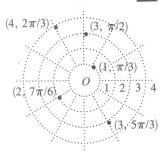

Figure 9.44

Solution: To locate $(1, \pi/3)$, we proceed counterclockwise around the unit circle through an angle of $\pi/3$. To locate the point $(3, \pi/2)$, we proceed counterclockwise from the polar axis and draw an angle of $\pi/2$ radians. On the terminal side of this angle we mark a point 3 units from the origin. The other points are plotted in a similar manner (Figure 9.44).

Polar graph paper makes it very easy to plot points specified by polar coordinates. The grid on polar graph paper consists of concentric circles and rays originating from a common center.

Practice Exercise 1

Use polar coordinates to plot the following points: $(2, \pi/4)$, $(1, 5\pi/6)$, $(3, 4\pi/3)$, $(2, 3\pi/2)$, $(1, 7\pi/4)$.

Answer:

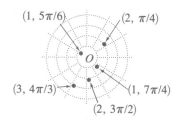

The Nonuniqueness of Polar Coordinates

In a Cartesian coordinate system, each point has a corresponding *unique* pair of coordinates. This is no longer true in a polar coordinate system. To any point, we can assign an *infinite* set of polar coordinates. For example, referring to Figure 9.44, you can see that if we add or subtract integer multiples of 2π to the value $\pi/3$, then the point $(1, \pi/3)$ also has coordinates $(1, 7\pi/3)$, $(1, 13\pi/3)$, $(1, -5\pi/3)$, and so on. As we already know, all angles of the form $\theta + 2k\pi$, $k = 0, \pm 1$,

$\pm 2, \ldots$, are coterminal; thus if (r, θ) is a pair of coordinates of P (Figure 9.43), then $(r, \theta + 2k\pi)$, $k = 0, \pm 1, \pm 2, \ldots$, are also coordinates of the same point.

So far in our discussion on polar coordinates, the number r has represented the distance from a point to the pole, therefore r is a positive number. On many occasions, particularly when sketching graphs in polar coordinates, it is convenient to allow r to be a negative number, which allows us to consider *signed* distances from the pole. For example, to locate the point $(3, \pi/4)$, we proceed counter-clockwise, starting from the polar axis, through an angle of $\pi/4$. The desired point lies on the terminal side of the angle, 3 units away from the origin (Figure 9.45a). To locate the point $(-3, \pi/4)$, we draw the angle $\pi/4$ as before, but the point will be $|-3| = 3$ units from the origin in the *opposite* direction (Figure 9.45b).

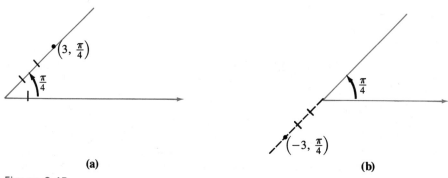

(a)   **(b)**

Figure 9.45

EXAMPLE **2**   Plot the points with coordinates $P(-1, 4\pi/3)$, $Q(-2, -\pi/4)$.

Solution:

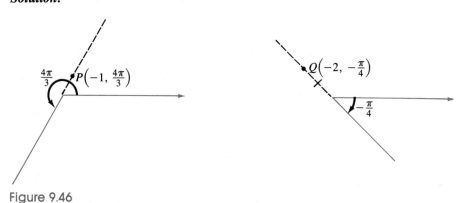

Figure 9.46

Note that another set of coordinates for P is $(1, \pi/3)$ and for Q it is $(2, 3\pi/4)$.

Practice Exercise 2   Plot the points $A(-3, -3\pi/4)$ and $B(-2, 2\pi/3)$.

Answer:

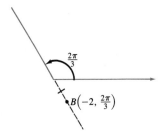

The Relationship between Cartesian and Polar Coordinates

Suppose that the polar axis in a polar coordinate system coincides with the positive *x*-axis of a Cartesian coordinate system. Then, as illustrated in Figure 9.47, the Cartesian coordinates (x, y) of a point *P* and the polar coordinates (r, θ) of the same point are related by the following equations:

$$x = r \cos \theta, \qquad y = r \sin \theta; \tag{9.34}$$

$$r^2 = x^2 + y^2, \qquad \theta = \arctan\left(\frac{y}{x}\right). \tag{9.35}$$

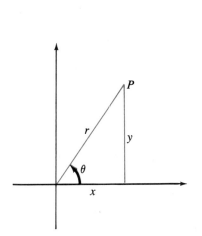

Figure 9.47

EXAMPLE **3**

Find **(a)** the Cartesian coordinates corresponding to $(4, 2\pi/3)$, and **(b)** polar coordinates corresponding to $\left(\sqrt{2}, \sqrt{2}\right)$.

Solution: **(a)** If $r, \theta) = (4, 2\pi/3)$, then by (9.34), we have

$$x = 4 \cos\left(\frac{2\pi}{3}\right) = 4\left(-\frac{1}{2}\right) = -2,$$

$$y = 4 \sin\left(\frac{2\pi}{3}\right) = 4\left(\frac{\sqrt{3}}{2}\right) = 2\sqrt{3}.$$

Thus, the corresponding pair of Cartesian coordinates is $\left(-2, 2\sqrt{3}\right)$.
(b) If $(x, y) = \left(\sqrt{2}, \sqrt{2}\right)$, then by (9.35),

$$r^2 = 4 \qquad \text{and} \qquad \theta = \arctan 1.$$

It follows that one possible pair of polar coordinates is $(2, \pi/4)$. Other pairs of polar coordinates are $(-2, 5\pi/4)$ or $(-2, -3\pi/4)$ (see Figure 9.48).

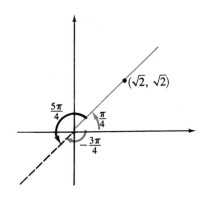

Figure 9.48

Practice Exercise 3

(a) Find the pair of Cartesian coordinates corresponding to $(3, \pi/4)$.
(b) Find two pairs of polar coordinates corresponding to $\left(-\sqrt{3}, 1\right)$.
Answer: **(a)** $\left(3\sqrt{2}/2, 3\sqrt{2}/2\right)$ **(b)** $(2, 5\pi/6)$ or $(-2, -\pi/6)$

Graphing Polar Equations

Any relation of the form $r = f(\theta)$, where *f* is a function of θ, is called a *polar equation*. For example,

$$r = 4 \sin \theta, \qquad r = 2 \cos \theta, \qquad \text{and} \qquad r = 2$$

are all polar equations. The graph of a polar equation consists of all points P that have a pair of polar coordinates (r, θ) satisfying the given equation.

The most basic way of graphing a polar equation is by tabulating some values and then plotting the corresponding points.

EXAMPLE 4

Sketch the graph of the polar equation $r = 4 \sin \theta$.

Solution: Figure 9.49 shows a table of values and the corresponding points plotted on a polar coordinate grid.

θ	r
0	0
$\pi/6$	2
$\pi/3$	$2\sqrt{3}$
$\pi/2$	4
$2\pi/3$	$2\sqrt{3}$
$5\pi/6$	2
π	0
$7\pi/6$	-2
$4\pi/3$	$-2\sqrt{3}$
$3\pi/2$	-4

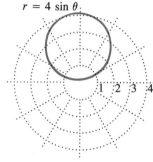

Figure 9.49

Note that the complete curve is obtained as θ varies from 0 to π. When θ varies from π to 2π, the values of $\sin \theta$ (and so the values of r) are negative; according to our convention of plotting points, these values will correspond to points we have already plotted. For example, $(2, \pi/6)$ and $(-2, 7\pi/6)$ are polar coordinates of the same point.

Practice Exercise 4

Graph the polar equation $r = -2 \sin \theta$.

Answer:

θ	r
0	0
$\pi/6$	-1
$\pi/4$	$-\sqrt{2}$
$\pi/2$	-2
$2\pi/3$	$-\sqrt{3}$
π	0

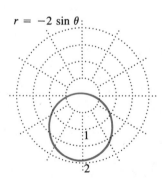

EXAMPLE 5

Graph $r = 1 - \cos \theta$.

Solution: We make a table of values and plot the corresponding points. In our table, we have selected values of θ such that the corresponding values of r are rational numbers.

θ	r
0	0
$\pi/3$	1/2
$\pi/2$	1
$2\pi/3$	3/2
π	2
$4\pi/3$	3/2
$3\pi/2$	1
$5\pi/3$	1/2
2π	0

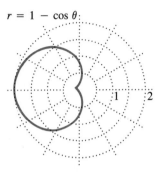

$r = 1 - \cos\theta$

Figure 9.50

Practice Exercise 5

Graph $r = 1 + \cos\theta$.

Answer:

$r = 1 + \cos\theta$

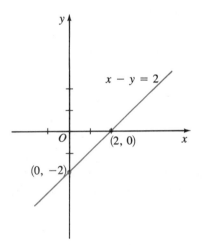

$x - y = 2$

$(2, 0)$

$(0, -2)$

Figure 9.51

On certain occasions, it is convenient to convert a polar equation to one in rectangular coordinates; sometimes it is easier to sketch.

EXAMPLE 6

Graph the polar equation $r = \dfrac{2}{\cos\theta - \sin\theta}$.

Solution: We rewrite the given equation as follows:

$$r(\cos\theta - \sin\theta) = 2$$
$$r\cos\theta - r\sin\theta = 2.$$

Now, using the conversion formulas (9.34), we can write the equation in rectangular coordinates:

$$x - y = 2.$$

This equation represents a straight line (Figure 9.51).

Practice Exercise 6   Convert the equation

$$r = \frac{1}{2\cos\theta + \sin\theta}$$

to rectangular coordinates and graph it.

Answer:

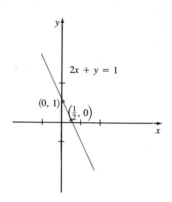

EXERCISES 9.10

1. Plot the points with given polar coordinates.
   (a) $(2, \pi/3)$ (b) $(-3, \pi/6)$ (c) $(4, -\pi/2)$
   (d) $(-1, -2\pi/3)$
2. Plot the points with the following polar coordinates:
   (a) $(4, \pi/2)$ (b) $(-2, 2\pi/3)$ (c) $(2, -5\pi/6)$
   (d) $(-2, -3\pi/2)$
3. Find Cartesian coordinates corresponding to the following polar coordinates.
   (a) $(2, \pi/3)$ (b) $(-2, \pi/4)$ (c) $(-3, -\pi/6)$
   (d) $(4, -\pi/4)$
4. Find rectangular coordinates for the points with the following polar coordinates.
   (a) $(4, \pi/2)$ (b) $(-3, 5\pi/6)$ (c) $(-2, -2\pi/3)$
   (d) $(5, -2\pi/3)$
5. For the given rectangular coordinates, find the corresponding polar coordinates, with $r > 0$ and $0 \le \theta < 2\pi$.
   (a) $(1, 1)$ (b) $\left(-\sqrt{3}, 1\right)$ (c) $(0, -5)$
   (d) $(-3, 3)$
6. Find polar coordinates (with $r > 0$ and $0 \le \theta < 2\pi$) for each of the following points.
   (a) $(-1, 1)$ (b) $\left(-2, 2\sqrt{3}\right)$ (c) $(3, 0)$
   (d) $(4, 4)$

In Exercises 7–16, graph each of the following equations.

7. $r = 2\cos\theta$ 8. $r = 3\cos\theta$
9. $r = 1 + \sin\theta$ 10. $r = 1 - \sin\theta$
11. $r = 2 - 2\cos\theta$ 12. $r = 3 + 3\cos\theta$
13. $r = 2 - \cos\theta$ 14. $r = 3 + \cos\theta$
15. $r = \sin 2\theta$ 16. $r = \cos 2\theta$

For each of the equations in Exercises 17–30, find a corresponding equation in Cartesian coordinates and graph it.

17. $r = 4$ 18. $r = 9$
19. $r = 4\sin\theta$ 20. $r = -2\sin\theta$
21. $r = 3\sec\theta$ 22. $r = 2\csc\theta$
23. $r = \dfrac{2}{3\sin\theta - \cos\theta}$ 24. $r = \dfrac{3}{\sin\theta - 2\cos\theta}$
25. $r\cos\theta - 2 = 0$ 26. $r\sin\theta + 1 = 0$
27. $r(2\sin\theta + \cos\theta) = 1$ 28. $r(\sin\theta - 3\cos\theta) = 4$
29. $r = \dfrac{2}{1 - \sin\theta}$ 30. $r = \dfrac{1}{1 - \cos\theta}$

In Exercises 31–40, convert each of the following equations into polar coordinates.

31. $x^2 + y^2 = 16$ 32. $x^2 + y^2 = 25$
33. $x - 5 = 0$ 34. $y - 3 = 0$
35. $x + 2y = 4$ 36. $2x - y = 3$
37. $x^2 + y^2 - 2y = 0$ 38. $x^2 + y^2 + 6x = 0$
39. $x^2 = 2y$ 40. $y^2 = 4x$

CHAPTER SUMMARY

A *trigonometric identity* is a relationship, involving trigonometric functions, that is satisfied for all values of the variable. From the Pythagorean trigonometric identity,

$$\sin^2 x + \cos^2 x = 1,$$

and from the definitions of tangent, cotangent, secant, and cosecant, other trigonometric identities can be derived. When working with trigonometric identities you should follow the Hints for Verifying Trigonometric Identities at the end of Section 9.1.

The following identities are extremely useful in many applications.

Addition and Subtraction Identities

$$\sin(\alpha \pm \beta) = \sin \alpha \cos \beta \pm \cos \alpha \sin \beta$$

$$\cos(\alpha \pm \beta) = \cos \alpha \cos \beta \mp \sin \alpha \sin \beta$$

$$\tan(\alpha \pm \beta) = \frac{\tan \alpha \pm \tan \beta}{1 \mp \tan \alpha \tan \beta}$$

Double-Angle Identities

$$\sin 2\alpha = 2 \sin \alpha \cos \alpha$$

$$\cos 2\alpha = \cos^2 \alpha - \sin^2 \alpha$$

$$= 1 - 2 \sin^2 \alpha$$

$$= 2 \cos^2 \alpha - 1$$

$$\tan 2\alpha = \frac{2 \tan \alpha}{1 - \tan^2 \alpha}$$

Half-Angle Identities

$$\sin \frac{\alpha}{2} = \pm \sqrt{\frac{1 - \cos \alpha}{2}}$$

$$\cos \frac{\alpha}{2} = \pm \sqrt{\frac{1 + \cos \alpha}{2}}$$

$$\tan \frac{\alpha}{2} = \pm \sqrt{\frac{1 - \cos \alpha}{1 + \cos \alpha}}$$

A *trigonometric equation* is an equation involving trigonometric functions. The difference between a trigonometric equation and a trigonometric identity is that an equation may be satisfied for *some* values of the variable, called *solutions* of the equation, while an identity must be satisfied for *all* values of the variable.

When solving *oblique triangles,* you will frequently use

The Law of Sines

$$\frac{a}{\sin \alpha} = \frac{b}{\sin \beta} = \frac{c}{\sin \gamma}$$

and/or

The Law of Cosines

$$a^2 = b^2 + c^2 - 2bc \cos \alpha.$$

A *vector* **v** *in the plane* can be written as an ordered pair,

$$\mathbf{v} = \langle a, b \rangle,$$

or as a *linear combination* of the vectors **i** (unit vector pointing in the positive *x*-direction) and **j** (unit vector pointing in the positive *y*-direction),

$$\mathbf{v} = a\mathbf{i} + b\mathbf{j}.$$

The numbers *a* and *b* are the *components* of **v**. The *length* or *magnitude* of **v** is the number

$$|\mathbf{v}| = \sqrt{a^2 + b^2}.$$

A *unit vector* is a vector of magnitude 1.

The *inner* or *(dot) product* of two vectors $\mathbf{u} = \langle u_1, u_2 \rangle$ and $\mathbf{v} = \langle v_1, v_2 \rangle$ is defined to be

$$\mathbf{u} \cdot \mathbf{v} = u_1 v_1 + u_2 v_2.$$

The following property of the inner product,

$$\mathbf{u} \cdot \mathbf{v} = |\mathbf{u}| \, |\mathbf{v}| \cos \theta,$$

involves the *angle θ between* two vectors, where

$$\cos \theta = \frac{\mathbf{u} \cdot \mathbf{v}}{|\mathbf{u}| \, |\mathbf{v}|}, \qquad \text{hence} \qquad \theta = \cos^{-1}\left(\frac{\mathbf{u} \cdot \mathbf{v}}{|\mathbf{u}| \, |\mathbf{v}|}\right).$$

Two vectors are *orthogonal* (or *perpendicular*) exactly when their inner product is zero. Every complex number $z = x + iy$ can be represented in *trigonometric* or *polar form*:

$$z = r(\cos \theta + i \sin \theta),$$

$$\text{where}$$

$$r^2 = x^2 + y^2 \qquad \text{and} \qquad \tan \theta = \frac{y}{x}.$$

De Moivre's formula,

$$z^n = r^n(\cos n\theta + i \sin n\theta),$$

allows the computation, in a simple manner, of powers of a complex number in polar form.

The *nth root formula,*

$$w = r^{1/n}\left[\cos\left(\frac{\theta + 2k\pi}{n}\right) + i \sin\left(\frac{\theta + 2k\pi}{n}\right)\right],$$

$$k = 0, 1, \ldots, n - 1$$

gives the n distinct nth roots of the number $z = r(\cos \theta + i \sin \theta)$. As an application, this formula can be used to find the *nth roots of unity*.

In a *polar coordinate system*, the coordinates of a point P are given by an ordered pair (r, θ), where r is the distance from P to the origin O and θ is one of the angles determined by the *polar axis* and the segment OP. Cartesian and polar coordinates are related by the following equations:

$$x = r \cos \theta, \qquad y = r \sin \theta,$$

$$r^2 = x^2 + y^2, \qquad \theta = \arctan\left(\frac{y}{x}\right).$$

CHAPTER TEST

1. Write $\sin 42° \cos 34° + \cos 42° \sin 34°$ as one trigonometric function of a single angle.
2. Express $2 \sin b \cos 4b$ as a sum or difference of trigonometric functions.
3. Let a be a second-quadrant angle such that $\sin a = 2/3$, and let b be an angle such that $0 < b < \pi/2$ and $\cos b = 3/5$. Find $\cos(a + b)$.
4. Let $\cos \alpha = -3/8$, where $180° < \alpha < 270°$. Calculate $\sin 2\alpha$.
5. Find all solutions of the trigonometric equation $\sin 2x - \sqrt{2} \sin x = 0$ in the interval $[0, 2\pi)$.
6. Solve the triangle ABC, given that $\alpha = 65°$, $\beta = 60°$, and $b = 14.5$.
7. Points A and B are 120 and 140 yards away from point C, respectively. If angle ACB measures $48°$, approximate the distance from A to B.
8. Write the complex number $80i$ in trigonometric form.
9. What is the product of the complex numbers $z_1 = 5(\cos 120° - i \sin 120°)$ and $z_2 = 3(\cos 55° + i \sin 55°)$?
10. Find all cube roots of $27(\cos 30° + i \sin 30°)$.

REVIEW EXERCISES

Verify the trigonometric identities in Exercises 1–28.

1. $\cos^2 x(\sec^2 x - 1) = \sin^2 x$ **2.** $\dfrac{\cot^2 u - 1}{1 - \tan^2 u} = \cot^2 u$

3. $\dfrac{1 + \cos u}{\sin u} + \dfrac{\sin u}{1 + \cos u} = 2 \csc u$

4. $\dfrac{1 - \cos^2 x}{\cos^2 x} = \tan^2 x$

5. $\dfrac{\cos u}{1 - \sin u} + \dfrac{\cos u}{1 + \sin u} = 2 \sec u$

6. $\dfrac{\sec x + 1}{\sin x + \tan x} = \csc x$

7. $\dfrac{\tan \theta}{\sec \theta - 1} = \dfrac{\sec \theta + 1}{\tan \theta}$

8. $\dfrac{1}{\csc \alpha + \tan \alpha} = \dfrac{\cos \alpha}{\cot \alpha + \sin \alpha}$

9. $\dfrac{\tan a}{\sin a - 2 \tan a} = \dfrac{1}{\cos a - 2}$

10. $2 \sec^2 u + \dfrac{\csc u}{1 - \csc u} = \dfrac{\csc u}{1 + \csc u}$

11. $\dfrac{\cot x - \cos x}{\cot x + \cos x} = \dfrac{\csc x - 1}{\csc x + 1}$

12. $\dfrac{\cot x + \cos x}{\cos x \cot x} = \dfrac{\cos x \cot x}{\cot x - \cos x}$

13. $\dfrac{\sin a \cos b - \cos a \sin b}{\cos a \cos b + \sin a \sin b} = \dfrac{\tan a - \tan b}{1 + \tan a \tan b}$

14. $\dfrac{\tan a - \tan b}{1 + \tan a \tan b} = \dfrac{\cot b - \cot a}{\cot a \cot b + 1}$

15. $\cot^2 x + 2 = \dfrac{\csc^4 x - 1}{\cot^2 x}$

16. $\sin^4 x - \cos^4 x = 1 - 2 \cos^2 x$

17. $2 - \csc^2 \theta = \dfrac{1 - \cot^4 \theta}{\csc^2 \theta}$

18. $\tan^4 \alpha + \sec^2 \alpha = \sec^4 \alpha - \tan^2 \alpha$

19. $\dfrac{\sin^3 x + \cos^3 x}{\sin x + \cos x} = 1 - \sin x \cos x$

20. $\dfrac{\sec^2 x + \tan^2 x}{1 - \sin^4 x} = \sec^4 x$

21. $(\sin \theta + \cos \theta)^2 = 1 + \sin 2\theta$

22. $1 - \cos 2\theta = \tan \theta \sin 2\theta$

23. $2 \tan \alpha \csc 2\alpha = 1 + \tan^2 \alpha$

24. $\sin 2\alpha = (1 - \cos 2\alpha) \cot \alpha$

25. $2 \tan u \cot 2u = 1 - \tan^2 u$

26. $\sec 2u = \dfrac{\csc^2 u}{\csc^2 u - 2}$

27. $\tan \dfrac{\theta}{2} = \dfrac{\sin \theta}{1 + \cos \theta}$

28. $\cot \dfrac{\theta}{2} = \dfrac{\sin \theta}{1 - \cos \theta}$

C 29. Approximate $\tan 2\alpha$ if $\cos \alpha = -0.5736$ and α is a second-quadrant angle.

C 30. Approximate $\tan \dfrac{\theta}{2}$ if $\sin \theta = 0.9659$ and θ is an acute angle.

In Exercises 31–34, solve each of the following equations over the given interval.

31. $2 \sin \theta + 1 = 0$, $[0, 2\pi)$

32. $\sqrt{3} \cot \theta - 1 = 0$, $[0, 2\pi)$

33. $\tan 2\theta + \sqrt{3} = 0$, $\left[-\dfrac{\pi}{4}, \dfrac{\pi}{4}\right]$

34. $\cot\left(\dfrac{\theta}{2}\right) - \sqrt{3} = 0$, $[0, \pi]$

In Exercises 35–40, find all solutions of the given equations.

35. $\sin \alpha - \cos \alpha = 1$ **36.** $\sec \alpha + \tan \alpha = 1$

37. $4 \sin^2 x - 1 = 0$

38. $(\sin x - 1)(2 \sin x + \sqrt{2}) = 0$

39. $\csc^2 u = \cot u + 1$ **40.** $2 \sin^2 u = 2 - \cos u$

In Exercises 41–44, use the law of sines to solve each triangle. Whenever necessary, approximate angles to the nearest ten minutes and side lengths to the nearest tenth.

41. $\alpha = 28°$, $\beta = 65°$, $a = 15$
42. $\alpha = 78°$, $\beta = 42°$, $b = 12$
43. $\beta = 120°$, $\gamma = 35°$, $a = 18$
44. $\beta = 105°$, $\gamma = 15°$, $c = 10$

In Exercises 45–48, use the law of cosines to solve each triangle. Approximate angles to the nearest ten minutes and side lengths to the nearest tenth.

45. $a = 9$, $b = 12$, $c = 15$
46. $a = 10$, $b = 24$, $c = 26$
47. $a = 4$, $b = 5$, $c = 6$
48. $a = 7$, $b = 9$, $c = 11$

In Exercises 49–74, solve the given problem.

49. A pole is tilted 8° away from the vertical. A guy wire 56 ft long is attached to the top of the pole and forms a 65° angle with the ground. Find the length of the pole to the nearest foot.

50. An antenna 50 m high is situated at the top of a hill. At a point 120 m down the hill, the angle between the surface of the hill and the line of sight to the top of the antenna is 18°30′. Find the angle between the surface of the hill and level ground to the nearest minute.

51. Two men 1.85 km apart observe a balloon in the sky between them. The angles of elevation of the balloon from each of the two men are 68° and 42°. If the two men and the balloon are in the same vertical plane, find the height of the balloon above the ground.

52. From a point on the ground, an observer finds that the angle of elevation of the top of a building is 72°. He moves 120 ft away from the building and finds that the angle of elevation is now 58°. How high is the building?

53. Find to the nearest minute the greatest angle of a triangle whose sides are 10.25, 8.18, and 6.72 units of length.

54. Find to the nearest minute the smallest angle of the triangle considered in Exercise 53.

☐ **55.** Prove that the area of any triangle is half the product of any two sides and the sine of the included angle.

☐ **56.** Let ABC be a triangle with sides a, b, and c facing angles α, β, and γ. Prove that its area is given by

$$\frac{a^2 \sin \beta \sin \gamma}{2 \sin \alpha}.$$

☐ **57.** Let a, b, and c be the lengths of the sides of a triangle, and let α be the angle opposite to side a. Prove each of the following statements.
  (a) If α is acute, then $a^2 < b^2 + c^2$.
  (b) If α is a right angle, then $a^2 = b^2 + c^2$.
  (c) If α is obtuse, then $a^2 > b^2 + c^2$.

☐ **58.** Prove that for every triangle ABC,

$$a^2 + b^2 + c^2 = 2(bc \cos \alpha + ac \cos \beta + ab \cos \gamma).$$

☐ **59.** Show that in any triangle ABC,

$$b^2 - a^2 = c(b \cos \alpha - a \cos \beta).$$

☐ **60.** Prove that for every triangle ABC, the following relation holds:

$$\frac{a^2 + b^2 + c^2}{2abc} = \frac{\cos \alpha}{a} + \frac{\cos \beta}{b} + \frac{\cos \gamma}{c}.$$

61. If $\mathbf{v} = \langle -5, 4 \rangle$ and $A(-3, 4)$ is the initial point of $\mathbf{v}$, what are the coordinates of the terminal point?

62. Find the terminal point of a vector $\mathbf{u} = \langle 4, -8 \rangle$ whose initial point is $P(-4, -3)$.

63. Vectors $\mathbf{u}$ and $\mathbf{v}$ of magnitudes 5 and 8, respectively, make an angle of 120°. Find $|\mathbf{u} + \mathbf{v}|$ and $|\mathbf{u} - \mathbf{v}|$.

64. Let $|\mathbf{PQ}| = 4$ and $|\mathbf{PR}| = 6$ and assume that $\mathbf{PQ}$ and $\mathbf{PR}$ make an angle of 60°. Find $|\mathbf{PQ} + \mathbf{PR}|$ and $|\mathbf{PQ} - \mathbf{PR}|$.

65. What is the unit vector in the direction of the vector $\mathbf{v} = \langle 7, -12 \rangle$?

66. Find the unit vector in the opposite direction of the vector $\mathbf{a} = \langle -3, 8 \rangle$.

67. The magnitude of a vector is 8 units and its direction angle is 45°. Write the vector as a sum of its horizontal and vertical components.

68. Let 10 units and 150° be the length and direction angle of a vector. Find the horizontal and vertical components of the vector.

69. What is the angle between each of the following pairs of vectors?
  (a) $\mathbf{a} = 3\mathbf{i} - 4\mathbf{j}$, $\mathbf{b} = -6\mathbf{i} + 3\mathbf{j}$
  (b) $\mathbf{r} = \langle 8, 3 \rangle$, $\mathbf{s} = \langle -8, 4 \rangle$

70. Find the angle between each of the following pairs of vectors.
  (a) $\mathbf{p} = -4\mathbf{i} + 2\mathbf{j}$, $\mathbf{q} = 5\mathbf{i} - 3\mathbf{j}$
  (b) $\mathbf{u} = \langle -2, 6 \rangle$, $\mathbf{v} = \langle -4, -4 \rangle$

71. What force is required to keep a 4000-lb vehicle parked on a hill that makes a 30° angle with the horizontal?

72. Find the force necessary to push a 560-kg box up a ramp inclined 25° with the horizontal. Assume that friction is neglected.

73. Two forces acting on a point and making an angle of 60° have a resultant of 252 lb. If one of the forces is twice the other, find the magnitude of each force.

74. Two forces of equal magnitude act on a point and make an angle of 120°. If the resultant has a magnitude of 372 N, what is the magnitude of each force?

In Exercises 75–80, write each of the given numbers in trigonometric form.

75. $1 + i$

76. $5 - 5i$

77. $3\sqrt{3} + 3i$

78. $\frac{1}{2} + \frac{1}{2}i$

79. $-4i$

80. $6i$

In Exercises 81–84, find $z_1 z_2$ and z_1/z_2 for each of the following pairs of numbers.

81. $z_1 = 2[\cos(\pi/6) + i \sin(\pi/6)]$,
  $z_2 = 3[\cos(\pi/3) + i \sin(\pi/3)]$

82. $z_1 = -3(\cos 15° + i \sin 15°)$,
  $z_2 = \cos 120° + i \sin 120°$

83. $z_1 = \sqrt{2}(\cos 45° + i \sin 45°)$,
  $z_2 = \sqrt{3}(\cos 285° + i \sin 285°)$

84. $z_1 = 3[\cos(7\pi/6) + i \sin(7\pi/6)]$,
  $z_2 = -2[\cos(\pi/3) + i \sin(\pi/3)]$

In Exercises 85–90, use De Moivre's formula to find each of the following complex numbers.

85. $(1 + i)^4$ **86.** $(\sqrt{3} + i)^5$

87. $(1 - i\sqrt{3})^6$ **88.** $(-1 + i)^8$

89. $[3(\cos 25° + i \sin 25°)]^4$

90. $[\sqrt{3}(\cos 150° + i \sin 150°)]^3$

In Exercises 91–96, find all the nth roots of z, for n and z as indicated. Whenever possible, write your answers in rectangular form.

91. $z = 1 - i\sqrt{3}$, $n = 2$ **92.** $z = -16i$, $n = 4$

93. $z = -8 - 8i$, $n = 3$

94. $z = 16 + 16i\sqrt{3}$, $n = 5$

95. $z = 81(\cos 220° + i \sin 220°)$, $n = 4$

96. $z = 64(\cos 270° + i \sin 270°)$, $n = 3$

In Exercises 97–100, find all solutions of the given equations.

97. $x^3 + 27 = 0$ **98.** $x^4 + 4 = 0$

99. $x^4 - 16i = 0$ **100.** $x^3 + 8i = 0$

101. Find the rectangular coordinates of the points whose polar coordinates are as given.

   (a) $(3, \pi/4)$ **(b)** $(-2, \pi/3)$ **(c)** $(4, -5\pi/6)$

102. Find Cartesian coordinates corresponding to the following polar coordinates.

   (a) $(5, \pi/6)$ **(b)** $(3, -\pi/4)$ **(c)** $(-3, 2\pi/3)$

Graph each of the following equations in Exercises 103 and 104.

103. $r = 3 \sin \theta$ **104.** $r = -4 \sin \theta$

For each of the following equations in Exercises 105–108, find a corresponding equation in Cartesian coordinates and graph it.

105. $r = 4 \csc \theta$ **106.** $r = 2 \sec \theta$

107. $r = \dfrac{3}{2 \sin \theta + \cos \theta}$ **108.** $r = \dfrac{-2}{\sin \theta - 3 \cos \theta}$

☐ **109.** Show that $\left(\sqrt{2}/2\right) + \left(\sqrt{2}/2\right)i$ is a solution of $x^4 + x^2 + 1 = i$.

☐ **110.** Show that $\cos(\pi/4) - i \sin(\pi/4)$ is a solution of $x^6 + x^4 + x^2 + 1 = 0$.

CHAPTER 10

■

SYSTEMS

OF

EQUATIONS

In most of this chapter we will discuss various methods of solving systems of linear equations. We begin with a review of the methods of substitution and elimination for solving two-by-two linear systems. Next, we will explain the powerful and simple Gaussian elimination method of solving general systems of linear equations. Matrices and their properties are discussed in Section 10.3. Every system of linear equations can be written in matrix form, and if the matrix of the coefficients is invertible, then the system is solvable by matrix algebra. In Section 10.4 we define determinants, describe their properties, and discuss Cramer's rule for solving linear systems. Section 10.5 contains a brief discussion on nonlinear systems of equations. Finally, in the last section, we study systems of linear inequalities and discuss linear programming.

10.1 SYSTEMS OF TWO LINEAR EQUATIONS

Suppose that a manufacturer produces two types of toys, cars and trucks, and that each requires wheels and axles according to the following table:

	Cars	Trucks
Wheels	4	10
Axles	2	3

454

Such a table is called a *production matrix*. How many cars and trucks can be produced if the manufacturer has a stock of 300 wheels and 110 axles and uses his entire stock? To solve this problem, let x be the number of cars and y the number of trucks to be produced. According to the table, four wheels are needed for each car and ten wheels for each truck, so the equation describing the use of the wheels is $4x + 10y = 300$. Also, two axles are needed for each car and three axles for each truck, so we get the equation $2x + 3y = 110$. Thus, we obtain the system of two linear equations in two *unknowns, x and y:*

$$\begin{cases} 4x + 10y = 300 \\ 2x + 3y = 110. \end{cases}$$

Such a system is also referred to as a *two-by-two linear system*. To *solve the system* means to find two numbers that, when substituted for x and y, satisfy both equations. You can check, by direct substitution, that the solution of the system is $x = 25$ and $y = 20$. Thus, if the entire stock of wheels and axles is to be used, the manufacturer must produce 25 cars and 20 trucks.

We frequently encounter systems of linear equations in the course of solving problems in mathematics or the applied sciences. Depending on the particular problem, the system we want to solve may have two, three, or more linear equations in several variables.

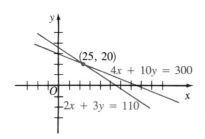

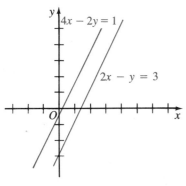

Figure 10.1

Geometric Interpretation

Consider the problem of finding all real numbers x and y satisfying the preceding system of equations. Each equation represents a line in the Cartesian plane, as illustrated in Figure 10.1. Since the two lines are neither parallel nor identical, they have a unique point of intersection whose coordinates are obtained by solving the system of linear equations. Since the solution of the system is $x = 25$ and $y = 20$, the coordinates of the point of intersection are (25, 20). Conversely, if we can find the coordinates of the point of intersection by geometric means, then we will also have the solution of the two-by-two system.

Consistent, Inconsistent, and Dependent Systems

There are systems of equations that do not have solutions. For example, the two equations in the system

$$\begin{cases} 2x - y = 3 \\ 4x - 2y = 1 \end{cases}$$

represent parallel lines (Figure 10.2). Since these lines are distinct, they do not intersect. Thus, the system does not have any solutions. We can also reason by contradiction. Multiplying both sides of the first equation by 2, we get

$$\begin{cases} 4x - 2y = 6 \\ 4x - 2y = 1. \end{cases}$$

But this implies that $6 = 1$, which is absurd. Thus, there cannot be any pair of values for x and y that make both equations true simultaneously.

There are systems of equations that have infinitely many solutions. For example, consider the system

Figure 10.2

$$\begin{cases} 2x + 3y = 1 \\ 4x + 6y = 2. \end{cases}$$

Although the two equations are formally different, they represent the same line (Figure 10.3). In fact, the second equation is a multiple of the first. Every point on the line—such as $(0, 1/3)$, $(1/2, 0)$, $(2, -1)$, and so on—is a solution of the system. Since the line contains infinitely many points, the system has infinitely many solutions.

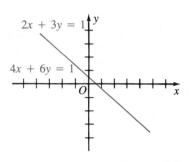

$2x + 3y = 1$

$4x + 6y = 1$

Figure 10.3

Summary

These examples show that a two-by-two linear system may have a *unique solution, no solutions,* or an *infinite number of solutions.* A system that has a unique solution is called *consistent;* a system with no solutions is called *inconsistent;* and a system with infinitely many solutions is called *dependent.*

Now we review the methods of substitution and elimination for solving two-by-two linear systems.

Solving by Substitution

With this method we solve one of the equations for one of the variables. Then we substitute the expression just obtained into the second equation and solve for the remaining variable.

EXAMPLE 1   Solve the system

$$\begin{cases} 2x + 4y = 10 \\ x - 2y = -7. \end{cases}$$

Solution: Solving the second equation for x, we get

$$x = 2y - 7. \tag{10.1}$$

Next we substitute $2y - 7$ for x in the first equation, obtaining an equation in the variable y that we solve for y:

$$2(2y - 7) + 4y = 10$$
$$4y - 14 + 4y = 10$$
$$8y = 24$$
$$y = 3.$$

Substituting 3 for y in (10.1), we obtain x:

$$x = 2(3) - 7 = -1.$$

We could instead have substituted 3 for y in either of the two equations of the given system to obtain the same $x = -1$.

Now to check that $x = -1$ and $y = 3$ are solutions of the given system, we substitute -1 for x and 3 for y into the two equations and verify that the result is an identity in each case:

$$2(-1) + 4(3) = 10$$
$$(-1) - 2(3) = -7.$$

Practice Exercise 1   Find the solution of the system

$$\begin{cases} 3x - 2y = 7 \\ 4x - y = 6. \end{cases}$$

Answer: $x = 1, y = -2$

Solving by Elimination

With this method we eliminate one of the variables between the two equations, obtaining a single equation in the remaining variable. This equation is then solved for the remaining variable.

EXAMPLE **2**   Solve the system

$$\begin{cases} x - y = 1 \\ 2x + y = 5. \end{cases}$$

Solution: By adding the two equations, we eliminate the variable y:

$$\begin{array}{r} x - y = 1 \\ +\ \underline{2x + y = 5} \\ 3x\quad\ \ = 6 \end{array}$$

so

$$x = 2.$$

Substituting 2 for x in the first equation, say, and solving for y, we get

$$2 - y = 1$$
$$y = 1.$$

You can check that $(2, 1)$ is the solution of the system.

Practice Exercise 2   Solve the two-by-two system

$$\begin{cases} 4x - 3y = 9 \\ x + 3y = 21. \end{cases}$$

Answer: $x = 6, y = 5$

Sometimes it is necessary to multiply one equation, or both, by constants before eliminating one of the variables.

EXAMPLE 3

Solve the system

$$\begin{cases} 4x + 10y = 300 \\ 2x + 3y = 110. \end{cases}$$

Solution: If a linear system has a solution, then multiplying one equation, or both, by nonzero numbers does not affect the solution of the system. Multiplying the second equation by -2 and leaving the first equation unchanged, we obtain the new system

$$\begin{cases} 4x + 10y = 300 \\ -4x - 6y = -220. \end{cases}$$

Adding the two equations, we eliminate x and get

$$4y = 80$$
$$y = 20.$$

Substituting 20 for y in the first equation gives us

$$4x + 10 \cdot 20 = 300$$
$$4x = 100$$
$$x = 25.$$

The solution of the system is then $x = 25$ and $y = 20$.

HISTORICAL NOTE

LINEAR SYSTEMS

Linear systems have a long history in East Asia. The consideration of systems of linear equations that are defined by the square array of their coefficients and are solved by means of what are now called matrix transformations is found in the Chinese classic, *Nine Chapters on the Mathematical Art*, which dates from the period of the Han dynasty (206 B.C.–A.D. 220).

Practice Exercise 3

Solve the system

$$\begin{cases} 3x - 5y = 11 \\ 4x + 3y = 5. \end{cases}$$

Answer: $x = 2, y = -1$

Note that in Example 2, we *added* the two equations in order to *eliminate* one of the variables, while in Example 3, we *multiplied* the second equation by a *nonzero constant* before adding the two equations. The method described in both examples is a particular case of a general method for solving systems of linear equations called *Gaussian elimination*, which will be discussed in detail in the next section.

Many word problems can be solved by using two-by-two linear systems.

EXAMPLE 4

A silversmith has two bars of silver alloy in stock. One alloy is 50% silver and the other is 75% silver. How many grams of each alloy should he use to obtain 40 g of a 60% silver alloy?

Solution: If x denotes the number of grams of the 50% silver alloy and y the number of grams of the 75% silver alloy to be used, then

$$x + y = 40.$$

The amount of silver in x grams of the 50% alloy is $0.5x$, the amount of silver in y grams of the 75% alloy is $0.75y$, and the amount of silver in the 40 g of the 60% alloy to be obtained is $0.6 \times 40 = 24$ g. Thus

$$0.5x + 0.75y = 24.$$

We must now solve the system

$$\begin{cases} x + \quad y = 40 \\ 0.5x + 0.75y = 24. \end{cases}$$

Proceeding by substitution, we get

$$y = 40 - x.$$

Next,

$$0.5x + 0.75(40 - x) = 24$$
$$0.5x + 30 - 0.75x = 24$$
$$-0.25x + 30 = 24$$
$$-0.25x = -6$$
$$x = \frac{-6.00}{-0.25}$$
$$x = 24\,\text{g}.$$

Then

$$y = 40 - 24 = 16\,\text{g}.$$

Thus he should use 24 grams of the 50% silver alloy and 16 grams of the 75% silver alloy.

Practice Exercise 4   Find the dimensions of a rectangle whose perimeter measures 54 ft, if its length is 3 ft longer than its width.

Answer: 15 ft by 12 ft

EXERCISES 10.1

Solve each of Exercises 1–22 by substitution or elimination.

1. $\begin{cases} 4x - 3y = 7 \\ \quad 2y = 6 \end{cases}$

2. $\begin{cases} 3x \quad\quad = 15 \\ 5x - 4y = \ 1 \end{cases}$

3. $\begin{cases} 3x - 8y = -1 \\ \quad 2x = \ 5y \end{cases}$

4. $\begin{cases} 4x - 3y = \quad 0 \\ 2x - 5y = -14 \end{cases}$

5. $\begin{cases} 3x - 2y = \ 7 \\ \ x + 4y = 21 \end{cases}$

6. $\begin{cases} 2x - 4y = 10 \\ 5x + 3y = 12 \end{cases}$

7. $\begin{cases} 4x - 3y = \ 6 \\ 2x + 5y = 16 \end{cases}$

8. $\begin{cases} \ x + 2y = 5 \\ 5x + \ y = 2 \end{cases}$

9. $\begin{cases} 5x + 7y = \ 9 \\ 2x + 6y = -6 \end{cases}$

10. $\begin{cases} 2x - 4y = \ 8 \\ \ x + 5y = -17 \end{cases}$

11. $\begin{cases} 8x + \ 6y = -1 \\ 4x + 10y = -4 \end{cases}$

12. $\begin{cases} 9x - \ 5y = 1 \\ 6x + 15y = 8 \end{cases}$

13. $\begin{cases} 5x - 6y = 1 \\ 3x - 2y = 7 \end{cases}$

14. $\begin{cases} \ 2x + 5y = \ 7 \\ -4x + 3y = 25 \end{cases}$

15. $\begin{cases} 6x - 4y = 2 \\ 9x - 2y = 5 \end{cases}$

16. $\begin{cases} 3x - 10y = -1 \\ 6x + \ 5y = \ 3 \end{cases}$

17. $\begin{cases} 0.5x - 0.25y = 3.5 \\ 1.2x - 1.5y \ = 3 \end{cases}$

18. $\begin{cases} 0.8x + \ 4y = \ 10 \\ \ 2x - 10y = -9 \end{cases}$

19. $\begin{cases} \dfrac{2}{3}x - \dfrac{1}{2}y = 2 \\[2mm] \dfrac{3}{4}x - \dfrac{2}{3}y = \dfrac{11}{6} \end{cases}$

20. $\begin{cases} \dfrac{5}{6}x - \dfrac{3}{8}y = 2 \\[2mm] \dfrac{1}{2}x + \dfrac{3}{4}y = 9 \end{cases}$

21. $\begin{cases} \dfrac{x}{5} + \dfrac{y}{3} = \dfrac{1}{15} \\ -\dfrac{x}{2} + 2y = \dfrac{11}{2} \end{cases}$ **22.** $\begin{cases} 2x - \dfrac{1}{2}\,y = \dfrac{13}{2} \\ \dfrac{x}{2} + \dfrac{y}{3} = -\dfrac{2}{3} \end{cases}$

In Exercises 23–28, solve each system. [*Hint:* For the first four, set $u = 1/x$ and $v = 1/y$ and then solve the corresponding system. What change of variables should you use to solve the last two?)

23. $\begin{cases} \dfrac{5}{x} - \dfrac{1}{y} = 7 \\ \dfrac{3}{x} + \dfrac{2}{y} = -1 \end{cases}$ **24.** $\begin{cases} \dfrac{1}{x} - \dfrac{3}{y} = 9 \\ -\dfrac{2}{x} - \dfrac{4}{y} = 2 \end{cases}$

25. $\begin{cases} \dfrac{4}{x} - \dfrac{3}{y} = 5 \\ -\dfrac{3}{x} + \dfrac{7}{y} = 1 \end{cases}$ **26.** $\begin{cases} \dfrac{5}{x} + \dfrac{2}{y} = 4 \\ -\dfrac{3}{x} - \dfrac{5}{y} = 9 \end{cases}$

27. $\begin{cases} \dfrac{2}{x-1} - \dfrac{3}{y-2} = 3 \\ \dfrac{1}{x-1} + \dfrac{4}{y-2} = 7 \end{cases}$ **28.** $\begin{cases} \dfrac{3}{x+2} + \dfrac{2}{y-3} = 4 \\ \dfrac{1}{x+2} - \dfrac{3}{y-3} = 5 \end{cases}$

In Exercises 29–40, write the two-by-two linear system corresponding to each problem and then solve the system.

29. Two zinc alloys contain 20% and 50% of zinc. How many kilograms of each alloy should be mixed to obtain 30 kg of a 45% alloy?

30. A chemical manufacturer wishes to prepare 600 gallons of a 50% acid solution by mixing a 60% acid solution with a 36% acid solution. How many gallons of each solution should he mix?

31. On a river, a boat took 3 h to travel 24 miles upstream and $1\frac{1}{2}$ h to return. Find the speed of the boat in still water and the speed of the current.

32. With the aid of a tail wind, an airplane travels 1050 miles in 3 h and 30 min. Flying against the same wind, the airplane takes 4 h and 12 min to travel the same distance. Find the speed of the airplane in still air and the speed of the wind.

33. A girl has 30 coins consisting of nickels and dimes. If the total amount is $2.10, how many nickels and dimes does the girl have?

34. Peter spends $6.32 buying 11-cent and 18-cent stamps. If he buys ten more 11-cent stamps than 18-cent stamps, how many stamps of each denomination does he buy?

35. On a 200-mile-long highway, two cars start at the same time from opposite ends of the highway and travel towards each other with constant speed. The first car travels 10 mph faster than the other car. If the two cars pass each other after 2 h, find the speed of each car.

36. Two trains leave towns 120 km apart and travel toward each other at constant speed. If they pass each other after 45 min and one train travels 20 km/h faster than the other, find the speed of each train in kilometers per hour.

37. If a rectangle has perimeter 140 cm and its width is equal to 3/4 of its length, what are its dimensions?

38. The perimeter of a rectangle is 36 ft long. Find the dimension of the rectangle if the length is 3 ft longer than twice the width.

39. The following table shows the percentage of zinc and copper in two alloys, A and B.

	Zinc	Copper
Alloy A	30	50
Alloy B	20	60

How many kilograms of each alloy should be mixed to obtain 100 kg of a new alloy containing 26% zinc?

40. Use the table in Exercise 39 to find how many grams of each alloy should be combined to obtain 60 g of a new alloy containing 56% copper.

10.2 SYSTEMS OF LINEAR EQUATIONS IN MORE THAN TWO VARIABLES; GAUSSIAN ELIMINATION

A *linear equation in n unknowns* $x_1, x_2, \ldots, x_n$ is an equation of the form

$$a_1 x_1 + a_2 x_2 + \cdots + a_n x_n = b,$$

where $a_1, a_2, \ldots, a_n$ and b are real numbers, and $x_1, x_2, \ldots, x_n$ (also called *variables*) are quantities to be determined. The number a_i is the *coefficient* of x_i in the equation. A *solution* of the equation is an *n*-tuple of real numbers $(\alpha_1, \alpha_2, \ldots, a_n)$ such that

$$a_1\alpha_1 + a_2\alpha_2 + \cdots + a_n\alpha_n = b.$$

For example, $3x_1 - x_2 + 2x_3 = 5$ is a linear equation in three unknowns. The numbers 3, -1, and 2 are respectively the coefficients of x_1, x_2, and x_3. The triples $(1, 0, 1)$, $(2, 1, 0)$, and $(0, -1, 2)$ are examples of solutions of this equation.

A collection of linear equations is called a *system of linear equations*. For example,

$$\begin{cases} 2x_1 - 3x_2 + x_3 = 1 \\ x_1 + 2x_2 - x_3 = 3 \end{cases}$$

is a system of two linear equations in three unknowns.

A *solution* of a system in *n* unknowns is an *n*-tuple of real numbers that satisfies each equation of the system. For example, $(1, 2, 3)$ is a solution of the system

$$\begin{cases} 2x - y + z = 3 \\ x + 2y - z = 2 \\ 3x - 2y + 2z = 5. \end{cases}$$

Here, for simplicity of notation, the variables are denoted by x, y, and z instead of x_1, x_2, and x_3.

The General Form of Three-by-Three Equations

The general form of a system of three linear equations in three unknowns is

$$\begin{cases} a_1x + b_1y + c_1z = d_1 \\ a_2x + b_2y + c_2z = d_2 \\ a_3x + b_3y + c_3z = d_3. \end{cases} \qquad (10.2)$$

Consistent, Inconsistent, and Dependent Systems

It can be shown that a system of three linear equations in three unknowns has either a *unique solution*, *no solution*, or *infinitely many solutions*. If the solution is unique, the system is said to be *consistent*; if no solutions exist, the system is said to be *inconsistent*; and if there is an infinite number of solutions, the system is said to be *dependent*. The same definitions apply to the case of a system of *n* linear equations in *n* unknowns.

A Geometric Interpretation of Solutions

We already know that a Cartesian system of coordinates can be defined on a plane. Points are represented by ordered pairs of real numbers and each ordered pair of real numbers determines a point. In a similar way, a Cartesian system of coordinates can be defined in a three-dimensional Euclidean space. Points are now represented by *ordered triples* (x, y, z) of real numbers, and each ordered triple of real numbers determines a unique point in the Euclidean space.

If a Cartesian coordinate system is defined on a three-dimensional space, then it can be shown that a linear equation

$$ax + by + cz = d$$

represents a *plane*. Thus finding the solution of the system (10.2) corresponds to locating the points of intersection of three planes.

If the system has a unique solution, then the three planes intersect at a single point P (Figure 10.4).

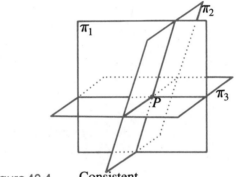

Figure 10.4   Consistent

If the system has no solution, then the three planes have no point in common. In this case, the three planes may be parallel (Figure 10.5a); or two planes may be parallel and the third plane either coincides with one of the two or crosses them (Figure 10.5b); or the three planes may intersect along three parallel lines (Figure 10.5c).

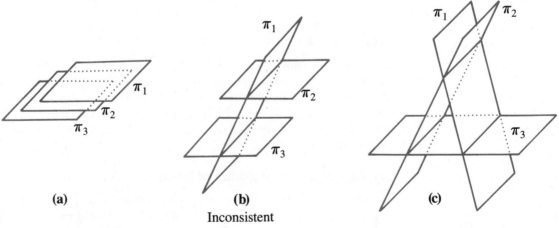

(a)

(b)

Inconsistent

(c)

Figure 10.5

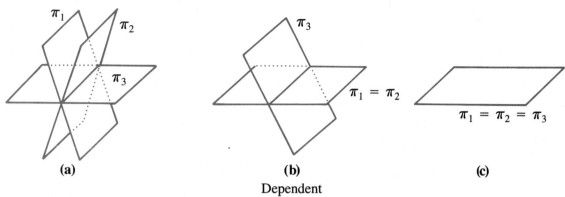

(a)   **(b)**   **(c)**

Dependent

Figure 10.6

If the system has an infinite number of solutions, then either the three planes have a common line of intersection (Figure 10.6a), or two planes coincide and the third plane is different (Figure 10.6b), or all three planes coincide (Figure 10.6c).

Solving by Substitution

Systems of linear equations such as (10.2) can be solved by the method of substitution: we solve any one of the equations for one of the variables; replace it in the other two equations to obtain a two-by-two linear system; and then solve the new system using one of the methods described in the previous section.

EXAMPLE 1   Solve the system

$$\begin{cases} 2x - y + z = 3 \\ x + 2y - z = 2 \\ 3x - 2y + 2z = 5. \end{cases} \tag{10.3}$$

Solution: Solving the second equation for z, we get

$$z = -2 + x + 2y.$$

Substituting for z in the first and third equations, we obtain

$$\begin{cases} 2x - y + (-2 + x + 2y) = 3 \\ 3x - 2y + 2(-2 + x + 2y) = 5, \end{cases}$$

which simplifies to the two-by-two linear system

$$\begin{cases} 3x + y = 5 \\ 5x + 2y = 9. \end{cases}$$

Solving the first equation for y gives us

$$y = 5 - 3x.$$

Substituting for y in the second equation and simplifying, we get

$$5x + 2(5 - 3x) = 9,$$

or

$$-x = -1,$$

hence $x = 1$. To find y we substitute 1 for x in the equation $y = 5 - 3x$, obtaining

$$y = 5 - 3(1) = 2.$$

Finally, to find z, we substitute for x and y in the equation $z = -2 + x + 2y$, and obtain

$$z = -2 + 1 + 2(2) = 3.$$

The ordered triple $(1, 2, 3)$ is then the unique solution of the system. We leave it to you to check that the triple $(1, 2, 3)$ satisfies each equation of the system.

Remember that to check the solution you must substitute 1 for x, 2 for y, and 3 for z into *all three* equations of the system (10.3) and verify that the result is an identity in each case.

Practice Exercise 1   Solve the three-by-three linear system

$$\begin{cases} x - 2y + z = -2 \\ 3x + y - z = -3 \\ -x + 2y + 3z = 6. \end{cases}$$

Answer: $x = -1, y = 1, z = 1$

The method of substitution extends without difficulty to any system of linear equations. However, this method has several drawbacks: it can be too lengthy, time consuming, and tedious. Instead, it is preferable to use the *Gaussian elimination method*, which allows us to transform the given system into a simpler *equivalent* system.

Equivalent Systems

Two systems of linear equations in n unknowns are said to be *equivalent* if they have the same solutions.

For example, the systems

$$\begin{cases} x - y = 1 \\ 2x + y = 5 \end{cases} \quad \text{and} \quad \begin{cases} 2x - 2y = 2 \\ 4x - y = 7 \end{cases}$$

are equivalent since they have the same solution, $(2, 1)$. Also, the system (10.3) is equivalent to the system

$$\begin{cases} x + 2y - z = 2 \\ -5y + 3z = -1 \\ z = 3. \end{cases} \tag{10.4}$$

The linear system (10.4) is said to be in *upper-triangular form*. To solve it, notice that the last equation is already solved for z. Substituting 3 for z in the second equation and solving for y, we obtain $y = 2$. Finally, setting $y = 2$ and $z = 3$ in the first equation and solving for x, we get $x = 1$.

Gaussian Elimination

Any operation that transforms a system of linear equations into an equivalent one is called an *admissible operation*. In the Gaussian elimination method, a given system is transformed into an equivalent one in upper- (or lower-) triangular form by means of admissible operations. The solution of the triangular system, which is easily obtained, is the solution of the original system.

The following are three admissible operations that may be applied to a system to get an equivalent one.

I. Changing the order in which the equations are listed.

II. Multiplying any equation by a nonzero number.

For example, the systems

$$\begin{cases} x - y = 1 \\ 2x + y = 5 \end{cases} \quad \text{and} \quad \begin{cases} 2x - 2y = 2 \\ 2x + y = 5 \end{cases}$$

are equivalent. The first equation in the second system is twice the first equation in the first system.

III. Replacing any equation by the sum of that equation and another equation in the system.

For example, the systems

$$\begin{cases} 2x - 2y = 2 \\ 2x + y = 5 \end{cases} \quad \text{and} \quad \begin{cases} 2x - 2y = 2 \\ 4x - y = 7 \end{cases}$$

are equivalent. The second equation in the second system is obtained by adding the first and second equations in the first system.

When transforming a system into an equivalent one, we often combine the admissible operations II and III. For example, the two systems

$$\begin{cases} x - y = 1 \\ 2x + y = 5 \end{cases} \quad \text{and} \quad \begin{cases} x - y = 1 \\ 4x - y = 7 \end{cases}$$

are equivalent since they have the same solution, $(2, 1)$. The second system is obtained from the first one by adding 2 times the first equation to the second.

In the following example, we show how the three-by-three linear system (10.3) of Example 1 can be transformed into the upper-triangular system (10.4).

EXAMPLE 2   Solve the system

$$\begin{cases} 2x - y + z = 3 \\ x + 2y - z = 2 \\ 3x - 2y + 2z = 5. \end{cases}$$

Solution: First, we interchange the first and second equations (admissible operation I) and write the system as follows:

$$\begin{cases} x + 2y - z = 2 \\ 2x - y + z = 3 \\ 3x - 2y + 2z = 5. \end{cases}$$

By doing this, we obtain a system in which the coefficient of x in the first equation is 1. Next, using admissible operations II and III, we can eliminate the terms containing x in the second and third equations. Adding -3 times the first equation to the third (this is indicated below with the notation $-3R_1 + R_3$), we obtain the equivalent system

$$\begin{cases} x + 2y - z = 2 \\ 2x - y + z = 3 \qquad -3R_1 + R_3 \\ - 8y + 5z = -1. \end{cases}$$

Adding -2 times the first equation to the second, we get the system

$$\begin{cases} x + 2y - z = 2 \\ -5y + 3z = -1 \qquad -2R_1 + R_2 \\ -8y + 5z = -1. \end{cases}$$

We now work with the second and third equations, seeking to eliminate the term containing y in the third equation. Adding $-8/5$ times the second equation to the third, we get the upper-triangular system

$$\begin{cases} x + 2y - z = 2 \\ -5y + 3z = -1 \qquad -\dfrac{8}{5}R_2 + R_3 \\ \dfrac{1}{5}z = \dfrac{3}{5}. \end{cases}$$

Finally, if we multiply the last equation by 5, we obtain the system (10.4),

$$\begin{cases} x + 2y - z = 2 \\ -5y + 3z = -1 \qquad 5R_3 \\ z = 3, \end{cases}$$

whose solution is $x = 1$, $y = 2$, and $z = 3$.

Practice Exercise 2   Using the Gaussian elimination method, transform the linear system

$$\begin{cases} x - 2y + z = -2 \\ 3x + y - z = -3 \\ -x + 2y + 3z = 6 \end{cases}$$

into an equivalent system in upper-triangular form and solve it.

Answer: $x = -1, y = 1, z = 1$

The method of elimination for two-by-two systems described in Section 10.1 is a variant of the Gaussian elimination method. To see this, consider the two-by-two linear system

$$\begin{cases} 4x + 10y = 300 \\ 2x + 3y = 110, \end{cases}$$

discussed in Example 3 of Section 10.1. Multiplying the second equation by -2 (admissible operation II), we get the equivalent system

$$\begin{cases} 4x + 10y = 300 \\ -4x - 6y = -220. \end{cases}$$

Adding the first equation to the second while leaving the first equation unchanged (admissible operation III), we obtain the upper-triangular system

$$\begin{cases} 4x + 10y = 300 \\ 4y = 80. \end{cases}$$

Solving the last equation for y, we get $y = 20$. Substituting 20 for y in the first equation yields $x = 25$.

The Matrix of Coefficients and the Augmented Matrix

It is useful to observe that when we solve a system of linear equations by Gaussian elimination, the computations involve only the coefficients and constant terms of the equations. Since the variables play no role in the calculations, the process can be speeded up by omitting them. Referring to the system considered in Example 2, the array of numbers

$$\begin{bmatrix} 2 & -1 & 1 \\ 1 & 2 & -1 \\ 3 & -2 & 2 \end{bmatrix},$$

consisting of the coefficients of x, y, and z, is called the *matrix of coefficients* of the system. If we add a column consisting of the constant terms 3, 2, and 5, we obtain the *augmented matrix* of the system,

$$\begin{bmatrix} 2 & -1 & 1 & | & 3 \\ 1 & 2 & -1 & | & 2 \\ 3 & -2 & 2 & | & 5 \end{bmatrix}.$$

Matrices will be studied in detail in the next section.

Two augmented matrices are said to be *row-equivalent* if they are augmented matrices of two equivalent systems of equations. For instance, the matrices

$$\begin{bmatrix} 2 & -1 & 1 & | & 3 \\ 1 & 2 & -1 & | & 2 \\ 3 & -2 & 2 & | & 5 \end{bmatrix} \quad \text{and} \quad \begin{bmatrix} 1 & 2 & -1 & | & 2 \\ 0 & -5 & 3 & | & -1 \\ 0 & 0 & 1 & | & 3 \end{bmatrix}$$

are row-equivalent, since they are augmented matrices of the two equivalent systems considered in Example 2.

Admissible Row Operations

An augmented matrix can be transformed into a row-equivalent matrix by performing *admissible row operations*. Such operations are the following:

  I. Interchanging two rows;
  II. Multiplying any row by a nonzero number;
  III. Replacing any row by the sum of its elements and the elements of another row.

Observe that these admissible row operations correspond to the admissible operations for systems of equations listed previously.

In the example that follows, we solve a system of linear equations by the augmented matrix method.

EXAMPLE 3   Solve the linear system

$$\begin{cases} 2x - y + 3z = 4 \\ x \quad\;\; - 4z = 6 \\ 5x - y + 2z = 11. \end{cases}$$

Solution: We first write the augmented matrix of the system:

$$\begin{bmatrix} 2 & -1 & 3 & \vdots & 4 \\ 1 & 0 & -4 & \vdots & 6 \\ 5 & -1 & 2 & \vdots & 11 \end{bmatrix}.$$

Interchanging the first and second rows (which we indicate by $R_1 \leftrightarrow R_2$) according to row operation I, we obtain the row-equivalent matrix

$$\begin{bmatrix} 1 & 0 & -4 & \vdots & 6 \\ 2 & -1 & 3 & \vdots & 4 \\ 5 & -1 & 2 & \vdots & 11 \end{bmatrix}. \quad R_1 \leftrightarrow R_2$$

This matrix corresponds to interchanging the first and second equations of the given system of equations. Next we want to obtain zeros in the first entries of the second and third rows. This is the same as eliminating the terms containing x in the second and third equations of the corresponding system. Adding -5 times the first row to the third one, we get

$$\begin{bmatrix} 1 & 0 & -4 & \vdots & 6 \\ 2 & -1 & 3 & \vdots & 4 \\ 0 & -1 & 22 & \vdots & -19 \end{bmatrix}. \quad -5R_1 + R_3$$

Adding -2 times the first row to the second row, we have

$$\begin{bmatrix} 1 & 0 & -4 & \vdots & 6 \\ 0 & -1 & 11 & \vdots & -8 \\ 0 & -1 & 22 & \vdots & -19 \end{bmatrix}. \quad -2R_1 + R_2$$

Next, without changing the first column, we want to obtain a zero in the second entry of the third row. To do this, we add -1 times the second row to the third:

$$\begin{bmatrix} 1 & 0 & -4 & | & 6 \\ 0 & -1 & 11 & | & -8 \\ 0 & 0 & 11 & | & -11 \end{bmatrix}. \quad -R_2 + R_3$$

This is the augmented matrix of the upper triangular system

$$\begin{cases} x & -4z = & 6 \\ -y + 11z = & -8 \\ 11z = & -11, \end{cases}$$

which is equivalent to the given system. Solving the last system, we obtain the solution $x = 2$, $y = -3$, and $z = -1$.

Practice Exercise 3   Solve the following system by the augmented matrix method.

$$\begin{cases} 3x - 2y + 2z = 0 \\ -x - y + 3z = 9 \\ -2x \quad - z = 2 \end{cases}$$

Answer: $x = -2$, $y = -1$, $z = 2$

Observe that our objective in Example 3 was to transform the original augmented matrix into a matrix of the form

$$\begin{bmatrix} a & b & c & | & l \\ 0 & d & e & | & m \\ 0 & 0 & f & | & n \end{bmatrix},$$

where the matrix of coefficients is in *upper-triangular form,* that is, it contains zeros below the *main diagonal*. Once this is achieved, we write the corresponding system of linear equations and solve it.

Similarly, by performing row-admissible operations, we could have obtained a matrix such as

$$\begin{bmatrix} a' & 0 & 0 & | & l' \\ b' & d' & 0 & | & m' \\ c' & e' & f' & | & n' \end{bmatrix},$$

where the matrix of coefficients is now in *lower-triangular* form, that is, it contains zeros *above* the main diagonal. The corresponding system is readily solvable and the solution, of course, would be the same.

A System with Infinitely Many Solutions

The method of Gaussian elimination applies in general to any system of linear equations. If the system has solutions, they can be obtained by this method. If no solutions exist, then we obtain a contradiction.

EXAMPLE 4

Solve the system

$$\begin{cases} x - 3y + z = 2 \\ 2x - 5y + z = 5. \end{cases}$$

Solution: This is a system of two equations in three unknowns, and we cannot expect it to have a unique solution. As we already said, each equation represents a plane; two planes may intersect or coincide—in which cases there will be infinitely many solutions—or they may be parallel, in which case there will be no solution. Adding -2 times the first equation to the second, we eliminate x from the second equation and obtain the equivalent system

$$\begin{cases} x - 3y + z = 2 \\ y - z = 1. \end{cases} \quad -2R_1 + R_2$$

It is not possible to eliminate y or z from the second equation. You should check that any attempt to do so would reintroduce x in the second equation. Now we solve the second equation $y - z = 1$ for one of the variables, say y, obtaining

$$y = 1 + z.$$

(We could have solved it for z. The conclusion we are about to reach would be the same.) Substituting for y in the first equation and solving for x, we get

$$x - 3(1 + z) + z = 2,$$

or

$$x = 5 + 2z.$$

The given system has an infinite number of solutions: to each value assigned to z there corresponds a value of y given by $y = 1 + z$ and a value of x given by $x = 5 + 2z$. Thus, the solutions of the system consist of all ordered triples of the form $(5 + 2z, 1 + z, z)$, where z is the arbitrary number. For example, if $z = -1$, we obtain the solution $(3, 0, -1)$; if $z = 0$, we get $(5, 1, 0)$; if $z = 1/2$, we have $(6, 3/2, 1/2)$; and so on.

Practice Exercise 4

Use the augmented matrix method to solve the system

$$\begin{cases} 3x - y + 2z = 1 \\ x + y - z = 0. \end{cases}$$

Answer: Infinitely many solutions: $(x, 1 - 5x, 1 - 4x)$, with x any number

A System with No Solution

EXAMPLE 5

Solve the system

$$\begin{cases} x - 2y - z = 1 \\ 3x - 5y - 6z = 5 \\ x - y - 4z = 6. \end{cases}$$

Solution: Adding -3 times the first equation to the second, and then adding -1 times the first equation to the third leads us to the equivalent system

$$\begin{cases} x - 2y - z = 1 \\ y - 3z = 2 \\ y - 3z = 5. \end{cases} \quad -3R_1 + R_2 \text{ and } -R_1 + R_3$$

Since the last two equations cannot be solved for y and z, the given system does not have solutions. This can also be seen by continuing with the Gaussian elimination and trying to eliminate y from the third equation. Adding -1 times the second equation to the third gives us

$$\begin{cases} x - 2y - z = 1 \\ y - 3z = 2 \\ 0 = 3. \end{cases} \quad -R_2 + R_3$$

We have reached a contradiction: $0 = 3$! Therefore, the given system has no solution. We have an inconsistent system of equations.

Practice Exercise 5   Use Gaussian elimination or the augmented matrix method to solve the system

$$\begin{cases} x - y + 3z = 2 \\ 3x - z = 1 \\ 2x + y - 4z = 1 \end{cases}$$

Answer: No solution

Homogeneous Systems

If all the numbers appearing on the right-hand side of a system of linear equations are zero, the system is said to be *homogeneous*. The general form of a homogeneous system of three linear equations in three unknowns is

$$\begin{cases} a_1x + b_1y + c_1z = 0 \\ a_2x + b_2y + c_2z = 0 \\ a_3x + b_3y + c_3z = 0. \end{cases} \tag{10.5}$$

Every homogeneous system has at least one solution, namely, the triple $(0, 0, 0)$. This is called the *trivial solution*. A homogeneous system may have nontrivial solutions. They can be found by the same methods used for nonhomogeneous systems.

EXAMPLE 6   Find nontrivial solutions of the system

$$\begin{cases} x + y + z = 0 \\ 3x + y - z = 0 \\ x + 2y + 3z = 0. \end{cases}$$

Solution: If we add -3 times the first equation to the second and then -1 times the first equation to the third, we obtain the equivalent system

$$\begin{cases} x + y + z = 0 \\ - 2y - 4z = 0 \\ y + 2z = 0. \end{cases} \quad -3R_1 + R_2 \text{ and } -R_1 + R_3$$

In the new system, the second equation, $-2y - 4z = 0$, is equal to the third equation, $y + 2z = 0$, multiplied by -2. Thus the given system is equivalent to

$$\begin{cases} x + y + z = 0 \\ y + 2z = 0. \end{cases}$$

Solving the last equation for y, we get $y = -2z$. Substituting for y into the first equation and solving for x gives us $x = z$. Thus the variables x and y depend on the variable z. To each value assigned to z there corresponds a value of x and a value of y. The ordered triple obtained in this way is a solution of the original system. For example, if $z = 1$, then $x = 1$ and $y = -2$, and we obtain the triple $(1, -2, 1)$, which is a solution of the given system. Similarly, $(-2, 4, -2)$, $(1/2, -1, 1/2)$ are also solutions. We conclude that the system has infinitely many solutions given by the ordered triples $(z, -2z, z)$, where z is an arbitrary number.

Also, notice that the solutions can be represented in a different manner. Since $x = z$, we can write $y = -2z$ as $y = -2x$. By doing so, the solutions are now given by the ordered triples $(x, -2x, x)$, where x is an arbitrary number.

As an exercise, you can check that the solutions can also be given by the ordered triples $(-y/2, y, -y/2)$, where y is an arbitrary number.

Practice Exercise 6   Find all the solutions of the homogeneous system

$$\begin{cases} 2x + 4z = 0 \\ y + 2z = 0 \\ x - y = 0. \end{cases}$$

Answer: $(x, x, -x/2)$, with x any number

An Application

We end this section by discussing a word problem that leads us to a system of linear equations.

EXAMPLE 7   A manufacturer produces three kinds of equipment, A, B, and C, that have to be transported by truck. He has three types of trucks, F, G, and H, that can carry various units of equipment according to the following table:

Equipment	Trucks		
	F	G	H
A	2	1	1
B	1	3	3
C	3	2	1

If the manufacturer has to deliver 15 units of equipment A, 20 units of equipment B, and 22 units of equipment C, how many trucks of each type should he use so that all trucks operate at full capacity?

Solution: Let x be the number of trucks of type F, y the number of trucks of type G, and z the number of trucks of type H. According to the table, truck F carries

2 units of equipment A, truck G carries 1 unit of equipment A, and truck H carries 1 unit of equipment A. Together the three trucks must carry 15 units of equipment A, so we get the equation $2x + y + z = 15$. With similar reasoning for equipments B and C, we obtain two other equations: $x + 3y + 3z = 20$ and $3x + 2y + z = 22$. We thus obtain the three-by-three system

$$\begin{cases} 2x + y + z = 15 \\ x + 3y + 3z = 20 \\ 3x + 2y + z = 22. \end{cases}$$

Using Gaussian elimination or the augmented matrix method, you can check that the solution of this system is $x = 5$, $y = 2$, and $z = 3$. Thus the manufacturer should use 5 trucks of type F, 2 trucks of type G, and 3 trucks of type H.

Practice Exercise 7   Judy spent \$6.66 buying 11-, 20-, and 40-cent stamps. She bought twice as many 20-cent stamps as 11-cent stamps and half as many 40-cent stamps as the sum of 11- and 20-cent stamps. How many stamps of each denomination did she buy?

Answer: six 11-cent stamps, twelve 20-cent stamps, and nine 40-cent stamps

EXERCISES 10.2

In Exercises 1–8, solve each system by substitution.

1. $\begin{cases} 3x - y - z = 2 \\ 2y - z = 5 \\ 2z = 6 \end{cases}$

2. $\begin{cases} 4x = 2 \\ 2x - y = 0 \\ 6x - 2y + z = 3 \end{cases}$

3. $\begin{cases} 3x - y + 2z = 1 \\ 2y = 0 \\ 4x - 5y = 4 \end{cases}$

4. $\begin{cases} 2y = 4 \\ y - 3z = 5 \\ x - 2y - 3z = 2 \end{cases}$

5. $\begin{cases} y + z = 5 \\ x + z = 4 \\ x + y = 3 \end{cases}$

6. $\begin{cases} x - 2y + z = 4 \\ 2x - y + 2z = 2 \\ x - 2y = 5 \end{cases}$

7. $\begin{cases} 2x - y = 3 \\ y - 3z = -8 \\ x + 2z = 8 \end{cases}$

8. $\begin{cases} x - 2y - 3z = 0 \\ y - 2z = 4 \\ x + 2y + 3z = 2 \end{cases}$

Solve each of the systems in Exercises 9–18 by Gaussian elimination.

9. $\begin{cases} x - 2y + z = 0 \\ y + 2z = -2 \\ 2x - 3y + 3z = -1 \end{cases}$

10. $\begin{cases} x + 2y - 3z = 1 \\ -2x + y - z = -7 \\ y + z = -1 \end{cases}$

11. $\begin{cases} y - 2z = 5 \\ x + z = 0 \\ 2x - y = 3 \end{cases}$

12. $\begin{cases} x - y = -1 \\ x + z = 4 \\ y - 3z = -3 \end{cases}$

13. $\begin{cases} 2x + y - 3z = 3 \\ 4x - 2y - 3z = 1 \\ 2x - y + 3z = -1 \end{cases}$

14. $\begin{cases} 3x + 4y - z = 2 \\ 3x - 2y - 2z = -3 \\ 6x - 4y - 3z = -4 \end{cases}$

15. $\begin{cases} x + 2y + 3z = 6 \\ 2x + z = -1 \\ 3x + y + z = 0 \end{cases}$

16. $\begin{cases} 3x + 2y + z = 3 \\ x + 2z = 4 \\ 2y + z = -3 \end{cases}$

17. $\begin{cases} x + y - z - w = 1 \\ 2x - y + 2z + w = 2 \\ x - 2y + z + 3w = 2 \\ 2x + y - z - 2w = 1 \end{cases}$

18. $\begin{cases} x - 3y + 4w = 0 \\ 3x - 2z = 1 \\ y + 2z - w = 2 \\ 2x - 3w = 5 \end{cases}$

In Exercises 19–26, use admissible row operations on the corresponding augmented matrix to solve each system.

19. $\begin{cases} 3x + 2y = 2 \\ 9x - 4y = 1 \end{cases}$ **20.** $\begin{cases} 3x - 4y = -1 \\ 6x + 8y = 10 \end{cases}$

21. $\begin{cases} x + 2y = 2 \\ 3x - 4y = 1 \end{cases}$ **22.** $\begin{cases} x + 4y = 1 \\ 2x + 2y = -1 \end{cases}$

23. $\begin{cases} x - 2y - 3z = 4 \\ 2x + y - 2z = 4 \\ 3x + 2y + z = 2 \end{cases}$ **24.** $\begin{cases} 2x - y - z = 4 \\ x + 2y - z = 0 \\ x + y + 3z = -2 \end{cases}$

25. $\begin{cases} x + 2y = -1 \\ 2x + z = 8 \\ y + 2z = -7 \end{cases}$ **26.** $\begin{cases} x + y + z = 4 \\ 2x - 2z = -4 \\ 3x - 2y = 17 \end{cases}$

In Exercises 27–34, solve each system in terms of an arbitrary variable.

27. $\begin{cases} 2x - y + 4z = 1 \\ x + 3z = 0 \end{cases}$ **28.** $\begin{cases} x + 2y = 3 \\ 2x - 3y + z = 1 \end{cases}$

29. $\begin{cases} 2x + 3y - z = 4 \\ x - 2y + 3z = 2 \end{cases}$ **30.** $\begin{cases} x - y + 3z = 1 \\ 3x + 2y - z = 8 \end{cases}$

31. $\begin{cases} x + z = 3 \\ 3x - 2y = 1 \end{cases}$ **32.** $\begin{cases} 2x - z = 4 \\ 4y - 3z = 0 \end{cases}$

33. $\begin{cases} x - y + z = 0 \\ 2x + y - 3z = 0 \\ 4x - y - z = 0 \end{cases}$ **34.** $\begin{cases} 3x - 4y + z = 0 \\ x + y = 0 \\ 2x - 5y + z = 0 \end{cases}$

In Exercises 35–40, use any of the methods described in this section to find the solution (if any) of each system.

35. $\begin{cases} 2x - 3y = 5 \\ -3x + y = 3 \\ x - y = 1 \end{cases}$ **36.** $\begin{cases} 2x - 3y = 1 \\ 3x + 2y = 8 \\ x + y = 2 \end{cases}$

37. $\begin{cases} x + 2y = -1 \\ 5x - y = 6 \\ 2x - y = 0 \end{cases}$ **38.** $\begin{cases} x + y = 1 \\ 2x + 4y = 3 \\ 6x - 2y = 2 \end{cases}$

39. $\begin{cases} x + y + 2z = 0 \\ y + z = 0 \\ 3x + z = 0 \end{cases}$ **40.** $\begin{cases} 2x - y + z = 0 \\ x + 2y - z = 0 \\ 3x - y + 2z = 0 \end{cases}$

In Exercises 41–50, solve the given problem.

41. Mary spends $8.86 on 11-, 18-, and 40-cent stamps. The total number of 11- and 18-cent stamps she buys is the same as the number of 40-cent stamps, and there are four more 18-cent stamps than 11-cent stamps. How many stamps of each type does she buy?

42. A boy has $5.00 in nickels, dimes, and quarters. The total number of nickels and quarters is twice the number of dimes, and the number of nickels is twice the difference between the numbers of dimes and quarters. Find how many nickels, dimes, and quarters the boy has.

43. The following table shows the percentages of copper, zinc, and tin that are used to produce alloys A, B, and C.

	A	B	C
Copper	50	60	40
Zinc	30	15	30
Tin	20	25	30

How many kilograms of each alloy must be used to produce 100 kg of a new alloy containing 52% copper and 24% tin?

44. Use the table of Exercise 43 to find how many kilograms of each alloy must be combined to produce 120 kg of an alloy that contains 52.5% copper and 17.5% zinc.

45. The perimeter of a triangle measures 42 m. Find the dimensions of the triangle if one side is 6 m longer than the shortest side and is half as long as the longest side.

46. The longest side of a triangle is 20 cm longer than one of the sides and 7 cm longer than the other side. Find the length of each side if the perimeter of the triangle measures 138 cm.

47. A grocer blends three types of coffee that sell for $2.60, $2.80, and $3.20 per pound so as to obtain 150 lb of coffee worth $3.00/lb. If he uses equal amounts of the two lower-priced coffees, how many pounds of each type of coffee should he blend?

48. A merchant makes 90 lb of a mixture of peanuts, cashews, and hazelnuts that can be sold for $3.80/lb. Peanuts sell for $2.70, cashews for $3.60, and hazelnuts for $4.50/lb. She mixes ten more pounds of cashews than peanuts. How many pounds of each nut does she mix?

49. Charlie won $18,000 in the New Jersey Lottery and has invested it in three ways. The first part he invested in a money market fund paying 8.5% per year. The second part, $2,000 less than the first, is invested in certificates of deposit paying 7.5% per year. Finally, the rest he has invested in municipal bonds earning 6.5% per year. At the end of the first year he receives $1,390 in interest. How much is invested at each rate?

50. In a given year, an investor received $515 in dividends corresponding to an investment in three types of bonds yielding 7.5%, 8.5%, and 9% per year. If the amount invested at 8.5% is twice the amount invested at 7.5%, and the amount invested at 9% equals the sum of the amounts invested at 7.5% and 8.5%, find how much money is invested in each type of bond.

MATRICES

In the previous section we introduced the notion of matrices when we were solving a system of linear equations. A matrix is an important concept that has many applications in mathematics and the applied sciences.

Matrices

A rectangular array of numbers, written within brackets, is called a *matrix*.

$$\begin{bmatrix} a_{11} & a_{12} & \cdots & a_{1n} \\ a_{21} & 2_{22} & \cdots & a_{2n} \\ \cdots & \cdots & \cdots & \cdots \\ a_{m1} & a_{m2} & \cdots & a_{mn} \end{bmatrix} \qquad (10.6)$$

The numbers are called the *elements* or *entries* of the matrix.

For example,

$$\begin{bmatrix} 2 & 1 \\ 3 & 2 \end{bmatrix} \quad \begin{bmatrix} 1 & 3 & 5 \end{bmatrix} \quad \begin{bmatrix} 2 \\ 4 \\ 6 \end{bmatrix}$$

$$\begin{bmatrix} 3 & 0 & -3 \\ 1 & 4 & -1 \end{bmatrix} \quad \begin{bmatrix} 2 & 3 & 1 \\ 0 & -1 & 4 \\ -2 & 5 & 2 \end{bmatrix}$$

are matrices.

It is common practice to denote each entry a_{ij} of a matrix with two subscripts i and j. The first subscript, i, identifies the *row*, while the second, j, identifies the *column* where the element is located. For example, in the matrix

$$\begin{bmatrix} 2 & 3 & 1 \\ 0 & -1 & 4 \\ -2 & 5 & 2 \end{bmatrix},$$

the element a_{32}, located in the third row and second column, is 5. Also, $a_{13} = 1$, $a_{21} = 0$, and so on.

We often use the short notation $[a_{ij}]_{\substack{1 \le i \le m \\ 1 \le j \le n}}$ or simply $[a_{ij}]$, when no confusion is possible, to denote the matrix (10.6).

A matrix with m rows and n columns is said to be an $m \times n$ *matrix* or an m-by-n *matrix*. For example, $\begin{bmatrix} 1 & 3 & 5 \end{bmatrix}$ is a 1-by-3 matrix, and $\begin{bmatrix} 3 & 0 & -3 \\ 1 & 4 & -1 \end{bmatrix}$ is a 2×3 matrix.

When $m = n$, we say that the matrix is a *square matrix*. An $n \times n$ matrix is also called a matrix of *order n*. For example,

$$\begin{bmatrix} 2 & 1 \\ 3 & 2 \end{bmatrix} \quad \text{and} \quad \begin{bmatrix} 2 & 3 & 1 \\ 0 & -1 & 4 \\ -2 & 5 & 2 \end{bmatrix}$$

are square matrices of order 2 and 3, respectively. A matrix of order 1 consists of a single element.

Equality of Matrices

Two matrices are *equal* if and only if they have the same number of rows and columns and the same entry in each position.

In other words $[a_{ij}]_{\substack{1 \le i \le m \\ 1 \le j \le n}} = [b_{ij}]_{\substack{1 \le i \le p \\ 1 \le j \le q}}$ if and only if $m = p$, $n = q$, and $a_{ij} = b_{ij}$ for all indices i and j.

The Sum of Two Matrices

If $A = [a_{ij}]$ and $B = [b_{ij}]$ are two $m \times n$ matrices, then their sum $A + B$ is defined to be the matrix $[a_{ij} + b_{ij}]$.

EXAMPLE 1 Find the sum of the matrices

$$A = \begin{bmatrix} 3 & 0 & -3 \\ 1 & 4 & -1 \end{bmatrix} \quad \text{and} \quad B = \begin{bmatrix} -1 & 2 & 3 \\ -2 & -4 & 5 \end{bmatrix}.$$

Solution: We have

$$A + B = \begin{bmatrix} 3 & 0 & -3 \\ 1 & 4 & -1 \end{bmatrix} + \begin{bmatrix} -1 & 2 & 3 \\ -2 & -4 & 5 \end{bmatrix}$$

$$= \begin{bmatrix} 2 & 2 & 0 \\ -1 & 0 & 4 \end{bmatrix}.$$

Practice Exercise 1 Find the sum of the matrices

$$M = \begin{bmatrix} 1 & -1 \\ -3 & 4 \\ 2 & -5 \end{bmatrix} \quad \text{and} \quad N = \begin{bmatrix} 3 & 2 \\ 0 & -2 \\ -4 & 4 \end{bmatrix}.$$

Answer: $M + N = \begin{bmatrix} 4 & 1 \\ -3 & 2 \\ -2 & -1 \end{bmatrix}$

Matrix addition satisfies the commutative and associative properties. If $A = [a_{ij}]$, $B = [b_{ij}]$, and $C = [c_{ij}]$, are $m \times n$ matrices, then

$$A + B = B + A \quad \text{Commutativity}$$

and

$$(A + B) + C = A + (B + C) \quad \text{Associativity}$$

We can also write

$$[a_{ij}] + [b_{ij}] = [b_{ij}] + [a_{ij}]$$

and

$$([a_{ij}] + [b_{ij}]) + [c_{ij}] = [a_{ij}] + ([b_{ij}] + [c_{ij}]).$$

The Zero Matrix

A matrix whose entries are all zeros is called a *zero matrix*.

If $A = [a_{ij}]$ is an $m \times n$ matrix and O is the $m \times n$ zero matrix, then $A + O = O + A = A$.

The Additive Inverse of a Matrix

If $A = [a_{ij}]$ is an $m \times n$ matrix, there is always an $m \times n$ matrix $B = [b_{ij}]$ such that $A + B = O$. The matrix B is called the *additive inverse* of A.

If B is the additive inverse of A, then its entries b_{ij} are such that $b_{ij} = -a_{ij}$, and we write $B = -A$.

The Difference of Two Matrices

If $A = [a_{ij}]$ and $B = [b_{ij}]$ are two $m \times n$ matrices, then their difference $A - B$ is defined by

$$A - B = A + (-B).$$

EXAMPLE 2   Find $A - B$ if

$$A = \begin{bmatrix} 2 & -3 & 6 \\ -4 & 1 & -2 \end{bmatrix} \quad \text{and} \quad B = \begin{bmatrix} 4 & 2 & 3 \\ -1 & 3 & -5 \end{bmatrix}.$$

Solution: We have

$$A - B = \begin{bmatrix} 2 & -3 & 6 \\ -4 & 1 & -2 \end{bmatrix} - \begin{bmatrix} 4 & 2 & 3 \\ -1 & 3 & -5 \end{bmatrix}$$

$$= \begin{bmatrix} 2 & -3 & 6 \\ -4 & 1 & -2 \end{bmatrix} + \begin{bmatrix} -4 & -2 & -3 \\ 1 & -3 & 5 \end{bmatrix}$$

$$= \begin{bmatrix} -2 & -5 & 3 \\ -3 & -2 & 3 \end{bmatrix}.$$

Practice Exercise 2   For the matrices of Example 2, find $B - A$.

Answer: $B - A = \begin{bmatrix} 2 & 5 & -3 \\ 3 & 2 & -3 \end{bmatrix}$

Scalar Multiplication

Let $A = [a_{ij}]$ be an $m \times n$ matrix and let α be a number. The product of α by A is the matrix αA defined by $[\alpha a_{ij}]$. We say that αA is the *scalar product* of α by A.

EXAMPLE 3   Multiply 3 by the matrix $A = \begin{bmatrix} 5 & 0 & 5 \\ -2 & 1 & -4 \end{bmatrix}$.

Solution: We have

$$3A = 3 \begin{bmatrix} 5 & 0 & 5 \\ -2 & 1 & -4 \end{bmatrix} = \begin{bmatrix} 15 & 0 & 15 \\ -6 & 3 & -12 \end{bmatrix}.$$

Practice Exercise 3   Let $\alpha = \dfrac{1}{2}$ and $A = \begin{bmatrix} 4 & -2 \\ 3 & 8 \end{bmatrix}$. Find αA.

Answer: $\dfrac{1}{2}A = \begin{bmatrix} 2 & -1 \\ \dfrac{3}{2} & 4 \end{bmatrix}$

From the definitions of additive inverse and scalar product, it follows that $-A = (-1)A$; that is, the additive inverse of a matrix is equal to the scalar product of -1 by the matrix.

Properties of the Scalar Product

If α and β are numbers and A and B are $m \times n$ matrices, then the following properties of scalar multiplication are easy to check.

$$\alpha(A + B) = \alpha A + \alpha B$$

$$(\alpha + \beta)A = \alpha A + \beta A$$

$$\alpha(\beta A) = (\alpha\beta)A$$

Matrix Multiplication

Let $A = [a_{ij}]$ be an $m \times p$ matrix and let $B = [b_{ij}]$ be a $p \times n$ matrix. The product (row by column) of A and B is the $m \times n$ matrix C whose entries are defined by

$$c_{ij} = a_{i1}b_{1j} + a_{i2}b_{2j} + \cdots + a_{ip}b_{pj}. \tag{10.7}$$

We write $C = AB$.

Observe that to find the entry c_{ij} of the product AB, we multiply each element in the ith row of A by the corresponding element in the jth column of B, and add the products:

$$i\begin{bmatrix} a_{i1} & a_{i2} & \cdots & a_{ip} \\ & & \vdots & \end{bmatrix}\begin{bmatrix} & b_{1j} & \\ \cdots & b_{2j} & \cdots \\ & \vdots & \\ & b_{pj} & \end{bmatrix} = \begin{bmatrix} & & \\ \cdots & c_{ij} & \cdots \\ & & \end{bmatrix}i,$$

with c_{ij} given by (10.7).

It is important to observe that the product (row by column) of A and B is defined only if *the number of columns of A is the same as the number of rows of B*. Then the product AB has *the same number of rows as A* and *the same number of columns as B*.

EXAMPLE **4**   Find AB if

$$A = \begin{bmatrix} 1 & -1 & 2 \\ 3 & 0 & 1 \end{bmatrix} \quad \text{and} \quad B = \begin{bmatrix} 2 & 0 \\ 1 & -1 \\ -2 & 3 \end{bmatrix}.$$

Solution: According to (10.7), we have

$$AB = \begin{bmatrix} 1 & -1 & 2 \\ 3 & 0 & 1 \end{bmatrix}\begin{bmatrix} 2 & 0 \\ 1 & -1 \\ -2 & 3 \end{bmatrix}$$

$$= \begin{bmatrix} 1 \cdot 2 + (-1) \cdot 1 + 2 \cdot (-2) & 1 \cdot 0 + (-1)(-1) + 2 \cdot 3 \\ 3 \cdot 2 + 0 \cdot 1 + 1 \cdot (-2) & 3 \cdot 0 + 0 \cdot (-1) + 1 \cdot 3 \end{bmatrix}$$

$$= \begin{bmatrix} -3 & 7 \\ 4 & 3 \end{bmatrix}.$$

Practice Exercise 4   If A and B are the matrices of Example 4, find BA.

$$\textbf{\textit{Answer:}}\quad BA = \begin{bmatrix} 2 & -2 & 4 \\ -2 & -1 & 1 \\ 7 & 2 & -1 \end{bmatrix}$$

Properties of Matrix Multiplication

Whenever it is defined, matrix multiplication satisfies the following properties:

$$A(BC) = (AB)C \quad \text{Associativity}$$

$$\left.\begin{array}{l} A(B + C) = AB + AC \\ (B + C)A = BA + CA \end{array}\right\} \quad \text{Distributivity}$$

In general, the multiplication of matrices is a *noncommutative* operation; that is, $AB \neq BA$ for most matrices. Indeed, if A and B are not square matrices, one of the two products may be undefined.

EXAMPLE 5   If $A = \begin{bmatrix} 1 & 2 \\ 0 & 1 \end{bmatrix}$ and $B = \begin{bmatrix} 0 & -1 \\ -2 & 2 \end{bmatrix}$, verify that $AB \neq BA$.

Solution: We have

$$AB = \begin{bmatrix} 1 & 2 \\ 0 & 1 \end{bmatrix}\begin{bmatrix} 0 & -1 \\ -2 & 2 \end{bmatrix} = \begin{bmatrix} -4 & 3 \\ -2 & 2 \end{bmatrix}$$

and

$$BA = \begin{bmatrix} 0 & -1 \\ -2 & 2 \end{bmatrix}\begin{bmatrix} 1 & 2 \\ 0 & 1 \end{bmatrix} = \begin{bmatrix} 0 & -1 \\ -2 & -2 \end{bmatrix}.$$

As we can see, $AB \neq BA$.

Practice Exercise 5   Let $A = \begin{bmatrix} 1 & 1 \\ 0 & 1 \end{bmatrix}$ and $B = \begin{bmatrix} 0 & 0 \\ 2 & 0 \end{bmatrix}$. Find AB and BA.

$$\textbf{\textit{Answer:}}\quad AB = \begin{bmatrix} 2 & 0 \\ 2 & 0 \end{bmatrix}, BA = \begin{bmatrix} 0 & 0 \\ 2 & 2 \end{bmatrix}.$$

The Product of Square Matrices

Of special interest is the multiplication of square matrices. If A and B are square matrices of order n, then the products AA, AB, BA, and BB are defined and are all matrices of order n.

> ### MATHEMATICAL VIGNETTE
>
> ## COUNTING THE NUMBER OF ROUTES BETWEEN CITIES
>
> The vertices 1, 2, and 3 of the equilateral triangle
>
>
>
> represent three cities equidistant from each other by 1 mile. In the matrix
>
> $$A = \begin{bmatrix} 0 & 1 & 1 \\ 1 & 0 & 1 \\ 1 & 1 & 0 \end{bmatrix},$$
>
> each entry a_{ij}, $1 \le i \le 3$, $1 \le j \le 3$, gives the number of routes *one mile long* between the cities i and j. For example, $a_{12} = 1$ indicates that there is only one route one mile long joining cities 1 and 2.
>
> The square of the matrix A (defined as the product of A by itself and denoted by A^2) is
>
> $$A^2 = \begin{bmatrix} 2 & 1 & 1 \\ 1 & 2 & 1 \\ 1 & 1 & 2 \end{bmatrix}.$$
>
> Each entry a_{ij} denotes the number of routes two miles long joining the cities i and j. For example, $a_{11} = 2$ indicates that there are two routes two miles long starting and ending at city 1: from 1 to 2 and back to 1, and from 1 to 3 and back to 1. Also, there is only one route 2 miles long joining 1 and 3, so $a_{13} = a_{31} = 1$.
>
> Suppose that you have four cities equidistant from each other by 1 mile. Write the four-by-four matrix showing the number of routes one mile long between these cities. Square the matrix and obtain a new matrix giving the number of routes 2 miles long between the four cities.

> ### Identity Matrix
>
> The $n \times n$ matrix $I_n = [\delta_{ij}]$ whose entries are
>
> $$\delta_{ij} = 1 \quad \text{if} \quad i = j$$
> $$\delta_{ij} = 0 \quad \text{if} \quad i \ne j$$
>
> is called the *identity matrix of order n.*

Such a matrix has *ones* along the main diagonal and *zeros* elsewhere. For example,

$$I_2 = \begin{bmatrix} 1 & 0 \\ 0 & 1 \end{bmatrix} \quad \text{and} \quad I_3 = \begin{bmatrix} 1 & 0 & 0 \\ 0 & 1 & 0 \\ 0 & 0 & 1 \end{bmatrix}$$

are the identity matrices of order 2 and 3, respectively.

It can be shown that if A is an $n \times n$ matrix, then $AI_n = I_nA = A$. For example, it is easy to apply the definition of matrix multiplication to see that

$$\begin{bmatrix} a & b \\ c & d \end{bmatrix}\begin{bmatrix} 1 & 0 \\ 0 & 1 \end{bmatrix} = \begin{bmatrix} 1 & 0 \\ 0 & 1 \end{bmatrix}\begin{bmatrix} a & b \\ c & d \end{bmatrix} = \begin{bmatrix} a & b \\ c & d \end{bmatrix}.$$

The Multiplicative Inverse of a Matrix

If A is an $n \times n$ matrix, and if there is a matrix B of order n such that

$$AB = BA = I_n,$$

then A is said to be *invertible* and B is the *multiplicative inverse* or, simply, the *inverse* of A.

For example, the matrix $B = \begin{bmatrix} 3 & -4 \\ -2 & 3 \end{bmatrix}$ is the inverse of $A = \begin{bmatrix} 3 & 4 \\ 2 & 3 \end{bmatrix}$. To verify this, we compute the products

$$\begin{aligned} AB &= \begin{bmatrix} 3 & 4 \\ 2 & 3 \end{bmatrix}\begin{bmatrix} 3 & -4 \\ -2 & 3 \end{bmatrix} \\ &= \begin{bmatrix} 3\cdot 3 + 4\cdot(-2) & 3\cdot(-4) + 4\cdot 3 \\ 2\cdot 3 + 3\cdot(-2) & 2\cdot(-4) + 3\cdot 3 \end{bmatrix} \\ &= \begin{bmatrix} 1 & 0 \\ 0 & 1 \end{bmatrix} = I_2 \end{aligned}$$

and

$$\begin{aligned} BA &= \begin{bmatrix} 3 & -4 \\ -2 & 3 \end{bmatrix}\begin{bmatrix} 3 & 4 \\ 2 & 3 \end{bmatrix} \\ &= \begin{bmatrix} 3\cdot 3 + (-4)\cdot 2 & 3\cdot 4 + (-4)\cdot 3 \\ (-2)\cdot 3 + 3\cdot 2 & (-2)\cdot 4 + 3\cdot 3 \end{bmatrix} \\ &= \begin{bmatrix} 1 & 0 \\ 0 & 1 \end{bmatrix} = I_2. \end{aligned}$$

■ **Important remark**   *Not every matrix has an inverse.* However, if a matrix has an inverse, it can be proved that the inverse is *unique*. In Section 10.4 on *determinants,* we shall state a necessary and sufficient condition for a matrix to have an inverse. If A is an invertible matrix, its inverse is denoted by A^{-1}.

EXAMPLE 6   Find the inverse of the matrix $A = \begin{bmatrix} 3 & 4 \\ 2 & 3 \end{bmatrix}$.

Solution: We wish to find a 2×2 matrix B such that $AB = BA = I_2$. If we set

$$B = \begin{bmatrix} x & u \\ y & v \end{bmatrix},$$

then we must have

$$\begin{bmatrix} 3 & 4 \\ 2 & 3 \end{bmatrix} \begin{bmatrix} x & u \\ y & v \end{bmatrix} = \begin{bmatrix} 1 & 0 \\ 0 & 1 \end{bmatrix}.$$

Computing the product, we get

$$\begin{bmatrix} 3x + 4y & 3u + 4v \\ 2x + 3y & 2u + 3v \end{bmatrix} = \begin{bmatrix} 1 & 0 \\ 0 & 1 \end{bmatrix},$$

from which we obtain two systems of linear equations:

$$\begin{cases} 3x + 4y = 1 \\ 2x + 3y = 0 \end{cases} \quad \text{and} \quad \begin{cases} 3u + 4v = 0 \\ 2u + 3v = 1. \end{cases}$$

Solving these two systems, we obtain

$$x = 3, y = -2 \quad \text{and} \quad u = -4, v = 3.$$

Thus

$$B = \begin{bmatrix} 3 & -4 \\ -2 & 3 \end{bmatrix}.$$

We already checked, just before this example, that $AB = BA = I_2$; therefore B is the multiplicative inverse of matrix A.

Practice Exercise 6   What is the inverse of the matrix $A = \begin{bmatrix} 5 & 3 \\ 3 & 2 \end{bmatrix}$?

Answer: $A^{-1} = \begin{bmatrix} 2 & -3 \\ -3 & 5 \end{bmatrix}$

Systems of Linear Equations in Matrix Form

Every $n \times n$ system of linear equations,

$$\begin{cases} a_{11}x_1 + a_{12}x_2 + \cdots + a_{1n}x_n = b_1 \\ a_{21}x_1 + a_{22}x_2 + \cdots + a_{2n}x_n = b_2 \\ \qquad\qquad\qquad \vdots \\ a_{n1}x_1 + a_{n2}x_2 + \cdots + a_{nn}x_n = b_n, \end{cases} \qquad (10.8)$$

can be expressed in compact form using matrix notation. In fact, if we set $A = [a_{ij}]_{\substack{1 \le i \le n \\ 1 \le j \le n}}$, $X = [x_j]_{1 \le j \le n}$, and $B = [b_i]_{1 \le i \le n}$, then we can write the system (10.8) as

$$AX = B. \qquad (10.9)$$

The matrix A is called the *matrix of coefficients.*
    For example, the two-by-two system

$$\begin{cases} 3x + 4y = 1 \\ 2x + 3y = -2 \end{cases}$$

can be written in matrix notation as

$$\begin{bmatrix} 3 & 4 \\ 2 & 3 \end{bmatrix}\begin{bmatrix} x \\ y \end{bmatrix} = \begin{bmatrix} 1 \\ -2 \end{bmatrix}.$$

Using the Inverse Matrix to Solve a System

If A^{-1} is the inverse of A and $B \neq 0$, then the system (10.8) has a unique solution given by

$$X = A^{-1}B. \tag{10.10}$$

Indeed, multiplying both sides of (10.9) by A^{-1}, we have

$$A^{-1}(AX) = A^{-1}B.$$

Using the associativity of the matrix product, we get

$$(A^{-1}A)X = A^{-1}B.$$

Since $AA^{-1} = A^{-1}A = I_n$, it follows that

$$I_nX = A^{-1}B,$$

from which we obtain (10.10).

EXAMPLE 7

Solve the given system by using the inverse matrix.

$$\begin{cases} 3x + 4y = 1 \\ 2x + 3y = -2 \end{cases}$$

Solution: We know from Example 6 that the matrix of coefficients $A = \begin{bmatrix} 3 & 4 \\ 2 & 3 \end{bmatrix}$ has inverse $A^{-1} = \begin{bmatrix} 3 & -4 \\ -2 & 3 \end{bmatrix}$. Thus, according to (10.10), we have

$$\begin{bmatrix} x \\ y \end{bmatrix} = \begin{bmatrix} 3 & -4 \\ -2 & 3 \end{bmatrix}\begin{bmatrix} 1 \\ -2 \end{bmatrix}$$

$$= \begin{bmatrix} 3 \cdot 1 + (-4) \cdot (-2) \\ (-2) \cdot 1 + 3 \cdot (-2) \end{bmatrix}$$

$$= \begin{bmatrix} 11 \\ -8 \end{bmatrix}.$$

The solution of the given system is thus $(11, -8)$.

> **MATHEMATICAL VIGNETTE**
>
> INVERSE MATRIX
>
> Efficient computer programs have been written to find the inverse of a matrix and to solve a system of n equations in n unknowns by using exactly the method of Example 7.

Practice Exercise 7

Use the inverse of the matrix of coefficients to solve the system

$$\begin{cases} 5x + 3y = 1 \\ 3x + 2y = 0. \end{cases}$$

Answer: $x = 2, y = -3$

Advantages of the Method

Formula (10.10) is important from a theoretical viewpoint. In computations, the use of (10.10) may require a lot of work, especially for higher-order systems.

However, if the matrix of coefficients A remains the same and $b_1, b_2, \ldots, b_n$ are given many different values, the use of formula (10.10) will save much work in solving the resulting systems.

Finding the Inverse of a Matrix by Row Operations

As we already remarked, if the matrix of coefficients of an $n \times n$ nonhomogeneous system of linear equations has a multiplicative inverse, then we can solve the system. Thus, it is important to have methods to find the inverse of a matrix. In Example 6, we described a way of finding the inverse of a 2×2 matrix. The method extends to higher-order matrices, but it has a serious drawback: the number of systems of linear equations to be solved increases with the order of the matrix.

A more effective method, which we now discuss in the particular case of 3×3 matrices, uses the admissible row operations of Section 10.2. Suppose that we want to find the inverse of the matrix

$$A = \begin{bmatrix} a & b & c \\ d & e & f \\ g & h & i \end{bmatrix}.$$

We form an *augmented matrix*

$$[A \,|\, I_3] = \begin{bmatrix} a & b & c & | & 1 & 0 & 0 \\ d & e & f & | & 0 & 1 & 0 \\ g & h & i & | & 0 & 0 & 1 \end{bmatrix}$$

by placing the identity matrix I_3 to the right of the matrix A. If the inverse A^{-1} exists, then by performing as many admissible row operations as necessary, it is possible to transform the augmented matrix into a matrix of the form

$$[I_3 \,|\, B] = \begin{bmatrix} 1 & 0 & 0 & | & a' & b' & c' \\ 0 & 1 & 0 & | & d' & e' & f' \\ 0 & 0 & 1 & | & g' & h' & i' \end{bmatrix},$$

where I_3 now appears to the left of a 3×3 matrix. It can be proved that the matrix

$$B = \begin{bmatrix} a' & b' & c' \\ d' & e' & f' \\ g' & h' & i' \end{bmatrix}$$

is the desired multiplicative inverse A^{-1}. If it is not possible to transform the augmented matrix $[A \,|\, I_3]$ into a matrix of the form $[I_3 \,|\, B]$, then the multiplicative inverse of A does not exist.

This method generalizes to the case of any $n \times n$ matrices.

EXAMPLE 8   Find the inverse of the matrix

$$A = \begin{bmatrix} 2 & 0 & 3 \\ 1 & -1 & 0 \\ 0 & 1 & -1 \end{bmatrix}$$

by transforming the augmented matrix $[A \,|\, I_3]$ to the form $[I_3 \,|\, B]$.

Solution: First we form the matrix $[A \mid I_3]$:

$$\begin{bmatrix} 2 & 0 & 3 & \vline & 1 & 0 & 0 \\ 1 & -1 & 0 & \vline & 0 & 1 & 0 \\ 0 & 1 & -1 & \vline & 0 & 0 & 1 \end{bmatrix}.$$

Next we perform admissible row operations until the left matrix is transformed into the identity matrix.

$$\begin{bmatrix} 2 & 0 & 3 & \vline & 1 & 0 & 0 \\ 1 & -1 & 0 & \vline & 0 & 1 & 0 \\ 0 & 1 & -1 & \vline & 0 & 0 & 1 \end{bmatrix}$$

$$\downarrow \quad R_1 \leftrightarrow R_2$$

$$\begin{bmatrix} 1 & -1 & 0 & \vline & 0 & 1 & 0 \\ 2 & 0 & 3 & \vline & 1 & 0 & 0 \\ 0 & 1 & -1 & \vline & 0 & 0 & 1 \end{bmatrix}$$

$$\downarrow \quad -2R_1 + R_2$$

$$\begin{bmatrix} 1 & -1 & 0 & \vline & 0 & 1 & 0 \\ 0 & 2 & 3 & \vline & 1 & -2 & 0 \\ 0 & 1 & -1 & \vline & 0 & 0 & 1 \end{bmatrix}$$

$$\downarrow \quad \tfrac{1}{2} R_2$$

$$\begin{bmatrix} 1 & -1 & 0 & \vline & 0 & 1 & 0 \\ 0 & 1 & \frac{3}{2} & \vline & \frac{1}{2} & -1 & 0 \\ 0 & 1 & -1 & \vline & 0 & 0 & 1 \end{bmatrix}$$

$$\downarrow \quad -R_2 + R_3$$

$$\begin{bmatrix} 1 & -1 & 0 & \vline & 0 & 1 & 0 \\ 0 & 1 & \frac{3}{2} & \vline & \frac{1}{2} & -1 & 0 \\ 0 & 0 & -\frac{5}{2} & \vline & -\frac{1}{2} & 1 & 1 \end{bmatrix}$$

$$\downarrow \quad -\tfrac{2}{5} R_3$$

$$\begin{bmatrix} 1 & -1 & 0 & \vline & 0 & 1 & 0 \\ 0 & 1 & \frac{3}{2} & \vline & \frac{1}{2} & -1 & 0 \\ 0 & 0 & 1 & \vline & \frac{1}{5} & -\frac{2}{5} & -\frac{2}{5} \end{bmatrix}$$

$$\downarrow \quad -\tfrac{3}{2} R_3 + R_2$$

$$\begin{bmatrix} 1 & -1 & 0 & \vline & 0 & 1 & 0 \\ 0 & 1 & 0 & \vline & \frac{1}{5} & -\frac{2}{5} & \frac{3}{5} \\ 0 & 0 & 1 & \vline & \frac{1}{5} & -\frac{2}{5} & -\frac{2}{5} \end{bmatrix}$$

$$\downarrow R_2 + R_1$$

$$\begin{bmatrix} 1 & 0 & 0 & \vdots & \dfrac{1}{5} & \dfrac{3}{5} & \dfrac{3}{5} \\ 0 & 1 & 0 & \vdots & \dfrac{1}{5} & -\dfrac{2}{5} & \dfrac{3}{5} \\ 0 & 0 & 1 & \vdots & \dfrac{1}{5} & -\dfrac{2}{5} & -\dfrac{2}{5} \end{bmatrix}$$

Thus the inverse of the given matrix is

$$A^{-1} = \begin{bmatrix} \dfrac{1}{5} & \dfrac{3}{5} & \dfrac{3}{5} \\ \dfrac{1}{5} & -\dfrac{2}{5} & \dfrac{3}{5} \\ \dfrac{1}{5} & -\dfrac{2}{5} & -\dfrac{2}{5} \end{bmatrix} = \dfrac{1}{5}\begin{bmatrix} 1 & 3 & 3 \\ 1 & -2 & 3 \\ 1 & -2 & -2 \end{bmatrix}.$$

Practice Exercise 8   Find the inverse of the matrix

$$A = \begin{bmatrix} 0 & 1 & 1 \\ 1 & 1 & 0 \\ 1 & 0 & -2 \end{bmatrix}$$

by admissible row operations.

Answer: $A^{-1} = \begin{bmatrix} -2 & 2 & -1 \\ 2 & -1 & 1 \\ -1 & 1 & -1 \end{bmatrix}$

EXAMPLE 9

Solve the following three-by-three system by using the inverse of the matrix of coefficients.

$$\begin{cases} 2x & + 3z = 2 \\ x - y & = 5 \\ y - z = 1 \end{cases}$$

Solution: In matrix form, the given system can be written as

$$AX = B,$$

where

$$A = \begin{bmatrix} 2 & 0 & 3 \\ 1 & -1 & 0 \\ 0 & 1 & -1 \end{bmatrix}, \quad X = \begin{bmatrix} x \\ y \\ z \end{bmatrix}, \quad \text{and} \quad b = \begin{bmatrix} 2 \\ 5 \\ 1 \end{bmatrix}.$$

Now the solution can be found by formula (10.10),

$$X = A^{-1}B,$$

where we know from Example 8 that

$$A^{-1} = \frac{1}{5} \begin{bmatrix} 1 & 3 & 3 \\ 1 & -2 & 3 \\ 1 & -2 & -2 \end{bmatrix}$$

is the inverse of the matrix A. Note that our calculations will be easier if we leave the scalar $1/5$ outside the matrix. We have

$$\begin{bmatrix} x \\ y \\ z \end{bmatrix} = \frac{1}{5} \begin{bmatrix} 1 & 3 & 3 \\ 1 & -2 & 3 \\ 1 & -2 & -2 \end{bmatrix} \begin{bmatrix} 2 \\ 5 \\ 1 \end{bmatrix}$$

$$= \frac{1}{5} \begin{bmatrix} 2 + 15 + 3 \\ 2 - 10 + 3 \\ 2 - 10 - 2 \end{bmatrix}$$

$$= \frac{1}{5} \begin{bmatrix} 20 \\ -5 \\ -10 \end{bmatrix} = \begin{bmatrix} 4 \\ -1 \\ -2 \end{bmatrix}.$$

The solution is then $x = 4$, $y = -1$, and $z = -2$.

Practice Exercise 9   Use the inverse of the matrix of coefficients to solve the system

$$\begin{cases} y + z = -1 \\ x + y = 6 \\ x - 2z = 10. \end{cases}$$

Answer: $x = 4$, $y = 2$, $z = -3$

MATHEMATICAL VIGNETTE

MATRICES AND ROTATION OF AXES

On an x, y-coordinate plane, the *counterclockwise rotation about the origin by an angle of* 90° can be described by the matrix

$$M = \begin{bmatrix} 0 & -1 \\ 1 & 0 \end{bmatrix}.$$

For if (a, b) is an ordered pair of numbers, then the product

$$\begin{bmatrix} 0 & -1 \\ 1 & 0 \end{bmatrix} \begin{bmatrix} a \\ b \end{bmatrix} = \begin{bmatrix} -b \\ a \end{bmatrix}$$

indicates that the matrix M *transforms* any point $P(a, b)$ into the point $Q(-b, a)$. By carefully plotting these points in a Cartesian coordinate system, you can check that Q is obtained from P by a counterclockwise rotation of 90° about the origin.

EXERCISES 10.3

In Exercises 1–8, perform the indicated operations.

1. $\begin{bmatrix} 2 & 1 \\ 3 & -2 \end{bmatrix} + \begin{bmatrix} 4 & 0 \\ -5 & 3 \end{bmatrix}$

2. $\begin{bmatrix} 1 & -5 \\ -2 & 4 \end{bmatrix} - \begin{bmatrix} 3 & -2 \\ -2 & -4 \end{bmatrix}$

3. $\begin{bmatrix} 2 \\ -4 \\ 5 \end{bmatrix} + \begin{bmatrix} -3 \\ 2 \\ -1 \end{bmatrix}$

4. $\begin{bmatrix} 3 & -2 & 4 \end{bmatrix} - \begin{bmatrix} 1 & -3 & 2 \end{bmatrix}$

5. $\begin{bmatrix} 3 & 1 & -2 \\ 0 & 1 & 4 \end{bmatrix} - \begin{bmatrix} 1 & -3 & 4 \\ -3 & 1 & -2 \end{bmatrix}$

6. $\begin{bmatrix} 4 & -2 & 6 \\ -3 & 5 & -4 \end{bmatrix} + \begin{bmatrix} -3 & 3 & -3 \\ 3 & -2 & 6 \end{bmatrix}$

7. $\begin{bmatrix} 1 & -2 \\ 2 & -3 \\ 0 & -1 \\ 2 & 1 \end{bmatrix} - \begin{bmatrix} 2 & 3 \\ -1 & 0 \\ -2 & 2 \\ 0 & 2 \end{bmatrix}$

8. $\begin{bmatrix} 1 & -1 & 0 & 1 \\ -2 & 0 & 2 & -1 \end{bmatrix} + \begin{bmatrix} -2 & 1 & -1 & 2 \\ 1 & -2 & 3 & 0 \end{bmatrix}$

In Exercises 9–14, use the definition of matrix equality to find the value of each variable.

9. $\begin{bmatrix} x & y \\ 1 & z \end{bmatrix} = \begin{bmatrix} 3 & -2 \\ 1 & 4 \end{bmatrix}$

10. $\begin{bmatrix} 6 & 2 \\ -4 & -5 \end{bmatrix} = \begin{bmatrix} 2u & v \\ \dfrac{w}{3} & -5 \end{bmatrix}$

11. $\begin{bmatrix} 6 & 2 \\ -3 & -2 \\ 8 & -1 \end{bmatrix} = \begin{bmatrix} 3r & 2 \\ -3 & \dfrac{s}{2} \\ 2t & -1 \end{bmatrix}$

12. $\begin{bmatrix} 5m & -n & -2 \\ 1 & 4 & -\dfrac{p}{3} \end{bmatrix} = \begin{bmatrix} 15 & 3 & -2 \\ 1 & 4 & -3 \end{bmatrix}$

13. $\begin{bmatrix} x-2 & y & 3 \\ 3z & 2v & 5 \end{bmatrix} + \begin{bmatrix} 4 & 2y-1 & 2 \\ -2 & 3-v & 7 \end{bmatrix} = \begin{bmatrix} 6 & -1 & 5 \\ 4 & 1 & 2w \end{bmatrix}$

14. $\begin{bmatrix} a+4 & b \\ -5 & 2c \\ 3+d & 10 \end{bmatrix} + \begin{bmatrix} -3 & b-1 \\ 7 & 3-c \\ 6 & -7 \end{bmatrix} = \begin{bmatrix} 5 & 3 \\ 2 & 1 \\ 4d & 3 \end{bmatrix}$

In Exercises 15–22, find $A + B$, $A - B$, $3A$, and $2A - 3B$.

15. $A = \begin{bmatrix} 2 & 1 \\ 3 & -2 \end{bmatrix}$, $B = \begin{bmatrix} 1 & -2 \\ 2 & -3 \end{bmatrix}$

16. $A = \begin{bmatrix} 3 & 0 \\ 0 & -3 \end{bmatrix}$, $B = \begin{bmatrix} 2 & -2 \\ -2 & 2 \end{bmatrix}$

17. $A = \begin{bmatrix} 2 & -1 & 5 \\ -1 & 3 & -2 \end{bmatrix}$, $B = \begin{bmatrix} -2 & 1 & -2 \\ 1 & -3 & 4 \end{bmatrix}$

18. $A = \begin{bmatrix} 3 & 2 \\ -1 & -2 \\ -3 & 0 \end{bmatrix}$, $B = \begin{bmatrix} -3 & 5 \\ 1 & 4 \\ 5 & -2 \end{bmatrix}$

19. $A = \begin{bmatrix} 1 \\ -2 \\ 1 \end{bmatrix}$, $B = \begin{bmatrix} -2 \\ 3 \\ 4 \end{bmatrix}$

20. $A = \begin{bmatrix} 1 & -1 \\ 2 & -2 \\ 3 & -3 \end{bmatrix}$, $B = \begin{bmatrix} -2 & 1 \\ 0 & 0 \\ 1 & -2 \end{bmatrix}$

21. $A = \begin{bmatrix} 1 \\ 2 \\ 0 \\ 2 \end{bmatrix}$, $B = \begin{bmatrix} 3 \\ 0 \\ 2 \\ 2 \end{bmatrix}$

22. $A = \begin{bmatrix} 1 & -1 & 0 & 1 \end{bmatrix}$, $B = \begin{bmatrix} -2 & 1 & -1 & 2 \end{bmatrix}$

In Exercises 23–32, find AB and BA, whenever possible.

23. $A = \begin{bmatrix} 1 & -1 \\ 3 & 0 \end{bmatrix}$, $B = \begin{bmatrix} 2 & 0 \\ 1 & -1 \end{bmatrix}$

24. $A = \begin{bmatrix} -1 & 2 \\ 0 & 1 \end{bmatrix}$, $B = \begin{bmatrix} 1 & -1 \\ -2 & 3 \end{bmatrix}$

25. $A = \begin{bmatrix} 3 \\ -2 \\ 4 \end{bmatrix}$, $B = \begin{bmatrix} 1 & -3 & 2 \end{bmatrix}$

26. $A = \begin{bmatrix} 4 & 0 & -1 \end{bmatrix}$, $B = \begin{bmatrix} 0 \\ -2 \\ 3 \end{bmatrix}$

27. $A = \begin{bmatrix} 2 & -2 & 1 \\ 3 & 0 & 1 \end{bmatrix}$, $B = \begin{bmatrix} 0 & -2 \\ -1 & -1 \\ 3 & 2 \end{bmatrix}$

28. $A = \begin{bmatrix} 1 & -3 & 1 \\ 2 & -1 & 2 \end{bmatrix}$, $B = \begin{bmatrix} 2 & 1 \\ 1 & 0 \\ -4 & -2 \end{bmatrix}$

29. $A = \begin{bmatrix} 2 & 0 & -2 \\ 0 & 2 & 2 \\ -2 & 0 & 2 \end{bmatrix}$, $B = \begin{bmatrix} 0 & 1 & 0 \\ 1 & 0 & 1 \\ 0 & 1 & 0 \end{bmatrix}$

30. $A = \begin{bmatrix} 1 & 2 & 3 \\ 0 & 1 & 2 \\ 0 & 0 & 1 \end{bmatrix}$, $B = \begin{bmatrix} 1 & 2 & 1 \\ 0 & 1 & 2 \\ 0 & 0 & 1 \end{bmatrix}$

31. $A = \begin{bmatrix} 4 \\ 1 \\ 0 \\ 2 \end{bmatrix}$, $B = \begin{bmatrix} 2 & 0 & -1 & 3 \end{bmatrix}$

32. $A = \begin{bmatrix} 1 & -2 & 0 & 0 \end{bmatrix}$, $B = \begin{bmatrix} 0 \\ -1 \\ 3 \\ 2 \end{bmatrix}$

In Exercises 33–38, use the method of Example 6 to find the inverse of each matrix, if the inverse exists.

33. $\begin{bmatrix} 1 & 1 \\ 0 & 1 \end{bmatrix}$ **34.** $\begin{bmatrix} 1 & 0 \\ 1 & 1 \end{bmatrix}$ **35.** $\begin{bmatrix} 2 & -1 \\ 1 & 1 \end{bmatrix}$

36. $\begin{bmatrix} 1 & -1 \\ 1 & 1 \end{bmatrix}$ **37.** $\begin{bmatrix} 2 & 2 \\ 0 & 2 \end{bmatrix}$ **38.** $\begin{bmatrix} 2 & 0 \\ -1 & -3 \end{bmatrix}$

In Exercises 39–50, find the inverse (if it exists) of each matrix by row operation.

39. $\begin{bmatrix} 1 & 0 \\ 1 & 1 \end{bmatrix}$ **40.** $\begin{bmatrix} 2 & -1 \\ 1 & 1 \end{bmatrix}$ **41.** $\begin{bmatrix} 1 & 2 \\ 2 & 4 \end{bmatrix}$

42. $\begin{bmatrix} 2 & -3 \\ 4 & -6 \end{bmatrix}$ **43.** $\begin{bmatrix} 1 & 0 & 1 \\ 0 & 1 & 0 \\ 0 & 0 & 1 \end{bmatrix}$ **44.** $\begin{bmatrix} 1 & 0 & 1 \\ 0 & 1 & 0 \\ 1 & 0 & 0 \end{bmatrix}$

45. $\begin{bmatrix} 1 & 2 & 3 \\ 0 & 1 & 2 \\ 0 & 0 & 1 \end{bmatrix}$ **46.** $\begin{bmatrix} 2 & 0 & 0 \\ 1 & 2 & 0 \\ 1 & 1 & 2 \end{bmatrix}$ **47.** $\begin{bmatrix} 2 & 1 & 1 \\ 0 & 0 & 1 \\ 0 & 0 & 2 \end{bmatrix}$

48. $\begin{bmatrix} 1 & 0 & 0 \\ 2 & 1 & 0 \\ 3 & 2 & 0 \end{bmatrix}$ **49.** $\begin{bmatrix} 1 & 0 & 0 \\ 0 & 2 & 0 \\ 0 & 0 & 3 \end{bmatrix}$ **50.** $\begin{bmatrix} 0 & 0 & 1 \\ 0 & 2 & 0 \\ 3 & 0 & 0 \end{bmatrix}$

In Exercises 51–60, solve each linear system by using formula (10.10).

51. $\begin{cases} 2x - y = 2 \\ 4x - y = 1 \end{cases}$ **52.** $\begin{cases} 3x - 2y = 4 \\ 2x + 6y = -1 \end{cases}$

53. $\begin{cases} x - y = 5 \\ 2x - 3y = 12 \end{cases}$ **54.** $\begin{cases} -x + 3y = 1 \\ x - 2y = -2 \end{cases}$

55. $\begin{cases} 2y = 4 \\ y - 3z = 5 \\ x - 2y - 3z = 2 \end{cases}$ **56.** $\begin{cases} x - y = -1 \\ x + z = 4 \\ y - 3z = -3 \end{cases}$

57. $\begin{cases} x - z = -2 \\ 2y + z = 13 \\ y + z = 9 \end{cases}$ **58.** $\begin{cases} 2x - z = 3 \\ y - 3z = 8 \\ x + z = 0 \end{cases}$

59. $\begin{cases} 2x - y = 7 \\ 2y + z = -10 \\ x + 3y + 2z = 15 \end{cases}$ **60.** $\begin{cases} x + y = 0 \\ x + 2y + z = 2 \\ y + 3z = 8 \end{cases}$

10.4   DETERMINANTS AND CRAMER'S RULE

With every *square* matrix A we can associate a number, called the *determinant* of A and denoted by det A. The notion of determinants is important in both applied and theoretical mathematics. In this section, we define determinants and discuss several of their properties. Also, we will study *Cramer's rule,* which uses determinants to solve systems of linear equations.

Determinants will be defined in an *inductive way.* First determinants of 1×1 matrices are defined. (This is the trivial case.) Next we define determinants of 2×2 matrices. Using this definition, we define determinants of 3×3 matrices, and so on. Assuming that the definition of determinants of $(n - 1) \times (n - 1)$ matrices is known, we can define determinants of $n \times n$ matrices. This way of defining determinants uses the *principle of mathematical induction,* a method of proof discussed in Chapter 12.

The Definition of Determinants

If $A = [a]$ is a 1×1 matrix, then by definition we set

$$\det A = a.$$

If $A = \begin{bmatrix} a_{11} & a_{12} \\ a_{21} & a_{22} \end{bmatrix}$ is a 2 × 2 matrix, then its determinant is defined by

$$\det A = a_{11}a_{22} - a_{21}a_{12}. \tag{10.11}$$

The determinant of A is also denoted by

$$\det A = \begin{vmatrix} a_{11} & a_{12} \\ a_{21} & a_{22} \end{vmatrix}.$$

EXAMPLE 1   If $A = \begin{bmatrix} 2 & 1 \\ 4 & -5 \end{bmatrix}$, find det A.

Solution: According to (10.11), we have

$$\det A = \begin{vmatrix} 2 & 1 \\ 4 & -5 \end{vmatrix} = 2 \cdot (-5) - (1 \cdot 4) = -10 - 4 = -14.$$

Practice Exercise 1   Find the determinant of the matrix

$$B = \begin{bmatrix} -1 & 2 \\ -3 & -4 \end{bmatrix}.$$

Answer: det $B = 10$

In order to extend the definition of determinant to $n \times n$ matrices, with $n \geq 3$, we have to introduce the notions of *minors* and *cofactors*.

Minors and Cofactors

If $A = [a_{ij}]$ is an $n \times n$ matrix, we denote by M_{ij} the $(n-1) \times (n-1)$ matrix obtained by deleting the ith row and jth column from the matrix A. The matrix M_{ij} is called the *complementary matrix* of the element a_{ij}.

Next, assuming that determinants of $(n-1) \times (n-1)$ matrices are known, we define the *minor* of the element a_{ij} to be the determinant of the matrix M_{ij}. Also, the number $A_{ij} = (-1)^{i+j} \det M_{ij}$ is said to be the *cofactor* of the element a_{ij}. Thus the *cofactor of a_{ij} is the determinant of the complementary matrix M_{ij} multiplied by 1 if the sum $i + j$ is even, or by -1 if the sum $i + j$ is odd.*

EXAMPLE 2   Find the minor and cofactor of the element a_{23} in the matrix

$$A = \begin{bmatrix} 3 & 0 & -2 \\ 0 & 1 & 4 \\ 1 & -2 & 1 \end{bmatrix}.$$

Solution: By deleting the second row and third column of the given matrix, as shown,

$$\begin{bmatrix} 3 & 0 & -2 \\ 0 & 1 & 4 \\ 1 & -2 & 1 \end{bmatrix},$$

we obtain the 2×2 complementary matrix

$$M_{23} = \begin{bmatrix} 3 & 0 \\ 1 & -2 \end{bmatrix}.$$

The minor of the element $a_{23} = 4$ of A is then

$$\det M_{23} = \begin{vmatrix} 3 & 0 \\ 1 & -2 \end{vmatrix} = 3(-2) - 1 \cdot 0 = -6.$$

The cofactor of a_{23} is

$$\begin{aligned} A_{23} &= (-1)^{2+3} \det M_{23} \\ &= (-1)^5(-6) \\ &= (-1)(-6) = 6. \end{aligned}$$

Practice Exercise 2   For the matrix in Example 2, find the minor and cofactor of the element a_{32}.

Answer: $\det M_{32} = \begin{vmatrix} 3 & -2 \\ 0 & 4 \end{vmatrix} = 12, A_{32} = -12$

By using the definition of cofactors, we may redefine the determinant of a 2×2 matrix as follows:

$$\det A = a_{11}A_{11} + a_{12}A_{12}. \tag{10.12}$$

Indeed, the cofactor of a_{11} is

$$A_{11} = (-1)^{1+1}\det [a_{22}] = a_{22},$$

while the cofactor of a_{12} is

$$A_{12} = (-1)^{1+2}\det [a_{21}] = -a_{21}.$$

Thus, formula (10.12) follows from formula (10.11). We conclude that, to find the determinant of a 2×2 matrix, *we multiply each element in the first row by its corresponding cofactor, and add the products.*

The Determinant of a Three-by-Three Matrix

We can now extend the definition of determinants to 3×3 matrices. If

$$A = \begin{bmatrix} a_{11} & a_{12} & a_{13} \\ a_{21} & a_{22} & a_{23} \\ a_{31} & a_{32} & a_{33} \end{bmatrix}$$

is a 3×3 matrix, its determinant is defined by

$$\det A = a_{11}A_{11} + a_{12}A_{12} + a_{13}A_{13}. \tag{10.13}$$

That is, *the determinant of A is the sum of all the products of the elements in the first row by their corresponding cofactors.*

EXAMPLE 3   If

$$A = \begin{bmatrix} 3 & 0 & -2 \\ 0 & 1 & 4 \\ 1 & -2 & 1 \end{bmatrix}.$$

find det A.

Solution: We have

$$M_{11} = \begin{bmatrix} 1 & 4 \\ -2 & 1 \end{bmatrix}, \quad M_{12} = \begin{bmatrix} 0 & 4 \\ 1 & 1 \end{bmatrix}, \quad M_{13} = \begin{bmatrix} 0 & 1 \\ 1 & -2 \end{bmatrix}.$$

Hence

$$A_{11} = (-1)^{1+1} \begin{vmatrix} 1 & 4 \\ -2 & 1 \end{vmatrix}, \quad A_{12} = (-1)^{1+2} \begin{vmatrix} 0 & 4 \\ 1 & 1 \end{vmatrix}$$

and

$$A_{13} = (-1)^{1+3} \begin{vmatrix} 0 & 1 \\ 1 & -2 \end{vmatrix}.$$

Thus according to (10.13), we have

$$
\begin{aligned}
\det A &= (3)(-1)^{1+1} \begin{vmatrix} 1 & 4 \\ -2 & 1 \end{vmatrix} + (0)(-1)^{1+2} \begin{vmatrix} 0 & 4 \\ 1 & 1 \end{vmatrix} \\
&\quad + (-2)(-1)^{1+3} \begin{vmatrix} 0 & 1 \\ 1 & -2 \end{vmatrix} \\
&= (3) \begin{vmatrix} 1 & 4 \\ -2 & 1 \end{vmatrix} + (0) \begin{vmatrix} 0 & 4 \\ 1 & 1 \end{vmatrix} + (-2) \begin{vmatrix} 0 & 1 \\ 1 & -2 \end{vmatrix} \\
&= (3)[(1)(1) - (-2)(4)] + (-2)[(0)(-2) - (1)(1)] \\
&= (3)(9) + (-2)(-1) \\
&= 27 + 2 \\
&= 29.
\end{aligned}
$$

Practice Exercise 3   Find the determinant of the matrix

$$B = \begin{bmatrix} 1 & -1 & 0 \\ 1 & 0 & 1 \\ 0 & 1 & -3 \end{bmatrix}.$$

Answer: $\det B = -4$.

The Determinant of an $n \times n$ Matrix

In general, if

$$A = \begin{bmatrix} a_{11} & a_{12} & \cdots & a_{1n} \\ a_{21} & a_{22} & \cdots & a_{2n} \\ \cdots & \cdots & \cdots & \cdots \\ a_{n1} & a_{n2} & \cdots & a_{nn} \end{bmatrix}$$

is an $n \times n$ matrix and we assume that determinants of $(n - 1) \times (n - 1)$ matrices are defined, then

$$\det A = a_{11}A_{11} + a_{12}A_{12} + \cdots + a_{1n}A_{1n}. \tag{10.14}$$

This formula is similar to formulas (10.12) and (10.13). We shall refer to it as the *cofactor expansion along the first row* of the determinant of the matrix A.

The determinant of A is also denoted by

$$\det A = \begin{vmatrix} a_{11} & a_{12} & \cdots & a_{1n} \\ a_{21} & a_{22} & \cdots & a_{2n} \\ \cdots & \cdots & \cdots & \cdots \\ a_{n1} & a_{n2} & \cdots & a_{nn} \end{vmatrix}.$$

In what follows we shall refer to *elements, rows,* and *columns of a determinant,* instead of elements, rows, and columns of the matrix corresponding to the determinant.

An important theorem that we quote without proof states that the determinant of an $n \times n$ matrix can be obtained as the *cofactor expansion along any of its rows or columns.* Thus, definition (10.14) is equivalent to

$$\det A = a_{i1}A_{i1} + a_{i2}A_{i2} + \cdots + a_{in}A_{in}, \tag{10.15}$$

which is *the cofactor expansion along the ith row of the determinant of the matrix A.* It is also equivalent to the formula

$$\det A = a_{1j}A_{1j} + a_{2j}A_{2j} + \cdots + a_{nj}A_{nj},$$

which is *the cofactor expansion along the jth column of the determinant of the matrix A.*

Hints for Computing a Determinant

The determinant of any matrix can be computed by cofactor expansion along any row or column that you may choose. In order to shorten your calculations, choose a row or column with the largest number of zeros. Notice also that if *all* entries of a row or column are zero, then the determinant is equal to zero.

EXAMPLE 4   Find det A if

$$A = \begin{bmatrix} 3 & -1 & 2 & 0 \\ 2 & 1 & -3 & 1 \\ 1 & 0 & -1 & 0 \\ 4 & 1 & -2 & 0 \end{bmatrix}.$$

Solution: Since the last column has three zeros, we use the cofactor expansion along the fourth column to get

$$\det A = (1)(-1)^{2+4} \begin{vmatrix} 3 & -1 & 2 \\ 1 & 0 & -1 \\ 4 & 1 & -2 \end{vmatrix} = \begin{vmatrix} 3 & -1 & 2 \\ 1 & 0 & -1 \\ 4 & 1 & -2 \end{vmatrix}.$$

We now expand the 3×3 determinant by the elements of the second row or column. Choosing the second row, we have

$$\det A = \begin{vmatrix} 3 & -1 & 2 \\ 1 & 0 & -1 \\ 4 & 1 & -2 \end{vmatrix} = (1)(-1)^{2+1} \begin{vmatrix} -1 & 2 \\ 1 & -2 \end{vmatrix} + (-1)(-1)^{2+3} \begin{vmatrix} 3 & -1 \\ 4 & 1 \end{vmatrix}$$

$$= -[(-1) \cdot (-2) - 1 \cdot 2] + [3 \cdot 1 - 4 \cdot (-1)]$$
$$= 0 + 7 = 7.$$

Practice Exercise 4   Find det A if

$$A = \begin{bmatrix} 2 & 0 & 0 & 0 \\ 1 & 2 & 0 & -1 \\ -1 & 3 & 0 & 2 \\ 3 & -1 & 1 & 2 \end{bmatrix}.$$

Answer: det $A = -14$

Computing Third-Order Determinants

The determinant of a three-by-three matrix can be computed according to the following scheme:

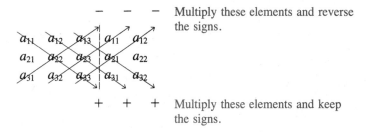

That is,

$$\begin{vmatrix} a_{11} & a_{12} & a_{13} \\ a_{21} & a_{22} & a_{23} \\ a_{31} & a_{32} & a_{33} \end{vmatrix} = a_{11}a_{22}a_{33} + a_{12}a_{23}a_{31} + a_{13}a_{21}a_{32}$$
$$- a_{31}a_{22}a_{13} - a_{32}a_{23}a_{11} - a_{33}a_{21}a_{12}.$$

This scheme can be checked by computing the right-hand side of (10.13). Note that we have six terms, three with a plus sign and three with a minus sign, as the diagram indicates.

EXAMPLE 5   Use the determinant scheme to evaluate the determinant

$$\begin{vmatrix} 3 & 0 & -2 \\ 0 & 1 & 4 \\ 1 & -2 & 1 \end{vmatrix}.$$

Solution: This determinant was computed in Example 3 by the cofactor expansion along the first row. Now, we compute it by using the scheme just described.

$$\begin{vmatrix} 3 & 0 & -2 \\ 0 & 1 & 4 \\ 1 & -2 & 1 \end{vmatrix} = 3 \cdot 1 \cdot 1 + 0 \cdot 4 \cdot 1 + (-2)(-2) \cdot 0$$
$$- 1 \cdot 1 \cdot (-2) - (-2) \cdot 4 \cdot 3 - 1 \cdot 0 \cdot 0$$
$$= 3 + 24 + 2$$
$$= 29$$

■ A word of caution The student should be very much aware that this scheme applies *only* to 3 × 3 determinants. *The method does not apply to higher-order determinants.*

Practice Exercise 5   Use the determinant scheme to compute the determinant of the matrix

$$B = \begin{vmatrix} 1 & -1 & 0 \\ 1 & 0 & 1 \\ 0 & 1 & -3 \end{vmatrix}.$$

Answer: $\det B = -4$

Properties of Determinants

The properties that follow greatly simplify the task of evaluating determinants of matrices of order 3 or greater.

I. If we interchange two rows (or columns) of a determinant, the sign of the determinant changes.

For example,

$$\begin{vmatrix} a_{21} & a_{22} \\ a_{11} & a_{12} \end{vmatrix} = a_{21}a_{12} - a_{11}a_{22}$$

$$= -(a_{11}a_{22} - a_{21}a_{12})$$

$$= -\begin{vmatrix} a_{11} & a_{12} \\ a_{21} & a_{22} \end{vmatrix}.$$

II. If we multiply a row (or column) of a determinant by a constant, the new determinant is equal to the original determinant times the constant.

For example,

$$\begin{vmatrix} ca_{11} & ca_{12} \\ a_{21} & a_{22} \end{vmatrix} = (ca_{11})a_{22} - a_{21}(ca_{12})$$

$$= c(a_{11}a_{22} - a_{21}a_{12})$$

$$= c\begin{vmatrix} a_{11} & a_{12} \\ a_{21} & a_{22} \end{vmatrix}.$$

III. If we add a multiple of a row (or column) of a determinant to another row (or column), the value of the determinant is unchanged.

For example,

$$\begin{vmatrix} a_{11} & a_{12} \\ a_{21} + ca_{11} & a_{22} + ca_{12} \end{vmatrix} = a_{11}(a_{22} + ca_{12}) - a_{12}(a_{21} + ca_{11})$$

$$= a_{11}a_{22} + ca_{11}a_{12} - a_{12}a_{21} - ca_{12}a_{11}$$

$$= a_{11}a_{22} - a_{12}a_{21}$$

$$= \begin{vmatrix} a_{11} & a_{12} \\ a_{21} & a_{22} \end{vmatrix}.$$

Properties I, II, and III, which we proved for 2×2 determinants, extend to the general case of $n \times n$ determinants. The proofs, which are beyond the scope of this book, will not be given here.

Property I implies that if the corresponding elements in two rows (or columns) of a matrix are equal, then its determinant is zero. To see this, note that if a matrix A has two equal rows, then interchanging the rows does not change the matrix. But, by property I, we have

$$\det A = -\det A,$$

which implies that $\det A = 0$.

EXAMPLE 6   Without computing the determinant, verify that

$$\begin{vmatrix} 1 & 2 & -1 \\ 2 & 1 & 2 \\ 5 & 7 & -1 \end{vmatrix} = 0.$$

Solution: By using properties I, II, and III, we have

$$\begin{vmatrix} 1 & 2 & -1 \\ 2 & 1 & 2 \\ 5 & 7 & -1 \end{vmatrix}$$

$$\downarrow \qquad -R_2 + R_3$$

$$= \begin{vmatrix} 1 & 2 & -1 \\ 2 & 1 & 2 \\ 3 & 6 & -3 \end{vmatrix}$$

$$\downarrow \qquad \text{Use property II}$$

$$= 3\begin{vmatrix} 1 & 2 & -1 \\ 2 & 1 & 2 \\ 1 & 2 & -1 \end{vmatrix} = 0,$$

since the first and third rows are equal.

Practice Exercise 6   Using properties I, II, and III of determinants, show that

$$\begin{vmatrix} 1 & 0 & -2 \\ 3 & 1 & -4 \\ 5 & 1 & -8 \end{vmatrix} = 0$$

without actually expanding the determinant.

EXAMPLE **7**   Show that the determinant equation

$$\begin{vmatrix} x & y & 1 \\ x_1 & y_1 & 1 \\ x_2 & y_2 & 1 \end{vmatrix} = 0$$

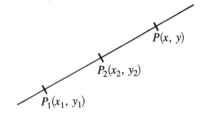

represents an equation of the line through the points $P_1(x_1, y_1)$ and $P_2(x_2, y_2)$, where $P_1 \neq P_2$.

Solution: Expanding the determinant by the cofactors of the elements of the first row and equating to the result zero, we get

$$\begin{vmatrix} y_1 & 1 \\ y_2 & 1 \end{vmatrix} x - \begin{vmatrix} x_1 & 1 \\ x_2 & 1 \end{vmatrix} y + \begin{vmatrix} x_1 & y_1 \\ x_2 & y_2 \end{vmatrix} = 0,$$

or

$$(y_1 - y_2)x + (x_2 - x_1)y + (x_1 y_2 - x_2 y_1) = 0.$$

The latter is an equation of the form $Ax + By + C = 0$, where $A = y_1 - y_2$, $B = x_2 - x_1$, and $C = x_1 y_2 - x_2 y_1$. The coefficients A and B are not both zero (because $P_1 \neq P_2$), so the equation represents a line (Section 3.2). Substituting (x_1, y_1) for (x, y) in the determinant makes the first two rows equal, so the determinant is zero. Thus the point $P_1(x_1, y_1)$ belongs to the line. Similarly, $P_2(x_2, y_2)$ also belongs to the line.

Practice Exercise 7   Use the method described in Example 7 to find an equation of a line through the points $P_1(2, -3)$ and $P_2(-1, 4)$.

Answer: $\begin{vmatrix} x & y & 1 \\ 2 & -3 & 1 \\ -1 & 4 & 1 \end{vmatrix} = 0$ or $7x + 3y - 5 = 0$

EXAMPLE **8**   Evaluate det A if

$$A = \begin{bmatrix} 1 & 0 & -2 & 1 \\ 2 & -1 & 3 & 0 \\ -2 & 0 & 2 & 0 \\ -3 & 2 & -1 & 1 \end{bmatrix}.$$

Solution: Noticing that the last column of A has two zeros, we try to produce another zero in it. This can be achieved by adding -1 times the fourth row to the first:

$$\det A = \begin{vmatrix} 4 & -2 & -1 & 0 \\ 2 & -1 & 3 & 0 \\ -2 & 0 & 2 & 0 \\ -3 & 2 & -1 & 1 \end{vmatrix}. \qquad -R_4 + R_1$$

Now, we use the cofactor expansion along the fourth column:

$$\det A = \begin{vmatrix} 4 & -2 & -1 & 0 \\ 2 & -1 & 3 & 0 \\ -2 & 0 & 2 & 0 \\ -3 & 2 & -1 & 1 \end{vmatrix}$$

$$= (1)(-1)^{4+4} \begin{vmatrix} 4 & -2 & -1 \\ 2 & -1 & 3 \\ -2 & 0 & 2 \end{vmatrix} = \begin{vmatrix} 4 & -2 & -1 \\ 2 & -1 & 3 \\ -2 & 0 & 2 \end{vmatrix}.$$

There is a zero in the third row of the last determinant. We can produce another zero by adding the first column to the third (which we indicate by $C_1 + C_3$), obtaining

$$\det A = \begin{vmatrix} 4 & -2 & 3 \\ 2 & -1 & 5 \\ -2 & 0 & 0 \end{vmatrix}. \quad C_1 + C_3$$

By the cofactor expansion along the last row, we get

$$\det A = (-2)(-1)^{1+3} \begin{vmatrix} -2 & 3 \\ -1 & 5 \end{vmatrix}$$

$$= (-2)[(-2) \cdot 5 - (-1) \cdot 3]$$

$$= (-2)(-10 + 3) = (-2)(-7) = 14.$$

Practice Exercise 8   Find the determinant

$$\begin{vmatrix} 1 & 0 & 2 & 0 \\ -2 & 1 & -4 & 3 \\ 3 & 4 & 6 & 1 \\ -1 & 2 & -1 & 2 \end{vmatrix}$$

by first producing as many zeros as possible along a row or column and then expanding the determinant along that row or column.

Answer: 11

Cramer's Rule

Determinants provide a general method of solving systems of linear equations. Consider, for example, the two-by-two system

$$\begin{cases} a_{11}x_1 + a_{12}x_2 = b_1 \\ a_{21}x_1 + a_{22}x_2 = b_2 \end{cases} \tag{10.16}$$

and assume that the determinant of the matrix of coefficients is nonzero, that is,

$$D = \begin{vmatrix} a_{11} & a_{12} \\ a_{21} & a_{22} \end{vmatrix} = a_{11}a_{22} - a_{21}a_{12} \neq 0.$$

Next, we solve the system (10.16) by elimination. Multiplying the first equation by a_{22} and the second by a_{12}, we obtain

$$\begin{cases} a_{11}a_{22}x_1 + a_{12}a_{22}x_2 = b_1 a_{22} \\ a_{21}a_{12}x_1 + a_{22}a_{12}x_2 = b_2 a_{12}. \end{cases}$$

Subtracting the second equation from the first one, we eliminate the terms containing x_2:

$$(a_{11}a_{22} - a_{21}a_{12})x_1 = b_1 a_{22} - b_2 a_{12};$$

hence

$$x_1 = \frac{b_1 a_{22} - b_2 a_{12}}{a_{11}a_{22} - a_{21}a_{12}}.$$

Proceeding in a similar manner with system (10.16) and eliminating the terms containing x_1, we obtain

$$x_2 = \frac{b_2 a_{11} - b_1 a_{21}}{a_{11}a_{22} - a_{21}a_{22}}.$$

The solution (x_1, x_2) of system (10.16) can be expressed in terms of determinants as follows:

$$x_1 = \frac{\begin{vmatrix} b_1 & a_{12} \\ b_2 & a_{22} \end{vmatrix}}{|D|} \tag{10.17}$$

and

$$x_2 = \frac{\begin{vmatrix} a_{11} & b_1 \\ a_{21} & b_2 \end{vmatrix}}{|D|}. \tag{10.18}$$

Formulas (10.17) and (10.18) together are called *Cramer's rule* for a 2×2 linear system.

EXAMPLE 9

Using Cramer's rule, solve the system

$$\begin{cases} 2x - y = 8 \\ 3x + 2y = -2. \end{cases}$$

Solution: First, we must make sure that the determinant of the matrix of coefficients is different from zero:

$$\begin{vmatrix} 2 & -1 \\ 3 & 2 \end{vmatrix} = 2 \cdot 2 - (-1) \cdot 3 = 4 + 3 = 7 \neq 0.$$

Next, using (10.17) and (10.18), we obtain the solution of the system:

$$x = \frac{\begin{vmatrix} 8 & -1 \\ -2 & 2 \end{vmatrix}}{7} = \frac{16 - 2}{7} = \frac{14}{7} = 2$$

and

$$y = \frac{\begin{vmatrix} 2 & 8 \\ 3 & -2 \end{vmatrix}}{7} = \frac{-4 - 24}{7} = \frac{-28}{7} = -4.$$

Practice Exercise 9   Use Cramer's rule to solve the system

$$\begin{cases} 2x - y = 1 \\ x + 2y = -12 \end{cases}.$$

Answer: $x = -2, y = -5$

The General Case

Formulas (10.17) and (10.18) generalize to $n \times n$ linear systems. Let

$$\begin{cases} a_{11}x_1 + a_{12}x_2 + \cdots + a_{1n}x_n = b_1 \\ a_{21}x_1 + a_{22}x_2 + \cdots + a_{2n}x_n = b_2 \\ \phantom{a_{21}x_1 + a_{22}x_2} \vdots \\ a_{n1}x_1 + a_{n2}x_2 + \cdots + a_{nn}x_n = b_n \end{cases} \qquad (10.19)$$

be an $n \times n$ linear system and assume that the determinant D of the matrix of coefficients is nonzero, that is,

$$D = \begin{vmatrix} a_{11} & a_{12} & \cdots & a_{1n} \\ a_{21} & a_{22} & \cdots & a_{2n} \\ \cdots & \cdots & \cdots & \cdots \\ a_{n1} & a_{n2} & \cdots & a_{nn} \end{vmatrix} \neq 0.$$

Then, the jth component of the solution $(x_1, x_2, \cdots, x_n)$ of the system (10.19) is given by the formula

$$x_j = \frac{\begin{vmatrix} a_{11} & \cdots & b_1 & \cdots & a_{1n} \\ a_{21} & \cdots & b_2 & \cdots & a_{2n} \\ \cdots & \cdots & \cdots & \cdots & \cdots \\ a_{n1} & \cdots & b_n & \cdots & a_{nn} \end{vmatrix}}{D}, \text{ for } 1 \le j \le n. \qquad (10.20)$$

where the j column replacement occurs at position j.

In other words, the jth component x_j of the solution is the quotient of two determinants. The denominator is the determinant of the matrix of coefficients. The numerator is the determinant of the matrix obtained from the matrix of coefficients by replacing the jth column with the constants $b_1, b_2, \ldots, b_n$.

Except for 2×2 and 3×3 systems, Cramer's rule is not very useful in practice. In most cases, the Gaussian elimination method is more efficient. However, Cramer's rule is interesting for historical reasons, besides being a valuable theoretical tool.

EXAMPLE 10   Use Cramer's rule to solve the system

$$\begin{cases} x & + 2z = 1 \\ 3x - y & = -3 \\ 2y + z = 1. \end{cases}$$

Solution: We have

$$D = \begin{vmatrix} 1 & 0 & 2 \\ 3 & -1 & 0 \\ 0 & 2 & 1 \end{vmatrix} = 1 \begin{vmatrix} -1 & 0 \\ 2 & 1 \end{vmatrix} - 0 \begin{vmatrix} 3 & 0 \\ 0 & 1 \end{vmatrix} + 2 \begin{vmatrix} 3 & -1 \\ 0 & 2 \end{vmatrix}$$

$$= 1[(-1)(1) - 2(0)] + 2[3(2) - 0(-1)]$$
$$= 1(-1) + 2(6)$$
$$= 11 \neq 0.$$

According to formula (10.20), we get

$$x = \frac{\begin{vmatrix} 1 & 0 & 2 \\ -3 & -1 & 0 \\ 1 & 2 & 1 \end{vmatrix}}{11} = \frac{-11}{11} = -1,$$

$$y = \frac{\begin{vmatrix} 1 & 1 & 2 \\ 3 & -3 & 0 \\ 0 & 1 & 1 \end{vmatrix}}{11} = \frac{0}{11} = 0,$$

and

$$z = \frac{\begin{vmatrix} 1 & 0 & 1 \\ 3 & -1 & -3 \\ 0 & 2 & 1 \end{vmatrix}}{11} = \frac{11}{11} = 1,$$

where the computation of each determinant is left as an exercise for the reader.

Practice Exercise 10   Solve by Cramer's rule the system

$$\begin{cases} x + y & = 0 \\ x + 2y + z = 2 \\ y + 3z = 8. \end{cases}$$

Answer: $x = 1, y = -1, z = 3$

An Application of Determinants

Recall that the law of cosines was discussed in Section 9.6. Now we show how to derive the law of cosines by setting up and solving a 3 × 3 linear system.

Let *ABC* be the triangle with sides *a*, *b*, and *c* and opposite angles α, β, and γ, respectively (Figure 10.7). Using the trigonometric definitions of Chapter 7, we get

$$c \cos \beta + b \cos \gamma = a.$$

Reasoning the same way with the other two sides, *b* and *c*, we get two more equations:

$$c \cos \alpha + a \cos \gamma = b \quad \text{and} \quad b \cos \alpha + a \cos \beta = c.$$

The three equations form a linear system,

$$\begin{cases} \quad\quad c \cos \beta + b \cos \gamma = a \\ c \cos \alpha \quad\quad\quad + a \cos \gamma = b \\ b \cos \alpha + a \cos \beta \quad\quad\quad = c, \end{cases}$$

that we wish to solve for $\cos \alpha$, $\cos \beta$, and $\cos \gamma$. The determinant of the matrix of coefficients is

$$\begin{vmatrix} 0 & c & b \\ c & 0 & a \\ b & a & 0 \end{vmatrix} = 2abc.$$

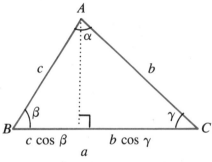

Figure 10.7

This determinant is nonzero, since $a \neq 0$, $b \neq 0$, and $c \neq 0$. We thus can apply Cramer's rule to get $\cos \alpha$:

$$\cos \alpha = \frac{\begin{vmatrix} a & c & b \\ b & 0 & a \\ c & a & 0 \end{vmatrix}}{2abc} = \frac{ac^2 + ab^2 - a^3}{2abc}$$

$$= \frac{c^2 + b^2 - a^2}{2bc}.$$

Hence

$$a^2 = b^2 + c^2 - 2bc \cos \alpha.$$

Solving the system for $\cos \beta$ and $\cos \gamma$, we obtain

$$b^2 = a^2 + c^2 - 2ac \cos \beta$$

and

$$c^2 = a^2 + b^2 - 2ab \cos \gamma.$$

EXERCISES 10.4

For each matrix in Exercises 1–6, write in determinant form the complementary matrix, the minor, and the cofactor of the given entries.

1. $\begin{bmatrix} 2 & -1 \\ 3 & -2 \end{bmatrix}$; a_{11}, a_{21}　　2. $\begin{bmatrix} -3 & 1 \\ -2 & 2 \end{bmatrix}$; a_{12}, a_{22}

3. $\begin{bmatrix} 1 & -2 & 0 \\ 3 & -1 & 2 \\ 0 & -2 & 1 \end{bmatrix}$; a_{11}, a_{32}, a_{13}　4. $\begin{bmatrix} 0 & 1 & 2 \\ 1 & 0 & 1 \\ 2 & 1 & 0 \end{bmatrix}$; a_{13}, a_{22}, a_{31}

5. $\begin{bmatrix} 1 & 0 & 0 & 2 \\ 2 & 1 & 0 & 3 \\ -2 & 0 & 0 & -1 \\ -3 & 0 & 1 & 2 \end{bmatrix}$; a_{11}, a_{13}, a_{43}

6. $\begin{bmatrix} -2 & 0 & 1 & 2 \\ -1 & 3 & 0 & 1 \\ 2 & 2 & -1 & 1 \\ 1 & -1 & 3 & -2 \end{bmatrix}$; a_{21}, a_{33}, a_{42}

In Exercises 7–22, evaluate each determinant.

7. $\begin{vmatrix} 2 & -1 \\ 3 & -2 \end{vmatrix}$　　8. $\begin{vmatrix} -3 & 1 \\ -2 & 2 \end{vmatrix}$

9. $\begin{vmatrix} 2 & -3 \\ -1 & -2 \end{vmatrix}$　　10. $\begin{vmatrix} -2 & -1 \\ 3 & -1 \end{vmatrix}$

11. $\begin{vmatrix} 0 & 1 \\ 2 & 0 \end{vmatrix}$　　12. $\begin{vmatrix} 2 & 0 \\ 0 & 3 \end{vmatrix}$

13. $\begin{vmatrix} 1 & -2 & 3 \\ 3 & -1 & 2 \\ 0 & -2 & 1 \end{vmatrix}$　　14. $\begin{vmatrix} 0 & 1 & 2 \\ 1 & 0 & 1 \\ 2 & 1 & 0 \end{vmatrix}$

15. $\begin{vmatrix} 1 & 2 & 3 \\ 0 & -1 & -2 \\ -3 & -2 & 1 \end{vmatrix}$　　16. $\begin{vmatrix} 2 & 0 & 0 \\ 1 & 3 & 4 \\ -2 & -1 & -2 \end{vmatrix}$

17. $\begin{vmatrix} 2 & 1 & 2 \\ 0 & 3 & 1 \\ 0 & 0 & 4 \end{vmatrix}$　　18. $\begin{vmatrix} 1 & 2 & -2 \\ -2 & 3 & 0 \\ 4 & 0 & 0 \end{vmatrix}$

19. $\begin{vmatrix} 2 & -1 & 3 \\ 1 & 2 & 1 \\ 4 & -2 & 6 \end{vmatrix}$　　20. $\begin{vmatrix} 1 & 0 & -2 \\ 2 & 1 & 3 \\ 4 & 1 & -1 \end{vmatrix}$

21. $\begin{vmatrix} 1 & 0 & 0 & 2 \\ 2 & 1 & 0 & 3 \\ -2 & 0 & 0 & -1 \\ -3 & 0 & 1 & 2 \end{vmatrix}$　　22. $\begin{vmatrix} -2 & 0 & 1 & 2 \\ -1 & 3 & 0 & 1 \\ 2 & 2 & -1 & 1 \\ 1 & -1 & 3 & -2 \end{vmatrix}$

23. Write in determinant form an equation of the line through the points $(2, 3)$ and $(-1, -2)$.

24. Write in determinant form an equation of the line through the points $(-3, 1)$ and $(2, -4)$.

In Exercises 25–28, compute the 3×3 determinants by using the scheme illustrated in Example 5.

25. $\begin{vmatrix} 1 & 2 & -1 \\ 0 & -2 & 1 \\ -1 & 2 & 3 \end{vmatrix}$　　26. $\begin{vmatrix} 2 & 0 & -2 \\ 1 & -1 & 1 \\ 3 & 0 & -3 \end{vmatrix}$

27. $\begin{vmatrix} 1 & 2 & 3 \\ 0 & -1 & -2 \\ -3 & -2 & 1 \end{vmatrix}$　　28. $\begin{vmatrix} 2 & 0 & 0 \\ 1 & 3 & 4 \\ -2 & -1 & -2 \end{vmatrix}$

In Exercises 29–34, verify each statement without expanding the determinant. Use only properties I, II, and III of determinants.

29. $\begin{vmatrix} 1 & 2 & 3 \\ 0 & -1 & -2 \\ 3 & 6 & 9 \end{vmatrix} = 0$　30. $\begin{vmatrix} -2 & -1 & -2 \\ 3 & 2 & 1 \\ 4 & 2 & 4 \end{vmatrix} = 0$

31. $\begin{vmatrix} 1 & 2 & -1 \\ 0 & -2 & 1 \\ 2 & 2 & -1 \end{vmatrix} = 0$　32. $\begin{vmatrix} 2 & 0 & -2 \\ 1 & -1 & 1 \\ -1 & 3 & -5 \end{vmatrix} = 0$

☐ 33. $\begin{vmatrix} 2 & 1 & 1 & 3 \\ -1 & -2 & 2 & 1 \\ 1 & 0 & 0 & -1 \\ 5 & 2 & 2 & 5 \end{vmatrix} = 0$

☐ 34. $\begin{vmatrix} 0 & 0 & 0 & 3 \\ 1 & 2 & -2 & 1 \\ -2 & 5 & -5 & 4 \\ -1 & 1 & -1 & 1 \end{vmatrix} = 0$

In Exercises 35–44, solve each system by Cramer's rule.

35. $\begin{cases} 2x - y = 4 \\ 3x - 2y = 3 \end{cases}$　　36. $\begin{cases} 2x + 2y = -1 \\ 2x + 3y = 6 \end{cases}$

37. $\begin{cases} 10x - 5y = -1 \\ 9x + 6y = 3 \end{cases}$　38. $\begin{cases} x - y = 1 \\ x - 4y = 2 \end{cases}$

39. $\begin{cases} 3x - 2y = 5 \\ 9x - 6y = 1 \end{cases}$　40. $\begin{cases} 5x + 4y = -3 \\ 10x + 8y = 7 \end{cases}$

41. $\begin{cases} y - 2z = 5 \\ x + z = 0 \\ 2x - y = 3 \end{cases}$　42. $\begin{cases} x - y = -1 \\ x + z = 4 \\ y - 3z = -3 \end{cases}$

43. $\begin{cases} 3x - 2y + z = 1 \\ x + y - 4z = 9 \\ 2x - y - z = 4 \end{cases}$　44. $\begin{cases} x + 4y - z = 2 \\ 3x - y + 2z = 3 \\ x + 3y - z = 1 \end{cases}$

In each of Exercises 45–50, find the corresponding linear system and then solve it by Cramer's rule.

45. Find a and b so that the line with equation $ax + by + 1 = 0$ passes through the points $(1, -1)$ and $(2, -3)$.

46. The points $(-2, 1)$ and $(3, 4)$ lie on the line $ax + by - 2 = 0$. What are the values of a and b?

47. How many grams each of 12-carat gold (12/24 pure gold) and 20-carat gold (20/24 pure) should a goldsmith mix to get 60 g of 18-carat gold?

48. A lab assistant has 30% and 60% acid solutions. How many cc of each should she mix to get 180 cc of a 50% solution?

49. In a family of three, the father is three times as old as the daughter. Four years from now the mother will be twice as old as the daughter. How old is each individual if the sum of their present ages is 112 years?

50. A manufacturer employs 90 workers and pays $7, $9, and $12 an hour. Twice as many workers are paid $7 an hour as are paid $9 an hour. If the total of the hourly wages is $820, find how many workers are each paid $7, $9, and $12.

10.5   NONLINEAR SYSTEMS OF EQUATIONS

In the previous sections we described several methods of solving systems of linear equations, such as substitution, Gaussian elimination, the inverse matrix method, and Cramer's rule. These are general methods that can be applied to solve any given linear system. In this section we shall study certain *nonlinear systems of equations*. A system is said to be *nonlinear* if it contains at least one equation that is not a linear equation. Nonlinear systems are in most cases more difficult to solve than linear systems. Moreover, there is no general method that can be applied to solve any given nonlinear system. In what follows we shall discuss only examples of nonlinear systems in two unknowns that can be solved by either substitution or elimination.

EXAMPLE 1   Find all solutions of the system

$$\begin{cases} 2x + y = -1 \\ x^2 - y = 4. \end{cases}$$

Solution: We solve the second equation for y and get

$$y = x^2 - 4.$$

Substituting for y in the first equation, we obtain

$$2x + x^2 - 4 = -1$$
$$x^2 + 2x - 3 = 0$$
$$(x + 3)(x - 1) = 0.$$

Hence, $x = -3$ and $x = 1$ are solutions. Now we substitute these two values into either of the two equations. Setting $x = -3$ into $y = x^2 - 4$, we obtain

$$y = (-3)^2 - 4 = 9 - 4 = 5.$$

Similarly, substituting 1 for x in the same equation, we get

$$y = 1^2 - 4 = 1 - 4 = -3.$$

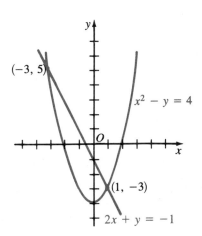

Figure 10.8

As an exercise, you can check that the same values for y can be obtained by substituting $x = -3$ and $x = 1$ into the equation $2x + y = -1$. Thus the two pairs $(-3, 5)$ and $(1, -3)$ are the solutions of the system.

We can now give a geometrical interpretation of this result. The equation $2x + y = -1$ represents a line, while the equation $x^2 - y = 4$ represents a parabola (Figure 10.8 on page 505). The two pairs $(-3, 5)$ and $(1, -3)$ are the coordinates of the points of intersection of the line and the parabola.

Practice Exercise 1 Find the intersections of the line $2x - y = -1$ and the parabola $x^2 + y = 1$.

Answer: $(-2, -3)$, $(0, 1)$

EXAMPLE 2 Find the solution of the system

$$\begin{cases} -x + 6y = 1 \\ 3xy = 1. \end{cases}$$

Solution: We solve the first equation for x and obtain

$$x = 6y - 1.$$

We substitute for x into the second and get

$$3(6y - 1)y = 1$$
$$18y^2 - 3y - 1 = 0$$
$$(6y + 1)(3y - 1) = 0.$$

Thus $y = -1/6$ or $y = 1/3$. We can now substitute each of these values into either of the two equations of the system and solve for x. Working with the second equation, we obtain

$$3x\left(-\frac{1}{6}\right) = 1, \qquad \text{so} \qquad x = -2.$$

Similarly,

$$3x\left(\frac{1}{3}\right) = 1, \qquad \text{so} \qquad x = 1.$$

Thus, $(1, 1/3)$ and $(-2, -1/6)$ are the two solutions of the given system. Figure 10.9 shows the graphs of the equations $-x + 6y = 1$ and $3xy = 1$ and the two points of intersection of the graphs.

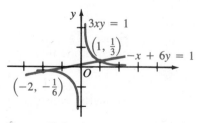

Figure 10.9

Practice Exercise 2   Find all solutions of the system

$$\begin{cases} xy + 1 = 0 \\ 2x + y = -1. \end{cases}$$

Answer: $(1/2, -2), (-1, 1)$

EXAMPLE 3   Find the points of intersection of the parabola $x = y^2$ and the circle $x^2 + y^2 = 6$.

Solution: The two curves are sketched in Figure 10.10, where we can see that there are two points of intersection. The coordinates of these points are the solutions of the nonlinear system

$$\begin{cases} x = y^2 \\ x^2 + y^2 = 6. \end{cases}$$

Substituting x for the term y^2 in the second equation, we obtain a quadratic equation in x:

$$x^2 + x = 6$$
$$x^2 + x - 6 = 0$$
$$(x - 2)(x + 3) = 0,$$

which has solutions $x = 2$ and $x = -3$. Now, the first equation of the system, $x = y^2$, tells us that x is always nonnegative. Thus the solution $x = -3$ of the quadratic equation $x^2 + x - 6 = 0$ must be discarded. Next, substituting $x = 2$ into the first equation gives us

$$2 = y^2$$
$$y = \pm\sqrt{2}.$$

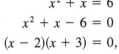

Figure 10.10   Thus the points of intersection of the given curves are $\left(2, \sqrt{2}\right)$ and $\left(2, -\sqrt{2}\right)$.

Practice Exercise 3   Find all solutions of the system

$$\begin{cases} 4x + y^2 = 0 \\ x^2 + y^2 = 5. \end{cases}$$

What is the geometrical interpretation of these solutions?

Answer: $(-1, 2)$ and $(-1, -2)$ are the points of intersection of the parabola $4x + y^2 = 0$ and the circle $x^2 + y^2 = 5$.

EXAMPLE 4   Solve the system

$$\begin{cases} x^2 + y^2 = 1 \\ \dfrac{x^2}{4} + 4y^2 = 1. \end{cases}$$

Solution: The first equation represents a circle and the second an ellipse (Figure 10.11). The two curves may intersect in at most four points, which correspond to the solutions of the system. These can be found by elimination. We multiply the first equation by -4 and add the result to the second.

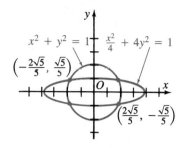

$x^2 + y^2 = 1$ $\frac{x^2}{4} + 4y^2 = 1$

$\left(-\frac{2\sqrt{5}}{5}, \frac{\sqrt{5}}{5}\right)$

$\left(\frac{2\sqrt{5}}{5}, -\frac{\sqrt{5}}{5}\right)$

Figure 10.11

$$-4x^2 - 4y^2 = -4$$

$$\frac{x^2}{4} + 4y^2 = 1$$

$$\overline{\qquad\qquad\qquad\qquad}$$

$$-4x^2 + \frac{x^2}{4} = -3$$

$$-16x^2 + x^2 = -12$$

$$-15x^2 = -12$$

$$x^2 = \frac{12}{15} = \frac{4}{5}$$

$$x = \pm\frac{2}{\sqrt{5}} = \pm\frac{2\sqrt{5}}{5}$$

Now we substitute $2/\sqrt{5}$ for x into either one of the original equations. Working with the first equation, we have

$$\left(\frac{2}{\sqrt{5}}\right)^2 + y^2 = 1$$

$$\frac{4}{5} + y^2 = 1$$

$$y^2 = \frac{1}{5}$$

$$y = \pm\frac{1}{\sqrt{5}} = \pm\frac{\sqrt{5}}{5}.$$

Similarly, substituting $-2\sqrt{5}/5$ for x into the first equation also yields $y = \pm\sqrt{5}/5$. Thus we have four solutions: $(2\sqrt{5}/5, \sqrt{5}/5)$, $(2\sqrt{5}/5, -\sqrt{5}/5)$, $(-2\sqrt{5}/5, \sqrt{5}/5)$, and $(-2\sqrt{5}/5, -\sqrt{5}/5)$. You should check that these are actually solutions of the given system.

Practice Exercise 4   The two ellipses $2x^2 + 3y^2 = 1$ and $3x^2 + 2y^2 = 1$ have four points of intersection. Find the coordinates of these points.

Answer: $(\sqrt{5}/5, \sqrt{5}/5)$, $(\sqrt{5}/5, -\sqrt{5}/5)$, $(-\sqrt{5}/5, \sqrt{5}/5)$, $(-\sqrt{5}/5, -\sqrt{5}/5)$

EXERCISES 10.5

In Exercises 1–28, find all solutions of each of the given systems.

1. $\begin{cases} x = y^2 \\ x - 3y = 0 \end{cases}$

2. $\begin{cases} x^2 + y = 0 \\ 2x - y = 0 \end{cases}$

3. $\begin{cases} y = x - 2 \\ x^2 + y = 0 \end{cases}$

4. $\begin{cases} y + x = 6 \\ y^2 - x = 0 \end{cases}$

5. $\begin{cases} x^2 - 4y = 0 \\ 3x - 4y = -4 \end{cases}$

6. $\begin{cases} 4x^2 - y = 0 \\ 4x + y = 0 \end{cases}$

7. $\begin{cases} 3x - y = 0 \\ xy = 3 \end{cases}$

8. $\begin{cases} 4xy = 3 \\ y = \frac{3}{4}x \end{cases}$

9. $\begin{cases} 2xy = 1 \\ 2x - 4y = 3 \end{cases}$

10. $\begin{cases} x - 2y = 6 \\ xy + 4 = 0 \end{cases}$

11. $\begin{cases} xy = -1 \\ y + x = 1 \end{cases}$

12. $\begin{cases} xy = 2 \\ y - x = 4 \end{cases}$

13. $\begin{cases} x^2 - 2y = 0 \\ x^2 + y^2 = 8 \end{cases}$

14. $\begin{cases} x^2 + 3y = 0 \\ x^2 + y^2 = 18 \end{cases}$

15. $\begin{cases} y = x^2 - 1 \\ x^2 + y^2 = 13 \end{cases}$

16. $\begin{cases} x - y^2 = 1 \\ x^2 + y^2 = 5 \end{cases}$

17. $\begin{cases} x^2 + y^2 = 7 \\ x - y^2 = -1 \end{cases}$

18. $\begin{cases} x^2 + y = 1 \\ x^2 + y^2 = 3 \end{cases}$

19. $\begin{cases} 2x^2 + 2y^2 = 1 \\ x^2 + 4y^2 = 1 \end{cases}$

20. $\begin{cases} x^2 + y^2 = 1 \\ 5x^2 + y^2 = 1 \end{cases}$

21. $\begin{cases} x^2 + 4y^2 = 1 \\ 2x^2 + y^2 = 1 \end{cases}$

22. $\begin{cases} 3x^2 + y^2 = 1 \\ 4x^2 + y^2 = 1 \end{cases}$

23. $\begin{cases} x^2 - 2x + y^2 = 0 \\ x^2 + y^2 = 1 \end{cases}$

24. $\begin{cases} x^2 + y^2 - 4y = 0 \\ x^2 + y^2 = 1 \end{cases}$

25. $\begin{cases} 2x^2 - xy + y^2 = 8 \\ x^2 + xy - y^2 = 4 \end{cases}$

26. $\begin{cases} x^2 + xy - 2y^2 = 1 \\ -3x^2 + xy - 2y^2 = -3 \end{cases}$

27. $\begin{cases} x^2 + 2xy - y^2 = 7 \\ x^2 \quad - y^2 = -5 \end{cases}$

28. $\begin{cases} 2x^2 - xy + 2y^2 = 12 \\ x^2 \quad + y^2 = 5 \end{cases}$

In Exercises 29–32, use properties of exponential and logarithmic functions to solve each system.

29. $\begin{cases} 3^x - y = 0 \\ 9^x - y = 6 \end{cases}$

30. $\begin{cases} x - 16^y = -12 \\ x - 4^y = 0 \end{cases}$

31. $\begin{cases} y - \log_2(x + 2) = 3 \\ y + \log_2(x + 3) = 4 \end{cases}$

32. $\begin{cases} y - \log_3(x + 3) = 6 \\ y + \log_3(x + 5) = 7 \end{cases}$

In Exercises 33–40, solve the given problem.

33. The difference of two positive numbers is 8 and their product is 240. What are the numbers?

34. Find two numbers whose sum is 4 and whose product is -96.

35. The product of two positive numbers is 384 and their ratio is 2/3. Find the numbers.

36. The ratio of two natural numbers is 3 to 7 and their product is 525. What are the numbers?

37. What are the dimensions of a rectangle whose perimeter is 36 m and whose area is 80 m^2?

38. Find the dimensions of a rectangle with an area of 75 ft^2 and perimeter of 40 ft.

39. The area of a right triangle is 150 m^2 and the hypotenuse is 25 m. What are the dimensions of the two legs?

40. The sum of the squares of the length and width of a rectangle is 74 cm^2 and the area of the rectangle is 35 cm^2. Find the dimensions of the rectangle.

10.6   SYSTEMS OF INEQUALITIES: LINEAR PROGRAMMING

Linear Inequalities

A line lying in a plane divides the plane into two parts, each of which is called a *half-plane*. Each half-plane may or may not contain the line. If a Cartesian coordinate system is defined in the plane and the line is represented by a linear equation, then we may ask what can be said about the coordinates of points on each of the half-planes. To answer this question consider, as an example, the line whose equation is $2x - y + 1 = 0$ (Figure 10.12). As we already know, points such as $(1, 3)$, $(0, 1)$, and $(-1/2, 0)$ that satisfy the equation $2x - y + 1 = 0$ lie on the line. If a point $P(x, y)$ does not lie on the line, then $2x - y + 1 \neq 0$. Consequently, *either* $2x - y + 1 > 0$ *or* $2x - y + 1 < 0$. For example, the point $(1, 4)$ does not lie on the line because $2 \cdot 1 - 4 + 1 = -1 < 0$. Similarly, the point $(1, 2)$ does not belong to the line because $2 \cdot 1 - 2 + 1 = 1 > 0$. Moreover, it can be shown that all points (x, y) on the half-plane that contains $(1, 4)$ satisfy the inequality $2x - y + 1 < 0$, and all points (x, y) on the same half-plane as the point $(1, 2)$ satisfy the inequality $2x - y + 1 > 0$.

Thus each half-plane is characterized as the set of points whose coordinates satisfy one of the strict inequalities

$$2x - y + 1 > 0 \quad \text{or} \quad 2x - y + 1 < 0.$$

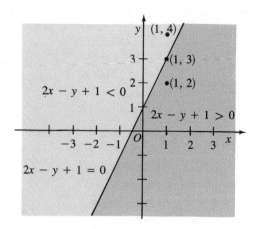

Figure 10.12

If the half-plane contains the line itself, then it is the set of points whose coordinates satisfy

$$2x - y + 1 \geq 0 \quad or \quad 2x - y + 1 \leq 0.$$

The following is a general result about linear inequalities in two variables x and y.

Linear Inequalities

A line $Ax + Bx + C = 0$, with A and B not both equal to zero, divides the coordinate plane into two half-planes. If a point $P(x, y)$ not on the line, belongs to one of the half-planes, then the coordinates of P satisfy one of the *strict linear inequalities*

$$Ax + By + C > 0 \quad or \quad Ax + By + C < 0.$$

Each inequality

$$Ax + By + C \geq 0 \quad or \quad Ax + By + C \leq 0$$

represents one of the half-planes and the line itself.

EXAMPLE **1**   Graph the inequality $4x - 2y + 6 \leq 0$.

Solution: **First method.** The graph consists of the line $4x - 2y + 6 = 0$ and one of the half-planes determined by this line. To find the half-plane, we must solve the given inequality for y:

$$4x - 2y + 6 \leq 0$$
$$-2y \leq -4x - 6 \quad \text{Multiply by } -1/2. \text{ Note that the sense}$$
$$y \geq 2x + 3 \quad \text{of the inequality is reversed.}$$

Any point with first coordinate x lies *on* the line if its second coordinate is *equal to* $2x + 3$. Any point with first coordinate x and second coordinate y *greater than*

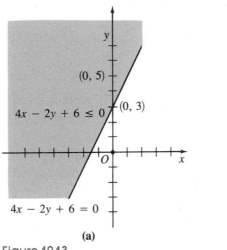
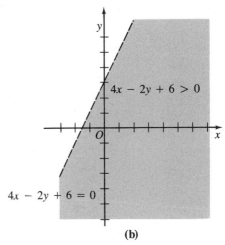

(a)   **(b)**

Figure 10.13

$2x + 3$ lies *above* the line. The graph, shown in Figure 10.13a, consists of the line and the region above the line. A solid line is used to indicate that the line is included in the graph.

Second method. Here we choose a *test point* and check to see if the coordinates of the point satisfy the given inequality. For example, the point (0, 5) is such that

$$4 \cdot 0 - 2 \cdot 5 + 6 = -10 + 6 = -4 < 0.$$

Thus it satisfies the inequality $4x - 2y + 6 \leq 0$. It follows that *all* points on the same side of the line as (0, 5) also satisfy the same inequality (Figure 10.13a).

Notice that the origin $O = (0, 0)$ is such that

$$4 \cdot 0 - 2 \cdot 0 + 6 = 6 > 0.$$

Thus (0, 0) *does not* satisfy the inequality $4x - 2y + 6 \leq 0$. It follows that the origin belongs to the region defined by the inequality $4x - 2y + 6 > 0$. The graph of this inequality is illustrated in Figure 10.13b. The dashed line indicates that the line $4x - 2y + 6 = 0$ is not included in the graph.

Practice Exercise 1   Graph the inequality $2x + 3y - 6 > 0$.

Answer:

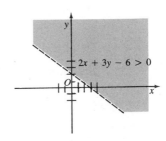

Nonlinear Inequalities

The graph of a function may also divide the plane into two regions.

Nonlinear Inequalities

Let $y = f(x)$ be a function with domain D. If $x \in D$, all points (x, y) such that $y > f(x)$ are said to be *above* the graph of $y = f(x)$, and all points (x, y) such that $y < f(x)$ are said to be *below* the graph of $y = f(x)$.

The *graph of the inequality $y > f(x)$ $[y < f(x)]$* consists of all points above [below] the graph of $y = f(x)$.

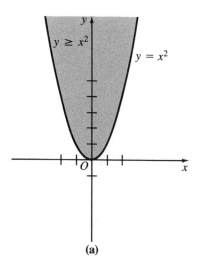
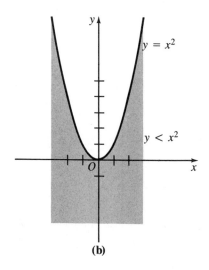

(a)   (b)

Figure 10.14

Consider, for example, the function $y = x^2$. Figure 10.14a illustrates the graph of the inequality $y \geq x^2$. The solid line indicates that the parabola $y = x^2$ is included in the graph. Figure 10.14b shows the graph of the inequality $y < x^2$. The dotted line indicates that the curve $y = x^2$ is not included in the graph.

EXAMPLE 2   Graph the inequality $y < \sqrt{1 - x}$.

Solution: The domain of the function defined by $y = \sqrt{1 - x}$ consists of all points such that $1 - x \geq 0$, that is, $x \leq 1$. The graph of this function, as well as the graph of the given inequality, are shown in Figure 10.15.

Practice Exercise 2   Graph the inequality $y \geq \sqrt{x + 4}$.

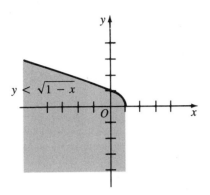

Figure 10.15

Answer:

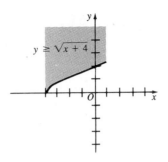

Systems of Inequalities in Two Variables

Sometimes we may have to deal simultaneously with several linear and/or non-linear inequalities, instead of a single inequality as discussed in Examples 1 and 2. Two or more inequalities form a *system of inequalities*. The *solution set* of a system consists of all ordered pairs (x, y) that satisfy all the inequalities of the system. To graph the solution set, we just graph each inequality and then find the region common to all of the graphs.

EXAMPLE 3   Graph the system of linear inequalities

$$\begin{cases} x + y < 4 \\ 2x - y + 3 > 0. \end{cases}$$

Solution: Solving each inequality for y, we obtain the equivalent system

$$\begin{cases} y < -x + 4 \\ y < 2x + 3. \end{cases}$$

Now we graph each inequality, as shown in Figure 10.16 in the margin below. The region common to the two graphs is the graph of the given system.

Practice Exercise 3   Graph the system of linear inequalities

$$\begin{cases} -x + y < 4 \\ 2x + y - 2 > 0. \end{cases}$$

Answer:

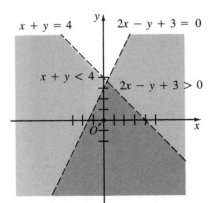

Figure 10.16

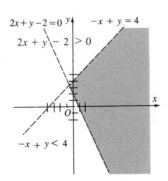

EXAMPLE **4**   In the next example we consider a *nonlinear* system of inequalities.

Graph the system of inequalities

$$\begin{cases} y > x^2 - 1 \\ y \le x + 1. \end{cases}$$

Solution: The graphs of these inequalities are the lightly shaded areas shown in Figure 10.17. Notice that the graph of the inequality $y \le x + 1$ contains the line $y = x + 1$. The region common to both graphs (the heavily shaded area) is the graph of the given system of inequalities.

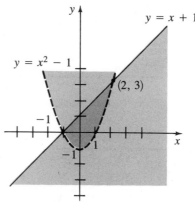

Figure 10.17

Practice Exercise 4   Graph the system of inequalities

$$\begin{cases} y \le 3 - x^2 \\ y < -x + 1. \end{cases}$$

Answer:

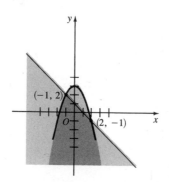

Linear Programming

An *optimization problem* has the following form: given a function, and given some subset of its domain, find the points in the subset where the function takes its largest (smallest) value and find the largest (smallest) value. We have already solved some simple optimization problems. For example, to find the vertex of a parabola, we had to find a point at which a quadratic function attained its maximum (minimum) on $\mathbb{R}$.

The theory of *linear programming* studies optimization problems in which the given function is a *linear function of two variables x and y* (or, in more advanced settings, of more than two variables). The given subset of the domain is *the set of solutions of a system of linear inequalities*. The given function is called the *objective function;* the set of solutions of the system of linear inequalities is called the *feasible set;* and the inequalities are called the *constraints* of the problem. The solution of the optimization problem may be called the *maximum (minimum) feasible* solution or *optimal* solution. (Linear programming has innumerable applications in management and economics, and the language of the theory—words like *feasible, objective*—reflects its business orientation.)

We shall use the technique of graphing systems of linear inequalities to solve linear programming problems. As an example, consider the following linear programming problem, which arises in attempting to manage a business efficiently.

EXAMPLE 5

A manufacturer of furniture can produce at most 30 desks and at most 40 chairs per week. Each desk requires 4 hours of labor and each chair requires 2 hours of labor. The manufacturer has a maximum of 160 hours of labor available per week. If the profit on each desk is $40 and the profit on each chair is $30, find the number of desks and chairs the manufacturer should produce per week for maximum profit.

Solution: Let x represent the number of desks and y the number of chairs produced per week. Since the number of desks produced per week is at most 30, it follows that the inequality $0 \le x \le 30$ is a constraint on the number of desks produced in a week. Similarly, the inequality $0 \le y \le 40$ is a constraint on the number of chairs produced weekly. In addition, another constraint on desks and chairs is derived from the maximum number of hours of labor available per week. If each desk requires 4 hours of labor and each chair requires 2 hours of labor, then x desks and y chairs require $4x + 2y$ hours of labor per week. Since this quantity cannot exceed 160 hours, the inequality $4x + 2y \le 160$ gives us a third constraint to our problem. Thus the set of constraints on the number of desks and chairs is given by the following system of linear inequalities:

$$\begin{cases} 0 \le x \le 30 \\ 0 \le y \le 40 \\ 4x + 2y \le 160. \end{cases}$$

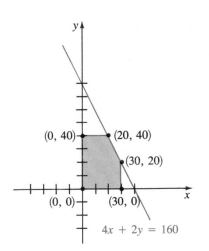

$4x + 2y = 160$

Figure 10.18

The graph of this system (the set of all feasible solutions) is the shaded region in Figure 10.18.

Since the profit on each desk is $40 and the profit on each chair is $30, it follows that the profit on x desks and y chairs is $P = 40x + 30y$. This is the objective function that we have to maximize on the set of all feasible solutions.

We now analyze the values of P on the set of constraints. For convenience, this set is again illustrated in Figure 10.19. Suppose that no desks or chairs were produced. Then, the profit would be zero and the equation $40x + 30y = 0$ would represent the line of zero profit.

Now, if 30 desks and no chairs were produced, then the profit would be $P = 40 \cdot 30 + 30 \cdot 0 = 1200$, and the equation $40x + 30y = 1200$ would repre-

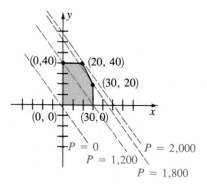

Figure 10.19

sent the line along which profit is $1,200 (Figure 10.19). Notice that the point (0, 40) corresponding to a production of no desks and 40 chairs also lies on this line. Thus, the same profit of $1,200 would be achieved by producing a total of 40 chairs and no desks. Consider, now, a production of 30 desks and 20 chairs. This yields a profit $P = 40 \cdot 30 + 30 \cdot 20 = 1200 + 600 = 1800$. The equation $40x + 30y = 1800$ represents the line along which profit is $1,800. All lines of constant profit are parallel to each other (they all have slope $-4/3$), and as they move away from the origin they represent higher profit. We can see (Figure 10.19) that the maximum profit is attained at the point (20, 40). When 20 desks and 40 chairs are produced, the profit is $P = 40 \cdot 20 + 30 \cdot 40 = 800 + 1200 = 2000$, and this is the maximum profit on our set of constraints. We conclude that the manufacturer should produce 20 desks and 40 chairs per week for a maximum profit of $2,000.

The Linear Programming Theorem

Example 5 shows that the optimal solution of the problem occurred at a *vertex* of the set of constraints. In fact, there is a theorem that asserts that *if the optimal solution of a linear programming problem exists, then it occurs at a vertex of the set of feasible solutions.*

The Linear Programming Theorem

Consider a linear function $L = Ax + By + C$, where x and y are subject to constraints given by a linear system of inequalities. If L has a maximum (or a minimum) on the solution set of the system, then it occurs at a vertex of the set.

In view of this theorem, we may now go back to Example 5 and make the following table to find the optimal solution of the problem.

Vertex	$P = 40x + 30y$
(0, 0)	$40 \cdot 0 + 30 \cdot 0 = 0$
(30, 0)	$40 \cdot 30 + 30 \cdot 0 = 1200$
(0, 40)	$40 \cdot 0 + 30 \cdot 40 = 1200$
(30, 20)	$40 \cdot 30 + 30 \cdot 20 = 1800$
(20, 40)	**$40 \cdot 20 + 30 \cdot 40 = 2000$**

In this table we are computing the values of $P = 40x + 30y$ only on the vertices of the set of all feasible solutions. As we already know, the largest value of P is $2,000 and the corresponding optimal solution is (20, 40).

■ Important remark   The importance of the linear programming theorem is that it reduces the problem of finding the maximum or minimum of the objective function over an *infinite* set (the feasible set) to a finite calculation: we have to check the values of the objective function at only *finitely many points* (the vertices of the feasible set).

Practice Exercise 5

Under the same assumptions of Example 5, suppose that the profit on each desk is $50 and the profit on each chair is $20. What is the number of desks and chairs the manufacturer should produce for maximum profit? How much is his profit?

Answer: 30 desks and 20 chairs; $1,900

EXAMPLE   6

An advertising agency is comparing ad costs for two monthly magazines. A half-page ad costs $240 in magazine X and $320 in magazine Y. According to a survey, during a given month a given advertisement will be noticed by 7200 readers of magazine X and 6000 readers of magazine Y. Also, 300 readers of magazine X and 500 readers of magazine Y are likely to complete an attached questionnaire card asking for additional information about a given ad. The agency estimates that to profit from the advertising campaign, at least 72,000 readers should be reached and at least 4500 should answer the questionnaire. How many monthly ads should be placed with each magazine in order to minimize the cost? What is the minimum cost?

Solution: If x is the number of half-page ads with magazine X and y the number with magazine Y, then the monthly cost to run the advertising campaign is $C = 240x + 320y$. We have to minimize this function on the set of constraints for this problem. The constraints are the following:

$$x \geq 0, \quad y \geq 0$$

Readers noticing the ads	$7200x + 6000y \geq 72{,}000$
Readers returning the cards	$300x + 500y \geq 4{,}500$

After simplification we obtain the following system of linear inequalities:

$$x \geq 0$$
$$y \geq 0$$
$$6x + 5y \geq 60$$
$$3x + 5y \geq 45.$$

The set of feasible solutions (i.e., the solution set of this system) is the shaded area in Figure 10.20. The vertices of this set are $(0, 12)$, $(5, 6)$, and $(15, 0)$.

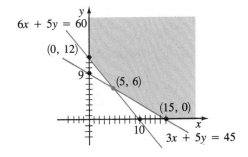

Figure 10.20

According to the linear programming theorem, the optimal solution (i.e., the minimum cost) occurs at one of these vertices. We may now arrange our work in tabular form as follows.

Vertex	$C = 240x + 320y$
$(0, 12)$	$240 \cdot 0 + 320 \cdot 12 = 0 + 3840 = 3840$
$(5, 6)$	**$240 \cdot 5 + 320 \cdot 6 = 1200 + 1920 = 3120$**
$(15, 0)$	$240 \cdot 15 + 320 \cdot 0 = 3600 + 0 = 3600$

Thus to minimize cost, five ads should be placed with magazine X and six ads with magazine Y. The minimum cost is $3120.

Practice Exercise 6   Under the same assumptions as Example 6, suppose that the advertising agency estimates that at least 108,000 readers should be reached and at least 6000 should answer the cards. How many monthly ads should now be placed with each magazine to minimize the cost? Find the minimum cost.

Answer: Ten ads with magazine X, six with magazine Y; \$4,320

EXERCISES 10.6

In Exercises 1–4, decide whether each of the ordered pairs satisfies the given inequality.

1. $3x - 2y < 1$
$(1, -3), (-2, 3)$

2. $x + 5y > 2$
$(1, -4), (3, 1)$

3. $3x - 4y + 1 \geq 0$
$\left(\frac{1}{3}, \frac{1}{2}\right), (-2, 3)$

4. $-2x + 5y - 3 \leq 0$
$(-2, 3), \left(\frac{1}{2}, \frac{4}{5}\right)$

In Exercises 5–14, graph each of the following inequalities in a Cartesian plane.

5. $2x + 1 > 0$
6. $-3x + 1 < 0$
7. $x + 5y \geq 2$
8. $3x - 2y \leq 1$
9. $-2x + 4y - 3 < 0$
10. $3x - 6y + 9 > 0$
11. $y \leq \sqrt{x - 4}$
12. $y \geq \sqrt{9 + x}$
13. $y > x^2 - 1$
14. $y \leq x^2 - 4$

In Exercises 15–24, graph each nonlinear system of inequalities.

15. $\begin{cases} y > x^2 \\ y < x + 2 \end{cases}$

16. $\begin{cases} y < 1 - x^2 \\ y > -6(x + 1) \end{cases}$

17. $\begin{cases} x^2 + y^2 < 1 \\ x \leq y \end{cases}$

18. $\begin{cases} x^2 + y^2 < 5 \\ 2x \geq y \end{cases}$

19. $\begin{cases} 4x^2 + y^2 < 4 \\ 4x < y \end{cases}$

20. $\begin{cases} x^2 + 9y^2 < 1 \\ x < 2y \end{cases}$

21. $\begin{cases} y \geq x^2 - 1 \\ x^2 + y^2 < 3 \end{cases}$

22. $\begin{cases} y < 4 - x^2 \\ x^2 + y^2 \geq 6 \end{cases}$

23. $\begin{cases} x^2 + y^2 < 2 \\ \dfrac{x^2}{9} + y^2 < 1 \end{cases}$

24. $\begin{cases} x^2 + y^2 \geq 2 \\ \dfrac{x^2}{4} + y^2 < 1 \end{cases}$

In Exercises 25–34, graph each system of linear inequalities. In each case specify the vertices.

25. $\begin{cases} y \geq 3x \\ y < -x + 4 \end{cases}$

26. $\begin{cases} 4y > -x \\ y \geq 2x + 6 \end{cases}$

27. $\begin{cases} x - y \geq 2 \\ 3x + y < 0 \end{cases}$

28. $\begin{cases} 2x - 4y < 6 \\ x - 3y \geq 0 \end{cases}$

29. $\begin{cases} 3x - 4y - 1 > 0 \\ 5x + 8y + 2 \leq 0 \end{cases}$

30. $\begin{cases} 6x - 3y + 2 \geq 0 \\ 2x + 8y - 3 < 0 \end{cases}$

31. $\begin{cases} 3x + 2y \geq 6 \\ 3x + 2y \leq 12 \\ x \geq 0 \\ y \leq 0 \end{cases}$

32. $\begin{cases} x - 4y < 6 \\ 2x - 8y > 16 \\ x \leq 0 \\ y \leq 0 \end{cases}$

33. $\begin{cases} 3x + 4y \leq 24 \\ x - 4y \geq -8 \\ x \leq 6 \\ y \geq 0 \end{cases}$

34. $\begin{cases} 2x - y + 4 > 0 \\ 3x + 2y - 15 < 0 \\ x \leq 3 \\ y \geq 0 \end{cases}$

In Exercises 35–40, an objective function L and a set of constraints are given. Find the maximum and minimum values of the objective function on the set of constraints. Specify the corresponding optimal solutions in each case.

35. $L = 2x + 3y - 5$
$$\begin{cases} x + y \leq 4 \\ 2x - y \geq -3 \\ 0 \leq x \leq 3 \\ 0 \leq y \leq 3 \end{cases}$$

36. $L = x - 2y$
$$\begin{cases} x - y \leq 4 \\ 2x + y \leq 2 \\ -3 \leq x \leq 1 \\ 0 \leq y \leq 2 \end{cases}$$

37. $L = 4x - 2y + 1$
$$\begin{cases} 3x - y \geq 0 \\ x + y \geq 0 \\ 1 \leq x \leq 3 \\ y \geq 0 \end{cases}$$

38. $L = 3x + 4y - 2$
$$\begin{cases} x + y \geq -6 \\ y - 2x \geq 0 \\ -4 \leq x \leq 0 \\ y \leq 0 \end{cases}$$

39. $L = 5x + 4y$
$$\begin{cases} 2x + 3y \geq 12 \\ 2x + y \geq 8 \\ x \geq 0 \\ y \geq 0 \end{cases}$$

40. $L = 7x + 4y - 2$
$$\begin{cases} 3x + y \geq 6 \\ 3x + 4y \geq 15 \\ x \geq 0 \\ y \geq 0 \end{cases}$$

In Exercises 41–50, solve each given problem.

41. A refinery with the capacity of refining at most 2500 barrels of oil per day produces gasoline and heating oil. The profit on each barrel of gasoline is $4, and the profit on each barrel of heating oil is $3. If at least 300 barrels of gasoline and 400 barrels of heating oil must be produced each day, find how many barrels of gasoline and heating oil should be produced for maximum profit.

42. A merchant has a roasting plant that can produce 1800 lb of roasted coffee per day. A shipment of Brazilian and Colombian coffee has just arrived, and she has to roast at least 200 lb of Brazilian coffee and 350 lb of Colombian coffee per day. The profit on each pound of Brazilian coffee is 50 cents, and that on each pound of Colombian coffee is 40 cents. How many pounds of each coffee should she produce for maximum profit?

43. The assembly line of a manufacturer produces two models of golf carts, Birdie Custom and Bogey De Luxe. It takes 10 hours of labor to produce a Birdie Custom and 15 hours of labor to produce a Bogey De Luxe, and the manufacturer has up to 90 hours of labor available per day. The assembly line can produce as many as 6 Birdie Customs and as many as 4 Bogey De Luxe per day. The profit on each Birdie Custom is $150, and that on each Bogey De Luxe is $240. How many of each model should the manufacturer assemble per day to have maximum profit? What is the maximum profit?

44. A factory assembles two types of motors, small and large. One hour of labor is required to assemble a small motor and four hours are needed for a large motor, and there are 80 hours of labor available each day. The maximum number

of small motors that can be assembled in a day is 60 and the maximum number of large motors is 15. The owner can make a profit of $120 on each small motor and $250 on each large motor. What is the maximum daily profit that can be realized?

45. The following table gives the number of units of protein and fiber in one gram of each of two ingredients, A and B, used in the preparation of a certain brand of cat food.

	A	B
Protein	3	2
Fiber	3	4

Each gram of ingredient A costs 8¢ and each gram of ingredient B costs 6¢. If each can of food has to contain at least 180 units of protein and 240 units of fiber, find the number of grams of each ingredient that should be used to minimize the cost.

46. The owner of a dog kennel prepares her own brand of dog food in such a way that her dogs receive at least 6 oz of protein and 4 oz of fat per day. For this she mixes two brand-name dog foods: Dog's Delight and Pet's Treat. Each brand contain protein and fat as follows.

	Protein (percentage)	*Fat* (percentage)
Dog's Delight	40	20
Pet's Treat	25	25

Dog's Delight costs 5¢/oz and Pet's Treat 4¢/oz. What is the least expensive mix of the two brand-name dog foods?

47. A farmer has 360 acres available for planting barley and corn. It takes 1/2 hour of labor per acre to plant barley and 3/4 hour of labor per acre to plant corn, and the farmer has a maximum of 210 hours available for this job. He expects a profit of $35 per acre from barley and $45 per acre from corn. How many acres of each crop should he plant to maximize his profit? What is this maximum profit?

48. Find the maximum daily profit that can be realized on manufacturing x tables and y bookcases under the following assumptions.
   (a) Each table requires 4 hours of labor and each bookcase requires 2 hours.
   (b) There are 90 hours of labor available each day.
   (c) At most 15 tables and 25 bookcases can be produced each day.
   (d) The profit on each table is $35 and on each bookcase, $20.

49. A town operates two centers, A and B, that collect paper, glass, and aluminum for recycling. Center A receives an average of 360 lb of paper, 180 lb of glass, and 120 lb of aluminum per day. Center B receives an average of 120 lb of paper, 540 lb of glass, and 120 lb of aluminum per day. The operating costs for centers A and B are \$30 and \$20 per day, respectively. The town has agreed to supply at least 1800 lb of paper, 2700 lb of glass, and 1080 lb of aluminum per week to a recycling company. To minimize the operating cost, how many days a week should each center remain open?

50. A farmer wishes to prepare a supply of cattle food by mixing two commercially available foods, A and B. Food A costs 40¢/lb and B costs 50¢/lb. The farmer mixes the two products so that the mixture contains at least 25 units of protein, at least 25 units of carbohydrates, and at least 20 units of fat per pound. The specifications for the two commercially available foods are listed in the following table.

Food	Protein (units/lb)	Carbohydrates (units/lb)	Fat (units/lb)
A	3	2	2
B	2	3	2

How many pounds of each food should be mixed to minimize the cost?

CHAPTER SUMMARY

In this chapter you have primarily studied two-by-two and three-by-three linear systems. Such systems can be classified as follows.

Consistent	Unique solution
Inconsistent	No solutions
Dependent	Infinitely many solutions

A linear system can always be solved by the method of *substitution*. This simple method may become too lengthy and tedious when used to solve linear systems of more than two equations and two unknowns. It is preferable to employ *Gaussian elimination*, a method of great power and flexibility. By performing *admissible operations* (see Section 10.2), you may be able to transform a three-by-three system,

$$\begin{cases} a_1x + b_1y + c_1z = d_1 \\ a_2x + b_2y + c_2z = d_2 \\ a_3x + b_3y + c_3z = d_3, \end{cases}$$

into an equivalent system in *upper-triangular form*,

$$\begin{cases} ax + by + cz = l \\ dy + ez = m \\ fz = n, \end{cases}$$

which can be solved easily.

When using Gaussian elimination, you can work with the system of equations or just the *augmented matrix*:

$$\begin{bmatrix} a_1 & b_1 & c_1 & d_1 \\ a_2 & b_2 & c_2 & d_2 \\ a_3 & b_3 & c_3 & d_3 \end{bmatrix}$$

Admissible row operations (see Section 10.2) will allow you to transform this matrix into an *upper-triangular matrix*:

$$\begin{bmatrix} a & b & c & l \\ 0 & d & e & m \\ 0 & 0 & f & n \end{bmatrix}$$

Another method of solving an $n \times n$ system uses the *algebra of matrices*. An *m-by-n matrix* is an array of $m \cdot n$ numbers written in m rows and n columns. Matrices can be added and subtracted, and multiplied by a number (*scalar multiplication*). In certain cases, the product of two matrices can be defined. Then if

$$A = \begin{bmatrix} a_1 & b_1 & c_1 \\ a_2 & b_2 & c_2 \\ a_3 & b_3 & c_3 \end{bmatrix}, \quad X = \begin{bmatrix} x \\ y \\ z \end{bmatrix}, \quad \text{and} \quad D = \begin{bmatrix} d_1 \\ d_2 \\ d_3 \end{bmatrix},$$

the three-by-three system considered before can be written as a *matrix equation:*

$$AX = D.$$

If the matrix A has an *inverse* A^{-1}, the solution is given by the formula

$$X = A^{-1}D.$$

Hence the importance of methods to find the inverse of a matrix (such as the ones described in Section 10.3), when such an inverse exists.

A third method of solving an $n \times n$ system uses *Cramer's rule*, which is based upon the notion of the *determinant* of a matrix. The computation of determinants of two-by-two or three-by-three matrices is a simple task. It becomes more complex as the order of the matrices increases. You should refer to Section 10.4, where determinants and Cramer's rule were discussed in detail.

Nonlinear systems are in most cases more difficult to solve. However, in special cases, they can be solved by *substitution* or *elimination*.

In a *linear programming problem*, the aim is to maximize or minimize a linear function, called the *objective function*, on a set called the *feasible set*. This is the solution set of a *system of linear inequalities*. The *constraints* of the problem are the linear inequalities that define the feasible set.

The *linear programming theorem* states that the maximum or minimum of the objective function occurs at a vertex of the feasible set. Linear programming has innumerable applications in management sciences and economics.

CHAPTER TEST

1. Solve the linear system

$$\begin{cases} 3x - 5y = 3 \\ 4x + 2y = -4. \end{cases}$$

2. Solve the following three-by-three system by Gaussian elimination:

$$\begin{cases} x - y + z = 4 \\ x + 2z = 3 \\ y - 2z = -2. \end{cases}$$

3. The perimeter of a triangle is 47 cm. If one side is 10 cm longer than another side and 6 cm shorter than the third side, what are the dimensions of the triangle?

4. Find $3A - 2B$, given

$$A = \begin{bmatrix} -2 & 3 \\ -1 & -1 \end{bmatrix} \quad \text{and} \quad B = \begin{bmatrix} 2 & -1 \\ -3 & 1 \end{bmatrix}.$$

5. Find the inverse of the matrix

$$\begin{bmatrix} -2 & -1 & -1 \\ 0 & -3 & -3 \\ 0 & 0 & -1 \end{bmatrix}.$$

6. Find the minor and cofactor of the element a_{23} of the matrix

$$\begin{bmatrix} 1 & -1 & -1 \\ 0 & -1 & 0 \\ 3 & 0 & 1 \end{bmatrix}.$$

7. Solve the nonlinear system

$$\begin{cases} 2x^2 + y = 0 \\ y - x = -6. \end{cases}$$

8. Graph the system of linear inequalities

$$\begin{cases} 2x + 4y < -2 \\ x - 2y \geq 1. \end{cases}$$

9. Find the maximum and minimum values of the objective function $L = x + y + 2$ on the following set of constraints:

$$\begin{cases} x - y \leq 4 \\ 2x + y \leq 2 \\ -3 \leq x \leq 1 \\ 0 \leq y \leq 2 \end{cases}$$

10. A refinery produces gasoline and heating oil by refining at most 3000 barrels of crude oil per day. The profit on each barrel of gasoline is $3, and the profit on each barrel of heating oil is $4. What is the maximum profit if at least 800 barrels of gasoline and 300 barrels of heating oil are produced each day?

REVIEW EXERCISES

In Exercises 1–4, find $2A + B$, $A - 3B$, and $5B$.

1. $A = \begin{bmatrix} 2 & 1 \\ 2 & -3 \end{bmatrix}, \quad B = \begin{bmatrix} 1 & -2 \\ 3 & 2 \end{bmatrix}$

2. $A = \begin{bmatrix} 1 & -1 \\ 2 & 3 \\ -1 & 1 \end{bmatrix}, \quad B = \begin{bmatrix} 4 & -2 \\ 3 & -1 \\ 2 & 0 \end{bmatrix}$

3. $A = \begin{bmatrix} 5 & 0 \\ 0 & -5 \end{bmatrix}$, $B = \begin{bmatrix} 0 & 2 \\ -2 & 0 \end{bmatrix}$

4. $A = \begin{bmatrix} 1 & 3 & -1 \\ -3 & 1 & 3 \end{bmatrix}$, $B = \begin{bmatrix} 2 & 0 & -2 \\ 0 & 3 & 0 \end{bmatrix}$

In Exercises 5–8, find AB and BA.

5. $A = \begin{bmatrix} 0 & 2 & -1 & -2 \end{bmatrix}$, $B = \begin{bmatrix} 2 \\ 1 \\ -3 \\ 2 \end{bmatrix}$

6. $A = \begin{bmatrix} -2 & 4 \\ 0 & 2 \end{bmatrix}$, $B = \begin{bmatrix} 2 & -1 \\ -2 & 4 \end{bmatrix}$

7. $A = \begin{bmatrix} 1 & -1 & 2 \\ 4 & 0 & 3 \end{bmatrix}$, $B = \begin{bmatrix} 3 & 2 \\ -2 & 0 \\ 1 & -2 \end{bmatrix}$

8. $A = \begin{bmatrix} 1 & 1 & 1 \\ 0 & 1 & 1 \\ 0 & 0 & 1 \end{bmatrix}$, $B = \begin{bmatrix} 1 & 0 & 0 \\ 1 & 1 & 0 \\ 1 & 1 & 1 \end{bmatrix}$

In Exercises 9–14, find the inverse of each matrix.

9. $\begin{bmatrix} 2 & -1 \\ 3 & 2 \end{bmatrix}$

10. $\begin{bmatrix} 3 & 4 \\ 1 & 2 \end{bmatrix}$

11. $\begin{bmatrix} 1 & 2 & 3 \\ 0 & 2 & 3 \\ 0 & 0 & 3 \end{bmatrix}$

12. $\begin{bmatrix} 1 & 1 & 2 \\ 1 & 2 & 0 \\ 2 & 0 & 0 \end{bmatrix}$

☐ 13. $\begin{bmatrix} 1 & 0 & 0 & 0 \\ 2 & 1 & 0 & 0 \\ 3 & 3 & 1 & 0 \\ 4 & 4 & 4 & 1 \end{bmatrix}$

☐ 14. $\begin{bmatrix} 0 & -1 & -1 & -1 \\ 1 & 2 & 1 & 1 \\ 1 & 2 & 2 & 1 \\ 1 & 2 & 2 & 2 \end{bmatrix}$

In Exercises 15–20, evaluate the determinants.

15. $\begin{vmatrix} 0 & 2 & 2 \\ 3 & 0 & 2 \\ 3 & 3 & 0 \end{vmatrix}$

16. $\begin{vmatrix} 3 & 1 & 0 \\ 1 & 0 & 1 \\ 0 & 1 & 3 \end{vmatrix}$

17. $\begin{vmatrix} 3 & 2 & 0 \\ 1 & 0 & 2 \\ 0 & 1 & 2 \end{vmatrix}$

18. $\begin{vmatrix} 2 & 0 & 1 \\ 0 & -2 & 0 \\ -1 & 0 & 2 \end{vmatrix}$

19. $\begin{vmatrix} 1 & 2 & 0 & 3 \\ -1 & -2 & 0 & 1 \\ -1 & 2 & 1 & 1 \\ -3 & 2 & 1 & 2 \end{vmatrix}$

20. $\begin{vmatrix} 1 & 0 & 1 & 0 \\ 0 & 1 & 0 & 1 \\ 1 & 0 & 1 & 1 \\ 0 & 1 & 1 & 1 \end{vmatrix}$

☐ 21. If $ab \neq 0$, find the inverse of the matrix $\begin{bmatrix} a & 0 \\ 0 & b \end{bmatrix}$.

☐ 22. If $abc \neq 0$, find the inverse of the matrix $\begin{bmatrix} a & 0 & 0 \\ 0 & b & 0 \\ 0 & 0 & c \end{bmatrix}$.

In Exercises 23–28, solve each system.

23. $\begin{cases} x - 3y = 7 \\ 2x - y = -4 \end{cases}$

24. $\begin{cases} 6x + 5y = 4 \\ 3x - 15y = -5 \end{cases}$

25. $\begin{cases} x + y - z = 0 \\ 2x - 3y - 4z = 6 \\ 2x - y - z = 8 \end{cases}$

26. $\begin{cases} x + y - 2z = -1 \\ 3x - 2y - z = 2 \\ 2x - 3y - 3z = -5 \end{cases}$

27. $\begin{cases} x + y = 1 \\ 2y + z = 0 \\ 4x - 3z = 0 \end{cases}$

28. $\begin{cases} y - z = 2 \\ x + 2z = 11 \\ x - y = 0 \end{cases}$

Solve for x in Exercises 29–32.

29. $\begin{vmatrix} x & 2 \\ -3 & -2 \end{vmatrix} = 0$

30. $\begin{vmatrix} 1 & x \\ -2 & 4 \end{vmatrix} = 6$

31. $\begin{vmatrix} 1 & x & -1 \\ 3 & 0 & 2 \\ 5 & 0 & 4 \end{vmatrix} = -8$

32. $\begin{vmatrix} 1 & 2 & -1 \\ x & 0 & 1 \\ 3 & 1 & -2 \end{vmatrix} = 14$

In Exercises 33–60, solve the given problem.

33. Find a and b so that the line of the equation $ax + by + 1 = 0$ passes through the points $(1, 2)$ and $(-2, -3)$.

34. Find a and b so that the line of the equation $ax + by - 1 = 0$ contains the points $(6, -1)$ and $(1/2, 5/6)$.

35. Find a and b so that the lines $ax + by + 1 = 0$ and $bx - ay - 8 = 0$ intersect at the point $(3, -2)$.

36. Find a and b so that the lines $ax + by + 1 = 0$ and $bx - ay - 5 = 0$ intersect at the point $(1, -1)$.

37. John has $500 more invested at 8% per year than he has invested at 7.5% per year. Together the two investments provide him with an annual income of $425. How much does John have invested at each rate?

38. A man has $6000 invested in two types of bonds paying 9% and 8% per year. If the annual income from his investments is $515, how much does he have invested in each type of bond?

39. A boy has four times more nickels than dimes, for a total of $4.80. How many nickels and dimes does he have?

40. Ann has $2.40 in nickels and dimes. If she had half as many nickels and twice as many dimes as she does, the new total would be $3.90. How many nickels and dimes does Ann have?

41. A company pays its salespeople on a basis of a percentage of the first $15,000 in sales, plus another percentage of any amount over $15,000. If a salesperson earns $1908 on sales of $27,000 and $2611 on sales of $36,500, find the two percentages.

42. In a biological experiment, two species of fish, A_1 and A_2, are kept in the same environment and are fed two types of food, F_1 and F_2. The following table gives the number of units of each type of food that each species consumes daily.

	A_1	A_2
F_1	2	2
F_2	3	1

If 74 units of food F_1 and 67 units of food F_2 are available per day, find the population size of each species that would exactly consume all the available food.

43. A manufacturer uses two machines, M and N, to produce two products, P and Q. The amount of time, in hours, that each machine is used to produce one unit of each product is given by the following table.

	P	Q
M	1	$\dfrac{1}{2}$
N	$\dfrac{3}{4}$	$\dfrac{1}{4}$

Machine M is available 16 hours a day, and machine N is available 11 hours a day. Find how many units of P and Q should be produced daily so that M and N are fully in operation.

44. Determine the values for a, b, and c so that the graph of the parabola $y = ax^2 + bx + c$ contains the points $(-1, -3)$, $(1, 3)$, and $(-2, -3)$.

45. Find a, b, and c so that the graph of $y = ax^2 + bx + c$ contains the points $(0, 1)$, $(1, 0)$, and $(2, 3)$.

46. The perimeter of a triangle is 110 m. Side c is 30 m longer than the sum of sides a and b, and side a is $1/3$ the difference between sides b and a. Find the length of each side.

47. A merchant wants to mix peanuts costing \$2.10/lb, almonds costing \$4.20/lb, and cashews costing \$4.00/lb to obtain 100 lb of a mixture costing \$2.90/lb. He also wants the amount of peanuts to be three times the amount of almonds. How many pounds of each variety should he mix?

48. A family consists of a father, a mother, and a son. The mother is three times as old as the son. Two years from now the father will be three times as old as the son. Eight years ago, the age of the father was the sum of the ages of the mother and the son. Find the present age of each member of the family.

49. A man has \$425 in one-dollar, five-dollar, and ten-dollar bills. He has 75 bills in all, and 5 more ten-dollar bills than one-dollar bills. How many bills of each kind does the man have?

50. A lab technician mixes three different sulfuric acid solutions that have concentrations of 20%, 45%, and 50%, to obtain 180 liters of a 42% solution. If he uses twice as much of the 50% solution as of the 20% solution, how many liters of each solution does he use?

51. A company employs 120 workers and pays \$6, \$8, and \$10 an hour. The total of the hourly wages is \$840. Three times as many workers are paid \$8 an hour as are paid \$10 an hour. How many workers are paid \$6? \$8? \$10?

52. Copper, zinc, and lead are mixed in different amounts according to the given table to produce three types of alloys, A, B, and C.

	A	B	C
Copper	45%	55%	60%
Zinc	30%	20%	25%
Lead	25%	25%	15%

How many pounds of each alloy should be mixed to obtain 200 lb of a new alloy that is 25.25% zinc and 22.50% lead?

☐ **53.** Show that

$$\begin{vmatrix} 1 & 1 & 1 \\ x & y & z \\ x^2 & y^2 & z^2 \end{vmatrix} = (x - y)(y - z)(z - x).$$

☐ **54.** Show that

$$\begin{vmatrix} a_{11} & a_{12} & a_{13} \\ 0 & a_{22} & a_{23} \\ 0 & 0 & a_{33} \end{vmatrix} = a_{11}a_{22}a_{33}.$$

☐ **55.** Without expanding the determinant, show that

$$\begin{vmatrix} a & b + c & 1 \\ b & c + a & 1 \\ c & a + b & 1 \end{vmatrix} = 0.$$

☐ **56.** Show that

$$\begin{vmatrix} a_{11} & a_{12} & a_{13} & a_{14} \\ 0 & a_{22} & a_{23} & a_{24} \\ 0 & 0 & a_{33} & a_{34} \\ 0 & 0 & 0 & a_{44} \end{vmatrix} = a_{11}a_{22}a_{33}a_{44}.$$

57. Show that

$$\begin{vmatrix} \cos\theta & -\sin\theta \\ \sin\theta & \cos\theta \end{vmatrix} = 1.$$

58. Find all the roots of the following equation:

$$\begin{vmatrix} 1 & x_1 & x_2 \\ 1 & x & x_2 \\ 1 & x_1 & x \end{vmatrix} = 0.$$

59. Find all the roots of the equation

$$\begin{vmatrix} 1 & x_1 & x_2 & x_3 \\ 1 & x & x_2 & x_3 \\ 1 & x_1 & x & x_3 \\ 1 & x_1 & x_2 & x \end{vmatrix} = 0.$$

60. Show that

$$\begin{vmatrix} a & b & 0 & 0 \\ c & d & 0 & 0 \\ 0 & 0 & e & f \\ 0 & 0 & g & h \end{vmatrix} = \begin{vmatrix} a & b \\ c & d \end{vmatrix} \cdot \begin{vmatrix} e & f \\ g & h \end{vmatrix}.$$

In Exercises 61–64, solve each nonlinear system of equations by algebraic methods.

61. $\begin{cases} y = x^2 + 3 \\ x^2 + y^2 = 9 \end{cases}$ **62.** $\begin{cases} xy = 3 \\ 2y + x = -5 \end{cases}$

63. $\begin{cases} 3x^2 - 2y = 0 \\ 5x - 2y = -2 \end{cases}$ **64.** $\begin{cases} x^2 + y = 2 \\ x^2 + y^2 = 8 \end{cases}$

In Exercises 65–68, graph each nonlinear system of inequalities.

65. $\begin{cases} 3y > x \\ x - y^2 > 0 \end{cases}$ **66.** $\begin{cases} y - 2x > 0 \\ x^2 + y^2 < 5 \end{cases}$

67. $\begin{cases} x^2 + y^2 < 8 \\ x^2 + y < 2 \end{cases}$ **68.** $\begin{cases} x^2 - 2x + y^2 < 0 \\ x^2 + y^2 < 1 \end{cases}$

In Exercises 69–72, graph each of the following systems of linear inequalities.

69. $\begin{cases} 2x - 5y < 9 \\ 3x + y > 5 \end{cases}$ **70.** $\begin{cases} 6x - 2y - 8 > 0 \\ 3x + y - 10 < 0 \end{cases}$

71. $\begin{cases} 3x + 4y \geq 10 \\ x + 2y \geq 4 \\ x \geq 0 \\ y \geq 0 \end{cases}$ **72.** $\begin{cases} 2x + 3y \leq -8 \\ x + y \leq -3 \\ x \leq 0 \\ y \leq 0 \end{cases}$

In Exercises 73–76, find the maximum and the minimum values of the objective function L on the given set of constraints. Specify the optimal solution in each case.

73. $L = 3x + 4y$
$$\begin{cases} x + y \leq 4 \\ 2x - y \geq -3 \\ 0 \leq x \leq 3 \\ 0 \leq y \leq 3 \end{cases}$$

74. $L = 2x - 4y + 1$
$$\begin{cases} x - y \leq 4 \\ 2x + y \leq 2 \\ -3 \leq x \leq 1 \\ 0 \leq y \leq 2 \end{cases}$$

75. $L = x + 3y - 4$
$$\begin{cases} x + y \geq -6 \\ x - 2y \geq 0 \\ -4 \leq x \leq 0 \\ y \leq 0 \end{cases}$$

76. $L = 2x + 5y$
$$\begin{cases} 3x + y \geq 6 \\ 3x + 4y \geq 15 \\ x \geq 0 \\ y \geq 0 \end{cases}$$

In Exercises 77–80, solve the given problem.

77. The difference of two positive numbers is 7 and the product is 450. Find the two numbers.

78. The perimeter of a rectangle is 46 m and its area is 120 m². What are the dimensions of the rectangle?

79. A manufacturer assembles two models of bicycles: Standard and De Luxe. It takes 2 hours of labor to assemble a Standard model and 4 hours to assemble a De Luxe, and the manufacturer has up to 40 hours of labor available per day. The assembly line can produce at most 10 Standard and 8 De Luxe per day. The profit on each Standard model is $60, and that on each De Luxe is $80. Find how many bicycles of each model the manufacturer should assemble per day to realize maximum profit. What is the maximum profit?

80. The following table gives the number of units of fat and protein contained in one pound of two ingredients, A and B, used in the preparation of a brand of dog food.

	A	B
Fat	3	2
Protein	3	4

Each pound of ingredient A costs 50¢ and each pound of B costs 40¢. If each bag of dog food has to contain at least 12 units of fat and 18 units of protein, find the number of pounds of each ingredient that should be used to minimize the cost.

CHAPTER 11

ZEROS

OF

POLYNOMIALS

The present chapter begins with a brief review of the definition of polynomials and the operations of addition and multiplication of polynomials. Next, we describe the division algorithm for polynomials, the division of a polynomial by $x - a$, and the remainder and factor theorems. We define the notion of multiplicity of a zero, and then we prove the theorem about the number of zeros of a polynomial. If a polynomial has integer coefficients, then its rational zeros, if they exist, can be found with the help of a simple test explained in Section 11.4. In the last section of this chapter, we discuss the method of partial fraction decomposition.

11.1 PRELIMINARIES

Polynomials were defined in Section 1.4 of Chapter 1. Recall that if n is a positive integer, a *polynomial of degree n* is an expression of the form

$$a_n x^n + a_{n-1} x^{n-1} + \cdots + a_1 x + a_0,$$

where $a_0, a_1, \ldots, a_n$ are real numbers and $a_n \neq 0$. The term $a_k x^k$ in a given polynomial is called the *term of degree k* or the *k*th-degree term. If $a_n \neq 0$, then $a_n x^n$ is the *leading term* and a_n is the *leading coefficient*. The term a_0 is the *constant term* of the polynomial.

As we already know, the polynomial function defined by

$$A(x) = a_n x^n + \cdots + a_1 x + a_0$$

has for its domain of definition the set $\mathbb{R}$ of all real numbers. If $b \in \mathbb{R}$, the function value of $A(x)$ at $x = b$ is defined by

$$A(b) = a_n b^n + \cdots + a_1 b + a_0.$$

EXAMPLE 1

If $A(x) = 3x^2 + 5x + 3$, find $A(-1)$, $A(2)$, and $A(-1)/A(2)$.

Solution: We have

$$A(-1) = 3(-1)^2 + 5(-1) + 3 = 3 - 5 + 3 = 1,$$
$$A(2) = 3(2)^2 + 5(2) + 3 = 12 + 10 + 3 = 25,$$
$$\frac{A(-1)}{A(2)} = \frac{1}{25}.$$

Practice Exercise 1

Let $P(x) = 4x^3 - 5x + 6$. Find $P(-2)$, $P(1)$, and $3P(1) + P(-2)$.

Answer: $P(-2) = -16$, $P(1) = 5$, $3P(1) + P(-2) = -1$

EXAMPLE 2

If $P(x) = x^3 + ax^2 - 4x + 1$ and $P(-2) = -11$, find a.

Solution: We have

$$-11 = P(-2) = (-2)^3 + a(-2)^2 - 4(-2) + 1$$
$$-11 = -8 + 4a + 8 + 1.$$

Thus

$$-12 = 4a$$

and

$$a = \frac{-12}{4} = -3.$$

Practice Exercise 2

If $A(x) = x^3 + bx - 4$ and $A(-1) = -3$, find b.

Answer: $b = -2$

Constant Polynomials

A constant function $A(x) = a_0$, with $a_0 \neq 0$, can be regarded as a polynomial function of degree zero. The constant function $A(x) = 0$ is sometimes called the *zero polynomial* and its degree is not defined.

Equality

Two polynomials $A(x)$ and $B(x)$ are *identically equal* if they are equal as functions, that is, if $A(x) = B(x)$ *for all values of x.*

When no confusion is possible, we simply say of two polynomials that they are *equal* instead of saying that they are *identically equal*. The following theorem, stated without proof, tells us exactly when two given polynomials are equal.

Identically Equal Polynomials

Two polynomials $A(x)$ and $B(x)$ are equal if and only if they have the same degree and the coefficients of terms of same degree are equal.

For example, let

$$A(x) = 2x^3 - 4x^2 + 7$$

and

$$B(x) = b_4 x^4 + b_3 x^3 + b_2 x^2 + b_1 x + b_0.$$

If $A(x)$ is identically equal to $B(x)$, then the theorem implies that

$$b_4 = 0, \quad b_3 = 2, \quad b_2 = -4, \quad b_1 = 0, \quad b_0 = 7.$$

At this point we suggest that you review the examples and exercises about the sum and product of polynomials in Section 1.4 of Chapter 1. They may help you understand the following formal definitions of sum and product.

Let

$$B(x) = b_m x^m + b_{m-1} x^{m-1} + \cdots + b_1 x + b_0$$

be a polynomial of degree m, and let $A(x)$ be the polynomial of degree n that was defined at the beginning of this section.

The Sum of Two Polynomials

The sum $A(x) + B(x)$ is defined by

$$A(x) + B(x) = (a_0 + b_0) + (a_1 + b_1)x + (a_2 + b_2)x^2 + \cdots.$$

That is, to add polynomials, we add the coefficients of terms of the same degree.

The Product of Two Polynomials

The product $A(x) \cdot B(x)$ is defined by

$$
\begin{aligned}
A(x) \cdot B(x) &= (a_n x^n + \cdots + a_1 x + a_0)(b_m x^m + \cdots + b_1 x + b_0) \\
&= a_n b_m x^{n+m} + \cdots + (a_2 b_0 + a_1 b_1 + a_0 b_2)x^2 \\
&\quad + (a_1 b_0 + a_0 b_1)x + a_0 b_0.
\end{aligned}
$$

The definitions of sum and product of two polynomials extend to the case of any *finite* number of polynomials. The degree of a product of polynomials is equal to the *sum* of the degrees of the individual polynomials. Moreover, the zero polynomial, 0, is the identity for the sum, while the constant polynomial 1 is the identity for the product.

The Division Algorithm

If a and b are two nonnegative integers, with $b \neq 0$, then there is a unique pair of nonnegative integers, q and r, such that

$$a = bq + r, \quad \text{with } 0 \le r < b.$$

The number a is called the *dividend*, b is the *divisor*, q is the *quotient*, and r is the *remainder* in the division of a by b. When $r = 0$, we say that a is *divisible* by b, and that b is a *factor* of a.

The following theorem, known as *the division algorithm for polynomials*, is analogous to the preceding result for integers.

The Division Algorithm

Let $A(x)$ be a polynomial of degree n and let $B(x) \ne 0$ be a polynomial of degree m, such that $n \ge m$. Then there are two polynomials $Q(x)$ and $R(x)$ such that

$$A(x) = B(x) \cdot Q(x) + R(x), \qquad (11.1)$$

where the degree of $Q(x)$ is $n - m$, and either $R(x) = 0$ or the degree of $R(x)$ is less than m.

The polynomial $A(x)$ is called the *dividend*, $B(x)$ is the *divisor*, $Q(x)$ is the *quotient*, and $R(x)$ is the *remainder* in the division of $A(x)$ by $B(x)$. It is important to observe that the degree of $Q(x)$ is the *difference* of the degrees of $A(x)$ and $B(x)$, and the degree of $R(x)$ is always *smaller* than the degree of $B(x)$, unless $R(x) = 0$.

EXAMPLE 3

Find the quotient and remainder in the division of $4x^3 - 8x^2 + 5x + 1$ by $2x - 1$.

Solution: We use the method of long division as follows:

$$
\begin{array}{r}
2x^2 - 3x + 1 \quad \longleftarrow \text{Quotient} \\
2x - 1 \overline{\smash{)}\, 4x^3 - 8x^2 + 5x + 1} \\
\underline{4x^3 - 2x^2} \qquad\qquad\qquad \\
-6x^2 + 5x \qquad\quad \\
\underline{-6x^2 + 3x} \qquad\quad \\
2x + 1 \\
\underline{2x - 1} \\
2 \quad \longleftarrow \text{Remainder}
\end{array}
$$

Thus the *quotient* is $2x^2 - 3x + 1$ and the *remainder* is 2. By performing the algebraic operations, you can check that the following relation holds:

$$4x^3 - 8x^2 + 5x + 1 = (2x - 1)(2x^2 - 3x + 1) + 2.$$

This is exactly relation (11.1).

Practice Exercise 3

Divide $6x^3 + 19x^2 + x + 3$ by $3x + 2$.

Answer: Quotient: $2x^2 + 5x - 3$, remainder: 9
$6x^3 + 19x^2 + x + 3 = (3x + 2)(2x^2 + 5x - 3) + 9$

EXAMPLE 4

Use the method of long division to find the quotient and remainder in the division of $x^4 - x^2 + 5$ by $x^2 + 2x + 1$.

Solution: The terms of degree 3 and 1 are missing from the dividend $x^4 - x^2 + 5$. In such cases it is helpful to allow spaces for the two terms; we then work as follows.

$$
\begin{array}{r}
x^2 - 2x + 2 \\
x^2 + 2x + 1\overline{\smash{)}x^4 \qquad\quad - x^2 \qquad + 5} \\
\underline{x^4 + 2x^3 + x^2} \\
-2x^3 - 2x^2 \\
\underline{-2x^3 - 4x^2 - 2x} \\
2x^2 + 2x + 5 \\
\underline{2x^2 + 4x + 2} \\
-2x + 3
\end{array}
$$

Thus the quotient is $x^2 - 2x + 2$ and the remainder is $-2x + 3$. We may now write

$$x^4 - x^2 + 5 = (x^2 + 2x + 1)(x^2 - 2x + 2) + (-2x + 3)$$

Practice Exercise 4   Find the quotient and remainder in the division of $x^5 - 3x^4 + 5x^2$ by $x^2 - 3x + 2$.

Answer: Quotient: $x^3 - 2x - 1$, remainder: $x + 2$
$$x^5 - 3x^4 + 5x^2 = (x^2 - 3x + 2)(x^3 - 2x - 1) + (x + 2)$$

Exact Division

If $R(x) = 0$, the expression (11.1) becomes

$$A(x) = B(x)Q(x). \tag{11.2}$$

In this case, the division is *exact*, and $A(x)$ is said to be *divisible by $B(x)$*. Also $B(x)$ is called a *factor* of $A(x)$.

Notice that this is exactly what happens with integers: if one integer is a factor of another, then the division is exact.

EXAMPLE **5**   Divide $x^5 + x + 1$ by $x^2 + x + 1$.

Solution: Again we proceed by long division.

$$
\begin{array}{r}
x^3 - x^2 + 1 \\
x^2 + x + 1\overline{\smash{)}x^5 \qquad\qquad\qquad\quad x + 1} \\
\underline{x^5 + x^4 + x^3} \\
-x^4 - x^3 \\
\underline{-x^4 - x^3 - x^2} \\
x^2 + x + 1 \\
\underline{x^2 + x + 1} \\
0
\end{array}
$$

Since the remainder is 0, it follows that $x^5 + x + 1$ is divisible by $x^2 + x + 1$, and we have

$$x^5 + x + 1 = (x^2 + x + 1)(x^3 - x^2 + 1).$$

Practice Exercise 5   Find the quotient and remainder in the division of $x^6 + x^4 - 5x^2 + 3$ by $x^2 + 3$.

Answer: Quotient: $x^4 - 2x^2 + 1$, remainder: 0
$$x^6 + x^4 - 5x^2 + 3 = (x^2 + 3)(x^4 - 2x^2 + 1)$$

EXERCISES 11.1

1. If $P(x) = 2x^3 - x^2 + 2x - 1$, find the following function values.
   (a) $P(1)$ **(b)** $P(2)$ **(c)** $P(1)/P(2)$
   (d) $P(1/2)$ **(e)** $2P(1) + 3P(2)$

2. If $Q(x) = x^4 - 16$, find the following function values.
   (a) $Q(2)$ **(b)** $Q(0)$ **(c)** $Q(1/2)$
   (d) $1/Q(0)$ **(e)** $Q(2) \cdot Q(1/2)$

3. If $A(x) = x^3 - 2x^2 + ax - 1$ and $A(1) = 4$, find a.

4. If $B(x) = ax^4 + 2x - 3$ and $B(2) = 33$, find a.

In Exercises 5–10, find constants a, b, c and d so that $P(x) = Q(x)$.

5. $P(x) = 4x^2 - 3x + 2$,
   $Q(x) = (a + 1)x^2 + (b + 1)x + (c - 3)$

6. $P(x) = x^2 - 4x + 8$,
   $Q(x) = (2a)x^2 + (b - 2)x + 4c$

7. $P(x) = 3x^2 - 4x + 6$,
   $Q(x) = ax^2 + (b - 2a)x + (c + 1)$

8. $P(x) = 4x^2 - x - 3$,
   $Q(x) = (a + b)x^2 + (b + c)x + c$

9. $P(x) = x^3 - 2x + 1$,
   $Q(x) = (a - b)x^3 + (b + c)x^2 + (c - 2d)x + d$

10. $P(x) = 2x^3 - 1$,
   $Q(x) = ax^3 + (b - 2a)x^2 + (a - 2c)x + (3a - d)$

In Exercises 11–20, use long division to find the quotient and the remainder.

11. $(6x^2 + 11x + 4) \div (2x + 5)$

12. $(8x^2 - 8x + 6) \div (2x - 3)$

13. $(x^3 - 2x^2 + 4x - 1) \div (3x - 4)$

14. $(2x^3 - 7x^2 - 7x + 12) \div (2x + 3)$

15. $(x^4 + 2x^3 - 2x^2 + 7x + 2) \div (x^2 - x + 2)$

16. $(x^5 - 3x^4 + x^3 + 4x^2 - x + 2) \div (x^2 - 3x - 1)$

17. $(8x^6 - 3x^2 + 1) \div (4x^3 - x - 2)$

18. $(2x^5 - x) \div (2x^2 - 1)$

19. $(x^6 - x^2 - 1) \div (2x^3 - 4)$

20. $(x^4 - 3x^3 + 2x^2 + x + 1) \div (2x^2 - 1)$

Formula (11.1) can be written

$$\frac{A(x)}{B(x)} = Q(x) + \frac{R(x)}{B(x)},$$

where the polynomial $Q(x)$ is the quotient of $A(x)$ by $B(x)$ and $R(x)/B(x)$ is a *proper rational expression*, that is, a rational expression where the degree of the numerator is smaller than the degree of the denominator. In Exercises 21–30, write each rational expression as a sum of a polynomial and a proper rational expression.

21. $\dfrac{x^2 - 4x + 5}{x - 1}$

22. $\dfrac{2x^2 + x - 3}{x + 2}$

23. $\dfrac{2x^3 - 4x - 2}{2x + 4}$

24. $\dfrac{3x^3 - 7x^2 + 11x + 4}{3x - 1}$

25. $\dfrac{x^4 - 5x^2 + 3}{2x - 1}$

26. $\dfrac{6x^4 + 2x + 1}{2x + 1}$

27. $\dfrac{10x^6 + 5x^3 - x}{2x^4 - x^2 - 1}$

28. $\dfrac{x^6 - 1}{x^4 + 1}$

29. $\dfrac{3x^4 - 2x^3 + x^2 + 2x + 3}{x^2 - x - 1}$

30. $\dfrac{2x^5 - 4x^3 + x - 1}{3x^2 - x - 1}$

11.2   DIVISION BY $x - a$

The division algorithm yields interesting consequences when the divisor is a first-degree polynomial of the form

$$x - a,$$

where the leading coefficient is 1 and a is an arbitrary number.

    If the dividend $A(x)$ has degree n, then the quotient $Q(x)$ is a polynomial of degree $n - 1$. The remainder is either 0 or a polynomial of degree 0, that is, a constant. We then have

$$A(x) = (x - a)Q(x) + R, \qquad (11.3)$$

where R is a number.

If in the last formula a is substituted for x, we get

$$
\begin{aligned}
A(a) &= (a - a)Q(a) + R \\
&= 0 \cdot Q(a) + R \\
&= R.
\end{aligned}
$$

Thus,

$$R = A(a) \qquad (11.4)$$

and we have proved the following theorem.

The Remainder Theorem

The remainder in the division of a polynomial $A(x)$ by $x - a$ is $A(a)$.

This result is extremely useful in applications. Among other things, *it allows us to find the remainder when $A(x)$ is divided by $(x - a)$, without performing the division.*

EXAMPLE 1

Find the remainder in the division of $A(x) = 2x^4 - 4x^3 - 4x^2 + 13x - 7$ by $B(x) = x + 2$.

Solution: First we write

$$B(x) = x + 2 = x - (-2).$$

Thus, in this example, $a = -2$. To find the remainder, we compute the function value of the polynomial $A(x)$ at $x = -2$:

$$
\begin{aligned}
A(-2) &= 2 \cdot (-2)^4 - 4 \cdot (-2)^3 - 4 \cdot (-2)^2 + 13(-2) - 7 \\
&= 32 + 32 - 16 - 26 - 7 \\
&= 15.
\end{aligned}
$$

By the remainder theorem, 15 is the remainder in the division of the given polynomial by $x + 2$. As an exercise, you can check this result by carrying out the actual division.

Practice Exercise 1

Find the remainder in the division of $A(x) = 2x^4 - 4x^3 - 4x^2 + 13x - 7$ by $B(x) = x - 2$.

Answer: $R = 3$.

When $R = A(a) = 0$, formula (11.3) becomes

$$A(x) = (x - a)Q(x). \qquad (11.5)$$

Thus, when $A(a) = 0$, $A(x)$ is *divisible* by $x - a$, and $x - a$ is a factor of $A(x)$. Conversely, if $A(x)$ is divisible by $x - a$, then the remainder R is 0. Since

$R = A(a)$ according to (11.4), we get $A(a) = 0$. Thus, we have proved the following theorem.

The Factor Theorem

A linear polynomial $x - a$ is a factor of $A(x)$ if and only if $A(a) = 0$.

EXAMPLE 2 Verify that $P(x) = x^3 - 3x^2 + 6x - 8$ is divisible by $x - 2$.

Solution: According to the remainder theorem, it suffices to compute the function value of $P(x)$ at $x = 2$:

$$R = P(2)$$
$$= 2^3 - 3 \cdot (2)^2 + 6 \cdot 2 - 8$$
$$= 8 - 12 + 12 - 8$$
$$= 0.$$

As an exercise, you can check this result by long division.

Practice Exercise 2 Is $x^4 - 9x^2 + x$ divisible by $x + 3$?

Answer: No: remainder upon division is -3.

Horner's Method

The numerical evaluation of a polynomial can be greatly simplified by proceeding as follows. Suppose that we want to compute the function value of $P(x) = x^3 - 3x^2 + 6x - 8$ at $x = 2$. First, we write the polynomial in nested form as follows:

$$P(x) = x^3 - 3x^2 + 6x - 8$$
$$= [x^2 - 3x + 6]x - 8 \qquad \text{Factor } x \text{ from the first three terms}$$
$$= [(x - 3)x + 6]x - 8. \qquad \text{Factor } x \text{ from the first two terms inside the square brackets}$$

Next, we evaluate $P(x)$ at $x = 2$:

$$P(2) = [(2 - 3) \cdot 2 + 6] \cdot 2 - 8$$
$$= [-2 + 6] \cdot 2 - 8 \qquad \text{Eliminate parentheses}$$
$$= 4 \cdot 2 - 8 \qquad\qquad \text{Eliminate brackets}$$
$$= 0.$$

This procedure, known as *Horner's method,* applies to any polynomial

$$P(x) = a_n x^n + a_{n-1} x^{n-1} + \cdots + a_1 x + a_0$$

by writing it in nested form,

$$P(x) = \{ \cdots [(a_n x + a_{n-1})x + a_{n-2}]x + \cdots + a_1 \}x + a_0.$$

This method is extremely useful if you are computing function values of a polynomial with the help of a calculator, particularly if your calculator has a memory. Storing the numerical value of x at the beginning of the computation, and recalling it whenever necessary, will speed up your calculations.

EXAMPLE 3

Find the remainder in the division of $P(x) = 0.5x^3 - 0.2x^2 + 0.3x - 0.4$ by $x - 1.5$.

Solution: First, we write

$$P(x) = [(0.5x - 0.2)x + 0.3]x - 0.4.$$

Since, by the remainder theorem, $R = P(1.5)$, we compute the function value of $P(x)$ at $x = 1.5$:

$$
\begin{aligned}
P(1.5) &= [(0.5 \times 1.5 - 0.2) \times 1.5 + 0.3] \times 1.5 - 0.4 \\
&= [0.55 \times 1.5 + 0.3] \times 1.5 - 0.4 \\
&= 1.125 \times 1.5 - 0.4 \\
&= 1.6875 - 0.4 \\
&= 1.2875.
\end{aligned}
$$

The following keystrokes were performed on a TI-30-II in order to obtain the function value of $P(x)$ at $x = 1.5$.

1.5 $\boxed{\text{STO}}$ 0.5 $\boxed{\times}$ $\boxed{\text{RCL}}$ $\boxed{-}$ 0.2 $\boxed{=}$ $\boxed{\times}$ $\boxed{\text{RCL}}$ $\boxed{+}$ 0.3 $\boxed{=}$ $\boxed{\times}$ $\boxed{\text{RCL}}$ $\boxed{-}$ 0.4 $\boxed{=}$

Practice Exercise 3

Find the function value of $A(x) = 0.3x^4 - 1.5x^3 - 0.4x^2 + 2.1x - 1.3$ at $x = 0.16$.

Answer: $A(0.16) = -0.9801874$

Synthetic Division

When dividing a polynomial by $x - a$, it is possible to avoid long division by using a shorthand method called *synthetic division,* which we now explain.

EXAMPLE 4

Divide $A(x) = 4x^3 - x^2 - 8x + 5$ by $x - 2$.

Solution: First, let us perform the long division for comparison.

$$
\begin{array}{r}
4x^2 + 7x + 6 \\
x - 2 \overline{\smash{)}4x^3 - x^2 - 8x + 5} \\
\underline{4x^3 - 8x^2} \\
7x^2 - 8x \\
\underline{7x^2 - 14x} \\
6x + 5 \\
\underline{6x - 12} \\
17
\end{array}
$$

We obtain $Q(x) = 4x^2 + 7x + 6$ as quotient and $R = 17$ as remainder.

Next, we describe the method of *synthetic division*.

(a) We write the coefficients of the dividend $A(x)$, and we write the number $a (= 2$ in this example) in the upper corner to the left of the coefficients.

$$2 \ \bigg|\ \ 4 \qquad -1 \qquad -8 \qquad 5$$

(b) We bring down the first coefficient of the dividend.

$$2 \ \bigg|\ \ 4 \qquad -1 \qquad -8 \qquad 5$$
$$\downarrow$$
$$4$$

This is the *first coefficient of the quotient*.

(c) We multiply the first coefficient of the quotient by 2 (the number in the upper left corner) and add it to -1, the second coefficient of the dividend.

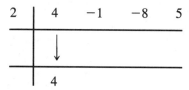

$$2 \ \bigg|\ \ 4 \qquad -1 \qquad -8 \qquad 5$$

This is the *second coefficient of the quotient*.

(d) We proceed in a similar manner until all coefficients of the dividend have been used.

$$2 \ \bigg|\ \ 4 \qquad -1 \qquad -8 \qquad 5$$

The circled number is the *remainder*. The numbers to the left are the coefficients of the quotient. As before, we get

$$Q(x) = 4x^2 + 7x + 6 \qquad \text{and} \qquad R = 17.$$

Practice Exercise 4   Use the method of synthetic division to find the quotient and remainder in the division of $-4x^3 + 3x^2 - x + 1$ by $x - 1$.

Answer: $Q(x) = -4x^2 - x - 2$, $R = -1$

EXAMPLE **5**   Use the method of synthetic division to divide $x^4 + 1$ by $x + 1$.

Solution: First, notice that $x + 1 = x - (-1)$, so $a = -1$. Next, the missing terms of degrees 3, 2, and 1 in the dividend correspond to coefficients equal to 0. We write the synthetic division in the following format.

$$
\begin{array}{c|ccccc}
-1 & 1 & 0 & 0 & 0 & 1 \\
   &   & -1 & 1 & -1 & 1 \\
\hline
   & 1 & -1 & 1 & -1 & ②
\end{array}
$$

Thus,

$$Q(x) = x^3 - x^2 + x - 1 \quad \text{and} \quad R = 2.$$

Practice Exercise 5   Use synthetic division to divide $x^5 - 2x^3 - 7x$ by $x + 2$.

Answer: $Q(x) = x^4 - 2x^3 + 2x^2 - 4x + 1$, $R = -2$.

The method of synthetic division can also be used to evaluate polynomial functions.

EXAMPLE 6   If $P(x) = 2x^3 - 5x + 6$, find $P(3)$.

Solution:

$$
\begin{array}{c|cccc}
3 & 2 & 0 & -5 & 6 \\
  &   & 6 & 18 & 39 \\
\hline
  & 2 & 6 & 13 & 45
\end{array}
$$

Thus, according to (11.4), we have $R = P(3) = 45$.

Of course, the same result could have been obtained directly:

$$
\begin{aligned}
P(3) &= 2(3)^3 - 5(3) + 6 \\
     &= 54 - 15 + 6 \\
     &= 45.
\end{aligned}
$$

Practice Exercise 6   Let $P(x) = 4x^3 - x^2 - 2x + 12$. Use synthetic division to find $P(-2)$.

Answer: $R = P(-2) = -20$

■ **Important note**   Synthetic division is done only with a linear divisor of the form $x - a$.

EXERCISES 11.2

Without performing the division, find the remainder in the following divisions in Exercises 1–10.

1. $(2x^2 - 6x - 4) \div (x + 2)$

2. $(-x^2 - x - 1) \div (x - 2)$

3. $(-x^3 - 2x^2 + x + 1) \div \left(x + \dfrac{1}{2}\right)$

4. $(5x^3 - x - 6) \div \left(x - \dfrac{2}{3}\right)$

5. $(x^5 - x^3 - 2x + 2) \div (x - 2)$

6. $(x^6 - 3x^4 + 4x^2 - 2) \div (x - 1)$

7. $(x^8 - 3x^6 - 4x^5 - 2x^4 + x^3 - 2x^2 + 3x - 4) \div x$

8. $(x^{10} - 5x^5 - 8x + 1) \div x$

[C] **9.** $(0.2x^3 - 0.03x^2 + 0.21x - 0.18) \div (x + 0.5)$

[C] **10.** $(0.5x^4 - 0.12x^2 + 0.03) \div (x - 0.02)$

In Exercises 11–22, find the quotient and remainder by synthetic division.

11. $(2x^2 - 6x - 4) \div (x - 2)$

12. $(-x^2 - x - 1) \div (x - 2)$

13. $(-x^3 - 2x^2 + x + 1) \div \left(x + \dfrac{1}{2}\right)$

14. $(5x^3 - x - 6) \div \left(x - \dfrac{2}{3}\right)$

15. $(4x^3 - 3x + 4) \div (x - 4)$

16. $(x^4 + x^2 + 1) \div (x + 2)$

17. $(x^5 - x^3 - 2x + 2) \div (x + 2)$

18. $(x^6 - 3x^4 + 4x^2 - 2) \div (x - 1)$

19. $(x^4 - 1) \div (x + 1)$

20. $(x^5 + 1) \div (x + 1)$

[C] **21.** $(0.2x^3 - 0.03x^2 + 0.21x - 0.18) \div (x + 0.5)$

[C] **22.** $(0.5x^4 - 0.12x^2 + 0.03) \div (x - 0.02)$

In Exercises 23–26, factor as directed.

23. Show that $x - 3$ is a factor of the polynomial $x^3 - 6x^2 + 11x - 6$. Factor this polynomial completely.

24. Show that $x + 4$ is a factor of the polynomial $x^3 + 4x^2 - x - 4$. Determine the other factors.

25. Factor the polynomial $2x^3 - x^2 - 5x - 2$ completely.

26. Factor the polynomial $x^4 + 3x^3 + x^2 - 3x - 2$ completely.

In Exercises 27–32, apply Horner's method to find the indicated function values. Use your calculator if you wish.

27. $P(x) = 4x^3 - 3x^2 + 5x - 1$, $P(2)$

28. $Q(t) = t^4 - 3t^2 + 5$, $Q(-3)$

[C] **29.** $A(x) = x^3 - 2x^2 + 5x - 2$, $A(0.18)$

[C] **30.** $B(x) = 2x^3 - 5x + 3$, $B(1.25)$

[C] **31.** $A(t) = 0.1t^4 - 0.4t^3 + 0.3t^2 - 0.5$, $A(1.2)$

[C] **32.** $B(s) = 1.3s^4 + 2.5s^2 - 0.31$, $B(0.02)$

33. Find a value of a such that $x + 4$ is a factor of $x^3 + x^2 - 17x + a$.

34. For what value of a is the remainder of $(x^4 - 3x^2 + 2a) \div (x - 2)$ equal to -3?

35. For what value of a is $x - 3$ a factor of $x^3 - 7x^2 + 3ax - 18$?

36. Find a value of a such that $2x^3 - 3x^2 + 2ax + 4$ is divisible by $x + 1/2$.

37. Find a value of a such that the remainder of $(x^3 - ax^2 + 5) \div (x - 2)$ is -3.

38. For what value of a is the remainder of $(x^4 + ax + 10) \div (x + 3)$ equal to -2?

39. Use synthetic division to obtain the quotient and the remainder of the division of the polynomial $P(x) = 3x^2 - 2x + 1$ by $Q(x) = x - a$.

40. Determine by synthetic division the quotient of $ax^2 + bx + c$ divided by $x - 1$.

11.3   **THE ZEROS OF A POLYNOMIAL**

The Zero or Root of a Polynomial

A number r is said to be a *zero* or a *root* of a polynomial $P(x)$ if the function value of the polynomial at $x = r$ is equal to zero, that is, if $P(r) = 0$. In other words, r is a *solution* of the equation $P(x) = 0$.

If r is a zero of $P(x)$ then, by the factor theorem, the polynomial $P(x)$ is divisible by $x - r$, and vice versa. For example, 1 is a root of $x^2 + x - 2$, since $1^2 + 1 - 2 = 0$. The polynomial $x^2 + x - 2$ is then divisible by $x - 1$ and we have the factorization

$$x^2 + x - 2 = (x - 1)(x + 2).$$

From the last expression, we see that -2 is the other solution of the equation $x^2 + x - 2 = 0$.

EXAMPLE 1   Is 3 a root of $P(x) = x^3 - 2x^2 - 2x - 3$?

Solution: To answer this question, we evaluate $P(x)$ at $x = 3$:

$$\begin{aligned} P(3) &= (3)^3 - 2(3)^2 - 2(3) - 3 \\ &= 27 - 18 - 6 - 3 \\ &= 0. \end{aligned}$$

Thus 3 is a zero of $P(x)$.

Practice Exercise 1   Is -2 a zero of the polynomial $A(x) = 2x^4 - 3x^2 - x - 6$?

Answer: No: $A(-2) = 16 \neq 0$.

EXAMPLE 2   Show that $x + 2$ is a factor of the polynomial $2x^3 + x^2 - 5x + 2$.

Solution: We can proceed in two different ways. One way is by evaluating the given polynomial at $x = -2$, as we did in the previous example. The other way is by using synthetic division.

$$\begin{array}{r|rrrr} -2 & 2 & 1 & -5 & 2 \\ & & -4 & 6 & -2 \\ \hline & 2 & -3 & 1 & ⓪ \end{array}$$

The remainder is 0, so -2 is a solution of the equation $2x^3 + x^2 - 5x + 2 = 0$. It then follows that $x + 2$ is a factor of the polynomial $2x^3 + x^2 - 5x + 2$. The method of synthetic division also gives the other factor:

$$2x^3 + x^2 - 5x + 2 = (x + 2)(2x^2 - 3x + 1).$$

Notice that the quadratic polynomial $2x^2 - 3x + 1$ can be factored into $(2x - 1)(x - 1)$. Thus, we obtain the complete factorization of the given polynomial:

$$2x^3 + x^2 - 5x + 2 = (x + 2)(2x - 1)(x - 1).$$

Practice Exercise 2   Is $x - 3$ a factor of the polynomial $x^3 - 4x^2 + 4x - 3$?

Answer: Yes: $x^3 - 4x^2 + 4x - 3 = (x - 3)(x^2 - x + 1)$. (Can you factor $x^2 - x + 1$ any further?)

Complex Zeros

Zeros of polynomials can be real or complex numbers. Even polynomials with real coefficients may have complex (nonreal) zeros. For example, the quadratic polynomial $x^2 + 1$, which led us to the definition of complex numbers, has zeros i and $-i$. Notice that the remainder and factor theorems are also true in the case of complex zeros. For example, $x^2 + 1 = (x + i)(x - i)$.   Similarly,

$x^3 - 2x^2 + x - 2 = (x - 2)(x + i)(x - i)$ since 2, i, and $-i$ are the three roots of the polynomial $x^3 - 2x^2 + x - 2$. Both theorems also remain true in the case of polynomials with complex coefficients. Throughout this book we deal mostly with polynomials with real coefficients. However, in some instances we will have to consider polynomials with complex coefficients. The computations with such polynomials are no more difficult than the computations with real-coefficient polynomials.

The Fundamental Theorem of Algebra

When dealing with polynomials it is important to know how to find their zeros. It is very simple to find the zero of a linear polynomial. Quadratic equations can be solved by using the quadratic formula discussed in Section 2.5. As the degree of the polynomial increases, so does the difficulty in solving the corresponding polynomial equation.

It appears that the Babylonians (about 1950 B.C.) had already considered and solved quite general quadratic equations, seeking only positive roots. They must have known an equivalent version of the quadratic formula at least in certain special cases. The Hindu mathematician Bhaskara (about A.D. 1150) investigated negative roots, possibly under Chinese influence, and observed that positive numbers had two real roots while negative numbers had none.

Efforts to find formulas to solve equations of higher degrees continued for many centuries without success. In 1545, Hieronimo Cardano (1501–1576), an Italian mathematician, published in his *Ars Magna* formulas to solve equations of degrees three and four. These formulas are known as *Cardano's formulas*, although they were not discovered by him. Scipio del Ferro (1465–1526), professor at the University of Bologna, classified and solved all the cases of the general third-degree equation. Del Ferro did not publish, but his methods were rediscovered about 1535 by Niccolo Tartaglia (1500–1577), an Italian engineer and mathematician. It was Tartaglia who revealed the methods to Cardano. Also, the solution of the general fourth-degree equation was found by Cardano's young student Ludovico Ferrari (1527–1565).

Cardano's formulas involving rational combinations of the coefficients and root extraction are too complicated and almost never used in practice. For more than two centuries mathematicians tried unsuccessfully to find formulas involving rational combinations of the coefficients and root extractions to solve equations of degree $n > 4$. However, the Norwegian mathematician Niels Henrik Abel (1802–1829) proved this to be impossible. In other words, equations of degree higher than four cannot be solved by root extractions except for special values of the coefficients.

Even if a polynomial equation of degree higher than four cannot be solved by root extractions, it is conceivable that the equation has solutions. In 1799, Carl Friedrich Gauss (1777–1855), a German mathematician, proved the following important theorem.

The Fundamental Theorem of Algebra

Every polynomial of degree n with real (or complex) coefficients has at least one zero.

The proof of this theorem involves techniques beyond the scope of this book. We point out that the fundamental theorem of algebra is an *existential* theorem: it only asserts the *existence* of a zero without giving any indication of *how* to find it. Also, the zero may be a real or a complex number.

Complex zeros of polynomials with *real* coefficients always occur in *conjugate pairs*. Let us state this more precisely.

Complex Zeros of Polynomials with Real Coefficients

If r is a complex zero of a polynomial $P(x) = a_n x^n + \cdots + a_1 x + a_0$ with real coefficients, then $\bar{r}$, the complex conjugate of r, is also a zero of $P(x)$.

Indeed, if r is a zero of $P(x)$, then

$$P(r) = a_n r^n + \cdots + a_1 r + a_0 = 0.$$

Taking complex conjugates, we have

$$\overline{a_n r^n + \cdots + a_1 r + a_0} = 0.$$

Since the complex conjugate of a sum or product is the sum or product of complex conjugates (Section 1.9), we obtain

$$\bar{a}_n \bar{r}^n + \cdots + \bar{a}_1 \bar{r} + \bar{a}_0 = 0.$$

But the complex conjugate of a real number is the real number itself, so

$$a_n \bar{r}^n + \cdots + a_1 \bar{r} + a_0 = 0,$$

which proves that $\bar{r}$ is also a zero of $P(x)$.

The Multiplicity of a Zero

Let r be a zero of a polynomial of degree $n \geq 1$. Then, by the factor theorem, the linear polynomial $x - r$ is a factor of $P(x)$ and we have

$$P(x) = (x - r)Q_1(x),$$

where $Q_1(x)$ is a polynomial of degree $n - 1$. Either r is also a root of $Q_1(x)$ or it is not. If it is, then we can apply the factor theorem again and thus obtain

$$Q_1(x) = (x - r)Q_2(x).$$

This gives us the factorization

$$\begin{aligned} P(x) &= (x - r)Q_1(x) \\ &= (x - r)^2 Q_2(x), \end{aligned}$$

where the degree of $Q_2(x)$ is $n - 2$. This process can be continued until we reach a stage where

$$P(x) = (x - r)^k Q_k(x)$$

and r is not a root of $Q_k(x)$. (We must always reach that stage, because no such factorization can exist with $k > n$.) It is possible to show that the number k and the polynomial $Q_k(x)$, which has degree $n - k$, are uniquely determined by r. The number k is called the *multiplicity* of the root r of $P(x)$. This means that if r is a root of $P(x)$ of multiplicity k, then $(x - r)^k$ *is a factor of* $P(x)$, *while* $(x - r)^{k+1}$ *is not*. In other words, k is the largest of the positive integers m for which $(x - r)^m$ is a factor of $P(x)$.

For example, 1 is a zero of multiplicity 2 of $P(x) = x^3 + x^2 - 5x + 3$, because $P(x) = (x - 1)^2(x + 3)$. The number -3 is a zero of multiplicity 1 of $P(x)$.

A zero of multiplicity 2 is also called a *double zero*, and a zero of multiplicity 3 is called a *triple zero*.

When counting the number of zeros of a polynomial, *a zero of multiplicity k is counted as k zeros*.

The Number of Zeros of a Polynomial

By the fundamental theorem of algebra, if $P(x)$ is a polynomial of degree $n \geq 1$, then it has at least one zero, r_1. By the factor theorem, $P(x)$ is divisible by $x - r_1$, and we can write

$$P(x) = (x - r_1)Q_1(x),$$

where $Q_1(x)$ is a polynomial of degree $n - 1$. If $n = 1$, then the degree of $Q_1(x)$ is zero, so $Q_1(x) = a_1$ is a constant. If $n > 1$, then we reason with $Q_1(x)$ in the same way we did with $P(x)$, to obtain

$$Q_1(x) = (x - r_2)Q_2(x),$$

where r_2 is a zero of $Q_1(x)$ and where $Q_2(x)$ is a polynomial of degree $n - 2$. Note that r_2 may be equal to r_1. Even if this is the case, we continue to use different notations for the same root. Replacing $Q_1(x)$ in the expression for $P(x)$ yields

$$P(x) = (x - r_1)[(x - r_2)Q_2(x)]$$
$$= (x - r_1)(x - r_2)Q_2(x).$$

Either $n = 2$, in which case $Q_2(x) = a_2$, a constant, or $n > 2$, in which case we can again apply the fundamental theorem of algebra. After a finite number of steps, we obtain the factorization

$$P(x) = (x - r_1)(x - r_2) \cdots (x - r_n)a_n, \tag{11.6}$$

where $a_n \neq 0$ is the leading coefficient of $P(x)$, and $r_1, r_2, \cdots, r_n$ are the zeros (not necessarily distinct) of $P(x)$. Thus, every polynomial of degree n has *at least* n zeros.

Now, let r be a number different from $r_1, \cdots, r_n$. Then we have

$$P(r) = a_n(r - r_1)(r - r_2) \cdots (r - r_n) \neq 0$$

since all factors are different from zero. We conclude that $r_1, r_2, \cdots, r_n$ are the only zeros of $P(x)$ and, therefore, that $P(x)$ has *at most* n zeros. Thus, we have proved the following theorem.

The Number of Zeros of a Polynomial

Every polynomial of degree $n \geq 1$ has exactly n zeros, counting multiplicities.

EXAMPLE **3**

Verify that -2 is a double zero of $P(x) = x^4 + 2x^3 + x^2 + 12x + 20$, and find all the roots of the polynomial.

Solution: We divide $P(x)$ by $x + 2 = x - (-2)$.

$$
\begin{array}{r|rrrrr}
-2 & 1 & 2 & 1 & 12 & 20 \\
   &   & -2 & 0 & -2 & -20 \\
\hline
   & 1 & 0 & 1 & 10 & ⓪
\end{array}
$$

The remainder is zero, so -2 is a root of $P(x)$. The quotient in the division of $P(x)$ by $x + 2$ is $Q_1(x) = x^3 + x + 10$, and we have

$$P(x) = (x + 2)(x^3 + x + 10).$$

Next, we divide $Q_1(x)$ by $x + 2$.

$$
\begin{array}{r|rrrr}
-2 & 1 & 0 & 1 & 10 \\
   &   & -2 & 4 & -10 \\
\hline
   & 1 & -2 & 5 & ⓪
\end{array}
$$

The remainder is zero, so -2 is a root of $Q_1(x)$, and we have

$$Q_1(x) = (x + 2)(x^2 - 2x + 5).$$

Replacing $Q_1(x)$ in the expression for $P(x)$, we obtain

$$
\begin{aligned}
P(x) &= (x + 2)(x + 2)(x^2 - 2x + 5) \\
     &= (x + 2)^2(x^2 - 2x + 5).
\end{aligned}
$$

Setting $Q_2(x) = x^2 - 2x + 5$, we can check that $Q_2(-2) = 13 \neq 0$. Thus, -2 is indeed a double root of $P(x)$. To find the other roots, we must solve the quadratic equation

$$x^2 - 2x + 5 = 0.$$

Using the quadratic formula, we obtain two complex zeros:

$$x = \frac{2 \pm \sqrt{4 - 20}}{2} = \frac{2 \pm 4i}{2} = 1 \pm 2i.$$

The four zeros of the given polynomial are: -2 with *multiplicity two*, $1 + 2i$, and $1 - 2i$.

Practice Exercise 3

Verify that 2 is a double root of the polynomial $x^4 - 4x^3 + 5x^2 - 4x + 4$. Find all roots of the polynomial.

Answer: 2 (double root), i, $-i$

EXAMPLE **4**

Factor the polynomial $P(x) = x^6 - 5x^5 + 6x^4$ and determine its zeros.

Solution: We have

$$P(x) = x^4(x^2 - 5x + 6)$$
$$= x^4(x - 2)(x - 3).$$

This shows that the given polynomial has six zeros: 0 with multiplicity four, 2, and 3.

Practice Exercise 4   Determine all zeros of the polynomial $x^5 + 3x^4 - 4x^3$. Factor this polynomial completely.

Answer: Five zeros: 0 with multiplicity three, 1, -4
$$x^5 + 3x^4 - 4x^3 = x^3(x^2 + 3x - 4) = x^3(x - 1)(x + 4)$$

Polynomials with Given Zeros

We can use the results of the preceding examples to "build up" polynomials with given zeros, as shown in the next example.

EXAMPLE 5   **(a)** Find a polynomial of degree 3 with zeros -1, 1, and 2.
(b) Find a polynomial of degree 3 having the same zeros and taking the value 6 at $x = 0$.

Solution: **(a)** If a polynomial $P(x)$ has zeros -1, 1, and 2, then $x + 1$, $x - 1$, and $x - 2$ are factors of $P(x)$. Since we want $P(x)$ to have degree 3, it suffices to write

$$P(x) = (x - 1)(x + 1)(x - 2)$$
$$= (x^2 - 1)(x - 2)$$
$$= x^3 - 2x^2 - x + 2.$$

Note that the polynomial $P(x)$ is not *uniquely* determined. Any polynomial of the form

$$Q(x) = a(x - 1)(x + 1)(x - 2),$$

where a is a constant, also has zeros -1, 1, and 2. In order to determine a, we need an additional condition, as in part b of this example.
(b) According to (11.6), we write

$$P(x) = a_3(x - 1)(x + 1)(x - 2).$$

When $x = 0$, we want $P(0) = 6$. Thus

$$P(0) = a_3(0 - 1)(0 + 1)(0 - 2) = 6$$
$$2a_3 = 6$$
$$a_3 = 3.$$

Thus we obtain

$$P(x) = 3(x - 1)(x + 1)(x - 2)$$
$$= 3(x^2 - 1)(x - 2)$$
$$= 3(x^3 - 2x^2 - x + 2)$$
$$= 3x^3 - 6x^2 - 3x + 6,$$

which is the desired polynomial.

Practice Exercise 5   Find a polynomial $P(x)$ of lowest degree with zeros -1, 0, and -2 and such that $P(1) = 12$.

Answer: $P(x) = 2x(x + 1)(x + 2)$

EXAMPLE **6**   Find a third-degree polynomial with real coefficients and having zeros 1 and $2 + 3i$.

Solution: If a polynomial has real coefficients and $2 + 3i$ is a zero, then the complex conjugate $2 - 3i$ is also a zero. The three zeros of the third-degree polynomial we are looking for are 1, $2 + 3i$, and $2 - 3i$. Therefore,

$$\begin{aligned}
P(x) &= (x - 1)[x - (2 + 3i)][x - (2 - 3i)] \\
&= (x - 1)[(x - 2) - 3i][(x - 2) + 3i] \\
&= (x - 1)[(x - 2)^2 + 9] \\
&= (x - 1)(x^2 - 4x + 13) \\
&= x^3 - 5x^2 + 17x - 13.
\end{aligned}$$

Practice Exercise 6   Write a third-degree polynomial $P(x)$ with real coefficients and having leading coefficient 2 and zeros $1 - i$ and -3.

Answer: $P(x) = 2x^3 + 2x^2 - 8x + 12$

HISTORICAL NOTE

CARL FRIEDRICH GAUSS

Carl Friedrich Gauss (1777–1855) was a German mathematician and astronomer who made notable contributions in several branches of mathematics and physics. He investigated the geometry of curved surfaces, was an authority in geodesy, and applied mathematical theory to electricity and magnetism. The fundamental theorem of algebra, proved in 1799, was part of Gauss's doctoral dissertation, "A New Proof that Every Rational Integral Function of One Variable Can Be Resolved into Real Factors of the First and Second Degree." The mathematical work of Gauss is so vast and important that he is often referred to as the Prince of Mathematicians. Gauss himself is placed in the same class as Archimedes and Newton.

EXERCISES 11.3

In Exercises 1–10, find whether the given number is a zero of the given polynomial.

1. $x^2 - 3x + 3$, 1

2. $2x^2 - 4x - 6$, -1

3. $3x^3 - 7x + 10$, -2

4. $x^3 - 4x^2 + 9$, 3

5. $2x^2 - 5x - 3$, $-\dfrac{1}{2}$

6. $3x^2 - 7x - 20$, $\dfrac{5}{3}$

7. $x^4 - 2x^3 + 3x^2 - 3$, -2

8. $x^5 - 3x^2 - 6$, 2

9. $x^2 - 2x + 5$, $1 - 2i$

10. $x^2 - 6x + 13$, $3 + 2i$

In Exercises 11–14, find whether the linear polynomial is a factor of the given polynomial. Whenever possible, find the other factors.

11. $2x^3 - 4x^2 + 3x - 6, \ x - 2$

12. $3x^3 + 14x^2 + 8x, \ x + 4$

13. $2x^4 + 6x^3 + 5x^2 + 14x - 3, \ x + 3$

14. $x^4 + 4x^3 + 4x^2 - 4x + 8, \ x - 2$

In Exercises 15–18, a zero of $P(x)$ is given. Find another zero.

15. $P(x) = x^2 - 4x + 5, \ 2 - i$

16. $P(x) = x^2 - 2x + 10, \ 1 + 3i$

17. $P(x) = 2x^3 - 7x^2 + 22x + 13, \ 2 + 3i$

18. $P(x) = x^3 + 8, \ 1 - i\sqrt{3}$

In Exercises 19–24, a polynomial and one of its roots are given. Find the other roots.

19. $x^3 - x^2 - 9x - 12, \ 4$

20. $2x^3 - x^2 - 15x + 18, \ 2$

21. $x^4 + 3x^2 - 4, \ 2i$

22. $x^3 + 2x^2 + 5x - 26, \ -2 + 3i$

23. Show that 3 is a double root of $x^4 - 5x^3 + 4x^2 + 3x + 9$ and find the other roots.

24. Show that -1 is a double root of $2x^4 + 4x^3 + 3x^2 + 2x + 1$ and find the other roots.

In Exercises 25–32, find the zeros of the given polynomials and determine the multiplicity of each zero.

25. $P(x) = (x - 5)(3x + 2)$

26. $P(x) = (x + 2)^3(4x - 1)^2$

27. $P(x) = (2x + 3)^2(x - 1)^3(3x - 4)$

28. $P(x) = (4x^2 - 3)^2$

29. $P(x) = x^4 + 2x^3 + 2x^2$ **30.** $P(x) = (x^2 - 4x + 3)^2$

31. $P(x) = (x^2 - 2x + 4)^2$ **32.** $P(x) = x^3(x^2 - x + 1)$

In Exercises 33–40, find a polynomial $P(x)$ of lowest degree having the indicated zeros and satisfying the given condition.

33. $-2, 2, 3; \ P(1) = 6$ **34.** $-2, -1, 4; \ P(2) = 8$

35. $-1, 2, 4; \ P(0) = 8$ **36.** $-2, 0, 3; \ P(1) = -6$

37. -1 (multiplicity 2), 3 (multiplicity 2); leading coefficient 1

38. 2 (multiplicity 3), 1 (multiplicity 2); leading coefficient 2

39. $1 + 2i, \ 1 - 2i, \ 3; \ P(0) = 15$

40. $2 - 2i, \ 2 + 2i, \ 1$ (multiplicity 2); leading coefficient 1

11.4   A TEST FOR RATIONAL ZEROS

If a polynomial has *integral coefficients,* there is a very useful test that allows us to determine whether the polynomial has *rational zeros.*

Let

$$P(x) = a_n x^n + a_{n-1} x^{n-1} + \cdots + a_1 x + a_0$$

be a polynomial of degree n ($a_n \neq 0$) with integral coefficients and such that $a_0 \neq 0$. Assume that the rational number p/q, in its lowest terms, is a zero of $P(x)$. Then,

$$P\left(\frac{p}{q}\right) = a_n\left(\frac{p}{q}\right)^n + a_{n-1}\left(\frac{p}{q}\right)^{n-1} + \cdots + a_1\left(\frac{p}{q}\right) + a_0 = 0.$$

Multiplying by q^n, we obtain

$$a_n p^n + a_{n-1} p^{n-1} q + \cdots + a_1 p q^{n-1} + a_0 q^n = 0. \tag{11.7}$$

Solving for $a_0 q^n$, we get

$$a_0 q^n = -a_n p^n - a_{n-1} p^{n-1} q - \cdots - a_1 p q^{n-1},$$

which can be written

$$a_0 q^n = p(-a_n p^{n-1} - a_{n-1} p^{n-2} q - \cdots - a_1 q^{n-1}).$$

This shows that p is a factor of $a_0 q^n$. Since p does not divide q (the fraction p/q is in lowest terms), we see that p must divide a_0. Therefore, p is a factor of a_0.

Similarly, solving (11.7) for $a_n p^n$, we get

$$a_n p^n = q(-a_{n-1}p^{n-1} - \cdots - a_1 p q^{n-2} - a_0 q^{n-1}).$$

Arguing as before, we conclude that q is a factor of a_n.

Thus, we have proved the following theorem.

A Test for Rational Zeros

Let

$$P(x) = a_n x^n + a_{n-1}x^{n-1} + \cdots + a_1 x + a_0$$

be a polynomial with integral coefficients such that $a_n \neq 0$ and $a_0 \neq 0$. If a rational number p/q is a zero of $P(x)$, then p must divide the constant term a_0, and q must divide the leading coefficient a_n.

EXAMPLE 1   Find the rational roots of $P(x) = 2x^3 + 9x^2 + 3x - 4$ and factor the polynomial.

Solution: According to the rational root test, if a fraction p/q, reduced to lowest terms, is a zero of the given polynomial then p is a factor of -4 and q is a factor of 2. The choices for p are then

$$\pm 1, \ \pm 2, \quad \text{and} \quad \pm 4,$$

and those for q are

$$\pm 1 \quad \text{and} \quad \pm 2.$$

Thus, the possible rational roots are

$$\pm 1, \ \pm \frac{1}{2}, \ \pm 2, \quad \text{and} \quad \pm 4.$$

Every rational zero r of $P(x) = 2x^3 + 9x^2 + 3x - 4$ must belong to this set of rational numbers, so we next compute the function values $P(r)$ until a zero is obtained. Since

$$P(-1) = -2 + 9 - 3 - 4 = 0,$$

we see that -1 is a root and $x + 1$ is a factor of $P(x)$. Dividing $P(x)$ by $x + 1$, we obtain

$$2x^3 + 9x^2 + 3x - 4 = (x + 1)(2x^2 + 7x - 4).$$

The remaining zeros of $P(x)$ are the zeros of the quadratic polynomial $2x^2 + 7x - 4$. At this point, we could either use this polynomial to find the other zeros of $P(x)$, or we could solve the quadratic equation $2x^2 + 7x - 4 = 0$. In any case, it is easy to verify that $1/2$ and -4 are also roots of $P(x)$. Thus we obtain the factorization

$$2x^3 + 9x^2 + 3x - 4 = 2(x + 1)\left(x - \frac{1}{2}\right)(x + 4).$$

Practice Exercise 1   Find all the roots of $A(x) = 2x^3 - x^2 - 7x + 6$ and factor this polynomial completely.

Answer: Roots: $-2, 1, \dfrac{3}{2}$

$$2x^3 - x^2 - 7x + 6 = 2\left(x - \frac{3}{2}\right)(x - 1)(x + 2)$$

■ **A word of caution**   *It is important to observe that the test for rational roots does not say that every polynomial with integer coefficients has rational roots.* It only states that if a rational number (in its lowest terms) *is* a root, then it must satisfy the conditions of the theorem. *What the theorem does is enable us to list a finite set of rational numbers that includes every rational root of the polynomial.* This reduces the problem of finding all the rational roots of a polynomial to the mechanical task of checking the possibilities on the list. *There are no rational roots that are not on that list.*

In particular, if the leading coefficient of a polynomial with integer coefficients is 1 and the polynomial has rational roots, then these roots are integers.

EXAMPLE 2   Find the rational zeros of $P(x) = x^4 - x^3 - 4x^2 + 4x$ and factor the polynomial.

Solution: Since the constant term of $P(x)$ is zero, we see that $x = 0$ is a root of $P(x)$, and we can write

$$P(x) = x(x^3 - x^2 - 4x + 4).$$

The remaining zeros are the roots of the third-degree polynomial $Q(x) = x^3 - x^2 - 4x + 4$. Since the leading coefficient is 1, the only possible rational zeros are integers. According to the rational root test, they must be among the integral factors of 4, that is, $\pm 1, \pm 2, \pm 4$. Because

$$Q(-1) = (-1)^3 - (-1)^2 - 4(-1) + 4$$
$$= -1 - 1 + 4 + 4$$
$$= 6,$$

-1 is not a zero. On the other hand,

$$Q(1) = 1^3 - 1^2 - 4(1) + 4$$
$$= 1 - 1 - 4 + 4$$
$$= 0,$$

so 1 is a root. Dividing $Q(x)$ by $x - 1$, we have

$$x^3 - x^2 - 4x + 4 = (x - 1)(x^2 - 4).$$

Since -2 and 2 are the roots of the quadratic polynomial $x^2 - 4$, we can now write $P(x)$ as a product of linear factors:

$$x^4 - x^3 - 4x^2 + 4x = x(x - 1)(x - 2)(x + 2).$$

Practice Exercise 2   Find all the zeros of $x^4 - 2x^3 - 5x^2 + 6x$ and factor the polynomial.

Answer: Zeros: $-2, 0, 1, 3$
$$x^4 - 2x^3 - 5x^2 + 6x = x(x + 2)(x - 1)(x - 3).$$

The Case of Rational Coefficients

The test for rational zeros also applies to polynomials with *rational* coefficients. Indeed, if a polynomial equation has rational coefficients, then multiplying both sides by the least common denominator of the coefficients transforms the equation into an equivalent one with integral coefficients.

EXAMPLE 3   Find the zeros of $P(x) = \frac{1}{6}x^3 - x^2 + \frac{3}{2}x - \frac{1}{3}$.

Solution: We have to solve the equation

$$\frac{1}{6}x^3 - x^2 + \frac{3}{2}x - \frac{1}{3} = 0.$$

Multiplying both sides by 6, we obtain an equivalent equation with integral coefficients:

$$x^3 - 6x^2 + 9x - 2 = 0.$$

The leading coefficient is 1, so the only possible rational roots are the integral factors of -2, namely, ± 1 and ± 2. We use synthetic division,

$$
\begin{array}{r|rrrr}
2 & 1 & -6 & 9 & -2 \\
  &   & 2 & -8 & 2 \\
\hline
  & 1 & -4 & 1 & 0
\end{array}
$$

and find that 2 is a root. Hence $x - 2$ is a factor of $P(x)$ and we have

$$x^3 - 6x^2 + 9x - 2 = (x - 2)(x^2 - 4x + 1).$$

The other roots of $P(x)$ are the solutions of the equation $x^2 - 4x + 1 = 0$. Using the quadratic formula, we obtain the solutions $2 + \sqrt{3}$ and $2 - \sqrt{3}$. Therefore, the three zeros of $P(x)$ are 2, $2 + \sqrt{3}$, and $2 - \sqrt{3}$.

Practice Exercise 3   Find all zeros, real or complex, of the polynomial $\frac{x^3}{12} - \frac{x^2}{3} + \frac{x}{12} - \frac{1}{3} = 0$.

Answer: 4, $\pm i$

Guidelines for Finding Rational Zeros

1. To find *rational zeros* of a polynomial, first make sure that the polynomial has *integral* or *rational* coefficients.
2. If the coefficients are rational numbers, then clear all denominators from the polynomial equation.
3. Form a complete set of fractions p/q, where p divides the constant term and q divides the leading term of the polynomial equation.
4. For each of the fractions, compute the function value until you find a zero of the polynomial. Determine the quotient and apply the rational root test to the quotient.

EXERCISES 11.4

Find all roots of the polynomials in Exercises 1–14.

1. $x^3 + x^2 - 4x - 4$ **2.** $x^3 - 3x^2 - x + 3$

3. $x^3 - 3x^2 - 4x + 12$ **4.** $2x^3 - 6x^2 + 6x - 2$

5. $x^3 - x$ **6.** $x^3 + x^2 - 6x$

7. $x^3 + x^2 + x + 2$ **8.** $x^4 - 4x^2$

9. $2x^3 + x^2 - 2x - 1$ **10.** $3x^3 + 5x^2 - 11x + 3$

11. $4x^3 - x$ **12.** $25x^4 - x^2$

13. $x^3 - 4x^2 + 4x - 4$ **14.** $2x^3 + 2x^2 + 3x + 3$

Factor each polynomial in Exercises 15–32.

15. $x^3 - x$ **16.** $x^4 - 4x^2$

17. $\dfrac{1}{2}x^3 - 2x$ **18.** $3x^4 - 12x^2$

19. $x^3 + x^2 - 4x - 4$ **20.** $3x^3 - 9x^2 - 3x + 9$

21. $x^3 + x^2 - 6x$ **22.** $4x^3 - 13x - 6$

23. $x^4 - 16$ **24.** $3x^3 + 3x$

25. $x^3 - 4x^2 + x - 4$ **26.** $x^3 - x^2 + 2x - 2$

27. $2x^3 - 3x^2 + 4x - 6$ **28.** $8x^3 + 4x^2 - 2x - 1$

29. $\dfrac{3}{4}x^4 - \dfrac{3}{2}x^3 - \dfrac{3}{2}x^2 + 6x - 6$

30. $\dfrac{1}{3}x^5 - \dfrac{1}{2}x^4 - \dfrac{1}{3}x + \dfrac{1}{2}$

31. $\dfrac{1}{4}x^3 - \dfrac{5}{4}x^2 + \dfrac{3}{2}x - \dfrac{1}{2}$ **32.** $x^3 + \dfrac{2}{3}x^2 - \dfrac{1}{4}x - \dfrac{1}{6}$

Show that the equations in Exercises 33–40 have no rational roots.

33. $x^2 - 2 = 0$ **34.** $x^2 - 3 = 0$

35. $2x^2 - x + 3 = 0$ **36.** $4x^2 - 12x + 7 = 0$

37. $x^3 - 3x^2 + 4x - 6 = 0$

38. $3x^3 - 4x^2 + 7x + 5 = 0$

39. $x^4 + x^3 - x^2 - 2x - 2 = 0$

40. $2x^4 - x^3 + 4x^2 - 2x + 3 = 0$

11.5   PARTIAL FRACTION DECOMPOSITION

In Section 1.6, we learned how to combine two or more rational expressions into a single one by addition or subtraction. For example,

$$\frac{5}{x + 1} + \frac{x}{x^2 + 2} = \frac{5(x^2 + 2) + x(x + 1)}{(x + 1)(x^2 + 2)}$$

$$= \frac{6x^2 + x + 10}{x^3 + x^2 + 2x + 2}.$$

In many situations (such as in integral calculus), the reverse process is required: we have to express a quotient of two polynomials as the sum of two (or more) simpler quotients, called *partial fractions*. Before we describe the method of *partial fraction decomposition* of a rational function, a few remarks are in order.

Irreducible Polynomials

A polynomial with real coefficients is said to be *irreducible over* $\mathbb{R}$ if it cannot be expressed as a product of two polynomials of positive degree with real coefficients. Otherwise, the polynomial is said to be *reducible over* $\mathbb{R}$.

For example, $x + 1, 2x + 3, x^2 + 1,$ and $x^2 + x + 1$ are irreducible over $\mathbb{R}$. On the other hand, the polynomials $x^2 - 1, 2x^2 - 5x - 3,$ and $x^3 - 6x^2 + 11x - 6$ are reducible over $\mathbb{R}$, because

$$x^2 - 1 = (x - 1)(x + 1),$$
$$2x^2 - 5x - 3 = (2x + 1)(x - 3),$$
$$x^3 - 6x^2 + 11x - 6 = (x - 1)(x - 2)(x - 3).$$

Note that if all the roots of a polynomial are real, then the polynomial is reducible over $\mathbb{R}$, and it can be written as a product of linear factors.

A quadratic polynomial $ax^2 + bx + c$ with real coefficients is *irreducible* over $\mathbb{R}$ if and only if it has *complex roots*; that is, if and only if the *discriminant* $b^2 - 4ac$ is *negative*. For example, $x^2 + 1$, $x^2 + x + 1$, and $3x^2 - 2x + 1$ are irreducible over $\mathbb{R}$. Note that if a quadratic polynomial with real coefficients is irreducible over $\mathbb{R}$, then it has exactly two complex roots, which form a pair of complex conjugate numbers.

We now state without proof the following important theorem.

The Factorization Theorem

Every polynomial with real coefficients can always be expressed as a product of *linear polynomials* and/or *quadratic polynomials that have real coefficients* and are *irreducible over* $\mathbb{R}$.

For example,

$$2x^2 - 5x - 3 = (2x + 1)(x - 3),$$
$$x^3 - 6x^2 + 11x - 6 = (x - 1)(x - 2)(x - 3),$$
$$x^3 - 1 = (x - 1)(x^2 + x + 1),$$
$$x^4 - 2x^2 + 1 = (x + 1)(x + 1)(x - 1)(x - 1) = (x + 1)^2(x - 1)^2.$$

It follows from the factorization theorem that every polynomial with real coefficients has two types of factors; $(ax + b)^k$ and/or $(ax^2 + bx + c)^m$, where $ax^2 + bx + c$ is irreducible over $\mathbb{R}$. The first factor corresponds to the real zero $-b/a$ with multiplicity k, while the second factor corresponds to a pair of complex conjugate zeros, each of which has multiplicity m.

Partial Fractions

Let $A(x)$ and $B(x)$ be two polynomials with real coefficients, such that *the degree of $A(x)$ is smaller than the degree of $B(x)$*. Then it follows from a theorem in algebra that the quotient $A(x)/B(x)$ can be written as a sum of the form

$$\frac{A(x)}{B(x)} = F_1(x) + F_2(x) + \cdots + F_p(x),$$

where each $F_i(x)$ is a rational expression of either of the types

$$\frac{A_k}{(ax + b)^k} \quad \text{or} \quad \frac{A_m x + B_m}{(ax^2 + bx + c)^m};$$

here A_k, A_m, and B_m are real constants, k and m are suitable positive integers, and $ax^2 + bx + c$ is irreducible over $\mathbb{R}$. The sum $F_1(x) + F_2(x) + \cdots + F_p(x)$ is called the *partial fraction decomposition* of $A(x)/B(x)$ and each term $F_i(x)$ is called a *partial fraction*.

If the degree of $A(x)$ is *greater than or equal* to the degree of $B(x)$, then dividing $A(x)$ by $B(x)$ and using (11.1), we obtain

$$\frac{A(x)}{B(x)} = Q(x) + \frac{R(x)}{B(x)},$$

where $Q(x)$ is a polynomial and $R(x)/B(x)$ is a proper rational expression. Since the degree of $R(x)$ is less than the degree of $B(x)$, we can replace $R(x)/B(x)$ by its partial fraction decomposition. Thus we conclude that every rational function $A(x)/B(x)$ can be written

$$\frac{A(x)}{B(x)} = Q(x) + F_1(x) + F_2(x) + \cdots + F_p(x),$$

where $Q(x)$ is a polynomial (if the degree of $A(x)$ is greater than or equal to the degree of $B(x)$) and $F_1(x)$, $F_2(x)$, . . . , $F_p(x)$ are partial fractions.

From now on, we assume that $A(x)/B(x)$ is a *proper rational function*, that is, the degree of $A(x)$ is less than the degree of $B(x)$. In order to find the partial fraction decomposition of $A(x)/B(x)$, we must write the denominator $B(x)$ as a product of factors of the type $(ax + b)^k$ and $(ax^2 + bx + c)^m$, where $ax^2 + bx + c$ is irreducible. Next we proceed according to the following rules.

Rules for Partial Fraction Decomposition

1. For each factor of the type $(ax + b)^k$ with $k \geq 1$, the partial fraction decomposition of $A(x)/B(x)$ contains a sum of k terms,

$$\frac{A_1}{ax + b} + \frac{A_2}{(ax + b)^2} + \cdots + \frac{A_k}{(ax + b)^k},$$

where $A_1, A_2, \ldots, A_k$ are real constants.

2. For each factor of the type $(ax^2 + bx + c)^m$, where $m \geq 1$ and $ax^2 + bx + c$ is irreducible over $\mathbb{R}$, the partial fraction decomposition of $A(x)/B(x)$ contains a sum of m terms,

$$\frac{A_1x + B_1}{ax^2 + bx + c} + \frac{A_2x + B_2}{(ax^2 + bx + c)^2} + \cdots + \frac{A_mx + B_m}{(ax^2 + bx + c)^m},$$

where $A_1, B_1, \ldots, A_m, B_m$ are real constants.

EXAMPLE 1   Decompose $\dfrac{5x - 1}{2x^2 - 5x - 3}$ into partial fractions.

Solution: First, we factor the denominator and obtain

$$2x^2 - 5x - 3 = (2x + 1)(x - 3).$$

The denominator contains two linear factors: $2x + 1$ and $x - 3$. According to rule 1, the partial fraction decomposition of $(5x - 1)/(2x^2 - 5x - 3)$ will contain terms of the form $A/(2x + 1)$ and $B/(x - 3)$. Since $2x + 1$ and $x - 3$ are the only factors of the denominators, we can write

$$\frac{5x - 1}{2x^2 - 5x - 3} = \frac{A}{2x + 1} + \frac{B}{x - 3}.$$

Next, we have to determine the constants A and B. If we multiply both sides of the last expression by $2x^2 - 5x - 3 = (2x + 1)(x - 3)$, we obtain

$$5x - 1 = A(x - 3) + B(2x + 1).$$

Since the last expression is an identity, it must be satisfied for all values of x. If we set $x = 3$, then the first term on the right-hand side equals zero, and we can obtain B:

$$5(3) - 1 = A(3 - 3) + B[2(3) + 1]$$
$$14 = 7B$$
$$B = 2.$$

If we set $x = -1/2$, then the second term on the right-hand side equals zero, and we obtain the value of A:

$$5\left(-\frac{1}{2}\right) - 1 = A\left(-\frac{1}{2} - 3\right) + B\left[2\left(-\frac{1}{2}\right) + 1\right]$$
$$-\frac{7}{2} = A\left(-\frac{7}{2}\right)$$
$$A = 1.$$

Thus the given expression is decomposed into partial fractions as follows:

$$\frac{5x - 1}{2x^2 - 5x - 3} = \frac{1}{2x + 1} + \frac{2}{x - 3}.$$

Practice Exercise 1   Decompose the expression $\dfrac{5x - 5}{3x^2 - 7x + 2}$ into partial fractions.

Answer: $\dfrac{5x - 5}{3x^2 - 7x + 2} = \dfrac{1}{x - 2} + \dfrac{2}{3x - 1}$

EXAMPLE 2   Decompose $\dfrac{6x^3 - 19x^2 + 6x + 5}{2x^2 - 5x - 3}$ into partial fractions.

Solution: Since the degree of the numerator is greater than the degree of the denominator, we have to write the given expression as a sum of a polynomial and a proper rational expression. After performing a long division (whose computation we leave to the reader as an exercise), we obtain

$$\frac{6x^3 - 19x^2 + 6x + 5}{2x^2 - 5x - 3} = 3x - 2 + \frac{5x - 1}{2x^2 - 5x - 3}.$$

Next, we proceed to decompose into partial fractions the proper rational expression $(5x - 1)/(2x^2 - 5x - 3)$. But this is exactly the expression considered in Example 1. Thus the given expression can be written

$$\frac{6x^3 - 19x^2 + 6x + 5}{2x^2 - 5x - 3} = 3x - 2 + \frac{1}{2x + 1} + \frac{2}{x - 3}.$$

Practice Exercise 2   Decompose the rational expression $\dfrac{6x^3 - 11x^2 + 2x - 3}{3x^2 - 7x + 2}$ into partial fractions.

Answer: $\dfrac{6x^3 - 11x^2 + 2x - 3}{3x^2 - 7x + 2} = 2x + 1 + \dfrac{1}{x - 2} + \dfrac{2}{3x - 1}$

EXAMPLE 3   Decompose $\dfrac{5x^2 + 2x + 2}{x^3 - 1}$ into partial fractions.

Solution: Factoring the denominator, we have

$$x^3 - 1 = (x - 1)(x^2 + x + 1).$$

where $x^2 + x + 1$ is a quadratic polynomial that is irreducible over the reals. According to rules 1 and 2, the partial fraction decomposition of the given rational expression will contain only two terms: $A/(x - 1)$ and $(Bx + C)/(x^2 + x + 1)$. Writing

$$\frac{5x^2 + 2x + 2}{x^3 - 1} = \frac{A}{x - 1} + \frac{Bx + C}{x^2 + x + 1},$$

we now have to determine the constants A, B, and C. Multiplying both sides by $x^3 - 1$, we obtain

$$5x^2 + 2x + 2 = A(x^2 + x + 1) + (Bx + C)(x - 1).$$

We first set $x = 1$ and solve for A:

$$5 \cdot 1^2 + 2 \cdot 1 + 2 = A(1^2 + 1 + 1) + (B \cdot 1 + C)(1 - 1)$$
$$9 = 3A$$
$$A = 3.$$

Now we substitute 3 for A, set $x = 0$, and solve for C:

$$5 \cdot 0^2 + 2 \cdot 0 + 2 = 3(0^2 + 0 + 1) + (B \cdot 0 + C)(0 - 1)$$
$$2 = 3 + C(-1)$$
$$C = 1.$$

Finally, substituting 3 for A and 1 for C, we rewrite the expression for $5x^2 + 2x + 2$ as follows:

$$5x^2 + 2x + 2 = 3(x^2 + x + 1) + (Bx + 1)(x - 1).$$

Now, we set $x = -1$ and solve for B:

$$5 = 3 + (-B + 1)(-2)$$
$$B = 2.$$

Hence,

$$\frac{5x^2 + 2x + 2}{x^3 - 1} = \frac{3}{x - 1} + \frac{2x + 1}{x^2 + x + 1}.$$

Practice Exercise 3

Decompose the rational expression $\dfrac{5x^2 + 1}{x^3 + 1}$ into partial fractions.

Answer: $\dfrac{5x^2 + 1}{x^3 + 1} = \dfrac{2}{x + 1} + \dfrac{3x - 1}{x^2 - x + 1}$

EXAMPLE 4

Decompose $\dfrac{4x^2 - 3x + 5}{(x - 1)^2(x + 2)}$ into partial fractions.

Solution: The denominator contains two linear factors: $x - 1$ and $x + 2$. Since the factor $x - 1$ appears with the exponent 2, then according to rule 1, the partial fraction decomposition will contain terms of the form $A/(x - 1)$ and $B/(x - 1)^2$. Moreover, the contribution from the factor $x + 2$ will be of the form $C/(x + 2)$. Thus we can write

$$\frac{4x^2 - 3x + 5}{(x - 1)^2(x + 2)} = \frac{A}{x - 1} + \frac{B}{(x - 1)^2} + \frac{C}{x + 2}.$$

In order to determine the constants, we multiply both sides by $(x - 1)^2(x + 2)$ and get

$$4x^2 - 3x + 5 = A(x - 1)(x + 2) + B(x + 2) + C(x - 1)^2.$$

If we set $x = 1$, then

$$6 = 3B$$
$$B = 2.$$

If we set $x = -2$, then

$$27 = 9C$$
$$C = 3.$$

Finally, setting $x = 0$ and substituting 2 for B and 3 for C, we obtain

$$5 = -2A + 4 + 3$$
$$A = 1.$$

Therefore,

$$\frac{4x^2 - 3x + 5}{(x - 1)^2(x + 2)} = \frac{1}{x - 1} + \frac{2}{(x - 1)^2} + \frac{3}{x + 2}.$$

Practice Exercise 4

Decompose $\dfrac{5x^2 + 15x + 7}{(x - 1)(x + 2)^2}$ into partial fractions.

Answer: $\dfrac{5x^2 + 15x + 7}{(x - 1)(x + 2)^2} = \dfrac{3}{x - 1} + \dfrac{2}{x + 2} + \dfrac{1}{(x + 2)^2}$

EXAMPLE 5

Decompose $\dfrac{x^3 - 2x^2 + 2x - 4}{(x^2 + x + 1)^2}$ into partial fractions.

Solution: The quadratic polynomial $x^2 + x + 1$ is irreducible over $\mathbb{R}$ (why?). Thus, using rule 2, we can write

$$\frac{x^3 - 2x^2 + 2x - 4}{(x^2 + x + 1)^2} = \frac{Ax + B}{x^2 + x + 1} + \frac{Cx + D}{(x^2 + x + 1)^2}.$$

Multiplying both sides by $(x^2 + x + 1)^2$, we obtain

$$x^3 - 2x^2 + 2x - 4 = (Ax + B)(x^2 + x + 1) + (Cx + D).$$

Now we must use a different technique to determine the coefficients A, B, C, and D. Multiplying and rearranging the terms on the right-hand side, we get

$$x^3 - 2x^2 + 2x - 4 = Ax^3 + (A + B)x^2 + (A + B + C)x + (B + D).$$

According to the theorem about identically equal polynomials (Section 11.1), these two polynomials are identically equal if and only if the coefficients of the terms of the same degree are equal. Thus we obtain a system of linear equations in the unknowns A, B, C, and D:

$$\begin{aligned} A &= 1 \\ A + B &= -2 \\ A + B + C &= 2 \\ B + D &= -4. \end{aligned}$$

This system can be easily solved as follows. We already have $A = 1$. Substituting 1 for A into the second equation and solving for B, we get $B = -3$. Substituting 1 for A and -3 for B into the third equation and solving for C, we obtain $C = 4$. Finally, substituting -3 for B into the last equation gives us $D = -1$. Thus we have the partial fraction decomposition

$$\frac{x^3 - 2x^2 + 2x - 4}{(x^2 + x + 1)^2} = \frac{x - 3}{x^2 + x + 1} + \frac{4x - 1}{(x^2 + x + 1)^2}.$$

Practice Exercise 5   Decompose the expression $\dfrac{2x^2 - x + 5}{(x^2 + x + 2)^2}$ into partial fractions.

Answer: $\dfrac{2x^2 - x + 5}{(x^2 + x + 2)^2} = \dfrac{2}{x^2 + x + 2} - \dfrac{3x - 1}{(x^2 + x + 2)^2}$

EXERCISES 11.5

In Exercises 1–10, find the constants A, B, C, and D for which the left-hand side is equal to the right-hand side.

1. $\dfrac{4x - 14}{(x - 2)(x - 4)} = \dfrac{A}{x - 2} + \dfrac{B}{x - 4}$

2. $\dfrac{7x + 15}{(x + 3)(x - 4)} = \dfrac{A}{x + 3} + \dfrac{B}{x - 4}$

3. $\dfrac{x + 3}{(x - 4)(3x + 2)} = \dfrac{A}{x - 4} + \dfrac{B}{3x + 2}$

4. $\dfrac{2x + 7}{(x + 6)(2x - 3)} = \dfrac{A}{x + 6} + \dfrac{B}{2x - 3}$

5. $\dfrac{5x^2 + 9x + 3}{x(x + 1)^2} = \dfrac{A}{x} + \dfrac{B}{x + 1} + \dfrac{C}{(x + 1)^2}$

6. $\dfrac{7x + 11}{(x - 2)(x + 3)^2} = \dfrac{A}{x - 2} + \dfrac{B}{x + 3} + \dfrac{C}{(x + 3)^2}$

7. $\dfrac{2x + 3}{(x + 1)(x^2 + x + 1)} = \dfrac{A}{x + 1} + \dfrac{Bx + C}{x^2 + x + 1}$

8. $\dfrac{x^2 - 9x + 1}{(x - 4)(x^2 + 3)} = \dfrac{A}{x - 4} + \dfrac{Bx + C}{x^2 + 3}$

9. $\dfrac{2x^2 - x + 6}{(x^2 - x + 2)^2} = \dfrac{Ax + B}{x^2 - x + 2} + \dfrac{Cx + D}{(x^2 - x + 2)^2}$

10. $\dfrac{x^3 + x}{(x^2 - x + 1)^2} = \dfrac{Ax + B}{x^2 - x + 1} + \dfrac{Cx + D}{(x^2 - x + 1)^2}$

In Exercises 11–20, decompose the given rational expressions into partial fractions.

11. $\dfrac{x - 8}{x^2 - x - 12}$

12. $\dfrac{-x + 9}{x^2 + 5x - 6}$

13. $\dfrac{-7x + 5}{2x^2 + 5x - 12}$

14. $\dfrac{16x + 2}{6x^2 - x - 1}$

15. $\dfrac{9x^2 + 9x - 12}{2x^3 - x^2 - 6x}$

16. $\dfrac{3x^2 + 12x - 2}{3x^3 + 5x^2 - 2x}$

17. $\dfrac{4x^2 + x - 2}{x^3 - x^2}$

18. $\dfrac{3x^2 - 12x + 11}{x^3 - 5x^2 + 8x - 4}$

19. $\dfrac{2x^3 + 5x^2 + 8x + 1}{(x^2 + 2x + 2)^2}$

20. $\dfrac{x^3 + 4x^2 + 2x}{(x^2 + 1)^2}$

CHAPTER SUMMARY

A *polynomial function* with real coefficients and degree n is defined by

$$A(x) = a_n x^n + a_{n-1} x^{n-1} + \cdots + a_1 x + a_0, \qquad a_n \neq 0.$$

Two polynomials $A(x)$ and $B(x)$ are *identically equal* if

$$A(x) = B(x) \qquad \text{for all values of } x.$$

This means that $A(x)$ and $B(x)$ have the *same degree* and that the coefficients of terms of same degree are *equal*.

The Division Algorithm

If $A(x)$ (*dividend*) is a polynomial of degree n and $B(x) \neq 0$ (*divisor*) is a polynomial of degree m, with $n \geq m$, then there exist two polynomials $Q(x)$ (*quotient*) and $R(x)$ (*remainder*) such that

$$A(x) = B(x) \cdot Q(x) + R(x),$$

where the degree of $Q(x)$ is $n - m$, and either $R(x) = 0$ or its degree is less than m.

Exact Division

When $R(x) = 0$, then

$$A(x) = B(x) \cdot Q(x).$$

In this case $A(x)$ is said to be *divisible* by $B(x)$, and $B(x)$ is said to be a *factor* of $A(x)$.

Division by $x - a$

Of special interest is the case when the divisor $B(x)$ is a first-degree polynomial of the form $x - a$. Then

$$A(x) = (x - a)Q(x) + R,$$

where $Q(x)$ has degree $n - 1$ and R is a constant.

The Remainder Theorem

$$R = A(a)$$

The Factor Theorem

$x - a$ is a factor of $A(x)$ if and only if $A(a) = 0$.

In view of the remainder and factor theorems, it is useful to have efficient methods to compute function values of polynomials. One of them is *Horner's method*, explained in Section 11.2. The method of *synthetic division* allows you to find, *without* performing long division, the quotient and remainder in the division of $A(x)$ by $x - a$.

The Fundamental Theorem of Algebra

Every polynomial of degree $n \geq 1$ has *exactly* n (real or complex) *zeros* or *roots*, counting multiplicities.

A Test for Rational Zeros

Let

$$P(x) = a_nx^n + a_{n-1}x^{n-1} + \cdots + a_1x + a_0$$

be a polynomial with integral coefficients such that $a_n \neq 0$ and $a_0 \neq 0$. If a fraction p/q, in lowest terms, is a zero of $P(x)$, then p divides a_0 and q divides a_n.

The Factorization Theorem

Every polynomial with real coefficients can always be written as a product of *linear polynomials and/or quadratic polynomials with real coefficients.*

Partial Fraction Decomposition

Every rational expression $A(x)/B(x)$, where the degree of $A(x)$ is *smaller* than the degree of $B(x)$, can be written as a finite sum:

$$\frac{A(x)}{B(x)} = F_1(x) + F_2(x) + \cdots + F_p(x),$$

where each $F_i(x)$ is of the form

$$\frac{A_k}{(ax + b)^k} \quad \text{or} \quad \frac{A_mx + B_m}{(ax^2 + bx + c)^m}$$

with A_k, A_m, and B_m real numbers, k and m positive integers, and $ax^2 + bx + c$ *irreducible* over $\mathbb{R}$.

CHAPTER TEST

1. Let $P(x) = x^3 + x^2 - x + 3$. Find $P(1)$, $P(-1)$, and $P(1) \cdot P(-1)$.

2. If $P(x) = -5x^2 - 1$ and $Q(x) = (a - 1)x^2 + (b + 1)x + (c - 2)$, what are the constants a, b, and c so that $P(x) = Q(x)$?

3. Use synthetic division to find the quotient and remainder of $4x^3 + 4x^2 - 3x$ divided by $x + 1$.

4. Show that $x - 4$ is a factor of the polynomial $x^3 - x^2 - 10x - 8$, and factor the polynomial completely.

5. Find k so that $x - 4$ is a factor of the polynomial $x^3 + 3x^2 + 5x + k$.

6. Find the zeros and determine the multiplicity of each zero of the polynomial

$$P(x) = (2x - 3)(2x - 4)^3(2x + 1)^2.$$

7. What polynomial of degree 3 with real coefficients and leading coefficient 1 has zeros -3 and $1 + i$?

8. Determine all rational solutions of the polynomial equation $9x^3 - 12x^2 + x + 2 = 0$.

9. For what constants A and B does the following decomposition into partial fractions hold?

$$\frac{-8x - 17}{(x + 4)(x - 1)} = \frac{A}{x + 4} + \frac{B}{x - 1}$$

10. Decompose into partial fractions:

$$\frac{3x - 1}{x^2 - 4x + 4}.$$

REVIEW EXERCISES

Use long division to find the quotient and remainder in Exercises 1–4.

1. $(6x^3 - 4x^2 + 5x - 2) \div (2x^2 - 1)$

2. $(9x^3 - 6x^2 - 4x - 2) \div (3x^2 + x)$

3. $(2x^6 - 3x^5 + x^4 - 5x^3 - 4x^2 + 2x - 3) \div (x^2 - 3x + 1)$

4. $(3x^6 - 4x^5 + 2x^4 - 2x^3 + 3x^2 + 4x - 1) \div (x^2 + 2x + 2)$

Find the quotient and the remainder by synthetic division in Exercises 5–10.

5. $(2x^3 - 3) \div (x - 4)$

6. $(4x^3 - 2x) \div (x + 5)$

7. $(2x^4 - 3x^2 + 1) \div (x + 3)$

8. $(3x^5 - 4x + 1) \div (x - 2)$

c 9. $(0.3x^3 - 0.2x^2 + 0.1x + 1) \div (x + 0.4)$

c 10. $(2.1x^3 + 3.2x^2 - 1.5x - 1.3) \div (x - 1.2)$

In Exercises 11–16, show that the linear polynomial is a factor of the given polynomial. Factor the polynomial.

11. $2x^3 - 13x^2 + 16x - 5; \; x - 5$

12. $3x^3 + 25x^2 + 36x - 36; \; x + 6$

13. $2x^3 + 3x^2 - 8x + 3; \; 2x - 1$

14. $3x^3 - 2x^2 - 12x + 8$; $3x - 2$

C **15.** $0.5x^3 + 0.8x^2 - 0.7x - 0.6$; $x + 2$

C **16.** $x^3 + 2.6x^2 - 5.2x + 1.6$; $x - 0.4$

In Exercises 17–20, verify that the given number is a zero of the given polynomial and find the other zeros. Express the polynomial as a product of linear factors.

17. $5x^2 + 13x - 6, \dfrac{2}{5}$

18. $x^2 - 4x + 1, 2 - \sqrt{3}$

19. $x^4 - 2x^3 + 4x^2 + 2x - 5, 1 - 2i$

20. $x^4 - 4x^3 + x^2 + 16x - 20, 2 + i$

21. Find a third-degree polynomial with real coefficients and having zeros -1 and $3 - 2i$.

22. Find a third-degree polynomial with real coefficients and having zeros $1 + i\sqrt{3}$ and 0.

23. Show that 1 is a root of $x^4 - 6x^3 + 12x^2 - 10x + 3$ and find the other roots.

24. Show that -1 is a double root of $x^4 - 2x^2 + 1$ and find the other roots.

In Exercises 25–28, find a polynomial $P(x)$ of lowest degree with the indicated zeros and satisfying the given conditions.

25. $-1, -2, -3, P(1) = 12$

26. 2 (multiplicity 1), 3 (multiplicity 2), leading coefficient 4

27. $2 + i, 2 - i,$ 3 (multiplicity 2), leading coefficient 1

28. $1 + i\sqrt{2}, 1 - i\sqrt{2},$ 0 (multiplicity 3), leading coefficient 2

Find all rational roots of the following polynomials in Exercises 29–32.

29. $x^3 - 5x^2 + 3x + 9$

30. $x^3 - 9x^2 + 26x - 24$

31. $2x^3 - 7x^2 + 4x + 4$

32. $3x^3 - 17x^2 + 28x - 12$

Show that the equations in Exercises 33–36 have no rational roots.

33. $2x^2 - 3 = 0$

34. $4x^2 - 3 = 0$

35. $x^2 - x - 3 = 0$

36. $3x^2 - x - 1 = 0$

Decompose the rational expressions given in Exercises 37–40 into partial fractions.

37. $\dfrac{2x^2 - x}{(x - 1)(x^2 - x + 1)}$

38. $\dfrac{4x^2 + 10x + 14}{(x + 3)(x^2 + x + 4)}$

39. $\dfrac{3x^2 + 7x + 1}{x(x + 1)^2}$

40. $\dfrac{x^3 - x^2 + 4x - 3}{(x^2 + 2)^2}$

41. Find all fifth roots of 2. (*Hint:* Solve the equation $x^5 - 2 = 0$ using the method discussed in Section 9.9 of Chapter 9.)

42. Find all fourth roots of 3.

43. Verify that i is a zero of the polynomial $x^2 - 3ix - 2$ but that the complex conjugate $-i$ of i is not a zero. Why doesn't this contradict the second theorem of Section 11.3 (page 539)? Find the other root.

44. Verify that $1 + i$ is a zero of the polynomial $x^2 - (1 + 3i)x + 2(i - 1)$, but that $1 - i$ is not a zero. Find the other zero.

☐ **45.** Prove that every polynomial of odd degree with real coefficients has at least one real root.

☐ **46.** Prove that if n is even, then $x + a$ is a factor of $x^n - a^n$.

☐ **47.** Prove that if n is odd, then $x + a$ is a factor of $x^n + a^n$.

☐ **48.** Prove that $x - a$ is a factor of $x^n - a^n$ for all positive integers n.

☐ **49.** Prove that if $r_1, r_2, \ldots, r_n$ are the zeros of the polynomial

$$P(x) = x^n + a_{n-1}x^{n-1} + \cdots + a_1x + a_0,$$

then $r_1 + r_2 + \cdots + r_n = -a_{n-1}$.

☐ **50.** Under the assumptions of Exercise 49, prove that $r_1r_2\cdots r_n = (-1)^n a_0$.

CHAPTER 12

MATHEMATICAL INDUCTION, SEQUENCES, SERIES, AND PROBABILITY

In this chapter, we state the principle of mathematical induction and discuss the method of proof by mathematical induction. Next, we derive the binomial theorem and some of its applications. We also discuss the notions of sequences and series, in particular, arithmetic and geometric series. Finally, permutations, combinations, set partitioning, and elementary probability are discussed in the last two sections of the chapter.

12.1 THE PRINCIPLE OF MATHEMATICAL INDUCTION

Suppose that we want to prove the validity of the formula

$$1 + 2 + 3 + \cdots + n = \frac{n(n + 1)}{2} \qquad (12.1)$$

for every natural number n.

As a first attempt, we could check this formula for several values of n. For example, if $n = 1, 2,$ and 3, we have

$$1 = \frac{1(1 + 1)}{2},$$

$$1 + 2 = 3 = \frac{2(2 + 1)}{2},$$

$$1 + 2 + 3 = 6 = \frac{3(3 + 1)}{2}.$$

Thus formula (12.1) is true for $n = 1, 2,$ and 3. As an exercise, you could also check the validity of the formula for $n = 4, 5,$ and 6.

At this point, we might ask ourselves whether this method of checking, sometimes called *experimental induction*, constitutes a mathematical proof. In other words, from the fact that (12.1) is true whenever we choose a particular n, can we conclude that (12.1) is true for *all* n? The answer is negative for the following reasons. First, this method of checking *never ends*. Second, even if we had checked the validity of formula (12.1) for all values of n from 1 to 1000, there is no guarantee that the formula would be true for $n = 1001$.

To see that experimental induction is an unreliable and misleading method of proof, we consider a different example. Let

$$p(n) = n^2 - n + 11$$

be a quadratic function defined over the set of natural numbers. Is $p(n)$ a prime number for every natural number n? Let us try experimental induction. If $n = 1$, then $p(1) = 11$ is a prime number. If $n = 2$ and 3, then $p(2) = 13$ and $p(3) = 17$, which are also prime numbers. Continuing the checking, we can verify that $p(n)$ is a prime number for $n = 1, 2, 3, \ldots, 10$. These checks might lead us to believe that $p(n)$ is a prime number for all n. However,

$$p(11) = 11^2 - 11 + 11 = 11^2$$

is not a prime number! Although $p(n)$ is a prime for the first 10 choices of the natural number n, it fails to be a prime when $n = 11$.

This shows that experimental induction cannot be used to prove statements or theorems. Thus, if we want to prove formula (12.1), a more rigorous method, such as the method of *mathematical induction,* must be used.

The Axiom of Induction

Let S be a set of natural numbers satisfying the following properties:

(i) 1 is an element of S;
(ii) if the natural number k is an element of S, the successor $k + 1$ is also an element of S.

Then S is the set of all natural numbers.

This is a basic axiom of the natural numbers system. Every mathematical proof based upon the axiom of induction is said to be a proof by mathematical induction.

EXAMPLE 1   Prove the formula

$$1 + 2 + 3 + \cdots + n = \frac{n(n + 1)}{2}$$

by mathematical induction.

Solution: Let S be the set of all natural numbers for which the given formula is true. As we already mentioned, the formula is true for $n = 1$. Thus 1 is an element of S, which is part i of the axiom of induction. Next, to show that S satisfies part ii of the axiom, we must show that whenever we assume that a certain number k belongs to S—that is, whenever we assume that $1 + 2 + \cdots + k = k(k + 1)/2$ is true—we can deduce from that fact that $k + 1$ must also belong to S, that is, that $1 + 2 + \cdots + (k + 1) = (k + 1)[(k + 1) + 1]/2$ must also be true. To do this, we write

$$1 + 2 + \cdots + (k + 1) = 1 + 2 + \cdots + k + (k + 1).$$

Since formula (12.1) is true when we substitute k for n, we have

$$
\begin{aligned}
1 + 2 + \cdots + (k + 1) &= [1 + 2 + \cdots + k] + (k + 1) \\
&= \frac{k(k + 1)}{2} + (k + 1) \\
&= \frac{k(k + 1) + 2(k + 1)}{2} \quad \text{Write both terms with the same denominator} \\
&= \frac{(k + 1)(k + 2)}{2} \quad \text{Factor} \\
&= \frac{(k + 1)[(k + 1) + 1]}{2},
\end{aligned}
$$

and this is formula (12.1) with $n = k + 1$. Thus $k + 1$ is an element of S. We conclude that the set S satisfies properties i and ii of the axiom of induction. Therefore S coincides with the set $\mathbb{N}$ of all natural numbers and, consequently, the statement (12.1) is true for every natural number.

Practice Exercise 1   Using induction, prove the formula

$$1 + 3 + 5 + \cdots + (2n - 1) = n^2.$$

(*Hint*: Check the formula for $n = 1$. Assume it to be true for k and prove it for $k + 1$.)

In what follows, the notation $P(n)$ will stand for a statement about natural numbers, such as "Every even number is of the form $2n$," "The sum of the first n natural numbers is $n(n + 1)/2$," and so on. The following is an equivalent version of the axiom of induction.

The Principle of Mathematical Induction

Let $P(n)$ express a statement about natural numbers, such that

(i) $P(1)$ is a true statement;
(ii) for each natural number k, the truth of $P(k)$ implies the truth of $P(k + 1)$.

Then $P(n)$ is a true statement for all natural numbers.

EXAMPLE 2   Prove by induction that $5^n - 2^n$ is divisible by 3 for any natural number n.

Solution: Let $P(n)$ be the statement "$5^n - 2^n$ is divisible by 3." First we claim that $P(1)$ is true. Indeed, if $n = 1$, then $5 - 2 = 3$ is divisible by 3. Next let us assume that $P(k)$ is true and then prove that $P(k + 1)$ is true. We start by writing

$$5^{k+1} - 2^{k+1} = 5^k \cdot 5 - 2^k \cdot 2$$
$$= 5^k \cdot (3 + 2) - 2^k \cdot 2$$
$$= 5^k \cdot 3 + 5^k \cdot 2 - 2^k \cdot 2$$
$$= 5^k \cdot 3 + (5^k - 2^k) \cdot 2.$$

We now examine each term on the right-hand side. The first term, $5^k \cdot 3$, is clearly divisible by 3. The second one, $(5^k - 2^k) \cdot 2$, is also divisible by 3, since we are assuming that $P(k)$ is a true statement. Since both terms are divisible by 3, their sum, $5^{k+1} - 2^{k+1}$, is also divisible by 3. Thus, we have proved that $P(k + 1)$ is true. From the principle of induction, it follows that $5^n - 2^n$ is divisible by 3 for all natural numbers.

Practice Exercise 2   Using induction, prove that 2 is a factor of $5^n - 3^n$ for all n.

On many occasions, it is useful to have the following version of the principle of induction.

The Modified Principle of Induction

Let $P(n)$ express a statement about a natural number n, and let m be a fixed natural number such that

(i) $P(m)$ is a true statement;
(ii) for each natural number $k \geq m$, the truth of $P(k)$ implies the truth of $P(k + 1)$.

Then $P(n)$ is a true statement for all natural numbers $n \geq m$.

Notice that when $m = 1$, the modified principle of induction coincides with the principle of induction.

EXAMPLE 3   Prove by induction that $2^n < n!$ for all $n \geq 4$.

Solution: Recall that $n!$, read "n factorial," is defined by $n! = 1 \cdot 2 \ldots n$. Let $P(n)$ be the statement "$2^n < n!$" and let $m = 4$. Since $2^4 = 16 < 4! = 24$, it follows that $P(4)$ is true. Suppose now that $P(k)$ is true for $k \geq 4$, and let us prove that $P(k + 1)$ is true. We have

$$2^{k+1} = 2^k \cdot 2 < k! \cdot 2,$$

because we have assumed that $2^k < k!$. Since $k \geq 4$, we see that $k + 1 > 2$. Thus we get

$$2^{k+1} < k! \cdot 2 < k! \cdot (k + 1) = (k + 1)!.$$

Hence

$$2^{k+1} < (k + 1)!,$$

that is, $P(k + 1)$ is true. By the modified principle of induction, it follows that $2^n < n!$ for all $n \geq 4$.

You should check that the inequality of Example 3 is false for $n = 1, 2,$ and 3.

Practice Exercise 3   Using induction, prove that $2^{n+1} < n!$ for all $n \geq 5$.

The Summation Notation

A sum such as $1 + 2 + 3 + \cdots + n$ is often represented in shorthand notation as follows:

$$1 + 2 + 3 + \cdots + n = \sum_{j=1}^{n} j.$$

The symbol Σ is the Greek capital letter *sigma*, and it stands for *sum*. The letter j under the sigma is called the *index of summation*, and the numbers 1 and n are called the *limits of summation*. The symbol $\sum_{j=1}^{n} j$ is read, "the sum of j from 1 to n." For example,

$$\sum_{j=1}^{6} j = 1 + 2 + 3 + 4 + 5 + 6.$$

Other examples are

$$\sum_{j=1}^{4} (2j - 1) = 1 + 3 + 5 + 7$$

and

$$\sum_{j=1}^{5} j^2 = 1^2 + 2^2 + 3^2 + 4^2 + 5^2.$$

More generally, if $a_1, a_2, \ldots, a_n$ are n numbers, their sum is represented by

$$a_1 + a_2 + \cdots + a_n = \sum_{j=1}^{n} a_j.$$

EXAMPLE 4   Evaluate each of the following sums.

(a) $\sum_{j=1}^{5} (2j + 3)$ (b) $\sum_{j=3}^{6} \dfrac{2j + 1}{j - 1}$

Solution: (a) In the expression $2j + 3$, we first replace j by 1, then by 2, and so on, until finally j is replaced by 5; we compute the numbers and add.

$$\sum_{j=1}^{5} (2j + 3) = (2 \cdot 1 + 3) + (2 \cdot 2 + 3) + (2 \cdot 3 + 3)$$
$$+ (2 \cdot 4 + 3) + (2 \cdot 5 + 3)$$
$$= 5 + 7 + 9 + 11 + 13$$
$$= 45$$

(b) We have

$$\sum_{j=3}^{6} \frac{2j + 1}{j - 1} = \frac{2 \cdot 3 + 1}{3 - 1} + \frac{2 \cdot 4 + 1}{4 - 1} + \frac{2 \cdot 5 + 1}{5 - 1} + \frac{2 \cdot 6 + 1}{6 - 1}$$
$$= \frac{7}{2} + \frac{9}{3} + \frac{11}{4} + \frac{13}{5}$$
$$= \frac{237}{20}.$$

Practice Exercise 4   Find the following sums.

(a) $\displaystyle\sum_{j=1}^{5} (2j - 1)$ (b) $\displaystyle\sum_{j=1}^{4} \frac{j - 1}{2j}$

Answer: **(a)** 25   **(b)** $\dfrac{23}{24}$

Properties of Sums

Let n be a natural number, let a_j and b_j be real numbers for $j = 1, 2, \ldots,$ n, and let c be any real number. The following are useful and important properties of sums.

1. $\displaystyle\sum_{j=1}^{n} c = n \cdot c$

2. $\displaystyle\sum_{j=1}^{n} (a_j + b_j) = \sum_{j=1}^{n} a_j + \sum_{j=1}^{n} b_j$

3. $\displaystyle\sum_{j=1}^{n} ca_j = c \sum_{j=1}^{n} a_j$

Each one of these three formulas can be proved by induction. As an illustration, we prove formula 1, which says that if we add n terms equal to a constant c, then the sum equals the product $n \cdot c$. First, by definition,

$$\sum_{j=1}^{1} a_j = a_1.$$

Thus, if $n = 1$, we have

$$\sum_{j=1}^{1} c = c = 1 \cdot c,$$

and formula 1 holds for $n = 1$. Assuming it to be true for k, let us prove it for $k + 1$. We have

$$\sum_{j=1}^{k+1} c = \left(\sum_{j=1}^{k} c\right) + c$$

$$= k \cdot c + c \qquad \text{Induction assumption}$$

$$= (k + 1) \cdot c. \qquad \text{The distributive law}$$

Thus, formula 1 holds for $k + 1$. By the principle of induction, formula 1 is true for every n.

The proofs of formulas 2 and 3 are left as exercises.

EXAMPLE 5

Find the sum $\displaystyle\sum_{j=1}^{5} (2j + 3)$.

Solution: This is the sum we calculated in Example 4a; we now compute it using formulas 1, 2, and 3. We have

$$\sum_{j=1}^{5} (2j + 3) = \sum_{j=1}^{5} 2j + \sum_{j=1}^{5} 3$$

$$= 2\left(\sum_{j=1}^{5} j\right) + 5 \cdot 3$$

$$= 2(1 + 2 + 3 + 4 + 5) + 15$$

$$= 2 \cdot 15 + 15$$

$$= 45.$$

Practice Exercise 5

Find the sum $\displaystyle\sum_{j=1}^{5}(2j + 1)$.

Answer: 35

EXERCISES 12.1

In Exercises 1–10, use mathematical induction to prove each formula for all natural numbers.

1. $2 + 4 + 6 + \cdots + 2n = n(n + 1)$
2. $1 + 3 + 5 + \cdots + (2n - 1) = n^2$
3. $1 \cdot 2 + 2 \cdot 3 + 3 \cdot 4 + \cdots + n(n + 1)$
$$= \frac{n(n + 1)(n + 2)}{3}$$
4. $\dfrac{1}{1 \cdot 2} + \dfrac{1}{2 \cdot 3} + \dfrac{1}{3 \cdot 4} + \cdots + \dfrac{1}{n(n + 1)} = \dfrac{n}{n + 1}$
5. $1^2 + 2^2 + 3^2 + \cdots + j^2 = \dfrac{j(j + 1)(2j + 1)}{6}$
6. $1^3 + 2^3 + 3^3 + \cdots + k^3 = \left[\dfrac{k(k + 1)}{2}\right]^2$

7. $1^3 + 3^3 + 5^3 + \cdots + (2j - 1)^3 = j^2(2j^2 - 1)$
8. $1^2 + 3^2 + 5^2 + \cdots + (2j - 1)^2 = \dfrac{j(4j^2 - 1)}{3}$
9. $\dfrac{1}{2} + 1 + \dfrac{3}{2} + 2 + \cdots + \dfrac{k}{2} = \dfrac{k(k + 1)}{4}$
10. $1 + 8 + 16 + \cdots + 8(k - 1) = (2k - 1)^2$

Use mathematical induction to prove each of the statements in Exercises 11–20.

11. 4 is a factor of $7^k - 3^k$ for all natural numbers k.
12. $9^k - 5^k$ is divisible by 4 for all $k \in \mathbb{N}$.
13. $4^{2n} - 1$ is divisible by 5 for all $n \in \mathbb{N}$.
14. 7 is a factor of $8^n - 1$ for all natural numbers n.
15. $m^3 - m + 3$ is a multiple of 3 for all $m \in \mathbb{N}$.
16. $m^2 + m$ is an even number for all $m \in \mathbb{N}$.

17. $3^{2j} - 1$ is divisible by 8 for all natural numbers j.

18. $j^3 + 2j$ is divisible by 3 for all natural numbers j.

19. $x^n - a^n$ is divisible by $x - a$ for all natural numbers n.
(*Hint:* $x^{n+1} - a^{n+1} = (x^n - a^n)x + a^n(x - a)$.)

20. $x + a$ is a factor of $x^{2n-1} + a^{2n-1}$ for all natural numbers n.

Find the sums given in Exercises 21–30.

21. $\displaystyle\sum_{j=1}^{6}(j + 1)$

22. $\displaystyle\sum_{j=1}^{5} 3j$

23. $\displaystyle\sum_{j=1}^{5}(2j - 1)$

24. $\displaystyle\sum_{j=1}^{6}(3j - 2)$

25. $\displaystyle\sum_{j=1}^{4}\frac{2j + 1}{2}$

26. $\displaystyle\sum_{j=1}^{5}\frac{3j - 3}{4}$

27. $\displaystyle\sum_{j=1}^{6} j^2$

28. $\displaystyle\sum_{j=1}^{5}(2j - 1)^2$

29. $\displaystyle\sum_{j=3}^{5}(j^2 - 2j)$

30. $\displaystyle\sum_{j=2}^{4}(2^j + 1)$

In Exercises 31–40, prove each inequality by mathematical induction.

31. $2n > n$ for all natural numbers n.

32. $3^n \geq 1 + 2n$ for all natural numbers n.

33. $2^{k+3} < (k + 3)!$ for all $k \in \mathbb{N}$.
(*Hint:* $n! = 1 \cdot 2 \cdots (n - 1) \cdot n$)

34. $k^2 \geq 2k - 1$ for all $k \in \mathbb{N}$.

35. $j^2 > 2j$ for all $j \geq 3$.

36. $2^j > j^2$ for all $j \geq 5$.

37. $2^{m-1} < m!$ for all $m > 2$.

38. If $0 < a < 1$, then $0 < a^n < 1$ for all natural numbers n.

39. If $a > 1$, then $a^n > 1$ for all natural numbers n.

40. If $0 < b < a$, then $b^n < a^n$ for all natural numbers n.

41. Prove formula 2 of the properties of sums. (*Hint:* Use mathematical induction to prove that

$$(a_1 + b_1) + (a_2 + b_2) + \cdots + (a_n + b_n)$$
$$= (a_1 + a_2 + \cdots + a_n) + (b_1 + b_2 + \cdots + b_n).)$$

42. Prove formula 3 of the properties of sums.

43. Let a and b be real numbers. Use mathematical induction to prove that $(ab)^n = a^n b^n$ for every natural number n.

44. Let a be a real number. Prove that $a^n \cdot a^m = a^{n+m}$ for every pair of natural numbers n and m. (*Hint:* First use induction to show that $a^n \cdot a = a^{n+1}$.)

45. Let a be a real number. Prove that $(a^n)^m = a^{nm}$ for every pair of natural numbers n and m.

46. If z is a complex number, let $\bar{z}$ denote its complex conjugate. Prove that $\overline{z_1 + z_2 + \cdots + z_n} = \bar{z}_1 + \bar{z}_2 + \cdots + \bar{z}_n$.

47. Prove that $\overline{z_1 \cdot z_2 \cdots z_n} = \bar{z}_1 \cdot \bar{z}_2 \cdots \bar{z}_n$.

48. Prove that $\cos(n\pi) = (-1)^n$ for each natural number n.

49. Prove that $\sin(\alpha + n\pi) = (-1)^n \sin \alpha$ for each natural number n.

50. Let a and r be real numbers with $r \neq 1$. Prove that

$$a + ar + ar^2 + \cdots + ar^{n-1} = \frac{a(1 - r^n)}{1 - r},$$

where n is any natural number.

12.2   THE BINOMIAL THEOREM

Consider the binomial expression

$$(a + b)^n,$$

where n is a natural number. Using the principle of mathematical induction, we shall prove a formula that gives the expansion of $(a + b)^n$ without having to compute the corresponding products of the binomial $a + b$. This expansion is a sum of products of powers of a and b multiplied by certain coefficients, which are called *binomial coefficients*. These coefficients have interesting properties and play an important role in finite probability.

Before explaining how the expansion formula for $(a + b)^n$ can be obtained, we first discuss some properties of the binomial coefficients. We assume that you already know the notation

$$n! = 1 \cdot 2 \cdots n \tag{12.2}$$

(see the solution to Example 3). We recall also the special definitions $1! = 1$ and $0! = 1$.

The Binomial Coefficient

If k and n are natural numbers, with $k \leq n$, the *binomial coefficient* $\binom{n}{k}$ is defined by

$$\binom{n}{k} = \frac{n!}{k! \, (n - k)!}. \tag{12.3}$$

Since $n! = n(n - 1) \cdots (n - k + 1)(n - k)!$, it follows that

$$\binom{n}{k} = \frac{n!}{k! \, (n - k)!} = \frac{n(n - 1) \cdots (n - k + 1)(n - k)!}{k! \, (n - k)!}$$

Thus, we can also write the binomial coefficient $\binom{n}{k}$ as follows:

$$\binom{n}{k} = \frac{n(n - 1) \cdots (n - k + 1)}{k!}. \tag{12.4}$$

EXAMPLE 1   Evaluate $\binom{5}{2}$, $\binom{6}{0}$, and $\binom{7}{1}$.

Solution: According to formula (12.3), we have:

$$\binom{5}{2} = \frac{5!}{2! \, (5 - 2)!} = \frac{5!}{2! \, 3!} = \frac{5 \cdot 4 \cdot 3!}{2! \, 3!} = 10,$$

$$\binom{6}{0} = \frac{6!}{0! \, (6 - 0)!} = \frac{6!}{6!} = 1,$$

$$\binom{7}{1} = \frac{7!}{1! \, (7 - 1)!} = \frac{7!}{6!} = \frac{7 \cdot 6!}{6!} = 7.$$

Practice Exercise 1   Find the numbers $\binom{7}{2}$, $\binom{6}{3}$, and $\binom{5}{5}$.

Answer: 21, 20, and 1

Properties of Binomial Coefficients

The following are properties of the binomial coefficients.

$$\textbf{1.} \quad \binom{n}{k} = \binom{n}{n - k}$$

This can be seen as follows:

$$\binom{n}{n-k} = \frac{n!}{(n-k)! \, [n-(n-k)]!}$$

$$= \frac{n!}{(n-k)! \, k!} = \binom{n}{k}$$

---

2. $\displaystyle \binom{n}{0} = \binom{n}{n} = 1$

The first equality is obtained from property 1 by letting $k = 0$. To show the second equality, we write

$$\binom{n}{n} = \frac{n!}{n! \, (n-n)!} = \frac{n!}{n! \, 0!} = 1.$$

---

3. $\displaystyle \binom{n}{k-1} + \binom{n}{k} = \binom{n+1}{k}$

This is proved by first writing

$$\binom{n}{k-1} + \binom{n}{k} = \frac{n!}{(k-1)! \, (n-k+1)!} + \frac{n!}{k! \, (n-k)!}$$

Since the least common denominator for the last two fractions is $k! \, (n-k+1)!$, we have

$$\binom{n}{k-1} + \binom{n}{k} = \frac{n! \, k + n! \, (n-k+1)}{k! \, (n-k+1)!}$$

$$= \frac{n! \, (k+n-k+1)}{k! \, (n-k+1)!} \qquad \text{The distributive law}$$

$$= \frac{n! \, (n+1)}{k! \, (n-k+1)!}$$

$$= \frac{(n+1)!}{k! \, (n+1-k)!}$$

$$= \binom{n+1}{k}.$$

Powers of a Binomial

We already know from Section 1.5 that

$$(a+b)^2 = a^2 + 2ab + b^2.$$

This formula tells us that the square of the binomial $a + b$ has three terms:

$$a^2 = a^2b^0, \quad 2ab, \quad \text{and} \quad b^2 = a^0b^2.$$

As we move from left to right, the powers of a decrease by 1, while the powers of b increase by 1. Moreover, the coefficients of the expansion are, respectively,

$$\binom{2}{0} = 1, \quad \binom{2}{1} = 2, \quad \text{and} \quad \binom{2}{2} = 1.$$

Similarly, if we compute the cube of $a + b$, we get

$$(a + b)^3 = a^3 + 3a^2b + 3ab^2 + b^3.$$

This expansion has four terms:

$$a^3 = a^3b^0, \quad 3a^2b, \quad 3ab^2, \quad \text{and} \quad a^0b^3 = b^3.$$

From one term to the next, the powers of a decrease by 1 while the powers of b increase by 1. From left to right the coefficients are

$$\binom{3}{0} = 1, \quad \binom{3}{1} = 3, \quad \binom{3}{2} = 3, \quad \text{and} \quad \binom{3}{3} = 1.$$

In view of these two observations, we conjecture that the following formula is true in general.

The Binomial Theorem

If a and b are real numbers and n is a natural number, then

$$(a + b)^n = \binom{n}{0}a^n + \binom{n}{1}a^{n-1}b + \binom{n}{2}a^{n-2}b^2$$
$$+ \cdots + \binom{n}{n-1}ab^{n-1} + \binom{n}{n}b^n \qquad (12.5)$$

or, in summation notation,

$$(a + b)^n = \sum_{k=0}^{n}\binom{n}{k}a^{n-k}b^k. \qquad (12.6)$$

The proof is done by mathematical induction. First, we prove that (12.6) is true for $n = 1$:

$$(a + b)^1 = \sum_{k=0}^{1}\binom{1}{k}a^{1-k}b^k$$
$$= \binom{1}{0}a^1b^0 + \binom{1}{1}a^0b^1$$
$$= a + b.$$

Next, assuming that (12.6) is true for some natural number N, let us prove that it is true for $N + 1$. We have

$$(a + b)^{N+1} = (a + b)(a + b)^N$$

$$= (a + b) \sum_{k=0}^{N} \binom{N}{k} a^{N-k} b^k$$

$$= a \sum_{k=0}^{N} \binom{N}{k} a^{N-k} b^k + b \sum_{k=0}^{N} \binom{N}{k} a^{N-k} b^k$$

$$= \left[\binom{N}{0} a^{N+1} + \binom{N}{1} a^N b + \binom{N}{2} a^{N-1} b^2 + \cdots + \binom{N}{N} a b^N \right]$$

$$+ \left[\binom{N}{0} a^N b + \binom{N}{1} a^{N-1} b^2 + \cdots + \binom{N}{N-1} a b^N + \binom{N}{N} b^{N+1} \right]$$

$$= \binom{N}{0} a^{N+1} + \left[\binom{N}{0} + \binom{N}{1} \right] a^N b + \left[\binom{N}{1} + \binom{N}{2} \right] a^{N-1} b^2$$

$$+ \cdots + \left[\binom{N}{N-1} + \binom{N}{N} \right] a b^N + \binom{N}{N} b^{N+1}.$$

Using property 3 and noticing that

$$\binom{N}{0} = \binom{N+1}{0} \qquad \text{and} \qquad \binom{N}{N} = \binom{N+1}{N+1},$$

we can rewrite the right-hand side of the last expression and get

$$(a + b)^{N+1} = \binom{N+1}{0} a^{N+1} + \binom{N+1}{1} a^N b + \binom{N+1}{2} a^{N-1} b^2$$

$$+ \cdots + \binom{N+1}{N+1} b^{N+1}$$

$$= \sum_{k=0}^{N+1} \binom{N+1}{k} a^{N+1-k} b^k,$$

which means that (12.6) is true for $N + 1$. By the principle of induction, the formula is true for every natural number n.

Formulas (12.5) and (12.6) are the *binomial expansion* of $(a + b)^n$.

EXAMPLE 2   Use the binomial formula to expand $(a + b)^5$.

Solution: According to (12.6) with $n = 5$, we have

$$(a + b)^5 = \sum_{k=0}^{5} \binom{5}{k} a^{5-k} b^k$$

$$= \binom{5}{0} a^5 + \binom{5}{1} a^4 b + \binom{5}{2} a^3 b^2 + \binom{5}{3} a^2 b^3 + \binom{5}{4} a b^4 + \binom{5}{5} b^5.$$

We now compute the binomial coefficients:

$$\binom{5}{0} = 1.$$

$$\binom{5}{1} = \frac{5!}{1! \, 4!} = 5,$$

$$\binom{5}{2} = \frac{5!}{2! \, 3!} = 10.$$

By property 2, we have

$$\binom{5}{3} = \binom{5}{2}, \quad \binom{5}{4} = \binom{5}{1} \quad \text{and} \quad \binom{5}{5} = \binom{5}{0}.$$

Thus,

$$(a + b)^5 = a^5 + 5a^4b + 10a^3b^2 + 10a^2b^3 + 5ab^4 + b^5.$$

Practice Exercise 2   Expand $(x + y)^4$ by the binomial theorem.

Answer: $(x + y)^4 = x^4 + 4x^3y + 6x^2y^2 + 4xy^3 + y^4$

EXAMPLE **3**   Evaluate $(x - 2y)^4$.

Solution: The power $(x - 2y)^4$ is of the form $(a + b)^4$ with $a = x$ and $b = -2y$. By the binomial theorem, we have

$$(x - 2y)^4 = [x + (-2y)]^4$$
$$= \binom{4}{0}x^4 + \binom{4}{1}x^3(-2y) + \binom{4}{2}x^2(-2y)^2$$
$$+ \binom{4}{3}x(-2y)^3 + \binom{4}{4}(-2y)^4$$
$$= x^4 - \binom{4}{1}2x^3y + \binom{4}{2}4x^2y^2 - \binom{4}{3}8xy^3 + \binom{4}{4}16y^4$$
$$= x^4 - 8x^3y + 24x^2y^2 - 32xy^3 + 16y^4,$$

after computing the binomial coefficients.

Practice Exercise 3   Find the binomial expansion of $(2a - b)^5$.

Answer: $(2a - b)^5 = 32a^5 - 80a^4b + 80a^3b^2 - 40a^2b^3 + 10ab^4 - b^5$

The $(k + 1)$st Term in the Binomial Expansion

The $(k + 1)$st term in the expansion of $(a + b)^n$ is

$$\binom{n}{k}a^{n-k}b^k \quad \text{for} \quad 0 \le k \le n.$$

Note that *the exponent of b in the $(k + 1)$st term is k.*

EXAMPLE **4**   Find the sixth term in the expansion of $(3u - 2v)^8$.

Solution: Let $a = 3u$ and $b = -2v$. Since the exponent of b in the sixth term is 5, we see that the sixth term is

$$\binom{8}{5}(3u)^{8-5}(-2v)^5 = \frac{8!}{5!\ 3!}(3u)^3(-2v)^5$$

$$= \frac{8!}{5!\ 3!}27u^3(-32v^5)$$

$$= -56 \cdot 27 \cdot 32u^3v^5$$

$$= -48384u^3v^5.$$

Practice Exercise 4   Find the fifth term of the expansion of $(2a + 3b)^7$.

Answer: $\binom{7}{4}(2a)^3(3b)^4 = 22{,}680a^3b^4.$

Pascal's Triangle

The following array of numbers

$$
\begin{array}{ccccccccccccc}
 & & & & & & 1 & & & & & & \\
 & & & & & 1 & & 1 & & & & & \\
 & & & & 1 & & 2 & & 1 & & & & \\
 & & & 1 & & 3 & & 3 & & 1 & & & \\
 & & 1 & & 4 & & 6 & & 4 & & 1 & & \\
 & 1 & & 5 & & 10 & & 10 & & 5 & & 1 & \\
1 & & 6 & & 15 & & 20 & & 15 & & 6 & & 1 \\
\end{array}
$$

. . . .

obtained by writing the binomial coefficients in the expansion of $(a + b)^n$, where $n = 0, 1, 2, \ldots$, is called *Pascal's triangle*. Note that the numbers 1, 2, and 1 in the third row from the top are the binomial coefficients in the expansion of $(a + b)^2$. Similarly, the numbers 1, 5, 10, 10, 5, and 1 in the sixth row are the binomial coefficients of $(a + b)^5$. Also observe that in the fourth row from the top, 6 is the sum of 3 and 3, the two numbers above it. In the fifth row, 10 is the sum of 6 and 4, the two numbers above it, and so on. Can you write the next two lines of Pascal's triangle?

HISTORICAL NOTE

PASCAL'S TRIANGLE

The triangular array of numbers was named after Blaise Pascal (1623–1662), a French scientist and religious philosopher, who wrote a treatise on the triangular array. However, Pascal's triangle was known to mathematicians even earlier. In 1303, the Chinese mathematician Chu-Shi-kie wrote of these triangular arrays.

EXERCISES 12.2

Evaluate each binomial coefficient in Exercises 1–4.

1. $\binom{9}{6}$ **2.** $\binom{8}{3}$ **3.** $\binom{12}{9}$ **4.** $\binom{11}{7}$

In Exercises 5–14, expand each binomial using the binomial theorem.

5. $(a + b)^4$

6. $(a + b)^7$

7. $(x + y)^6$

8. $(x + y)^8$

9. $(2a - b)^5$

10. $(3a + 2b)^5$

11. $(x - 2)^5$

12. $(2x - 5)^4$

13. $\left(a - \dfrac{1}{a}\right)^6$

14. $\left(3m - \dfrac{2}{m}\right)^5$

In Exercises 15–28, find the required quantity.

15. Find the fifth term in the expansion of $(2x - a)^7$.

16. Find the seventh term in the expansion of $(3x + 4a)^9$.

17. Find the eighth term in the expansion of $\left(\dfrac{u}{2} - 2\right)^{10}$.

18. Find the tenth term in the expansion of $\left(2a + \dfrac{b}{2}\right)^{12}$.

19. Find the term independent of y in the expansion of $\left(2y - \dfrac{1}{2y}\right)^8$.

20. Find the term independent of x in the expansion of $\left(\dfrac{3x}{4} - \dfrac{4}{3x}\right)^{10}$.

21. Find the middle term in the expansion of $(a^{1/4} + b^{1/4})^8$.

22. Find the middle term in the expansion of $(u^{1/2} - \sqrt{2})^{12}$.

23. Find the two middle terms in the expansion of $(xy - a)^7$.

24. Find the two middle terms in the expansion of $(3x - 2)^5$.

25. Find the term involving b^6 in the expansion of $(a - 3b^2)^5$.

26. Find the term involving x^3 in the expansion of $\left(x^2 - \dfrac{1}{x}\right)^6$.

27. Use the first five terms in the binomial expansion of $(1 + 0.01)^8$ to approximate $(1.01)^8$. Use your calculator to compute $(1.01)^8$ and compare the answers.

28. Approximate $(0.99)^{10}$ using the first five terms in the binomial expansion of $(1 - 0.01)^{10}$. Compute $(0.99)^{10}$ by using your calculator and compare your results.

In Exercises 29–32, use the binomial expansion to approximate each number to three decimal places.

29. $(1.99)^6$

30. $(2.01)^8$

31. $(1.02)^{10}$

32. $(0.98)^{10}$

In Exercises 33–40, solve the given problem.

33. Find all positive integers for which the following relation holds:

$$12n! = (n + 2)!.$$

34. For what positive integers n does the following relation hold?

$$30(n - 2)! = n!$$

35. Find integral solutions of the equation

$$\frac{(n + 4)!}{(n + 2)!} = 30.$$

36. Solve the equation

$$\frac{(n - 5)!}{(n - 3)!} = \frac{1}{42},$$

where n is an integer.

37. Show the formula $(n + 1)! = n! \, (n + 1)$.

38. Show the formula $(n + 1)! = (n - 1)! \, (n^2 + n)$.

39. Show that $\binom{n}{0} = \binom{n + 1}{0}$.

40. Show that $\binom{n}{n} = \binom{n + 1}{n + 1}$.

12.3   ## SEQUENCES AND SERIES

As we discussed in Chapter 4, a function is a rule that assigns to each element x of a set X a unique element y of a set Y. A *sequence* is a particular kind of function that we now define.

Sequences

Let Y be an arbitrary set. A *sequence of elements of Y* is a function $f\colon \mathbb{N} \to Y$ whose domain is the set $\mathbb{N}$ of all natural numbers.

The element $f(1)$ of Y is called the *first term* of the sequence, $f(2)$ is the *second term*, and so on. The element $f(n)$ is said to be the *nth term* of the sequence.

It is customary to denote $f(n)$ (the image of n under f) by a letter with a subscript. Thus we may write $f(n) = a_n$, where the subscript n is the argument of the function f. By doing this, we can use any one of the following notations to indicate the sequence f:

$$a_1, a_2, \ldots, a_n, \ldots,$$

$$\{a_n\}_{n=1}^{\infty}, \quad \text{or simply} \quad \{a_n\}.$$

We also say that a_n is the *general term* of the sequence $\{a_n\}$.

The set Y where f takes its values is arbitrary. However, we shall consider only *sequences of real numbers*. These correspond to the case when $Y = \mathbb{R}$.

EXAMPLE 1

Write the first five terms of each sequence.
(a) $f(n) = n^2$ **(b)** $\{(-1)^n 2^n\}$

Solution: As $n = 1, 2, 3, 4$, and 5, we get
(a) $1, 4, 9, 16, 25$; **(b)** $-2, 4, -8, 16, -32$.

Practice Exercise 1

Write the first four terms of each sequence.

(a) $\left\{\dfrac{n}{n+1}\right\}$ **(b)** $f(n) = \dfrac{n}{n^2 + 1}$

Answer: **(a)** $\dfrac{1}{2}, \dfrac{2}{3}, \dfrac{3}{4}, \dfrac{4}{5}$ **(b)** $\dfrac{1}{2}, \dfrac{2}{5}, \dfrac{3}{10}, \dfrac{4}{17}$

Recurrence Relation

In Example 1 and Practice Exercise 1, the sequences were given by an explicit formula. This may not always be the case, as the next example shows.

EXAMPLE 2

Write the first five terms of the sequence defined by

$$a_1 = 3, \quad a_k = a_{k-1} + 5, \quad \text{for all } k \geq 2.$$

Solution: The expression $a_k = a_{k-1} + 5$, $k \geq 2$, is called a *recurrence relation*. It allows us to determine the kth term of the sequence when the $(k-1)$st term is known. Since $a_1 = 3$, then

$$a_2 = a_1 + 5 = 3 + 5 = 8.$$

Also,

$$a_3 = a_2 + 5 = 8 + 5 = 13,$$

$$a_4 = a_3 + 5 = 13 + 5 = 18,$$
$$a_5 = a_4 + 5 = 18 + 5 = 23.$$

Thus the five terms of the sequence are 3, 8, 13, 18, and 23.

In this example it is possible to find an explicit expression for a_n, the nth term of the sequence. To see this, we observe that if $n = 2$, 3, and 4, then

$$a_2 = a_1 + 5,$$
$$a_3 = a_2 + 5 = a_1 + 5 + 5 = a_1 + 2 \cdot 5,$$
$$a_4 = a_3 + 5 = a_1 + 2 \cdot 5 + 5 = a_1 + 3 \cdot 5.$$

These relations seem to indicate that

$$a_n = a_1 + (n - 1) \cdot 5, \quad \text{for all } n \geq 2.$$

Or, since $a_1 = 3$,

$$a_n = 3 + (n - 1) \cdot 5, \quad \text{for all } n \geq 2.$$

This result can be proved by induction, and the proof is left as an exercise.

Practice Exercise 2   Write the first five terms of the sequence defined by

$$a_1 = 1, \quad a_k = \frac{1}{2} a_{k-1}, \quad k \geq 2.$$

Write the nth term of this sequence.

Answer: $a_1 = 1, a_2 = \dfrac{1}{2}, a_3 = \dfrac{1}{4}, a_4 = \dfrac{1}{8}, a_5 = \dfrac{1}{16}; a_n = \dfrac{1}{2^{n-1}}$, for all $n \geq 1$

MATHEMATICAL VIGNETTE

DAISIES AND FIBONACCI NUMBERS

The following is an interesting recursion formula:

$$x_n = x_{n-1} + x_{n-2}, \quad \text{for } n \geq 2.$$

If $x_0 = 1$ and $x_1 = 1$, then this recursion formula gives us the sequence of numbers

$$1, 1, 2, 3, 5, 8, 13, 21, 34, 55, 89, \ldots,$$

called *Fibonacci numbers,* after Leonardo Fibonacci, an Italian mathematician of the late twelfth and early thirteenth century.

The discovery of Fibonacci numbers throughout nature is a fascinating subject. Botanists have long observed that flower heads of daisies and sunflowers are composed of hundreds of tiny *florets,* arranged in superimposed rows of clockwise and counterclockwise spirals. The numbers of spirals going in each direction are, in almost every case, adjacent Fibonacci numbers. For example, it is very common to find daisies with 21 spirals going one way and 34 going the other. In many varieties of pine cones, the numbers of spirals of scales exhibit, in pairs, the Fibonacci numbers 3, 5, 8, and 13. The eyes of pineapples, also distributed in sets of spirals, are found in arrangements of the Fibonacci numbers 5, 8, 13, 21, and 34.

Adapted from Stephen R. Braun, "Botany With a Twist," *Science 86,* (May 1986).

Newton's Method for Approximating Square Roots

An important example of a sequence defined by a recurrence relation is provided by *Newton's method for approximating square roots* of a nonnegative real number. It is proved in calculus that the recurrence relation

$$x_n = \frac{1}{2}\left(x_{n-1} + \frac{a}{x_{n-1}}\right), \; n \geq 2. \tag{12.7}$$

defines a sequence of approximations for the square root of a nonnegative number a. We start the sequence of approximations by choosing for x_1 any reasonable approximation of the square root of a. The number x_1 is our *initial guess*. Once x_1 is selected, formula (12.7) gives us x_2, x_3, and so on. These numbers will be better approximations for the square root of a.

The method, named in honor of Isaac Newton (1642–1727), an English mathematician, is very effective and quickly yields decimal approximations to any desired accuracy.

◨ EXAMPLE 3   Using Newton's method, find four approximations of $\sqrt{2}$, taking $x_1 = 1.5$ as an initial guess.

Solution: If $x_1 = 1.5$, then using (12.7) we get

$$x_2 = \frac{1}{2}\left(1.5 + \frac{2}{1.5}\right) = 1.4166667.$$

Next,

$$x_3 = \frac{1}{2}\left(1.4166667 + \frac{2}{1.4166667}\right) = 1.4142157$$

and, finally

$$x_4 = \frac{1}{2}\left(1.4142157 + \frac{2}{1.4142157}\right) = 1.4142136.$$

This result already coincides with the approximation for $\sqrt{2}$ given by our calculator. A better approximation is 1.414213562. By comparing these approximations we can see how effective Newton's method is: with only three computations we have obtained an approximation for $\sqrt{2}$ accurate to six decimal places.

Using Your Calculator

If your calculator has a memory, you can generate a sequence of approximations for the square root of 2 (or any other positive number) in a very simple way. First, store the initial guess in the memory and, by recalling it whenever necessary, perform the computations indicated in formula (12.7) to obtain the second approximation of $\sqrt{2}$. Next, store the obtained result and repeat the same operations to obtain the third approximation, and so on. The following keystrokes were implemented in a TI-30-II (a similar procedure could be adapted to other calculators). After entering the initial guess of 1.5, storing it, and pressing the following keys:

$$\boxed{\text{RCL}}\,\boxed{+}\,2\,\boxed{\div}\,\boxed{\text{RCL}}\,\boxed{=}\,\boxed{\div}\,2\,\boxed{=}\,\boxed{\text{STO}},$$

we obtained the display 1.4166667, which was stored in the memory. Repeating the same keystrokes, we obtained the third element of the approximating sequence, 1.4142157, and so on.

Practice Exercise 3   Use Newton's method to find the first four approximations for $\sqrt{3}$. Take as your first guess $x_1 = 1.6$.

Answer: $x_1 = 1.6$, $x_2 = 1.7375$, $x_3 = 1.7320594$, $x_4 = 1.7320508$

HISTORICAL NOTE

SIR ISAAC NEWTON

Sir Isaac Newton (1642–1727) was an English mathematician, physicist, natural philosopher, and one of the greatest figures in the entire history of science. His scientific work contained major contributions to mathematics and experimental and theoretical physics, and it profoundly influenced eighteenth-century thought.

Newton's most important discoveries—including the development of differential calculus, the formulation of the laws of gravitation, and the discovery that white light is made up of the colors of the spectrum—were made in the two-year period of 1665 and 1666.

Newton also formulated the laws of motion, invented the reflecting telescope, and made important contributions to optics and astronomy.

Series

With every sequence $\{a_n\}$ we can associate another sequence $\{s_n\}$, whose general term is defined by the sum

$$s_n = a_1 + a_2 + \cdots + a_n = \sum_{j=1}^{n} a_j.$$

Conversely, given a sequence $\{s_n\}$, if we set

$$a_1 = s_1 \quad \text{and} \quad a_n = s_n - s_{n-1} \quad \text{for } n \geq 2,$$

we obtain a new sequence $\{a_n\}$ such that $s_n = a_1 + a_2 + \cdots + a_n$.

Series

A *series* is a pair of sequences $\{a_n\}$, $\{s_n\}$ whose elements a_n and s_n are linked by the relations $s_n = a_1 + a_2 + \cdots + a_n$, for all n. The element a_n is called the *nth term* and the element s_n the *nth partial sum* of the series. The sequence $\{a_n\}$ is the *generating sequence* and $\{s_n\}$ is called the *sequence of partial sums* of the series.

Very often when no confusion is possible, instead of referring to a series as a pair of sequences, we will simply say the *series of general term a_n*, or the *series whose terms are $a_1, a_2, \ldots, a_n, \ldots$*.

EXAMPLE **4**   Find the first four partial sums of the series whose terms are $1, 2, \ldots, n, \ldots$ (the sequence of all natural numbers). Find a general expression for the nth partial sum.

Solution: We have

$$s_1 = 1,$$
$$s_2 = 1 + 2 = 3,$$
$$s_3 = 1 + 2 + 3 = 6,$$
$$s_4 = 1 + 2 + 3 + 4 = 10.$$

The nth partial sum is

$$s_n = 1 + 2 + \cdots + n,$$

which, from formula (12.1), is equal to

$$s_n = \frac{n(n+1)}{2}.$$

Practice Exercise 4   Find the first four partial sums of the series whose terms are $2, 4, 6, \ldots, 2n, \ldots$ (the sequence of all even numbers). Also find s_n, the nth partial sum.

Answer: $s_1 = 2$, $s_2 = 6$, $s_3 = 12$, $s_4 = 20$; $s_n = n(n+1)$.

EXAMPLE 5   Find a general expression for the nth partial sum of the series with general term $\dfrac{1}{n(n+1)}$.

Solution: We have

$$s_n = \frac{1}{1 \cdot 2} + \frac{1}{2 \cdot 3} + \cdots + \frac{1}{(n-1)n} + \frac{1}{n(n+1)}.$$

Since

$$\frac{1}{n(n+1)} = \frac{1}{n} - \frac{1}{n+1},$$

we can rewrite s_n as follows:

$$s_n = \left(\frac{1}{1} - \frac{1}{2}\right) + \left(\frac{1}{2} - \frac{1}{3}\right) + \cdots + \left(\frac{1}{n-1} - \frac{1}{n}\right) + \left(\frac{1}{n} - \frac{1}{n+1}\right).$$

Simplifying, we obtain

$$s_n = 1 - \frac{1}{n+1} = \frac{n}{n+1}.$$

Practice Exercise 5   Find s_n the nth partial sum of the series with general term $n(n+1)$. (*Hint:* See Exercise 3 of Section 12.1.)

Answer: $s_n = \dfrac{n(n+1)(n+2)}{3}$

Explicit Expression for the Partial Sums of a Series

Whenever possible, it is very convenient to have an explicit expression for the nth partial sum of a series. For example, we saw in Example 4 that

$$s_n = \frac{n(n + 1)}{2}$$

is the nth partial sum of the series with general term n. Thus if we wish to find the partial sum

$$1 + 2 + 3 + \cdots + 25,$$

we use the formula for s_n with $n = 25$, which reduces our computations to a minimum:

$$1 + 2 + 3 + \cdots + 25 = s_{25} = \frac{25 \cdot 26}{2} = 325.$$

As another example, suppose that we want to compute the sum

$$\sum_{j=1}^{19} \frac{1}{j(j + 1)}.$$

According to Example 5, we have

$$s_n = \sum_{j=1}^{n} \frac{1}{j(j + 1)} = \frac{n}{n + 1}.$$

Thus

$$s_{19} = \frac{19}{20} = 0.95.$$

■ **Important remark**   Unfortunately, in most cases it is difficult, if not impossible, to find an explicit formula for the nth partial sum of a series. However, for two important series, *arithmetic* and *geometric series*, it is always possible to find explicit formulas for the nth partial sum. Arithmetic and geometric sequences and series will be discussed in detail in the next two sections.

EXERCISES 12.3

Write the first five terms of each of the sequences in Exercises 1–10.

1. $a_n = 5 - 2n$

2. $a_n = 3n - 7$

3. $a_n = (-1)^n n^2$

4. $a_n = \frac{(-1)^n n}{n + 1}$

5. $a_n = (-3)^n$

6. $a_n = 3(2^{-n})$

7. $a_n = \left(1 + \frac{1}{n}\right)^n$

8. $a_n = \frac{(-1)^{n+1}}{n^2}$

9. $a_n = \frac{(-1)^n 2^n}{2^n - 1}$

10. $a_n = 1 + (-1)^n$

In Exercises 11–16, find the general term and write the first five terms of each sequence.

11. All positive even numbers

12. All negative odd numbers

13. All negative integral multiples of 4

14. All positive integers divisible by 5

15. All positive integers that, when divided by 3, have remainder 2

16. All positive integers with remainder 1 when divided by 4

The sequences in Exercises 17–28 are defined by recurrence relations. Find the first five terms of each sequence.

17. $a_1 = -5, \ a_k = 2a_{k-1} + 7$ **18.** $a_1 = 2, \ a_k = a_{k-1} + 3$

19. $a_1 = 0, \ a_k = 4 - 3a_{k-1}$ **20.** $a_1 = 2, \ a_k = \frac{a_{k-1}}{4}$

21. $a_1 = 3, \ a_k = \frac{1}{a_{k-1}^2}$ **22.** $a_1 = 7, \ a_k = \frac{k}{a_{k-1}}$

23. $a_1 = 3$, $a_k = (k - 1)a_{k-1}$ **24.** $a_1 = 1$, $a_k = 2a_{k-1}$

25. $a_1 = 2$, $a_k = \dfrac{3}{4} a_{k-1}$ **26.** $a_1 = 3$, $a_k = \dfrac{1}{3} a_{k-1}$

27. $a_1 = 1$, $a_k = a_{k-1} + 4$ **28.** $a_1 = 3$, $a_k = 5 + a_{k-1}$

In Exercises 29–36, the nth partial sum s_n of a series is given. Find a_n, the nth term of the series.

29. $s_n = 2^{n+1} - 2$ **30.** $s_n = 1 - \left(\dfrac{1}{2}\right)^n$

31. $s_n = n(n + 1)$ **32.** $s_n = n^2$

33. $s_n = \dfrac{n(n + 1)}{4}$ **34.** $s_n = (2n - 1)^2$

35. $s_n = 3^n - 1$ **36.** $s_n = \dfrac{n(5n - 1)}{2}$

C In Exercises 37–40, use Newton's method together with the given initial guess to find four approximations for each square root.

37. $\sqrt{5}$, $x_1 = 2$ **38.** $\sqrt{6}$, $x_1 = 3$

39. $\sqrt{10}$, $x_1 = 3.2$ **40.** $\sqrt{11}$, $x_1 = 3.3$

12.4   ## ARITHMETIC SEQUENCES AND SERIES

Arithmetic Sequences

A sequence $\{a_n\}$ is called an *arithmetic sequence* if there is a number d such that

$$a_{n+1} - a_n = d$$

for every $n \in \mathbb{N}$.

The number d is called the *common difference* of the sequence. Arithmetic sequences are also called *arithmetic progressions*.

The simplest example of an arithmetic sequence is $1, 2, \ldots, n, \ldots$, the sequence of all natural numbers, where $d = 1$. The sequence

$$13, 18, 23, \ldots, 5n - 2, \ldots, \quad \text{for } n \geq 3,$$

(see Example 2 of Section 12.3) is an arithmetic sequence with $d = 5$.

If $\{a_n\}$ is an arithmetic sequence whose common difference is d, then

$$a_{n+1} = a_n + d, \quad \text{for all } n \in \mathbb{N}.$$

This is a recurrence relation that yields the successive elements of the sequence, given a_1, such as

$$a_2 = a_1 + d,$$
$$a_3 = a_2 + d = a_1 + 2d,$$
$$a_4 = a_3 + d = a_1 + 3d.$$

It can be proved, by mathematical induction, that the general term of an arithmetic sequence is given by

$$a_n = a_1 + (n - 1)d. \tag{12.8}$$

EXAMPLE 1

If 9 and 12 are the fifth and the sixth terms of an arithmetic progression, find the first and the tenth terms of the progression.

Solution: We have $a_5 = 9$ and $a_6 = 12$, so

$$d = a_6 - a_5 = 12 - 9 = 3.$$

From (12.8) with $n = 5$, $a_5 = 9$, and $d = 3$, we have

$$9 = a_1 + (5 - 1) \cdot 3$$
$$9 = a_1 + (4) \cdot 3$$
$$a_1 = -12 + 9 = -3.$$

Next, we find a_{10}:

$$\begin{aligned} a_{10} &= a_1 + (10 - 1)d \\ &= (-3) + (10 - 1)3 \\ &= -3 + 9 \cdot 3 \\ &= 24. \end{aligned}$$

Practice Exercise 1

Suppose that 43 and 40 are the fourth and fifth terms of an arithmetic sequence, respectively. Find the first and the fifteenth terms of the sequence.

Answer: $a_1 = 52$, $a_{15} = 10$

EXAMPLE 2

If the third and seventh terms of an arithmetic sequence are 6 and -2, respectively, find the twentieth term of the sequence.

Solution: We have $a_3 = 6$ and $a_7 = -2$. Using (12.8), we get a linear system of equations in a_1 and d:

$$\begin{cases} 6 = a_1 + (3 - 1)d \\ -2 = a_1 + (7 - 1)d \end{cases}$$

The solution is $a_1 = 10$ and $d = -2$. We can now find the term a_{20}:

$$\begin{aligned} a_{20} &= 10 + (20 - 1) \cdot (-2) \\ &= 10 + 19 \cdot (-2) \\ &= -28. \end{aligned}$$

Practice Exercise 2

Let 23 and 33 be the fifth and tenth terms of an arithmetic sequence, respectively. Find the eighteenth term of the sequence.

Answer: $a_{18} = 49$

A Property of Arithmetic Sequences

The following property of the terms of an arithmetic sequence will be used later. If $n > 1$ is a fixed natural number and r is such that $1 \leq r \leq n - 1$ and $2r < n$, write the first n elements of the sequence as follows:

$$\underbrace{a_1, \ldots, a_r}_{r \text{ terms}}, a_{r+1}, \ldots, a_{n-r}, \underbrace{a_{n-r+1}, \ldots, a_n}_{r \text{ terms}}.$$

There are r terms to the left of a_{r+1} and r terms to the right of a_{n-r}. We want to show that

$$a_{r+1} + a_{n-r} = a_1 + a_n, \quad \text{for all } 1 \le r \le n - 1. \qquad (12.9)$$

According to (12.8), we have

$$a_{r+1} = a_1 + rd,$$
$$a_{n-r} = a_1 + (n - r - 1)d.$$

By adding both expressions and simplifying, we get

$$
\begin{aligned}
a_{r+1} + a_{n-r} &= (a_1 + rd) + [a_1 + (n - r - 1)d] \\
&= a_1 + rd + a_1 + (n - 1)d - rd \\
&= a_1 + [a_1 + (n - 1)d] \\
&= a_1 + a_n,
\end{aligned}
$$

which is formula (12.9).

Arithmetic Series

An *arithmetic series* is a series whose generating sequence is an arithmetic sequence. If $a_1, a_2, \ldots, a_n, \ldots$ is an arithmetic sequence with common difference d, then the nth partial sum of the series is given by

$$s_n = \frac{(a_1 + a_n)n}{2}. \qquad (12.10)$$

To prove formula (12.10), we write s_n in two different ways:

$$s_n = a_1 + a_2 + \cdots + a_{n-1} + a_n$$

and

$$s_n = a_n + a_{n-1} + \cdots + a_2 + a_1.$$

By adding term by term and using the commutativity and associativity of addition, we obtain

$$2s_n = (a_1 + a_n) + (a_2 + a_{n-1}) + \cdots + (a_{n-1} + a_2) + (a_n + a_1).$$

By virtue of (12.9), each term between parentheses on the right-hand side is equal to $a_1 + a_n$. Since there are n such terms, we have

$$2s_n = (a_1 + a_n) \cdot n;$$

hence (12.10),

$$s_n = \frac{(a_1 + a_n) \cdot n}{2}.$$

Alternatively, substituting $a_n = a_1 + (n - 1)d$ into the last expression yields the formula

$$s_n = [2a_1 + (n - 1)d]\frac{n}{2}. \qquad (12.11)$$

EXAMPLE 3

Find the sum of all the odd numbers between 25 and 87, inclusive.

Solution: The odd numbers from 25 to 87 form an arithmetic progression with $a_1 = 25$ and $d = 2$. We use (12.8) to find the place of 87 in such progression:

$$87 = 25 + (n - 1)2,$$

so

$$n = 32.$$

Next, we use (12.10) to find s_{32}:

$$s_{32} = \frac{(25 + 87) \cdot 32}{2} = 1792.$$

Practice Exercise 3

Find the sum of all even numbers from 18 to 42 inclusive.

Answer: 390

EXAMPLE 4

Find the sum of the first 28 terms of an arithmetic progression if the first term is 105 and the common difference is -3.

Solution: Use (12.11) with $n = 28$, $a_1 = 105$, and $d = -3$ to obtain

$$s_{28} = [2 \cdot 105 + (28 - 1)(-3)]\frac{28}{2}$$
$$= (210 - 81) \cdot 14$$
$$= 1806.$$

Practice Exercise 4

The first term of an arithmetic sequence is 25 and the common difference is 4. Find the sum of the first 30 terms of the sequence.

Answer: 2490

EXERCISES 12.4

In Exercises 1–6, find the common difference and determine the next two terms of each arithmetic progression.

1. 2, 4, 6, . . .

2. 3, 6, 9, . . .

3. 2, 0.5, -1, . . .

4. $\frac{1}{3}, \frac{5}{6}, \frac{4}{3}, \ldots$

5. log 2, log 4, log 8, . . .

6. $\log 10^{-1}$, $\log 10^{-2}$, $\log 10^{-3}$, ...

For each of the arithmetic sequences in Exercises 7–12, find d and a_n.

7. 15, 19, 23, ...

8. 21, 28, 35, ...

9. -13, -2, 9, ...

10. 6, -6, -18, ...

11. a, $a + 2r$, $a + 4r$, ...

12. a, $a - r^2$, $a - 2r^2$, ...

In Exercises 13–16, find the required term(s).

13. If 4 and 7 are the first two terms of an arithmetic sequence, find the fifteenth term.

14. Find the tenth term of an arithmetic progression whose first and second terms are -3 and 5, respectively.

15. If 15 and 6 are the second and fifth terms of an arithmetic progression, find the first and the ninth terms.

16. The third and the fifth terms of an arithmetic sequence are 12 and 20, respectively. Find the eleventh term.

In Exercises 17–20, a_1 and d are the first term and the common difference of an arithmetic progression. Find s_n for each given n.

17. $a_1 = -10$, $d = 4$, $n = 12$

18. $a_1 = 42$, $d = -3$, $n = 15$

19. $a_1 = \frac{1}{2}$, $d = \frac{1}{3}$, $n = 16$

20. $a_1 = \frac{2}{3}$, $d = -\frac{3}{4}$, $n = 21$

In Exercises 21–26, find each sum.

21. $\displaystyle\sum_{n=1}^{25} (2n + 3)$

22. $\displaystyle\sum_{n=1}^{30} (3n + 5)$

23. $\displaystyle\sum_{n=1}^{30} (4 - 3n)$

24. $\displaystyle\sum_{n=1}^{22} (12 - 4n)$

25. $\displaystyle\sum_{n=1}^{15} \left(\frac{n}{2} + 3\right)$

26. $\displaystyle\sum_{n=1}^{21} \left(\frac{3n}{4} - 2\right)$

In Exercises 27–40, solve the given problem.

27. Find the sum of all even numbers between 21 and 95.

28. Find the sum of all odd numbers between 36 and 112.

29. How many multiples of 4 are there between 15 and 94? Find their sum.

30. How many natural numbers divisible by 5 are there between 27 and 108? Find their sum.

31. A boy saves 10¢ on July 1, 20¢ on July 2, 30¢ on July 3, and so on. How much does he save during July?

32. A clock strikes once at 1:00, twice at 2:00, and so on. Assuming that the clock does not strike on the half-hour, how many times does the clock strike between 12:30 A.M. and 12:30 P.M.?

33. A pile of logs has 20 logs in the bottom layer, 19 in the second layer, and so on, until there is one log in the last layer. How many logs are there in the pile?

34. A stack of pipes has 35 pipes in the bottom row, 34 in the second, and so on, with one pipe in the top row. How many pipes are there in the stack?

35. The sum of three numbers in arithmetic sequence is 60, and the quotient of the third by the first is 7. Find the three numbers.

36. Find three numbers in arithmetic progression such that the sum of the first and the third is 40 and the product of the first and the second is 300.

37. The sum of three numbers in arithmetic progression is 33. The sum of squares of the first and third numbers is 274. Find the numbers.

38. The sum of three terms of an arithmetic sequence is 66 and the difference of squares of the third and first terms is 1056. What are the numbers?

39. Using formula (12.10), show that the sum of the first n odd natural numbers is n^2.

40. Using formula (12.10), show that the sum of the first n even natural numbers is $n(n + 1)$.

12.5   ## GEOMETRIC SEQUENCES AND SERIES

Geometric Sequences

A sequence $\{a_n\}$ is called a *geometric sequence* if there is a number $r \neq 0$ such that

$$\frac{a_{n+1}}{a_n} = r, \qquad (12.12)$$

for every $n \in \mathbb{N}$.

The number r is called the *common ratio* of the sequence. Geometric sequences are also called *geometric progressions*.

The relation (12.12) can also be written

$$a_{n+1} = a_n r, \quad n \in \mathbb{N}.$$

Setting $n = 1, 2, \ldots$, we obtain the successive elements of the sequence, given the first element a_1:

$$a_2 = a_1 r,$$
$$a_3 = a_2 r = a_1 r^2,$$
$$a_4 = a_3 r = a_1 r^3,$$

and so on. This indicates that the general term of the sequence is given by

$$a_n = a_1 r^{n-1}. \tag{12.13}$$

This formula can be proved by induction.

As an example, suppose that n_0 denotes the initial number of bacteria in a culture where the number of bacteria is doubling every hour. Then the numbers

$$n_0, \quad 2n_0, \quad 2^2 n_0, \quad 2^3 n_0, \ldots$$

represent the initial number of bacteria, the number of bacteria after 1 hour, after 2 hours, and so on. These numbers form a geometric sequence of ratio 2.

Similarly, the numbers

$$1, \frac{1}{2}, \frac{1}{2^2}, \frac{1}{2^3}, \ldots$$

form a geometric sequence of ratio $1/2$.

EXAMPLE 1   If the first term of a geometric progression is 2 and the ratio is 3, find the first four terms of the progression.

Solution: We have $a_1 = 2$ and $r = 3$. Thus,

$$a_1 = 2,$$
$$a_2 = 2 \cdot 3 = 6,$$
$$a_3 = 2 \cdot 3^2 = 18,$$
$$a_4 = 2 \cdot 3^3 = 54.$$

Practice Exercise 1   The common ratio of a geometric sequence is $1/4$. If the first term is 32, find the next four terms of the sequence.

Answer: $8, 2, \dfrac{1}{2}, \dfrac{1}{8}$

EXAMPLE 2   If the fourth and the ninth terms of a geometric sequence are $-1/8$ and $1/256$, find the common ratio r.

Solution: By using formula (12.13) with $n = 4$ and $n = 9$, respectively, we have:

$$-\frac{1}{8} = a_1 r^3$$

and

$$\frac{1}{256} = a_1 r^8.$$

Dividing the first equation by the second, we get

$$\frac{-1/8}{1/256} = \frac{a_1 r^3}{a_1 r^8}$$

$$-32 = \frac{1}{r^5}$$

$$r^5 = -\frac{1}{32}.$$

Therefore

$$r = -\frac{1}{2}.$$

Practice Exercise 2   The seventh term of a geometric sequence is 405 and the tenth term is 10935. Find the common ratio of the sequence.

Answer: 3

Geometric Series

A *geometric series* is a series whose generating sequence is a geometric sequence. If $a_1, a_2, \ldots, a_n, \ldots$ is a geometric sequence with common ratio $r \neq 1$, then the nth partial sum of the series is given by

$$s_n = \frac{a_1(1 - r^n)}{1 - r}. \qquad (12.14)$$

This formula can be proved by induction (Exercise 50 of Section 12.1) or as follows. Let us write

$$s_n = a_1 + a_1 r + a_1 r^2 + \cdots + a_1 r^{n-1} = \sum_{k=1}^{n} a_1 r^{k-1}.$$

Multiplying both sides by r, we obtain

$$r s_n = a_1 r + a_1 r^2 + \cdots + a_1 r^n.$$

Subtracting this equation from the previous one and simplifying, we obtain

$$s_n - r s_n = (a_1 + a_1 r + \cdots + a_1 r^{n-1}) - (a_1 r + \cdots + a_1 r^{n-1} + a_1 r^n)$$
$$= a_1 - a_1 r^n$$

or

$$s_n(1 - r) = a_1(1 - r^n).$$

Hence

$$s_n = \frac{a_1(1 - r^n)}{1 - r},$$

which is formula (12.14).

◯ EXAMPLE 3   Find the sum of the first 15 terms of the geometric progression whose first term is 1 and whose ratio is 2.

Solution: Apply formula (12.14) with $n = 15$, $a_1 = 1$, and $r = 2$, to obtain

$$s_{15} = \frac{1 \cdot (1 - 2^{15})}{1 - 2} = 2^{15} - 1 = 32{,}767.$$

◯ Practice Exercise 3   The first term of a geometric sequence is 3 and the common ratio is 0.2. Find the sum of the first five terms of the sequence.

Answer: 3.7488

The Sum of a Geometric Series

About 450 B.C. the Greek philosopher Zeno of Elea (495–435 B.C.) published a book of paradoxes as a challenge to the philosophers and mathematicians of his time. One of them can be paraphrased as follows. In order for a runner to go a certain distance, he must first run half of the distance; in order to go half the distance, he must first run a quarter of the distance; in order to go a quarter of the distance, he must first run one eighth of the distance, and so on. Since to go any distance at all the runner has an infinite number of distances to run, it appears impossible ever to begin running. Thus, concluded Zeno, motion itself is impossible!

This is clearly a paradox, since we know that motion *is* possible and that once started, the runner will eventually reach his destination. The difficulty with Zeno's paradox results, perhaps, from our perception that it is impossible to do infinitely many things in a finite length of time.

From a mathematical viewpoint, we can say that the distance of 1 mile is the "sum" of half a mile plus a quarter of a mile plus an eighth of a mile, and so on, *ad infinitum* (Figure 12.1, from right to left). Thus, we may write

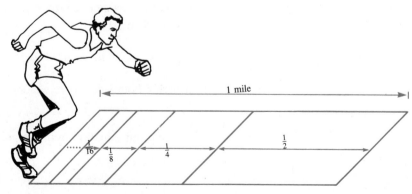

Figure 12.1

$$\frac{1}{2} + \frac{1}{4} + \frac{1}{8} + \cdots = 1,$$

where the dots indicate that we are computing a "sum" containing infinitely many terms.

Note that the addition operation is defined for only a finite number of terms, so it is necessary to define the concept of an *infinite sum*, that is, a *sum containing infinitely many terms*. The precise definition of infinite sums requires the definition of *limit of a sequence*, a topic beyond the scope of this book. However, we will justify it by using the example that we have discussed so far. Consider the geometric series whose terms are $1/2$, $1/2^2$, $1/2^3$, ... According to formula (12.14), with $a_1 = 1/2$ and $r = 1/2$, the nth partial sum of the series is

$$s_n = \frac{\frac{1}{2}\left(1 - \frac{1}{2^n}\right)}{1 - \frac{1}{2}}.$$

Simplifying, we obtain

$$s_n = \frac{\frac{1}{2}\left(1 - \frac{1}{2^n}\right)}{\frac{1}{2}} = 1 - \frac{1}{2^n},$$

for every natural number n. Now as n increases without bound, 2^n increases without bound, so that $1/2^n$ decreases to zero. We conclude that s_n approaches 1 as n increases without bound. On the other hand,

$$s_n = \frac{1}{2} + \frac{1}{4} + \frac{1}{8} + \cdots + \frac{1}{2^n}$$

and, as n increases without bound, the number of terms of this sum increases without bound. Thus we can write

$$\frac{1}{2} + \frac{1}{4} + \frac{1}{8} + \cdots + \frac{1}{2^n} + \cdots = 1 \quad \text{or} \quad \sum_{j=1}^{\infty} \frac{1}{2^j} = 1.$$

We say that 1 is *the sum of the geometric sequence whose terms are* $\{1/2^n\}$.

The same reasoning applies to any geometric sequence whose ratio is less than 1 in absolute value. In fact, if the ratio r of a geometric sequence $\{a_n\}$ is such that $|r| < 1$, then the sequence of partial sums $\{s_n\}$ approaches the number

$$\frac{a_1}{1 - r}$$

as n increases without bound.

To see this, we write (12.14) as follows:

$$s_n = \frac{a_1}{1 - r} - \frac{a_1}{1 - r} r^n.$$

Now, it can be proved that if $|r| < 1$, then r^n *approaches* 0 *as n increases without bound*. Thus the second term of the difference approaches 0 and, consequently, s_n approaches $a_1/(1 - r)$, as n increases without bound.

The Sum of a Geometric Series

Let $\{a_n\}$ be a geometric sequence whose ratio r is such that $|r| < 1$. The number

$$s = \frac{a_1}{1 - r} \qquad (12.15)$$

is called *the sum of the geometric series of general term a_n*.

Alternative ways of writing the sum of the geometric series are

$$a_1 + a_1 r + a_1 r^2 + \cdots + a_1 r^{n-1} + \cdots = \frac{a_1}{1 - r} \qquad (12.16)$$

or, using the summation notation,

$$\sum_{k=1}^{\infty} a_1 r^{k-1} = \frac{a_1}{1 - r}. \qquad (12.17)$$

The expression on the left side of (12.17) is called an *infinite sum*.

We also note that if $|r| \geq 1$, the geometric series has no sum because the term $[a_1/(1 - r)]r^n$ in the expression for s_n does not exhibit any regular behavior as n becomes large.

EXAMPLE 4   Find the infinite sum $\displaystyle\sum_{n=1}^{\infty} \left(\frac{3}{4}\right)^{n-1}$.

Solution: This is an infinite series with first term $a_1 = 1$ and ratio $r = 3/4 < 1$. According to (12.17), we obtain

$$\sum_{n=1}^{\infty} \left(\frac{3}{4}\right)^{n-1} = \frac{1}{1 - \dfrac{3}{4}} = \frac{1}{\dfrac{1}{4}} = 4.$$

Practice Exercise 4   Find the sum of the series $\displaystyle\sum_{n=1}^{\infty} \left(\frac{2}{3}\right)^n$.

Answer: 2

EXAMPLE 5   A Ping-Pong ball dropped from a height of 1 m rebounds one-half of the distance after each fall. Find the total distance the ball travels before coming to rest.

Solution: The ball travels 1 m when dropped. After the first rebound the ball travels $1/2$ m up and $1/2$ m down, that is, 1 m; after the second rebound it travels $2 \cdot 1/4$ m $= 1/2$ m; after the third rebound it travels $2 \cdot 1/8 = 1/4$ m, and so on. The total distance traveled is then given by the infinite series

$$1 + 1 + \frac{1}{2} + \frac{1}{4} + \cdots = 1 + \frac{1}{1 - \dfrac{1}{2}} \qquad \begin{matrix} \text{use (12.15) with } a_1 = 1 \\ \text{and } r = 1/2 \end{matrix}$$

$$= 1 + 2 = 3 \text{ m}.$$

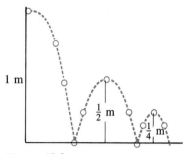

Figure 12.2

Practice Exercise 5

A soccer ball dropped from a height of 1 m rebounds 3/4 of the distance after each fall. What is the total distance traveled by the ball before coming to rest?

Answer: 7 m

Decimal Numbers as Fractions

Infinite geometric series can also be used to convert repeating decimals into rational numbers; in fact, a repeating decimal can be viewed as an expression for a certain kind of geometric series.

EXAMPLE 6

Write the repeating decimal $0.\overline{36}$ as a fraction.

Solution: Since

$$0.\overline{36} = 0.363636 \ldots$$
$$= 0.36 + 0.0036 + 0.000036 + \cdots,$$

the right-hand side is the sum of an infinite geometric series whose first term is 0.36 and whose ratio is 0.01. Using formula (12.15), we get

$$0.\overline{36} = \frac{0.36}{1 - 0.01} = \frac{0.36}{0.99} = \frac{4}{11}.$$

You can check this result by dividing 4 by 11.

Practice Exercise 6

Write the repeating decimal $0.\overline{45}$ as a fraction.

Answer: $\dfrac{5}{11}$

EXERCISES 12.5

In Exercises 1–6, find the ratio and determine the next two terms of each geometric progression.

1. 2, 4, 8, . . .

2. 3, 9, 27, . . .

3. $1, -\dfrac{1}{2}, \dfrac{1}{4}, \ldots$

4. $8, -2, \dfrac{1}{2}, \ldots$

5. $\dfrac{1}{2}, \dfrac{1}{3}, \dfrac{2}{9}, \ldots$

6. $\dfrac{2}{3}, -\dfrac{1}{2}, \dfrac{3}{8}, \ldots$

In Exercises 7–10, the first three terms of a geometric sequence are given. Find the ratio and the general term.

7. log 3, log 9, log 81, . . .

8. log 2, log 8, log 512, . . .

9. log 16, log 4, log 2, . . .

10. log 1296, log 36, log 6, . . .

In Exercises 11–18, find the required term(s).

11. If the first term of a geometric sequence is 5 and the ratio is 4, find the first four terms of the sequence.

12. The first term of a geometric sequence is -2 and the ratio is $-3/5$. Find the first four terms of the progression.

13. Find the fifth term of a geometric progression whose first two terms are 3/8 and 1/4.

14. Find the seventh term of a geometric progression whose first two terms are 6 and 9.

15. The third and sixth terms of a geometric sequence are 2 and $-1/4$, respectively. Find the first term.

16. The second and sixth terms of a geometric sequence with a positive ratio are 4 and 1/64. Find the ninth term.

C 17. Find the sum of the first ten terms of a geometric sequence whose first term is 6 and whose ratio is -2.

C 18. Find the sum of the first four terms of a geometric sequence whose first term is 100 and whose ratio is 3/4.

In Exercises 19–22, find each sum.

19. $\sum_{k=1}^{12} (-2)^{k-1}$

20. $\sum_{k=1}^{8} \left(\frac{1}{3}\right)^{k-1}$

21. $\sum_{k=1}^{10} 3^{k-1}$

22. $\sum_{k=1}^{15} \left(\frac{1}{2}\right)^{k-1}$

In Exercises 23–32, find the sum of each geometric series.

23. $1 - \frac{1}{2} + \frac{1}{4} - \frac{1}{8} + \cdots$

24. $\frac{2}{3} + \frac{4}{9} + \frac{8}{27} + \frac{16}{81} + \cdots$

25. $3 + \frac{3}{4} + \frac{3}{16} + \frac{3}{64} + \cdots$

26. $2 - \frac{2}{5} + \frac{2}{25} - \frac{2}{125} + \cdots$

27. $27 + 9 + 3 + 1 + \cdots$

28. $625 + 125 + 25 + 5 + \cdots$

29. $\sum_{n=1}^{\infty} \left(\frac{5}{6}\right)^{n-1}$

30. $\sum_{n=1}^{\infty} 2\left(-\frac{3}{4}\right)^{n-1}$

31. $\sum_{n=1}^{\infty} \left(-\frac{4}{5}\right)^{n}$

32. $\sum_{n=1}^{\infty} \left(\frac{2}{3}\right)^{n}$

In Exercises 33–40, write each repeating decimal as a fraction.

33. $0.\overline{18}$

34. $0.\overline{15}$

35. $0.\overline{63}$

36. $0.\overline{12}$

37. $0.3\overline{18}$

38. $0.4\overline{06}$

39. $0.12\overline{42}$

40. $0.25\overline{45}$

41. A tennis ball dropped from a height of 6 ft rebounds 0.8 of the distance after each fall. Determine the total distance covered by the ball before coming to rest.

42. The first swing of the bob of a pendulum is 2 ft. In each subsequent swing the bob travels 3/4 of the preceding swing. Find how far the bob will travel before coming to rest.

43. Assume that each new generation of a bug population is 1.2 times as large as the last generation. If there were 200 bugs in the first generation, how many bugs would there be in the fourth generation?

44. After being exposed to a new brand of insecticide, an insect population is decaying so that each generation is 0.6 times as large as the last generation. Suppose that there were 300 insects in the first generation. How many insects will there be in the fourth generation?

45. In a biology lab, a population of fruit flies is increasing at a rate of 30% per week. What is the population after 10 weeks if in the first week the population was 100 fruit flies?

46. If a population p_0 grows at the constant rate of $100r$ percent per year, show that the population $p(t)$ after t years is given by $p(t) = p_0(1 + r)^t$. How long will it take for the population of a certain country to double if it is increasing at a rate of 3% per year?

47. A wheel rotating at a rate of 1200 rpm is slowing down. If in each minute it rotates 1/3 as many times as in the preceding minute, find how many revolutions the wheel will make before stopping.

48. Attached at the end of a spring, a weight bobs up and down at a rate of 150 oscillations per minute. Due to air resistance, the oscillations are slowing down in such a way that in each minute, the weight oscillates only 3/4 as many times as in the previous minute. How many times will the weight oscillate before coming to a complete stop?

49. Suppose that each New Yorker regularly spends 3/4 of his or her income in New York City, and saves or spends the rest elsewhere. Suppose also that during a convention, out-of-town conventioneers spent a total of $200,000 in New York City. So, 3/4 of the total of $200,000 will be spent in New York City, another 3/4 of the spent amount will in turn be spent in New York City, and so on. What is the total amount of money spent in New York City as a result of the convention? (This is called the *multiplier effect* in economics.)

50. A company estimates that as a result of advertising, 1,000,000 people will buy a new product being marketed. It also estimates that satisfied buyers will induce 50% more people to buy the product, and that the new buyers will induce another 50% to buy the product, and so on. How many people are estimated to buy the product?

PERMUTATIONS, COMBINATIONS, AND
SET PARTITIONING

Suppose that four people are members of a school committee. In how many ways can a president and vice president be chosen? To answer this question, let us label the people as A, B, C, and D. If A is chosen to be the president, then the vice president can be chosen in *three* different ways: either B, or C, or D. If B is the president, then there are also three possible choices for a vice president: A, or C, or D, and so on. To help us count the number of all possibilities, we form a *tree diagram* as follows.

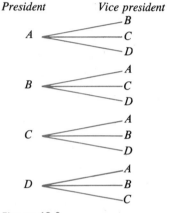

Figure 12.3

Therefore, we have a total of 12 possibilities.

This number is obtained as follows. There are four possible choices for a committee president. Once a president is selected, there are three possible choices for a vice president. Altogether the total number of choices is $4 \cdot 3 = 12$.

This example illustrates the *fundamental principle of counting*, which we now state.

The Fundamental Principle of Counting

Suppose that two operations, O_1 and O_2, are performed in order, with n_1 possible outcomes for the first operation and n_2 possible outcomes for the second operation. Then, there are $n_1 n_2$ possible combined outcomes of the first and second operations.

Suppose that k operations, $O_1, O_2, \ldots, O_k$, are performed in order, with n_1 the number of outcomes for O_1, n_2 the number for O_2, $\ldots$, n_k the number for O_k. Then,

$$n_1 \cdot n_2 \cdots n_k$$

is the total number of outcomes of the operations performed in that order.

EXAMPLE 1

Five horses are competing in a race for first or second place. In how many different ways can the race be decided, assuming that there are no ties for either first or second place?

Solution: There are five possible outcomes for first place. Once the first place is decided, there are four possible outcomes for second place. Thus, according to the fundamental principle of counting, the first and second places can be decided in $5 \cdot 4 = 20$ different ways.

Practice Exercise 1

In how many ways can we choose a chairperson and a vice chairperson from a committee of six people?

Answer: 30

EXAMPLE 2

Ten people form a committee. In how many different ways can a chairperson, a vice chairperson, and a secretary be selected?

Solution: Here we have three operations, O_1, O_2, and O_3. Operation O_1 consists of selecting a chairperson, O_2 a vice chairperson, and O_3 a secretary. There are ten different ways to select the chairperson. Once the chairperson is selected, there are nine ways that a vice chairperson can be selected. Finally, after a chairperson and vice chairperson are selected, there are eight ways to select a secretary. Thus, the total number of selections for a chairperson, vice chairperson, and secretary is $10 \cdot 9 \cdot 8 = 720$.

Practice Exercise 2

Eight athletes are competing for a gold, silver, or bronze medal in an Olympic event. How many different outcomes can the competition have?

Answer: 336

Permutations

In all the examples we have discussed so far, we have considered only *ordered arrangements* of objects. For example, if person A is elected president of the school committee and B vice president, then we have an arrangement that we may denote by (A, B). Such an arrangement is, of course, different from the arrangement (B, A), where B would be the president and A the vice president. Arrangements where the *order* distinguishes one from another are called *permutations*.

Suppose that we have a collection of n distinct objects. How many permutations of k objects $(k \le n)$ can we obtain from the original collection of n objects? The number of such permutations will be denoted by $P(n, k)$. How can we determine this number? Notice that we are performing k operations in order on a set of n objects. The number of outcomes of the first operation is n, of the second operation, $n - 1$, of the third operation, $n - 2, \ldots$, of the kth operation, $n - (k + 1)$. By the fundamental principle of counting, we obtain

$$P(n, k) = n(n - 1) \cdots (n - k + 1). \tag{12.18}$$

At this point, we recall that $(n - k)! = 1 \cdot 2 \cdots (n - k)$. Multiplying and dividing the right side of the equation in (12.18) by $(n - k)!$, we obtain

$$P(n, k) = n(n - 1) \cdots (n - k + 1)\left[\frac{(n - k)!}{(n - k)!}\right]$$

$$= \frac{n(n - 1) \cdots (n - k + 1)(n - k) \cdots 2 \cdot 1}{(n - k)!}$$

$$= \frac{n!}{(n - k)!}.$$

Thus we obtain the following formula.

Permutations of *k* Objects from a Set of *n* Objects

$$P(n, k) = \frac{n!}{(n - k)!}, \qquad 0 \le k \le n \qquad\qquad (12.19))$$

© EXAMPLE 3   Find the number of permutations of four objects from a set of 30 objects.

Solution: We have

$$P(30, 4) = \frac{30!}{(30 - 4)!}$$

$$= \frac{30!}{26!}$$

$$= \frac{30 \cdot 29 \cdot 28 \cdot 27 \cdot 26!}{26!}$$

$$= 657{,}720.$$

© **Practice Exercise 3**   What is the number of permutations of three objects from a set of 80 objects?

Answer: $\dfrac{80!}{77!} = 492{,}960$

Permutations of *n* Objects

If in formulas (12.18) and (12.19) the number k is equal to n, we obtain the number of permutations of n objects from a set of n objects:

$$P(n, n) = n(n - 1) \cdots 2 \cdot 1 = n!. \qquad\qquad (12.20)$$

Permutations of n objects from a set of n objects are called, for simplicity, *permutations of n objects*. Formula (12.20) tells us that the number of permutations of n objects is *n factorial*.

© EXAMPLE 4   Find the number of permutations of eight objects.

Solution: We have $P(8, 8) = 8! = 8 \cdot 7 \cdot 6 \cdot 5 \cdot 4 \cdot 3 \cdot 2 \cdot 1 = 40{,}320$. Note that you may obtain this number by multiplying 8 by 7, the result by 6, the result by

5, and so on until you reach 1. However, if you use a scientific calculator you can find this result with a few keystrokes. For example, the keystrokes 8 $\boxed{x!}$ will give you 40,320.

◰ Practice Exercise 4   What is the number of permutations of ten objects?

Answer: 3,628,800

Combinations

Suppose that after choosing the president and vice president of the school committee (discussed at the beginning of this section), the four committee members go out for dinner. Upon arriving at the restaurant, they are told that the only table available is a table that seats only two people. They decide to wait for a larger table. While they are waiting, one of the committee members figures out that the number of seating arrangements for four people at a table for two is 6. How was this number determined? First, notice that if A and B form a seating arrangement denoted by AB, this arrangement is the same as BA. In other words, the order of the arrangement is now immaterial. Referring to the tree diagram of Figure 12.3, you can see that the following list contains all possible seating arrangements.

$$\begin{array}{lll} AB & & \\ AC & BC & \\ AD & BD & CD \end{array}$$

Figure 12.4

There is a total of six different arrangements.

Arrangements where the order of the objects is immaterial are called *combinations*. The problem of finding the number of seating arrangements for the school committee members is the same as the problem of finding the *number of combinations of two objects from a set of four objects*. This number is denoted by $C(4, 2)$.

In general, the number of combinations of k objects from a set of n objects is denoted by $C(n, k)$. How can we find this number? First let us consider the seating arrangement BC of Figure 12.4. If we arrange the two people B and C in a given order, there are then two permutations, BC and CB. We see that taking all combinations of two objects from a set of four objects and then taking all permutations of two objects gives us the set of *all* permutations of two objects from a set of four objects. Thus

$$P(4, 2) = C(4, 2) \cdot P(2, 2),$$

so

$$C(4, 2) = \frac{P(4, 2)}{P(2, 2)} = \frac{4!}{2! \; 2!} = 6.$$

If $C(n, k)$ denotes the number of combinations of k objects from a set of n objects, and we multiply this number by $P(k, k) = k!$, the number of permutations of k objects, we obtain $P(n, k)$, the number of permutations of k objects from a set of n objects. That is,

$$P(n, k) = C(n, k) \cdot P(k, k).$$

Hence, we derive the formula

$$C(n, k) = \frac{P(n, k)}{P(k, k)}. \qquad (12.21)$$

If we replace $P(n, k)$ with $n!/(n - k)!$ and $P(k, k)$ with $k!$, formula (12.21) can be written as follows.

Combinations of k Objects from a Set of n Objects

$$C(n, k) = \frac{n!}{(n - k)! \, k!} \qquad (12.22)$$

We have seen these numbers before: in Section 12.2, page 566, we defined the binomial coefficient

$$\binom{n}{k} = \frac{n!}{k! \, (n - k)!}$$

which is the same fraction (except for the order of the factors in the denominator). Thus we have the important relation

$$C(n, k) = \binom{n}{k}.$$

EXAMPLE 5   A teacher allows each student to choose any three problems from a list of eight exam problems to work at home. How many possible choices does a student have?

Solution: Each student chooses a set of three problems (in any order) from a list of eight problems. Thus, we have to find the number of combinations of three objects from a set of eight objects:

$$C(8, 3) = \frac{8!}{(8 - 3)! \, 3!}$$

$$= \frac{8!}{5! \, 3!} = \frac{8 \cdot 7 \cdot 6 \cdot 5!}{5! \, 3!} = \frac{8 \cdot 7 \cdot 6}{3 \cdot 2} = 56.$$

Practice Exercise 5   How many four-person subcommittees can be formed from a committee of ten people?

Answer: $C(10, 4) = \dfrac{10!}{6! \, 4!} = 210$

Partitioning n Objects in k Cells

Suppose that twelve congressmen are to be assigned to three committees of three, four, and five people in such a way that each of them will be assigned to only one committee. In how many ways can this be done? Let us perform three operations in the following order. In the first operation, we form committees of three from the twelve congressmen. There are $C(12, 3)$ possible outcomes for the first operation. Once the committee of three is formed, there will be nine congressmen left. In the second operation we form committees of four from nine congressmen, obtaining $C(9, 4)$ committees. Once a committee of three and a committee of four are set up, there will be five congressmen left. Our third operation will be to form committees of five members. Since there are only five congressmen left, we can form only one committee. Note that $1 = C(5, 5)$. Now, by the fundamental principle of counting, the total number of committees of three, four, and five congressmen is

$$C(12, 3) \cdot C(9, 4) \cdot C(5, 5) = \frac{12!}{3! \; 9!} \cdot \frac{9!}{4! \; 5!} \cdot 1$$

$$= \frac{12!}{3! \; 4! \; 5!} = 27{,}720.$$

What we just did was to partition 12 objects into three cells, with one cell each of three, four, and five distinct objects. (It is important to note that $3 + 4 + 5 = 12$.) The number of all such partitions is denoted by $\Gamma(12, 3, 4, 5)$, and we have $\Gamma(12, 3, 4, 5) = 12!/3! \; 4! \; 5!$. This discussion justifies the following result, which can be derived from the fundamental principle of counting.

Partitions of n Objects in k Cells

If n objects are to be partitioned into k cells such that

1. each object belongs to only one cell;
2. the first cell contains n_1 objects, the second cell n_2 objects, . . . , the kth cell n_k objects;
3. $n_1 + n_2 + \cdots + n_k = n$;

then the number of all such partitions is

$$\Gamma(n, n_1, \ldots, n_k) = \frac{n!}{n_1! \; n_2! \ldots n_k!}. \qquad (12.23)$$

◨ EXAMPLE 6   Two people are playing a card game that uses a 40-card deck. Four cards are turned face up and each person is dealt three cards. Find how many deals are possible.

Solution: After four cards are turned face up, the remaining 36 cards are to be partitioned into three cells. Two cells correspond to the two players receiving three cards each. The 30 cards that remain correspond to the third cell. Thus we have

$$\Gamma(36, 3, 3, 30) = \frac{36!}{3! \; 3! \; 30!}$$

$$= 38{,}955{,}840.$$

ⓒ Practice Exercise 6   Four people are playing the same card game of Exercise 5. How many deals are possible if each person receives three cards?

Answer: $\Gamma(36, 3, 3, 3, 3, 24) \simeq 4.6262008 \times 10^{14}$

Distinguishable Permutations

An *anagram* is a transposition of the letters of a word to form another word. How many anagrams can be obtained from the word BOOK? First, notice that if we interchange the two letters O and keep fixed the letters B and K, a permutation of the letters B, O, O, K has been performed without changing the given word. Such a permutation is called a *nondistinguishable permutation*. However, by interchanging the last two letters, a *distinguishable permutation* is performed, giving us the anagram BOKO. Now, if all the letters were different, the number of all anagrams would be 4! = 24, the number of permutations of four objects. Since two letters are the same, it follows that the number n of anagrams that can be obtained from the word BOOK must be less than 24. To find this number, notice that to each distinguishable permutation there correspond 2! nondistinguishable permutations that are obtained by interchanging the letter O. Thus if we multiply n by 2!, we obtain the number of permutations of four objects; that is, $2! \, n = 4!$, or

$$n = \frac{4!}{2!} = 12.$$

The same reasoning applies to other cases. Suppose that we want to find the number of anagrams obtained from the word ANACONDA. There are 3! arrangements of the letter A and 2! arrangements of the letter N (a total of 3! 2! arrangements) that do not change any given anagram. Thus, if n denotes the number of all anagrams obtained from the word ANACONDA, then 3! 2! n is equal to 8!, the number of permutations of eight different objects. It follows that

$$n = \frac{8!}{3! \, 2!}.$$

These two examples illustrate the following general result.

The Number of Distinguishable Permutations

The number of distinguishable permutations of n objects, of which n_1 are equal, n_2 are equal of another kind, . . . , n_k are equal of a further kind, and such that $n = n_1 + n_2 + \cdots + n_k$, is

$$\Gamma(n, n_1, \ldots, n_k) = \frac{n!}{n_1! \, n^2! \ldots n_k!}. \qquad (12.24)$$

EXAMPLE 7   How many distinguishable permutations can be obtained with the letters of the word INDIANA?

Solution: The given word has seven letters, of which the letter A is repeated twice, the letter I is repeated twice, and the letter N is repeated twice. According to formula (12.24), the number of distinguishable permutations is

$$\Gamma(7, 2, 2, 2, 1) = \frac{7!}{2!\ 2!\ 2!\ 1!} = 630.$$

Practice Exercise 7   Find the number of anagrams of the word ARKANSAS.

Answer: $\Gamma(8, 3, 2, 1, 1, 1) = 3360$

EXERCISES 12.6

In Exercises 1–24, evaluate each number.

1. $P(6, 2)$
2. $P(8, 2)$
3. $P(8, 3)$
4. $P(6, 3)$
5. $P(6, 6)$
6. $P(7, 0)$
© 7. $P(40, 4)$
© 8. $P(36, 5)$
© 9. $P(80, 5)$
© 10. $P(120, 4)$
11. $C(5, 2)$
12. $C(7, 3)$
13. $C(9, 3)$
14. $C(12, 8)$
15. $C(10, 6)$
16. $C(8, 4)$
17. $C(40, 2)$
18. $C(40, 3)$
© 19. $C(52, 13)$
© 20. $C(48, 10)$
21. $\Gamma(10, 3, 3, 4)$
22. $\Gamma(12, 2, 2, 8)$
© 23. $\Gamma(52, 5, 5, 5, 5, 32)$
© 24. $\Gamma(52, 11, 11, 30)$

In Exercises 25–50, solve the given problem.

25. In how many ways can a president and a vice president be elected from a committee of ten people?
26. In how many ways can a secretary and a treasurer be elected from a committee of 12 people?
27. Eight people form a committee. In how many ways can a chairperson, vice chairperson, and secretary be selected?
28. Twelve people form a committee. In how many different ways can a chairperson, a vice chairperson, a secretary, and a treasurer be selected?
29. Ten athletes are competing for a gold, silver, or bronze medal in the high jump. How many outcomes can this competition have?
30. Eight horses are running for first, second, or third place. Find the number of different ways in which the first, second, and third place can be decided, assuming that there are no ties.
31. In how many ways can five students be seated in a row of five seats?
32. In how many ways can seven cashiers be assigned to seven different cash registers?
33. How many three-digit numbers can be obtained from the digits 1, 2, 3, and 4? How many four-digit numbers? (No digit can be used more than once.)
34. How many ordered arrangements containing three letters can you form by using the vowels a, e, i, o, u? How many containing five letters? (No repetitions of letters in an arrangement are allowed.)

35. If the letters in the word SMILEY are used to form a three-letter code word, find how many code words can be obtained.
36. The numbers 5, 6, 7, 8, and 9 are being used to form three- and four-digit numbers in which no digit is repeated. How many three-digit numbers can be formed? How many four-digit numbers?
37. How many five-person subcommittees can be formed from a committee of twelve people?
38. A box has space available for four books. In how many ways can eight books be arranged in the box?
39. How many seven-card hands are possible from a standard 52-card deck?
40. How many nine-card hands are possible from a standard 52-card deck?
41. A baseball league is organized with eight teams. If each team is to play every other team exactly once, find how many games must be scheduled.
42. How many line segments can you draw joining two points at a time if you are given nine points, no three of which lie on a straight line?
43. Ten members of a committee are to be assigned to three subcommittees of three, three, and four people. In how many ways can this be done if each person is to be assigned to only one subcommittee?
44. In how many ways can fifteen people be divided into three committees of 4, 5, and 6 people?
45. Four people are playing bridge, a card game where each player is dealt 13 cards from a 52-card deck. How many different deals are possible?
46. If four people are playing poker and each is dealt seven cards from a 52-card deck, find how many different deals are possible.
47. Find the number of distinguishable permutations obtained with the letters of the word MISSISSIPPI.
48. How many anagrams can you form with the letters of the word TENNESSEE?
49. Eight boys and six girls were elected to a student committee. How many four-student subcommittees can be formed? How many subcommittees consisting of two boys and two girls can be formed?
50. A student council consists of ten girls and six boys. How many six-student committees are possible? How many committees consisting of four girls and two boys can be formed?

12.7   ELEMENTARY PROBABILITY

Probability is a mathematical theory whose applications touch nearly every area of human activity. The theory of probability is used in actuarial science, medical statistics, physics and astronomy, meteorology, game theory, decision making in the business world, and so on. It deals with situations that involve an uncertain future.

In probability, an *experiment* is any process that generates well-defined *outcomes*. For example, the experiment of flipping a coin has two possible outcomes: the coin can land heads up (H) or tails up (T). The tossing of a die has six possible outcomes: the number appearing on the upper face can be 1, 2, 3, 4, 5, or 6.

In probability theory we seek answers to questions of the following type. If a coin is flipped, what are the chances of obtaining heads? If a die is tossed, what is the likelihood of rolling a 5? What is the probability of obtaining four aces if you are dealt five cards from a deck of 52 playing cards?

Throughout this section we shall discuss a few elementary ideas in probability and restrict our discussion to experiments whose outcomes are *equally likely*. That is, if a coin is flipped, the chance that it lands heads up is the same as that of landing tails up. In the experiment of tossing a die, we assume that the probability of any of the six outcomes is exactly the same. The coin and the die are then said to be *fair*.

The *sample space* of an experiment is the set of all possible outcomes. For example, the flipping of a coin has two outcomes, H and T, so that the sample space of this experiment is

$$S = \{H, T\}.$$

Analogously, the sample space for the experiment of tossing a die is

$$S = \{1, 2, 3, 4, 5, 6\}.$$

We shall consider only *finite* sample spaces.

An *event* is any subset of the sample space. For example, $E = \{H\}$, a subset of $S = \{H, T\}$, is the event of obtaining heads after flipping a coin. If you toss a die and observe a 5 appearing on the upper face, this is an event denoted by $E_1 = \{5\}$. The event $E_2 = \{1, 2, 3, 4\}$ could be interpreted as observing a number less than 5.

We now give the following definition.

Probability of an Event

Let S be a sample space and let E be an event. If $n(S)$ and $n(E)$ denote the number of elements of S and E, respectively, then the *probability of an event* E is defined by the ratio

$$P(E) = \frac{n(E)}{n(S)}. \qquad\qquad \textbf{(12.25)}$$

Since E is always a subset of S, it follows that $n(E) \leq n(S)$, so $0 \leq P(E) \leq 1$. Thus, the probability of any event is always a number less than or equal to 1.

For instance, the sample space of the experiment of flipping a coin is, as we already know, $S = \{H, T\}$, so $n(S) = 2$. If $E = \{H\}$, then $n(E) = 1$. Thus in the flipping of a coin, the probability that heads will turn up is $P(E) = 1/2$. Suppose that we are tossing a die. What is the probability that a number less than 5 will appear on the upper face of the die? As we said before, the subset $E_2 = \{1, 2, 3, 4\}$ represents the event of observing a number less than 5. Since $S = \{1, 2, 3, 4, 5, 6\}$, it follows that $P(E_2) = n(E_2)/n(S) = 4/6 = 2/3$.

EXAMPLE 1

Two dice are tossed. What is the probability that **(a)** the sum of the numbers is 5? **(b)** The sum of the numbers is 7?

Solution: Suppose that one die is white and the other is yellow, and we use ordered pairs to represent the outcome of each toss as follows. The pair $(2, 3)$ indicates that a 2 has appeared on the upper face of the white die and a 3 on the yellow die. Now, if 1 appears on the white die, we may get one of the six pairs

$$(1, 1), (1, 2) (1, 3), (1, 4), (1, 5), \text{ and } (1, 6).$$

Similarly, if a 2 shows up on the white die, there are six possibilities for the number on the yellow die:

$$(2, 1), (2, 2), (2, 3), (2, 4), (2, 5), \text{ and } (2, 6).$$

Reasoning in the same way with the numbers 3, 4, 5, and 6, we conclude that the total number of ordered pairs is 36. Thus if S is the sample space of this experiment, then $n(S) = 36$.

(a) Let E_1 be the event that the sum of the numbers is 5. Then E_1 is given by the subset

$$E_1 = \{(1, 4), (2, 3), (3, 2), (4, 1)\}.$$

Since $n(E_1) = 4$, we see that

$$P(E_1) = \frac{4}{36} = \frac{1}{9}.$$

This is the probability that the sum of the numbers is 5.

(b) The following set contains all pairs whose sum is 7:

$$E_2 = \{(1, 6), (2, 5), (3, 4), (4, 3), (5, 2), (6, 1)\}.$$

We have

$$P(E_2) = \frac{n(E_2)}{n(S)} = \frac{6}{36} = \frac{1}{6}.$$

That is, the probability is $1/6$ that the sum of the numbers is 7.

Practice Exercise 1

If two coins are flipped, find the probability that **(a)** both coins will turn up tails; **(b)** one coin will turn up heads and the other tails.

Answer: **(a)** $\dfrac{1}{4}$ **(b)** $\dfrac{1}{2}$

⊡ EXAMPLE 2

You are dealt five cards from a deck of 52 playing cards. What is the probability of getting four aces?

Solution: Denote by S the sample space of all five-card hands from the 52-card deck. Since the order of the cards in a hand is immaterial, it follows that $n(S) = C(52, 5)$. There are four aces in a 52-card deck. If the four aces are dealt, then the fifth card can be any one of the 48 remaining cards. Since the event E is getting a five-card hand with four aces, it follows that $n(E) = 48$. Thus

$$P(E) = \frac{n(E)}{n(S)} = \frac{48}{C(52, 5)}$$

$$= \frac{48 \cdot (5! \; 47!)}{52!} \approx 1.85 \times 10^{-5}.$$

⧉ Practice Exercise 2   An experiment consists of drawing five cards from a deck of 52 playing cards. What is the probability of getting three jacks?

Answer: $\dfrac{C(4, 3)C(48, 2)}{C(52, 5)} \approx 1.736 \times 10^{-3}$

Mutually Exclusive Events

Suppose that you are tossing two dice. Let E_1 be the event of obtaining 5 as the sum of the numbers, and let E_2 be that of obtaining the sum 7. These two events are *mutually exclusive:* you cannot obtain the sum 5 *and* the sum 7 at the same time. As we saw in Example 1,

$$E_1 = \{(1, 4), (2, 3), (3, 2), (4, 1)\},$$
$$E_2 = \{(1, 6), (2, 5), (3, 4), (4, 3), (5, 2), (6, 1)\},$$

and these two sets have no elements in common.

In general, let S be the sample space of an experiment and let E_1 and E_2 be two events associated with the experiment. The events E_1 and E_2 are said to be *mutually exclusive* if the sets E_1 and E_2 have no elements in common. We also say that the sets E_1 and E_2 are *disjoint*.

Let E be the *union* of these sets, denoted by $E = E_1 \cup E_2$. Since $n(E) = n(E_1) + n(E_2)$, it follows that

$$P(E) = \frac{n(E)}{n(S)} = \frac{n(E_1) + n(E_2)}{n(S)} = \frac{n(E_1)}{n(S)} + \frac{n(E_2)}{n(S)},$$

so

$$P(E) = P(E_1) + P(E_2).$$

Thus the probability of the union of two mutually exclusive (or disjoint) events is the sum of their respective probabilities. Since this result depends upon an obvious fact about counting—that the number of elements in the union of several disjoint sets is the sum of the number of elements in the respective sets—it generalizes to any finite family of mutually exclusive events, as follows.

The Probability of a Union of Mutually Exclusive Events

Let $E_1, E_2, \ldots, E_k$ be mutually exclusive events associated with a certain experiment. If

$$E = E_1 \cup E_2 \cup \cdots \cup E_k,$$

then

$$P(E) = P(E_1) + P(E_2) + \cdots + P(E_k). \qquad (12.26)$$

EXAMPLE 3

Two dice are tossed. What is the probability that the sum of the numbers is 5 or 7?

Solution: As we have already observed, the two events

$$E_1 = \{(1, 4), (2, 3), (3, 2), (4, 1)\}$$

and

$$E_2 = \{(1, 6), (2, 5), (3, 4), (4, 3), (5, 2), (6, 1)\}$$

are mutually exclusive. The event of obtaining the sum 5 or 7 is the union $E = E_1 \cup E_2$ of these events. Since $n(E_1) = 4$, $n(E_2) = 6$, and $n(S) = 36$ (see Example 1), we obtain the probability of E from (12.26)

$$P(E) = P(E_1) + P(E_2)$$
$$= \frac{4}{36} + \frac{6}{36}$$
$$= \frac{10}{36} = \frac{5}{18},$$

which is the probability that the sum of the numbers is 5 or 7.

Practice Exercise 3

One card is drawn from a 52-card deck. What is the probability of obtaining an ace or a king?

Answer: $\dfrac{4}{52} + \dfrac{4}{52} = \dfrac{2}{13}$

Odds

The probability of an event is sometimes expressed in terms of *odds*, either *odds in favor* or *odds against* the event. If p is the probability that a certain event E occurs, then $p' = 1 - p$ is the probability that the event *does not* occur.

The Odds of an Event

The odds in favor of an event E are defined by the ratio

$$\frac{p}{1 - p}.$$

The odds against E are defined by the ratio

$$\frac{p'}{1 - p'}.$$

EXAMPLE 4

What are the odds in favor of rolling a 4 with a fair die?

Solution: Let E be the event of observing a 4 on the upper face of a die. The probability that E occurs is clearly $p = p(E) = 1/6$. Now, the probability that E

does not occur is $p' = 1 - p = 5/6$. (Can you see why?) Thus the odds in favor of rolling a 4 are

$$\frac{1/6}{5/6} = \frac{1}{5}.$$

Alternatively, we can say that the odds in favor of rolling a 4 with a fair die are *one to five*. This means that over the long run, rolling a 4 will occur once for every five times it does not occur.

Practice Exercise 4   Two dice are being tossed. What are the odds in favor of the sum of the numbers being 5?

Answer: 1 to 8

EXAMPLE 5   British bookmakers posted the odds as 1 to 4 in favor of Argentina winning the 1986 World Soccer Cup in Mexico City. According to the bookmakers, what was the probability of Argentina's being the champion?

Solution: If p denotes the probability of Argentina's being the champion, then $1 - p$ is the probability against it. By the definition of odds in favor, we have

$$\frac{p}{1 - p} = \frac{1}{4}.$$

Solving this equation for p, we get

$$4p = 1 - p$$
$$5p = 1$$
$$p = \frac{1}{5} = 0.2.$$

Thus, the probability was 0.2 or 20%.

Practice Exercise 5   According to the same British bookmakers, the odds in favor of England's winning the World Soccer Cup in 1986 were 1 to 12. Find the corresponding probability.

Answer: $p = \dfrac{1}{13} \simeq 7.7\%$

EXERCISES 12.7

A six-sided die is tossed. Find the probability of each of the following events in Exercises 1–6.

1. The number showing is odd.
2. The number showing is prime.
3. The number showing is less than 5.
4. The number showing is greater than or equal to 3.

5. The number showing is less than 7.
6. The number showing is greater than 7.

Assume that two dice are tossed and we observe the sum of the numbers showing on the upper faces. Find the probability that the sum satisfies the condition in each of Exercises 7–14.

7. Equal to 6

8. Equal to 7

9. Less than or equal to 7

10. Greater than 6

11. Equal to a prime number

12. Equal to a multiple of 4

13. Equal to 7 and at least 11

14. At most 5 or equal to 10

15. If two coins are flipped, what is the probability that both coins turn up heads?

16. If two coins are flipped, what is the probability that one coin turns up heads and the other tails?

Exercises 17–20 refer to the following experiment. Marbles marked with the numbers 1, 2, 3, 4, and 5 are placed in a box. After the contents are mixed well, two marbles are simultaneously selected from the box. Find a sample space S with equally likely outcomes for this experiment. Also, find the probability of each of the following events.

17. One marble is marked with an odd number and the other with an even number.

18. Both marbles are marked with an odd number.

19. The sum of the numbers is at least 9.

20. The sum of the numbers is at most 9.

In Exercises 21–24, consider the following situation. You are taking a three-question true/false quiz, and you are guessing all the answers. Produce a sample space for all your possible answers, and find the probability of each of the following events.

21. Your three guesses are right.

22. Only one of your guesses is right.

23. At least two of your guesses are right.

24. Exactly two of your guesses are right.

25. There are five questions on a true-false test. If a student guesses the answer for each question, find the probability that **(a)** five answers are correct; **(b)** three answers are correct and two are incorrect.

26. Refer to Exercise 25. Find the probability that **(a)** two answers are correct and three are incorrect; **(b)** at least three answers are correct.

In Exercises 27–32, our experiment consists of dealing five cards from a deck of 52 playing cards.

© 27. Find the probability of being dealt five cards of the same suit.

© 28. What is the probability of being dealt four cards of the same suit?

© 29. What is the probability of being dealt three aces and two jacks?

© 30. Find the probability of being dealt two aces and three kings.

© 31. In poker, a *straight flush* is a five-card hand of the same suit and in consecutive sequence, such as the 6, 7, 8, 9, and 10

of hearts. What is the probability of being dealt a straight flush?

© 32. Referring to Exercise 31, a straight flush 10, J, Q, K, A of one suit is called a *royal flush*. Find the probability of being dealt a royal flush.

In Exercises 33–50, solve the given problems.

33. One card is drawn from a 52-card deck. What is the probability that the card is **(a)** a 5 or a jack? **(b)** a diamond or a spade?

34. Two dice are rolled. Find the probability that **(a)** the sum of the numbers is 6 or 10; **(b)** the sum of the numbers is no less than 11 or no more than 3.

35. A box contains four white, six red, and eight blue marbles. If a marble is drawn at random from the box, find the probability that **(a)** a red marble is drawn; **(b)** a white or a blue marble is drawn.

36. In Exercise 35, what is the probability of drawing **(a)** a blue marble? **(b)** a white or a red marble?

37. Denise and Pat are members of an eight-person committee. If a two-person subcommittee is to be selected at random, what is the probability that Denise and Pat will both be selected?

38. Two convention officers are to be selected from a group of five to form a subcommittee to check delegate credentials. Find the probability that Judy and Lynn (both convention officers) will be selected.

39. Three coffee varieties, Java, Mocha, and Santos, are being tasted by a person who has no ability to distinguish a difference between coffees. If the person is asked to rank the three varieties according to taste preference, what is the probability that the person will rank Santos as best?

40. In Exercise 39, what is the probability that the person will rank Santos as best and Mocha as second best?

41. Find the odds in favor of an event E, given that **(a)** $P(E) = 5/8$; **(b)** $P(E) = 0.3$.

42. What are the odds in favor of an event if its probability is **(a)** $3/5$? **(b)** 0.25?

43. Find the probability of an event E given that **(a)** the odds in favor of E are 3 to 2; **(b)** the odds against E are 5 to 3.

44. Find the probability of an event if **(a)** the odds in favor of it are 11 to 4; **(b)** the odds against it are 7 to 3.

45. Find the odds in favor of **(a)** rolling a number greater than 4 with a fair die; **(b)** drawing a diamond from a standard deck of 52 cards.

46. Find the odds against **(a)** rolling a number less than 3 with a fair die; **(b)** drawing an ace from a standard deck of 52 cards.

47. A batch of ten batteries contains three defective ones. **(a)** What is the probability of selecting a defective battery? **(b)** What are the odds in favor of selecting a good one?

48. An urn contains four black balls and eight white balls. **(a)** What is the probability of drawing a black ball? **(b)** What are the odds in favor of selecting a white ball?

49. In a wine cellar there are 15 bottles of red Burgundy, 18 of white Burgundy, 10 of red Bordeaux, and 12 of white Bordeaux. If a bottle is selected at random, what are the odds in favor of selecting a white Burgundy?

50. Given the conditions of Exercise 49, what are the odds in favor of selecting a bottle of red wine?

CHAPTER SUMMARY

Proof by induction is based upon the *principle of mathematical induction;* it is a method that you will frequently use to prove statements about natural numbers.

Principle of Mathematical Induction

Let $P(n)$ be a statement about natural numbers. If you can show that (i) $P(1)$ is a true statement, and (ii) for each natural number k, the truth of $P(k)$ implies the truth of $P(k + 1)$, then $P(n)$ is a true statement for *all* natural numbers.

Mathematical induction is used to prove the *binomial theorem.*

The Binomial Theorem

$$(a + b)^n = \sum_{k=0}^{n} \binom{n}{k} a^{n-k} b^k,$$

where the numbers

$$\binom{n}{k} = \frac{n!}{k!\,(n-k)!}, \qquad k = 0, 1, \ldots, n,$$

are the *binomial coefficients.*

The array of numbers

```
            1
          1   1
        1   2   1
      1   3   3   1
    1   4   6   4   1
  1   5  10  10   5   1
```
. . . .

obtained by writing all the binomial coefficients corresponding to $n = 0, 1, 2, \ldots$, is called *Pascal's triangle.*

Sequences are represented by $\{a_n\}$, $n = 1, 2, 3, \ldots$, where a_n is the *general term* of the sequence. Of particular importance are *arithmetic* and *geometric* sequences.

Arithmetic Sequences

$d = a_{n+1} - a_n$
(*d is the common difference*)

$a_n = a_1 + (n - 1)d$
(*formula for the general term*)

$s_n = \dfrac{(a_1 + a_n)n}{2}$
(*sum of the first n terms*)

$s_n = [2a_1 + (n - 1)d]n/2$
(*sum of the first n terms*)

Geometric Sequences

$\dfrac{a_{n+1}}{a_n} = r \neq 0$
(*r is the common ratio*)

$a_n = a_1 r^{n-1}$
(*formula for the general term*)

$s_n = \dfrac{a_1(1 - r^n)}{1 - r}$
(*sum of the first n terms*)

$s = \displaystyle\sum_{n=1}^{\infty} a_n = \dfrac{a_1}{1 - r}$
(*sum of an infinite geometric series when* $|r| < 1$)

There are many different ways to arrange n objects into sets of k objects. The following are the formulas giving the number of the different arrangements.

Permutations of *k* Objects from a Set of *n* Objects

$$P(n, k) = n(n - 1) \cdots (n - k + 1)$$

$$= \frac{n!}{(n - k)!}, \qquad 0 \leq k \leq n$$

Permutations of *n* Objects

$$P(n, n) = n(n - 1) \cdots 2 \cdot 1 = n!$$

Combinations of *k* Objects from a Set of *n* Objects

$$C(n, k) = \binom{n}{k} = \frac{n!}{(n - k)! \, k!}$$

Partitioning *n* Objects in *k* Cells; Distinguishable Permutations

$$\Gamma(n, n_1, \ldots, n_k) = \frac{n!}{n_1! \, n_2! \cdots n_k!},$$

$$(n = n_1 + n_2 + \cdots n_k)$$

In the *theory of probability*, we deal with *experiments* that have different *outcomes*. A *sample space* is the set of all possible outcomes; an *event* is a subset of a sample space.

The Probability of an Event

$$P(E) = \frac{n(E)}{n(S)}$$

$n(E)$ = number of elements of the event E

$n(S)$ = number of elements of the sample space S

Probability of Mutually Exclusive Events

$$P(E) = P(E_1) + P(E_2)$$

E_1, E_2 mutually exclusive events; $E = E_1 \cup E_2$

Odds in Favor and Odds against an Event

If p is the probability that a certain event E *occurs*, then $p' = 1 - p$ is the probability that the event *does not occur.*

Odds in favor of E: $\quad \dfrac{p}{1 - p}$

Odds against E: $\quad \dfrac{p'}{1 - p'}$

CHAPTER TEST

1. Use mathematical induction to prove that 5 is a factor of $7^n - 2^n$, for all n.
2. What is the fourth term in the expansion of $(2a + b)^7$?
3. Write the first five terms of the sequence defined by the following recurrence relation:

$$a_1 = -1, \quad a_k = a_{k-1} + 8, \quad k \geq 2.$$

4. Find the tenth term of an arithmetic sequence whose first and second terms are 5 and -4.
5. The first term of an arithmetic sequence is 8 and the common difference is -4. Find s_8, the sum of the first eight terms of the sequence.

6. What are the first four terms of a geometric sequence whose first term is -3 and whose ratio is 2?
7. Find the sum of the infinite series $\sum_{n=1}^{\infty} 3\left(\frac{1}{2}\right)^{n-1}$.
8. How many three-person subcommittees can be formed from a committee of 14 people?
9. How many four-digit numbers can be obtained from the digits 2, 3, 4, 5, 6, and 7 if no digit is used more than once?
10. A box contains eight white, four red, and eight blue marbles. If a marble is drawn at random, what is the probability that a blue or a white marble is drawn?

REVIEW EXERCISES

Evaluate the following numbers.

1. $\dbinom{18}{15}$

2. $\dbinom{25}{24}$

Expand using the binomial theorem.

3. $\left(\dfrac{a}{2} + 2b\right)^5$

4. $\left(3x - \dfrac{y}{3}\right)^6$

5. Find the sixth term in the expansion of $(2u - b)^7$.
6. Find the eighth term in the expansion of $(3a - 4b)^{10}$.
7. Find the middle term in the expansion of $(3x - 4)^6$.
8. Find the two middle terms in the expansion of $\left(2x + \dfrac{a}{2}\right)^7$.
9. Find the term containing a^2 in the expansion of $(a^{1/2} + b^{1/2})^8$.
10. Find the term containing u in the expansion of $(u^{1/3} - 2)^7$.
11. Use the binomial theorem to evaluate $(1.02)^6$.
12. Use the binomial theorem to evaluate $(0.99)^5$.

In Exercises 13–18, write the first five terms of each sequence.

13. $a_n = 5 - \dfrac{n}{2}$

14. $a_n = \dfrac{n}{3} - 7$

15. $a_n = \dfrac{(-1)^{n-1}}{n!}$

16. $a_n = (-1)^{n+1} n^3$

17. $a_n = \dfrac{(-1)^n (n - 1)}{n + 1}$

18. $a_n = \dfrac{1}{(2n + 1)!}$

In Exercises 19–22, find the general term a_n of a sequence, given the general partial sum s_n of the series whose terms are the a_k.

19. $s_n = 4^n + 1$

20. $s_n = \dfrac{n(3n + 1)}{2}$

21. $s_n = \dfrac{n(n + 1)(n + 2)}{3}$

22. $s_n = \dfrac{n(n + 1)(2n + 1)}{6}$

23. Use Newton's method to find four decimal approximations for $\sqrt{7}$. Take as initial guess $x_1 = 2.1$.
24. Use Newton's method to find four decimal approximations for $\sqrt{8}$. Take $x_1 = 3.1$ as initial guess.

In Exercises 25–28, find the common difference and determine the next two terms of each arithmetic progression.

25. $10, 7, 4, \ldots$

26. $2, 0.5, -1, \ldots$

27. $\ln 3, \ln 1, \ln 3^{-1}$
28. $\log 10, \log 100, \log 1000, \ldots$
29. If 6 and 11 are the third and fifth terms of an arithmetic sequence, find the tenth term.
30. If 3.25 and 2.5 are the fourth and seventh terms of an arithmetic sequence, find the second term.
31. How many natural numbers divisible by 6 are there between 15 and 81? Find their sum.
32. Find the sum of all the numbers divisible by 8 that are between 50 and 122.

In Exercises 33–36, find the ratio and determine the next two terms of the geometric progression.

33. $\dfrac{3}{5}, \dfrac{1}{2}, \dfrac{5}{12}, \ldots$

34. $\dfrac{4}{5}, \dfrac{7}{10}, \dfrac{49}{80}, \ldots$

35. $5, 7.5, 11.25, \ldots$

36. $5, 1, 0.2, \ldots$

37. The third and fifth terms of a geometric progression with positive terms are 4/9 and 16/81. Find the second term.

38. The second and fourth terms of a geometric progression are 80 and 5. Find the sixth term.

Find the sum of each series in Exercises 39 and 40.

39. $\displaystyle\sum_{n=1}^{\infty} 3\left(-\frac{1}{3}\right)^{n-1}$ **40.** $\displaystyle\sum_{n=1}^{\infty} 5\left(\frac{2}{5}\right)^{n-1}$

In Exercises 41–44, write the repeating decimals as fractions.

41. $0.\overline{27}$ **42.** $0.\overline{54}$

43. $0.0\overline{24}$ **44.** $0.1\overline{72}$

In Exercises 45–52, prove each proposition by mathematical induction.

45. $4 + 8 + 12 + \cdots + 4n = 2n(n + 1)$ for all $n \in \mathbb{N}$.

46. $\displaystyle\sum_{j=1}^{\infty} \frac{1}{(2j - 1)(2j + 1)} = \frac{n}{2n + 1}$ for all $n \in \mathbb{N}$.

47. $a(b_1 + \cdots + b_n) = ab_1 + \cdots + ab_n$ for all $n \geq 2$.

48. $\log(a_1 \cdot a_2 \cdots a_n) = \log a_1 + \log a_2 + \cdots + \log a_n$, $n \geq 2$.

49. $(1 + x)^n \geq 1 + nx$ if $x > -1$, for all natural numbers n.

50. De Moivre's formula: $z^n = [r(\cos \theta + i \sin \theta)]^n = r^n(\cos n\theta + i \sin n\theta)$.

51. $(a + b)^n > a^n + b^n$, a and b positive, $n > 1$.

52. $x^{2n} - a^{2n}$ is divisible by $x + a$.

53. Prove that $\dbinom{n}{0} + \dbinom{n}{1} + \dbinom{n}{2} + \cdots + \dbinom{n}{n} = 2^n$ for all n.
(*Hint*: Write $2^n = (1 + 1)^n$ and expand, using the binomial theorem.)

54. Prove that $\displaystyle\sum_{k=0}^{n} (-1)^k \dbinom{n}{k} = 0$ for all n.

In Exercises 55–70, solve the given problem.

55. A free-falling object travels 16 ft during the first second, 48 ft during the second second, and so on. How many feet will the object fall during the eleventh second? How many feet will it fall in 11 seconds?

56. An auditorium has 15 rows of seats. If there are 20 seats in the first row, 22 in the second row, 24 in the third row, and so on, find the total number of seats.

57. In a tapered ladder each rung, from bottom to top, is 1/8 in. shorter than the previous one. If the bottom rung is 12 in. long and the ladder has 11 rungs, find the length of the top rung. Assuming no waste, how many inches of rung material would be needed to build the ladder?

58. The number of bacteria in a colony doubles each hour. If the initial number of bacteria is 1800, what is the number of bacteria in the colony after 6 hours? After $7\frac{1}{2}$ hours?

59. A weight at the end of a line swings through an arc 2 m long in its first swing. On each subsequent swing, the weight travels 0.8 the length of the preceding swing. How far will the weight travel before coming to rest?

60. To stimulate the economy of a depressed area, the government distributes $2,000,000 through a subsidy program. Assuming that each institution or individual will spend 80% of the amount received, and that 80% of the spent amount will in turn be spent, and so on, find the total amount of money that will be spent as a result of the program.

61. Evaluate the following numbers.
(a) $P(8, 3)$ (b) $P(10, 6)$ (c) $C(10, 5)$
(d) $C(12, 4)$ (e) $\Gamma(8, 4, 3, 1)$ (f) $\Gamma(10, 6, 2, 2)$

62. Evaluate each of the following numbers.
(a) $P(30, 4)$ (b) $P(96, 3)$ (c) $C(15, 8)$
(d) $C(20, 10)$ (e) $\Gamma(18, 8, 6, 4)$ (f) $\Gamma(20, 9, 5, 3, 3)$

63. In how many ways can four flags of different colors be arranged on a pole?

64. In how many ways can five candidates be listed on a ballot?

65. How many distinct lines can be drawn through eight points, no three of which lie on the same line?

66. Three noncollinear points determine a circle. How many distinct circles can be drawn through eight points, no three of which are collinear?

67. A junior soccer league has eight teams. How many games must be scheduled if each team has to play every other team twice?

68. The technique of paired comparisons is often used in panel tests of consumer products. In this technique each product is paired with each of the other products and the comparison is then made. If a sample consists of eight products, find how many paired comparisons can be made.

69. In how many ways can a jury of 12 people be selected from a group of 18 prospective jurors? (Assume that the order of selection is immaterial.)

70. In several states the automobile license plate consists of three letters followed by three digits. If no repetitions are permitted, find the total number of distinct license plates.

In Exercises 71 and 72, two dice are rolled. Find the probability of each event.

71. The sum of the numbers is greater than or equal to 8.

72. The sum of the numbers is equal to 7 or at least 10.

In Exercises 73 and 74, six marbles numbered 1, 2, 3, 4, 5, and 6 are placed in a box. After the marbles are mixed, two are simultaneously picked up. Find the probability of each event.

73. Both marbles are marked with an even number.

74. The sum of the numbers is at most 8.

In Exercises 75–80, solve the given problem.

75. Jack and Jill belong to a 20-member hiking club. Every year two people are selected at random from the club to serve as secretary and treasurer. What is the probability that Jack is selected as secretary and Jill as treasurer?

76. Find the probability of guessing ten correct answers on a ten-question true/false test.

77. Find the probability of being dealt a five-card hand of three diamonds and two spades from a 52-card deck.

78. Five cards are dealt from a standard 52-card deck. What is the probability of obtaining three jacks?

79. A subcommittee of five people is to be chosen from a student committee that consists of five girls and five boys. What is the probability that there will be three girls and two boys in the subcommittee?

80. A *full house* in poker consists of three cards of one rank (such as three eights or three queens) and two cards of another rank (such as two fours or two kings). What is the probability of being dealt a full house from a standard 52-card deck in five-card poker?

■

LINEAR INTERPOLATION

When the logarithm or antilogarithm of a number is not found in a table, it is possible to approximate the logarithm or antilogarithm by the method of linear interpolation, which we now describe. The same method applies to computations involving the trigonometric values of angles, and most other tabulated data.

Linear interpolation is based upon the following general property. Let $y = f(x)$ be a given function and suppose that for two distinct values a and b of x, the function values $f(a)$ and $f(b)$ are known. If c is a number between a and b, the function value $f(c)$ can be approximated by the ordinate of the point P that lies on the line segment joining the points $A(a, f(a))$ and $B(b, f(b))$ and has abscissa c. (Figure A1). Notice that the point C on the graph of f has coordinates $(c, f(c))$,

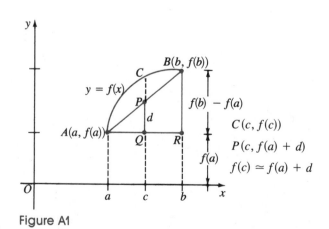

Figure A1

while the point P on the line segment AB has coordinates $(c, f(a) + d)$. Since we do not know the function value $f(c)$, we are approximating it by $f(a) + d$. The quantity d can be easily computed by using the similarity of the right triangles AQP and ARB:

$$\frac{d}{f(b) - f(a)} = \frac{c - a}{b - a}.$$

Thus,

$$d = \frac{f(b) - f(a)}{b - a}(c - a).$$

Substituting for d, we have

$$f(c) \simeq f(a) + \frac{f(b) - f(a)}{b - a}(c - a). \tag{A1}$$

EXAMPLE **1**   Approximate log 5.814.

Solution: The number 5.814 is not found in Table 2, but it lies between the numbers 5.81 and 5.82, whose logarithms are given by Table 2. Since the common logarithm is an increasing function, we have

$$\log 5.81 < \log 5.814 < \log 5.82$$

or

$$0.7642 < \log 5.814 < 0.7649.$$

According to Figure A2, we determine d as follows:

$$\frac{d}{0.0007} = \frac{0.004}{0.01} = \frac{4}{10} = \frac{2}{5}.$$

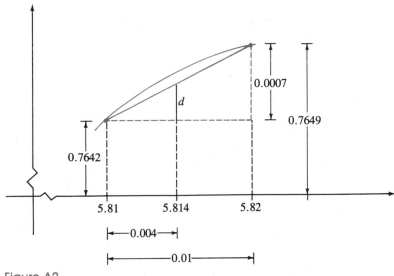

Figure A2

Hence,

$$d = \frac{2(0.0007)}{5}$$

$$= \frac{0.0014}{5}$$

$$= 0.00028.$$

Now, according to formula (A1), we obtain

$$\log 5.814 \simeq 0.7642 + 0.00028$$

$$\simeq 0.7645.$$

When using the method of linear interpolation, it is helpful to organize our work as follows:

$$0.01 \left\{ 0.004 \left\{ \begin{array}{cc} x & \log x \\ 5.81 & 0.7642 \\ \\ 5.814 & ? \\ 5.82 & 0.7649 \end{array} \right\} d \right\} 0.0007$$

where the numbers next to the braces are the differences of the corresponding pairs. Now, we write the ratios

$$\frac{d}{0.0007} = \frac{0.004}{0.01} = \frac{2}{5}$$

and compute the value of d as we did before.

Practice Exercise 1   Approximate the number of log 8.742.

Answer: 0.9416

EXAMPLE **2**   Approximate log 5814.

Solution: First, we write 5814 in scientific notation: 5.814×10^3. Next, we take logarithms:

$$\log 5814 = \log(5.814 \times 10^3)$$

$$= \log 5.814 + \log 10^3$$

$$= \log 5.814 + 3.$$

Now we use linear interpolation to compute log 5.814 (as in Example 1). Thus,

$$\log 5814 = \log 5.814 + 3$$

$$\simeq 0.7645 + 3$$

$$= 3.7645.$$

Practice Exercise 2   Find an approximation for log 874.2.

Answer: 2.9416

EXAMPLE **3** Approximate x if $\log x = 2.5116$.

Solution: We write

$$\log x = 0.5116 + 2,$$

where 0.5116 is the mantissa and 2 is the characteristic of $\log x$. Next, we observe that the number 0.5116 is not found in Table 2 but lies between 0.5105 and 0.5119, the common logarithms of the numbers 3.24 and 3.25, respectively. Now we arrange our work as follows:

$$0.01\left\{d\left\{\begin{matrix} x & \log x \\ 3.24 & 0.5105 \\ \\ ? & 0.5116 \\ 3.25 & 0.5119 \end{matrix}\right\}0.0011\right\}0.0014$$

and write the ratio

$$\frac{d}{0.01} = \frac{0.0011}{0.0014} = \frac{11}{14}.$$

Solving for d, we get

$$d = \frac{11(0.01)}{14} \simeq 0.008.$$

By adding this number to 3.24, we obtain 3.248. Thus,

$$\log 3.248 \simeq 0.5116.$$

Next, we write

$$\log x = 0.5116 + 2$$
$$\simeq \log 3.248 + 2$$
$$= \log(3.248 \times 10^2)$$
$$= \log(324.8).$$

Hence, $x \simeq 324.8$.

Practice Exercise 3 Find an approximation for x if $\log x = 0.8794$.

Answer: 7.575

Now, we discuss some examples involving trigonometric functions.

EXAMPLE **4** Approximate the value $\sin 31°12'$.

Solution: The angle of measure $31°12'$ lies between the angles of measure $31°10'$ and $31°20'$, whose trigonometric values are found in Table 4. We interpolate as follows:

$$10'\left\{2'\left\{\begin{matrix} \theta & \sin\theta \\ 31°10' & 0.5175 \\ \\ 31°12' & ? \\ 31°20' & 0.5200 \end{matrix}\right\}d\right\}0.0025$$

Thus,

$$\frac{d}{0.0025} = \frac{2}{10} = \frac{1}{5}$$

and

$$d = \frac{0.0025}{5} = 0.0005.$$

Hence,

$$\sin 31°12' \simeq \sin 31°10' + 0.0005$$
$$\simeq 0.5175 + 0.0005$$
$$= 0.5180$$

Practice Exercise 4   Approximate the number tan 23°15′.

Answer: 0.4297

EXAMPLE 5   Find, to the nearest minute, a third-quadrant angle θ such that $\sin \theta = -0.5246$.

Solution: If α denotes the reference angle of θ, then $\sin \alpha = 0.5246$. This number is not in the sine column of Table 4, but lies between the values $\sin 31°30' = 0.5225$ and $\sin 31°40' = 0.5250$. Now, we interpolate:

$$10'\left\{ d\left\{ \begin{array}{cc} x & \sin x \\ 31°30' & \left. \begin{array}{c} 0.5225 \end{array} \right\} \\ \alpha & \left. \begin{array}{c} 0.5246 \end{array} \right\}0.0021 \\ 31°40' & 0.5250 \end{array} \right. \right\}0.0025$$

Thus,

$$\frac{d}{10} = \frac{0.0021}{0.0025} = \frac{21}{25}$$

and

$$d = \frac{21(10)}{25} = 8.4' \simeq 8'.$$

It follows that

$$\alpha = 31°30' + 8' = 31°38'.$$

The corresponding third-quadrant angle is then

$$\theta = 180° + 31°38' = 211°38'.$$

Practice Exercise 5   Let α be a second-quadrant angle such that $\cos \alpha = -0.8611$. Approximate α to the nearest minute.

Answer: 149° 28′

In Examples 4 and 5, we have dealt with angles measured in degrees and minutes. The same method applies to angles measured in radians.

EXERCISES

Use linear interpolation to approximate the common logarithms of the numbers in Exercises 1–6. If a calculator is available, compare your approximations with the values obtained with the calculator.

1. 32.56

2. 125.4

3. 1382

4. 3261

5. 0.002351

6. 0.0005321

In Exercises 7–14, use linear interpolation to approximate x. If a calculator is available, compare your answers with those obtained with the calculator.

7. $\log x = 1.3145$

8. $\log x = 2.3126$

9. $\log x = -0.4138$

10. $\log x = -2.1456$

11. $\log x = 3.0856$

12. $\log x = -4.1378$

13. $\log x = 6.5614 - 8$

14. $\log x = 2.4156 - 5$

Use linear interpolation to approximate the given numbers in Exercises 15–22.

15. $\cos 15°36'$

16. $\tan 26°48'$

17. $\sec 62°15'$

18. $\csc 45°32'$

19. $\sin 0.7600$

20. $\tan 1.324$

21. $\cos 1.4678$

22. $\sin 38°45'$

In Exercises 23–30, use linear interpolation to approximate, to the nearest minute, the degree measures of the angles θ such that $0° \le \theta \le 360°$.

23. $\cos \theta = 0.3261$

24. $\tan \theta = 1.3216$

25. $\sec \theta = 2.0018$

26. $\sin \theta = 0.8108$

27. $\sin \theta = 0.1758$

28. $\cot \theta = 4.128$

29. $\cos \theta = -0.8669$

30. $\sin \theta = -0.5856$

TABLES

TABLE 1 Values of e^x and e^{-x}

x	e^x	e^{-x}	x	e^x	e^{-x}	x	e^x	e^{-x}
0.00	1.000	1.000	0.35	1.419	0.705	0.70	2.014	0.497
0.01	1.010	0.990	0.36	1.433	0.698	0.71	2.034	0.492
0.02	1.020	0.980	0.37	1.448	0.691	0.72	2.054	0.487
0.03	1.031	0.970	0.38	1.462	0.684	0.73	2.075	0.482
0.04	1.041	0.961	0.39	1.477	0.677	0.74	2.096	0.477
0.05	1.051	0.951	0.40	1.492	0.670	0.75	2.117	0.472
0.06	1.062	0.942	0.41	1.507	0.664	0.76	2.138	0.468
0.07	1.073	0.932	0.42	1.522	0.657	0.77	2.160	0.463
0.08	1.083	0.923	0.43	1.537	0.651	0.78	2.182	0.458
0.09	1.094	0.914	0.44	1.553	0.644	0.79	2.203	0.454
0.10	1.105	0.905	0.45	1.568	0.638	0.80	2.226	0.449
0.11	1.116	0.896	0.46	1.584	0.631	0.81	2.248	0.445
0.12	1.127	0.887	0.47	1.600	0.625	0.82	2.270	0.440
0.13	1.139	0.878	0.48	1.616	0.619	0.83	2.293	0.436
0.14	1.150	0.869	0.49	1.632	0.613	0.84	2.316	0.432
0.15	1.162	0.861	0.50	1.649	0.607	0.85	2.340	0.427
0.16	1.174	0.852	0.51	1.665	0.600	0.86	2.363	0.423
0.17	1.185	0.844	0.52	1.682	0.595	0.87	2.387	0.419
0.18	1.197	0.835	0.53	1.699	0.589	0.88	2.411	0.415
0.19	1.209	0.827	0.54	1.716	0.583	0.89	2.435	0.411
0.20	1.221	0.819	0.55	1.733	0.577	0.90	2.460	0.407
0.21	1.234	0.811	0.56	1.751	0.571	0.91	2.484	0.403
0.22	1.246	0.803	0.57	1.768	0.566	0.92	2.509	0.399
0.23	1.259	0.795	0.58	1.786	0.560	0.93	2.535	0.395
0.24	1.271	0.787	0.59	1.804	0.554	0.94	2.560	0.391

TABLE 1 *(continued)*

x	e^x	e^{-x}	x	e^x	e^{-x}	x	e^x	e^{-x}
0.25	1.284	0.779	0.60	1.822	0.549	0.95	2.586	0.387
0.26	1.297	0.771	0.61	1.840	0.543	0.96	2.612	0.383
0.27	1.310	0.763	0.62	1.859	0.538	0.97	2.638	0.379
0.28	1.323	0.756	0.63	1.878	0.533	0.98	2.664	0.375
0.29	1.336	0.748	0.64	1.896	0.527	0.99	2.691	0.372
0.30	1.350	0.741	0.65	1.916	0.522	1.00	2.718	0.368
0.31	1.363	0.733	0.66	1.935	0.517	1.01	2.746	0.364
0.32	1.377	0.726	0.67	1.954	0.512	1.02	2.773	0.361
0.33	1.391	0.719	0.68	1.974	0.507	1.03	2.801	0.357
0.34	1.405	0.712	0.69	1.994	0.502	1.04	2.829	0.353
1.05	2.858	0.350	1.40	4.055	0.247	1.75	5.755	0.174
1.06	2.886	0.346	1.41	4.096	0.244	1.76	5.812	0.172
1.07	2.915	0.343	1.42	4.137	0.242	1.77	5.871	0.170
1.08	2.945	0.340	1.43	4.179	0.239	1.78	5.930	0.169
1.09	2.974	0.336	1.44	4.221	0.237	1.79	5.989	0.167
1.10	3.004	0.333	1.45	4.263	0.235	1.80	6.050	0.165
1.11	3.034	0.330	1.46	4.306	0.232	1.81	6.110	0.164
1.12	3.065	0.326	1.47	4.349	0.230	1.82	6.172	0.162
1.13	3.096	0.323	1.48	4.393	0.228	1.83	6.234	0.160
1.14	3.127	0.320	1.49	4.437	0.225	1.84	6.297	0.159
1.15	3.158	0.317	1.50	4.482	0.223	1.85	6.360	0.157
1.16	3.190	0.313	1.51	4.527	0.221	1.86	6.424	0.156
1.17	3.222	0.310	1.52	4.572	0.219	1.87	6.488	0.154
1.18	3.254	0.307	1.53	4.618	0.217	1.88	6.553	0.153
1.19	3.287	0.304	1.54	4.665	0.214	1.89	6.619	0.151
1.20	3.320	0.301	1.55	4.712	0.212	1.90	6.686	0.150
1.21	3.353	0.298	1.56	4.759	0.210	1.91	6.753	0.148
1.22	3.387	0.295	1.57	4.807	0.208	1.92	6.821	0.147
1.23	3.421	0.292	1.58	4.855	0.206	1.93	6.890	0.145
1.24	3.456	0.289	1.59	4.904	0.204	1.94	6.959	0.144
1.25	3.490	0.287	1.60	4.953	0.202	1.95	7.029	0.142
1.26	3.525	0.284	1.61	5.003	0.200	1.96	7.099	0.141
1.27	3.561	0.281	1.62	5.053	0.198	1.97	7.171	0.139
1.28	3.597	0.278	1.63	5.104	0.196	1.98	7.243	0.138
1.29	3.633	0.275	1.64	5.155	0.194	1.99	7.316	0.137
1.30	3.669	0.273	1.65	5.207	0.192	2.00	7.389	0.135
1.31	3.706	0.270	1.66	5.259	0.190	2.01	7.463	0.134
1.32	3.743	0.267	1.67	5.312	0.188	2.02	7.538	0.133
1.33	3.781	0.264	1.68	5.366	0.186	2.03	7.614	0.131
1.34	3.819	0.262	1.69	5.420	0.185	2.04	7.691	0.130
1.35	3.857	0.259	1.70	5.474	0.183	2.05	7.768	0.129
1.36	3.896	0.257	1.71	5.529	0.181	2.06	7.846	0.127
1.37	3.935	0.254	1.72	5.585	0.179	2.07	7.925	0.126
1.38	3.975	0.252	1.73	5.641	0.177	2.08	8.004	0.125
1.39	4.015	0.249	1.74	5.697	0.176	2.09	8.085	0.124

TABLE 1 *(continued*

x	e^x	e^{-x}	x	e^x	e^{-x}	x	x^e	e^{-x}
2.10	8.166	0.122	2.40	11.023	0.091	2.70	14.880	0.067
2.11	8.248	0.121	2.41	11.134	0.090	2.71	15.029	0.067
2.12	8.331	0.120	2.42	11.246	0.089	2.72	15.180	0.066
2.13	8.415	0.119	2.43	11.359	0.088	2.73	15.333	0.065
2.14	8.499	0.118	2.44	11.473	0.087	2.74	15.487	0.065
2.15	8.585	0.116	2.45	11.588	0.086	2.75	15.643	0.064
2.16	8.671	0.115	2.46	11.705	0.085	2.76	15.800	0.063
2.17	8.758	0.114	2.47	11.822	0.085	2.77	15.959	0.063
2.18	8.846	0.113	2.48	11.941	0.084	2.78	16.119	0.062
2.19	8.935	0.112	2.49	12.061	0.083	2.79	16.281	0.061
2.20	9.025	0.111	2.50	12.182	0.082	2.80	16.445	0.061
2.21	9.116	0.110	2.51	12.305	0.081	2.81	16.610	0.060
2.22	9.207	0.109	2.52	12.429	0.080	2.82	16.777	0.060
2.23	9.300	0.108	2.53	12.554	0.080	2.83	16.945	0.059
2.24	9.393	0.106	2.54	12.680	0.079	2.84	17.116	0.058
2.25	9.488	0.105	2.55	12.807	0.078	2.85	17.288	0.058
2.26	9.583	0.104	2.56	12.936	0.077	2.86	17.462	0.057
2.27	9.679	0.103	2.57	13.066	0.077	2.87	17.637	0.057
2.28	9.777	0.102	2.58	13.197	0.076	2.88	17.814	0.056
2.29	9.875	0.101	2.59	13.330	0.075	2.89	17.993	0.056
2.30	9.974	0.100	2.60	13.464	0.074	2.90	18.174	0.055
2.31	10.074	0.099	2.61	13.599	0.074	2.91	18.357	0.054
2.32	10.176	0.098	2.62	13.736	0.073	2.92	18.541	0.054
2.33	10.278	0.097	2.63	13.874	0.072	2.93	18.728	0.053
2.34	10.381	0.096	2.64	14.013	0.071	2.94	18.916	0.053
2.35	10.486	0.095	2.65	14.154	0.071	2.95	19.106	0.052
2.36	10.591	0.094	2.66	14.296	0.070	2.96	19.298	0.052
2.37	10.697	0.093	2.67	14.440	0.069	2.97	19.492	0.051
2.38	10.805	0.093	2.68	14.585	0.069	2.98	19.688	0.051
2.39	10.913	0.092	2.69	14.732	0.068	2.99	19.886	0.050
						3.00	20.086	0.050

TABLE 2 Common Logarithms

x	.00	.01	.02	.03	.04	.05	.06	.07	.08	.09
1.0	.0000	.0043	.0086	.0128	.0170	.0212	.0253	.0294	.0334	.0374
1.1	.0414	.0453	.0492	.0531	.0569	.0607	.0645	.0682	.0719	.0755
1.2	.0792	.0828	.0864	.0899	.0934	.0969	.1004	.1038	.1072	.1106
1.3	.1139	.1173	.1206	.1239	.1271	.1303	.1335	.1367	.1399	.1430
1.4	.1461	.1492	.1523	.1553	.1584	.1614	.1644	.1673	.1703	.1732
1.5	.1761	.1790	.1818	.1847	.1875	.1903	.1913	.1959	.1987	.2014
1.6	.2041	.2068	.2095	.2122	.2148	.2175	.2201	.2227	.2253	.2279
1.7	.2304	.2330	.2355	.2380	.2405	.2430	.2455	.2480	.2504	.2529
1.8	.2553	.2577	.2601	.2625	.2648	.2672	.2695	.2718	.2742	.2765
1.9	.2788	.2810	.2833	.2856	.2878	.2900	.2923	.2945	.2967	.2989
2.0	.3010	.3032	.3054	.3075	.3096	.3118	.3139	.3160	.3181	.3201
2.1	.3222	.3243	.3263	.3284	.3304	.3324	.3345	.3365	.3385	.3404
2.2	.3424	.3444	.3464	.3483	.3502	.3522	.3541	.3560	.3579	.3598
2.3	.3617	.3636	.3655	.3674	.3692	.3711	.3729	.3747	.3766	.3784
2.4	.3802	.3820	.3838	.3856	.3874	.3892	.3909	.3927	.3945	.3962
2.5	.3979	.3997	.4014	.4031	.4048	.4065	.4082	.4099	.4116	.4133
2.6	.4150	.4166	.4183	.4200	.4216	.4232	.4249	.4265	.4281	.4298
2.7	.4314	.4330	.4346	.4362	.4378	.4393	.4409	.4425	.4440	.4456
2.8	.4472	.4487	.4502	.4518	.4533	.4548	.4564	.4579	.4594	.4609
2.9	.4624	.4639	.4654	.4669	.4683	.4698	.4713	.4728	.4742	.4757
3.0	.4771	.4786	.4800	.4814	.4829	.4843	.4857	.4871	.4886	.4900
3.1	.4914	.4928	.4942	.4955	.4969	.4983	.4997	.5011	.5024	.5038
3.2	.5051	.5065	.5079	.5092	.5105	.5119	.5132	.5145	.5159	.5172
3.3	.5185	.5198	.5211	.5224	.5237	.5250	.5263	.5276	.5289	.5302
3.4	.5315	.5328	.5340	.5353	.5366	.5378	.5391	.5403	.5416	.5428
3.5	.5441	.5453	.5465	.5478	.5490	.5502	.5514	.5527	.5539	.5551
3.6	.5563	.5575	.5587	.5599	.5611	.5623	.5635	.5647	.5658	.5670
3.7	.5682	.5694	.5705	.5717	.5729	.5740	.5752	.5763	.5775	.5786
3.8	.5798	.5809	.5821	.5832	.5843	.5855	.5866	.5877	.5888	.5899
3.9	.5911	.5922	.5933	.5944	.5955	.5966	.5977	.5988	.5999	.6010
4.0	.6021	.6031	.6042	.6053	.6064	.6075	.6085	.6096	.6107	.6117
4.1	.6128	.6138	.6149	.6160	.6170	.6180	.6191	.6201	.6212	.6222
4.2	.6232	.6243	.6253	.6263	.6274	.6284	.6294	.6304	.6314	.6325
4.3	.6335	.6345	.6355	.6365	.6375	.6385	.6395	.6405	.6415	.6425
4.4	.6435	.6444	.6454	.6464	.6474	.6484	.6493	.6503	.6513	.6522
4.5	.6532	.6542	.6551	.6561	.6571	.6580	.6590	.6599	.6609	.6618
4.6	.6628	.6637	.6646	.6656	.6665	.6675	.6684	.6693	.6702	.6712
4.7	.6721	.6730	.6739	.6749	.6758	.6767	.6776	.6785	.6794	.6803
4.8	.6812	.6821	.6830	.6839	.6848	.6857	.6866	.6875	.6884	.6893
4.9	.6902	.6911	.6920	.6928	.6937	.6946	.6955	.6964	.6972	.6981
5.0	.6990	.6998	.7007	.7016	.7024	.7033	.7042	.7050	.7059	.7067
5.1	.7076	.7084	.7093	.7101	.7110	.7118	.7126	.7135	.7143	.7152
5.2	.7160	.7168	.7177	.7185	.7193	.7202	.7210	.7218	.7226	.7235
5.3	.7243	.7251	.7259	.7267	.7275	.7284	.7292	.7300	.7308.	.7316
5.4	.7324	.7332	.7340	.7348	.7356	.7364	.7372	.7380	.7388	.7396

TABLE 2 *(continued)*

x	.00	.01	.02	.03	.04	.05	.06	.07	.08	.09
5.5	.7404	.7412	.7419	.7427	.7435	.7443	.7451	.7459	.7466	.7474
5.6	.7482	.7490	.7497	.7505	.7513	.7520	.7528	.7536	.7543	.7551
5.7	.7559	.7566	.7574	.7582	.7589	.7597	.7604	.7612	.7619	.7627
5.8	.7634	.7642	.7649	.7657	.7664	.7672	.7679	.7686	.7694	.7701
5.9	.7709	.7716	.7723	.7731	.7738	.7745	.7752	.7760	.7767	.7774
6.0	.7782	.7789	.7796	.7803	.7810	.7818	.7825	.7832	.7839	.7846
6.1	.7853	.7860	.7868	.7875	.7882	.7889	.7896	.7903	.7910	.7917
6.2	.7924	.7931	.7938	.7945	.7952	.7959	.7966	.7973	.7980	.7987
6.3	.7993	.8000	.8007	.8014	.8021	.8028	.8035	.8041	.8048	.8055
6.4	.8062	.8069	.8075	.8082	.8089	.8096	.8102	.8109	.8116	.8122
6.5	.8129	.8136	.8142	.8149	.8156	.8162	.8169	.8176	.8182	.8189
6.6	.8195	.8202	.8209	.8215	.8222	.8228	.8235	.8241	.8248	.8254
6.7	.8261	.8267	.8274	.8280	.8287	.8293	.8299	.8306	.8312	.8319
6.8	.8325	.8331	.8338	.8344	.8351	.8357	.8363	.8370	.8376	.8382
6.9	.8388	.8395	.8401	.8407	.8414	.8420	.8426	.8432	.8439	.8445
7.0	.8451	.8457	.8463	.8470	.8476	.8482	.8488	.8494	.8500	.8506
7.1	.8513	.8519	.8525	.8531	.8537	.8543	.8549	.8555	.8561	.8567
7.2	.8573	.8579	.8585	.8591	.8597	.8603	.8609	.8615	.8621	.8627
7.3	.8633	.8639	.8645	.8651	.8657	.8663	.8669	.8675	.8681	.8686
7.4	.8692	.8698	.8704	.8710	.8716	.8722	.8727	.8733	.8739	.8745
7.5	.8751	.8756	.8762	.8768	.8774	.8779	.8785	.8791	.8797	.8802
7.6	.8808	.8814	.8820	.8825	.8831	.8837	.8842	.8848	.8854	.8859
7.7	.8865	.8871	.8876	.8882	.8887	.8893	.8899	.8904	.8910	.8915
7.8	.8921	.8927	.8932	.8938	.8943	.8949	.8954	.8960	.8965	.8971
7.9	.8976	.8982	.8987	.8993	.8998	.9004	.9009	.9015	.9020	.9025
8.0	.9031	.9036	.9042	.9047	.9053	.9058	.9063	.9069	.9074	.9079
8.1	.9085	.9090	.9096	.9101	.9106	.9112	.9117	.9122	.9128	.9133
8.2	.9138	.9143	.9149	.9154	.9159	.9165	.9170	.9175	.9180	.9186
8.3	.9191	.9196	.9201	.9206	.9212	.9217	.9222	.9227	.9232	.9238
8.4	.9243	.9248	.9253	.9258	.9263	.9269	.9274	.9279	.9284	.9289
8.5	.9294	.9299	.9304	.9309	.9315	.9320	.9325	.9330	.9335	.9340
8.6	.9345	.9350	.9355	.9360	.9365	.9370	.9375	.9380	.9385	.9390
8.7	.9395	.9400	.9405	.9410	.9415	.9420	.9425	.9430	.9435	.9440
8.8	.9445	.9450	.9455	.9460	.9465	.9469	.9474	.9479	.9484	.9489
8.9	.9494	.9499	.9504	.9509	.9513	.9518	.9523	.9528	.9533	.9538
9.0	.9542	.9547	.9552	.9557	.9562	.9566	.9571	.9576	.9581	.9586
9.1	.9590	.9595	.9600	.9605	.9609	.9614	.9619	.9624	.9628	.9633
9.2	.9638	.9643	.9647	.9652	.9657	.9661	.9666	.9671	.9675	.9680
9.3	.9685	.9689	.9694	.9699	.9703	.9708	.9713	.9717	.9722	.9727
9.4	.9731	.9736	.9741	.9745	.9750	.9754	.9759	.9763	.9768	.9773
9.5	.9777	.9782	.9786	.9791	.9795	.9800	.9805	.9809	.9814	.9818
9.6	.9823	.9827	.9832	.9836	.9841	.9845	.9850	.9854	.9859	.9863
9.7	.9868	.9872	.9877	.9881	.9886	.9890	.9894	.9899	.9903	.9908
9.8	.9912	.9917	.9921	.9926	.9930	.9934	.9939	.9943	.9948	.9952
9.9	.9956	.9961	.9965	.9969	.9974	.9978	.9983	.9987	.9991	.9996

TABLE 3 Natural Logarithms

x	.00	.01	02	.03	.04	.05	.06	.07	.08	.09
1.0	0.0000	0.0100	0.0198	0.0296	0.0392	0.0488	0.0583	0.0677	0.0770	0.0862
1.1	0.0953	0.1044	0.1133	0.1222	0.1310	0.1398	0.1484	0.1570	0.1655	0.1740
1.2	0.1823	0.1906	0.1989	0.2070	0.2151	0.2231	0.2311	0.2390	0.2469	0.2546
1.3	0.2624	0.2700	0.2776	0.2852	0.2927	0.3001	0.3075	0.3148	0.3221	0.3293
1.4	0.3365	0.3436	0.3507	0.3577	0.3646	0.3716	0.3784	0.3853	0.3920	0.3988
1.5	0.4055	0.4121	0.4187	0.4253	0.4318	0.4383	0.4447	0.4511	0.4574	0.4637
1.6	0.4700	0.4762	0.4824	0.4886	0.4947	0.5008	0.5068	0.5128	0.5188	0.5247
1.7	0.5306	0.5365	0.5423	0.5481	0.5539	0.5596	0.5653	0.5710	0.5766	0.5822
1.8	0.5878	0.5933	0.5988	0.6043	0.6098	0.6152	0.6206	0.6259	0.6313	0.6366
1.9	0.6419	0.6471	0.6523	0.6575	0.6627	0.6678	0.6729	0.6780	0.6831	0.6881
2.0	0.6931	0.6981	0.7031	0.7080	0.7130	0.7178	0.7227	0.7275	0.7324	0.7372
2.1	0.7419	0.7467	0.7514	0.7561	0.7608	0.7655	0.7701	0.7747	0.7793	0.7839
2.2	0.7885	0.7930	0.7975	0.8020	0.8065	0.8109	0.8154	0.8198	0.8242	0.8286
2.3	0.8329	0.8372	0.8416	0.8459	0.8502	0.8544	0.8587	0.8629	0.8671	0.8713
2.4	0.8755	0.8796	0.8838	0.8879	0.8920	0.8961	0.9002	0.9042	0.9083	0.9123
2.5	0.9163	0.9203	0.9243	0.9282	0.9322	0.9361	0.9400	0.9439	0.9478	0.9517
2.6	0.9555	0.9594	0.9632	0.9670	0.9708	0.9746	0.9783	0.9821	0.9858	0.9895
2.7	0.9933	0.9969	1.0006	1.0043	1.0080	1.0116	1.0152	1.0188	1.0225	1.0260
2.8	1.0296	1.0332	1.0367	1.0403	1.0438	1.0473	1.0508	1.0543	1.0578	1.0613
2.9	1.0647	1.0682	1.0716	1.0750	1.0784	1.0818	1.0852	1.0886	1.0919	1.0953
3.0	1.0986	1.1019	1.1053	1.1086	1.1119	1.1151	1.1184	1.1217	1.1249	1.1282
3.1	1.1314	1.1346	1.1378	1.1410	1.1442	1.1474	1.1506	1.1537	1.1569	1.1600
3.2	1.1632	1.1663	1.1694	1.1725	1.1756	1.1787	1.1817	1.1848	1.1878	1.1909
3.3	1.1939	1.1970	1.2000	1.2030	1.2060	1.2090	1.2119	1.2149	1.2179	1.2208
3.4	1.2238	1.2267	1.2296	1.2326	1.2355	1.2384	1.2413	1.2442	1.2470	1.2499
3.5	1.2528	1.2556	1.2585	1.2613	1.2641	1.2669	1.2698	1.2726	1.2754	1.2782
3.6	1.2809	1.2837	1.2865	1.2892	1.2920	1.2947	1.2975	1.3002	1.3029	1.3056
3.7	1.3083	1.3110	1.3137	1.3164	1.3191	1.3218	1.3244	1.3271	1.3297	1.3324
3.8	1.3350	1.3376	1.3403	1.3429	1.3455	1.3481	1.3507	1.3533	1.3558	1.3584
3.9	1.3610	1.3635	1.3661	1.3686	1.3712	1.3737	1.3762	1.3788	1.3813	1.3838
4.0	1.3863	1.3888	1.3913	1.3938	1.3962	1.3987	1.4012	1.4036	1.4061	1.4085
4.1	1.4110	1.4134	1.4159	1.4183	1.4207	1.4231	1.4255	1.4279	1.4303	1.4327
4.2	1.4351	1.4375	1.4398	1.4422	1.4446	1.4469	1.4493	1.4516	1.4540	1.4563
4.3	1.4586	1.4609	1.4633	1.4656	1.4679	1.4702	1.4725	1.4748	1.4770	1.4793
4.4	1.4816	1.4839	1.4861	1.4884	1.4907	1.4929	1.4952	1.4974	1.4996	1.5019
4.5	1.5041	1.5063	1.5085	1.5107	1.5129	1.5151	1.5173	1.5195	1.5217	1.5239
4.6	1.5261	1.5282	1.5304	1.5326	1.5347	1.5369	1.5390	1.5412	1.5433	1.5454
4.7	1.5476	1.5497	1.5518	1.5539	1.5560	1.5581	1.5602	1.5623	1.5644	1.5665
4.8	1.5686	1.5707	1.5728	1.5748	1.5769	1.5790	1.5810	1.5831	1.5851	1.5872
4.9	1.5892	1.5913	1.5933	1.5953	1.5974	1.5994	1.6014	1.6034	1.6054	1.6074
5.0	1.6094	1.6114	1.6134	1.6154	1.6174	1.6194	1.6214	1.6233	1.6253	1.6273
5.1	1.6292	1.6312	1.6332	1.6351	1.6371	1.6390	1.6409	1.6429	1.6448	1.6467
5.2	1.6487	1.6506	1.6525	1.6544	1.6563	1.6582	1.6601	1.6620	1.6639	1.6658
5.3	1.6677	1.6696	1.6715	1.6734	1.6753	1.6771	1.6790	1.6808	1.6827	1.6845
5.4	1.6864	1.6882	1.6901	1.6919	1.6938	1.6956	1.6974	1.6993	1.7011	1.7029

$$\ln(N \cdot 10^m) = \ln N + m \ln 10, \quad \ln 10 = 2.3026$$

TABLE 3 *(continued)*

x	.00	.01	.02	.03	.04	.05	.06	.07	.08	.09
5.5	1.7047	1.7066	1.7084	1.7102	1.7120	1.7138	1.7156	1.7174	1.7192	1.7210
5.6	1.7228	1.7246	1.7263	1.7281	1.7299	1.7317	1.7334	1.7352	1.7370	1.7387
5.7	1.7405	1.7422	1.7440	1.7457	1.7475	1.7492	1.7509	1.7527	1.7544	1.7561
5.8	1.7579	1.7596	1.7613	1.7630	1.7647	1.7664	1.7682	1.7699	1.7716	1.7733
5.9	1.7750	1.7766	1.7783	1.7800	1.7817	1.7834	1.7851	1.7867	1.7884	1.7901
6.0	1.7918	1.7934	1.7951	1.7967	1.7984	1.8001	1.8017	1.8034	1.8050	1.8066
6.1	1.8083	1.8099	1.8116	1.8132	1.8148	1.8165	1.8181	1.8197	1.8213	1.8229
6.2	1.8245	1.8262	1.8278	1.8294	1.8310	1.8326	1.8342	1.8358	1.8374	1.8390
6.3	1.8406	1.8421	1.8437	1.8453	1.8469	1.8485	1.8500	1.8516	1.8532	1.8547
6.4	1.8563	1.8579	1.8594	1.8610	1.8625	1.8641	1.8656	1.8672	1.8687	1.8703
6.5	1.8718	1.8733	1.8749	1.8764	1.8779	1.8795	1.8810	1.8825	1.8840	1.8856
6.6	1.8871	1.8886	1.8901	1.8916	1.8931	1.8946	1.8961	1.8976	1.8991	1.9006
6.7	1.9021	1.9036	1.9051	1.9066	1.9081	1.9095	1.9110	1.9125	1.9140	1.9155
6.8	1.9169	1.9184	1.9199	1.9213	1.9228	1.9242	1.9257	1.9272	1.9286	1.9301
6.9	1.9315	1.9330	1.9344	1.9359	1.9373	1.9387	1.9402	1.9416	1.9430	1.9445
7.0	1.9459	1.9473	1.9488	1.9502	1.9516	1.9530	1.9544	1.9559	1.9573	1.9587
7.1	1.9601	1.9615	1.9629	1.9643	1.9657	1.9671	1.9685	1.9699	1.9713	1.9727
7.2	1.9741	1.9755	1.9769	1.9782	1.9796	1.9810	1.9824	1.9838	1.9851	1.9865
7.3	1.9879	1.9892	1.9906	1.9920	1.9933	1.9947	1.9961	1.9974	1.9988	2.0001
7.4	2.0015	2.0028	2.0042	2.0055	2.0069	2.0082	2.0096	2.0109	2.0122	2.0136
7.5	2.0149	2.0162	2.0176	2.0189	2.0202	2.0215	2.0229	2.0242	2.0255	2.0268
7.6	2.0282	2.0295	2.0308	2.0321	2.0334	2.0347	2.0360	2.0373	2.0386	2.0399
7.7	2.0412	2.0425	2.0438	2.0451	2.0464	2.0477	2.0490	2.0503	2.0516	2.0528
7.8	2.0541	2.0554	2.0567	2.0580	2.0592	2.0605	2.0618	2.0631	2.0643	2.0656
7.9	2.0669	2.0681	2.0694	2.0707	2.0719	2.0732	2.0744	2.0757	2.0769	2.0782
8.0	2.0794	2.0807	2.0819	2.0832	2.0844	2.0857	2.0869	2.0882	2.0894	2.0906
8.1	2.0919	2.0931	2.0943	2.0956	2.0968	2.0980	2.0992	2.1005	2.1017	2.1029
8.2	2.1041	2.1054	2.1066	2.1078	2.1090	2.1102	2.1114	2.1126	2.1138	2.1150
8.3	2.1163	2.1175	2.1187	2.1199	2.1211	2.1223	2.1235	2.1247	2.1258	2.1270
8.4	2.1282	2.1294	2.1306	2.1318	2.1330	2.1342	2.1353	2.1365	2.1377	2.1389
8.5	2.1401	2.1412	2.1424	2.1436	2.1448	2.1459	2.1471	2.1483	2.1494	2.1506
8.6	2.1518	2.1529	2.1541	2.1552	2.1564	2.1576	2.1587	2.1599	2.1610	2.1622
8.7	2.1633	2.1645	2.1656	2.1668	2.1679	2.1691	2.1702	2.1713	2.1725	2.1736
8.8	2.1748	2.1759	2.1770	2.1782	2.1793	2.1804	2.1815	2.1827	2.1838	2.1849
8.9	2.1861	2.1872	2.1883	2.1894	2.1905	2.1917	2.1928	2.1939	2.1950	2.1961
9.0	2.1972	2.1983	2.1994	2.2006	2.2017	2.2028	2.2039	2.2050	2.2061	2.2072
9.1	2.2083	2.2094	2.2105	2.2116	2.2127	2.2138	2.2148	2.2159	2.2170	2.2181
9.2	2.2192	2.2203	2.2214	2.2225	2.2235	2.2246	2.2257	2.2268	2.2279	2.2289
9.3	2.2300	2.2311	2.2322	2.2332	2.2343	2.2354	2.2364	2.2375	2.2386	2.2396
9.4	2.2407	2.2418	2.2428	2.2439	2.2450	2.2460	2.2471	2.2481	2.2492	2.2502
9.5	2.2513	2.2523	2.2534	2.2544	2.2555	2.2565	2.2576	2.2586	2.2597	2.2607
9.6	2.2618	2.2628	2.2638	2.2649	2.2659	2.2670	2.2680	2.2690	2.2701	2.2711
9.7	2.2721	2.2732	2.2742	2.2752	2.2762	2.2773	2.2783	2.2793	2.2803	2.2814
9.8	2.2824	2.2834	2.2844	2.2854	2.2865	2.2875	2.2885	2.2895	2.2905	2.2915
9.9	2.2925	2.2935	2.2946	2.2956	2.2966	2.2976	2.2986	2.2996	2.3006	2.3016

TABLE 4 Trigonometric Functions

Degrees	Radians	Sin	Cos	Tan	Cot	Sec	Csc		
0°00′	.0000	.0000	1.0000	.0000	—	1.000	—	1.5708	**90°00′**
10	029	029	000	029	343.8	000	343.8	679	50
20	058	058	000	058	171.9	000	171.9	650	40
30	.0087	.0087	1.0000	.0087	114.6	1.000	114.6	1.5621	30
40	116	116	.9999	116	85.94	000	85.95	592	20
50	145	145	999	145	68.75	000	68.76	563	10
1°00′	.0175	.0175	.9998	.0175	57.29	1.000	57.30	1.5533	**89°00′**
10	204	204	998	204	49.10	000	49.11	504	50
20	233	233	997	233	42.96	000	42.98	475	40
30	0.0262	0.0262	.9997	.0262	38.19	1.000	38.20	1.5446	30
40	291	291	996	291	34.37	000	34.38	417	20
50	320	320	995	320	31.24	001	31.26	388	10
2°00′	.0349	.0349	.9994	.0349	28.64	1.001	28.65	1.5359	**88°00′**
10	378	378	993	378	26.43	001	26.45	330	50
20	407	407	992	407	24.54	001	24.56	301	40
30	.0436	.0436	.9990	.0437	22.90	1.001	22.93	1.5272	30
40	465	465	989	466	21.47	001	21.49	243	20
50	495	494	988	495	20.21	001	20.23	213	10
3°00′	.0524	.0523	.9986	.0524	19.08	1.001	19.11	1.5184	**87°00′**
10	553	552	985	553	18.07	002	18.10	155	50
20	582	581	983	582	17.17	002	17.20	126	40
30	.0611	.0610	.9981	.0612	16.35	1.002	16.38	1.5097	30
40	640	640	980	641	15.60	002	15.64	068	20
50	669	669	978	670	14.92	002	14.96	039	10
4°00′	.0698	.0698	.9976	.0699	14.30	1.002	14.34	1.5010	**86°00′**
10	727	727	974	729	13.73	003	13.76	981	50
20	756	756	971	758	13.20	003	13.23	952	40
30	.0785	.0785	.9969	.0787	12.71	1.003	12.75	1.4923	30
40	814	814	967	816	12.25	003	12.29	893	20
50	844	843	964	846	11.83	004	11.87	864	10
5°00′	.0873	.0872	.9962	.0875	11.43	1.004	11.47	1.4835	**85°00′**
10	902	901	959	904	11.06	004	11.10	806	50
20	931	929	957	934	10.71	004	10.76	777	40
30	.0960	.0958	.9954	.0963	10.39	1.005	10.43	1.4748	30
40	989	987	951	992	10.08	005	10.13	719	20
50	.1018	.1016	948	.1022	9.788	005	9.839	690	10
6°00′	.1047	.1045	.9945	.1051	9.514	1.006	9.567	1.4661	**84°00′**
10	076	074	942	080	9.255	006	9.309	632	50
20	105	103	939	110	9.010	006	9.065	603	40
30	.1134	.1132	.9936	.1139	8.777	1.006	8.834	1.4573	30
40	164	161	932	169	8.556	007	8.614	544	20
50	193	190	929	198	8.345	007	8.405	515	10
7°00′	.1222	.1219	.9925	.1228	8.144	1.008	8.206	1.4486	**83°00′**
10	251	248	922	257	7.953	008	8.016	457	50
20	280	276	918	287	7.770	008	7.834	428	40
30	.1309	.1305	.9914	.1317	7.596	1.009	7.661	1.4399	30
40	338	334	911	346	7.429	009	7.496	370	20
50	367	363	907	376	7.269	009	7.337	341	10
		Cos	**Sin**	**Cot**	**Tan**	**Csc**	**Sec**	**Radians**	**Degrees**

TABLE 4 *(continued)*

Degrees	Radians	Sin	Cos	Tan	Cot	Sec	Csc		
8°00′	.1396	.1392	.9903	.1405	7.115	1.010	7.185	1.4312	**82°00′**
10	425	421	899	435	6.968	010	7.040	283	50
20	454	449	894	465	6.827	011	6.900	254	40
30	.1484	.1478	.9890	.1495	6.691	1.011	6.765	1.4224	30
40	513	507	886	524	6.561	012	6.636	195	20
50	542	536	881	554	6.435	012	6.512	166	10
9°00′	.1571	.1564	.9877	.1584	6.314	1.012	6.392	1.4137	**81°00′**
10	600	593	872	614	197	013	277	108	50
20	629	622	868	644	084	013	166	079	40
30	.1658	.1650	.9863	.1673	5.976	1.014	6.059	1.4050	30
40	687	679	858	703	871	014	5.955	1.4021	20
50	716	708	853	733	769	015	855	992	10
10°00′	.1745	.1736	.9848	.1763	5.671	1.015	5.759	1.3963	**80°00′**
10	774	765	843	793	576	016	665	934	50
20	804	794	838	823	485	016	575	904	40
30	.1833	.1822	.9833	.1853	5.396	1.017	5.487	1.3875	30
40	862	851	827	883	309	018	403	846	20
50	891	880	822	914	226	018	320	817	10
11°00′	.1920	.1908	.9816	.1944	5.145	1.019	5.241	1.3788	**79°00′**
10	949	937	811	974	066	019	164	759	50
20	978	965	805	.2004	4.989	020	089	730	40
30	.2007	.1994	.9799	.2035	4.915	1.020	5.016	1.3701	30
40	036	.2022	793	065	843	021	4.945	672	20
50	065	051	787	095	773	022	876	643	10
12°00′	.2094	.2079	.9781	.2126	4.705	1.022	4.810	1.3614	**78°00′**
10	123	108	775	156	638	023	745	584	50
20	153	136	769	186	574	024	682	555	40
30	.2182	.2164	.9763	.2217	4.511	1.024	4.620	1.3526	30
40	211	193	757	247	449	025	560	497	20
50	240	221	750	278	390	026	502	468	10
13°00′	.2269	.2250	.9744	.2309	4.331	1.026	4.445	1.3439	**77°00′**
10	298	278	737	339	275	027	390	410	50
20	327	306	730	370	219	028	336	381	40
30	.2356	.2334	.9724	.2401	4.165	1.028	4.284	1.3352	30
40	385	363	717	432	113	029	232	323	20
50	414	391	710	462	061	030	182	294	10
14°00′	.2443	.2419	.9703	.2493	4.011	1.031	4.134	1.3265	**76°00′**
10	473	447	696	524	3.962	031	086	235	50
20	502	476	689	555	914	032	039	206	40
30	.2531	.2504	.9681	.2586	3.867	1.033	3.994	1.3177	30
40	560	532	674	617	821	034	950	148	20
50	589	560	667	648	776	034	906	119	10
15°00′	.2618	.2588	.9659	.2679	3.732	1.035	3.864	1.3090	**75°00′**
10	647	616	652	711	689	036	·822	061	50
20	676	644	644	742	647	037	782	032	40
30	.2705	.2672	.9636	.2773	3.606	1.038	3.742	1.3003	30
40	734	700	628	805	566	039	703	974	20
50	763	728	621	836	526	039	665	945	10
		Cos	**Sin**	**Cot**	**Tan**	**Csc**	**Sec**	**Radians**	**Degrees**

TABLE 4 (*continued*)

Degrees	Radians	Sin	Cos	Tan	Cot	Sec	Csc		
16°00′	.2793	.2756	.9613	.2867	3.487	1.040	3.628	1.2915	**74°00′**
10	822	784	605	899	450	041	592	886	50
20	851	812	596	931	412	042	556	857	40
30	.2880	.2840	.9588	.2962	3.376	1.043	3.521	1.2828	30
40	909	868	580	994	340	044	487	799	20
50	938	896	572	.3026	305	045	453	770	10
17°00′	.2967	.2924	.9563	.3057	3.271	1.046	3.420	1.2741	**73°00′**
10	996	952	555	089	237	047	388	712	50
20	.3025	979	546	121	204	048	356	683	40
30	.3054	.3007	.9537	.3153	3.172	1.049	3.326	1.2654	30
40	083	035	528	185	140	049	295	625	20
50	113	062	520	217	108	050	265	595	10
18°00′	.3142	.3090	.9511	.3249	3.078	1.051	3.236	1.2566	**72°00′**
10	171	118	502	281	047	052	207	537	50
20	200	145	492	314	018	053	179	508	40
30	.3229	.3173	.9483	.3346	2.989	1.054	3.152	1.2479	30
40	258	201	474	378	960	056	124	450	20
50	287	228	465	411	932	057	098	421	10
19°00′	.3316	.3256	.9455	.3443	2.904	1.058	3.072	1.2392	**71°00′**
10	345	283	446	476	877	059	046	363	50
20	374	311	436	508	850	060	021	334	40
30	.3403	.3338	.9426	.3541	2.824	1.061	2.996	1.2305	30
40	432	365	417	574	798	062	971	275	20
50	462	393	407	607	773	063	947	246	10
20°00′	.3491	.3420	.9397	.3640	2.747	1.064	2.924	1.2217	**70°00′**
10	520	448	387	673	723	065	901	188	50
20	549	475	377	706	699	066	878	159	40
30	.3578	.3502	.9367	.3739	2.675	1.068	2.855	1.2130	30
40	607	529	356	772	651	069	833	101	20
50	636	557	346	805	628	070	812	072	10
21°00′	.3665	.3584	.9336	.3839	2.605	1.071	2.790	1.2043	**69°00′**
10	694	611	325	872	583	072	769	1.2014	50
20	723	638	315	906	560	074	749	985	40
30	.3752	.3665	.9304	.3939	2.539	1.075	2.729	1.1956	30
40	782	692	293	973	517	076	709	926	20
50	811	719	283	.4006	496	077	689	897	10
22°00′	.3840	.3746	.9272	.4040	2.475	1.079	2.669	1.1868	**68°00′**
10	869	773	261	074	455	080	650	839	50
20	898	800	250	108	434	081	632	810	40
30	.3927	.3827	.9239	.4142	2.414	1.082	2.613	1.1781	30
40	956	854	228	176	394	084	595	752	20
50	985	881	216	210	375	085	577	723	10
23°00′	.4014	.3907	.9205	.4245	2.356	1.086	2.559	1.1694	**67°00′**
10	043	934	194	279	337	088	542	665	50
20	072	961	182	314	318	089	525	636	40
30	.4102	.3987	.9171	.4348	2.300	1.090	2.508	1.1606	30
40	131	.4014	159	383	282	092	491	577	20
50	160	041	147	417	264	093	475	548	10
		Cos	Sin	Cot	Tan	Csc	Sec	Radians	Degrees

TABLE 4 (*continued*)

Degrees	Radians	Sin	Cos	Tan	Cot	Sec	Csc		
24°00′	.4189	.4067	.9135	.4452	2.246	1.095	2.459	1.1519	**66°00′**
10	218	094	124	487	229	096	443	490	50
20	247	120	112	522	211	097	427	461	40
30	.4276	.4147	.9100	.4557	2.194	1.099	2.411	1.1432	30
40	305	173	088	592	177	100	396	403	20
50	334	200	075	628	161	102	381	374	10
25°00′	.4363	.4226	.9063	.4663	2.145	1.103	2.366	1.1345	**65°00′**
10	392	253	051	699	128	105	352	316	50
20	422	279	038	734	112	106	337	286	40
30	.4451	.4305	.9026	.4770	2.097	1.108	2.323	1.1257	30
40	480	331	013	806	081	109	309	228	20
50	509	358	001	841	066	111	295	199	10
26°00′	.4538	.4384	.8988	.4877	2.050	1.113	2.281	1.1170	**64°00′**
10	567	410	975	913	035	114	268	141	50
20	596	436	962	950	020	116	254	112	40
30	.4625	.4462	.8949	.4986	2.006	1.117	2.241	1.1083	30
40	654	488	936	.5022	1.991	119	228	054	20
50	683	514	923	059	977	121	215	1.1025	10
27°00′	.4712	.4540	.8910	.5095	1.963	1.122	2.203	1.0996	**63°00′**
10	741	566	897	132	949	124	190	966	50
20	771	592	884	169	935	126	178	937	40
30	.4800	.4617	.8870	.5206	1.921	1.127	2.166	1.0908	30
40	829	643	857	243	907	129	154	879	20
50	858	669	843	280	894	131	142	850	10
28°00′	.4887	.4695	.8829	.5317	1.881	1.133	2.130	1.0821	**62°00′**
10	916	720	816	354	868	134	118	792	50
20	945	746	802	392	855	136	107	763	40
30	.4974	.4772	.8788	.5430	1.842	1.138	2.096	1.0734	30
40	.5003	797	774	467	829	140	085	705	20
50	032	823	760	505	816	142	074	676	10
29°00′	.5061	.4848	.8746	.5543	1.804	1.143	2.063	1.0647	**61°00′**
10	091	874	732	581	792	145	052	617	50
20	120	899	718	619	780	147	041	588	40
30	.5149	.4924	.8704	.5658	1.767	1.149	2.031	1.0559	30
40	178	950	689	696	756	151	020	530	20
50	207	975	675	735	744	153	010	501	10
30°00′	.5236	.5000	.8660	.5774	1.732	1.155	2.000	1.0472	**60°00′**
10	265	025	646	812	720	157	1.990	443	50
20	294	050	631	851	709	159	980	414	40
30	.5323	.5075	.8616	.5890	1.698	1.161	1.970	1.0385	30
40	352	100	601	930	686	163	961	356	20
50	381	125	587	969	675	165	951	327	10
31°00′	.5411	.5150	.8572	.6009	1.664	1.167	1.942	1.0297	**59°00′**
10	440	175	557	048	653	169	932	268	50
20	469	200	542	088	643	171	923	239	40
30	.5498	.5225	.8526	.6128	1.632	1.173	1.914	1.0210	30
40	527	250	511	168	621	175	905	181	20
50	556	275	496	208	611	177	896	152	10
		Cos	**Sin**	**Cot**	**Tan**	**Csc**	**Sec**	**Radians**	**Degrees**

TABLE 4 (*continued*)

Degrees	Radians	Sin	Cos	Tan	Cot	Sec	Csc		
32°00′	.5585	.5299	.8480	.6249	1.600	1.179	1.887	1.0123	**58°00′**
10	614	324	465	289	590	181	878	094	50
20	643	348	450	330	580	184	870	065	40
30	.5672	.5373	.8434	.6371	1.570	1.186	1.861	1.0036	30
40	701	398	418	412	560	188	853	1.0007	20
50	730	422	403	453	550	190	844	977	10
33°00′	.5760	.5446	.8387	.6494	1.540	1.192	1.836	.9948	**57°00′**
10	789	471	371	536	530	195	828	919	50
20	818	495	355	577	520	197	820	890	40
30	.5847	.5519	.8339	.6619	1.511	1.199	1.812	.9861	30
40	876	544	323	661	501	202	804	832	20
50	905	568	307	703	1.492	204	796	803	10
34°00′	.5934	.5592	.8290	.6745	1.483	1.206	1.788	.9774	**56°00′**
10	963	616	274	787	473	209	781	745	50
20	992	640	258	830	464	211	773	716	40
30	.6021	.5664	.8241	.6873	1.455	1.213	1.766	.9687	30
40	050	688	225	916	446	216	758	657	20
50	080	712	208	959	437	218	751	628	10
35°00′	.6109	.5736	.8192	.7002	1.428	1.221	1.743	.9599	**55°00′**
10	138	760	175	046	419	223	736	570	50
20	167	783	158	089	411	226	729	541	40
30	.6196	.5807	.8141	.7133	1.402	1.228	1.722	.9512	30
40	225	831	124	177	393	231	715	483	20
50	254	854	107	221	385	233	708	454	10
36°00′	.6283	.5878	.8090	.7265	1.376	1.236	1.701	.9425	**54°00′**
10	312	901	073	310	368	239	695	396	50
20	341	925	056	355	360	241	688	367	40
30	.6370	.5948	.8039	.7400	1.351	1.244	1.681	.9338	30
40	400	972	021	445	343	247	675	308	20
50	429	995	004	490	335	249	668	279	10
37°00′	.6458	.6018	.7986	.7536	1.327	1.252	1.662	.9250	**53°00′**
10	487	041	969	581	319	255	655	221	50
20	516	065	951	627	311	258	649	192	40
30	.6545	.6088	.7934	.7673	1.303	1.260	1.643	.9163	30
40	574	111	916	720	295	263	636	134	20
50	603	134	898	766	288	266	630	105	10
38°00′	.6632	.6157	.7880	.7813	1.280	1.269	1.624	.9076	**52°00′**
10	661	180	862	860	272	272	618	047	50
20	690	202	844	907	265	275	612	.9018	40
30	.6720	.6225	.7826	.7954	1.257	1.278	1.606	.8988	30
40	749	248	808	.8002	250	281	601	959	20
50	778	271	790	050	242	284	595	930	10
39°00′	.6807	.6293	.7771	.8098	1.235	1.287	1.589	.8901	**51°00′**
10	836	316	753	146	228	290	583	872	50
20	865	338	735	195	220	293	578	843	40
30	.6894	.6361	.7716	.8243	1.213	1.296	1.572	.8814	30
40	923	383	698	292	206	299	567	785	20
50	952	406	679	342	199	302	561	756	10
		Cos	**Sin**	**Cot**	**Tan**	**Csc**	**Sec**	**Radians**	**Degrees**

TABLE 4 *(continued)*

Degrees	Radians	Sin	Cos	Tan	Cot	Sec	Csc		
40°00'	.6981	.6428	.7660	.8391	1.192	1.305	1.556	.8727	**50°00'**
10	.7010	450	642	441	185	309	550	698	50
20	039	472	623	491	178	312	545	668	40
30	.7069	.6494	.7604	.8541	1.171	1.315	1.540	.8639	30
40	098	517	585	591	164	318	535	610	20
50	127	539	566	642	157	322	529	581	10
41°00'	.7156	.6561	.7547	.8693	1.150	1.325	1.524	.8552	**49°00'**
10	185	583	528	744	144	328	519	523	50
20	214	604	509	796	137	332	514	494	40
30	.7243	.6626	.7490	.8847	1.130	1.335	1.509	.8465	30
40	272	648	470	899	124	339	504	436	20
50	301	670	451	952	117	342	499	407	10
42°00'	.7330	.6691	.7431	.9004	1.111	1.346	1.494	.8378	**48°00'**
10	359	713	412	057	104	349	490	348	50
20	389	734	392	110	098	353	485	319	40
30	.7418	.6756	.7373	.9163	1.091	1.356	1.480	.8290	30
40	447	777	353	217	085	360	476	261	20
50	476	799	333	271	079	364	471	232	10
43°00'	.7505	.6820	.7314	.9325	1.072	1.367	1.466	.8203	**47°00'**
10	534	841	294	380	066	371	462	174	50
20	563	862	274	435	060	375	457	145	40
30	.7592	.6884	.7254	.9490	1.054	1.379	1.453	.8116	30
40	621	905	234	545	048	382	448	087	20
50	650	926	214	601	042	386	444	058	10
44°00'	.7679	.6947	.7193	.9657	1.036	1.390	1.440	.8029	**46°00'**
10	709	967	173	713	030	394	435	999	50
20	738	988	153	770	024	398	431	970	40
30	.7767	.7009	.7133	.9827	1.018	1.402	1.427	.7941	30
40	796	030	112	884	012	406	423	912	20
50	825	050	092	942	006	410	418	883	10
45°00'	.7854	.7071	.7071	1.000	1.000	1.414	1.414	.7854	**45°00'**
		Cos	Sin	Cot	Tan	Csc	Sec	Radians	Degrees

ANSWERS TO
ODD-NUMBERED EXERCISES

■

CHAPTER 1

Exercises 1.1

1. $2 \cdot 2 \cdot 2 \cdot 2 \cdot 3 \cdot 5$ **3.** $2 \cdot 2 \cdot 2 \cdot 2 \cdot 3 \cdot 3 \cdot 3$
5. $2 \cdot 3 \cdot 5 \cdot 7 \cdot 11$ **7.** 11/15 **9.** 1/10 **11.** $0.\overline{428571}$
13. 0.525 **15.** $0.\overline{27}$ **17.** 0.125 **19.** 3/20
21. 41/333 **23.** 23/55 **25.** 73/66 **27.** $0.\overline{59}$
29. $0.3\overline{24415}$

Exercises 1.2

1. Associativity of addition **3.** Commutativity of multiplication **5.** Distributivity of the product over the sum
7. Distributive property and commutativity of multiplication
9. 11/40 **11.** $-1/18$ **13.** 4/135 **15.** 5/3 **17.** 21/2
19. $-5/14$ **21.** -2 **23.** 10/27 **25.** $-3/56$
27. 7/260 **29.** No; $\sqrt{2} + \left(-\sqrt{2}\right) = 0$, which is a rational number **35.** **(a)** rational **(b)** irrational **(c)** rational
(d) irrational **37.** $22/7 > \pi$

Exercises 1.3

1. 14641 **3.** 1/125 **5.** 16/9 **7.** -512 **9.** $-1/343$
11. $-3/64$ **13.** 13/36 **15.** 11/6 **17.** $1/4x^2y^6$
19. $1/x^{12}$ **21.** $(1 + x^4)/x^2$ **23.** $1/(x + y)^3$
25. $3x^2z^5/y^4$ **27.** $12/x^6$ **29.** $9x^4/y^2$ **31.** $c^8/256a^4b^8$
33. $1/4x^4y^6$ **35.** $3x^4y^3$ **37.** $1/3a^3b^4c^4$ **39.** x^4/a^8y^8
43. 5.15×10^9 **45.** 1.8×10^{-9} **47.** 5.142×10^{-10}
49. 2.13×10^6 **51.** 4.056×10^{-6} **53.** 5.55 to 5.65 ft
55. 15.755 to 15.765 km **57.** 21.3145 to 21.3155 kg
59. 3 **61.** 5 **63.** 4 **65.** 2 **67.** 2.0×10^{-3}
69. 5.6×10^3 **71.** 3.4×10^{-19} **73.** 3.5×10^{11}
75. 3.3×10^5 **77.** 3.5×10^{-12} **79.** 1.08×10^9 km
81. 5.983×10^{21} metric tons **83.** 7.347×10^{22} kg
85. 1.983×10^{20} N **87.** 1.863×10^5 miles/s
89. **(a)** 3.72×10^{-23} liter **(b)** 3.72×10^{-26} m³

Exercises 1.4

1. $x^2 - 2x - 3$ **3.** $\frac{1}{6}x^4 + 5x^2 - \frac{7}{10}$
5. $x^7 + 5x^5 + 4x^4 + 3x^3 + 2x^2 + x + 2$
7. $t^5 - t^4 + 6t^3 + 3t^2 + 3t + 1$
9. $3x^4 + 2x^3 - 4x^2 - x + 4$
11. $5x^3 - 7x^2y + 9xy^2 - 4y^3 - 4x^2 + 7xy - x + 5$
13. $8x^3 - 16x^2 + 10x - 2$
15. $3x^4 - 8x^3 + 14x^2 - 8x + 3$
17. $x^7 - 4x^6 - x^5 + 8x^4 - 2x^3$
19. $x^7 + x^6 - x^4 - x^3 + x^2 + x + 1$ **21.** $a^3 + b^3$
23. $x^2 + 2xy + y^2 - 1$ **25.** $4x^2 - 9y^2 + 24y - 16$
27. $x^4 - y^4 - x^2y^2 - 2xy^3$ **29.** $x^3 - 6x^2 + 11x - 6$
31. $27x^3 - 54x^2 + 33x - 6$ **33.** $8x^3 + 12x^2 + 6x + 1$
35. $a^4 - b^4$ **37.** $9x^4 - 13x^2 + 4$ **39.** $3x^3 + 4x - 5$
41. $3x^5 - 4x^3 - x^2 + 2x$ **43.** $4x^6 - 5x^4 + 3x^2$
45. $5x^2 - 3xy$ **47.** $x - 2xz^2 + 3x^2yz^2$
49. $4uv^3 - 2au^2v^2 + 3a^2u^3v$

Exercises 1.5

1. $4x^2 + 8x + 3$ **3.** $4x^2 + 4ax - 3a^2$
5. $4a^2x^2 - 20abx + 25b^2$ **7.** $16y^2 - 24by + 9b^2$
9. $x^4 - 25$ **11.** $x^2 - 2$ **13.** $a - b$
15. $x^2 - 2xy + y^2 + 2ax - 2ay + a^2$
17. $(x + 2)(x + 4)$ **19.** $(x - 3)(x - 8)$
21. $(x - 5)(x + 7)$ **23.** $(y - 6)(y + 2)$
25. $(p - 5)(p - 6)$ **27.** $(4x - 3)(x + 1)$
29. $2(x + 3)(x + 5)$ **31.** $5(x + 2)(x - 2)$
33. $3a(x - 1)(x + 3)$ **35.** $(8a + 3b)(8a - 3b)$
37. $(4x^2 + 9)(2x + 3)(2x - 3)$
39. $(x - 3)(x^2 + 3x + 9)$ **41.** $8(x + 2)(x^2 - 2x + 4)$
43. $2a(2x + 1)(4x^2 - 2x + 1)$ **45.** $(2x - 1)(x + 3)$
47. $(2x - 1)^2$ **49.** $(2x - 1)(3x - 1)$
51. $(x + 3)(3x + 2)$ **53.** $(c - 3d)(2a + b)$
55. $(2x - 3y)(3a - b)$ **57.** $(x^2 + 2a)(y + 3b)$

59. $(a + b - 2)(a^2 + 2ab + b^2 + 2a + 2b + 4)$
61. $(x + 2)(x^2 - 2x + 4)(x - 2)(x^2 + 2x + 4)$
63. $(x + 2y + 2z)(x + 2y - 2z)$
65. $(2x + 1)(4x^2 - 8x + 7)$ **67.** $4ab$
69. $\frac{1}{2}(u + 4)(u - 4)$ **71.** $\frac{1}{6}(3x + 1)(2x - 1)$
73. $\frac{1}{2}(x + 2)(x^2 - 2x + 4)$ **75.** 641.8

Exercises 1.6

1. $5ax^2$ **3.** $2x^5 + 3x^3 - x$ **5.** $(x - 3)/(x - 4)$
7. $2(2z + 1)/(3z - 1)$ **9.** $1/(x + 8)$
11. $(u - 3)/(u + 3)$ **13.** $2x/(2x - 1)$
15. $(4x^2 + 7x + 4)/(4x^2 + 11x + 6)$
17. $(4x + 5)/(x + 2)$ **19.** $(x^2 - 6x + 8)/(x - 3)$
21. $(-2x^2 + 4x + 12)/(2x - 1)(x + 4)$
23. $(13x + 12)/(x - 4)(x + 4)$
25. $8/(x + 4)$ **27.** $(x^2 + 5x - 1)/(x + 3)^2$
29. $(2x^2 - 4)/(2x + 1)(x + 1)(x - 1)$
31. $(5x + 1)/4(x - 1)(3x - 1)$
33. $(4x - 9)/(x + 1)(x + 2)(x - 3)$ **35.** $1/(x - 1)$
37. $y/2(y + 2)$ **39.** $1/4b^2$ **41.** $4/3(x - 1)$
43. $(x + 2)(x + 1)/(x - 1)(2x + 1)$
45. $(2x + 1)/(x - 1)$ **47.** $x^2/2$ **49.** $(4x - 2)/(2x + 1)$
51. $2x/3$ **53.** $x(x - 1)$ **55.** $x/2(2x - 1)$
57. $5(2x - 1)/(x + 3)$
59. $(3x + 1)(x + 1)/(2x - 1)(x + 4)$
61. $x^2/(x + 1)(x - 1)$ **63.** $(5 - x)/(3x - 7)$
65. $(2x^2 + 5x - 3)/(4x^2 - 6x - 2)$
67. $-(x + 3)/(x + 2)$ **69.** $(4 + x)/(x + 3)$

Exercises 1.7

1. 9 **3.** -2 **5.** $4/5$ **7.** -1 **9.** 29
11. $10 - 2\sqrt{21}$ **13.** $5\sqrt{6}$ **15.** $22\sqrt{3}$
17. $15 - 7\sqrt{3}$ **19.** 4 **21.** 12 **23.** $2a^2$
25. $-3a\sqrt[3]{a^2}$ **27.** $4x^2$ **29.** $3\sqrt{2}\,a^2b^2$ **31.** $4a^2x^3$
33. $2x^2y^3$ **35.** $2abc^2$ **37.** $4ab^2/c^3$
39. $2ab\sqrt[5]{ab^2}/c$ **41.** $\sqrt{2ab}/2a^2b$
43. $\sqrt[3]{12am^2n^2}/3n$ **45.** $2abx\sqrt[6]{2ab^2x^2}$ **47.** $m + n$
49. $5a\sqrt[4]{25a}$ **51.** $13\sqrt{2}$ **53.** $-10\sqrt{3}$
55. $10\sqrt{5} - 11\sqrt{2}$ **57.** $8\sqrt[3]{2}$ **59.** $3\sqrt{x}$
61. $10a\sqrt{2a}$ **63.** $(\sqrt{2} + \sqrt{6})/2$
65. $(a\sqrt{c} + \sqrt{bc})/c$ **67.** $-3 - 3\sqrt{2}$
69. $(\sqrt{15} + \sqrt{6})/3$ **71.** $(\sqrt{ac} - c)/c$ **73.** $1 + x$
75. $-2\sqrt{6} + \sqrt{30} - 4\sqrt{5} + 8$
77. $a(\sqrt{a} - 2)/(a - 4)$ **79.** $(x - \sqrt{xy})/(x - y)$
81. $(\sqrt{x - 1} + x - 1)/(2 - x)$
83. $(a - \sqrt{b})/(a^2 - b)$ **85.** $\sqrt{x^2 + 1} - x$

Exercises 1.8

1. $\sqrt{8^3}$ **3.** $1/\sqrt[4]{64}$ **5.** $\sqrt[5]{(a^2b^3)^3}$ **7.** $\sqrt{x^2 + y^2}$
9. $5^{2/3}$ **11.** $x^{3/4}$ **13.** $a^{5/2}$ **15.** $a^{17/4}$ **17.** 4

19. $1/9$ **21.** -27 **23.** $15u^2$ **25.** $25x^4$ **27.** $m^{1/3}n^{1/2}$
29. $25a^{1/2}/b$ **31.** $27/x^6y^3$ **33.** $1/6a^4b$
35. $1/16a^{4/3}b^2$ **37.** $z^{3/2}/6xy^2$ **39.** a^4b^3 **41.** $4a^3/5x^2$
43. $27a^{1/2}/8x^6$ **45.** $a^{1/2}/b^{3/4}$ **47.** $\sqrt[6]{72}$ **49.** $\sqrt[4]{2a^3}$
51. 1 **53.** $\sqrt[4]{2a^3}$ **55.** $\sqrt{2xy}$ **57.** $\sqrt[6]{mv}$ **59.** $\sqrt{2a}$

Exercises 1.9

1. $-1 + 5i$ **3.** $8 - 11i$ **5.** $2 - 13i$ **7.** $2 - 5i$
9. $7 - 9i$ **11.** $24 + 7i$ **13.** $-31 - 12i$ **15.** $\frac{15}{2} + \frac{5}{2}i$
17. 34 **19.** $145/144$ **21.** $-21 + 20i$ **23.** $-9 + 19i$
25. $30 - 15i$ **27.** $\frac{1}{2} - \frac{1}{2}i$ **29.** $-\frac{3}{4}i$ **31.** $-\frac{21}{29} + \frac{20}{29}i$
33. $\frac{1}{10} - \frac{7}{10}i$ **35.** $-\frac{4}{15} - \frac{7}{15}i$

Chapter Test

1. (a) rational (b) irrational **2.** p^6q^5r **3.** $8a^2 + 9a$
4. $3(-y + 1)/(y + 3)(y + 2)(y - 3)$
5. $3(4u + 5)(4u - 5)$
6. $-(\sqrt{10} - 5\sqrt{2} + 2\sqrt{5} - 10)/2$
7. $(5z - 2b)(4z + 3b)$ **8.** $6x - 3a^2x - 5a$
9. $(4v - 3w)(4n - 3q)$ **10.** $\frac{20}{17} - \frac{5}{17}i$

Review Exercises

1. 0.4375 **3.** $0.\overline{81}$ **5.** 0.234375 **7.** $259/50$
9. $47/333$ **11.** $29/22$ **13.** $1.\overline{4}$ **15.** $0.\overline{68}$
17. $15/2$ **19.** $1/4$ **21.** $-35/19$
23. No; for example $\sqrt{2} - \sqrt{2} = 0$
25. (a) rational (b) irrational (c) irrational (d) rational
27. $12x^2/a^4y^3$ **29.** $3a^2c^4/2b^5$ **31.** $3m/2y^5$
33. $m^6p^3q^{15}/8$ **35.** b^2x^6/a^4 **37.** $1/5a^3x^2$
39. $x^{1/2}y$ **41.** $2^{3/2}a^{7/2}/3^{3/2}x^2$ **43.** $b^{5/2}y^{3/4}$
45. 3.0×10^{-3} **47.** 9.3×10^{-18} **49.** 2.0×10^4
51. 5.73×10^{-22} **53.** 1.2×10^{11} **55.** 2.961×10^{12}
57. 9.468×10^{12} km **59.** 1.44×10^8
61. 6.4345×10^5 liters **63.** 1.318×10^{25} lb
65. $u^2 - 3u - 10$ **67.** $x^4 + x^2 - 6$ **69.** $9a^4 - b^4$
71. $x^2 - a$ **73.** $x - 4$ **75.** $(v - 5)^2$
77. $(3x - 2)(x + 4)$ **79.** $(2a + 5b)(x - 4y)$
81. $(2u - 3v)(b + a)$
83. $2x(2x + 3a + 3)(2x - 3a - 3)$
85. $(7x - 27)/(x - 6)(x - 3)$
87. $(6x + 2)/(x + 2)(x - 2)$
89. $(x + 7)/(x - 1)^2(x + 1)$ **91.** $y^2/(x^3 - y^3)$
93. $5/6(x - 1)$ **95.** $-5(x + 5)/x$ **97.** $(x + 1)/(x - 1)$
99. $4xy^3\sqrt{3x}$ **101.** $5ab\sqrt[4]{ab^3}$ **103.** $ab^4\sqrt{6a}$
105. $2\sqrt[5]{ab^3c^4}/c^2$ **107.** $2abc\sqrt[6]{b^2c^3}$ **109.** $9a^2x^4y^2$
111. $(4\sqrt{3} + \sqrt{15})/3$ **113.** $(a + \sqrt{ab})/a$
115. $2 + \sqrt{6} + \sqrt{10} + \sqrt{15}$
117. $(2x + 3\sqrt{xy} + y)/(x - y)$
119. $-\sqrt{x^2 - 1} - \sqrt{x^2 + x} - \sqrt{x^2 - x} - x$ **121.** 1
123. $8 - 13i$ **125.** $-31 - 12i$ **127.** 13 **129.** $34i$
131. $72 + 96i$ **133.** $\frac{4}{41} + \frac{5}{41}i$ **135.** $\frac{8}{89} + \frac{5}{89}i$

CHAPTER 2

Exercises 2.1

1. $\frac{10}{3}$ **3.** $-\frac{9}{8}$ **5.** $3\sqrt{2}/2$ **7.** -4 **9.** 1 **11.** 3
13. $3\sqrt{3}/2$ **15.** $8\sqrt{3} + 8\sqrt{2}$ **17.** 1 **19.** 1 **21.** $\frac{18}{11}$
23. $45/31$ **25.** No solution **27.** $\frac{7}{4}$ **29.** $\frac{3}{5}$ **31.** 4.17
33. 9.869 **35.** 6.29×10^{-14} **37.** $y = (x + 1)/(2x + 3)$
39. $u = (5v - 4)/(3v - 2)$ **41.** $y = (2x + 1)/(3 - 2x)$
43. $x = (2y + 5)/(1 - 3y)$ **45.** $v = (4u + 3)/(u - 2)$
47. $C = \frac{5}{9}(F - 32)$ **49. (a)** $T = PV/C$, **(b)** $V = CT/P$
51. (a) $y = (P/2) - x$, **(b)** $y = 100 \, \text{cm}$
53. $P_0 = P/(1 + rt)$, $r = (P - P_0)/P_0 t$ **55.** $r = S/2\pi h$
57. $m_1 = r^2 F/Gm_2$ **59.** $R = V/I$

Exercises 2.2

1. 111, 112, 113, 114 **3.** 158, 160, 162 **5.** 24, 25
7. 62, 64, 66 **9.** 31 cm, 18 cm **11.** 78 **13.** 250 dolls
15. \$125 **17.** \$80 **19.** 1320
21. 8 dimes, 12 quarters, 15 nickels **23.** \$192, \$2592
25. \$1183.60 **27.** \$22,000 at 7.1%, \$28,000 at 14.5%
29. 450 g of chemical A, 270 g of chemical B
31. 375 ft^3 of cement, 250 ft^3 of sand, 250 ft^3 of stone
33. 160 cm^3 **35.** 40 gal of A, 120 gal of B
37. 12 miles **39. (a)** 750 miles, **(b)** 3 h **41.** 36 mph
43. 2.5 h, 12:00 noon **45.** $\frac{12}{5}$ h **47.** $\frac{12}{7}$ h **49.** 20 min

Exercises 2.3

1. $-2 < \sqrt{3} < 2$ **3.** $-\sqrt{2} < -\frac{2}{3} < \frac{1}{2}$
5. $\frac{1}{4} < \sqrt{3} < \sqrt{5} < 4$ **7.** $-\frac{5}{2} < -\sqrt{2} < 1 < \sqrt{6}$
9. $-2.45 < -\sqrt{6} < \sqrt{2} < 17/12$
11. $\{x \in \mathbb{R}: -2 \leq x \leq 3\}$ **13.** $\{x \in \mathbb{R}: 2 \leq x < 8\}$

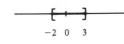

15. $\{x \in \mathbb{R}: x < 5\}$ **17.** $(-2, 7)$

19. $(-3, 3)$ **21.** $[-2, \infty)$

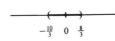

23. $5, -5$ **25.** $\frac{7}{2}, -\frac{7}{2}$ **27.** -2 **29.** $4, -2$
31. $-\frac{1}{2}, \frac{7}{2}$ **33.** $2, \frac{2}{3}$ **35.** $\frac{4}{3}, -\frac{2}{3}$ **37.** $-\frac{8}{5}, \frac{12}{5}$
39. $-\frac{3}{4}, \frac{9}{4}$ **41.** $9, -\frac{1}{5}$ **43.** $-\frac{1}{2}$ **45.** 15 **47.** 5
49. 14 **51.** $\pi - \frac{25}{8}$ **53.** $\sqrt{3} - \sqrt{2}$ **55.** $3 - a$
57. $10 - 2x$ **59.** 2, 5, 3 **61.** $\frac{5}{2}, 6, \frac{7}{2}$
63. 0.76, 0.42, 1.18

Exercises 2.4

1. $x > \frac{11}{6}$ **3.** $x \geq -1$

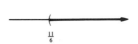

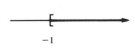

5. $x \geq \frac{28}{9}$ **7.** $x \geq \frac{6}{5}$

9. $x \leq -\frac{5}{12}$ **11.** $1 \leq x \leq \frac{3}{2}$

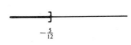

13. $\frac{11}{2} \leq x \leq \frac{23}{2}$ **15.** $-9 < x \leq 6$ **17.** $x > -\frac{2}{5}$
19. $x > \frac{1}{5}$ **21.** $x > -3$ **23.** $x < 4$ **25.** $x < -4$
27. $x < \frac{1}{3}$ **29.** $x > \frac{1}{4}$ **31.** $x > 4$ or $x < -4$

33. $-2 \leq x \leq 2$ **35.** $x < 3$ or $x > 5$

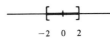

37. $-\frac{10}{3} < x < \frac{8}{3}$ **39.** $x > \frac{5}{6}$ or $x < \frac{1}{6}$

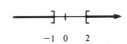

41. $x \geq 2$ or $x \leq -1$ **43.** $-1 < x < 7$

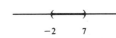

45. $-2 < x < 7$ **47.** $-\frac{11}{6} < x < \frac{13}{6}$

49. $x \leq -\frac{3}{2}$ or $x \geq \frac{9}{2}$ **51.** $x \leq -\frac{8}{9}$ or $x \geq \frac{16}{9}$

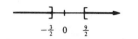

53. $x \neq -3$ **55.** $68° < F < 77°$ **57.** $10 < R < 20$
59. $50 \leq x \leq 70$ **61.** $4/3 \leq V \leq 4$
63. $4.5 \text{ sec} \leq t \leq 6.5 \text{ sec}$ **65.** $10.5 \leq MA \leq 21$
67. $50 \leq r \leq 60$ mph   **69.** 62 pairs of jeans

Exercises 2.5

1. ± 2 **3.** ± 4 **5.** $\pm 3\sqrt{2}/2$ **7.** $-1 \pm \sqrt{2}$
9. $(2 \pm \sqrt{11})/5$ **11.** $\pm i\sqrt{15}$ **13.** $-3 \pm i\sqrt{10}$
15. $(2 \pm 2i\sqrt{5})/3$ **17.** $2, 4$ **19.** $-5, -3$ **21.** $5, 1$
23. $0, 3$ **25.** $-\frac{3}{4}, 2$ **27.** $\frac{1}{3}, -\frac{3}{4}$ **29.** $-\frac{3}{4}, \frac{3}{2}$ **31.** $4, 0$
33. $0, -5$ **35.** $-2, 4$ **37.** $-5, -3$ **39.** $-8, -2$
41. $2, 4$ **43.** $2 \pm \sqrt{2}$ **45.** $\frac{1}{3}, \frac{2}{3}$ **47.** $(-3 \pm \sqrt{13})/4$
49. $(3 \pm i\sqrt{7})/4$ **51.** $(-2 \pm i\sqrt{2})/6$
53. $(-2 \pm \sqrt{14})/2$ **55.** $-2, 1$ **57.** $-\frac{3}{2}, 1$
59. $(-1 \pm \sqrt{7})/2$ **61.** $(1 \pm i\sqrt{19})/5$
63. $(1 \pm i\sqrt{7})/2$ **65.** $(1 \pm i\sqrt{35})/3$ **67.** $15, -4$
69. $\frac{1}{3}, -3$ **71.** $-1, 6$ **73.** $-\frac{1}{3}, 3$ **75.** $1, 2$
77. $0.25 \pm 0.32i$ **79.** $1.22 \pm 1.55i$
81. $2(x - 3)(x + 3)$ **83.** $(2x - 1)(x + 2)$
85. $(x - 1 + \sqrt{2})(x - 1 - \sqrt{2})$

Exercises 2.6

1. 144 **3.** -12 **5.** $3, 4$ **7.** $\frac{1}{3}, \frac{2}{3}$ **9.** 1 **11.** 11
13. $12, 9$ **15.** 3 **17.** 7 **19.** 2 **21.** $-3, 1$
23. $-2, 1$ **25.** 4 **27.** $(1 \pm i\sqrt{7})/2$ **29.** $0, 5$
31. $\frac{3}{2}, -2$ **33.** $\pm\sqrt{2}, \pm 2$ **35.** $-2, -\frac{1}{2}$
37. $\pm\sqrt{3 + \sqrt{17}}, \pm i\sqrt{\sqrt{17} - 3}$
39. $(\pm\sqrt{-3 + \sqrt{13}})/2, (\pm i\sqrt{3 + \sqrt{13}})/2$
41. $(12 \pm \sqrt{6})/6, 2 \pm \sqrt{2}$ **43.** $\pm\sqrt{2}, \pm i\sqrt{2}/2$
45. 4 **47.** $4, 169$ **49.** $\frac{1}{16}, \frac{1}{81}$ **51.** $8, 64$ **53.** $4, 1$
55. $-3, 0$ **57.** $-\sqrt[3]{3}, -\sqrt[3]{5}$ **59.** $2, 6$ **61.** $1, 2$
63. $r = \sqrt{A\pi}/\pi$ **65.** $d = \sqrt{GmMF}/F$
67. $t = \sqrt{2a(s - s_0)}/a$ **69.** $h = \sqrt{A^2 - \pi^2 r^4}/\pi r$

Exercises 2.7

1. $16, 17$ **3.** $24, 26$ **5.** $9, 18$ **7.** $5, \frac{1}{5}$ **9.** $6, 10$

11. $-15, 10$ **13.** $6, 9$ **15.** $8, 12$ **17.** $120, 240$
19. $5, 18$ **21.** $6, 8, 10$ **23.** $14, 14$ **25.** 1 m
27. 2 ft   **29.** 5% **31.** 60 **33.** 300 **35.** 2 mph
37. Peter takes 3 days, Mary takes 2 days   **39.** 50 and
60 mph   **41.** 150 m/min   **43.** **(a)** 10 s, 50 s   **(b)** 60 s
45. 80 in.   **47.** \$1.50   **49.** 1.619×10^5 m

Exercises 2.8

1. $(-\infty, -2) \cup (5, +\infty)$ **3.** $(-\infty, -2] \cup [2, +\infty)$
5. $(-\infty, -\frac{1}{2}] \cup [1, +\infty)$ **7.** $[0, 5]$
9. $(-\infty, -2) \cup (2, +\infty)$ **11.** $(-\infty, +\infty)$
13. No solution   **15.** $(-\infty, -3] \cup [0, +\infty)$
17. $(-\infty, -5) \cup (2, +\infty)$ **19.** $(-\infty, 2) \cup (5, +\infty)$
21. $(-\sqrt{6}/2, \sqrt{6}/2)$
23. $(-\infty, 2 - \sqrt{3}] \cup [2 + \sqrt{3}, +\infty)$
25. $(-1 - \sqrt{15})/2 < x < (-1 + \sqrt{15})/2$
27. $(-3, 5)$
29. $(-\infty, (1 - \sqrt{2})/2] \cup [(1 + \sqrt{2})/2, +\infty)$
31. $(-\infty, +\infty)$ **33.** $(-3, 1]$ **35.** $[0, 2] \cup (4, +\infty)$
37. $(-\infty, -3) \cup (5, +\infty)$ **39.** $(-\infty, -5) \cup [\frac{3}{2}, +\infty)$
41. $(-4, -1) \cup (2, +\infty)$ **43.** $(-\sqrt{3}, 0) \cup (0, \sqrt{3})$
45. $(-\infty, -3) \cup [1, 2]$ **47.** $[-2, -1) \cup (0, 4]$
49. $(-\infty, -3) \cup (-1, \frac{1}{2}) \cup (4, +\infty)$ **51.** $(-2, \frac{1}{2})$
53. **(a)** 5 s $< t < 55$ s   **(b)** 8.79 s $< t < 51.21$ s
(c) $0 \leq t < 6$ s or 54 s $< t < 60$ s   **55.** $0 < p < 150$
57. $t > 10$ h   **59.** $1.5 < x < 4.5$ **61.** width > 5 ft
63. $k < -4$ or $k > 4$ **65.** $20 \leq x \leq 50$
67. $a \leq x \leq b$

Chapter Test

1. 7 **2.** \$4,000 at 6%; \$6,000 at 8.5%   **3.** $-2, -8$
4. $\frac{13}{3} \leq x \leq 7$ **5.** $-\frac{1}{5}, -1$ **6.** $1, 8$ **7.** 20 m, 21 m
8. $-\frac{3}{2} < x < 1$ **9.** $(-\infty, 5) \cup (8, +\infty)$
10. $(2 - \sqrt{14})/2, (2 + \sqrt{14})/2$

Review Exercises

1. -1 **3.** $\frac{4}{13}$ **5.** 5 **7.** $-\frac{16}{5}$ **9.** 18 **11.** 1 **13.** $3, 4$
15. $-\frac{1}{2}, \frac{2}{3}$ **17.** $\frac{1}{3}$ **19.** 7 **21.** $-8 \pm \sqrt{70}$
23. $1 \pm \sqrt{3}/3$ **25.** $\frac{3}{4}$ **27.** 1 **29.** $-2, \frac{1}{2}$
31. $16, 81$ **33.** $2, 5$ **35.** $\frac{29}{7}, 5$ **37.** $b_1 = 2A/h - b_2$
39. $t = (l - l_0)/\alpha l_0$ **41.** $t = (-v_0 \pm \sqrt{v_0^2 + 2as})/a$
43. $r = (-h \pm \sqrt{h^2 + 2S/\pi})/2$ **45.** $x < -65$
47. $x \geq -41/23$ **49.** $x > -3$ **51.** $5 < x \leq \frac{15}{2}$
53. $\frac{1}{2} < x < \frac{5}{3}$ or $x > 3$ **55.** $-2 < x < \frac{1}{3}$ or $3 < x < 7$
57. 35, 36, 37, and 38   **59.** 1964   **61.** 216   **63.** 2 gal
65. \$400   **67.** 5.5 m, 8 m   **69.** 8 m   **71.** 1200
73. 5, 12, 13   **75.** 30 ft   **77.** 2.207
79. 162.7 foot-candles

CHAPTER 3

Exercises 3.1

1.

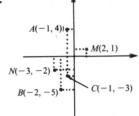

3. $\left(-\frac{3}{2}, \frac{9}{2}\right)$ **5.** $\left(-\frac{3}{8}, \frac{3}{4}\right)$

7. $((2 + a)/2, 0)$ **9.** $4\sqrt{2}$ **11.** $\sqrt{17}$ **13.** $\frac{7}{3}$
15. $\sqrt{26}$ **17.** $2\sqrt{1 + b^2}$ **19.** 5   **21.** 31.33
23. $d(A, B) = 5, d(A, C) = \sqrt{125}, d(B, C) = 10$
25. 25   **27.** $d(A, B) = d(B, C) = \sqrt{41}$
29. $d(A, B) = d(A, D) = d(C, D) = d(B, C) = 5\sqrt{2}$;
$d(B, D) = 10$ **31.** -2 **33.** 13, -22 **35.** $-\frac{1}{2}$, 16
37. $(4, 2)$ **39.** $d(A, P) = d(B, P) = \sqrt{26}$
41. $2x - 3y - 2 = 0$ **43.** $(4, 4), (1, 1)$
45. $(x - 2)^2 + (y + 4)^2 = 9$ **47.** On the line segment
49. Not on the line segment

Exercises 3.2

1. 1   **3.** $-1/12$ **5.** Points lie on the same line; common
slope: $3/2$ **7.** Common slope: $-1/2$; points are colinear
9. $x = -2$ **11.** $y = 3x - 2$ **13.** $y = (-2/5)x$
15. $y = -4x + 8$ **17.** $y = -2$ **19.** $2x - 3y + 6 = 0$
21. $y = (3/4)x + 5/4, m = 3/4, (0, 5/4), (-5/3, 0)$

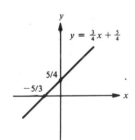

23. $y = (5/3)x - 5, m = 5/3, (0, -5), (3, 0)$

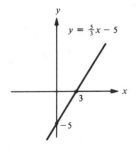

25. $y = -3/2, m = 0, (0, -3/2)$, no x-intercept

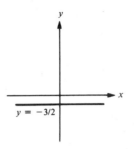

27. $y = (-3/2)x + 3, m = -3/2, (0, 3), (2, 0)$

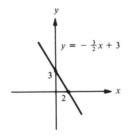

29. Perpendicular   **31.** Perpendicular

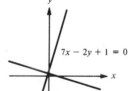
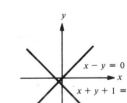

33. $y = 3x$ **35.** $x + 2y + 5 = 0$

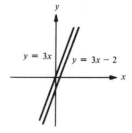
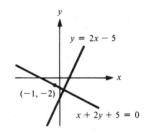

37. $x + y + 3 = 0$ **39.** $x - y + 10 = 0$

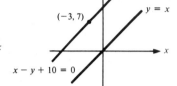

9.

No symmetry

11.

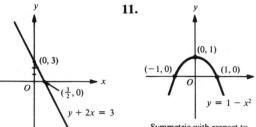

Symmetric with respect to the y-axis.

41. $3x + 4y + 3 = 0$ **43.** $-3/4$ **45.** 5

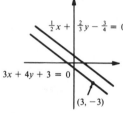

13.

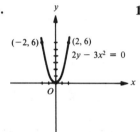

Symmetric with respect to the y-axis.

15.

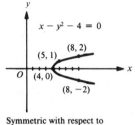

Symmetric with respect to the x-axis.

47. $5x - 19y - 85 = 0$ **49.** $m_{AB} = 2$, $m_{BC} = -1/2$
51. $7/3$ **53.** $2x + y - 6 = 0$ **55.** $v = (5/3)u + 5/3$
57. 2 **59.** (a) $s = w/5$, (b) 0.4 in.
61. $p = -200t + 1000$, $\$400$ **63.** 96 ft/s, 6 s

Exercises 3.3

1.

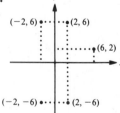

3.

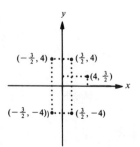

17.

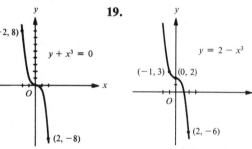

Symmetric with respect to the origin.

19.

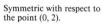

Symmetric with respect to the point $(0, 2)$.

5.

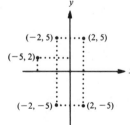

7.

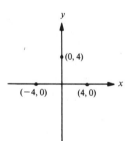

21.

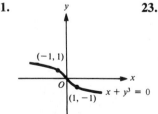

Symmetric with respect to the origin.

23.

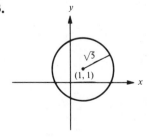

$(x - 1)^2 + (y - 1)^2 = 5$

25. $x^2 + y^2 = 8$

27. $(x - 2)^2 + (y + 1)^2 = 5$

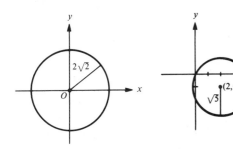

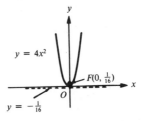

29. $(x - 3)^2 + (y - 5)^2 = 25$

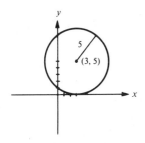

31. $(x - \frac{1}{2})^2 + (y - 4)^2 = \frac{13}{4}$

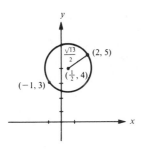

33. $(x - 3)^2 + (y - 3)^2 = 9$

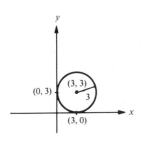

35. $x^2 + (y - 1)^2 = 9$, $C(0, 1)$, $r = 3$
37. $x^2 + y^2 = 9$, $C(0, 0)$, $r = 3$
39. $(x - 3)^2 + (y - 2)^2 = 4$, $C(3, 2)$, $r = 2$
41. $x^2 + (y - 2)^2 = 9$, $C(0, 2)$, $r = 3$
43. $(x + 3)^2 + (y + 2)^2 = 0$, $C(-3, -2)$, $r = 0$
45. $(x + (1/2))^2 + (y - (3/2))^2 = 5/2$, $C(-1/2, 3/2)$,
$r = \sqrt{10}/2$ **47.** $(x - 1)^2 + (y + 2)^2 = -1$; not a circle
49. $(x + \frac{3}{2})^2 + (y + 2)^2 = \frac{33}{4}$, $C(-\frac{3}{2}, -2)$, $r = \sqrt{33}/2$

Exercises 3.4

(*Note:* PA = principal axis, D = directrix)
1. $V(0, 0)$, $F(0, \frac{1}{16})$, PA: $x = 0$, D: $y = -\frac{1}{16}$

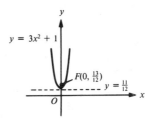

3. $V(0, 1)$, $F(0, \frac{13}{12})$; PA: $x = 0$, D: $y = \frac{11}{12}$

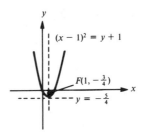

5. $V(0, 2)$, $F(0, 3)$, PA: $x = 0$, D: $y = 1$

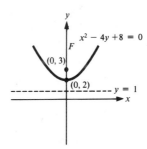

7. $V(1, -1)$, $F(1, -\frac{3}{4})$, PA: $x = 1$, D: $y = -\frac{5}{4}$

9. $y^2 = 6x$

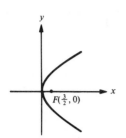

$F(\tfrac{3}{2}, 0)$

11. $(y - 2)^2 = 4(x - 1)$

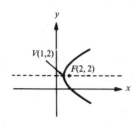

$V(1, 2)$ $F(2, 2)$

13. $y^2 = 8(x - 3)$

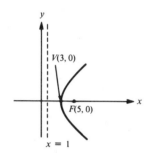

$V(3, 0)$ $F(5, 0)$ $x = 1$

15. $(x + 1)^2 = -2(y + 1)$

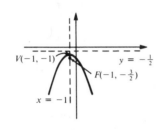

$V(-1, -1)$ $y = -\tfrac{1}{2}$ $F(-1, -\tfrac{3}{2})$ $x = -1$

17. $(x - 1)^2 = \tfrac{1}{4}(y - 2)$

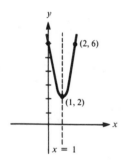

$(2, 6)$ $(1, 2)$ $x = 1$

19. Parabola opening upwards, $V(0, \tfrac{3}{4})$, $F(0, \tfrac{7}{4})$, PA: $x = 0$, D: $y = -\tfrac{1}{4}$ **21.** Parabola opening to the right, $V(-1, 1)$, $F(-\tfrac{3}{4}, 1)$, PA: $y = 1$, D: $x = -\tfrac{5}{4}$
23. $c < 4$ **25.** $(1, 3), (-2, -3)$
27. $C(0, 0)$, $V(\pm 4, 0)$, $F(\pm\sqrt{7}, 0)$, major axis: x-axis, minor axis: y-axis

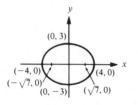

$(0, 3)$ $(-4, 0)$ $(4, 0)$ $(-\sqrt{7}, 0)$ $(0, -3)$ $(\sqrt{7}, 0)$

29. $C(0, 0)$, $V(\pm 3, 0)$, $F(\pm\sqrt{5}, 0)$, major axis: x-axis, minor axis: y-axis

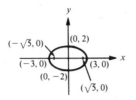

$(-\sqrt{5}, 0)$ $(0, 2)$ $(-3, 0)$ $(3, 0)$ $(0, -2)$ $(\sqrt{5}, 0)$

31. $C(1, -1)$, $V_1(-4, -1)$, $V_2(6, -1)$, $F_1(1 - \sqrt{21}, -1)$, $F_2(1 + \sqrt{21}, -1)$, major axis: $y = -1$, minor axis: $x = 1$

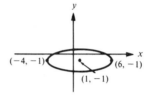

$(-4, -1)$ $(6, -1)$ $(1, -1)$

33. $x^2/16 + y^2/12 = 1$
35. $(x - 1)^2/25 + (y - 2)^2/16 = 1$
37. $(x - 3)^2/25 + y^2/16 = 1$ **39.** $x^2/225 + y^2/200 = 1$
41. $x^2/9 + y^2/5 = 1$ **43.** $x^2/25 + y^2/16 = 1$
45. $C(1, -1)$, major axis: $x = 1$, minor axis: $y = -1$
47. $C(2, -1)$, major axis: $y = -1$, minor axis; $x = 2$
49. 9.135×10^7 miles, 9.445×10^7 miles
51. $C(0, 0)$, $V(\pm 8, 0)$, $F(\pm 10, 0)$
53. $C(2, -3)$, $V_1(-7, -3)$, $V_2(11, -3)$, $F_1(-13, -3)$, $F_2(17, -3)$ **55.** $x^2/4 - y^2/12 = 1$
57. $(x - 2)^2/16 - (y - 1)^2/9 = 1$ **59.** $x^2 - y^2/4 = 1$
61. $(x - 1)^2/16 - (y - 1)^2/9 = 1$
63. $y^2/16 - x^2/9 = 1$
65. $C(0, -1)$, $V_1(0, 0)$, $V_2(0, -2)$
67. $C(-2, 1)$, $V_1(-3, 1)$, $V_2(-1, 1)$
69. $C(2, -1)$, $V_1(-1, -1)$, $V_2(5, -1)$

Chapter Test
1. $(0, \tfrac{3}{2})$, $\sqrt{89}$ **2.** $\tfrac{29}{3}$ **3.** $-\tfrac{13}{11}$ **4.** $y = -x + 4$
5. Perpendicular **6.** $(x - 1)^2 + (y + 2)^2 = 25$
7. $C(3, 4)$, $3\sqrt{3}$ **8.** $c > \tfrac{4}{3}$
9. $(x + 4)^2/36 + (y - 2)^2/27 = 1$ **10.** $x^2 - y^2/3 = 1$

Review Exercises
1. $(\tfrac{5}{2}, \tfrac{3}{2})$ **3.** $(\tfrac{13}{24}, -\tfrac{3}{20})$ **5.** $(6, 1)$ **7.** $(2 \pm \sqrt{11}, 0)$
9. $d(A, B) = \sqrt{125}$, $d(A, C) = 10$, $d(B, C) = 5$, area $= 25$
11. $d(A, B) = d(C, D) = 4\sqrt{2}$, $d(A, D) = d(B, C) = 6\sqrt{2}$
13. $7/3$ **15.** Symmetric with respect to the origin

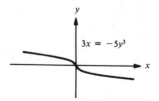

$3x = -5y^3$

17. Symmetric with respect to the line $x = 2$ **19.** $12/7$

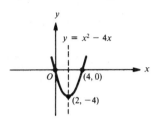

21. 0.041 **23.** $x + 3y + 20 = 0$
25. $10x - 6y + 9 = 0$ **27.** $10x + y - 6 = 0$
29. $3.95x + y - 2.08 = 0$ **31.** $3x + 2y - 15 = 0$
33. 8 **35.** $\approx 1.61, \approx 0.54$ **37.** $x + y - 8 = 0$
39. $v = 20t + 15, 3.5$ **41.** $V = 1200t + 20,000$

43. $x^2 + y^2 = 10$ **45.** $(x + 1)^2 + (y - 1)^2 = 25$
47. $(x + \sqrt{10})^2 + (y + \sqrt{10})^2 = 10$
49. $C(0, 0), r = 4$ **51.** $C(0, \frac{3}{2}), r = \sqrt{13}/2$
53. $C(-\frac{3}{2}, -\frac{5}{2}), r = 3$ **55.** $x^2 = 12y$ **57.** $y^2 = 8x$
59. $(x - 2)^2 = \frac{4}{3}(y - 1)$ **61.** $(1, 1)$ and $(\frac{1}{2}, \frac{1}{4})$
63. $(x + 1)^2/5 + y^2/9 = 1$
65. $(x - 1)^2/25 + (y - 3)^2/16 = 1$
67. $9x^2 + 16y^2 = 25$ **69.** $y^2/9 - x^2/16 = 1$
71. $x^2/4 - y^2/21 = 1$ **73.** $x^2/64 - y^2/36 = 1$
75. $C(2, 3), r = 3$
77. Ellipse, $C(-1, -1), a = \frac{5}{2}, b = \frac{5}{3}$
79. Parabola opening to the right, $V(-3, 3)$
81. Hyperbola, $C(1, -1), a = 3, b = \frac{3}{2}$ **83.** $\pm 2\sqrt{10}$
85. $(-1, 0), (3, 0)$ **87.** $(x + 2)^2 + (y - 3)^2 = 10$
89. 1.285×10^8 miles, 1.549×10^8 miles

CHAPTER 4

Exercises 4.1

1. $(-\infty, +\infty); (-\infty, +\infty)$ **3.** $(-\infty, +\infty); (-\infty, +\infty)$
5. $(-\infty, +\infty); (-\infty, 4]$ **7.** $(-\infty, +\infty); [-1, +\infty)$
9. $[0, +\infty); [-3, +\infty)$ **11.** $[0, +\infty); [1, +\infty)$
13. $[\frac{1}{2}, +\infty); [0, +\infty)$ **15.** $-10, -4, -2, \approx -0.33$
17. $-5, 0, 3, 1$ **19.** $\approx 1.59, -1, \approx 1.13$, not defined, $\frac{5}{2}$
21. not defined, $1, 2\sqrt{2}, \approx 1.80, \frac{1}{2}$
23. $(-\infty, +\infty)$ **25.** $(-\infty, +\infty)$

27. $(-\infty, +\infty)$ **29.** $(-\infty, +\infty)$

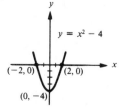

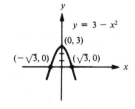

31. $[1, +\infty)$ **33.** $(-\infty, 4]$

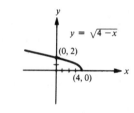

35. $[2, +\infty)$ **37.** $[0, +\infty)$

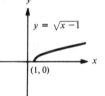

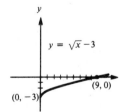

39. $[0, +\infty)$

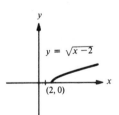

Exercises 4.2
1. $(-\infty, 2) \cup (2, +\infty)$ **3.** $(-\infty, \frac{4}{3}]$ **5.** $(-\infty, 1) \cup (1, +\infty)$
7. $(-\infty, 1) \cup (1, 2) \cup (2, +\infty)$ **9.** $[-1, 1]$
11. 9, 6, 8, ≈ 9.67 **13.** $-1, -1, -3, -(3 + \sqrt{2})$
15. $0, 4, \frac{2}{3}$, not defined **17.** $4, -2, 0, \approx 0.54$
19. $5 - 3a, 5 + 3a, 3a - 5, 10 - 3(a + b)$,
$5 - 3(a + b)$ **21.** $a^2 - 4, a^2 - 4, 4 - a^2$,
$a^2 + b^2 - 8, (a + b)^2 - 4$ **23.** $(a - 1)/(a + 1)$,
$(a + 1)/(a - 1), (-a + 1)/(a + 1), 2(ab - 1)/(a + 1)(b + 1)$,
$(a + b - 1)/(a + b + 1)$ **25.** $a/(1 - 5a), a - 5$,
$(a^2 - 10a + 1)/(a - 5)(1 - 5a), a/(a^2 - 5a + 1)$
27. $a/(2 - a), 2a - 1, (2a^2 - 2a + 2)/(2a - 1)(2 - a)$,
$a/(2a^2 + 2 - a)$ **29.** $(a^2 - 1)/a^2, 1/(1 - a^2)$,
$-(a^2 - 1)^2/a^2, (-a^4 - a^2 - 1)/a^2$ **31.** $x^2 + 3x + 2$,
$-x^2 + 3x + 8, 3x^3 + 5x^2 - 9x - 15, (3x + 5)/(x^2 - 3)$;
all domains are $(-\infty, +\infty)$ except that of f/g, which is
$(-\infty, -\sqrt{3}) \cup (-\sqrt{3}, \sqrt{3}) \cup (\sqrt{3}, +\infty)$
33. $(17x^2 + 10x - 1)/(2x - 1)(5x + 3)$,
$(13x^2 + 8x + 1)/(2x - 1)(5x + 3), (3x^2 + 3x)/(2x - 1)(5x + 3)$,
domains are $(-\infty, -\frac{3}{5}) \cup (-\frac{3}{5}, \frac{1}{2}) \cup (\frac{1}{2}, +\infty)$;
$3x(5x + 3)/(2x - 1)(x + 1)$,
domain is $(-\infty, -1) \cup (-1, \frac{1}{2}) \cup (\frac{1}{2}, +\infty)$
35. $2, -6/x, 1 - 9/x^2$, domains are $\{x \in \mathbb{R}: x \neq -3, x \neq 0\}$;
$(x - 3)/(x + 3)$, domain: $\{x \in \mathbb{R}: x \neq -3, x \neq 0\}$ **37.** 0
39. 3 **41.** $2a + h$ **43.** $6a + 3h - 1$ **45.** $4a + 2h - 3$
47. $A(x) = x(2x + 4)$ **49.** $A(p) = p^2/16$ **51.** $V(x) = 2x^3$
53. $V(h) = \pi h^3/16$ **55.** $L(h) = 8\pi\sqrt{h}$
57. $p(t) = 1000 + 60t$ **59.** $f = 5d_2/6$
61. $R = (250 + 10x)(100 - 2x)$ **63.** $A = 0.4x + 4.75$
65. $V = 2(2w - 4)(w - 4)$ **67.** (a) $L(h) = 2\pi h^2/3$,
(b) 0.13, 2.09, 3.27, 8.48 **69.** 1.40, 2.81, 4.44

9.
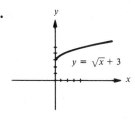

11.
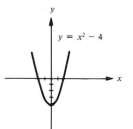

13.
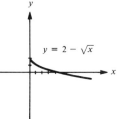

15.
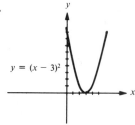

17.
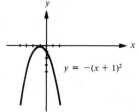

19.
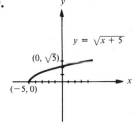

21.
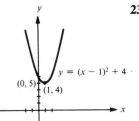

23.
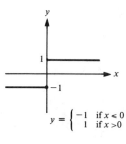

Exercises 4.3
1.
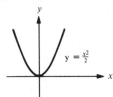

3.
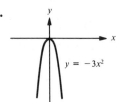

5.
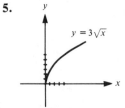

7.
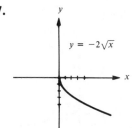

25.
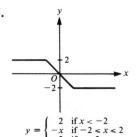

27.
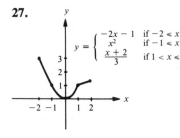

29.

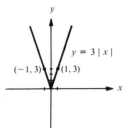

$y = 3|x|$

$(-1, 3)$ $(1, 3)$

31.

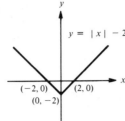

$y = |x| - 2$

$(-2, 0)$ $(2, 0)$

$(0, -2)$

63. Decreasing on $(-\infty, 0)$, increasing on $(0, +\infty)$

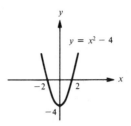

$y = x^2 - 4$

-2 2

-4

33.

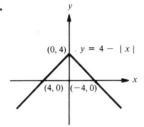

$(0, 4)$ $y = 4 - |x|$

$(4, 0)$ $(-4, 0)$

35.

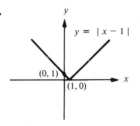

$y = |x - 1|$

$(0, 1)$

$(1, 0)$

65. Increasing on $(-\infty, 3)$, decreasing on $(3, +\infty)$

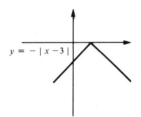

$y = -|x - 3|$

37.

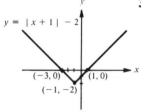

$y = |x + 1| - 2$

$(-3, 0)$ $(1, 0)$

$(-1, -2)$

39.

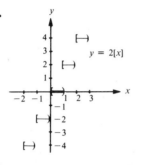

$y = 2[x]$

67. Increasing on $(-\infty, 0)$, neither increasing nor decreasing on $(0, +\infty)$

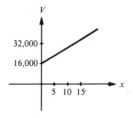

$y = \begin{cases} 0 & \text{if } x \geq 0 \\ 2x & \text{if } x < 0 \end{cases}$

41.

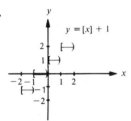

$y = [x] + 1$

43.

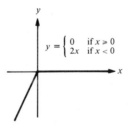

$y = [x - 1]$

45. Even   **47.** Odd   **49.** Even   **51.** Even   **53.** Even
55. Odd   **57.** Even   **59.** Odd
61. Decreasing on $(-\infty, +\infty)$

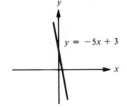

$y = -5x + 3$

69. $V(x) = 16000(1 + 0.08x)$

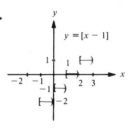

V

32,000

16,000

5   10   15

71.

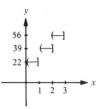

73.

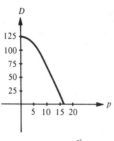

$$D(p) = 125 - \frac{p^2}{2}$$

43.

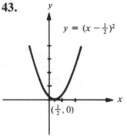

$$y = (x - \tfrac{1}{2})^2$$

45.

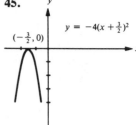

$$y = -4(x + \tfrac{3}{2})^2$$

Exercises 4.4

1. $u = 3v$ **3.** $y = 1/(3\sqrt{x})$ **5.** $y = 1/3x^2$ **7.** 3
9. 2/3   **11.** $\frac{28}{3}$ in.   **13.** 250 foot-candles   **15.** $E = kAT^4$
17. 180 ft   **19.** (a) $f = k\sqrt{T}/l$,   (b) Frequency remains
the same   **21.** $V = \frac{4}{3}\pi r^3$, 47.7 in.³   **23.** 162 lb
25. (a) $K = \frac{1}{2}mv^2$,   (b) 1.16×10^3
27. (a) $T = 2\pi\sqrt{l/g}$,   (b) 1.42 s   **29.** 11.9 years

47.

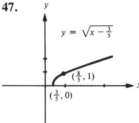

$$y = \sqrt{x - \tfrac{3}{5}}$$
$$(\tfrac{8}{5}, 1)$$
$$(\tfrac{3}{5}, 0)$$

49.

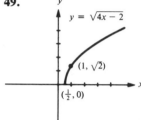

$$y = \sqrt{4x - 2}$$
$$(1, \sqrt{2})$$
$$(\tfrac{1}{2}, 0)$$

Chapter Test

1. $u \geq \frac{4}{3}$ **2.** 22, 22, 14
3. $(-\infty, -4) \cup (-4, -3) \cup (-3, +\infty)$
4. $-6 + 3h$ **5.** $A = 0.40x + 1.10$
6.   **7.** Neither   **8.**

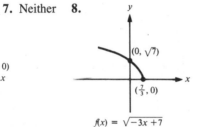

$$(3 - \sqrt{3}, 0) \qquad (3 + \sqrt{3}, 0)$$
$$(3, -3)$$
$$f(x) = (x - 3)^2 - 3$$

$$(0, \sqrt{7})$$
$$(\tfrac{7}{3}, 0)$$
$$f(x) = \sqrt{-3x + 7}$$

9. 3/4   **10.** $p = kx^2\sqrt[3]{y}/z^2$

Review Exercises

1. $(-\infty, \frac{5}{4}) \cup (\frac{5}{4}, +\infty)$ **3.** $(-\infty, \frac{2}{3}]$
5. $(-\infty, -\frac{5}{3}) \cup (-\frac{5}{3}, 2) \cup (2, +\infty)$
7. $(-\infty, -1) \cup (-1, 2) \cup (2, +\infty)$
9. $(-\infty, -3] \cup [3, +\infty)$ **11.** $(-\infty, -2) \cup [1, +\infty)$
13. $(-\infty, 0]$ **15.** $-\frac{9}{2}$, 1, 14, 14.6
17. $-47, 3, -47, -17$ **19.** $5, 7, \sqrt{5}/2$, not defined
21. $\frac{3}{5}, 0, -\frac{3}{5}, -0.45$ **23.** $0, \sqrt{11/13}, \sqrt{5}, 1.14$
25. -7 **27.** $10x + 5h - 1$ **29.** $-2/x(x + h)$
31. $2/(x + 1)(x + h + 1)$
33. $(2a + b)/b, 2(1 - a^3b)/(1 - a^2)(1 - ab)$,
$(a + b + a^2b)/(a + b - a^2b)$ **37.** $A(h) = h^2/6$
39. $L(r) = 10\pi r^2$ **41.**

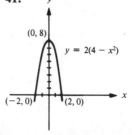

$$(0, 8)$$
$$y = 2(4 - x^2)$$
$$(-2, 0) \qquad (2, 0)$$

51.

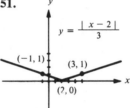

$$y = \frac{|x - 2|}{3}$$
$$(-1, 1)$$
$$(3, 1)$$
$$(2, 0)$$

53.

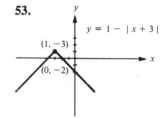

$$y = 1 - |x + 3|$$
$$(1, -3)$$
$$(0, -2)$$

55. Odd   **57.** Odd   **59.** Even   **61.** Even
63. Increasing on $(-\infty, +\infty)$

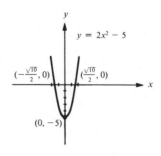

$$y = \tfrac{1}{2}x - 3$$
$$(6, 0)$$
$$(0, 3)$$

65. Decreasing on $(-\infty, 0)$, increasing on $(0 + \infty)$

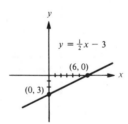

$$y = 2x^2 - 5$$
$$\left(-\tfrac{\sqrt{10}}{2}, 0\right) \qquad \left(\tfrac{\sqrt{10}}{2}, 0\right)$$
$$(0, -5)$$

67. Increasing on $(-\infty, -3)$, decreasing on $(-3, +\infty)$

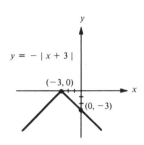

69. Increasing on $(4, +\infty)$ **71.** $P = kd$ **73.** $w = khr^2$

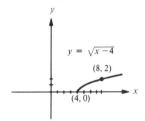

75. $3\sqrt{3}$ **77.** By a factor of 4 **79.** 13.5 gal

CHAPTER 5

Exercises 5.1

1. (a)

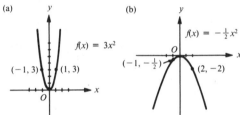

5. -12, no x-intercepts, opens upward **7.** -16, no x-intercepts, opens upward **9.** 81, two x-intercepts, opens downward **11.** 0, tangent to x-axis, opens upward **13.** -8, no x-intercepts, opens downward

15.

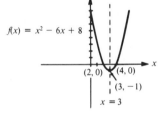

17.

19.

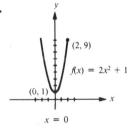

21.

3. (a)

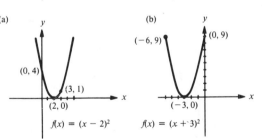

(c)

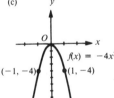

(d)

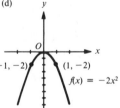

23.

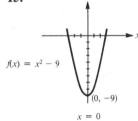

25.

(c)

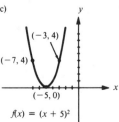

(d)

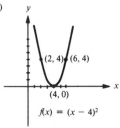

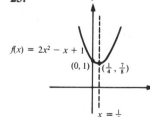

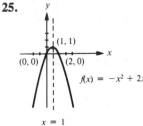

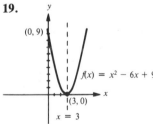

27.

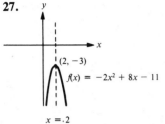

$f(x) = -2x^2 + 8x - 11$
$(2, -3)$
$x = 2$

29. 16

31. $-4 \le b \le 4$ **33.** 18, 18 **35.** 99225 m^2
37. 15 ft, 45 ft, 675 ft^2 **39.** 200, \$80 **41.** \$1600, 40¢
43. 121 ft, 5.5 s **45.** 5 days, 400 **47.** 300

Exercises 5.2

1.

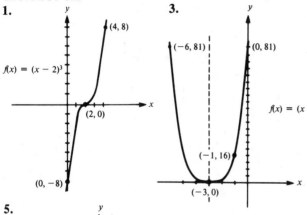

$f(x) = (x - 2)^3$
$(4, 8)$
$(2, 0)$
$(0, -8)$

3.

$f(x) = (x + 3)^4$
$(-6, 81)$
$(0, 81)$
$(-1, 16)$
$(-3, 0)$

5.

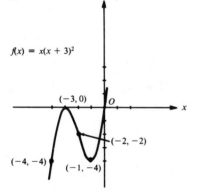

$f(x) = x(x + 3)^2$
$(-3, 0)$
O
$(-2, -2)$
$(-4, -4)$
$(-1, -4)$

7.

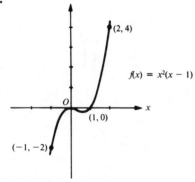

$f(x) = x^2(x - 1)$
$(2, 4)$
O
$(1, 0)$
$(-1, -2)$

9.

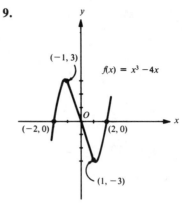

$f(x) = x^3 - 4x$
$(-1, 3)$
$(-2, 0)$
O
$(2, 0)$
$(1, -3)$

11.

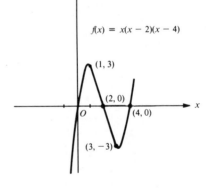

$f(x) = x(x - 2)(x - 4)$
$(1, 3)$
$(2, 0)$
O
$(4, 0)$
$(3, -3)$

13.

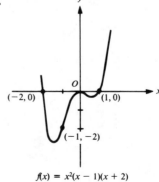

$(-2, 0)$
O
$(1, 0)$
$(-1, -2)$
$f(x) = x^2(x - 1)(x + 2)$

15.

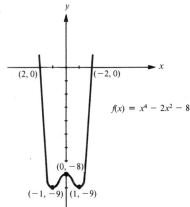

$f(x) = x^4 - 2x^2 - 8$

(2, 0) (−2, 0) (0, −8) (−1, −9) (1, −9)

21. Symmetric with respect to the point (0, 4)

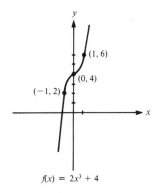

(1, 6) (0, 4) (−1, 2)

$f(x) = 2x^3 + 4$

17.

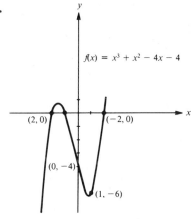

$f(x) = x^3 + x^2 - 4x - 4$

(2, 0) (−2, 0) (0, −4) (1, −6)

23. Symmetric with respect to the line $x = 0$

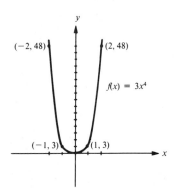

(−2, 48) (2, 48) $f(x) = 3x^4$ (−1, 3) (1, 3)

19.

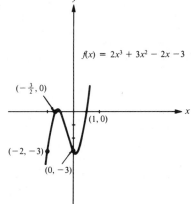

$f(x) = 2x^3 + 3x^2 - 2x - 3$

$\left(-\frac{3}{2}, 0\right)$ (1, 0) (−2, −3) (0, −3)

25. Symmetric with respect to $\left(0, -\frac{1}{3}\right)$

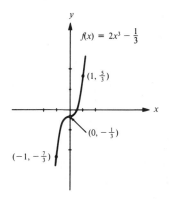

$f(x) = 2x^3 - \frac{1}{3}$ $\left(1, \frac{5}{3}\right)$ $\left(0, -\frac{1}{3}\right)$ $\left(-1, -\frac{7}{3}\right)$

27. Symmetric with respect to $(0, -4)$

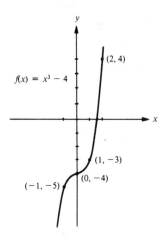

29. Symmetric with respect to $(-1, 0)$

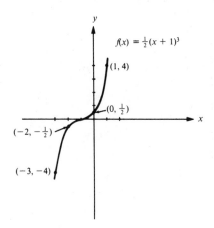

31. Symmetric with respect to $x = 2$

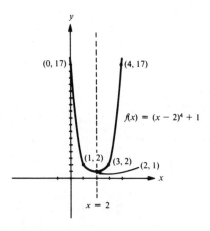

33. Symmetric with respect to $(\frac{1}{2}, 4)$

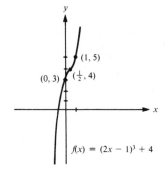

35.

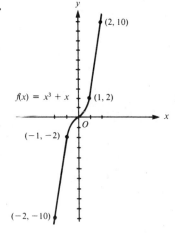

37.

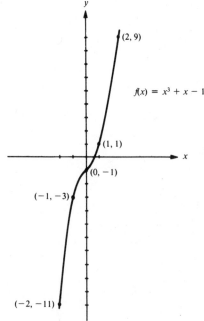

39.

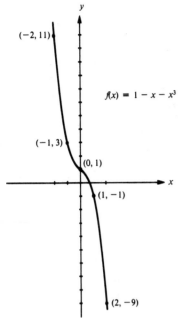

$f(x) = 1 - x - x^3$

$(-2, 11)$

$(-1, 3)$

$(0, 1)$

$(1, -1)$

$(2, -9)$

Exercises 5.3

(Note: VA = vertical asymptote, HA = horizontal asymptote)

1. VA: $x = -\frac{1}{2}$, HA: $y = 0$ **3.** VA: $x = -\frac{2}{3}$, HA: $y = \frac{2}{3}$
5. VA: $x = 2 \pm \sqrt{3}$, HA: $y = 2$ **7.** No VA; HA: $y = 0$
9. VA: $x = \frac{3}{5}$, no horizontal asymptote
11.

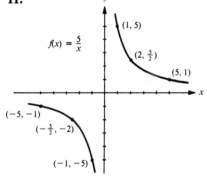

$f(x) = \frac{5}{x}$

$(1, 5)$

$(2, \frac{5}{2})$

$(5, 1)$

$(-5, -1)$

$(-\frac{5}{2}, -2)$

$(-1, -5)$

13.

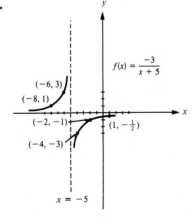

$f(x) = \frac{-3}{x + 5}$

$(-6, 3)$

$(-8, 1)$

$(-2, -1)$

$(1, -\frac{1}{2})$

$(-4, -3)$

$x = -5$

15.

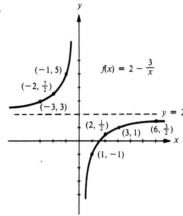

$(-1, 5)$

$(-2, \frac{7}{2})$

$f(x) = 2 - \frac{3}{x}$

$(-3, 3)$

$y = 2$

$(2, \frac{1}{2})$

$(3, 1)$ $(6, \frac{3}{2})$

$(1, -1)$

17.

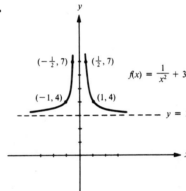

$(-\frac{1}{2}, 7)$ $(\frac{1}{2}, 7)$

$f(x) = \frac{1}{x^2} + 3$

$(-1, 4)$ $(1, 4)$

$y = 3$

19.

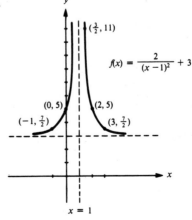

$(\frac{3}{2}, 11)$

$f(x) = \frac{2}{(x-1)^2} + 3$

$(0, 5)$ $(2, 5)$

$(-1, \frac{7}{2})$ $(3, \frac{7}{2})$

$x = 1$

21.

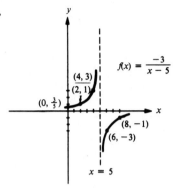

$$f(x) = \frac{-3}{x-5}$$

$(4, 3)$, $(2, 1)$, $(0, \frac{3}{5})$, $(8, -1)$, $(6, -3)$, $x = 5$

23.

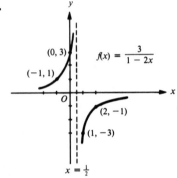

$$f(x) = \frac{3}{1-2x}$$

$(0, 3)$, $(-1, 1)$, $(2, -1)$, $(1, -3)$, $x = \frac{1}{2}$

25.

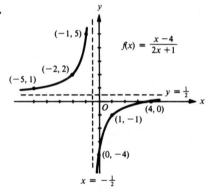

$$f(x) = \frac{x-4}{2x+1}$$

$(-1, 5)$, $(-2, 2)$, $(-5, 1)$, $y = \frac{1}{2}$, $(4, 0)$, $(1, -1)$, $(0, -4)$, $x = -\frac{1}{2}$

27.

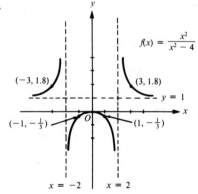

$$f(x) = \frac{x^2}{x^2 - 4}$$

$(-3, 1.8)$, $(3, 1.8)$, $y = 1$, $(-1, -\frac{1}{3})$, $(1, -\frac{1}{3})$, $x = -2$, $x = 2$

29.

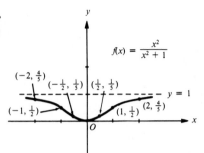

$$f(x) = \frac{x^2}{x^2 + 1}$$

$(-2, \frac{4}{5})$, $(-\frac{1}{2}, \frac{1}{5})$, $(\frac{1}{2}, \frac{1}{5})$, $y = 1$, $(-1, \frac{1}{2})$, $(1, \frac{1}{2})$, $(2, \frac{4}{5})$

31.

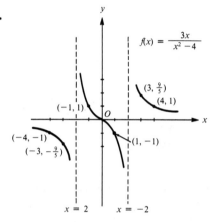

$$f(x) = \frac{3x}{x^2 - 4}$$

$(3, \frac{9}{5})$, $(4, 1)$, $(-1, 1)$, $(-4, -1)$, $(-3, -\frac{9}{5})$, $(1, -1)$, $x = 2$, $x = -2$

33.

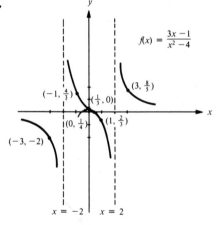

$$f(x) = \frac{3x-1}{x^2 - 4}$$

$(-1, \frac{4}{3})$, $(\frac{1}{3}, 0)$, $(3, \frac{8}{5})$, $(0, \frac{1}{4})$, $(1, \frac{2}{3})$, $(-3, -2)$, $x = -2$, $x = 2$

35.

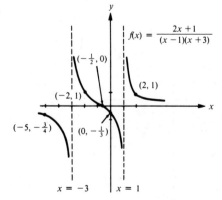

$$f(x) = \frac{2x+1}{(x-1)(x+3)}$$

$(-\frac{1}{2}, 0)$, $(2, 1)$, $(-2, 1)$, $(-5, -\frac{3}{4})$, $(0, -\frac{1}{3})$, $x = -3$, $x = 1$

37.

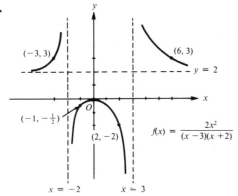

$$f(x) = \frac{2x^2}{(x-3)(x+2)}$$

39.

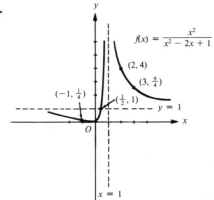

$$f(x) = \frac{x^2}{x^2 - 2x + 1}$$

41.

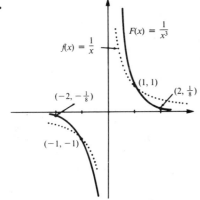

$$F(x) = \frac{1}{x^3}$$

$$f(x) = \frac{1}{x}$$

43. $y = \frac{3}{5}$ **45.** $y = 0$ **47.** $y = 0$
49. (a) $y = \frac{1}{3}$, (b) $y = 0$

Exercises 5.4
1. $6x - 5$, $6x + 5$, $4x + 9$ **3.** $3\sqrt{x + 5} + 1$, $\sqrt{3x + 6}$,
$9x + 4$ **5.** $-(5x^2 + 8)/(x^2 + 2)$, $1/(4x^2 - 20x + 27)$,
$4x - 15$ **7.** $x, x, x^9 + 3x^6 + 3x^3 + 2$ **9.** 2   **11.** 1.11
13. -0.197 **15.** 32   **17.** 2.6   **19.** 0   **21.** $F = f \circ h$

23. $H = g \circ h$ **25.** $V = g \circ g$
27. $f^{-1}(x) = (2x - 1)/x$, $x \neq 0$
29. $f^{-1}(x) = \sqrt{x + 4}$, $x \geq -4$
31. $f^{-1}(x) = (x^2 + 1)/4$, $x \geq 0$
33. $f^{-1}(x) = (x^3 + 4)/2$, $-\infty < x < +\infty$
35. Symmetric about the line $y = x$ **37.** Symmetric about
the line $y = x$
39.

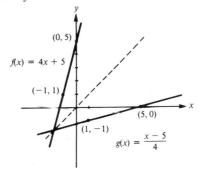

$$f(x) = 4x + 5$$

$$g(x) = \frac{x - 5}{4}$$

41.

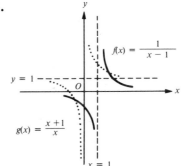

$$f(x) = \frac{1}{x - 1}$$

$$g(x) = \frac{x + 1}{x}$$

43.

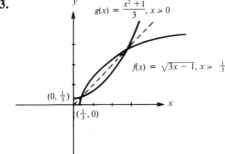

$$g(x) = \frac{x^2 + 1}{3}, x \geq 0$$

$$f(x) = \sqrt{3x - 1}, x \geq \frac{1}{3}$$

45. $(D \circ p)(c) = -0.018c^2 + 0.24c + 299.2$

Chapter Test
1. $(-4, 5)$, $x = -4$　　**2.** $(-2, -1)$

6. VA: $x = 3$, $x = 4$; HA: $y = 4$　**7.** $\sqrt{-3x - 6}$

8. -4　**9.** $(2x + 4)/(4 - x)$, $x \neq 4$

10. $f^{-1}(x) = \sqrt{(-6 - x)/5}$, $x \leq -6$

Review Exercises
1. -2 and 2, 8, $x = 0$, $(0, 8)$

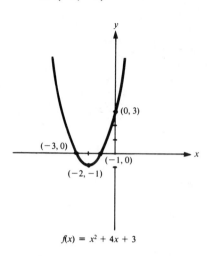

$(0, 3)$　$(-3, 0)$　$(-1, 0)$　$(-2, -1)$

$f(x) = x^2 + 4x + 3$

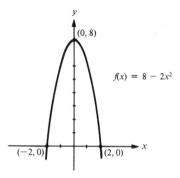

$(0, 8)$　$f(x) = 8 - 2x^2$　$(-2, 0)$　$(2, 0)$

3. $c > \frac{9}{8}$

4.

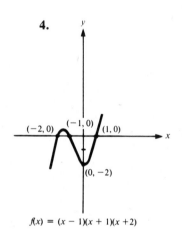

$(-2, 0)$　$(-1, 0)$　$(1, 0)$　$(0, -2)$

$f(x) = (x - 1)(x + 1)(x + 2)$

3. $-\frac{4}{3}$ and 1, -4, $x = -\frac{1}{6}$, $\left(-\frac{1}{6}, -\frac{49}{12}\right)$

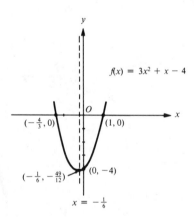

$f(x) = 3x^2 + x - 4$　O　$\left(-\frac{4}{3}, 0\right)$　$(1, 0)$　$\left(-\frac{1}{6}, -\frac{49}{12}\right)$　$(0, -4)$　$x = -\frac{1}{6}$

5.

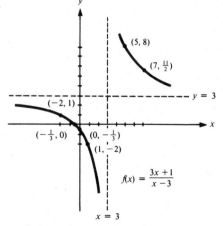

$(5, 8)$　$\left(7, \frac{11}{2}\right)$　$y = 3$　$(-2, 1)$　$\left(-\frac{1}{3}, 0\right)$　$\left(0, -\frac{1}{3}\right)$　$(1, -2)$

$f(x) = \dfrac{3x + 1}{x - 3}$

$x = 3$

5. no x-intercepts, -5, $x = 1$, $(1, -4)$

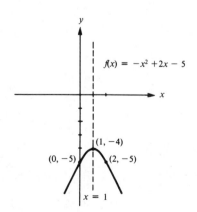

$f(x) = -x^2 + 2x - 5$　$(1, -4)$　$(0, -5)$　$(2, -5)$　$x = 1$

7.

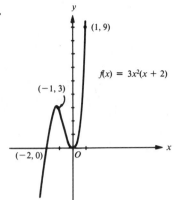

$f(x) = 3x^2(x + 2)$

(1, 9)

(−1, 3)

(−2, 0)   O

13.

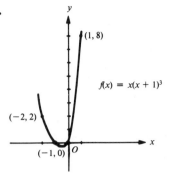

(1, 8)

$f(x) = x(x + 1)^3$

(−2, 2)

(−1, 0)   O

9.

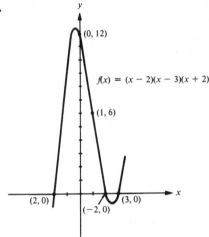

(0, 12)

$f(x) = (x - 2)(x - 3)(x + 2)$

(1, 6)

(2, 0)

(−2, 0)   (3, 0)

15.

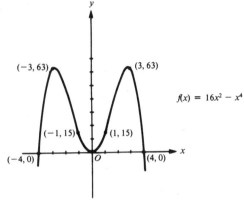

(−3, 63)   (3, 63)

$f(x) = 16x^2 - x^4$

(−1, 15)   (1, 15)

(−4, 0)   O   (4, 0)

11.

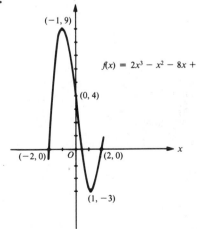

(−1, 9)

$f(x) = 2x^3 - x^2 - 8x + 4$

(0, 4)

(−2, 0)   O   (2, 0)

(1, −3)

17.

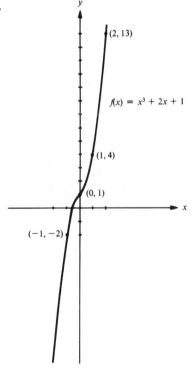

(2, 13)

$f(x) = x^3 + 2x + 1$

(1, 4)

(0, 1)

(−1, −2)

19.

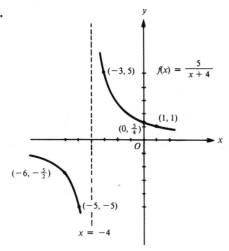

$f(x) = \dfrac{5}{x + 4}$

$(-3, 5)$, $(1, 1)$, $(0, \frac{5}{4})$, $(-6, -\frac{5}{2})$, $(-5, -5)$, $x = -4$

21.

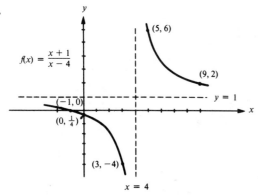

$f(x) = \dfrac{x + 1}{x - 4}$

$(5, 6)$, $(9, 2)$, $(-1, 0)$, $(0, \frac{1}{4})$, $(3, -4)$, $y = 1$, $x = 4$

23.

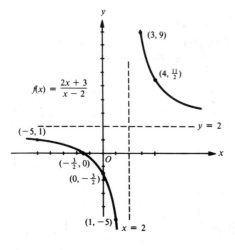

$f(x) = \dfrac{2x + 3}{x - 2}$

$(3, 9)$, $(4, \frac{11}{2})$, $(-5, 1)$, $(-\frac{3}{2}, 0)$, $(0, -\frac{3}{2})$, $(1, -5)$, $y = 2$, $x = 2$

25.

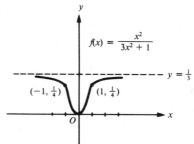

$f(x) = \dfrac{x^2}{3x^2 + 1}$

$y = \frac{1}{3}$, $(-1, \frac{1}{4})$, $(1, \frac{1}{4})$

27.

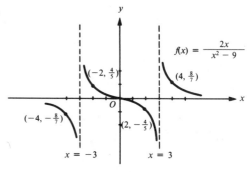

$f(x) = \dfrac{2x}{x^2 - 9}$

$(-2, \frac{4}{5})$, $(4, \frac{8}{7})$, $(-4, -\frac{8}{7})$, $(2, -\frac{4}{5})$, $x = -3$, $x = 3$

29.

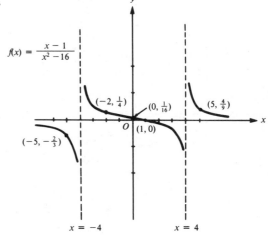

$f(x) = \dfrac{x - 1}{x^2 - 16}$

$(-2, \frac{1}{4})$, $(0, \frac{1}{16})$, $(5, \frac{4}{9})$, $(1, 0)$, $(-5, -\frac{2}{3})$, $x = -4$, $x = 4$

31.

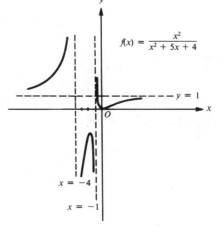

$f(x) = \dfrac{x^2}{x^2 + 5x + 4}$

$y = 1$, $x = -4$, $x = -1$

33. $y = \frac{3}{2}$ **35.** $y = \frac{4}{3}$ **37.** $y = 0$
39. $1/(3\sqrt{2x - 3} + 1)$, $\sqrt{-(9x + 1)/(3x + 1)}$,
$(3x + 1)/(3x + 4)$ **41.** $x, x, (x - 1)/(5 - x)$

43.

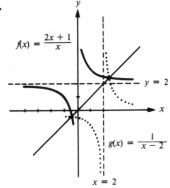

45.

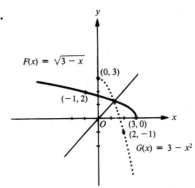

47. $f^{-1}(x) = (2 + 5x)/3x$, $x \neq 0$
49. $f^{-1}(x) = (x^2 + 4)/3$, $x \geq 0$
51. $f^{-1}(x) = -\sqrt{9 - x}$, $0 \leq x \leq 9$ **55.** $y = -1$
57. $P(t) = 55 - 60\sqrt{t} - 20t$

--- CHAPTER 6 ---

Exercises 6.1

1.

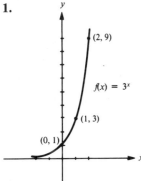

3.

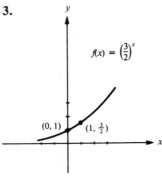

9.

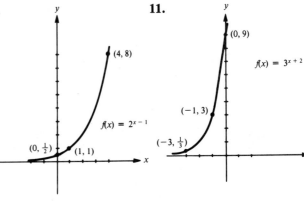

11.

5.

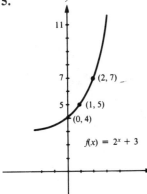

7.

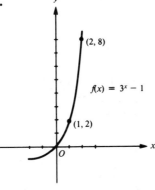

13.

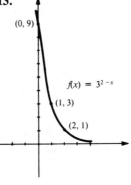

15.

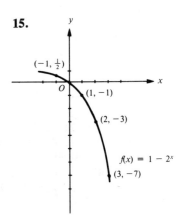

17. $16, \frac{1}{16}, 2$ **19.** $15, \frac{5}{3}, 45$ **21.** $19, \frac{13}{4}, 4$ **23.** 7.389
25. 1.649 **27.** 0.301 **29.** 1.051
31.

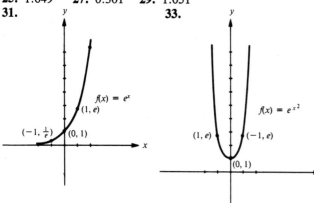

33.

35.

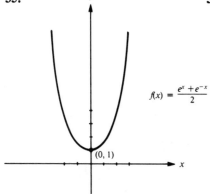

37. 4.729 **39.** 0.045

41. 1.633 **43.** 36.462 **45.** 7.767 **47.** −0.358, 0.01007, 0.726 **49.** 2.44141, 2.59374, 2.71692, 2.71827

Exercises 6.2
1. 7200, 14400 **3.** 15 million **5.** 15 g, 7.5 g
7. (a) \$10,771.36 **(b)** \$11,602.22; **(a)** \$10,778.84
(b) \$11,618.34 **9.** \$6,336.75 **11.** 180 million
13. 21,224 **15.** $n(t) = 15000 \cdot 2^t$; 120,000 **17.** 605,
429 **19.** $Q(t) = 100(1/2)^{t/1620}$; **(a)** 76 mg
(b) 25 mg **21.** 0.67 lm **23.** 23,000 years **25.** 30 s,
45 s **27.** 50% **29.** $3C_e/4$

Exercises 6.3
1. 4 **3.** −3 **5.** 4 **7.** −2 **9.** −3 **11.** $\log_2 32 = 5$
13. $\log_4(\frac{1}{16}) = -2$ **15.** $\log_{1/3}(\frac{1}{81}) = 4$ **17.** $\log_a c = b$
19. $2^7 = 128$ **21.** $2^{-10} = \frac{1}{1024}$ **23.** $10^2 = 100$
25. $10^r = q$ **27.** 1 **29.** 1.26 **31.** 1.30 **33.** −0.22
35. 0.52 **37.** 0.90 **39.** 1/2 **41.** 216 **43.** 81/16
45. $\sqrt[5]{4}$ **47.** 1 **49.** 2 **51.** 4 **53.** 7/16 **55.** $\log 4$
57. $\log_5 36$ **59.** $\ln(x^2 z^4/y^{1/4})$ **61.** $\log(\sqrt[3]{y}/x^2)$
63. $\log[(x + 1)x^2/(x - 2)]$ **65.** $\log[x^2(x + 2)^3/(x + 1)^4]$

67. $2\log x - 3\log(x - 2)$ **69.** $2\log x + 4\log(x - 2)$
71. $\frac{3}{2}\log(x - 2) + 2\log x$
73. $2\log x + 3\log(x + 1) - 5\log(x - 2)$
75. $\frac{1}{3}\log(x + 1) - \log x - \frac{2}{3}\log(x - 2)$

Exercises 6.4
1. 0.5705 **3.** 3.5119 **5.** 4.8531 **7.** −1.6478
9. 1.7093 **11.** 3460 **13.** 84,198 **15.** 5.45×10^{-3}
17. 20.8 **19.** 2.11×10^{-3} **21. (a)** 5.0, **(b)** 6.5
23. (a) 1.6×10^{-8}, **(b)** 10^{-5} **25.** $10^{-8} < [H^+] < 10^{-6}$
27. 120 dB; yes **29.** 6.5 **31.** $10^{5.6}$ **33.** First earthquake is 10 times more intense. **35.** 1.91 **37.** 3.53
39. (a) 65 **(b)** 58, 52

Exercises 6.5
1. 5 **3.** 3 **5.** −5/2 **7.** −15/2 **9.** 0.356
11. −1.893 **13.** 0.064 **15.** 0.4190 **17.** 20 **19.** 3
21. $(-3 + \sqrt{41})/2$ **23.** 1, 10^{-2} **25.** 10^{10} **27.** 0
29. 0.5493 **31.** 0.875 **33.** 1.26 **35.** 1.889
37. 1.185 **39.** 3.33 **41.** 2.2 **43.** 1.75 **45.** −1.33
47. 1.0686 **49.** 2.4330 **51.** 0.1884 **53.** −2.7210
55. 3.4 **57. (a)** 2.5 **(b)** 4 **59.** 83 min
61. -1.155×10^{-12} **63. (a)** 5.78 miles **(b)** 9.16 miles
65. (a) \$3,351.60 **(b)** 6.25% **67.** 4.4°C
69. 5.8 years

Chapter Test
1.

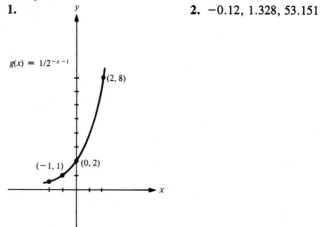

2. −0.12, 1.328, 53.151

3. 70, 35, 8.75 mg **4.** 31,128 **5.** $-3 + \log_2 5$
6. $\log[x^4(3x + 1)/\sqrt[3]{(x + 1)^2}]$ **7.** $10^{-10.5}$
8. 256, $9\sqrt{3}/16$ **9.** $10^{-0.6}$ **10.** 51/4

Review Exercises
1. (a) 8 **(b)** 7 **(c)** 25 **(d)** 2 **3. (a)** $\frac{1}{64}, \frac{1}{16}, 16$
(b) 0.3536, 9.931×10^{-5}, 0.1768 **5.** 10, 2, 80
7. (a) 16, 3 **(b)** 100

9. (a)

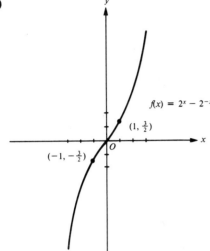

(b)

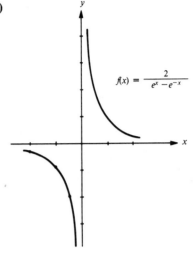

11. 14   **13.** 3/2   **15.** −7/6   **17.** 8   **19.** 16/15
21. 3/2   **23.** −11   **25.** 3/4   **27.** 3   **29.** 1.9734
31. 0   **33.** ln(84/5)   **35.** ln(32768/25)
37. $\log(a^7\sqrt[4]{c^3}/\sqrt{b})$ **39.** $\log[(x + 1)^2 x^4/(x + 2)^3]$
41. $\ln x + 2\ln(x - 1) - 3\ln(2x + 1)$

43. $\frac{1}{2}\ln x + \frac{3}{2}\ln(x - 1) - 4\ln(2x + 1)$ **45. (a)** −1.4930
(b) −2.5899   **(c)** −2.2038   **(d)** 6.6548   **47. (a)** 2.3
(b) 4.3   **49.** 80 dB   **51.** $I_1 = 3.16 \times 10^{-2} I_2$ **53.** 4.29
55. 33 years   **57.** 8.69×10^{-5} g,   **59.** $6,860.15
61. $237.04, $105.35   **63.** 114.7°F

CHAPTER 7

Exercises 7.1

1. 510°, −210°   **3.** 300°, −420°   **5.** $\pi/3$, $-5\pi/3$
7. Second   **9.** First   **11.** Second   **13.** $5\pi/6$
15. $-4\pi/3$ **17.** $-2\pi/3$ **19.** 150°   **21.** −30°
23. −405°   **25.** 5 rad   **27.** 25 rad   **29.** $\pi/3$ rad = 60°
31. $\pi/6$ rad = 30°   **33.** 60 cm   **35.** $72/\pi$ m
37. 2.4 cm   **39.** 30.73°   **41.** 85.51°   **43.** 229.18°
45. 85.94°   **47.** 286°28′44″   **49.** 68°45′18″
51. 8°35′40″   **53.** 0.7854   **55.** 0.6632   **57.** 2.1991
59. 0.6685   **61.** 1.1810   **63.** 0.2182
65. $\pi/6$ rad, $\pi/12$ rad, $5\pi/2$ rad
67. (a) $16\pi/15$ rad/s, $16\pi/15$ m/s   **(b)** $16\pi/3$ m
69. (a) $13\pi/5$ rad/s   **(b)** $78\pi/5$ in./s
71. (a) $4\pi/73$ rad   **(b)** $12\pi/73$ rad   **73.** 8 rad
75. (a) 3.336×10^3 km   **(b)** 6.672×10^3 km
(c) 6.371×10^3 km   **77.** 1.15 miles   **79.** 3662 miles

Exercises 7.2

In Exercises 1–9, answers are given in the order sine, cosine, tangent, cotangent, secant, and cosecant.

1. 4/5, 3/5, 4/3, 3/4, 5/3, 5/4   **3.** 8/17, 15/17, 8/15, 15/8, 17/15, 17/8   **5.** $3\sqrt{13}/13$, $2\sqrt{13}/13$, 3/2, 2/3, $\sqrt{13}/2$, $\sqrt{13}/3$ **7.** 0.5573, 0.8303, 0.6712, 1.4900, 1.2043, 1.7944   **9.** 0.556, 0.831, 0.670, 1.493, 1.204, 1.797   **11.** 20/3   **13.** 0.68   **15.** $\cos\alpha = \sqrt{39}/8$, $\tan\alpha = 5\sqrt{39}/39$, $\cot\alpha = \sqrt{39}/5$, $\sec\alpha = 8\sqrt{39}/39$, $\csc\alpha = 8/5$ **17.** $\sin\alpha = \sqrt{7}/4$, $\tan\alpha = \sqrt{7}/3$, $\cot\alpha = 3\sqrt{7}/7$, $\sec\alpha = 4/3$, $\csc\alpha = 4\sqrt{7}/7$
19. $\sin\alpha = 3\sqrt{34}/34$, $\cos\alpha = 5\sqrt{34}/34$, $\cot\alpha = 5/3$, $\sec\alpha = \sqrt{34}/5$, $\csc\alpha = \sqrt{34}/3$ **21.** $\sin\alpha = \sqrt{3}/3$, $\cos\alpha = \sqrt{6}/3$, $\tan\alpha = \sqrt{2}/2$, $\sec\alpha = \sqrt{6}/2$, $\csc\alpha = \sqrt{3}$ **23.** $\sin\alpha = 2/3$, $\cos\alpha = \sqrt{5}/3$, $\tan\alpha = 2\sqrt{5}/5$, $\cot\alpha = \sqrt{5}/2$, $\sec\alpha = 3\sqrt{5}/5$
25. $\cos 42°$ **27.** $\cot(\pi/12)$ **29.** $\csc 12°$ **31.** 0.5299
33. 3/5   **35.** $\sqrt{7}/3$ **37.** $3\sqrt{58}/58$ **39.** 0.2773
41. 0.5495   **43.** 0.1045   **45.** 0.4142   **47.** 0.3843
49. 0.2135   **51.** 0.3947   **53.** 1.1184

Exercises 7.3

1. $b = 6$, $c = 12$, $\beta = 30°$ **3.** $c = 3\sqrt{2}$, $\alpha = \beta = 45°$
5. $b = 3$, $c = 8$, $\alpha = 65°$ **7.** $b = 24$, $c = 28$,
$\beta = 57°50′$ **9.** $a = 46$, $c = 47$, $\beta = 14°40′$

11. $c = 13$, $\alpha = 22°37'11''$, $\beta = 67°22'49''$ **13.** $a = 6.1$, $c = 12.2$, $\beta = 59°40'$ **15.** $a = 5.6$, $b = 17.6$, $\alpha = 17°29'$ **17.** $b = 197.7$, $c = 234.2$, $\beta = 57°34'$ **19.** $a = 160.1$, $c = 270.7$, $\beta = 53°45'$ **21.** 920 ft   **23.** $51°20'$ **25.** $41°25'$ **27.** 112 ft   **29.** $96\sqrt{3}$ cm^2 **31.** 3549 m   **33.** 2 min   **35.** 41 ft, 121 ft   **37.** 42 m   **39.** 751 m

Chapter Test

1. $28\pi/15$ **2.** $48°$ **3.** $624/7\pi$ cm   **4.** 2π rad/s, 8π in/s   **5.** $\cos\alpha = 2\sqrt{14}/9$, $\tan\alpha = 5/2\sqrt{14}$, $\cot\alpha = 2\sqrt{14}/5$, $\sec\alpha = 9/2\sqrt{14}$, $\csc\alpha = 9/5$ **6.** $\sqrt{55}/8$ **7.** $b = 4.9$, $c = 9.7$, $\alpha = 59°30'$ **8.** $18.4°$ **9.** $3/\sqrt{34}$, $5/\sqrt{34}$ **10.** $4\pi/3$, $2\pi/3$

Review Exercises

1. $2\pi/5$, $\pi/12$, $-5\pi/18$, $3\pi/4$, $-17\pi/36$ **3.** $40°$, $252°$, $36°$, $70°$, $15°$ **5.** 0.4407, 2.1049, 1.3119, 1.1899, 3.6667 **7.** $229.18°$, $572.96°$, $85.94°$, $275.02°$, $128.92°$ **9.** 7.5 rad

11. 1.9 ft   **13.** 1.57 cm   **15.** (a) 80π rad/s   (b) 628.3 m   **17.** 17488 mph   **19.** 1106 in.

In Exercises 21–25, answers are given in the order sine, cosine, tangent, cotangent, secant, and cosecant.

21. $4\sqrt{65}/65$, $7\sqrt{65}/65$, $4/7$, $7/4$, $\sqrt{65}/7$, $\sqrt{65}/4$ **23.** $4/7$, $\sqrt{33}/7$, $4\sqrt{33}/33$, $\sqrt{33}/4$, $7\sqrt{33}/33$, $7/4$ **25.** $\sqrt{15}/4$, $1/4$, $\sqrt{15}$, $\sqrt{15}/15$, 4, $4\sqrt{15}/15$ **27.** $\cos x = \sqrt{55}/8$, $\tan x = 3\sqrt{55}/55$, $\cot x = \sqrt{55}/3$, $\sec x = 8\sqrt{55}/55$, $\csc x = 8/3$ **29.** $\sin x = 3\sqrt{10}/10$, $\cos x = \sqrt{10}/10$, $\cot x = 1/3$, $\sec x = \sqrt{10}$, $\csc x = \sqrt{10}/3$ **31.** $\tan x = \sqrt{21}/2$ **33.** $b = 6\sqrt{3} \approx 10.4$, $c = 12$, $\beta = 60°$ **35.** $a = 26$, $c = 28$, $\beta = 64°50'$ **37.** $c = 25$, $\alpha = 16°16'$, $\beta = 73°44'$ **39.** $a = 7.4$, $b = 7.3$, $\beta = 44°40'$ **41.** $2°23'$ **43.** 39 ft   **45.** 1039 m   **47.** 260 miles, 150 miles   **49.** 17,993 ft

---

CHAPTER 8

Exercises 8.1

1. $P(-\sqrt{2}/2, -\sqrt{2}/2)$, $\sin\theta = \cos\theta = -\sqrt{2}/2$, $\tan\theta = \cot\theta = 1$, $\sec\theta = \csc\theta = -\sqrt{2}$ **3.** $P(-\sqrt{2}/2, -\sqrt{2}/2)$, $\sin\alpha = \cos\alpha = -\sqrt{2}/2$, $\tan\alpha = \cot\alpha = 1$, $\sec\alpha = \csc\alpha = -\sqrt{2}$ **5.** $P(0, 1)$, $\sin\alpha = 1$, $\cos\alpha = 0$, $\tan\alpha$ not defined, $\cot\alpha = 0$, $\sec\alpha$ not defined, $\csc\alpha = 1$ **7.** $P(-\sqrt{2}/2, -\sqrt{2}/2)$, $\sin\theta = \cos\theta = -\sqrt{2}/2$, $\tan\theta = \cot\theta = 1$, $\sec\theta = \csc\theta = -\sqrt{2}$ **9.** $P(-1, 0)$, $\sin\theta = \tan\theta = 0$, $\cos\theta = \sec\theta = -1$, $\cot\theta$ and $\csc\theta$ not defined   **11.** $P(1, 0)$, $\sin\theta = \tan\theta = 0$, $\cos\theta = \sec\theta = 1$, $\cot\theta$ and $\csc\theta$ not defined   **13.** $P(-1, 0)$, $\sin\theta = \tan\theta = 0$, $\cos\theta = \sec\theta = -1$, $\cot\theta$ and $\csc\theta$ not defined   **15.** $\tan\theta = 2\sqrt{5}/5$, $\cot\theta = \sqrt{5}/2$, $\sec\theta = 3\sqrt{5}/5$, $\csc\theta = 3/2$ **17.** $\cos\theta = -\sqrt{15}/4$, $\tan\theta = -\sqrt{15}/15$, $\cot\theta = -\sqrt{15}$, $\csc\theta = 4$ **19.** $\cos\theta = -2\sqrt{2}/3$, $\cot\theta = 2\sqrt{2}$, $\sec\theta = -3\sqrt{2}/4$, $\csc\theta = -3$ **21.** $\cos\theta = \sqrt{7}/4$, $\tan\theta = -3\sqrt{7}/7$, $\cot\theta = -\sqrt{7}/3$, $\csc\theta = -4/3$ **23.** $\cos\theta = -\sqrt{7}/4$, $\tan\theta = -3\sqrt{7}/7$, $\cot\theta = -\sqrt{7}/3$, $\sec\theta = -4\sqrt{7}/7$, $\csc\theta = 4/3$ **25.** $\sin\theta = 3\sqrt{10}/10$, $\cos\theta = \sqrt{10}/10$, $\cot\theta = 1/3$, $\sec\theta = \sqrt{10}$, $\csc\theta = \sqrt{10}/3$ **27.** $\sin\theta = -2\sqrt{2}/3$, $\tan\theta = -2\sqrt{2}$, $\cot\theta = -\sqrt{2}/4$, $\sec\theta = 3$, $\csc\theta = -3\sqrt{2}/4$ **29.** $\sin\theta = \sqrt{2}/2$, $\cos\theta = -\sqrt{2}/2$, $\cot\theta = -1$, $\sec\theta = -\sqrt{2}$, $\csc\theta = \sqrt{2}$ **31.** $\cos\theta = -0.5735$, $\tan\theta = -1.4284$, $\cot\theta = -0.7001$,

$\sec\theta = -1.7437$, $\csc\theta = 1.2207$ **33.** $\sin\theta = -0.7662$, $\cos\theta = -0.6426$, $\cot\theta = 0.8386$, $\sec\theta = -1.5563$, $\csc\theta = -1.3051$

Exercises 8.2

In Exercises 1–19, answers are given in the order sine, cosine, tangent, cotangent, secant, and cosecant.

1. $5\sqrt{29}/29$, $2\sqrt{29}/29$, $5/2$, $2/5$, $\sqrt{29}/2$, $\sqrt{29}/5$ **3.** $3\sqrt{10}/10$, $-\sqrt{10}/10$, -3, $-1/3$, $-\sqrt{10}$, $\sqrt{10}/3$ **5.** $-7\sqrt{58}/58$, $-3\sqrt{58}/58$, $7/3$, $3/7$, $-\sqrt{58}/3$, $-\sqrt{58}/7$ **7.** $-2\sqrt{5}/5$, $\sqrt{5}/5$, -2, $-1/2$, $\sqrt{5}$, $-\sqrt{5}/2$ **9.** $-5\sqrt{41}/41$, $4\sqrt{41}/41$, $-5/4$, $-4/5$, $\sqrt{41}/4$, $-\sqrt{41}/5$ **11.** $2\sqrt{5}/5$, $\sqrt{5}/5$, 2, $1/2$, $\sqrt{5}$, $\sqrt{5}/2$ **13.** $3\sqrt{10}/10$, $-\sqrt{10}/10$, -3, $-1/3$, $-\sqrt{10}$, $\sqrt{10}/3$ **15.** $-\sqrt{2}/2$, $-\sqrt{2}/2$, 1, 1, $-\sqrt{2}$, $-\sqrt{2}$ **17.** $2\sqrt{13}/13$, $3\sqrt{13}/13$, $2/3$, $3/2$, $\sqrt{13}/3$, $\sqrt{13}/2$ **19.** $-3\sqrt{13}/13$, $-2\sqrt{13}/13$, $3/2$, $2/3$, $-\sqrt{13}/2$, $-\sqrt{13}/3$ **21.** $\sin\theta = 3\sqrt{34}/34$, $\cos\theta = 5\sqrt{34}/34$, $\cot\theta = 5/3$, $\sec\theta = \sqrt{34}/5$, $\csc\theta = \sqrt{34}/3$ **23.** $\sin\alpha = \sqrt{5}/5$, $\cos\alpha = -2\sqrt{5}/5$, $\tan\alpha = -1/2$, $\sec\alpha = -\sqrt{5}/2$, $\csc\alpha = \sqrt{5}$ **25.** $\cos\theta = -2\sqrt{2}/3$, $\tan\theta = -\sqrt{2}/4$, $\cot\theta = -2\sqrt{2}$, $\sec\theta = -3\sqrt{2}/4$, $\csc\theta = 3$ **27.** $\sin\theta = -\sqrt{21}/5$, $\tan\theta = \sqrt{21}/2$, $\cot\theta = 2\sqrt{21}/21$, $\sec\theta = -5/2$, $\csc\theta = -5\sqrt{21}/21$ **29.** $(4/5, -3/5)$

Exercises 8.3
1. 60°   **3.** 40°   **5.** $\pi/3$ **7.** $\pi/4$ **9.** 40°   **11.** $\pi/7$
13. $\sqrt{3}/2$ **15.** $-2\sqrt{3}/3$ **17.** 1   **19.** $\sqrt{3}/2$
21. $-1/2$ **23.** $1/2$ **25.** 0.2773   **27.** 1.820
29. 0.9644   **31.** 0.1045   **33.** 0.4142   **35.** -1.171
37. 0.3843   **39.** 0.3947   **41.** -2.2385 **43.** 0.6756
45. -0.7975 **47.** 0.6305   **49.** 1.0176

Exercises 8.4
1.

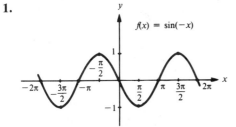

3.

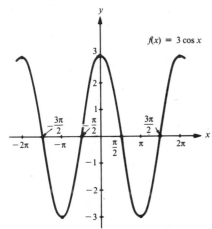

5.

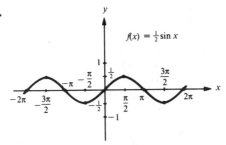

7.

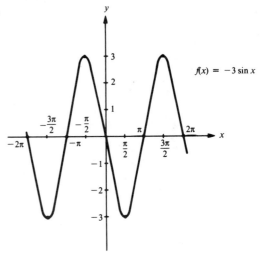

9.

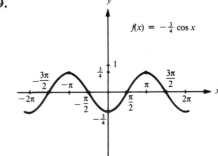

11. Amplitude: 1, period: $\pi/2$, phase shift: 0

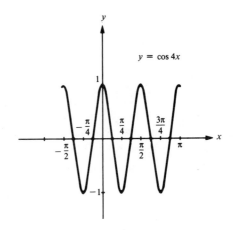

13. Amplitude: 1, period: 6π, phase shift: 0

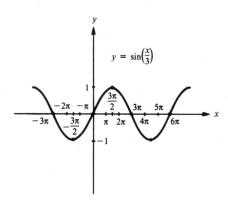

15. Amplitude: 3, period: π, phase shift: 0

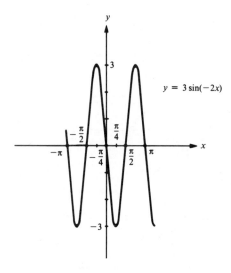

17. Amplitude: 3, period: 4, phase shift: 0

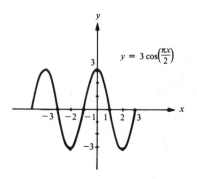

19. Amplitude: 1, period: 2π, phase shift: $-\pi/3$

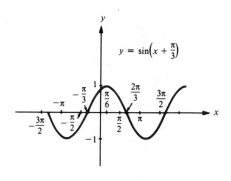

21. Amplitude: 1, period: 2, phase shift: 1

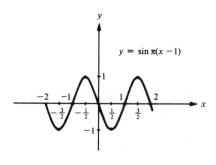

23. Amplitude: 2, period: 2π, phase shift: $\pi/3$

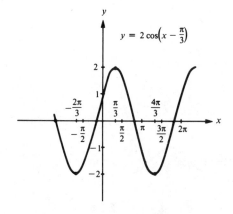

25. Amplitude: 3, period: π, phase shift: $\pi/2$

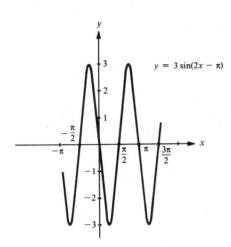

$y = 3 \sin(2x - \pi)$

27. Period: $\pi/3$

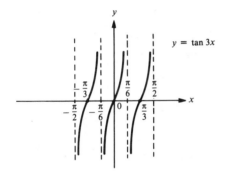

$y = \tan 3x$

29. Period: 2π

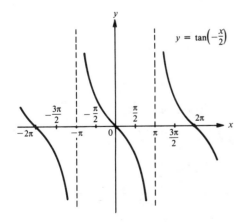

$y = \tan\left(-\dfrac{x}{2}\right)$

31.

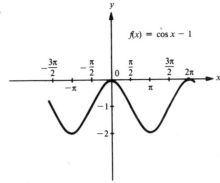

$f(x) = \cos x - 1$

33.

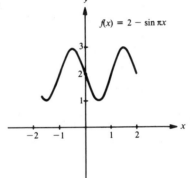

$f(x) = 2 - \sin \pi x$

35.

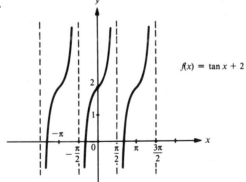

$f(x) = \tan x + 2$

37. Even   **39.** Odd

Exercises 8.5

1. (a) 5 (b) 1 (c) 2π (d) $1/2\pi$ (e) $-\pi/4$
3. (a) 4 (b) π (c) 2 (d) $1/2$ (e) -2
5. (a) $3/4$ (b) 2π (c) 1 (d) 1 (e) $\pi/4$
7. (a) 2 (b) $\pi/2$ (c) 4 (d) $1/4$ (e) $-\pi/3$
9. (a) 6 (b) π (c) 2 (d) $1/2$ (e) $\pi/2$
11. $x = 4\cos(60\pi t + \pi/2)$ **13.** (a) $0.25\,\text{ft}$ (b) $0.4\,\text{Hz}$
(c) $12.5\,\text{lb/ft}$ **15.** 6, $1/10\,\text{s}$, $10\,\text{Hz}$, -3 **17.** (a) 0
(b) $\pi/2$ (c) $-\pi/3$ **19.** (a) $1/60\,\text{s}$, $60\,\text{Hz}$ (b) 20 A
21. $y = 0.05\cos(8\sqrt{3}t/3)$, $1.36\,\text{s}$ **23.** 2
25. 1750, 6 years **27.** $4.7\,\text{s}$ **29.** $4.32\,\text{s}$

Exercises 8.6

1. $\pi/3$ **3.** $\pi/2$ **5.** $-\pi/6$ **7.** $\pi/4$ **9.** $\pi/4$
11. $\pi/6$ **13.** $11°30'$ **15.** $67°$ **17.** $43°30'$ **19.** $43°10'$
21. $\sqrt{2}/2$ **23.** $\sqrt{3}/3$ **25.** $\sqrt{3}/3$ **27.** $2\sqrt{3}/3$
29. $\sqrt{3}/2$ **31.** $\sqrt{7}/3$ **33.** $\sqrt{66}/3$ **35.** $\sqrt{5}/5$
37. $\sqrt{30}/6$ **39.** $-\sqrt{13}/2$ **41.** $-27°9'$ **43.** $-79°26'$
45. 0.7746 **47.** 0.3779 **49.** 0.9979 **51.** $x = \frac{1}{2}\arcsin y$
53. $x = \frac{1}{3}\sin^{-1}(y/2)$ **55.** $x = \arccos(3y) + 2$
57. $x = 2\cos(y/3)$ **59.** $x = \frac{1}{2}\tan(3y/2)$

Chapter Test

1. $(\sqrt{2}/2, \sqrt{2}/2)$ **2.** $\tan\alpha = -2\sqrt{5}$, $\csc\alpha = -3/2$,
$\sec\alpha = 3/\sqrt{5}$, $\cot\alpha = -\sqrt{5}/2$ **3.** $\sin\theta = 4/\sqrt{41}$,
$\cos\theta = 5/\sqrt{41}$, $\csc\theta = \sqrt{41}/4$, $\sec\theta = \sqrt{41}/5$,
$\cot\theta = 5/4$ **4.** $\sin\alpha = 2/\sqrt{5}$, $\cos\alpha = -1/\sqrt{5}$
5. $\sin\beta = -3/5$, $\cos\beta = -4/5$, $\tan\beta = 3/4$ **6.** $30°$,
$1/2$, $\sqrt{3}/2$ **7.** Amplitude: 1, period: π, phase shift: π

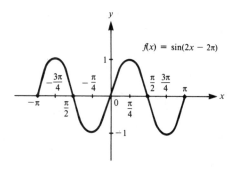

8. (a) 0.11 (b) $3.8/2\pi \approx 0.6\,\text{Hz}$ **9.** $4/\sqrt{17}$
10. $x = \frac{1}{7}\arcsin(y - 1)$

Review Exercises

Trigonometric values of an angle are given in the order sine, cosine, tangent, cotangent, secant, and cosecant.

1. $(-\sqrt{2}/2, \sqrt{2}/2)$, $\sqrt{2}/2$, $-\sqrt{2}/2$, -1, -1, $-\sqrt{2}$, $\sqrt{2}$
3. $(-1/2, -\sqrt{3}/2)$, $-\sqrt{3}/2$, $-1/2$, $\sqrt{3}$, $\sqrt{3}/3$, -2,
$-2\sqrt{3}/3$ **5.** $\sin\alpha = 1/3$, $\tan\alpha = \sqrt{2}/4$, $\cot\alpha = 2\sqrt{2}$,

$\sec\alpha = 3\sqrt{2}/4$ **7.** $\cos\beta = -\sqrt{21}/5$, $\cot\beta = -\sqrt{21}/2$,
$\sec\beta = -5\sqrt{21}/21$, $\csc\beta = 5/2$ **9.** $\sin\theta = -\sqrt{21}/5$,
$\tan\theta = -\sqrt{21}/2$, $\cot\theta = -2\sqrt{21}/21$, $\sec\theta = 5/2$,
$\csc\theta = -5\sqrt{21}/21$ **11.** $\sin\theta = \sqrt{10}/10$,
$\cos\theta = -3\sqrt{10}/10$, $\tan\theta = -1/3$, $\sec\theta = -\sqrt{10}/3$,
$\csc\theta = \sqrt{10}$ **13.** $\cos\theta = -0.8039$, $\tan\theta = -0.7399$,
$\cot\theta = -1.3515$, $\sec\theta = -1.2440$, $\csc\theta = 1.6812$
15. $-5\sqrt{34}/34$, $3\sqrt{34}/34$, $-5/3$, $-3/5$, $\sqrt{34}/3$, $-\sqrt{34}/5$
17. $2\sqrt{5}/5$, $-\sqrt{5}/5$, -2, $-1/2$, $-\sqrt{5}$, $\sqrt{5}/2$
19. $-2\sqrt{13}/13$, $-3\sqrt{13}/13$, $2/3$, $3/2$, $-\sqrt{13}/3$, $-\sqrt{13}/2$
21. $2\sqrt{5}/5$, $\sqrt{5}/5$, 2, $1/2$, $\sqrt{5}$, $\sqrt{5}/2$
23. $-3\sqrt{10}/10$, $-\sqrt{10}/10$, 3, $1/3$, $-\sqrt{10}$, $-\sqrt{10}/3$
25. (a) $50°$ (b) $40°$ (c) $\pi/2$ (d) $\pi/6$
(e) $30°15'$ (f) $24°30'$ **27.** (a) 0.7112 (b) 0.9636
(c) 1.108 (d) 1.202 (e) 0.9890 (f) 0.3987
29.

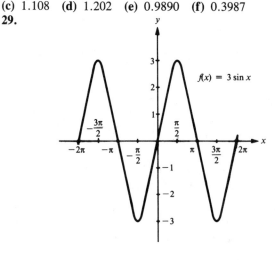

31.

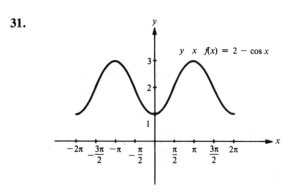

33.

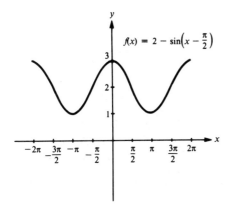

35. Amplitude: 2, period: 10π, phase shift: 0
37. Amplitude: 1, period: 2, phase shift: 2
39. Amplitude: 2, period: 2, phase shift: $4/\pi$
41.

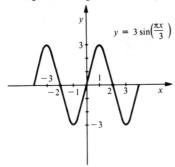

43.

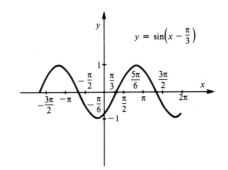

45.

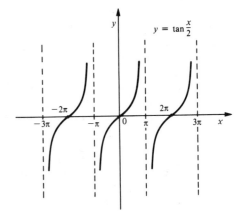

47. $16\sqrt{2}\,\text{Hz} \simeq 22.63\,\text{Hz}$
49. $y = 20\cos(4\pi t/3)$, 6.2 cm   **51.** $9.78\,\text{m/s}^2$,
$y = 0.5\cos(4.9t)$ **53.** $\sqrt{2}/2$ **55.** $5\pi/6$ **57.** $\sqrt{17}$
59. $56°13'$ **61.** $118°8'$ **63.** $45°$ **65.** 0.6712
67. 1.4945   **69.** 0.7071

CHAPTER 9

Exercises 9.1

1. $\cos\theta + 1$ **3.** 1   **5.** $\sin\alpha + \cos\alpha$
7. $1 + 2\sin u\cos u$ **9.** $2\sec^2\theta$

Exercises 9.2

1. $(\sqrt{6} - \sqrt{2})/4$ **3.** $2 - \sqrt{3}$ **5.** $(\sqrt{6} - \sqrt{2})/4$
7. $(\sqrt{6} + \sqrt{2})/4$ **9.** $-\sqrt{3}/2$ **11.** $(\sqrt{6} - \sqrt{2})/4$
13. $\sqrt{6} + \sqrt{2}$ **15.** $-\sqrt{2}/2$ **17.** $-\sqrt{3}/2$ **19.** -2
21. $\sin 60°$ **23.** $\cos 97°$ **25.** $\cos(9\pi/20)$
27. $\sin(3\pi/2)$ **29.** $\sin 2$ **31.** 1   **33.** $-33/65$
35. $(2 + 3\sqrt{5})/3\sqrt{10} = (2\sqrt{10} + 15\sqrt{2})/30$
37. $63/65$, $56/65$ **39.** $-16/65$, $-56/65$
41. $(2 - 2\sqrt{10})/9$, $-(4\sqrt{2} + \sqrt{5})/9$, quadrant III
43. $(8 - 3\sqrt{21})/25$, $(\sqrt{21} - 6)/6$, quadrant IV
45. $-(6 + \sqrt{35})/12$, $32\sqrt{5} - 27\sqrt{7}$, quadrant III
47. $\sin 8\theta + \sin 4\theta$ **49.** $(\sin 4\theta - \sin 2\theta)/2$
51. $\cos 3x + \cos 7x$ **53.** $(\cos 5x - \cos 11x)/2$

55. $2\sin(8a + 3b) + 2\sin(8a - 3b)$
57. $(3\cos 4a - 3\cos 6a)/2$

Exercises 9.3

1. $24/25$, $-7/25$, $-24/7$ **3.** $3\sqrt{7}/8$, $-1/8$, $-3\sqrt{7}$
5. $24/25$, $-7/25$, $-24/7$ **7.** $-4\sqrt{21}/25$, $-17/25$,
$4\sqrt{21}/17$ **9.** $\sqrt{2 + \sqrt{3}}/2$, $-(2 + \sqrt{3})$ **11.** $\sqrt{6}/4$,
$-\sqrt{15}/5$ **13.** $\sqrt{(10 - \sqrt{10})/20}$, $(\sqrt{10} - 1)/3$
15. $\sqrt{8 + 2\sqrt{15}}/4$, $-(4 + \sqrt{15})$ **17.** $2\sqrt{13}/13$
19. $3/5$ **21.** $\sqrt{2 - \sqrt{2}}/2$ **23.** $\sqrt{2 + \sqrt{3}}/2$
25. $\sqrt{2 + \sqrt{3}}/2$ **27.** $\sqrt{2 + \sqrt{2}}/2$ **29.** $24/25$
31. $119/169$ **33.** $\sqrt{10}/10$ **47.** -0.9397
49. -0.9659

Exercises 9.4

1. $\pi/6$ **3.** $\pi/6$ **5.** $120°$, $240°$ **7.** $30°$, $150°$
9. $3\pi/4$, $7\pi/4$ **11.** $45°$, $90°$, $225°$ **13.** $\pi/4$, $2\pi/3$, $3\pi/4$

15. 60°, 120°   **17.** 0, $\pi/4$ **19.** $2\pi/3$, π, $4\pi/3$
21. 270°   **23.** 0, $\pi/2$ **25.** 0°, 90°, 180°, 270°
27. 30°, 120°, 210°, 300°   **29.** $\pi/6$ **31.** 0, $\pi/3$, π
33. $\pi/4$, $\pi/2$ **35.** 60°, 300°   **37.** $2\pi/3$, π

In answers 39–53, $k = 0, \pm1, \pm2, \dots$.

39. $\pm\pi/6 + 2k\pi$ **41.** $\pi/6 + 2k\pi$, $5\pi/6 + 2k\pi$
43. $3\pi/4 + k\pi$ **45.** $\pi/4 + k\pi/2$ **47.** $\pi/2 + 2k\pi$,
$(2k + 1)\pi$ **49.** $\pi/2 + k\pi$, $7\pi/6 + 2k\pi$, $-\pi/6 + 2k\pi$
51. No solution   **53.** $k\pi$, $\pi/4 + k\pi$ **55.** 1.946 rad
57. 0.9389 rad, -0.3509 rad   **59.** 0.9553 rad, 2.1863 rad

Exercises 9.5

1. $b = 19$, $c = 18$, $\gamma = 67°$ **3.** $a = 10$, $b = 3$, $\gamma = 60°$
5. $b = 7$, $c = 6$, $\alpha = 75°$ **7.** $b = 4$, $c = 5$, $\alpha = 87°20'$
9. $c = 4\sqrt{3}$, $\beta = 90°$, $\gamma = 60°$ **11.** No solution
13. $a = 4.5$, $c = 3.2$, $\beta = 80°10'$ **15.** $c = 10.4$,
$\beta = 49°53'$, $\gamma = 95°7'$, or $c = 2.7$, $\beta = 130°7'$,
$\gamma = 14°53'$ **17.** $c = 16.1$, $\beta = 23°25'$, $\gamma = 126°25'$
19. No solution   **21.** $b = 3.1$, $c = 3.6$, $\alpha = 102°15'$
23. $a = 12.3$, $\alpha = 85°40'$, $\beta = 54°$, or $a = 2.9$,
$\alpha = 13°40'$, $\beta = 126°$ **25.** $c = 17.31$, $\beta = 63.15°$,
$\gamma = 74.85°$ **27.** $c = 7.22$, $\beta = 140.24°$, $\gamma = 11.09°$
29. $b = 11.47$, $c = 10.34$, $\gamma = 64°$ **31.** $b = 20.57$,
$c = 23.04$, $\gamma = 82.25°$ **33.** $a = 6.79$, $b = 27.63$,
$\beta = 61.7°$ **35.** 8 m^2 **37.** 161.8 cm^2 **39.** 253 yards
41. 14 miles, 21 miles   **43.** 90 ft   **45.** 5 km, 2 km
47. 25 miles

Exercises 9.6

1. $\alpha = 22°40'$, $\beta = 67°20'$, $\gamma = 90°$ **3.** $\alpha = 33°30'$,
$\beta = 50°40'$, $\gamma = 95°50'$ **5.** $b = 6$, $\alpha = 19°50'$,
$\gamma = 115°10'$ **7.** $a = 25$, $\beta = 27°$, $\gamma = 43°$
9. $\alpha = 29°14'$, $\beta = 51°13'$, $\gamma = 99°33'$ **11.** $\alpha = 88°20'$,
$\beta = 33°6'$, $\gamma = 58°34'$ **13.** $a = 8.53$, $\beta = 33°14'$,
$\gamma = 90°28'$ **15.** $b = 17.91$, $\alpha = 14°49'$, $\gamma = 59°46'$
17. $\alpha = 22.62°$, $\beta = 67.38°$, $\gamma = 90°$ **19.** $\alpha = 54.01°$,
$\beta = 45.55°$, $\gamma = 80.44°$ **21.** $b = 6.32$, $\alpha = 19.35°$,
$\gamma = 118.15°$ **23.** $c = 11.57$, $\alpha = 61.31°$, $\beta = 43.57°$
25. $a = 12.60$, $\beta = 26.18°$, $\gamma = 36.62°$ **27.** 7 in.
29. $15\sqrt{3}$ cm   **31.** 60.64°   **33.** 300 yards   **35.** 7 ft,
11.36 ft   **37.** 21 ft, 37 ft   **39.** 8.45 miles   **41.** 4.0 miles
43. 229 miles   **45.** 73.7 km   **47.** 180 m^2 **49.** 67.3 in.^{2}
51. 53.8 yd^2 **53.** 1500 m^2

Exercises 9.7

1. $\langle 2, -4 \rangle$ **3.** $\langle 4, 7 \rangle$ **5.** $(-3, 6)$ **7.** $(-1, -3)$
9. $(2, 1)$ **11.** $(3, -5)$ **13.** $\langle -8, -2 \rangle$ **15.** 3.23, 9.67
17. $\sqrt{34}$ **19.** $\sqrt{34}$ **21.** $\langle -9, 10 \rangle$ **23.** $\langle -19, 11 \rangle$
25. $\langle -\frac{5}{13}, \frac{12}{13} \rangle$; $\langle \frac{5}{13}, -\frac{12}{13} \rangle$ **27.** $-18\mathbf{i} + 19\mathbf{j}$ **29.** $\mathbf{i} + \sqrt{3}\,\mathbf{j}$
31. $-3\sqrt{3}\,\mathbf{i} + 3\mathbf{j}$ **33.** 45°   **35.** 300°   **37.** -26
39. -12 **41.** 169.70°   **43.** 90°   **45.** 4/3
47. 12.75 N   **49.** 90.6 kg   **51.** 124.4 lb, 46.3°
53. 400 lb   **55.** (a) $-375\mathbf{i} + 375\sqrt{3}\,\mathbf{j}$, (b) $750\sqrt{3}$ km

57. $\sqrt{10}$ m/s, 71.6°   **59.** 64 m   **61.** 500 lb on the
30°-angle cable, $500\sqrt{3}$ lb on the 60°-angle cable
63. 689 miles, 10° west of south   **65.** 93.3 lb, 71.5 lb

Exercises 9.8

1. $3\sqrt{2}[\cos(7\pi/4) + i\sin(7\pi/4)]$
3. $2[\cos(11\pi/6) + i\sin(11\pi/6)]$
5. $4[\cos(4\pi/3) + i\sin(4\pi/3)]$
7. $5[\cos(\pi/2) + i\sin(\pi/2)]$ **9.** $15(\cos\pi + i\sin\pi)$
11. $40[\cos(11\pi/12) + i\sin(11\pi/12)]$,
$\frac{5}{8}[\cos(5\pi/12) - i\sin(5\pi/12)]$
13. $54(\cos 345° + i\sin 345°)$, $\frac{2}{3}(\cos 135° + i\sin 135°)$
15. $8i$, $(\sqrt{3} - i)/4$ **17.** $15(1 - i)$, $(-5 + 5i)/3$

Exercises 9.9

1. $4 - 4i$ **3.** $-1024(\sqrt{3} - i)$ **5.** $64(1 + i\sqrt{3})$
7. 27 **9.** $-\frac{1}{32}(1 + i\sqrt{3})$ **11.** $(\sqrt{6}/2) + (\sqrt{2}/2)i$,
$(-\sqrt{6}/2) - (\sqrt{2}/2)i$ **13.** i, $(-\sqrt{3} - i)/2$, $(\sqrt{3} - i)/2$
15. $2^{1/5}(\cos 30° + i\sin 30°)$, $2^{1/5}(\cos 102° + i\sin 102°)$,
$2^{1/5}(\cos 174° + i\sin 174°)$, $2^{1/5}(\cos 246° + i\sin 246°)$,
$2^{1/5}(\cos 318° + i\sin 318°)$ **17.** $2^{7/6}(\cos 50° + i\sin 50°)$,
$2^{7/6}(\cos 110° + i\sin 110°)$, $2^{7/6}(\cos 170° + i\sin 170°)$,
$2^{7/6}(\cos 230° + i\sin 230°)$, $2^{7/6}(\cos 290° + i\sin 290°)$,
$2^{7/6}(\cos 350° + i\sin 350°)$ **19.** $\sqrt{3} + i$, $-1 + i\sqrt{3}$,
$-\sqrt{3} - i$, $1 - i\sqrt{3}$ **21.** $3(1 + i\sqrt{3})/2$, -3,
$3(1 - i\sqrt{3})/2$ **23.** 2, $2i$, -2, $-2i$ **25.** $(\sqrt{3} + i)/2$, i,
$(-\sqrt{3} + i)/2$, $(-\sqrt{3} - i)/2$, $-i$, $(\sqrt{3} - i)/2$
27. $\sqrt{3} + i$, $-\sqrt{3} + i$, $-2i$

Exercises 9.10

1.

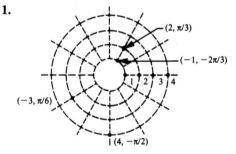

3. (a) $(1, \sqrt{3})$ (b) $(-\sqrt{2}, -\sqrt{2})$,   (c) $(-3\sqrt{3}/2, 3/2)$
(d) $(2\sqrt{2}, -2\sqrt{2})$ **5.** (a) $(\sqrt{2}, \pi/4)$,   (b) $(2, 5\pi/6)$
(c) $(5, 3\pi/2)$, (d) $(3\sqrt{2}, 3\pi/4)$
7.

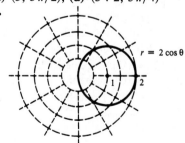

9.

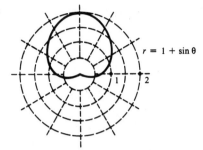

$r = 1 + \sin\theta$

11.

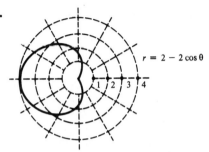

$r = 2 - 2\cos\theta$

13.

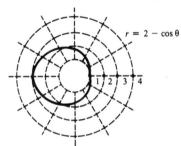

$r = 2 - \cos\theta$

15.

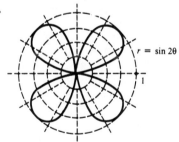

$r = \sin 2\theta$

17. $x^2 + y^2 = 16$

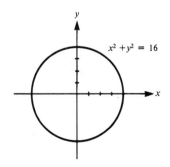

$x^2 + y^2 = 16$

19. $x^2 + (y - 2)^2 = 4$

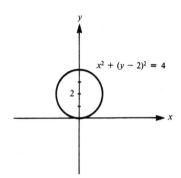

$x^2 + (y - 2)^2 = 4$

21. $x = 3$

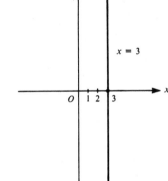

$x = 3$

23. $x - 3y + 2 = 0$

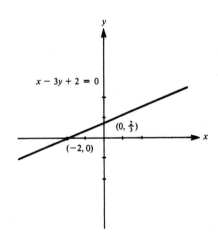

$x - 3y + 2 = 0$

$\left(0, \tfrac{2}{3}\right)$

$(-2, 0)$

25. $x = 2$

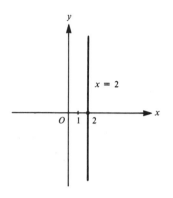

27. $x + 2y = 1$

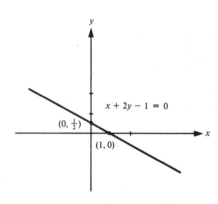

29. $4y = x^2 - 4$

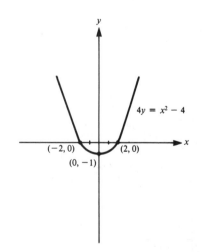

31. $r = 4$ **33.** $r = 5 \sec \theta$ **35.** $r = 4/(\cos \theta + 2 \sin \theta)$
37. $r = 2 \sin \theta$ **39.** $r = 2 \sin \theta \sec^2 \theta$

Chapter Test

1. $\sin 76°$ **2.** $\sin 5b - \sin 3b$ **3.** $-\sqrt{5}/5 - 8/15$
4. $3\sqrt{55}/32$ **5.** $0, \pi, \pi/4, 7\pi/4$ **6.** $a = 15.2$,
$c = 13.7, \gamma = 55°$ **7.** 107 yards
8. $80[\cos(\pi/2) + i\sin(\pi/2)]$ **9.** $15(\cos 65° - i\sin 65°)$
10. $3(\cos 10° + i\sin 10°), 3(\cos 130° + i\sin 130°)$,
$3(\cos 250° + i\sin 250°)$

Review Exercises

29. 2.7480 **31.** $7\pi/6, 11\pi/6$ **33.** $-\pi/6$
35. $(\pi/2) + 2k\pi, (2k + 1)\pi$ **37.** $\pm\pi/6 + k\pi$
39. $\pi/2 + k\pi, \pi/4 + k\pi$ **41.** $b = 29.0, c = 31.9$,
$\gamma = 87°$ **43.** $b = 36.9, c = 24.4, \alpha = 25°$
45. $\alpha = 36°50', \beta = 53°10', \gamma = 90°$ **47.** $\alpha = 41°20'$,
$\beta = 55°50', \gamma = 82°50'$ **49.** 51 ft **51.** 1.22 km
53. $86°21'$ **61.** $(-8, 8)$ **63.** $7, \sqrt{129}$
65. $\langle 7/\sqrt{193}, -12/\sqrt{193}\rangle$ **67.** $4\sqrt{2}\,\mathbf{i} + 4\sqrt{2}\,\mathbf{j}$
69. **(a)** $153.43°$ **(b)** $132.88°$ **71.** 2000 lb
73. $36\sqrt{7}$ lb, $72\sqrt{7}$ lb **75.** $\sqrt{2}(\cos 45° + i\sin 45°)$
77. $6(\cos 30° + i\sin 30°)$ **79.** $4(\cos 270° + i\sin 270°)$
81. $6i, (\sqrt{3} - i)/3$ **83.** $\sqrt{6}(\cos 330° + i\sin 330°)$,
$(\sqrt{6}/3)(\cos 240° - i\sin 240°)$ **85.** -4 **87.** 64
89. $81(\cos 100° + i\sin 100°)$ **91.** $-\sqrt{6}/2 + (\sqrt{2}/2)i$,
$\sqrt{6}/2 - (\sqrt{2}/2)i$ **93.** $2^{7/6}(\cos 75° + i\sin 75°)$,
$2^{7/6}(\cos 195° + i\sin 195°), 2^{7/6}(\cos 315° + i\sin 315°)$
95. $3(\cos 55° + i\sin 55°), 3(\cos 145° + i\sin 145°)$,
$3(\cos 235° + i\sin 235°), 3(\cos 325° + i\sin 325°)$
97. $(3 + 3i\sqrt{3})/2, -3, (3 - 3i\sqrt{3})/2$
99. $2[\cos(\pi/8) + i\sin(\pi/8)], 2[\cos(5\pi/8) + i\sin(5\pi/8)]$,
$2[\cos(9\pi/8) + i\sin(9\pi/8)], 2[\cos(13\pi/8) + i\sin(13\pi/8)]$
101. **(a)** $(3\sqrt{2}/2, 3\sqrt{2}/2)$ **(b)** $(-1, -\sqrt{3})$,
(c) $(-2\sqrt{3}, -2)$

103.

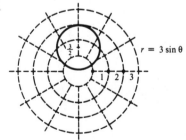

105. $y = 4$

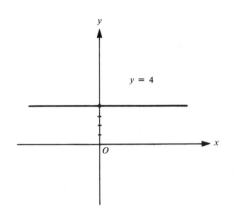

107. $x + 2y = 3$

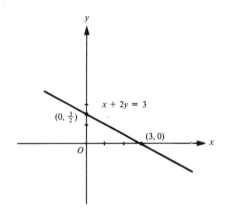

CHAPTER 10

Exercises 10.1

1. $(4, 3)$ **3.** $(5, 2)$ **5.** $(5, 4)$ **7.** $(3, 2)$ **9.** $(6, -3)$
11. $(\frac{1}{4}, -\frac{1}{2})$ **13.** $(5, 4)$ **15.** $(\frac{2}{3}, \frac{1}{2})$ **17.** $(10, 6)$
19. $(6, 4)$ **21.** $(-3, 2)$ **23.** $(1, -\frac{1}{2})$ **25.** $(\frac{1}{2}, 1)$
27. $(\frac{4}{3}, 3)$ **29.** 5 kg of 20% alloy, 25 kg of 50% alloy
31. boat: 12 mph, current: 4 mph   **33.** 18 nickels, 12
dimes   **35.** 55 mph, 45 mph   **37.** 40 cm, 30 cm
39. 60 kg of alloy A, 40 kg of alloy B

Exercises 10.2

1. $(3, 4, 3)$ **3.** $(1, 0, -1)$ **5.** $(1, 2, 3)$ **7.** $(2, 1, 3)$
9. $(1, 0, -1)$ **11.** $(2, 1, -2)$ **13.** $(\frac{1}{2}, 1, -\frac{1}{3})$
15. $(-1, 2, 1)$ **17.** $(1, 1, 0, 1)$ **19.** $(\frac{1}{3}, \frac{1}{2})$ **21.** $(1, \frac{1}{2})$
23. $(1, 0, -1)$ **25.** $(5, -3, -2)$
27. $(-3z, -1 - 2z, z)$ **29.** $(2 - z, z, z)$
31. $(x, (3x - 1)/2, 3 - x)$ **33.** $(2z/3, 5z/3, z)$
35. $(-2, -3)$ **37.** No solution   **39.** $(0, 0, 0)$
41. Six 11-cent, ten 18-cent, and sixteen 40-cent stamps
43. 40 kg of alloy A, 40 kg of alloy B, 20 kg of alloy C
45. 6, 12, 24   **47.** 30 lb of $2.60 coffee, 30 lb of $2.80
coffee, 90 lb of $3.20 coffee   **49.** $8,000 at 8.5%, $6,000
at 7.5%, $4,000 at 6.5%

Exercises 10.3

1. $\begin{bmatrix} 6 & 1 \\ -2 & 1 \end{bmatrix}$ **3.** $\begin{bmatrix} -1 \\ -2 \\ 4 \end{bmatrix}$ **5.** $\begin{bmatrix} 2 & 4 & -6 \\ 3 & 0 & 6 \end{bmatrix}$ **7.** $\begin{bmatrix} -1 & -5 \\ 3 & -3 \\ 2 & -3 \\ 2 & -1 \end{bmatrix}$

9. $x = 3, y = -2, z = 4$ **11.** $r = 2, s = -4, t = 4$
13. $x = 4, y = 0, z = 2, v = -2, w = 6$

15. $\begin{bmatrix} 3 & -1 \\ 5 & -5 \end{bmatrix}, \begin{bmatrix} 1 & 3 \\ 1 & 1 \end{bmatrix}, \begin{bmatrix} 6 & 3 \\ 9 & -6 \end{bmatrix}, \begin{bmatrix} 1 & 8 \\ 0 & 5 \end{bmatrix}$

17. $\begin{bmatrix} 0 & 0 & 3 \\ 0 & 0 & 2 \end{bmatrix}, \begin{bmatrix} 4 & -2 & 7 \\ -2 & 6 & -6 \end{bmatrix}, \begin{bmatrix} 6 & -3 & 15 \\ -3 & 9 & -6 \end{bmatrix},$
$\begin{bmatrix} 10 & -5 & 16 \\ -5 & 15 & -16 \end{bmatrix}$

19. $\begin{bmatrix} -1 \\ 1 \\ 5 \end{bmatrix}, \begin{bmatrix} 3 \\ -5 \\ -3 \end{bmatrix}, \begin{bmatrix} 3 \\ -6 \\ 3 \end{bmatrix}, \begin{bmatrix} 8 \\ -13 \\ -10 \end{bmatrix}$

21. $\begin{bmatrix} 4 \\ 2 \\ 2 \\ 4 \end{bmatrix}, \begin{bmatrix} -2 \\ 2 \\ -2 \\ 0 \end{bmatrix}, \begin{bmatrix} 3 \\ 6 \\ 0 \\ 6 \end{bmatrix}, \begin{bmatrix} -7 \\ 4 \\ -6 \\ -2 \end{bmatrix}$

23. $\begin{bmatrix} 1 & 1 \\ 6 & 0 \end{bmatrix}, \begin{bmatrix} 2 & -2 \\ -2 & -1 \end{bmatrix}$ **25.** $\begin{bmatrix} 3 & -9 & 6 \\ -2 & 6 & -4 \\ 4 & -12 & 8 \end{bmatrix}, [17]$

27. $\begin{bmatrix} 5 & 0 \\ 3 & -4 \end{bmatrix}, \begin{bmatrix} -6 & 0 & -2 \\ -5 & 2 & -2 \\ 12 & -6 & 5 \end{bmatrix}$

29. $\begin{bmatrix} 0 & 0 & 0 \\ 2 & 2 & 2 \\ 0 & 0 & 0 \end{bmatrix}, \begin{bmatrix} 0 & 2 & 2 \\ 0 & 0 & 0 \\ 0 & 2 & 2 \end{bmatrix}$

31. $\begin{bmatrix} 8 & 0 & -4 & 12 \\ 2 & 0 & -1 & 3 \\ 0 & 0 & 0 & 0 \\ 4 & 0 & -2 & 6 \end{bmatrix}, [14]$ **33.** $\begin{bmatrix} 1 & -1 \\ 0 & 1 \end{bmatrix}$

35. $\begin{bmatrix} \frac{1}{3} & \frac{1}{3} \\ -\frac{1}{3} & \frac{2}{3} \end{bmatrix}$ **37.** $\begin{bmatrix} \frac{1}{2} & -\frac{1}{2} \\ 0 & \frac{1}{2} \end{bmatrix}$ **39.** $\begin{bmatrix} 1 & 0 \\ -1 & 1 \end{bmatrix}$

41. No inverse

43. $\begin{bmatrix} 1 & 0 & -1 \\ 0 & 1 & 0 \\ 0 & 0 & 1 \end{bmatrix}$ **45.** $\begin{bmatrix} 1 & -2 & 1 \\ 0 & 1 & -2 \\ 0 & 0 & 1 \end{bmatrix}$ **47.** No inverse

49. $\begin{bmatrix} 1 & 0 & 0 \\ 0 & \frac{1}{2} & 0 \\ 0 & 0 & \frac{1}{3} \end{bmatrix}$ **51.** $(-\frac{1}{2}, -3)$ **53.** $(3, -2)$

55. $(3, 2, -1)$ **57.** $(3, 4, 5)$ **59.** $(-28, -63, 116)$

Exercises 10.4

1. $[-2], -2, -2; \ [-1], -1, 1$

3. $\begin{bmatrix} -1 & 2 \\ -2 & 1 \end{bmatrix}, 3, 3; \ \begin{bmatrix} 1 & 0 \\ 3 & 2 \end{bmatrix}, 2, -2; \ \begin{bmatrix} 3 & -1 \\ 0 & -2 \end{bmatrix}, -6, 6$

5. $\begin{bmatrix} 1 & 0 & 3 \\ 0 & 0 & -1 \\ 0 & 1 & 2 \end{bmatrix}, 1, 1; \ \begin{bmatrix} 2 & 1 & 3 \\ -2 & 0 & -1 \\ -3 & 0 & 2 \end{bmatrix}, 7, 7;$

$\begin{bmatrix} 1 & 0 & 2 \\ 2 & 1 & 3 \\ -2 & 0 & -1 \end{bmatrix}, 3, -3$ **7.** -1 **9.** -7 **11.** -2

13. -9 **15.** -2 **17.** 24 **19.** 0 **21.** -3

23. $\begin{vmatrix} x & y & 1 \\ 2 & 3 & 1 \\ -1 & -2 & 1 \end{vmatrix} = 0$

25. -8 **27.** -2 **35.** $(5, 6)$ **37.** $(\frac{3}{35}, \frac{13}{35})$
39. No solution **41.** $(2, 1, -2)$ **43.** $(1, 0, -2)$
45. $a = -2, b = -1$ **47.** 15 g of 12-carat, 45 g of
20-carat **49.** 54, 40, 18

Exercises 10.5
1. $(9, 3), (0, 0)$ **3.** $(-2, -4), (1, -1)$ **5.** $(4, 4),$
$(-1, \frac{1}{4})$ **7.** $(1, 3), (-1, -3)$ **9.** $(2, 1/4), (-1/2, -1)$
11. $((1 + \sqrt{5})/2, (1 - \sqrt{5})/2), ((1 - \sqrt{5})/2, (1 + \sqrt{5})/2)$
13. $(2, 2), (-2, 2)$ **15.** $(2, 3), (-2, 3)$
17. $(2, \sqrt{3}), (2, -\sqrt{3})$ **19.** $(\pm\sqrt{3}/3, \pm\sqrt{6}/6)$
21. $(\pm\sqrt{21}/7, \pm\sqrt{7}/7)$ **23.** $(1/2, \sqrt{3}/2),$
$(1/2, -\sqrt{3}/2)$ **25.** $(\pm2, 0), (2, 2), (-2, -2)$
27. $(2, 3), (-2, -3)$ **29.** $(1, 3)$ **31.** $(-1, 3)$
33. 12, 20 **35.** 16, 24 **37.** 8 m, 10 m **39.** 15 m, 20 m

Exercises 10.6
1. $(1, -3)$: no, $(-2, 3)$: yes **3.** $(\frac{1}{3}, \frac{1}{2})$: yes, $(-2, 3)$: no

5.

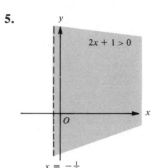

7.

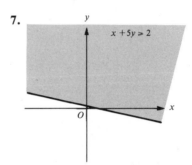

9.

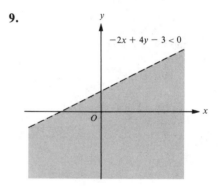

11.

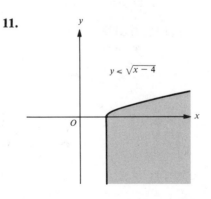

13.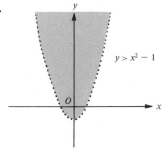

$y > x^2 - 1$

15.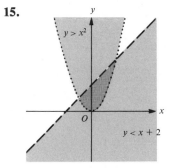

$y > x^2$

$y < x + 2$

17.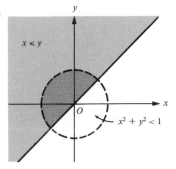

$x \leqslant y$

$x^2 + y^2 < 1$

19.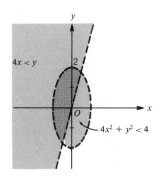

$4x < y$

2

$4x^2 + y^2 < 4$

21.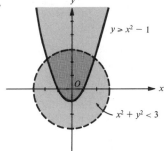

$y \geqslant x^2 - 1$

$x^2 + y^2 < 3$

23.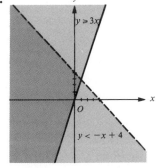

$x^2 + y^2 < 2$

$\dfrac{x^2}{9} + y^2 < 1$

25.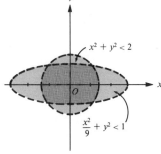

$y \geqslant 3x$

$y < -x + 4$

27.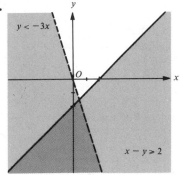

$y < -3x$

$x - y \geqslant 2$

29.

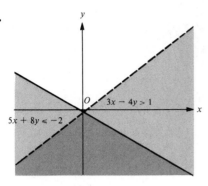

31.

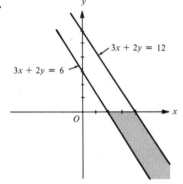

33.

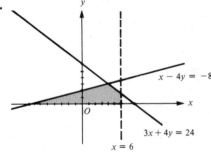

35. Maximum 6 at $(1, 3)$, minimum -5 at $(0, 0)$
37. Maximum 13 at $(3, 0)$, minimum -5 at $(3, 9)$
39. Minimum 23 at $(3, 2)$, no maximum   **41.** 400 barrels
of oil, 2100 barrels of gasoline   **43.** 3 Birdie Customs,
4 Bogey De Luxes, $1410   **45.** 40 g of A, 30 g of B
47. Barley: 240 acres, corn: 120 acres, $13,800
49. Center A: 3 days, center B: 6 days

Chapter Test
1. $\left(-\frac{7}{13}, -\frac{12}{13}\right)$ **2.** $\left(\frac{7}{3}, -\frac{4}{3}, \frac{1}{3}\right)$ **3.** $(7, 17, 23)$

4. $\begin{bmatrix} -10 & 11 \\ 3 & -5 \end{bmatrix}$ **5.** $\begin{bmatrix} -\frac{1}{2} & \frac{1}{6} & 0 \\ 0 & -\frac{1}{3} & 1 \\ 0 & 0 & -1 \end{bmatrix}$ **6.** $3, -3$

7. $\left(\frac{3}{2}, -\frac{9}{2}\right), (-2, -8)$
8.   **9.** $4, -1$ **10.** $11,200

Review Exercises
1. $\begin{bmatrix} 5 & 0 \\ 7 & -4 \end{bmatrix}$ $\begin{bmatrix} -1 & 7 \\ -7 & -9 \end{bmatrix}$ $\begin{bmatrix} 5 & -10 \\ 15 & 10 \end{bmatrix}$
3. $\begin{bmatrix} 10 & 2 \\ -2 & -10 \end{bmatrix}$ $\begin{bmatrix} 5 & -6 \\ 6 & -5 \end{bmatrix}$ $\begin{bmatrix} 0 & 10 \\ -10 & 0 \end{bmatrix}$

5. $[1]$ $\begin{bmatrix} 0 & 4 & -2 & -4 \\ 0 & 2 & -1 & -2 \\ 0 & -6 & 3 & 6 \\ 0 & 4 & -2 & -4 \end{bmatrix}$

7. $\begin{bmatrix} 7 & -2 \\ 15 & 2 \end{bmatrix}$ $\begin{bmatrix} 11 & -3 & 12 \\ -2 & 2 & -4 \\ -7 & -1 & -4 \end{bmatrix}$ **9.** $\begin{bmatrix} \frac{2}{7} & \frac{1}{7} \\ -\frac{3}{7} & \frac{2}{7} \end{bmatrix}$

11. $\begin{bmatrix} 1 & -1 & 0 \\ 0 & \frac{1}{2} & -\frac{1}{2} \\ 0 & 0 & \frac{1}{3} \end{bmatrix}$ **13.** $\begin{bmatrix} 1 & 0 & 0 & 0 \\ -2 & 1 & 0 & 0 \\ 3 & -3 & 1 & 0 \\ -8 & 8 & -4 & 1 \end{bmatrix}$

15. 30   **17.** -10 **19.** -16 **21.** $\begin{bmatrix} \frac{1}{a} & 0 \\ 0 & \frac{1}{b} \end{bmatrix}$

23. $(-19/5, -18/5)$
25. $(4, -2, 2)$ **27.** $(3, -2, 4)$ **29.** 3   **31.** 4
33. $(5, -3)$ **35.** $(1, 2)$ **37.** $2983.87 at 8%, $2483.87
at 7.5%   **39.** 16 dimes, 64 nickels   **41.** 6.8% on the first
$15,000, 7.4% over $15,000   **43.** 12 units of P, 8 units
of Q **45.** $(2, -3, 1)$ **47.** 60 lb peanuts, 20 lb almonds,
20 lb cashews   **49.** 25 one-dollar, 20 five-dollar, and 30
ten-dollar bills   **51.** 72 workers at $6.00; 36 workers at
$8.00; 12 workers at $10.00   **59.** x_1, x_2, x_3 **61.** $(0, 3)$
63. $\left(-\frac{1}{3}, \frac{1}{6}\right), (2, 6)$

65.

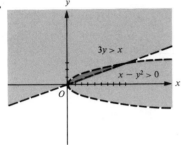
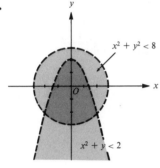

67.

69.

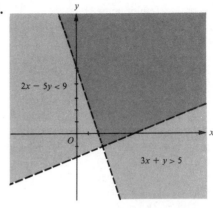
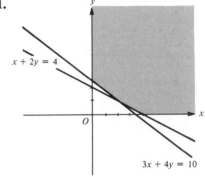

71.

73. Maximum 15 at $(1, 3)$, minimum 0 at $(0, 0)$
75. Maximum -4 at $(0, 0)$, minimum -22 at $(0, -6)$
77. 18, 25   **79.** 10 Standards, 5 De Luxes, $1,000

CHAPTER 11

Exercises 11.1

1. (a) 2 **(b)** 15 **(c)** 2/15 **(d)** 0 **(e)** 49   **3.** 6
5. $A = 3, B = -4, C = 5$ **7.** $A = 3, B = 2, C = 5$
9. $A = 1, B = 0, C = 0, D = 1$ **11.** $3x - 2; 14$
13. $\frac{1}{3}x^2 - \frac{2}{9}x + \frac{28}{27}; \frac{85}{27}$ **15.** $x^2 + 3x - 1; 4$
17. $2x^3 + \frac{1}{2}x + 1; -\frac{5}{2}x^2 + 2x + 3$
19. $\frac{1}{2}x^3 + 1; -x^2 + 3$ **21.** $x - 3 + 2/(x - 1)$
23. $x^2 - 2x + 2 - 10/(2x + 4)$
25. $\frac{1}{2}x^3 + \frac{1}{4}x^2 - \frac{19}{8}x - \frac{19}{16} + \frac{29}{16(2x - 1)}$
27. $5x^2 + \frac{5}{2} + (5x^3 + \frac{15}{2}x^2 - x + \frac{5}{2})/(2x^4 - x^2 - 1)$
29. $3x^2 + x + 5 + (8x + 8)/(x^2 - x - 1)$

Exercises 11.2

1. 16   **3.** 1/8   **5.** 22   **7.** -4 **9.** -0.3175

11. $2x - 2; -8$ **13.** $-x^2 - \frac{3}{2}x + \frac{7}{4}; \frac{1}{8}$
15. $4x^2 + 16x + 61; 248$
17. $x^4 - 2x^3 + 3x^2 - 6x + 10; -18$
19. $x^3 - x^2 + x - 1; 0$
21. $0.2x^2 - 0.13x + 0.275; -0.3175$
23. $(x - 3)(x - 2)(x - 1)$ **25.** $(x + 1)(x - 2)(2x + 1)$
27. 29   **29.** -1.158968 **31.** -0.55184 **33.** -20
35. 6   **37.** 4   **39.** $3x + 3a - 2; 3a^2 - 2a + 1$

Exercises 11.3

1. No   **3.** Yes   **5.** Yes   **7.** No   **9.** Yes
11. $(x - 2)(2x^2 + 3)$ **13.** $(x + 3)(2x^3 + 5x - 1)$
15. $2 + i$ **17.** $2 - 3i$ **19.** $(-3 \pm i\sqrt{3})/2$
21. $-2i, 1, -1$ **23.** $(-1 \pm i\sqrt{3})/2$

25. 5 (multiplicity 1), $-\frac{2}{3}$ (multiplicity 1)
27. $-\frac{3}{2}$ (multiplicity 2), 1 (multiplicity 3), $\frac{4}{3}$ (multiplicity 1)
29. $-1 \pm i$ (multiplicity 1), 0 (multiplicity 2)
31. $1 - i\sqrt{3}$ and $1 + i\sqrt{3}$, both with multiplicity 2
33. $x^3 - 3x^2 - 4x + 12$ **35.** $x^3 - 5x^2 + 2x + 8$
37. $x^4 - 4x^3 - 2x^2 + 12x + 9$
39. $-x^3 + 5x^2 - 11x + 15$

Exercises 11.4

1. $-2, -1, 2$ **3.** $-2, 2, 3$ **5.** $-1, 0, 1$ **7.** No rational roots   **9.** $-1, -\frac{1}{2}, 1$ **11.** $-\frac{1}{2}, 0, \frac{1}{2}$ **13.** No rational roots   **15.** $x(x + 1)(x - 1)$ **17.** $\frac{1}{2}x(x - 2)(x + 2)$
19. $(x + 1)(x + 2)(x - 2)$ **21.** $x(x + 3)(x - 2)$
23. $(x + 2i)(x - 2i)(x + 2)(x - 2)$
25. $(x - 4)(x + i)(x - i)$
27. $(2x - 3)(x + i\sqrt{2})(x - i\sqrt{2})$
29. $\frac{3}{4}(x + 2)(x - 2)(x - 1 - i)(x - 1 + i)$
31. $\frac{1}{4}(x - 1)(x - 2 - \sqrt{2})(x - 2 + \sqrt{2})$

Exercises 11.5

1. $A = 3, B = 1$ **3.** $A = 1/2, B = -1/2$ **5.** $A = 3,$ $B = 2, C = 1$ **7.** $A = 1, B = -1, C = 2$ **9.** $A = 0,$ $B = 2, C = 1, D = 2$ **11.** $11/7(x + 3) - 4/7(x - 4)$
13. $-1/(2x - 3) - 3/(x + 4)$
15. $2/x - 1/(2x + 3) + 3/(x - 2)$
17. $1/x + 2/x^2 + 3/(x - 1)$
19. $(2x + 1)/(x^2 + 2x + 2) + (2x - 1)/(x^2 + 2x + 2)^2$

Chapter Test

1. 4, 4, 16   **2.** $A = -4, B = -1, C = 1$ **3.** $4x^2 - 3; 3$
4. $(x - 4)(x + 1)(x + 2)$ **5.** -132 **6.** $\frac{3}{2}$ (multiplicity 1),
2 (multiplicity 3), $-\frac{1}{2}$ (multiplicity 2)
7. $x^3 + x^2 - 4x + 6$ **8.** $\frac{2}{3}, 1, -\frac{1}{3}$ **9.** $A = -3,$
$B = -5$ **10.** $3/(x - 2) + 5/(x - 2)^2$

Review Exercises

1. $3x - 2; (8x - 4)$ **3.** $2x^4 + 3x^3 + 8x^2 + 16x + 36;$
$94x - 39$ **5.** $2x^2 + 8x + 32; 125$
7. $2x^3 - 6x^2 + 15x - 45; 136$
9. $0.3x^2 - 0.32x + 0.228; 0.9088$
11. $(x - 5)(2x - 1)(x - 1)$ **13.** $(2x - 1)(x + 3)(x - 1)$
15. $0.1(x + 2)(5x + 3)(x - 1)$ **17.** $5(x - \frac{2}{5})(x + 3)$
19. $(x - 1 + 2i)(x - 1 - 2i)(x - 1)(x + 1)$
21. $x^3 - 5x^2 + 7x + 13$ **23.** 1 (multiplicity 3), 3
25. $\frac{1}{2}x^3 + 3x^2 + \frac{11}{2}x + 3$
27. $x^4 - 10x^3 + 38x^2 - 66x + 45$
29. $-1, 3$ (multiplicity 2)   **31.** $-\frac{1}{2}, 2$ (multiplicity 2)
37. $1/(x - 1) + (x + 1)/(x^2 - x + 1)$
39. $1/x + 2/(x + 1) + 3/(x + 1)^2$
41. $2^{1/5}, 2^{1/5}[\cos(2\pi/5) + i\sin(2\pi/5)],$
$2^{1/5}[\cos(4\pi/5) + i\sin(4\pi/5)],$
$2^{1/5}[\cos(6\pi/5) + i\sin(6\pi/5)],$
$2^{1/5}[\cos(8\pi/5) + i\sin(8\pi/5)]$

CHAPTER 12

Exercises 12.1

21. 27   **23.** 25   **25.** 12   **27.** 91   **29.** 26

Exercises 12.2

1. 84   **3.** 220   **5.** $a^4 + 4a^3b + 6a^2b^2 + 4ab^3 + b^4$
7. $x^6 + 6x^5y + 15x^4y^2 + 20x^3y^3 + 15x^2y^4 + 6xy^5 + y^6$
9. $32a^5 - 80a^4b + 80a^3b^2 - 40a^2b^3 + 10ab^4 - b^5$
11. $x^5 - 10x^4 + 40x^3 - 80x^2 + 80x - 32$
13. $a^6 - 6a^4 + 15a^2 - 20 + 15/a^2 - 6/a^4 + 1/a^6$
15. $280a^4x^3$ **17.** $-1920u^3$ **19.** 70   **21.** $70ab$
23. $-35x^4y^4a^3; 35x^3y^3a^4$ **25.** $-270a^2b^6$
27. 1.0828567   **29.** 62.104   **31.** 1.219   **33.** 2   **35.** 2

Exercises 12.3

1. $3, 1, -1, -3, -5$ **3.** $-1, 4, -9, 16, -25$ **5.** $-3,$
$9, -27, 81, -243$ **7.** 2, 9/4, 64/27, 625/256,
7776/3125   **9.** $-2, 4/3, -8/7, 16/15, -32/31$

11. $a_n = 2n, n \geq 1; 2, 4, 6, 8, 10$ **13.** $a_n = -4n,$
$n \geq 1; -4, -8, -12, -16, -20$ **15.** $a_n = 3n - 1,$
$n \geq 1; 2, 5, 8, 11, 14$ **17.** $-5, -3, 1, 9, 25$
19. $0, 4, -8, 28, -80$ **21.** 3; 1/9; 81; 1/6561;
43,046,721   **23.** 3, 3, 6, 18, 72   **25.** 2, 3/2, 9/8,
27/32, 81/128   **27.** 1, 5, 9, 13, 17   **29.** 2^n **31.** $2n$
33. $n/2$ **35.** $2(3^{n-1})$ **37.** $2, 2.25, 2.23\overline{61}, 2.236068$
39. $3.2, 3.1625, 3.16227767, 3.16227766$

Exercises 12.4

1. $d = 2; 8, 10$ **3.** $d = -1.5; -2.5, -4$ **5.** $d = \log 2;$
$\log 16, \log 32$ **7.** $d = 4; a_n = 15 + 4(n - 1)$
9. $d = 11; a_n = -13 + 11(n - 1)$ **11.** $d = 2r;$
$a_n = a + 2r(n - 1)$ **13.** 46   **15.** 18; -6 **17.** 144
19. 48   **21.** 725   **23.** -1275 **25.** 105   **27.** 2146
29. 1080   **31.** \$49.60   **33.** 210   **35.** 5, 20, 35
37. 7, 11, 15

Exercises 12.5

1. $r = 2$; 16, 32   **3.** $r = -1/2$; $-1/8$, 1/16
5. $r = 2/3$; 4/27, 8/81   **7.** $r = 2$; $a_n = (\log 3)2^{n-1}$,
$n \geq 1$ **9.** $r = 1/2$; $a_n = (\log 16)(1/2)^{n-1}$, $n \geq 1$
11. 5, 20, 80, 320   **13.** 2/27   **15.** 8   **17.** -2046
19. -1365 **21.** 29,524   **23.** 2/3   **25.** 4   **27.** 81/2
29. 6   **31.** $-4/9$ **33.** 2/11   **35.** 7/11   **37.** 7/22
39. 41/330   **41.** 54 ft   **43.** 346   **45.** 1379   **47.** 1800
49. \$800,000

Exercises 12.6

1. 30   **3.** 336   **5.** 720   **7.** 2,193,360
9. 2,884,801,920   **11.** 10   **13.** 84   **15.** 210   **17.** 780
19. 6.3501356×10^{11} **21.** 4200   **23.** 1.4782628×10^{24}
25. 90   **27.** 336   **29.** 720   **31.** 120   **33.** 24, 24
35. 120   **37.** 792   **39.** 133,784,560   **41.** 28
43. 4200   **45.** 5.3644738×10^{28} **47.** 34650
49. 1001; 420

Exercises 12.7

1. 1/2   **3.** 2/3   **5.** 1   **7.** 5/36   **9.** 7/12   **11.** 5/12
13. 0   **15.** 1/4   **17.** 3/5   **19.** 1/10   **21.** 1/8
23. 1/2   **25.** (a) 1/32   (b) 5/16   **27.** 1.98×10^{-3}
29. 9.23×10^{-6} **31.** 1.385×10^{-5} **33.** 2/13; 1/2
35. 1/3; 2/3   **37.** 1/28   **39.** 1/3   **41.** (a) 5 to 3
(b) 3 to 7   **43.** (a) 0.6   (b) 0.375
45. (a) 2 to 1   (b) 1 to 3   **47.** (a) 3/10   (b) 7 to 3
49. 18 to 37

Chapter Test

2. $560z^4b^3$ **3.** $-1, 7, 15, 23, 31$ **4.** -76 **5.** -48
6. $-3, -6, -12, -24$ **7.** 6   **8.** 364   **9.** 360
10. 4/5

Review Exercises

1. 816
3. $\frac{1}{32}a^5 + \frac{5}{8}a^4b + 5a^3b^2 + 20a^2b^3 + 40ab^4 + 32b^5$
5. $-84u^2b^5$ **7.** $-34560x^3$ **9.** $70a^2b^2$ **11.** 1.1261624
13. 9/2, 4, 7/2, 3, 5/2   **15.** 1, $-1/2$, 1/6, $-1/24$,
1/120   **17.** 0, 1/3, $-1/2$, 3/5, $-2/3$ **19.** $3 \cdot 4^{n-1}$
21. $n(n + 1)$ **23.** 2.1, 2.71$\overline{6}$, 2.6466769, 2.6457515
25. $d = -3$; 1, -2 **27.** $d = -\ln 3$; $\ln(1/9)$, $\ln(1/27)$
29. 47/2   **31.** 11; 528   **33.** $r = 5/6$; 25/72, 125/432
35. $r = 1.5$; 16.875, 25.3125   **37.** 2/3   **39.** 9/4
41. 3/11   **43.** 4/165   **55.** 336 ft, 1936 ft   **57.** 43/4 in.,
1001/8 in.   **59.** 10 m
61. (a) 336   (b) 151,200   (c) 252   (d) 495   (e) 280
(f) 1,260   **63.** 24   **65.** 28   **67.** 56   **69.** 18,564
71. 5/12   **73.** 1/5   **75.** 1/380   **77.** 0.0086
79. 0.3968

Appendix Exercises

1. 1.5127   **3.** 3.1405   **5.** -2.6287 **7.** 20.630
9. 0.386   **11.** 1217.8   **13.** 0.0364   **15.** 0.9631
17. 2.148   **19.** 0.6889   **21.** 0.1028   **23.** 70°58′
25. 60°1′   **27.** 10°8′   **29.** 150°6′

Graphs of Trigonometric Functions

$f(x) = \sin x$
Domain: all real numbers
Range: $[-1, 1]$
Odd function: $\sin(-x) = -\sin x$
Period $= 2\pi$, $\sin(x + 2\pi) = \sin x$

$f(x) = \cos x$
Domain: all real numbers
Range: $[-1, 1]$
Even function: $\cos(-x) = \cos x$
Period $= 2\pi$, $\cos(x + 2\pi) = \cos x$

$f(x) = \tan x$
Domain: all real numbers $x \neq (2k + 1)\pi/2$,
with k any integer
Range: $(-\infty, \infty)$
Odd function: $\tan(-x) = \tan x$
Period $= \pi$, $\tan(x + \pi) = \tan x$

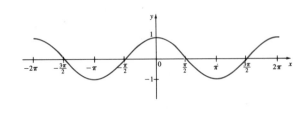

Trigonometric Functions of the Form

$y = A \sin(Bx + C)$ or $y = A \cos(Bx + C)$, $B > 0$
Amplitude: $|A|$
Period: $2\pi/B$
Phase shift: $-C/B$

Inverse Trigonometric Functions

$y = \sin^{-1} x$, $-1 \leq x \leq 1$, if and only if,
$\quad x = \sin y$, $-\pi/2 \leq y \leq \pi/2$
$y = \cos^{-1} x$, $-1 \leq x \leq 1$, if and only if,
$\quad x = \cos y$, $0 \leq y \leq \pi$
$y = \tan^{-1} x$, $-\infty < x < \infty$, if and only if,
$\quad x = \tan y$, $-\pi/2 < y < \pi/2$

Pythagorean Identities

$\sin^2 x + \cos^2 x = 1$
$\tan^2 x + 1 = \sec^2 x$
$\cot^2 x + 1 = \csc^2 x$

Trigonometric Identities

$\tan x = \sin x/\cos x \qquad \cot x = \cos x/\sin x$
$\sec x = 1/\cos x \qquad \csc x = 1/\sin x$
$\cot x = 1/\tan x$

Sum Identities

$\sin(x + y) = \sin x \cos y + \cos y \sin y$
$\cos(x + y) = \cos x \cos y - \sin x \sin y$
$\tan(x + y) = (\tan x + \tan y)/(1 - \tan x \tan y)$